AF478002

Acta Numerica 2019

Acta Numerica

Volume 28   2019

Published by the Press Syndicate of the University of Cambridge
The Pitt Building, Trumpington Street, Cambridge CB2 1RP
One Liberty Plaza, Floor 20, New York, NY 10006, USA
10 Stamford Road, Oakleigh, Melbourne 3166, Australia

© Cambridge University Press 2019

First published 2019

Printed in the UK by Bell & Bain

Library of Congress cataloguing in publication data available

A catalogue record for this book is available from the British Library

ISBN 978-1-108-478687
ISSN 0962-4929

Contents

Acta Numerica (2019), pp. 1–174
doi:10.1017/S0962492919000059

Solving inverse problems using data-driven models

Simon Arridge
Department of Computer Science,
University College London,
Gower Street, London WC1E 6BT, UK
E-mail: S.Arridge@cs.ucl.ac.uk

Peter Maass
Department of Mathematics,
University of Bremen, Postfach 330 440,
28344 Bremen, Germany
E-mail: pmaass@math.uni-bremen.de

Ozan Öktem
Department of Mathematics,
KTH – Royal Institute of Technology,
SE-100 44 Stockholm, Sweden
E-mail: ozan@kth.se

Carola-Bibiane Schönlieb
Department of Applied Mathematics and Theoretical Physics,
Cambridge University, Wilberforce Road,
Cambridge, CB3 0WA, UK
E-mail: C.B.Schoenlieb@damtp.cam.ac.uk

Recent research in inverse problems seeks to develop a mathematically coherent foundation for combining data-driven models, and in particular those based on deep learning, with domain-specific knowledge contained in physical–analytical models. The focus is on solving ill-posed inverse problems that are at the core of many challenging applications in the natural sciences, medicine and life sciences, as well as in engineering and industrial applications. This survey paper aims to give an account of some of the main contributions in data-driven inverse problems.

CONTENTS

1. Introduction

In several areas of science and industry there is a need to reliably recover a hidden multi-dimensional model parameter from noisy indirect observations. A typical example is when imaging/sensing technologies are used in medicine, engineering, astronomy and geophysics. These so-called inverse problems are often ill-posed, meaning that small errors in data may lead to large errors in the model parameter, or there are several possible model parameter values that are consistent with observations. Addressing ill-posedness is critical in applications where decision making is based on the recovered model parameter, for example in image-guided medical diagnostics. Furthermore, many highly relevant inverse problems are large-scale: they involve large amounts of data and the model parameter is high-dimensional.

Traditionally, an inverse problem is formalized as solving an equation of the form

$$g = \mathcal{A}(f) + e.$$

Here $g \in Y$ is the measured data, assumed to be given, and $f \in X$ is the model parameter we aim to reconstruct. In many applications, both g and f are elements in appropriate function spaces Y and X, respectively. The mapping $\mathcal{A} \colon X \to Y$ is the forward operator, which describes how the model parameter gives rise to data in the absence of noise and measurement errors, and $e \in Y$ is the observational noise that constitutes random corruptions in the data g. The above view constitutes a knowledge-driven approach, where the forward operator and the probability distribution of the observational noise are derived from first principles.

Classical research on inverse problems has focused on establishing conditions which guarantee that solutions to such ill-posed problems exist and on methods for approximating solutions in a stable way in the presence of noise (Engl, Hanke and Neubauer 2000, Benning and Burger 2018, Louis 1989, Kirsch 2011). Despite being very successful, such a knowledge-driven approach is also associated with some shortcomings. First, the forward model is always an approximate description of reality, and extending it might be challenging due to a limited understanding of the underlying physical or technical setting. It may also be limited due to computational complexity. Accurate analytical models, such as those based on systems of non-linear partial differential equations (PDEs), may reach a numerical complexity beyond any feasible real-time potential in the foreseeable future. Second, most applications will have inputs which do not cover the full model parameter space, but stem from an unknown subset or obey an unknown stochastic distribution. The latter shortcoming in particular has led to the advance of methods that incorporate information about the structure of the parameters to be determined in terms of sparsity assumptions (Daubechies, Defrise and De Mol 2004, Jin and Maass 2012*b*) or stochastic models (Kaipio and Somersalo 2007, Mueller and Siltanen 2012). While representing a significant advancement in the field of inverse problems, these models are, however, limited by their inability to capture very bespoke structures in data that vary in different applications.

At the same time, data-driven approaches as they appear in machine learning offer several methods for amending such analytical models and for tackling these shortcomings. In particular, deep learning (LeCun, Bengio and Hinton 2015), which has had a transformative impact on a wide range of tasks related to artificial intelligence, ranging from computer vision and speech recognition to playing games (Igami 2017), is starting to show its impact on inverse problems. A key feature in these methods is the use of *generic* models that are adapted to specific problems through learning against example data (training data). Furthermore, a common trait in the success stories for deep learning is the abundance of training data and the explicit agnosticism from *a priori* knowledge of how such data are generated. However, in many scientific applications, the solution method needs to be robust and there is insufficient training data to support an entirely data-driven approach. This seriously limits the use of entirely data-driven approaches for solving problems in the natural and engineering sciences, in particular for inverse problems.

A recent line of development in computational sciences combines the seemingly incompatible data- and knowledge-driven modelling paradigms. In the context of inverse problems, ideally one uses explicit knowledge-driven models when there are such available, and learns models from example data using data-driven methods only when this is necessary. Recently several algorithms have been proposed for this combination of model- and data-driven approaches for solving ill-posed inverse problems. These results are still primarily experimental and lack a thorough theoretical foundation; nevertheless, some mathematical concepts for treating data-driven approaches for inverse problems are emerging.

This survey attempts to provide an overview of methods for integrating data-driven concepts into the field of inverse problems. Particular emphasis is placed on techniques based on deep neural networks, and our aim is to pave the way for future research towards providing a solid mathematical theory. Some aspects of this development are covered in recent reviews of inverse problems and deep learning, for instance those of McCann, Jin and Unser (2017), Lucas, Iliadis, Molina and Katsaggelos (2018) and McCann and Unser (2019).

1.1. Overview

This survey investigates algorithms for combining model- and data-driven approaches for solving inverse problems. To do so, we start by reviewing some of the main ideas of knowledge-driven approaches to inverse problems, namely functional analytic inversion (Section 2) and Bayesian inversion (Section 3), respectively. These knowledge-driven inversion techniques are derived from first principles of knowledge we have about the data, the model parameter and their relationship to each other.

Knowledge- and data-driven approaches can now be combined in several different ways depending on the type of reconstruction one seeks to compute and the type of training data. Sections 4 and 5 represent the core of the survey and discuss a range of inverse problem approaches that introduce data-driven aspects in inverse problem solutions. Here, Section 4 is the data-driven sister section to functional analytic approaches in Section 2. These approaches are primarily designed to combine data-driven methods with functional analytic inversion. This is done either to make functional analytic approaches more data-driven by appropriate parametrization of these approaches and adapting these parametrizations to data, or to accelerate an otherwise costly functional analytic reconstruction method.

Many reconstruction methods, however, are not naturally formulated within the functional analytic view of inversion. An example is the posterior mean reconstruction, whose formulation requires adopting the Bayesian view of inversion. Section 5 is the data-driven companion to Bayesian inversion in Section 3, and surveys methods that combine data- and knowledge-driven methods in Bayesian inversion. The simplest is to apply data-driven post-processing of a reconstruction obtained via a knowledge-driven method. A more sophisticated approach is to use a learned iterative scheme that integrates a knowledge-driven model for how data are generated into a data-driven method for reconstruction. The latter is done by unrolling a knowledge-driven iterative scheme, and both approaches, which compute statistical estimators, can be combined with forward operators that are partially learned via a data-driven method.

The above approaches come with different trade-offs concerning demands on training data, statistical accuracy and robustness, functional complexity, stability and interpretability. They also impact the choice of machine learning methods and algorithms for training. Certain recent – and somewhat anecdotal – topics of data-driven inverse problems are discussed in Section 6, and exemplar practical inverse problems and their data-driven solutions are presented in Section 7.

Within data-driven approaches, deep neural networks will be a focus of this survey. For an introduction to deep neural networks the reader might find it helpful to consult some introductory literature on the topic. We recommend Courville, Goodfellow and Bengio (2017) and Higham and Higham (2018) for a general introduction to deep learning; see also Vidal, Bruna, Giryes and Soatto (2017) for a survey of work that aims to provide a mathematical justification for several properties of deep networks. Finally, the reader may also consult Ye, Han and Cha (2018), who give a nice survey of various types of deep neural network architectures.

Detailed structure of the paper. In Section 2 we discuss functional analytic inversion methods, and in particular the mathematical notion of ill-posedness (Section 2.3) and regularization (Section 2.4) as a means to counteract the latter. A special focus is on variational regularization methods (Sections 2.5–2.7), as those reappear in bilevel learning in Section 4.3 in the context of data-driven methods for inverse problems.

Statistical – and in particular Bayesian – approaches to inverse problems are described in Section 3. In contrast to functional analytic approaches (Section 2.4), in Bayesian inversion (Section 3.1) the model parameter is a random variable that follows a prior distribution. A key difference between Bayesian and functional analytic inversion is that in Bayesian inversion an approximation to the whole distribution of the model parameter conditioned on the measured data (posterior distribution) is computed, rather than a single model parameter as in functional analytic inversion. This means that reconstructed model parameters can be derived via different estimates of its posterior distribution (a concept that we will encounter again in Section 5, and in particular Section 5.1.2, where data-driven reconstructions are phrased as results of different Bayes estimators), but also that uncertainty of reconstructed model parameters can be quantified (Section 3.2.5). When evaluating different reconstructions of the model parameter – which is again important when defining learning, *i.e.* optimization criteria for inverse problem solutions – aspects of statistical decision theory can be used (Section 3.3). Also, the parallel concept of regularization, introduced in Section 3 for the functional analytic approach, is outlined in Section 3.2 for statistical approaches. The difficult problem of selecting a prior distribution for the model parameter is discussed in Section 3.4.

In Section 4 we present some central examples of machine learning combined with functional analytic inversion. These encompass classical parameter choice rules for inverse problems (Section 4.1) and bilevel learning (Section 4.3) for parameter learning in variational regularization methods. Moreover, dictionary learning is discussed in Section 4.4 as a companion to sparse reconstruction methods in Section 2.7, but with a data-driven dictionary. Also, the concept of a black-box denoiser, and its application to inverse problems by decoupling the regularization from the inversion of the data, is presented in Section 4.6. Two recent approaches that use deep neural network parametrizations for data-driven regularization in variational inversion models are investigated in Section 4.7. In Section 4.9 we discuss a range of learned optimization methods that use data-driven approximations as a means to speed up numerical computation. Finally, in Section 4.10 we introduce a new idea of using the recently introduced concept of deep inverse priors for solving inverse problems.

In Section 5 learning data-driven inversion models are phrased in the context of statistical regularization. Section 5.1.2 connects back to the difficulty in Bayesian inversion of choosing an appropriate prior (Section 3.4), and outlines how model learning can be used to compute various Bayes estimators. Here, in particular, fully learned inversion methods (Section 5.1.3), where the whole inversion model is data-driven, are put in context with learned iterative schemes (Section 5.1.4), in which data-driven components are interwoven with inverse model assumptions. In this context also we discuss post-processing methods in Section 5.1.5, where learned regularization together with simple knowledge-driven inversion methods are used sequentially. Section 5.2 addresses the computational bottleneck of Bayesian inversion methods by using learning, and shows how one can use learning to efficiently sample from the posterior.

Section 6 covers special topics of learning in inverse problems, and in Section 6.1 includes task-based reconstruction approaches that use ideas from learned iterative reconstruction (Section 5.1.4) and deep neural networks for segmentation and classification to solve joint reconstruction-segmentation problems, learning physics-based models via neural networks (Section 6.2.1), and learning corrections to forward operators by optimization methods that perform joint reconstruction-operator correction (Section 6.2).

Finally, Section 7 illustrates some of the data-driven inversion methods discussed in the paper by applying them to practical inverse problems. These include an introductory example on inversion of ill-conditioned linear systems to highlight the intricacy of using deep learning for inverse problems as a black-box approach (Section 7.1), bilevel optimization from Section 4.3 for parameter learning in TV-type regularized problems and variational models with mixed-noise data fidelity terms (Section 7.2), the application of learned iterative reconstruction from Section 5.1.4 to computed tomography (CT) and photoacoustic tomography (PAT) (Section 7.3), adversarial regularizers from Section 4.7 for CT reconstruction as an example of variational regularization with a trained neural network as a regularizer (Section 7.4), and the application of deep inverse priors from Section 4.10 to magnetic particle imaging (MPI) (Section 7.5).

In Section 8 we finish our discussion with a few concluding remarks and comments on future research directions.

2. Functional analytic regularization

Functional analysis has had a strong impact on the development of inverse problems. One of the first publications that can be attributed to the field of inverse problems is that of Radon (1917). This paper derived an explicit inversion formula for the so-called Radon transform, which was later identified as a key component in the mathematical model for X-ray CT. The derivation of the inversion formula, and its analysis concerning missing stability, makes use of operator formulations that are remarkably close to the functional analysis formulations that would be developed three decades later.

2.1. The inverse problem

There is no formal mathematical definition of an inverse problem, but from an applied viewpoint such problems are concerned with determining causes from desired or observed effects. It is common to formalize this as solving an operator equation.

Definition 2.1. An *inverse problem* is the task of recovering the model parameter $f_{\mathrm{true}} \in X$ from measured data $g \in Y$, where

$$g = \mathcal{A}(f_{\mathrm{true}}) + e. \tag{2.1}$$

Here, X (model parameter space) and Y (data space) are vector spaces with appropriate topologies and whose elements represent possible model parameters and data, respectively. Moreover, $\mathcal{A}\colon X \to Y$ (forward operator) is a known continuous operator that maps a model parameter to data in absence of observation noise and $e \in Y$ is a sample of a Y-valued random variable modelling the observation noise.

In most imaging applications, such as CT image reconstruction, elements in X are images represented by functions defined on a fixed domain $\Omega \subset \mathbb{R}^d$ and elements in Y represent imaging data by functions defined on a fixed manifold $\mathbb{M}$ that is given by the acquisition geometry associated with the measurements.

2.2. Introduction to some example problems

In the following, we briefly introduce some of the key inverse problems we consider later in this survey. All are from imaging, and we make a key distinction between (i) *image restoration* and (ii) *image reconstruction*. In the former, the data are a corrupted (*e.g.* noisy or blurry) realization of the model parameter (image) so the reconstruction and data spaces coincide, whereas in the latter the reconstruction space is the space of images but the data space has a definition that is problem-dependent. As we will see when discussing data-driven approaches to inverse problems in Sections 4 and 5, this differentiation is particularly crucial as the difference between image and data space poses additional challenges to the design of machine learning methods. Next, we describe very briefly some of the most common operators that we will refer to below. Here the inverse problems in Sections 2.2.1–2.2.3 are image restoration problems, while those in Sections 2.2.4 and 2.2.5 are examples of image reconstruction problems.

2.2.1. Image denoising

The observed data are the ideal solution corrupted by additive noise, so the forward operator in (2.1) is the identity transform $\mathcal{A} = \mathrm{id}$, and we get

$$g = f_{\mathrm{true}} + e, \tag{2.2}$$

In the simplest case the distribution of the observational noise is known. Furthermore, this distribution may in more advanced problems be correlated, spatially varying and of mixed type.

In Section 7.2 we will discuss bilevel learning of total variation (TV)-type variational models for denoising of data corrupted with mixed noise distributions.

2.2.2. Image deblurring

The observed data are given by convolution with a known filter function K together with additive noise, so (2.1) becomes

$$g = f_{\mathrm{true}} * K + e. \tag{2.3}$$

Any inverse problem of the type (2.1) with a linear forward operator that is translation-invariant will be of this form.

In the absence of noise, the inverse problem (*image deconvolution*) is exactly solvable by division in the Fourier domain, *i.e.* $f_{\mathrm{true}} = \mathcal{F}^{-1}[\mathcal{F}[g]/\mathcal{F}[K]]$, provided that $\mathcal{F}[K]$ has infinite support in the Fourier domain. In the presence of noise, the estimated solution is corrupted by noise whose frequency spectrum is the reciprocal of the spectrum of the filter K. The distribution of the observational also has the same considerations as in (2.2). Finally, extensions include the case of a spatially varying kernel and the case where K is unknown (*blind deconvolution*).

2.2.3. Image in-painting

Here, the observed data represents a noisy observation of the true model parameter $f_{\text{true}} \colon \Omega \to \mathbb{R}$ restricted to a fixed measurable set $\Omega_0 \subset \mathbb{R}^n$:

$$g = f_{\text{true}}\big|_{\Omega_0} + e. \tag{2.4}$$

In the above, $f_{\text{true}}\big|_{\Omega_0}$ is the restriction of f_{true} to Ω_0. Solutions take different forms depending on the size of connected components in Ω_0. Extensions include the case where Ω_0 is unknown or only partially known.

2.2.4. Computed tomography (CT)

The simplest physical model for CT assumes mono-energetic X-rays and disregards scattering phenomena. The model parameter is then a real-valued function $f \colon \Omega \to \mathbb{R}$ defined on a fixed domain $\Omega \subset \mathbb{R}^d$ ($d = 2$ for two-dimensional CT and $d = 3$ for three-dimensional CT) that has unit mass per volume. The forward operator is the one given by the Beer–Lambert law:

$$\mathcal{A}(f)(\omega, x) = \mathrm{e}^{-\mu \int_{-\infty}^{\infty} f(x+s\omega)\,\mathrm{d}s}. \tag{2.5}$$

Here, the unit vector $\omega \in S^{d-1}$ and $x \in \omega^{\perp}$ represent the line $\ell \colon s \mapsto x + s\omega$ along which the X-rays travel, and we also assume f decays fast enough for the integral to exist. In medical imaging, μ is usually set to a value that approximately corresponds to water at the X-ray energies used. The above represents pre-logarithm (or pre-log) data, and by taking the logarithm (or log) of data, one can recast the inverse problem in CT imaging to one where the forward model is the linear ray transform:

$$\mathcal{A}(f)(\omega, x) = \int_{-\infty}^{\infty} f(x + s\omega)\,\mathrm{d}s. \tag{2.6}$$

For low-dose imaging, pre-log data are Poisson-distributed with mean $\mathcal{A}(f_{\text{true}})$, where $\mathcal{A}$ is given as in (2.5), that is, $g \in Y$ is a sample of $\mathsf{g} \sim \text{Poisson}(\mathcal{A}(f_{\text{true}}))$. Thus, to get rid of the non-linear exponential in (2.5), it is common to take the log of data. With such post-log data the forward operator is linear and given as in (2.6). A complication with such post-log data is that the noise model becomes non-trivial, since one takes the log of a Poisson-distributed random variable (Fu *et al.* 2017). A common approximate noise model for post-log data is (2.1), with observational noise e which is a sample of a Gaussian or Laplace-distributed random variable.

In the case of *complete* data, that is, where a full angular set of data is measured, an exact inverse is obtained by the (Fourier-transformed) data backprojected on the same lines as used for the measurements and scaled by the absolute value of the spatial frequency, followed by the inverse Fourier transform. Thus, as in deblurring, the noise is amplified, but only linearly in spatial frequency, making the problem mildly ill-posed. Extensions include the *emission tomography* problem (single photon emission computed tomography (SPECT) and positron emission tomography (PET)) where the line integrals are exponentially attenuated by a function μ that may be unknown. A major challenge in tomography is to consider *incomplete* data, in particular the case where only a subset of lines is measured. This problem is much more ill-posed.

See Sections 7.3, 7.4 and 7.6 for instances of CT reconstruction that use deep neural networks in the solution of the inverse problem.

2.2.5. Magnetic resonance imaging (MRI)

The observed data are often considered to be samples of the Fourier transform of the ideal signal, so the MRI image reconstruction problem is an inverse problem of the type (2.1), where the forward operator is given as a discrete sampling operator concatenated with the Fourier transform. A correct description of the problem takes account of the complex-valued nature of the data, which implies that when e is normally distributed then the noise model of $|\mathcal{F}^{-1}[g]|$ is Rician. As in CT, the case of under-sampled data is of high practical importance. In MRI, the subsampling operator has to consist of connected trajectories in Fourier space but is not restricted to straight lines.

Extensions include the case of *parallel MRI* where the forward operator is combined with (several) spatial sensitivity functions. More exact forward operators take account of other non-linear physical effects and can reconstruct several functions in the solution space.

2.3. Notion of ill-posedness

A difficulty in solving (2.1) is that the solution is sensitive to variations in data, which is referred to as ill-posedness. More precisely, the notion of ill-posedness is usually attributed to Hadamard, who postulated that a well-posed problem must have three defining properties, namely that (i) it has a solution (existence) that is (ii) unique and that (iii) depends continuously on the data g (stability). Problems that do not fulfil these criteria are ill-posed and, according to Hadamard, should be modelled differently (Hadamard 1902, Hadamard 1923).

For example, instability arises when the forward operator $\mathcal{A}\colon X \to Y$ in (2.1) has an unbounded or discontinuous inverse. Hence, every non-degenerate compact operator between infinite-dimensional Hilbert spaces whose range is infinite naturally leads to ill-posed inverse problems. Slightly more generally, one can prove that continuous operators with non-closed range yield unbounded inverses and hence lead to ill-posed inverse problems. This class includes non-degenerate compact operators as well as convolution operators on unbounded domains.

Another way of describing ill-posedness is in terms of the set of $f \in X$ such that $\| \mathcal{A}(f) - g \| \leq \| e \|$ for given noise e in (2.1). This is an unbounded set for continuous operators with non-closed range. Finally, yet another way to understand the ill-posedness of a compact linear operator $\mathcal{A}$ is by means of its singular value decomposition. The decay of the spectrum $\{\sigma_k\}_{k \in \mathbb{N}}$ is strongly related to the ill-posedness: faster decay implies a more ill-posed problem. This allows us to determine the severity of the ill-posedness. More precisely, (2.1) is weakly ill-posed if σ_k decays with polynomial rate as $k \to \infty$ and strongly ill-posed if the decay is exponential: see Engl *et al.* (2000), Derevtsov, Efimov, Louis and Schuster (2011) and Louis (1989) for further details. As a final note, such classification is not possible when the forward operator is non-linear. In such cases, either linearized forward operators are analysed or a non-linear spectral analysis is considered for determining the degree of ill-posedness: see *e.g.* Hofmann (1994). Moreover, extensions to non-compact linear operators are considered by Hofmann *et al.* (2010), for example.

2.4. Regularization

Unfortunately, Hadamard's dogma stigmatized the study of ill-posed problems and thereby severely hampered the development of the field. Mathematicians' interest in studying ill-posed problems was revitalized by the pioneering works of Calderón and Zygmund (1952, 1956), Calderón (1958) and John (1955*a*, 1955*b*, 1959, 1960), who showed that instability is an intrinsic property in some of the most interesting and challenging problems in mathematical physics and applied analysis. To some extent, these papers constitute the origin of the modern theory of inverse problems and regularization.

The aim of functional analytic regularization theory is to develop stable schemes for estimating f_{true} from data g in (2.1) based on knowledge of $\mathcal{A}$, and to prove analytical results for the properties of the estimated solution. More precisely, a regularization of the inverse problem in (2.1) is formally a scheme that provides a well-defined parametrized mapping $\mathcal{R}_\theta \colon Y \to X$ (*existence*) that is continuous in Y for fixed θ (*stability*) and *convergent*. The latter means there is a way to select θ so that $\mathcal{R}_\theta(g) \to f_{\text{true}}$ as $g \to \mathcal{A}(f_{\text{true}})$.

Besides existence, stability and convergence, a complete mathematical analysis of a regularization method also includes proving *convergence rates* and *stability estimates*. Convergence rates provide an estimate of the difference between a regularized solution $\mathcal{R}_\theta(g)$ and the solution of (2.1) with $e = 0$ (provided it exists), whereas stability estimates provide a bound on the difference between $\mathcal{R}_\theta(g)$ and $\mathcal{R}_\theta(\mathcal{A}(f_{\text{true}}))$ depending on the error $\| e \|$. These theorems rely on 'source conditions': for example, convergence rate results are obtained under the assumption that the true solution f_{true} is in the range of $[\partial \mathcal{A}(f_{\mathrm{true}})]^* \colon Y \to X$. One difficulty is to formulate source conditions that are verifiable. This typically relates to a regularity assumption for f_{true} that ensures a certain convergence rate: see *e.g.* Engl, Kunisch and Neubauer (1989), Hofmann, Kaltenbacher, Pöschl and Scherzer (2007), Schuster, Kaltenbacher, Hofmann and Kazimierski (2012), Grasmair, Haltmeier and Scherzer (2008) and Hohage and Weidling (2016) for an example of this line of development.

From an algorithmic viewpoint, functional analytic regularization methods are subdivided into essentially four categories.

Approximate analytic inversion. These methods are based on stabilizing a closed-form expression for $\mathcal{A}^{-1}$. This is typically achieved by considering reconstruction operators that give a mollified solution, so the resulting approaches are highly problem-specific.

Analytic inversion has been hugely successful: for example, filtered backprojection (FBP) (Natterer 2001, Natterer and Wübbeling 2001) for inverting the ray transform is still the standard method for image reconstruction in CT used in clinical practice. Furthermore, the idea of recovering a mollified version can be stated in a less problem-specific manner, which leads to the method of approximate inverse (Louis 1996, Schuster 2007, Louis and Maass 1990).

Iterative methods with early stopping. Here one typically considers iteration methods based on gradient descent for the data misfit or discrepancy term $f \mapsto \| \mathcal{A}(f) - g \|^2$. The ill-posedness of the inverse problem leads to semiconvergent behaviour, meaning that the reconstruction error decreases until a certain data fit is achieved and then starts diverging. Hence a suitable stopping needs to be designed, which acts as a regularization.

Well-known examples are the iterative schemes of Kaczmarz and Landweber (Engl *et al.* 2000, Natterer and Wübbeling 2001, Kirsch 2011). There is also large body of literature addressing iteration schemes in Krylov spaces for inverse problems (conjugate gradient (CG) type methods) as well as accelerated and discretized versions thereof (*e.g.* CGLS, LSQR, GMRES): see Hanke-Bourgeois (1995), Hanke and Hansen (1993), Frommer and Maass (1999), Calvetti, Lewis and Reichel (2002) and Byrne (2008) for further reference. A different approach that has a statistical interpretation uses a fixed-point iteration for the maximum *a posteriori* (MAP) estimator leading to the maximum likelihood expectationmaximization (ML-EM) algorithm (Dempster *et al.* 1977).

Discretization as regularization. Projection or Galerkin methods, which search for an approximate solution of an inverse problems in a predefined subspace, are also a powerful tool for solving inverse problems. The level of discretization controls the approximation of the forward operator but it also stabilizes the inversion process: see Engl *et al.* (2000), Plato and

Vainikko (1990) and Natterer (1977). Such concepts have been discussed in the framework of parameter identification for partial differential equations (quasi-reversibility): see Lattès and Lions (1969) for an early reference and Hämarik, Kaltenbacher, Kangro and Resmerita (2016) and Kaltenbacher, Kirchner and Vexler (2011) for some recent developments.

Variational methods. The idea here is to minimize a measure of data misfit that is penalized using a regularizer (Kaltenbacher, Neubauer and Scherzer 2008, Scherzer *et al.* 2009):

$$\mathcal{R}_\theta(g) := \arg\min_{f \in X}\{\mathcal{L}(\mathcal{A}(f), g) + \mathcal{S}_\theta(f)\}, \tag{2.7}$$

where we make use of the notation in Definitions 2.2, 2.4 and 2.5. This is a generic, yet highly adaptable, framework for reconstruction with a natural plug-and-play structure where the forward operator $\mathcal{A}$, the data discrepancy $\mathcal{L}$ and the regularizer $\mathcal{S}_\theta$ are chosen to fit the specific aspects of the inverse problem. Well-known examples are classical Tikhonov regularization and TV regularization.

Sections 2.5 and 2.6 provide a closer look at the development of variational methods since these play an important role in Section 4, where data-driven methods are used in functional analytic regularization. To simplify these descriptions it is convenient to establish some key notions.

Definition 2.2. A *regularization functional* $\mathcal{S} : X \to \mathbb{R}_+$ quantifies how well a model parameter possesses desirable features: a larger value usually means less desirable properties.

In variational approaches to inverse problems, the value of $\mathcal{S}$ is considered as a *penalty term*, and in Bayesian approaches it is seen as the negative log of a *prior probability distribution*. Henceforth we will use $\mathcal{S}_\theta$ to denote a regularization term that depends on a parameter set $\theta \in \Theta$; in particular we will use θ as parameters that will be *learned*.

Remark 2.3. In some cases θ is a single scalar. We will use the notation $\lambda \mathcal{S}(f) \equiv \mathcal{S}_\theta(f)$ wherever such usage is unambiguous, and with the implication that $\theta = \lambda \in \mathbb{R}_+$. Furthermore, we will sometimes express the set θ explicitly, *e.g.* $\mathcal{S}_{\alpha,\beta}$, where the usage is unambiguous.

Definition 2.4. A *data discrepancy functional* $\mathcal{L}: Y \times Y \to \mathbb{R}$ is a scalar quantification of the similarity between two elements of data space Y.

The data discrepancy functional is considered to be a data fitting term in variational approaches to inverse problems. Although often taken to be a metric on data space, choosing it as an affine transformation of the negative log-likelihood of data allows for a statistical interpretation, since minimizing $f \mapsto \mathcal{L}(\mathcal{A}(f), g)$ amounts to finding a maximum likelihood solution. For Gaussian observational noise $e \sim N(0, \Gamma)$, the corresponding data discrepancy is then given by the Mahalanobis distance:

$$\mathcal{L}(g, v) := \|g - v\|_{\Gamma^{-1}}^2 \quad \text{for } g, v \in Y. \tag{2.8}$$

If data are Poisson-distributed, *i.e.* $g \sim \text{Poisson}(\mathcal{A}(f_{\text{true}}))$, then an appropriate data discrepancy functional is the Kullback–Leibler (KL) divergence. When elements in data space Y are real-valued functions on $\mathbb{M}$, then the KL divergence becomes

$$\mathcal{L}(g, v) := \int_{\mathbb{M}} [v(y) \log g(y) - \log v(y) - g(y) + v(y)] \, dy \quad \text{for } g, v \in Y. \tag{2.9}$$

Similarly, Laplace-distributed observational noise corresponds to a data discrepancy that is given by the 1-norm. One can also express the data log-likelihood for Poisson-distributed data with an additive observational noise term that is Gaussian, but the resulting expressions are quite complex (Benvenuto *et al.* 2008).

Definition 2.5. A *reconstruction operator* $\mathcal{R} \colon Y \to X$ is a mapping which gives a point estimate $\hat{f}$ as the solution to (2.1).

Henceforth we will use $\mathcal{R}_\theta$ to denote a reconstruction operator that depends on a parameter set $\theta \in \Theta$; in particular we will use θ as parameters that will be *learned*.

Remark 2.6. In variational approaches we will typically use the notation $\mathcal{R}_\lambda$ to denote an operator parametrized by a single scalar $\lambda > 0$ which corresponds (explicitly or implicitly) to optimizing a functional that includes a regularization penalty $\lambda \mathcal{S}$ as in Definition 2.2. More generally θ will also be used to define the parameters of an algorithm or neural network used to generate a solution f_θ that may or may not explicitly specify a regularization functional. Again we will assume the context provides an unambiguous justification for the choice between $\mathcal{R}_\theta$ and $\mathcal{R}_\lambda$. Furthermore, we will sometimes express the set θ explicitly, *e.g.* $\mathcal{R}_{\mathcal{W}, \psi}$, where the usage is unambiguous.

2.5. Classical Tikhonov regularization

Tikhonov (or Tikhonov–Phillips) regularization is arguably the most prominent technique for inverse problems. It was introduced by Tikhonov (1943, 1963), Phillips (1962) and Tikhonov and Arsenin (1977) for solving ill-posed inverse problems of the form (2.1), and can be stated in the form

$$\mathcal{R}_\lambda(g) := \underset{f \in X}{\arg\min} \left\{ \frac{1}{2} \|\mathcal{A}(f) - g\|^2 + \lambda \mathcal{S}(f) \right\}. \tag{2.10}$$

Bearing in mind the notation in Remarks 2.3 and 2.6, note that (2.10) has the form of (2.7) with $\mathcal{L}$ given by the squared L^2-distance. Here X and Y are Hilbert spaces (typically both are L^2 spaces). Moreover, the choice $\mathcal{S}(f) := \frac{1}{2}\|f\|^2$ is the most common one for the penalty term in (2.10). In fact, if $\mathcal{A}$ is linear, with $\mathcal{A}^*$ denoting its adjoint, then standard arguments for minimizing quadratic functionals yield

$$\mathcal{R}_\lambda = (\mathcal{A}^* \circ \mathcal{A} + \lambda\,\mathrm{id})^{-1} \circ \mathcal{A}^*. \qquad (2.11)$$

Until the late 1980s the analytical investigations of regularization schemes of the type in (2.10) were restricted either to linear operators or to rather specific approaches for selected non-linear problems. Many inverse problems, such as parameter identification in linear differential operators, lead to non-linear parameter-to-state maps, so the corresponding forward operator becomes non-linear (Arridge and Schotland 2009, Greenleaf, Kurylev, Lassas and Uhlmann 2007, Bal, Chung and Schotland 2016, Jin and Maass 2012b, Jin and Maass 2012a).

Analysis of (2.10) for non-linear forward operators is difficult, for example, singular value decompositions are not available. A major theoretical breakthrough came with the publications of Seidman and Vogel (1989) and Engl *et al.* (1989), which extended the theoretical investigation of Tikhonov regularization to the non-linear setting by introducing radically new concepts. Among others, it extended the notion of minimum norm solution used in theorems dealing with convergence rates to f_0-minimum norm solutions. This means one assumes the knowledge of some meaningful parameter $f_0 \in X$ and the aim of the regularization method is to approximate a solution that in the limit minimizes $\|f - f_0\|$ amongst all solutions of $\mathcal{A}(f) = g$. Hence, f_0 acts as a kind of prior. Among the main results of Engl *et al.* (1989) is a theorem that estimates the convergence rate assuming sufficient regularity of the forward operator $\mathcal{A}$ and a source condition that relates the penalty term to the functional $\mathcal{A}$ at f_{true}. This theorem, which is stated below, opened the path to many successful applications, particularly for parameter identification problems related to partial differential equations, and such assumptions occur in different variations in all theorems related to the variational approach.

Theorem 2.7. Consider the inverse problem in (2.1) where $\mathcal{A}\colon X \to Y$ is continuous, weakly sequentially closed and with a convex domain. Next, assume there exists a f_0-minimum norm solution f_{true} for some fixed $f_0 \in X$ and let data $g \in Y$ in (2.1) satisfy $\|\mathcal{A}(f_{\mathrm{true}}) - g\| \le \delta$. Also, let $f_\lambda^\delta \in X$ denote a minimizer of (2.10) with $\mathcal{S}(f) := \frac{1}{2}\|f\|^2$. Finally, assume the following.

(1) $\mathcal{A}$ has a continuous Fréchet derivative.

(2) There exists a $\gamma > 0$, such that

$$\|[\partial\mathcal{A}(f_{\text{true}})] - [\partial\mathcal{A}(f)]\|_{L(X,Y)} \leq \gamma\|f_{\text{true}} - f\|_X$$

for all $f \in \text{dom}(\mathcal{A}) \cap B_r(f_{\text{true}})$ with $r > 2\|f_{\text{true}} - f_0\|$. Here, $L(X,Y)$ is the vector space of Y-valued linear maps on X and $B_r(f_{\text{true}}) \subset X$ denotes a ball of radius r around f_{true}.

(3) There exists $v \in Y$ with $\gamma\|v\| < 1$ such that $f_{\text{true}} - f_0 = [\partial\mathcal{A}(f_{\text{true}})]^*(v)$.

Then, choosing $\lambda \propto \delta$ as $\delta \to 0$ yields

$$\|f_\lambda^\delta - f_{\text{true}}\| = O(\sqrt{\delta}) \quad \text{and} \quad \|\mathcal{A}(f_\lambda^\delta) - g\| = O(\delta).$$

2.6. Extension of classical Tikhonov regularization

Tikhonov regularization is a particular case of the more general variational regularization schemes (2.7). In fact, penalty terms for Tikhonov-type functionals, as well as suitable source conditions, have been generalized considerably in a series of papers. A key issue in solving an inverse problem is to use a forward operator $\mathcal{A}$ that is sufficiently accurate. It is also important to choose an appropriate data discrepancy $\mathcal{L}$, regularizer $\mathcal{S}_\theta$, and to have a parameter choice rule for setting θ. Depending on these choices, different reconstruction results are obtained.

The data discrepancy. Here the choice is ideally guided by statistical considerations for the observation noise (Bertero, Lantéri and Zanni 2008). Ideally one selects $\mathcal{L}$ as an appropriate affine transform of the negative log-likelihood of data, in which case minimizing $f \mapsto \mathcal{L}(\mathcal{A}(f), g)$ becomes the same as computing an maximum likelihood estimator. Hence, Poisson-distributed data that typically appear in photography (Costantini and Susstrunk 2004) and emission tomography applications (Vardi, Shepp and Kaufman 1985) lead to a data discrepancy given by the Kullback–Leibler divergence (Sawatzky, Brune, Müller and Burger 2009, Hohage and Werner 2016), while additive normally distributed data, as for Gaussian noise, result in a least-squares fit model.

The regularizer. As stated in Definition 2.2, $\mathcal{S}$ acts as a penalizer and is chosen to enforce stability by encoding *a priori* information about f_{true}. How to set the (regularization) parameter θ reflects noise level in data: see Section 4.1.

Classical Tikhonov regularization (2.10) uses Hilbert-space norms (or semi-norms) to regularize the inverse problems. In more recent years, Banach-space regularizers have become more popular in the context of sparsity-promoting and discontinuity-preserving regularization, which are revisited in Section 2.7. TV regularization was introduced by Rudin, Osher and Fatemi (1992) for image denoising due to its edge-preserving properties, favouring images f that have a sparse gradient. Here, the TV regularizer is given as

$$\mathcal{S}(f) := \mathrm{TV}(f) := |Df|(\Omega) = \int_\Omega \mathrm{d}|Df|, \qquad (2.12)$$

where $\Omega \subset \mathbb{R}^d$ is a fixed open and bounded set. The above functional (TV regularizer) uses the total variation measure of the distributional derivative of f defined on Ω (Ambrosio, Fusco and Pallara 2000). A drawback of using such a regularization procedure is apparent as soon as the true model parameter not only consists of constant regions and jumps but also possesses more complicated, higher-order structures, *e.g.* piecewise linear parts. In this case, TV introduces jumps that are not present in the true solution, which is referred to as staircasing (Ring 2000). Examples of generalizations of TV for addressing this drawback typically incorporate higher-order derivatives, *e.g.* total generalized variation (TGV) (Bredies, Kunisch and Pock 2011) and the infimal-convolution total variation (ICTV) model (Chambolle and Lions 1997). These read as

$$
\begin{aligned}
\mathcal{S}_{\alpha,\beta}(f) &:= \mathrm{ICTV}_{\alpha,\beta}(f) \\
&= \min_{\substack{v \in W^{1,1}(\Omega) \\ \nabla v \in BV(\Omega)}} \{\alpha \|Df - \nabla v\|_{\mathcal{M}(\Omega;\mathbb{R}^2)} + \beta \|D\nabla v\|_{\mathcal{M}(\Omega;\mathbb{R}^{2\times2})}\},
\end{aligned}
\qquad (2.13)
$$

and the second-order TGV (Bredies and Valkonen 2011, Bredies, Kunisch and Valkonen 2013) reads as

$$
\begin{aligned}
\mathcal{S}_{\alpha,\beta}(f) &:= \mathrm{TGV}^2_{\alpha,\beta}(f) \\
&= \min_{w \in BD(\Omega)} \{\alpha \|Df - w\|_{\mathcal{M}(\Omega;\mathbb{R}^2)} + \beta \|Ew\|_{\mathcal{M}(\Omega;\mathrm{Sym}^2(\mathbb{R}^2))}\}.
\end{aligned}
\qquad (2.14)
$$

Here

$$BD(\Omega) := \{w \in L^1(\Omega;\mathbb{R}^d) \mid \|Ew\|_{\mathcal{M}(\Omega;\mathbb{R}^{d\times d})} < \infty\}$$

is the space of vector fields of bounded deformation on Ω with E denoting the *symmetrized gradient* and $\mathrm{Sym}^2(\mathbb{R}^2)$ the space of symmetric tensors of order 2 with arguments in $\mathbb{R}^2$. The parameters α, β are fixed positive parameters. The main difference between (2.13) and (2.14) is that we do not generally have $w = \nabla v$ for any function v. That results in some qualitative differences of ICTV and TGV regularization: see *e.g.* Benning, Brune, Burger and Müller (2013). One may also consider Banach-space norms other than TV, such as Besov norms (Lassas, Saksman and Siltanen 2009), which behave more nicely with respect to discretization (see also Section 3.4). Different TV-type regularizers and their adaption to data by bilevel learning of parameters (*e.g.* α and β in ICTV and TGV) will be discussed in more detail in Section 4.3.1 and numerical results will be given in Section 7.2.

Finally, in applied harmonic analysis, ℓ_p-norms of wavelets have been proposed as regularizers (Daubechies *et al.* 1991, Mallat 2009, Unser and Blu 2000, Kutyniok and Labate 2012, Eldar and Kutyniok 2012, Foucart and Rauhut 2013). Other examples are non-local regularization (Gilboa and Osher 2008, Buades, Coll and Morel 2005), anisotropic regularizers (Weickert 1998) and, in the context of free discontinuity problems, the representation of images as a composition of smooth parts separated by edges (Blake and Zisserman 1987, Mumford and Shah 1989, Carriero, Leaci and Tomarelli 1996).

2.7. *Sparsity-promoting regularization*

Sparsity is an important concept in conceiving inversion models as well as learning parts of them. In what follows we review some of the main approaches to computing sparse solutions, and postpone learning sparse representations to Section 4.4.

2.7.1. *Notions of sparsity*

Let X be a separable Hilbert space, that is, we will assume that it has a countable orthonormal basis. A popular approach to sparse reconstruction uses the notion of a *dictionary* $\mathbb{D} := \{\phi_i\} \subset X$, whose elements are called atoms. Here, $\mathbb{D}$ is either given, *i.e.* knowledge-driven, or data-driven and derived from a set of realizations $f_i \in X$: see Bruckstein, Donoho and Elad (2009), Daubechies *et al.* (2004), Rubinstein, Bruckstein and Elad (2010), Lanusse, Starck, Woiselle and Fadili (2014) and Chen and Needell (2016).

A special class of dictionaries is that of *frames*. A dictionary $\mathbb{D} := \{\phi_i\}$ is a frame if there exists $C_1, C_2 > 0$ such that

$$C_1\|f\|^2 \le \sum_i |\langle f, \phi_i \rangle|^2 \le C_2\|f\|^2 \quad \text{for any } f \in X. \tag{2.15}$$

A frame is called *tight* if $C_1 = C_2 = 1$; it is called *over-complete* or *redundant* if $\mathbb{D}$ does not form a basis for X. Redundant dictionaries, *e.g.* translation-invariant wavelets, often work better than non-redundant dictionaries: see Peyré, Bougleux and Cohen (2011) and Elad (2010).

To construct sparse representations, *i.e.* sparsity-promoting regularizers from such a parametrization, we need the notions of an analysis and a synthesis operator. Given a dictionary $\mathbb{D}$, the analysis operator $\mathcal{E}_{\mathbb{D}} \colon X \to \Xi$ maps an element in X to a sequence in Ξ, which is typically in ℓ^2, such that

$$\boldsymbol{\xi} = \mathcal{E}_{\mathbb{D}}(f) \in \Xi \text{ has components } \xi_i = \langle f, \phi_i \rangle.$$

The corresponding synthesis operator $\mathcal{E}_{\mathbb{D}}^* \colon \Xi \to X$ is the adjoint of the analysis operator, that is,

$$\mathcal{E}_{\mathbb{D}}^*(\boldsymbol{\xi}) := \sum_i \xi_i \phi_i. \tag{2.16}$$

We further define the frame operator as $\mathcal{E}_{\mathbb{D}}^* \circ \mathcal{E}_{\mathbb{D}} \colon X \to X$, that is,

$$\mathcal{E}_{\mathbb{D}}^* \circ \mathcal{E}_{\mathbb{D}}(f) := \sum_i \langle f, \phi_i \rangle \phi_i. \tag{2.17}$$

Now, we can define an $f \in X$ to be s-sparse with respect to $\mathbb{D}$ if

$$\|\mathcal{E}_{\mathbb{D}}(f)\|_0 = \#\{i \,|\, \langle f, \phi_i \rangle \neq 0\} \leq s. \tag{2.18}$$

In most applications, the model parameter is not sparse in the strict sense, which leads to the weaker notion of compressibility. A model parameter $f \in X$ is compressible with respect to $\mathbb{D}$ if the following power decay law holds:

$$|\widetilde{\xi}_k| \leq C k^{-1/q} \text{ as } k \to \infty \text{ for some } C > 0 \text{ and } 0 < q < 1. \tag{2.19}$$

Here, $\widetilde{\boldsymbol{\xi}}$ is a non-increasing rearrangement of the sequence $\boldsymbol{\xi} = \{\langle f, \phi_i \rangle\} = \mathcal{E}_{\mathbb{D}}(f)$. Note that sparse signals are compressible, and in particular, if q is small in (2.19), then compressibility is equivalent to sparsity from any practical viewpoint. We further define $\boldsymbol{\xi}_s$ to consist of the s largest (in magnitude) coefficients of the sequence $\widetilde{\boldsymbol{\xi}}$.

2.7.2. Sparse recovery

Here we consider solving (2.1) when our prior model assumes f_{true} to be compressible with respect to a given dictionary $\mathbb{D} := \{\phi_i\}$. Then sparse recovery of f can be done by either the synthesis approach (*i.e.* sparse coding) or the analysis approach. In the synthesis approach the reconstruction operator $\mathcal{R}_\theta$ is given as

$$\mathcal{R}_\theta(g) := \mathcal{E}_{\mathbb{D}}^*(\hat{\xi}), \tag{2.20}$$

where

$$\hat{\xi} = \underset{\boldsymbol{\xi} \in \Xi}{\arg\min} \{\mathcal{L}(\mathcal{A}(\mathcal{E}_{\mathbb{D}}^*(\boldsymbol{\xi})), g) + \lambda \|\boldsymbol{\xi}\|_0\},$$

with $\theta = \{\mathbb{D}, \lambda\}$, *i.e.* θ is the scalar $\lambda > 0$ and the entire dictionary that defines the synthesis operator $\mathcal{E}_{\mathbb{D}}^*$. In the corresponding analysis approach, we get

$$\mathcal{R}_\theta(g) := \underset{f \in X}{\arg\min} \{\mathcal{L}(\mathcal{A}(f), g) + \lambda \|\mathcal{E}_{\mathbb{D}}(f)\|_0\}. \tag{2.21}$$

If $\mathbb{D}$ is an orthonormal basis then synthesis and analysis formulations are equivalent.

There are different strategies for numerically computing sparse representations that solve (2.20) and (2.21).

Greedy approaches. These build up an approximation in a greedy fashion, computing one non-zero entry of $\hat{\xi}$ at a time by making locally optimal choices at each step. One example of a greedy approach is iterative (hard) thresholding (Blumensath and Davies 2008, Blumensath 2013, Foucart 2016), where for an initial guess $\boldsymbol{\xi}^{(i0}$ and $i = 0, 1, \ldots$ one iterates

$$\boldsymbol{\xi}^{(i+1)} = T_s(\boldsymbol{\xi}^{(i)} - \mathcal{A}_{\mathbb{D}}^*(\mathcal{A}_{\mathbb{D}}(\boldsymbol{\xi}^{(i)}) - g)), \tag{2.22}$$

where $\mathcal{A}_{\mathbb{D}} := \mathcal{A} \circ \mathcal{E}_{\mathbb{D}}^*$. Here, $T_s(\boldsymbol{\xi})$ sets all but the largest (in magnitude) s elements of $\boldsymbol{\xi}$ to zero. This is therefore a proximal-gradient method with the proximal of the function being 0 at 0 and 1 everywhere else (see Section 8.2.7). Other examples are matching pursuit (MP) (Mallat and Zhang 1993), orthogonal matching pursuit (OMP) (Tropp and Gilbert 2007) and variants thereof such as StOMP (Donoho, Tsaig, Drori and Starck 2012), ROMP (Needell and Vershynin 2009) and CoSamp (Needell and Tropp 2009).

Convex relaxation. One of the most common approaches to solving sparse recovery is to replace the ℓ_0-(semi)norm with the ℓ_1-norm in (2.20) and (2.21). This leads to basis pursuit (Candès, Romberg and Tao 2006), also called Lasso in the statistics literature (Tibshirani 1996). The optimization literature for solving the resulting ℓ_1-type problems is vast. A few examples are interior-point methods (Candès *et al.* 2006, Kim *et al.* 2007), projected gradient methods (Figueiredo, Nowak and Wright 2007), iterative soft thresholding (see Section 8.2.7) (Daubechies *et al.* 2004, Fornasier and Rauhut 2008) and fast proximal gradient methods (FISTA and variants) (Bubeck 2015), to name just a few.

Combinatorial algorithms. A third class of approaches for sparse coding is that of combinatorial algorithms. They are particularly suitable when acquiring highly structured samples of the signal so that rapid reconstruction via group testing is efficient. This class of approaches includes Fourier sampling, chaining pursuit and HHS pursuit (Berinde *et al.* 2008).

The above approaches for solving (2.20) and (2.21) all have their advantages and disadvantages. First of all, greedy methods will generally not give the same solution as convex relaxation. However, if the restricted isometry property (RIP) from Section 2.7.3 holds, then both approaches have the same solution. Convex relaxation has the advantage that it succeeds with a very small number of possibly noisy measurements. However, their numerical solution tends to be computationally burdensome. Combinatorial algorithms, on the other hand, can be extremely fast (sublinear in the length of the target signal) but they require a very specific structure of the forward operator $\mathcal{A}$ and a large number of samples. The performance of greedy methods falls in between those of convex relaxation and combinatorial algorithms in their run-time and sampling efficiency.

2.7.3. Error estimates for sparse recovery

When the true model parameter is compressible, then it is possible to estimate the error committed by performing sparse recovery. Such estimates have been derived in the finite-dimensional setting when the matrix representing the linear forward operator satisfies the RIP:

$$(1 - \epsilon_s)\|f\|_2^2 \leq \|\mathcal{A}(f)\|_2^2 \leq (1 + \epsilon_s)\|f\|_2^2 \quad \text{for all } s\text{-sparse } f \in X, \quad (2.23)$$

for sufficiently small $\epsilon_s > 0$. Then we have the following error estimate for sparse recovery.

Theorem 2.8 (Candès, Romberg and Tao 2006). Let $\mathcal{A}\colon \mathbb{R}^n \to \mathbb{R}^m$ be a linear mapping whose matrix satisfies the RIP. If $g = \mathcal{A}(f_{\mathrm{true}}) + e$ with $\|e\| \leq \delta$ and

$$\hat{f}_\delta := \underset{f \in X}{\arg\min} \|f\|_1 \quad \text{subject to } \|\mathcal{A}(f) - g\|_2 \leq \delta, \quad (2.24)$$

then

$$\|\hat{f}_\delta - f_{\mathrm{true}}\|_2 \leq C\left[\delta + \frac{\|f_{\mathrm{true}} - f_{\mathrm{true}}^{(s)}\|_2}{\sqrt{s}}\right]. \quad (2.25)$$

In the above, $f_{\mathrm{true}}^{(s)} \in \mathbb{R}^n$ is a vector consisting of the s largest (in magnitude) coefficients of f_{true} and zeros otherwise.

Examples of matrices satisfying RIP are sub-Gaussian matrices and partial bounded orthogonal matrices (Chen and Needell 2016). Theorem 2.8 states that the reconstruction error is at most proportional to the norm of the noise in the data plus the *tail* $f_{\mathrm{true}} - f_{\mathrm{true}}^{(s)}$ of the signal. Cohen, Dahmen and DeVore (2009) show that this error bound is optimal (up to the precise value of C). Moreover, if f_{true} is s-sparse and $\delta = 0$ (noise-free data), then f_{true} can be reconstructed exactly. Furthermore, if f_{true} is compressible with (2.19), then

$$\|\hat{f}_\delta - f_{\mathrm{true}}\|_2 \leq C(\delta + C's^{1/2-1/q}). \quad (2.26)$$

Finally, error estimates of the above type have been extended to the infinite-dimensional setting in Adcock and Hansen (2016).

The choice of dictionary is clearly a central topic in sparsity-promoting regularization and, as outlined in Section 4.4, the dictionary can be learned beforehand or jointly alongside the signal recovery.

2.7.4. Error estimates for convex relaxation

Applying the convex relaxation to the analysis approach of (2.21) yields the regularized Tikhonov functional

$$\mathcal{R}_\theta(g) := \underset{f \in X}{\arg\min}\{\mathcal{L}(\mathcal{A}(f), g) + \lambda\|\mathcal{E}_{\mathbb{D}}(f)\|_p\} \quad (2.27)$$

with $1 \leq p < 2$. In the context of ill-posed inverse problems this functional was introduced and analysed in the ground-breaking paper by Daubechies *et al.* (2004). We emphasize that this analysis holds in infinite-dimensional function spaces, and does not depend on the finite-dimensional concepts used in compressive sampling or finite-dimensional sparse recovery concepts.

Since then, this model has been studied intensively, and in particular the case $p = 1$. Similar to Theorem 2.7, error estimates and regularizing properties, such as existence of minimizers, well-posedness, stability, convergence rates and error estimates, have been obtained for linear and non-linear operators: see *e.g.* Scherzer *et al.* (2009). Minimizers of this functional with $p = 1$ are indeed sparse even if the true solution is not. Using the notion of Bregman distances, such sparsity-promoting approaches have been extended to rather general choices of data discrepancy and regularizers. For a more complete introduction, see Chan and Shen (2006), Scherzer *et al.* (2009) and Bredies *et al.* (2011) and the recent survey by Benning and Burger (2018, Section 2).

3. Statistical regularization

Statistical regularization, and Bayesian inversion in particular, is a complete statistical inferential methodology for inverse problems. It offers a rich set of tools for incorporating data into the recovery of the model parameter, so it is a natural framework to consider when data-driven approaches from machine learning are to be used for solving ill-posed inverse problems.

A key element is to treat *both* the data and model parameter as realizations of certain random variables and phrase the inverse problem as a statistical inference question. In contrast, the functional analytic viewpoint (Section 2) allows for data to be interpreted as samples generated by a random variable, but there are no statistical assumptions on the model parameters.

Remark 3.1. In functional analytic regularization, a statistical model for data is mostly used to justify the choice of data discrepancy in a variational method and for selecting an appropriate regularization parameter. Within functional analytic regularization, one can more carefully account for statistical properties of data which can be useful for uncertainty quantification (Bissantz, Hohage, Munk and Ruymgaart 2007).

Bayesian statistics offers a natural setting for such a quest since it is natural to interpret measured data in an inverse problem as a sample of a random variable conditioned on data whose distribution is the data likelihood. The data likelihood can often be derived using knowledge-driven modelling. Solving an inverse problem can then be stated as finding the distribution of the model parameter conditioned on data (posterior distribution). The posterior describes all possible solutions given measured data, so in particular it provides an estimate of the statistical uncertainty of the solution that can be used for uncertainty quantification. *Many of the challenges in Bayesian inversion are associated with realizing these advantages without having access to the full posterior.* In particular, designing a 'good' prior and to have a computationally feasible means for exploring the posterior is essential for implementing and using Bayesian inversion. This, along with investigating its regularizing properties, drives much of the research in Bayesian inversion.

Motivated by the above, the focus of this brief survey is on Bayesian inversion. As one would expect, the theory was first developed in the finite-dimensional setting, *i.e.* when both model parameter and space and data spaces are finite-dimensional. Important early developments were made in the geophysics community (Tarantola and Valette 1982, Tarantola 2005), which had a great impact in the field; see also Calvetti and Somersalo (2017, Section 2) for a brief historical survey. Nice surveys of the finite-dimensional theory are given in Kaipio and Somersalo (2005) and Calvetti and Somersalo (2017), where many further references can be found.

Our focus is primarily on later developments that deal with Bayesian inversion in the infinite-dimensional (non-parametric) setting. Early work in this direction can be found in Mandelbaum (1984) and Lehtinen, Päivärinta and Somersalo (1989). Our brief survey in Sections 3.1, 3.2.1 and 3.2.2 is based on the excellent survey papers by Evans and Stark (2002), Stuart (2010) and Dashti and Stuart (2017). The sections concerning convergence (Section 3.2.3), convergence rates (Section 3.2.3) and characterization of the Bayesian posterior (Section 3.2.5) are based on the excellent short survey by Nickl (2017*b*).

3.1. Basic notions for Bayesian inversion

The vector spaces X and Y play the same role as in functional analytic regularization (Section 2), that is, elements in X represent model parameters and elements in Y represent data. For technical reasons, we also assume that both X and Y are separable Banach spaces. Both these spaces are also equipped with a Borel σ-algebra, and we let $\mathscr{P}_Y$ and $\mathscr{P}_Y$ denote the class of probability measures on X and Y, respectively. We also assume there exists a $(X \times Y)$-valued random variable $(\mathbb{f}, \mathbb{g}) \sim \mu$ that is distributed according to some joint law μ. Here, $\mathbb{f}$ generates elements in X (model parameters) and $\mathbb{g}$ generates elements in Y (data).

Remark 3.2. In this setting, integrals over X and/or Y, which are needed for defining expectation, are interpreted as a Bochner integral that extends the Lebesgue integral to functions that take values in a Banach space. See

Dashti and Stuart (2017, Section A.2) for a brief survey of Banach and Hilbert space-valued random variables.

3.1.1. The data model

A key assumption is that the conditional distribution of $(g \mid f = f) \sim \Pi^{f}_{\text{data}}$ exists (data likelihood) under the joint law μ for any $f \in X$. This allows us to define the *data model*, which is the statistical analogue of the forward operator, as the following $\mathscr{P}_{Y}$-valued mapping defined on X:

$$f \mapsto \Pi^{f}_{\text{data}} \quad \text{for any } f \in X. \tag{3.1}$$

Remark 3.3. The existence of regular conditional distributions can be ensured under very general assumptions. In particular, let f be an X-valued random variable and g a Y-valued random variable, where X is a Polish space, *i.e.* a complete and separable metric space, and Y is a general measurable space. Then there exists a regular conditional distribution of the conditional random variable $(f \mid g)$ (Kallenberg 2002, Theorem 6.3). In particular, when both X and X are Polish spaces, then both $(f \mid g)$ and $(g \mid f)$ exist.

The most common data model is the statistical analogue of (2.1) where the model parameter is allowed to be a random variable (Kallenberg 2002, Lemma 1.28 and Corollary 3.12):

$$g = \mathcal{A}(f) + e. \tag{3.2}$$

Here, $\mathcal{A} \colon X \to Y$ is the same forward operator as in (2.1), which models how data are generated in the absence of noise. Likewise, $e \sim \Pi_{\text{noise}}$, with $\Pi_{\text{noise}} \in \mathscr{P}_{Y}$ known, is the random variable that generates the observation noise. If e is independent from f, then (3.2) amounts to the data model

$$\Pi^{f}_{\text{data}} = \delta_{\mathcal{A}(f)} \circledast \Pi_{\text{noise}} = \Pi_{\text{noise}}(\cdot - \mathcal{A}(f)) \quad \text{for any } f \in X, \tag{3.3}$$

where $\circledast$ denotes convolution between measures.

Remark 3.4. Another common data model is when Π^{f}_{data} is a Poisson random measure on Y with mean equal to $\mathcal{A}(f)$ (Hohage and Werner 2016, Streit 2010, Vardi *et al.* 1985, Besag and Green 1993). This is suitable for modelling statistical properties of low-dose imaging data, such as data that is measured in line of response PET (Kadrmas 2004, Calvetti and Somersalo 2008, Section 3.2 of Natterer and Wübbeling 2001) and variants of fluorescence microscopy (Hell, Schönle and Van den Bos 2007, Diaspro *et al.* 2007).

3.1.2. The inverse problem

Following Evans and Stark (2002), a *(statistical) inverse problem* requires us to perform the task of recovering the posterior given a single sample (measured data) from the data model with unknown true model parameter. A more precise statement reads as follows.

Definition 3.5. A *statistical inverse problem* is the task of recovering the conditional distribution $\Pi^g_{\text{post}} \in \mathscr{P}_Y$ of $(\mathbb{f} \mid \mathbb{g} = g)$ under μ from measured data $g \in Y$, where

$$g \text{ is a single sample of } (\mathbb{g} \mid \mathbb{f} = f_{\text{true}}) \sim \Pi^{f_{\text{true}}}_{\text{data}}. \tag{3.4}$$

Here $f_{\text{true}} \in X$ is unknown while $f \mapsto \Pi^f_{\text{data}}$, which describes how data are generated, is known.

The conceptual difference that comes from adopting such a statistical view brings with it several potential advantages. The posterior, assuming it exists, describes all possible solutions, so recovering it represents a more complete solution to the inverse problem than recovering an approximation of f_{true}, which is the goal in functional analytic regularization. This is particularly the case when one seeks to quantify the uncertainty in the recovered model parameter in terms of statistical properties of the data. However, recovering the entire posterior is often not feasible, such as in inverse problems that arise in imaging. As an alternative, one can settle for exploring the posterior by computing suitable estimators (Section 3.3). Some may serve as approximations of f_{true} whereas others are designed for quantifying the uncertainty.

3.1.3. Bayes' theorem

A key part of solving the above inverse problem is to utilize a relation between the unknown posterior that one seeks to recover and the known data likelihood. Such a relation is given by Bayes' theorem.

A very general formulation of Bayes' theorem is given by Schervish (1995, Theorem 1.31). For simplicity we consider the formulation that holds for the special case where the data likelihood is given as in (3.2), with $\mathcal{A}\colon X \to Y$ a measurable map. Assume furthermore that $\mathbb{f} \sim \Pi_{\text{prior}}$ and $\mathbb{e} \sim \Pi_{\text{noise}}$, with $\mathbb{e}$ independent of $\mathbb{f}$. Then the data model is given as in (3.3), that is, at f it yields the translate of Π_{noise} by $\mathcal{A}(f)$.

Assume next that $\Pi^f_{\text{data}} \ll \Pi_{\text{noise}}$ (*i.e.* Π^f_{data} is absolutely continuous with respect to Π_{noise}) Π_{noise}-almost surely for all $f \in X$, so there exists some measurable map $\Phi\colon X \times Y \to \mathbb{R}$ (potential) such that

$$\frac{\mathrm{d}\Pi^f_{\text{data}}}{\mathrm{d}\Pi_{\text{noise}}}(g) = \exp(-\mathcal{L}(f, g)) \quad \text{for all } f \in X. \tag{3.5}$$

with

$$\mathbb{E}_{\mathbb{g} \sim \Pi_{\text{noise}}}[\exp(-\mathcal{L}(f, \mathbb{g}))] = 1.$$

The mapping $f \mapsto -\mathcal{L}(f, g)$ is called the (*data*) *log-likelihood* for the data $g \in Y$.

Remark 3.6. The equality for the Radon–Nikodym derivative in (3.5) means that

$$\mathbb{E}_{\mathrm{g}\sim\Pi_{\mathrm{data}}^{f}}[F(\mathrm{g})] = \mathbb{E}_{\mathrm{g}\sim\Pi_{\mathrm{noise}}}[\exp(-\mathcal{L}(f, \mathrm{g}))F(\mathrm{g})]$$

holds for any measurable $F \colon Y \to \mathbb{R}$.

Finally, assume $\mathcal{L}$ is μ_0-measurable, where $\mu_0 := \Pi_{\mathrm{prior}} \otimes \Pi_{\mathrm{noise}}$ and $\mu \ll \mu_0$, which means in particular that the joint law $(\mathbb{f}, \mathrm{g}) \sim \mu$ can be written as

$$\frac{\mathrm{d}\mu}{\mathrm{d}\mu_0}(f, g) = \exp(-\mathcal{L}(f, g)) \quad \text{for } (f, g) \in X \times Y.$$

Bearing in mind the above assumptions, we can now state Bayes' theorem (Dashti and Stuart 2017, Theorem 14), which expresses the posterior in terms of the data likelihood and the prior.

Theorem 3.7 (Bayes' theorem). The normalization constant $Z \colon Y \to \mathbb{R}$ is defined by

$$Z(g) := \mathbb{E}_{\mathbb{f}\sim\Pi_{\mathrm{prior}}}[\exp(-\mathcal{L}(\mathbb{f}, g))] \tag{3.6}$$

and we assume $Z(g) > 0$ holds Π_{noise}-almost surely for $g \in Y$. Then the posterior Π_{post}^{g}, which is the conditional distribution of $(\mathbb{f} \mid \mathrm{g} = g)$, exists under μ and it is absolutely continuous with respect to the prior, *i.e.* $\Pi_{\mathrm{post}}^{g} \ll \Pi_{\mathrm{prior}}$. Furthermore,

$$\frac{\mathrm{d}\Pi_{\mathrm{post}}^{g}}{\mathrm{d}\Pi_{\mathrm{prior}}}(f) = \frac{1}{Z(g)}\exp(-\mathcal{L}(f, g)) \tag{3.7}$$

holds μ_0-almost surely for $(f, g) \in X \times Y$.

Bayes' theorem is the basis for Bayesian inversion, where one seeks to solve the statistical inverse problem assuming access to *both* a prior and a data likelihood, whereas $f_{\mathrm{true}} \in X$ remains unknown. The data likelihood is given by the data model, which is in turn derived from knowledge about how data are generated. The choice of prior, however, is more subtle: it needs to act as a regularizer, and ideally it also encodes subjective prior beliefs about the unknown model parameter f_{true} by giving high probability to model parameters similar to f_{true} and low probability to other 'unnatural' model parameters. A brief survey of hand-crafted priors is provided in Section 3.4.

3.2. Regularization theory for Bayesian inversion

In functional analytic regularization, existence, stability and convergence are necessary if a reconstruction method is to be a regularization (see

Section 2.4). Moreover, a mathematical analysis also seeks to provide convergence rates and stability estimates. There is an ongoing effort to develop a similar theory for Bayesian inversion.

Methods for functional analytic regularization of ill-posed inverse problems typically regularize by a variational procedure or a spectral cut-off. In Bayesian inversion, regularization is mainly through the choice of an appropriate prior distribution. Since many different priors can serve as regularizers, a large portion of the theory seeks to characterize properties of Bayesian inversion methods that are independent of the prior. Much of the analysis in Sections 3.2.3–3.2.5 is therefore performed in the large-sample or small-noise limit, and under the assumption that data g is generated from $(\mathbf{g} \mid \mathbf{f} = f)$ where $f = f_{\mathrm{true}}$ (the true model parameter), instead of having f as a random sample of $\mathbf{f} \sim \Pi_{\mathrm{prior}}$ (prior).

Remark 3.8. The parametric setting refers to the case when the dimensions of X and Y are finite. In this context, a large-sample limit refers to an asymptotic analysis performed when the dimension of X is kept fixed and independent of the dimension of Y (sample size), which is allowed to grow. In the non-parametric setting, either both Y and X are infinite-dimensional from the outset, or one lets the dimension of X increase as the dimension of Y (sample size) increases.

3.2.1. Existence

Existence for Bayesian inversion follows when Bayes' theorem holds. Below we state the precise existence theorem (Dashti and Stuart 2017, Theorem 16) for the setting in Section 3.1.3, which covers the case when model parameter and data spaces are infinite-dimensional separable Banach spaces.

Theorem 3.9 (existence for Bayes inversion). Assume that $\mathcal{L}\colon X \times Y \to \mathbb{R}$ in (3.5) is continuously differentiable on some $X' \subset X$ that contains the support of the prior Π_{prior} and $\Pi_{\mathrm{prior}}(X' \cap B) > 0$ for some bounded set $B \subset X$. Also, assume there exists mappings $M_1, M_2\colon \mathbb{R}_+ \times \mathbb{R}_+ \to \mathbb{R}_+$ that are component-wise monotone, non-decreasing and where the following holds:

$$\begin{aligned}
-\mathcal{L}(f,g) &\le M_1(r, \|f\|), \\
|\mathcal{L}(f,g) - \mathcal{L}(f,v)| &\le M_2(r, \|f\|)\|g - v\|
\end{aligned} \tag{3.8}$$

for $f \in X$ and $g, v \in B_r(0) \subset Y$. Then, Z in (3.6) is finite, *i.e.* $0 < Z(g) < \infty$ for any $g \in Y$, and the posterior given by (3.7) yields a well-defined $\mathscr{P}_X$-valued mapping on $Y\colon g \mapsto \Pi^g_{\mathrm{post}}$.

Under certain circumstances it is possible to work with improper priors on X, for example by computing posterior distributions that approximate the posteriors one would have obtained using proper conjugate priors whose extreme values coincide with the improper prior.

3.2.2. Stability

One can show that small changes in the data lead to small changes in the posterior distribution (in Hellinger metric) on $\mathscr{P}_X$. The precise formulation given by Dashti and Stuart (2017, Theorem 16) reads as follows.

Theorem 3.10 (stability for Bayes inversion). Let the assumptions in Theorem 3.9 and assume in addition that

$$f \mapsto \exp(M_1(r, \|f\|))(1 + M_2(r, \|f\|))^2 \quad \text{is } \Pi_{\text{prior}}\text{-integrable on } X \quad (3.9)$$

for some fixed $r > 0$. Then

$$d_{\mathrm{H}}(\Pi_{\text{post}}^g, \Pi_{\text{post}}^v) \leq C(r)\|g + v\|$$

for some $C(r) > 0$ and $g, v \in B_r(0) \subset Y$. In the above, $d_{\mathrm{H}} \colon \mathscr{P}_X \times \mathscr{P}_X \to \mathbb{R}$ is the Hellinger metric (see Dashti and Stuart 2017, Definition 4).

Theorem 3.10 holds in particular when the negative log-likelihood of data is locally Hölder-continuous, which is the case for many standard probability distribution functions, for example when the negative log-likelihood is continuously differentiable.

By Theorem 3.10 the posterior is Lipschitz in the data with respect to the Hellinger metric, so rephrasing the inverse problem as the task of recovering the posterior instead of a model parameter acts as a regularization of an inverse problem that is ill-posed in the functional analytic sense. Note, however, that the above does not automatically imply that a particular estimator is continuous with respect to data. However, the Hellinger distance possesses the convenient property that continuity with respect to this metric implies the continuity of moments. Hence, a corollary to Theorem 3.10 is that *posterior moments, such as the mean and covariance, are continuous (Sprungk 2017, Corollary 3.23), that is, these estimators are regularizing.*

As a final note, one can also show that small changes in the data log-likelihood $\mathcal{L}$ in (3.5) lead to small changes in the posterior distribution, again in the Hellinger metric (Dashti and Stuart 2017, Theorem 18). This enables one to translate errors arising from inaccurate forward operator into errors in the Bayesian solution of the inverse problem, a topic that is also considered in Section 6.2.

3.2.3. Convergence

Posterior consistency is the Bayesian analogue of the notion of convergence in functional analytic regularization. More precisely, the requirement is that the posterior Π_{post}^g, where g is a sample of $(\mathsf{g} \mid \mathsf{f} = f_{\text{true}})$, concentrates in any small neighbourhood of the true model parameter $f_{\text{true}} \in X$ as information in data g increases indefinitely.[1] Intuitively, this means our knowledge about the model parameter becomes more accurate and precise as the amount of data increases indefinitely.

In the finite-dimensional setting, consistency of the posterior holds if and only if $f_{\text{true}} \in X$ is contained in the support of the prior Π_{prior} (Freedman 1963, Schwartz 1965) (provided the posterior is smooth with respect to the model parameter).

The situation is vastly more complex in the infinite-dimensional (non-parametric) setting. It is known that the posterior is consistent at every model parameter except possibly on a set of measure zero (Doob 1948, Breiman, Le Cam and Schwartz 1965), that is, Bayesian inversion is almost always consistent in the measure-theoretic sense. However, the situation changes if 'smallness' is measured in a topological sense, as shown by a classical counter-example in Freedman (1963) involving the simplest non-parametric problem where consistency fails. This is not a pathological counter-example: it is a generic property in the sense that most priors are 'bad' in a topological sense (Freedman 1965). In fact, consistency may fail for non-parametric models for very natural priors satisfying the support condition, which means even an infinite amount of data may not be sufficient to 'correct' for errors introduced by a prior. Hence, unlike the finite-dimensional setting, many priors do not 'wash out' as the information in the data increases indefinitely, so the prior may have a large influence on the corresponding posterior even in the asymptotic setting. Examples of posterior consistency results for Bayesian inversion are given in Ghosal, Ghosh and Ramamoorthi (1999), Ghosal, Ghosh and van der Vaart (2000), Neubauer and Pikkarainen (2008), Bochkina (2013), Agapiou, Larsson and Stuart (2013), Stuart and Teckentrup (2018), Kekkonen, Lassas and Siltanen (2016) and Kleijn and Zhao (2018).

To summarize, there are many reasonable priors for which posterior consistency holds at every point of the model parameter space X. A general class of such 'good' priors is that of tail-free priors (Freedman 1963) and neutral-to-the-right priors (Doksum 1974); see also Le Cam (1986, Section 17.7). The fact that there are too many 'bad' priors may therefore not be such a serious concern since there are also enough good priors approximating any given subjective belief. It is therefore important to have general results providing sufficient conditions for the consistency given a pair of model parameter and prior as developed, and Schwartz (1965) is an example in this direction.

---

[1] Increasing the 'information in data indefinitely' means $\mathrm{e} \to 0$ in (3.2), and if the data space Y is finite-dimensional, one also lets its dimension (sample size) increase. See Ghosal and van der Vaart (2017, Definition 6.1) for the precise definition.

3.2.4. Convergence rates

Posterior consistency is a weak property shared by many different choices of priors. More insight into the performance of Bayesian inversion under the choice of different priors requires characterizing other properties of the posterior, such as quantifying how quickly it converges to the true solution as the observational noise goes to 0. This leads to results about *contraction rates*, which is the Bayesian analogue of convergence rate theorems in functional analytic regularization.

Formally, consider the setting in (3.2) where the observational noise tends to zero as some scalar $\delta \to 0$, that is, for some fixed $f_{\text{true}} \in X$ we have

$$\mathsf{g} = \mathcal{A}(f_{\text{true}}) + \mathsf{e}_\delta \quad \text{where } \|\mathsf{e}_\delta\| \to 0 \text{ as } \delta \to 0. \tag{3.10}$$

A contraction rate theorem seeks to find the base rate for $\epsilon(\delta) \to 0$ such that

$$\Pi_{\text{post}}^{g^\delta}(\{f \in X : \ell_X(f, f_{\text{true}}) \geq \epsilon(\delta)\}) \to 0$$

$\Pi_{\text{post}}^{g_0}$-almost surely as $\epsilon \to 0$. Here, $g^\delta \in Y$ is a single sample of g in (3.10), $g_0 = \mathcal{A}(f_{\text{true}})$ is the corresponding ideal data, $\{\Pi_{\text{prior}}^\delta\}_\delta \subset \mathscr{P}_X$ is a sequence of prior distributions with $f \sim \Pi_{\text{prior}}^\delta$, and $\ell_X \colon X \times X \to \mathbb{R}$ is a (measurable) distance function on X. Research in this area has mainly focused on (a) obtaining contraction rates for Bayesian inversion where the prior is from some large class, or (b) improving the rates by changing the parameters of the prior depending on the level of noise (and even the data). Most of the work is done for additive observational white Gaussian noise, that is, the setting in (3.10) with $\mathsf{e}_\delta = \delta \mathbb{W}$, where $\mathbb{W}$ is a centred Gaussian white noise process that can be defined by its action on a separable Hilbert space Y. See Dashti and Stuart (2017, Sections A.3 and A.4) for a survey related to Gaussian measures and Wiener processes on separable Banach spaces.

As one might expect, initial results were for linear forward operators. The first results restricted attention to conjugate priors where the posterior has an explicit expression (it is in the same family as the prior) (Liese and Miescke 2008, Section 1.2). However, Bayesian non-parametric statistics often carries over to the inverse setting via the singular value decomposition of the linear forward operator. This allowed one to prove contraction rate results for Bayesian inversion with non-conjugate priors (Knapik, van der Vaart and van Zanten 2011, Knapik, van der Vaart and van Zanten 2013, Ray 2013, Agapiou *et al.* 2013, Agapiou, Stuart and Zhang 2014). In particular, if the Gaussian prior is non-diagonal for the singular value decomposition (SVD), then the posterior is still Gaussian and its contraction rate will be driven by the convergence rate of its posterior mean (since the variance does not depend on the data). Furthermore, in the linear setting, the posterior mean is the MAP, so the convergence rate of the MAP will be a posterior contraction rate.

Unfortunately, not many of the methods developed for proving contraction rates in the linear setting carry over to proving contraction rates for general, non-linear, inverse problems. A specific difficulty in the Bayesian setting is that the noise term often does not take values in the natural range of the forward operator. For example, consider the data model in (3.3) with observational white Gaussian noise, *i.e.* (3.10) with $e_\delta = \delta \mathbb{W}$, where $\mathbb{W}$ is a centred Gaussian white noise process that can be defined by its action on a separable Hilbert space Y. If we have a non-linear forward operator $\mathcal{A}$ given as the solution operator to an elliptic PDE, then the noise process $\mathbb{W}$ does not define a proper random element in $Y = L^2(\mathbb{M})$ for $\mathbb{M} \subset \mathbb{R}^d$: instead it defines a random variable only in a negative Sobolev space $W^{-\beta}$ with $\beta > d/2$. Nevertheless, there are also some results in the non-linear setting: for example, Kekkonen *et al.* (2016) derive contraction rate results for Bayesian inversion of inverse problems under Gaussian conjugate priors and where the forward operator is a linear hypoelliptic pseudodifferential operator. Recent developments that make use of techniques from concentration of measure theory have resulted in contraction rate theorems outside the conjugate setting. For example, Nickl and Söhl (2017) and Nickl (2017*a*) derive contraction rates for Bayesian inversion for parameter estimation involving certain classes of elliptic PDEs that are minimax-optimal in prediction loss: see *e.g.* Nickl (2017*a*, Theorem 28) for an example of a general contraction theorem. The case of a (possibly) non-linear forward operator and a Gaussian prior is considered in Nickl, van de Geer and Wang (2018), which studies properties of the MAP estimator. Finally, Gugushvili, van der Vaart and Yan (2018) derive contraction rates for 'general' priors expressed by scales of smoothness classes. The precise conditions are checked only for an elliptic example: it is not clear whether it works in other examples, such as the ray transform.

Remark 3.11. The proof techniques used for obtaining contraction rates in the non-linear setting depend on stability estimates for the forward problem that allow us to control $\| f - h \|$ in terms of $\| \mathcal{A}(f) - \mathcal{A}(h) \|$ in some suitable norm, and a dual form of the usual regularity estimates for solutions of PDEs that encodes the (functional analytic) ill-posedness of the problem. With estimates, one can use methods from non-parametric statistics to prove contraction rates for Bayesian inversion, with priors that do not require identifying a singular value-type basis underlying the forward operator.

3.2.5. Characterization of the posterior for uncertainty quantification
Posterior consistency and contraction rates are relevant results, but assessing the performance of Bayesian inversion methods in uncertainty quantification requires a more precise characterization of the posterior. The aim

is to characterize the fluctuations of $(\mathbb{f} \mid \mathbb{g} = g)$ near f_{true} when scaled by some inverse contraction rate.

One approach is to derive Bernstein–von Mises type theorems, which characterize the posterior distribution in terms of a canonical Gaussian distribution in the small-noise or large-sample limit. To better illustrate the role of such theorems, we consider the special case with Gaussian prior on the Hilbert space X and observational noise $\mathbb{e}_\delta = \delta \mathbb{W}$ in (3.10) (white noise model) where $\mathbb{W}$ denotes a Gaussian white noise process in the Hilbert space Y, that is,

$$\mathbb{g}^\delta = \mathcal{A}(f_{\text{true}}) + \delta \mathbb{W}.$$

Many of the results on contraction rates (Section 3.2.4) for Bayesian inversion in this setting are obtained for the distance function induced by the L^2-norm, so it is natural to initially consider the statistical fluctuations in L^2 of the random variable

$$\mathbb{z}_\delta := \frac{1}{\epsilon_\delta}(\mathbb{E}[\mathbb{f} \mid \mathbb{g} = g^\delta] - f_{\text{true}}).$$

Now, it turns out that there is no Bernstein–von Mises type asymptotics for $\mathbb{z}_\delta$ as the noise level δ tends to zero, that is, there is no Gaussian process $(\mathbb{G}(\phi))_{\phi \in C^\infty}$ such that

$$\left(\frac{1}{\delta}\langle \mathbb{z}_\delta - \mathbb{E}[\mathbb{z}_\delta], \phi\rangle_{L^2}\right)_{\phi \in C^\infty} \to (\mathbb{G}(\phi))_{\phi \in C^\infty} \quad \text{weakly as } \delta \to 0. \qquad (3.11)$$

To sidestep this difficulty, Castillo and Nickl (2013, 2014) seek to determine maximal families Ψ that replace C^∞ in (3.11) and where such an asymptotic characterization holds. This leads to non-parametric Bernstein–von Mises theorems, and while Castillo and Nickl (2013, 2014) considered 'direct' problems in non-parametric regression and probability density estimation, recent papers have obtained non-parametric Bernstein–von Mises theorems for certain classes of inverse problems. For example, Monard, Nickl and Paternain (2019) consider the case of inverting the (generalized) ray transform, whereas Nickl (2017a) considers PDE parameter estimation problems. The case with general linear forward problem is treated by Giordano and Kekkonen (2018), who build upon the techniques of Monard *et al.* (2019).

To give a flavour of the type of results obtained, we consider Theorem 2.5 of Monard *et al.* (2019), which is relevant to tomographic inverse problems involving inversion of the ray transform for recovering a function (images) defined on $\Omega \subset \mathbb{R}^d$, *i.e.* $X \subset L^2(\Omega)$. This theorem states that

$$\frac{1}{\delta}\langle (\mathbb{f} \mid \mathbb{g} = g^\delta) - \mathbb{E}[\mathbb{f} \mid \mathbb{g} = g^\delta], \phi\rangle_{L^2} \to \mathrm{N}(0, \|\mathcal{A} \circ (\mathcal{A}^* \circ \mathcal{A})^{-1}(\phi)\|_Y) \qquad (3.12)$$

as $\delta \to 0$ for any $\phi \in C^\infty(\Omega)$.

The convergence is in Π_{prior}-probability and the Y-norm on the right-hand side is a natural L^2-norm on the range of the ray transform. The limiting covariance is also shown to be minimal, that is, it attains the semi-parametric Cramér–Rao lower bound (or 'inverse Fisher information') for estimating $\langle f, \phi \rangle_{L^2}$ near f_{true}. A key step in the proof is to show smoothness of the 'Fisher information' operator $(\mathcal{A}^* \circ \mathcal{A})^{-1} \colon X \to X$, which is done using techniques from microlocal analysis (Monard *et al.* 2019, Theorem 2.2). The existence and mapping properties of this inverse Fisher information operator in (3.12) also plays a crucial role in proving that Bernstein–von Mises theorem for other inverse problems. For those with a non-linear forward operator, the information operator to be inverted is found after linearization as shown in Nickl (2017*a*) for parameter estimation in an elliptic PDE.

Relevance to applications. For large-scale problems, such as those arising in imaging, it is computationally challenging to even explore the posterior beyond computing point estimators (Section 3.5). This holds in particular for the task of computing Bayesian credible sets, which are relevant for uncertainty quantification. Hence, the theoretical results in Section 3.2.5 can be of interest in practical applications, since these provide good analytic approximations of the posterior that are much simpler to compute.

In particular, if a Bernstein–von Mises theorem holds, then the Bayesian and maximum likelihood estimators have the same asymptotic properties and the influence of the prior diminishes asymptotically as information in the data increases. Then, Bayesian credible sets are asymptotically equivalent to frequentist confidence regions,[2] so Bayesian inversion with 95% posterior credibility will have approximately 0.95 chance of returning the correct decision in repeated trials.

Many of the results, however, assume a Gaussian prior and data likelihood: for non-Gaussian problems in an infinite-dimensional setting the structure of the posterior can be very chaotic and difficult to characterize. Furthermore, the issue with any such asymptotic characterization of the posterior is that it is based on increasing information in data indefinitely. In reality data are fixed, and it is quite possible that the posterior is (very) non-normal and possibly multi-modal even though it behaves asymptotically like a Gaussian (or more generally the sampling distribution of a maximum likelihood estimator). Hence, a regularizing prior that provides 'best' contraction rates may not necessarily be the best when it comes to image quality for data with a given noise level. Here one would need to consider a prior that acts as more than just a regularizer. Finally, the assumptions

---

[2] A Bayesian credible set is a subset of the model parameter space X that contains a predefined fraction, say 95%, of the posterior mass. A frequentist confidence region is a subset of X that includes the unknown true model parameter with a predefined frequency as the experiment is repeated indefinitely.

of the Bernstein–von Mises theorem are fragile and easy to violate for a dataset or analysis, and it is difficult to know, without outside information, when this will occur. As nicely outlined in Nickl (2013, Section 2.25), the parametric (finite-dimensional) setting already requires many assumptions, such as a consistent maximum likelihood estimator, a true model parameter in the support of the prior, and a log-likelihood that is sufficiently regular. For example, data in an inverse problem are observational, and therefore it is unlikely that an estimator of f_{true}, such as maximum likelihood or the posterior mean, is consistent, in which case a Bernstein–von Mises theorem does not apply.

3.3. Reconstruction method as a point estimator

In most large-scale inverse problems it is computationally very challenging, if not impossible, to recover the entire posterior. However, a reconstruction method that seeks to compute an estimator for the posterior formally defines a mapping $\mathcal{R}\colon Y \to X$ (reconstruction operator). As such, it can be viewed as non-randomized decision rule (point estimator) in a statistical estimation problem: see Liese and Miescke (2008, Definition 3.1) for the formal definition. Computing a suitable point estimator is therefore an alternative to seeking to recover the posterior.

In the infinite-dimensional setting, the benefits of using one estimator rather than another (*e.g.* the conditional mean estimate rather than the MAP estimate) are not well understood. Statistical decision theory provides criteria for selecting and comparing decision rules, which can be used when selecting the estimator (reconstruction method). This requires phrasing the inverse problem as a statistical decision problem. More precisely, the tuple $((Y, \mathfrak{S}_Y), \{\Pi^f_{\mathrm{data}}\}_{f \in X})$ defines a statistical model which is parametrized by the Banach space X. The inverse problem is now a statistical decision problem (Liese and Miescke 2008, Definition 3.4) where the statistical model is parametrized by X, the decision space is $D := X$, and a given *loss function*:

$$\ell_X\colon X \times X \to \mathbb{R}. \tag{3.13}$$

The loss function is used to define the risk, which is given as the Π^f_{data}-expectation of the loss function. The risk seeks to quantify the downside that comes with using a particular reconstruction operator $\mathcal{R}\colon Y \to X$.

3.3.1. Some point estimators

Here we briefly define the most common point estimators that are commonly considered in solving many inverse problems. Overall, estimators that include an integration over X are computationally demanding. These include the conditional mean and many estimators associated with uncertainty quantification.

Maximum likelihood estimator. This estimator maximizes the data likelihood. Under the data model in (3.5), the (maximum likelihood) reconstruction operator $\mathcal{R}\colon Y \to X$ becomes

$$\mathcal{R}(g) \in \underset{f \in X}{\arg\max}\, \{\exp(-\mathcal{L}(f, g))\} = \underset{f \in X}{\arg\min}\, \mathcal{L}(f, g).$$

The advantage of the maximum likelihood estimator is that it does not involve any integration over X, so it is computationally feasible to use on large-scale problems. Furthermore, it only requires access to the data likelihood (no need to specify a prior), which is known. On the other hand, it does not act as a regularizer, so it is not suitable for ill-posed problems.

Maximum a posteriori (MAP) estimator. This estimator maximizes the posterior probability, that is, it is the 'most likely' model parameter given measured data g. In the finite-dimensional setting, the prior and posterior distribution can typically be described by densities with respect to the Lebesgue measure, and the MAP estimator is defined as the reconstruction operator $\mathcal{R}\colon Y \to X$ given by

$$\mathcal{R}(g) := \underset{f \in X}{\arg\max}\, \pi_{\mathrm{post}}(f \mid g) = \underset{f \in X}{\arg\min}\{\mathcal{L}(f, g) - \log \pi_{\mathrm{prior}}(f)\}.$$

The second equality above assumes a data model as in (3.5).

For many of the infinite-dimensional spaces there exists no analogue of the Lebesgue measure, which makes it difficult to define a MAP estimator through densities. One way to work around this technical problem is to replace the Lebesgue measure with a Gaussian measure on X. Hence, we assume the posterior and prior have densities with respect to some fixed centred (mean-zero) Gaussian measure μ_0 and E denotes its Cameron–Martin space.[3] Now, following Dashti, Law, Stuart and Voss (2013), we consider the centre of a small ball in X with maximal probability, and then study the limit of this centre as the radius of the ball shrinks to zero. Stated precisely, given fixed data $g \in Y$, assume there is a functional $\mathcal{J}\colon E \to \mathbb{R}$ that satisfies

$$\lim_{r \to 0} \frac{\Pi^g_{\mathrm{post}}(B_r(f_1))}{\Pi^g_{\mathrm{post}}(B_r(f_2))} = \exp(\mathcal{J}(f_1) - \mathcal{J}(f_2)).$$

Here, $B_r(f) \subset X$ is the open ball of radius $r > 0$ centred at $f \in X$ and $\mathcal{J}$ is the Onsager–Machlup functional (Ikeda and Watanabe 1989, p. 533).

---

[3] The Cameron–Martin space E associated with $\Pi \in \mathscr{P}_X$ consists of elements $f \in X$ such that $\delta_f \circledast \Pi \ll \Pi$, that is, the translated measure $B \mapsto \Pi(B - f)$ is absolutely continuous with respect to Π. The Cameron–Martin space is fundamental when dealing with the differential structure in X, mainly in connection with integration by parts formulas, and it inherits a natural Hilbert space structure from the space X^*.

For any fixed $f_1 \in X$, a model parameter $f_2 \in X$ for which the above limit is maximal is a natural candidate for the MAP estimator and is clearly given by minimizers of the Onsager–Machlup functional. The advantage of the MAP estimator is that it does not involve any integration over X. Furthermore, the prior often acts as a regularizer, so MAP can be useful for solving an ill-posed inverse problem. A disadvantage is that a MAP may not always exist (Section 3.3.2): one needs to provide an explicit prior, and many priors result in a non-smooth optimization problem. Due to the latter, MAP estimation is computationally more challenging than maximum likelihood estimation.

Conditional (posterior) mean. The reconstruction operator $\mathcal{R}\colon Y \to X$ is here defined as

$$\mathcal{R}(g) := \mathbb{E}[\mathbb{f} \mid \mathbb{g} = g] = \int_X f \, \mathrm{d}\Pi_{\mathrm{post}}^g(f).$$

This estimator involves integration over X, making it challenging to use on even small- to mid-scale problems. It also assumes access to the posterior, which in turn requires a prior. It does, however, act as a regularizer (Sprungk 2017, Corollary 3.23) and therefore it is suitable for ill-posed problems.

Bayes estimator. One starts by specifying a function $\ell_X\colon X \times X \to \mathbb{R}$ that quantifies proximity in X (it does not have to be a metric). Then, a Bayes estimator minimizes the expected loss with respect to the prior Π_{prior}, that is, $\mathcal{R}\colon Y \to X$ is defined as

$$\mathcal{R}(g) := \widehat{\mathcal{R}}(g), \quad \text{where } \widehat{\mathcal{R}} \in \operatorname*{arg\,min}_{\mathcal{R}\colon Y \to X} \mathbb{E}_{(\mathbb{f},\mathbb{g}) \sim \mu}[\ell_X(\mathcal{R}(\mathbb{g}), \mathbb{f})]. \tag{3.14}$$

This estimator is challenging to compute even for small- to mid-scale problems, due to the integration over $X \times Y$ and the minimization over all non-randomized decision rules. It also requires access to the joint distribution μ, which by the law of total probability is expressible as $\mu = \Pi_{\mathrm{prior}} \otimes \Pi_{\mathrm{data}}^f$ with known data likelihood Π_{data}^f. An important property of the Bayes estimator is that it is a regularizer, so it is suitable for ill-posed problems. It is also equivalent to the conditional mean when the loss is taken as the square of the L^2-norm (Helin and Burger 2015) and the conditional median when the loss is the L^1-norm. Furthermore, Theorem 1 of Burger and Lucka (2014) shows that a MAP estimator with a Gibbs-type prior where the energy functional is Lipschitz-continuous and convex equals a Bayes estimator where the loss is the Bregman distance of the aforementioned energy functional. Finally, in the finite-dimensional setting, one can show (Banerjee, Guo and Wang 2005) that the Bayes estimator is the same as the conditional mean if and only if ℓ_X is the Bregman distance of a strictly convex non-negative differentiable functional.[4]

Other estimators. Another important family of estimators in statistical decision theory is that of minimax estimators that minimize the maximum loss. Estimators for uncertainty quantification typically involve higher-order moments, such as the variance, or interval estimators that are given as a set of points in the model parameter space X.

3.3.2. Relation to functional analytic regularization

It is quite common to interpret a variational method as in (2.7) as a MAP estimators. This is especially the case when one chooses the data discrepancy $\mathcal{L}$ so that minimizing $f \mapsto \mathcal{L}(\mathcal{A}(f), g)$ corresponds to computing a maximum likelihood estimator.

Such an interpretation is almost always possible in the finite-dimensional setting since, as mentioned in Definition 2.2, the regularization functional $\mathcal{S}_\theta$ can be interpreted as a Gibbs-type prior $\rho_{\mathrm{prior}}(f) \propto \exp(-\mathcal{S}_\theta(f))$ (Kaipio and Somersalo 2005) where ρ_{prior} is the density for the prior. The situation is more complicated in the infinite-dimensional setting since a MAP estimator does not always exist, and if it exists then there is no general scheme for connecting the topological description of a MAP estimate to a variational problem. The main reason is that the posterior no longer has a natural density representation, which significantly complicates the definition and study of the underlying conditional probabilities.

Assume the prior measure is specified by a Gaussian random field, and the likelihood satisfies conditions in Theorem 3.9 that are necessary for the existence of a well-posed posterior measure. Then the MAP estimator is well-defined as the minimizer of an Onsager–Machlup functional defined on the Cameron–Martin space of the prior. If one has Gaussian noise, then this becomes a least-squares functional penalized by a regularization functional that is the Cameron–Martin norm of the Gaussian prior (Dashti *et al.* 2013); see also Dashti and Stuart (2017, Section 4.3). To handle the case with non-Gaussian priors, Helin and Burger (2015) introduce the notion of weak MAP estimate (wMAP) and show that a wMAP estimate can be connected to a variational formulation also for non-Gaussian priors. Furthermore, any MAP estimate in the sense of Dashti *et al.* (2013) is a wMAP estimate, so this is a generalization in the strict sense. The wMAP approach, however, fails when the prior does not admit continuous densities. In order to handle this, Clason, Helin, Kretschmann and Piiroinen (2018) introduce the notion of a generalized mode of a probability measure (Clason *et al.* 2018, Definition 2.3) and define a (generalized) MAP estimator as a generalized

---

[4] The Bregman distance for $x \mapsto \langle x, x \rangle$ gives the L^2-loss.

mode of the posterior measure. Generalized MAP reduces to the classical MAP in certain situations that include Gaussian priors but the former also exist for a more general class of priors, such as uniform priors. The main result in Clason *et al.* (2018) is that one can characterize a generalized MAP estimator as a minimizer of an Onsager–Machlup functional even in cases when the prior does not admit a continuous density.

As a final remark, while a MAP estimate with a Gaussian noise model does lead to an optimization problem with a quadratic data-fidelity term, Gribonval and Nikolova (2018) show via explicit examples that the converse is not true. They characterize those data models of the type in (3.2) where Bayes estimators can be expressed as the solution of a penalized least-squares optimization problem. One example is denoising in the presence of additive Gaussian noise and an arbitrary prior; another is a data model with (a variant of) Poisson noise and any prior probability on the unknowns. In these cases, the variational approach is rich enough to build all possible Bayes estimators via a well-chosen penalty.

3.4. *Explicit priors*

The choice of prior is important in Bayesian inversion, and much of the work has focused on characterizing families of priors that ensure posterior consistency holds and for which there are good convergence rates (Section 3.2.3).

From the general Bayesian statistics literature, much effort has gone into characterizing robust priors. These have limited influence (in the asymptotic regime) on the posterior (Bayesian robustness), and such non-informative priors are useful for Bayesian inversion when there is not enough information to choose a prior for the unknown model parameter, or when the information available is not easily translated into a probabilistic statement: see Berger (1985, Section 4.7.9) and Calvetti and Somersalo (2017). For example, hierarchical priors tend to be robust (Berger 1985, Section 4.7.9). Another class is that of conjugate priors, which are desirable from a computational perspective (Section 3.5.2) since they have tails that are typically of the same form as the likelihood function. Conjugate priors also remain influential when the likelihood function is concentrated in the (prior) tail, so natural conjugate priors are therefore not necessarily robust (Berger 1985, Section 4.7).

Priors on digitized images. A wide range of priors for Bayesian inversion have been suggested when the unknown model parameter represents a digitized image, the latter given by a real-valued function in some suitable class defined on a domain $\Omega \subset \mathbb{R}^n$.

One such class is that of *smoothness* priors, which reflect the belief that values of the model parameter at a point are close to the average of its values in a neighbourhood of that point. Another is *structural* priors, which allow for more abrupt (discontinuous) changes in the values of the unknown model parameter at specific locations. Yet another is *sparsity-promoting* priors (see Section 2.7), which encode the *a priori* belief that the unknown model parameter is compressible with respect to some underlying dictionary, that is, it can be transformed into a linear combination of dictionary elements where most coefficients vanish (or are small). Finally, there are hierarchical priors which are formed by combining other priors hierarchically into an overall prior. This is typically done in a two-step process where one first specifies some underlying priors, often taken as natural conjugate priors, and then mixes them in a second stage over hyper-parameters. Recently, Calvetti, Somersalo and Strang (2019) have reformulated the question of sparse recovery as an inverse problem in the Bayesian framework, and expressed the sparsity criteria by means of a hierarchical prior mode. More information and further examples of priors in the finite-dimensional setting are given by Kaipio and Somersalo (2005), Calvetti and Somersalo (2008, 2017) and Calvetti *et al.* (2019).

Priors on function spaces. Defining priors when X is an infinite-dimensional function space is somewhat involved. A common approach is to consider a convergent series expansion and then let the coefficients be generated by a random variable.

More precisely, consider the case when X is a Banach space of real-valued functions on some fixed domain $\Omega \subset \mathbb{R}^d$. Following Dashti and Stuart (2017, Section 2.1), let $\{\phi_i\}_i \subset X$ be a countable sequence whose elements are normalized, *i.e.* $\|\phi_i\|_X = 1$. Now consider model parameters $f \in X$ of the form

$$f = f_0 + \sum_i \alpha_i \phi_i,$$

where $f_0 \in X$ is not necessarily normalized to 1. A probability distribution on the coefficients α_i renders a real-valued random function on Ω: simply define the deterministic sequence $\{\gamma_i\}_i \subset \mathbb{R}$ and the i.i.d. random sequence $\{\xi_i\}_i \subset \mathbb{R}$ and set $\alpha_i = \gamma_i \xi_i$. This generates a probability measure Π_{prior} on X by taking the pushforward of the measure on the i.i.d. random sequence $\{\xi_i\}_i \subset \mathbb{R}$ under the map which takes the sequence into the random function.

Using the above technique, one can construct a uniform prior that can be shown to generate random functions that are all contained in a subset $X \subset L^\infty(\Omega)$, which can be characterized (Dashti and Stuart 2017, Theorem 2). It is likewise possible to define Gaussian priors where the random function exists as an L^2 limit in Sobolev spaces of sufficient smoothness (Dashti and Stuart 2017, Theorem 8). For imaging applications, much effort has been devoted to constructing edge-preserving priors. An obvious candidate is the TV-prior, but it is not 'discretization-invariant' as its edge-preserving property is lost when the discretization becomes finer. This prompted the development of other edge-preserving priors, such as Besov space priors, which are discretization-invariant (Lassas *et al.* 2009, Kolehmainen, Lassas, Niinimäki and Siltanen 2012). See also Dashti and Stuart (2017, Theorem 5), which characterizes Besov priors that generate random functions contained in separable Banach spaces. However, edge-preserving inversion using a Besov space prior often relies on the Haar wavelet basis. Due to the structure of the Haar basis, discontinuities are preferred on an underlying dyadic grid given by the discontinuities of the basis functions. As an example, on $[0, 1]$ a Besov space prior prefers a discontinuity at $x = 1/4$ over $x = 1/3$. Thus, in most practical cases, Besov priors rely on both a strong and unrealistic assumption. For this reason, Markkanen, Roininen, Huttunen and Lasanen (2019) propose another class of priors for edge-preserving Bayesian inversion, the Cauchy difference priors. Starting from continuous one-dimensional Cauchy motion, its discretized version, Cauchy random walk, can be used as a non-Gaussian prior for edge-preserving Bayesian inversion. As shown by Markkanen *et al.* (2019), one can also develop a suitable posterior distribution sampling algorithm for computing the conditional mean estimates with single-component Metropolis–Hastings. The approach is applied to CT reconstruction problems in materials science.

The above constructions of random functions through randomized series can be linked to each other through the notion of random fields as shown in Dashti and Stuart (2017, Section 2.5); see also Ghosal and van der Vaart (2017, Chapters 2 and 10) for further examples. These constructions also extend straightforwardly to $\mathbb{R}^n$- or $\mathbb{C}^n$-valued functions.

3.5. Challenges

The statistical view of an inverse problem in Bayesian inversion extends the functional analytic one, since the output is ideally the posterior that describes all possible solutions. This is very tractable and fits well within the scientific tradition of presenting data and inferred quantities with error bars.

Most priors are chosen to regularize the problem rather than improving the output quality. Next, algorithmic advances (Section 3.5.2) have resulted in methods that can sample in a computationally feasible manner from a posterior distribution in a high-dimensional setting, say up to 10^6 dimensions. This is still not sufficient for large-scale two-dimensional imaging or regular three-dimensional imaging applications. Furthermore, these methods require an explicit prior, which may not be feasible if one uses learning to obtain it. They may also make use of analytic approximations such as those given in Section 3.2.5, which restricts the priors that can come into question. For these reasons, most applications of Bayesian inversion on large-scale problems only compute a MAP estimator, whereas estimators requiring integration over the model parameter space remain computationally unfeasible. These include Bayes estimators and the conditional mean as well as estimators relevant for uncertainty quantification.

In conclusion, the above *difficulties in specifying a 'good' prior and in meeting the computational requirements have seriously limited the dissemination of Bayes inversion in large-scale inverse problems, such as those arising in imaging.* Before providing further remarks on this, let us mention that Section 5.1 shows how techniques from deep learning can be used to address the above challenges when computing a wide range of estimators. Likewise, in Section 5.2 we show how deep learning can be used to efficiently sample from the posterior.

3.5.1. Choosing a good prior

Current theory for choosing a 'good' prior mainly emphasizes the regularizing function of the prior (Section 3.4). In particular, one seeks a prior that ensures posterior consistency (Section 3.2.3) and good contraction rates (Section 3.2.4) and, if possible, also allows for an asymptotic characterization of the posterior (Section 3.2.5).

This theory, however, is for the asymptotic setting as the noise level in data tends to zero, whereas there seems to be no theory for Bayesian inversion that deals with the setting when the data have a fixed noise level. Here, the prior has a role that goes beyond acting as a regularizer, and its choice may have a large influence on the posterior. For example, it is far from clear whether a prior that provides 'optimal' contraction rates is the most suitable one when the data are fixed with a given noise level. Another difficulty is that norms used to quantify distance in the mathematical theorems have little to do with what is meant by a 'good' estimate of a model parameter. An archetypal example is the difficulty in quantifying the notion of 'image quality' in radiology. This is very difficult since the notion of image quality depends on the task motivating the imaging application. Hand-crafted priors surveyed in Section 3.4 have limitations in this regard, and in Sections 4.3, 4.4, 4.7 and 4.10 we survey work that uses data-driven modelling to obtain a prior.

3.5.2. Computational feasibility

The focus here is on techniques that are not based on deep learning for sampling from a very high-dimensional posterior distribution that lacks an explicit expression.

A well-known class of methods is based on Markov chain Monte Carlo (MCMC), where the aim is to define an iterative process whose stationary distribution coincides with the target distribution, which in Bayesian inversion is the posterior. MCMC techniques come in many variants, and one common variant is MCMC sampling with Metropolis–Hastings dynamics (Minh and Le Minh 2015), which generates a Markov chain with equilibrium distribution that coincides with the posterior in the limit. Other variants use Gibbs sampling, which reduces the autocorrelation between samples. Technically, Gibbs sampling can be seen as a special case of Metropolis–Hastings dynamics and it requires computation of conditional distributions. Further variants are auxiliary variable MCMC methods, such as slice sampling (Neal 2003), proximal MCMC (Green, Łatuszyński, Pereyra and Robert 2015, Durmus, Moulines and Pereyra 2018, Repetti, Pereyra and Wiaux 2019) and Hamiltonian Monte Carlo (Girolami and Calderhead 2011, Betancourt 2017). See also Dashti and Stuart (2017, Section 5) for a nice abstract description of MCMC in the context of infinite-dimensional Bayesian inversion.

An alternative approach to MCMC seeks to approximate the posterior with more tractable distributions (deterministic inference), for example in variational Bayes inference (Fox and Roberts 2012, Blei, Küçükelbir and McAuliffe 2017) and expectation propagation (Minka 2001). Variational Bayes inference has indeed emerged as a popular alternative to the classical MCMC methods for sampling from a difficult-to-compute probability distribution, which in Bayesian inversion is the posterior distribution. The idea is to start from a fixed family of probability distributions (variational family) and select the one that best approximates the target distribution under some similarity measure, such as the Kullback–Leibler divergence.

Blei *et al.* (2017, p. 860) try to provide some guidance on when to use MCMC and when to use variational Bayes. MCMC methods tend to be more computationally intensive than variational inference, but they also provide guarantees of producing (asymptotically) exact samples from the target density (Robert and Casella 2004). Variational inference does not enjoy such guarantees: it can only find a density close to the target but tends to be faster than MCMC. A recent development is the proof of a Bernstein–von Mises theorem (Wang and Blei 2017, Theorem 5), which shows that the variational Bayes posterior is asymptotically normal around the variational frequentist estimate. Hence, if the variational frequentist estimate is consistent, then the variational Bayes posterior converges to a Gaussian with a mean centred at the true model parameter. Furthermore, since variational Bayes rests on optimization, variational inference easily takes advantage of methods such as stochastic optimization (Robbins and Monro 1951, Kushner and Yin 1997) and distributed optimization (though some MCMC methods can also exploit these innovations (Welling and Teh 2011, Ahmed *et al.* 2012)). Thus, variational inference is suited to large data sets and scenarios where we want to quickly explore many models; MCMC is suited to smaller data sets and scenarios where we are happy to pay a heavier computational cost for more precise samples. Another factor is the geometry of the posterior distribution. For example, the posterior of a mixture model admits multiple modes, each corresponding to label permutations of the components. Gibbs sampling, if the model permits, is a powerful approach to sampling from such target distributions; it quickly focuses on one of the modes. For mixture models where Gibbs sampling is not an option, variational inference may perform better than a more general MCMC technique (*e.g.* Hamiltonian Monte Carlo), even for small datasets (Küçükelbir, Ranganath, Gelman and Blei 2017).

4. Learning in functional analytic regularization

There have been two main ways to incorporate learning into functional analytic regularization. The first relates to the 'evolution' of regularizing functionals, primarily within variational regularization (Sections 4.3, 4.4, 4.6, 4.7 and 4.10). Early approaches focused on using measured data to determine the regularization parameter(s), but as prior models became increasingly complex, this blended into approaches where one learns a highly parametrized regularizer from data. The second category uses learning to address the computational challenge associated with variational regularization. The idea is to 'learn how to optimize' in variational regularization given an *a priori* bound on the computational budget (Section 4.9).

4.1. Choosing the regularization parameter

To introduce this topic we first consider the simplified case where only one parameter is used for characterizing the scale of regularization as in (2.10) and in the functional analytic regularization methods encountered in Sections 2.5–2.7. Thus we use the notation of Remarks 2.3 and 2.6, that is, f_{true} denotes the true (unknown) model parameter and $f_\lambda := \mathcal{R}_\lambda(g)$ is the regularized solution obtained from applying a reconstruction operator $\mathcal{R}_\lambda \colon Y \to X$ on data g satisfying (2.1).

In this context, and recalling some historical classification, the three main types of parameter choice rules are characterized as *a posteriori*, *a priori* and error-free parameter choice rules: see Bertero and Boccacci (1998, Section 5.6) and Engl *et al.* (2000). With hindsight, many of these parameter choice rules can be seen as early attempts to use 'learning' from data in the context of reconstruction.

A posteriori rules. This class of rules is based on the assumption that a reasonably tight estimate of the data discrepancy and/or value of regularizer at the true solution can be accessed. That is, one knows $\epsilon > 0$ and/or $S > 0$ such that

$$\mathcal{L}(\mathcal{A}(f_{\text{true}}), g) \leq \epsilon \text{ for } g = \mathcal{A}(f_{\text{true}}) + e \quad \text{and/or} \quad \mathcal{S}(f_{\text{true}}) \leq S.$$

A prominent example of an *a posteriori* parameter choice rule is the Morozov discrepancy principle (Morozov 1966). Here, the regularization parameter

λ is chosen so that

$$\mathcal{L}(\mathcal{A}(f_\lambda), g) \leq \epsilon. \tag{4.1}$$

Another example is Miller's method (Miller 1970), where the regularization parameter λ is chosen so that

$$\mathcal{L}(\mathcal{A}(f_\lambda), g) \leq \epsilon \quad \text{and} \quad \mathcal{S}(f_\lambda) \leq S.$$

A priori **rules.** These methods combine an estimate of the noise level in the data with some knowledge of the smoothness of the solution as *a priori* information. Hence, the choice of regularization parameter can be made before computing f_λ. The choice of λ is ideally guided by a theorem that ensures the parameter choice rule yields an optimal convergence rate, for example as in Theorem 2.7 where the (scalar) regularization parameter is chosen in proportion to the noise level. For more detailed references, see Engl *et al.* (2000) and Kindermann (2011).

Error-free parameter choice rules. Use data to guide the choice of parameter, for example by balancing principles between the error in the fidelity and the regularization terms. An important example in this context is *generalized cross-validation* (Golub, Heat and Wahba 1979). Let $f_\lambda^{[k]} \in X$ denote the regularized solution obtained from data when we have removed the kth component g_k of g. Then the regularization parameter λ is chosen to predict the missing data values:

$$\text{minimize } \lambda \mapsto \sum_{k=1}^{m} |\mathcal{A}(f_\lambda^{[k]})_k - g_k| \quad \text{subject to } \mathcal{A}(f_\lambda^{[k]})_k \simeq g_k.$$

Another method within this class is the *L-curve* (Hansen 1992). Here the regularization parameter λ is chosen where the log-log plot of $\lambda \mapsto (\mathcal{L}(\mathcal{A}(f_\lambda), g) + \lambda \mathcal{S}(f_\lambda))$ has the highest curvature (*i.e.* it exhibits a corner).

Most of the work on parameter choice techniques addresses the case of a single scalar parameter. Much of the theory is developed for additive Gaussian noise, that is, when the data discrepancy $\mathcal{L}$ is a squared (possibly weighted) 2-norm. For error-free parameter choice rules, convergence $f_{\lambda(\delta)} \to f_{\text{true}}$ as $\delta \to 0$ cannot be guaranteed (Bakushinskii 1984). Error-free parameter choice rules are computationally very demanding as they require solutions for varying values of the regularization parameter. Although many rules have been proposed, very few of them are used in practice. In Section 4.3.1 we will encounter another instance of parameter choice rules for TV-type regularization problems via bilevel learning.

4.2. Learning in approximate analytic inversion

Early approaches to using learning in approximate analytic inversion (Section 2.4) mostly dealt with the FBP method for tomographic image reconstruction.

One of the first examples of the above is that of Floyd (1991), which learns the reconstruction filter of an analytic reconstruction algorithm. The same principle is considered by Würfl *et al.* (2018), who have designed a convolutional neural network (CNN) architecture specifically adapted to encode an analytic reconstruction algorithm for inverting the ray transform. A key observation is that the normal operator $\mathcal{A}^* \mathcal{A}$ is of convolutional type when $\mathcal{A}$ is the ray transform, and the reconstruction filter in FBP acts by a convolution as well. Hence, both of these are easily representable in a CNN. The paper gives explicit constructions of such CNNs for FBP in parallel-beam geometry and fan-beam geometry and the Feldkamp–Davis–Kress method in cone-beam geometry. Having the analytic reconstruction operator encoded as a CNN allows one to learn every other possible step in it, so the approach in Würfl *et al.* (2018) actually goes beyond learning the reconstruction filter of an analytic reconstruction algorithm. Finally, we mention Janssens *et al.* (2018), who consider a fan-beam reconstruction algorithm based on the Hilbert transform FBP (You and Zeng 2007) for which the filter is trained by a neural network.

4.3. Bilevel optimization

In variational methods (see Section 2.6), we define the reconstruction operator $\mathcal{R}_\theta \colon Y \to X$ by

$$\mathcal{R}_\theta(g) := \arg\min_{f \in X} \{ \mathcal{L}(\mathcal{A}(f), g) + \mathcal{S}_\theta(f) \} \quad \text{for } g \in Y. \tag{4.2}$$

As already mentioned, ideally $f \mapsto \mathcal{L}(\mathcal{A}(f), g)$ corresponds to an affine transformation of the negative log-likelihood of the data. However, it is less clear how to choose the regularizer $\mathcal{S}_\theta \colon X \to \mathbb{R}$ and/or the value for the parameter θ.

The focus here is on formulating a generic set-up for learning selected components of (4.2) from supervised training data given a loss function $\ell_X \colon X \times X \to \mathbb{R}$. This set-up can be tailored towards learning the regularization functional $\mathcal{S}$, or the data fidelity term $\mathcal{L}$, or an appropriate component in a forward operator $\mathcal{A}$, *e.g.* in blind image deconvolution (Hintermüller and Wu 2015). The starting point is to have access to supervised training data $(f_i, g_i) \in X \times Y$ that are generated by a $(X \times Y)$-valued random variable $(\mathbf{f}, \mathbf{g}) \sim \mu$. We can then form the following *bilevel optimization* formulation:

$$\begin{cases} \widehat{\theta} \in \arg\min_{\theta} \mathbb{E}_{(\mathsf{f},\mathsf{g})\sim\mu}[\ell_X(\mathcal{R}_\theta(\mathsf{g}),\mathsf{f})], \\ \mathcal{R}_\theta(g) := \arg\min_{f\in X}\{\mathcal{L}(\mathcal{A}(f),g) + \mathcal{S}_\theta(f)\}. \end{cases} \tag{4.3}$$

Note here that $\widehat{\theta}$ is a Bayes estimator, but μ is not fully known. Instead it is replaced by its empirical counterpart given by the supervised training data, in which case $\widehat{\theta}$ corresponds to *empirical risk minimization* Section 3.3.

In the bilevel optimization literature, as in the optimization literature as a whole, there are two main and mostly distinct approaches. The first one is the discrete approach that first discretizes the problem (4.2) and subsequently optimizes its parameters. In this way, optimality conditions and their well-posedness are derived in finite dimensions, which circumvents often difficult topological considerations related to convergence in infinite-dimensional function spaces, but also jeopardizes preservation of continuous structure (*i.e.* optimizing the discrete problem is not automatically equivalent to discretizing the optimality conditions of the continuous problem (De los Reyes 2015)) and dimension-invariant convergence properties.

Alternatively, (4.2) and its parameter θ are optimized in the continuum (*i.e.* appropriate infinite-dimensional function spaces) and then discretized. The resulting problems present several difficulties due to the frequent non-smoothness of the lower-level problem (think of TV regularization), which, in general, makes it impossible to verify Karush–Kuhn–Tucker constraint qualification conditions. This issue has led to the development of alternative analytical approaches in order to obtain first-order necessary optimality conditions (Bonnans and Tiba 1991, De los Reyes 2011, Hintermüller, Laurain, Löbhard, Rautenberg and Surowiec 2014). The bilevel problems under consideration are also related to generalized mathematical programs with equilibrium constraints in function spaces (Luo, Pang and Ralph 1996, Outrata 2000).

One of the first examples of the above is the paper by Haber and Tenorio (2003), who considered a regularization functional $\mathcal{S}_\theta\colon X \to \mathbb{R}$ that can depend on location and involves derivatives or other filters. Concrete examples are anisotropic weighted Dirichlet energy where θ is a function, that is,

$$\mathcal{S}_\theta(f) := \|\theta(\,\cdot\,)\nabla f(\,\cdot\,)\|_2^2 \quad \text{for } \theta\colon \Omega \to \mathbb{R},$$

and anisotropic weighted TV,

$$\mathcal{S}_\theta(f) := \|\theta(|\nabla f(\,\cdot\,)|)\|_1 \quad \text{for } \theta\colon \mathbb{R} \to \mathbb{R}.$$

The paper contains no formal mathematical statements or proofs, but there are many numerical examples showing how to use supervised learning techniques to determine a regularization functional given a training set of feasible solutions.

Another early example of learning a regularizer is the paper by Tappen (2007), who considered bilevel optimization for finding optimal regularizers parametrized by finite-dimensional Markov random field models. Bilevel optimization for optimal model design for inverse problems has also been discussed by Haber, Horesh and Tenorio (2010), Bui-Thanh, Willcox and Ghattas (2008) and Biegler *et al.* (2011).

A revival of bilevel learning in the context of non-smooth regularizers took place in 2013 with a series of papers: De los Reyes and Schönlieb (2013), Calatroni, De los Reyes and Schönlieb (2014), De los Reyes, Schönlieb and Valkonen (2016, 2017), Calatroni, De los Reyes and Schönlieb (2017), Van Chung, De los Reyes and Schönlieb (2017), Kunisch and Pock (2013), Chen, Pock and Bischof (2012), Chen, Yu and Pock (2015), Chung and Espanol (2017), Hintermüller and Wu (2015), Hintermüller and Rautenberg (2017), Hintermüller, Rautenberg, Wu and Langer (2017), Baus, Nikolova and Steidl (2014) and Schmidt and Roth (2014). A critical theoretical issue is the well-posedness of the learning; another is to derive a characterization of the optimal solutions that can be used in the design of computational methods. Such results were first derived by De los Reyes and Schönlieb (2013), Calatroni *et al.* (2014), De los Reyes, Schönlieb and Valkonen (2016, 2017), Van Chung *et al.* (2017) and Hintermüller and Rautenberg (2017), with applications to inverse problems and image processing (*e.g.* bilevel learning for image segmentation (Ranftl and Pock 2014)) as well as classification (*e.g.* learning an optimal set-up of support vector machines (Klatzer and Pock 2015)).

In what follows, we survey the main mathematical properties of bilevel learning and review the main parametrizations of regularizers in (4.2) that are considered in the literature.

4.3.1. Learning of TV-type regularizers and data fidelity terms

We start with a simple but theoretically and conceptually important example, namely the learning of total variation (TV)-type regularization models as proposed in De los Reyes and Schönlieb (2013), Kunisch and Pock (2013) and De los Reyes, Schönlieb and Valkonen (2017).

Let $X = BV(\Omega)$, where $\Omega \subset \mathbb{R}^n$ is a fixed open bounded set with Lipschitz boundary, and $Y = L^2(\mathbb{M}, \mathbb{R})$, where $\mathbb{M}$ is a manifold given by the data acquisition geometry. Next, let $\theta = (\lambda, \alpha)$, where $\lambda = (\lambda_1, \ldots, \lambda_M)$ and $\alpha = (\alpha_1, \ldots, \alpha_N)$ are non-negative scalar parameters. We also assume that the loss $\ell_X \colon X \times X \to \mathbb{R}$ is convex, proper and weak* lower semicontinuous. We next study the bilevel problem

$$\begin{cases} \widehat{\theta} \in \arg\min_{\theta}[\ell_X(\mathcal{R}_\theta(g), f)], \\ \mathcal{R}_\theta(g) = \arg\min_{f \in X}\{\mathcal{L}_\theta(\mathcal{A}(f), g) + \mathcal{S}_\theta(f)\}. \end{cases} \tag{4.4}$$

In the above, $\mathcal{L}_\theta \colon Y \times Y \to \mathbb{R}$ and $\mathcal{S}_\theta \colon X \to \mathbb{R}$ are defined for $\theta = (\lambda_i, \alpha_i)$ as

$$\mathcal{L}_\theta(\mathcal{A}(f), g) = \sum_{i=1}^{M} \lambda_i \mathcal{L}_i(\mathcal{A}(f), g) \quad \text{and} \quad \mathcal{S}_\theta(f) = \sum_{j=1}^{N} \alpha_j \| \mathcal{J}_j(f) \|_{\mathcal{M}(\Omega; \mathbb{R}^{m_j})}$$

$$(4.5)$$

where $\mathcal{L}_i \colon Y \times Y \to \mathbb{R}$ and $\mathcal{J}_j \in \mathcal{M}(\Omega; \mathbb{R}^{m_j})$ are given. Hence, in (4.4) the variational model is parametrized in terms of sums of different fidelity terms $\mathcal{L}_i$ and TV-type regularizers $\| \mathcal{J}_j(f) \|_{\mathcal{M}(\Omega; \mathbb{R}^{m_j})}$, weighted against each other with parameters λ_i and α_j (respectively). For $N = 1$ and $S_1 = D$ the distributional derivative, then

$$\| S_1(f) \|_{\mathcal{M}(\Omega; \mathbb{R}^{m_1})} = \mathrm{TV}(f).$$

This framework is the basis for the analysis of the learning model, in which convexity of the variational model and compactness properties in the space of functions of bounded variation are crucial for proving existence of an optimal solution: see De los Reyes, Schönlieb and Valkonen (2016). Richer parametrizations for bilevel learning are discussed in Chen *et al.* (2012, 2015), for example, where non-linear functions and convolution kernels are learned. Chen *et al.*, however, treat the learning model in finite dimensions, and a theoretical investigation of these more general bilevel learning models in a function space setting is a matter for future research.

In order to derive sharp optimality conditions for optimal parameters of (4.4) more regularity on the lower-level problem is needed. For shifting the problem (4.4) into a more regular setting, the Radon norms are regularized with Huber regularization and a convex, proper and weak* lower-semicontinuous smoothing functional $\mathrm{H} \colon X \to [0, \infty]$ is added to the lower-level problem, typically $\mathrm{H}(f) = \frac{1}{2} \| \nabla f \|^2$. In particular, the former is required for the single-valued differentiability of the solution map $(\lambda, \alpha) \mapsto f_{\alpha, \lambda}$, required by current numerical methods, irrespective of whether we are in a function space setting (see *e.g.* Rockafellar and Wets 1998, Theorem 9.56, for the finite-dimensional case). For parameters $\mu \geq 0$ and $\gamma \in (0, \infty]$, the lower-level problem in (4.4) is then replaced by

$$\mathcal{R}_\theta := \operatorname*{arg\,min}_{f \in X} \{ \mu \, \mathrm{H}(f) + \mathcal{L}(\mathcal{A}(f), g) + \mathcal{S}_\theta^\gamma(f) \}, \qquad (4.6)$$

with

$$\mathcal{S}_\theta^\gamma(f) := \sum_{j=1}^{N} \alpha_j \| \mathcal{J}_j(f) \|_{\mathcal{M}(\Omega; \mathbb{R}^{m_j})}^\gamma.$$

Here, $f \mapsto \| \mathcal{J}_j(f) \|_{\mathcal{M}(\Omega; \mathbb{R}^{m_j})}^\gamma$ is the Huberized TV measure, given as follows.

Definition 4.1. Given $\gamma \in (0, \infty]$, the Huber regularization for the norm $\| \cdot \|_2$ on $\mathbb{R}^n$ is defined by

$$
\|g\|_\gamma = \begin{cases} \|g\|_2 - \dfrac{1}{2\gamma}, & \|g\|_2 \geq \dfrac{1}{\gamma}, \\[2mm] \dfrac{\gamma}{2} \|g\|_2^2, & \|g\|_2 < \dfrac{1}{\gamma}. \end{cases}
$$

Then, for $\mu \in \mathcal{M}(\Omega; \mathbb{R}^{m_j})$ with Lebesgue decomposition $\mu = \nu L^n + \mu^s$ we have the Huber-regularized total variation measure,

$$
|\mu|_\gamma(V) := \int_v |\nu(x)|_\gamma \, \mathrm{d}x + |\mu^s|(V) \quad (V \subset \Omega \ \text{Borel-measurable}),
$$

and finally its Radon norm,

$$
\|\mu\|_{\mathcal{M}(\Omega; \mathbb{R}^{m_j})}^\gamma := \||\mu|_\gamma\|_{\mathcal{M}(\Omega; \mathbb{R}^{m_j})}.
$$

In all of these, we interpret the choice $\gamma = \infty$ to give back the standard unregularized total variation measure or norm. In this setting existence of optimal parameters and differentiability of the solution operator can be proved, and with this an optimality system can be derived: see De los Reyes *et al.* (2016). More precisely, for the special case of the TV-denoising model the following theorem holds.

Theorem 4.2 (TV denoising (De los Reyes *et al.* 2016)). Consider the denoising problem (2.2) where $g, f_{\text{true}} \in BV(\Omega) \bigcap L^2(\Omega)$, and assume $\mathrm{TV}(g) > \mathrm{TV}(f_{\text{true}})$. Also, let $\mathrm{TV}^\gamma(f) := \|Df\|_{\mathcal{M}(\Omega; \mathbb{R}^n)}^\gamma$. Then there exist $\bar{\mu}, \bar{\gamma} > 0$ such that any optimal solution $\alpha_{\gamma, \mu} \in [0, \infty]$ to the problem

$$
\begin{cases} \displaystyle\min_{\alpha \in [0, \infty]} \frac{1}{2} \|f_{\text{true}} - f_\alpha\|_{L^2(\Omega)}^2, \\[3mm] f_\alpha = \displaystyle\arg\min_{f \in BV(\Omega)} \left\{ \frac{1}{2} \|g - f\|_{L^2(\Omega)}^2 + \alpha \, \mathrm{TV}^\gamma(f) + \frac{\mu}{2} \|\nabla f\|_{L^2(\Omega; \mathbb{R}^n)}^2 \right\} \end{cases}
$$

satisfies $\alpha_{\gamma, \mu} > 0$ whenever $\mu \in [0, \bar{\mu}]$ and $\gamma \in [\bar{\gamma}, \infty]$.

Theorem 4.2 states that if g is a noisy image which oscillates more than the noise-free image f_{true}, then the optimal parameter is strictly positive, which is exactly what we would naturally expect. De los Reyes *et al.* (2016) proved a similar result for second-order TGV and ICTV regularization for the case when $X = Y$. The result was not extended to data with a general Y, but it is possible with additional assumptions on the parameter space.

Moreover, in much of the analysis for (4.4) we could allow for spatially dependent parameters α and λ. However, the parameters would then need to lie in a finite-dimensional subspace of $C_0(\Omega; \mathbb{R}^N)$: see De los Reyes and Schönlieb (2013) and Van Chung *et al.* (2017). Observe that Theorem 4.2 allows for infinite parameters α. Indeed, for regularization parameter learning it is important not to restrict the parameters to be finite, as this allows us to decide between TGV^2, TV and TV^2 regularization. De los Reyes *et al.* (2016) also prove a result on the approximation properties of the bilevel scheme with the smoothed variational model (4.6) as $\gamma \nearrow \infty$ and $\mu \searrow 0$. In particular, they prove that as the numerical regularization vanishes, any optimal parameters for the regularized models tend to optimal parameters of the original model (4.4) in the sense of Γ-convergence. Moreover, De los Reyes and Schönlieb (2013) take the limit as $\gamma \nearrow \infty$ but $\mu > 0$ fixed for the optimality system and derive an optimality system for the non-smooth problem. Recently Davoli and Liu (2018) expanded the analysis for bilevel optimization of total variation regularizers beyond (4.4) to a richer family of anisotropic total variation regularizers in which the parameter of the dual norm and the (fractional) order of derivatives becomes part of the parameter space that (4.4) is optimized over.

Remark 4.3. Let us note here that despite the apparent simplicity of only one parameter to optimize over in Theorem 4.2, even in the case of the forward operator $\mathcal{A} = \mathrm{id}$ being the identity, the bilevel optimization problem is non-convex in α (against common hypotheses previously stated in publications). Evidence for this provides the counter-example constructed by Pan Liu (private communication) in Figure 4.1. Here, the one-dimensional TV regularization problem for signal denoising is considered. Fed with a piecewise constant input g, the one-dimensional TV problem can be solved analytically (Strong *et al.* 1996) and its solution is denoted by $f(\alpha)$. Figure 4.1(b) shows the non-convexity of the associated L^2-loss function $I(\alpha) = \|f_{\mathrm{true}} - f(\alpha)\|^2$.

Computing optimal solutions to the bilevel learning problems requires a proper characterization of optimal solutions in terms of a first-order optimality system. Since (4.4)–(4.6) constitutes a PDE-constrained optimization problem, suitable techniques from this field may be utilized. For the limit cases, an additional asymptotic analysis needs to be performed in order to get a sharp characterization of the solutions as $\gamma \to \infty$ or $\mu \to 0$, or both.

Several instances of the abstract problem (4.4) have been considered in the literature. De los Reyes and Schönlieb (2013) considered the case with TV regularization in the presence of several noise models. They proved the Gâteaux differentiability of the solution operator, which led to the derivation of an optimality system. Thereafter they carried out an asymptotic analysis with respect to $\gamma \to \infty$ (with $\mu > 0$), obtaining an optimality system for the corresponding problem. In that case the optimization problem corresponds to one with variational inequality constraints and the characterization concerns C-stationary points. Also, De los Reyes *et al.* (2017) have investigated differentiability properties of higher-order regularization solution operators. They proved a stronger Fréchet-differentiability result

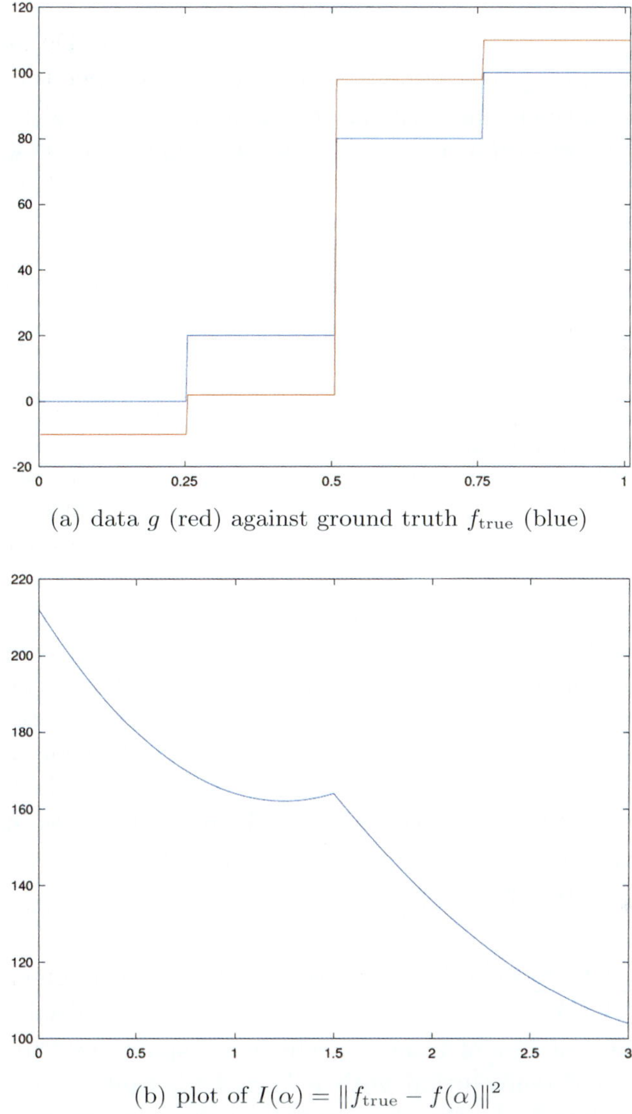

(a) data g (red) against ground truth f_{true} (blue)

(b) plot of $I(\alpha) = \|f_{\text{true}} - f(\alpha)\|^2$

Figure 4.1. Parameter optimality for TV denoising in Theorem 4.2. The non-convexity of the loss function, even for this one-parameter optimization problem, is clearly visible. Courtesy of Pan Liu.

for the TGV^2 case, which also holds for TV. In particular, these stronger results open the door to further necessary and sufficient optimality conditions. Further, using the adjoint optimality condition gradient, formulas for the reduced cost functional can be derived, which in turn feed into the design of numerical algorithms for solving (4.4): see Calatroni *et al.* (2016, Section 3).

4.3.2. From fields of experts to variational networks

The TV-type regularization approaches discussed above used popular hand-crafted variational regularizers in a first attempt to make them more data-driven, staying in the framework of optimizing proposed regularization models in infinite-dimensional function space. In what follows, we change our perspective to discrete variational regularization models (4.2) in which $X = \mathbb{R}^n$, and the functional $\mathcal{S}(\cdot)$ is equipped with parametrizations that go beyond TV-type regularizers; however, these are not covered by the theory in the previous section.

MRF parametrizations. A prominent example of such richer parametrizations of regularizers has its origins in MRF theory. MRFs, first introduced by Besag (1974) and then used by Cross and Jain (1983) for texture modelling, provide a probabilistic framework for learning image priors: see *e.g.* Zhu, Wu and Mumford (1998). Higher-order MRF models such as the celebrated Field of Experts (FoE) model (Roth and Black 2005) are the most popular ones, as they are able to express very complex image structures and yield good performance for various image processing tasks. The FoE model is learning a rich MRF image prior by using ideas from sparse coding (Olshausen and Field 1997), where image patches are approximated by a linear combination of learned filters, variations of which lead to the well-known principal component analysis (PCA) and independent component analysis (ICA), and the Product of Experts (PoE) model (Welling, Osindero and Hinton 2003).

In the context of bilevel learning, the FoE model has been used by Chen, Pock, Ranftl and Bischof (2013) and Chen, Ranftl and Pock (2014). Here, the regularizer is parametrized in the form

$$\mathcal{S}_\theta(f) := \sum_{i=1}^{N} \left[\alpha_i \sum_{k=1}^{n} \rho((J_i f)_k) \right] \quad \text{with } \theta = (\alpha_1, \ldots, \alpha_N, J_1, \ldots, J_N), \quad (4.7)$$

where N is the number of filters J_i which are sparse and implemented as a two-dimensional convolution with a kernel k_i, that is, $J_i f$ is the digitized version of $k_i * f$. In the FoE model we use the parametrization $J_i = \sum_{j=1}^{m} \beta_{i,j} B_j$, for a given set of basis filters $\{B_1, \ldots, B_m\}$ with $B_j \in \mathbb{R}^{n \times n}$. Moreover, the α_i are non-negative parameters and the non-linear function $\rho(z) = \log(1 + z^2)$. The shape of ρ is motivated by statistics over natural images that were shown to roughly follow a Student t-distribution (Huang and Mumford 1999). With these parametrizations, θ in (4.7) can be seen as $\theta = (\alpha_i, \beta_{i,j})$ for $i = 1, \ldots, n$ and $j = 1, \ldots, m$, and one learns the optimal θ from supervised training data (f_i, g_i) by minimizing the squared L^2-loss:

$$\ell_X(\mathcal{R}_\theta(g), f) := \| \mathcal{R}_\theta(g) - f \|_2^2 \quad \text{with } \mathcal{R}_\theta : Y \to X \text{ given by (4.4).}$$

Chen *et al.* (2014) investigated other non-linearities such as the non-smooth and non-convex $\rho(z) = \log(1 + |z|)$, and the non-smooth but convex ℓ_1-norm $\rho(z) = |z|$. Their experiments in particular suggested that, while the two log-type non-linearities gave very similar regularization performance, the convex ℓ_1-norm regularizer clearly did worse. Moreover, Chen *et al.* (2015) parametrized the non-linearities ρ with radial basis functions, whose coefficients are learned as well. Earlier MRF-based bilevel learning schemes exist: see *e.g.* Samuel and Tappen (2009).

A further development is the variational networks introduced by Kobler, Klatzer, Hammernik and Pock (2017) and Hammernik *et al.* (2018). The idea was to replace the above bilevel scheme with learning to optimize (Section 4.9) using a supervised loss, thereby leading to a learned iterative method. This will be discussed in more detail in Section 5.1.4, where the variational networks re-emerge in the framework of learned iterative schemes.

4.4. Dictionary learning in sparse models

In Section 2.7 we discussed sparsity as an important concept for modelling prior knowledge in inverse problems. Assuming that the model parameter possesses a sparse or compressible representation in a given dictionary $\mathbb{D}$ sparse recovery strategies, associated computational approaches and error estimates for the model parameter can be derived. In what follows, we turn to approaches which, rather than working with a given dictionary, learn the dictionary before or jointly with the recovery of the model parameter. Note that almost all work on dictionary learning in sparse models has been carried out in the context of denoising, *i.e.* $\mathcal{A} = \mathrm{id}$ only.

Patch-based local models. A particular class of sparse coding models are those which impose sparsity on patches (segmented) that are extracted from the signal. Let $g = \mathcal{A}(f_{\mathrm{true}}) + e$ and $\hat{f} \in X$. Further, let $\mathbb{D} = \{\phi_k\} \subset X$ be a dictionary, and $P_j : X \to X$ with $P_j(f) = f|_{\Omega_j}$ for $j = 1, \ldots N$ patches of f. Assume that

$$\hat{f} = \sum_{j=1}^{N} P_j(f_{\mathrm{true}}), \tag{4.8}$$

where $P_j(f_{\mathrm{true}}) \in X$ is compressible with respect to $\mathbb{D}$.

Some of the currently leading denoising methods are based on patch-based local models. Examples include K-SVD, which is a sparse coding approach on image patches (Elad and Aharon 2006), BM3D, which combines sparsity and self-similarity (Dabov, Foi, Katkovnik and Egiazarian 2007), and EPLL, which is a Gaussian mixture model of image patches (Zoran and Weiss 2011). Other examples are NCSR (Dong, Zhang, Shi and Li 2013), weighted nuclear norm regularization of image patches (WNNM) (Gu, Zhang, Zuo and Feng

2014) and SSC-GSM, a non-local sparsity approach with a Gaussian scale mixture (Dong, Shi, Ma and Li 2015).

Sparse-land models are a particular subclass of patch-based models. Here, each patch is sparse with respect to some global dictionary. Examples include sparse coding approaches that are applied patch-wise (Elad and Aharon 2006, Dong, Zhang, Shi and Wu 2011, Mairal and Ponce 2014, Romano and Elad 2015, Sulam and Elad 2015). In sparse-land models $\hat{f} \in X$ is reconstructed by solving

$$\min_{f \in X, \xi_i \in \Xi} \{ \mathcal{L}(\mathcal{A}(f), g) + \mathcal{S}_\theta(f, \xi_1, \ldots, \xi_N) \}, \tag{4.9}$$

where

$$\mathcal{S}_\theta(f, \xi_1, \ldots, \xi_N) := \sum_{j=1}^{N} [\lambda_j \| P_j(f) - \mathcal{E}_{\mathbb{D}}^*(\xi_j) \|_2^2 + \mu_j \| \xi_j \|_p^p], \tag{4.10}$$

with $\theta = (\lambda_j, \mu_j)_{j=1}^N \in (\mathbb{R}^2)^N$ and $\mathcal{E}_{\mathbb{D}}^* \colon \Xi \to X$ denoting the synthesis operator associated with the given dictionary $\mathbb{D}$. Bai *et al.* (2017) propose the following alternating scheme to solve the sparse-land reconstruction problem:

$$\begin{cases} f^{i+1} := \underset{f \in X}{\arg\min} \left\{ \mathcal{L}(\mathcal{A}(f), g) + \sum_{j=1}^{N} \lambda_j \| P_j(f) - \mathcal{E}_{\mathbb{D}}^*(\xi_j^j) \|_2^2 \right\} \\ \xi_j^{i+1} := \underset{\xi_i \in \Xi}{\arg\min} \{ \lambda_j \| P_j(f^{i+1}) - \mathcal{E}_{\mathbb{D}}^*(\xi_j) \|_2^2 + \mu_j \| \xi_j \|_p^p \} \quad \text{for } j = 1, \ldots, N. \end{cases} \tag{4.11}$$

The advantage of these approaches over plain-vanilla dictionary learning is that sparse-land models are computationally more feasible. Sparse-land models are one example of a dictionary learning approach, which will be discussed in the next section.

4.4.1. Reconstruction and dictionary learning

In sparse models, there are three ways to determine the dictionary. First, the dictionary is specified analytically. Second, the dictionary is determined from example data (dictionary learning). Third (as in the sparse-land models discussed above) the dictionary is determined jointly while performing reconstruction (joint reconstruction and dictionary learning). In what follows, we particularly focus on the second and third options for determining the dictionary.

Joint reconstruction and dictionary learning. The recent paper by Chambolle, Holler and Pock (2018) proposes a convex variational model for joint reconstruction and (convolutional) dictionary learning that applies to inverse problems where data are corrupted and/or incomplete. Chambolle *et al.* (2018) prove rigorous results on well-posedness and stability stated in the infinite-dimensional (non-digitized) setting.

Earlier approaches were less rigorous and considered the finite-dimensional setting with a patch-based dictionary (sparse-land model). One such example is adaptive dictionary-based statistical iterative reconstruction (ADSIR) (Zhang *et al.* 2016), which adds the dictionary $\mathbb{D} = \{\phi_i\} \subset X$ as an unknown to the sparse-land model as in global dictionary-based statistical iterative reconstruction (GDSIR) (Xu *et al.* 2012; see also Chun, Zheng, Long and Fessler 2017). This results in the following problem:

$$\min_{f \in X, \xi_i \in \Xi, \mathbb{D}} \{\mathcal{L}(\mathcal{A}(f), g) + \mathcal{S}_\theta(f, \xi_1, \ldots, \xi_N, \mathbb{D})\}, \tag{4.12}$$

where

$$\mathcal{S}_\theta(f, \xi_1, \ldots, \xi_N, \mathbb{D}) := \sum_{j=1}^{N} [\lambda_j \|P_j(f) - \mathcal{E}_{\mathbb{D}}^*(\xi_j)\|_2^2 + \mu_j \|\xi_j\|_p^p], \tag{4.13}$$

with $\theta = ((\lambda_j, \mu_j)_{j=1}^N) \in (\mathbb{R}^2)^N$, and $\mathcal{E}_{\mathbb{D}}^* : \Xi \to X$ denotes the synthesis operator associated with the dictionary $\mathbb{D}$. Usually an alternating minimization scheme is used to optimize over the three variables in (4.12).

Dictionary learning. Let $\ell_X : X \times X \to \mathbb{R}$ be a given loss function (*e.g.* the ℓ_2- or ℓ_1-norm). Further, let $f_1 \ldots f_N \in X$ be a given unsupervised training data, $\mathbb{D} = \{\phi_i\} \subset X$ a dictionary, and $\mathcal{E}_{\mathbb{D}}^* : \Xi \to X$ the synthesis operator given as $\mathcal{E}_{\mathbb{D}}^*(\xi) = \sum_i \xi_i \phi_i$ for $\xi \in \Xi$. One approach in dictionary learning is based on a sparsity assumption and solves

$$(\widehat{\mathbb{D}}, \hat{\xi}_i) \in \operatorname*{arg\,min}_{\xi_i \in \Xi, \mathbb{D} \subset X} \sum_{i=1}^{N} \ell_X(f_i, \mathcal{E}_{\mathbb{D}}^*(\xi_i)), \tag{4.14}$$

such that $\|\xi_i\|_0 \leq s$ for $i = 1, \ldots, N$, for a given sparsity level s. Alternatively, one looks for a dictionary that minimizes the total cost for representing signals in the training data while enforcing a constraint on the precision in the following way:

$$(\widehat{\mathbb{D}}, \hat{\xi}_i) = \operatorname*{arg\,min}_{\xi_i \in \Xi, \mathbb{D} \subset X} \sum_{i=1}^{N} \|\xi_i\|_0, \tag{4.15}$$

such that $\ell_X(f_i, \mathcal{E}_{\mathbb{D}}^*(\xi_i)) \leq \epsilon$ for $i = 1, \ldots, N$. A unified formulation is given by the unconstrained problem

$$(\widehat{\mathbb{D}}, \hat{\xi}_i) = \operatorname*{arg\,min}_{\xi_i \in \Xi, \mathbb{D} \subset X} \sum_{i=1}^{N} [\ell_X(f_i, \mathcal{E}_{\mathbb{D}}^*(\xi_i)) + \theta \|\xi_i\|_0]. \tag{4.16}$$

All three formulations are posed in terms of the ℓ_0-norm and are NP-hard to compute. This suggests the use of *convex relaxation*, by which (4.16) becomes

$$(\widehat{\mathbb{D}}, \hat{\xi}_i) = \underset{\xi_i \in \Xi, \mathbb{D} \subset X}{\arg\min} \sum_{i=1}^{N} [\ell_X(f_i, \mathcal{E}_{\mathbb{D}}^*(\xi_i)) + \theta\|\xi_i\|_1]. \tag{4.17}$$

If $\mathbb{D}$ is fixed then the sum in (4.17) decouples and leads to the convex relaxation of the *sparse coding* problem in (2.20) for $\mathcal{A} = \mathrm{id}$.

In the finite-dimensional setting $X = \mathbb{R}^m$ and Ξ is replaced by $\mathbb{R}^n$ for some n. Then the dictionary is $\mathbb{D} := \{\phi_k\}_{k=1\ldots n} \subset \mathbb{R}^m$ represented by an $n \times m$ matrix $\mathbf{D}$ leading to the synthesis operator becoming a mapping $\mathcal{E}_{\mathbf{D}}^* \colon \mathbb{R}^m \to \mathbb{R}^n$ with $\mathcal{E}_{\mathbf{D}}^*(\xi) = \mathbf{D} \cdot \xi$. In this setting (4.17) becomes

$$(\widehat{\mathbf{D}}, \hat{\xi}_i) := \underset{\xi_i \in \mathbb{R}^m, \mathbf{D} \in \mathbb{R}^{n \times m}}{\arg\min} \sum_{i=1}^{N} [\ell_X(f_i, \mathbf{D} \cdot \xi_i) + \theta\|\xi_i\|_1]. \tag{4.18}$$

Here, if $\mathbf{D}$ satisfies the RIP then the convex relaxation preserves the sparse solution (Candès *et al.* 2006). State-of-the-art dictionary learning algorithms are K-SVD (Aharon, Elad and Bruckstein 2006), geometric multi-resolution analysis (GRMA) (Allard, Chen and Maggioni 2012) and online dictionary learning (Mairal, Bach, Ponce and Sapiro 2010). Most work on dictionary learning to date has been done in the context of denoising, *i.e.* $\mathcal{A} = \mathrm{id}$; see also Rubinstein *et al.* (2010).

4.4.2. Convolutional sparse coding and convolutional dictionary learning

Dictionary learning in the context of sparse coding has been very popular and successful, with several seminal approaches arising in this field, as outlined in the previous section. However, there are still several issues with sparse coding related to the locality of learned structures and the computational effort needed. Sparse-land models (4.9), for instance, perform sparse coding over all the patches, and this tends to be a slow process.

The computational performance can be addressed by using learning to improve upon an optimization scheme (Section 4.9), for example the Learned Iterative Soft-Thresholding Algorithm (LISTA) (see Section 4.9.2) learns a finite number of unrolled Iterative Soft-Thresholding Algorithm (ISTA) iterates using unsupervised training data to match ISTA solutions (Gregor and LeCun 2010). Moreover, learning a dictionary over each patch independently as in (4.12) cannot account for global information, *e.g.* shift-invariance in images. What is needed is a computational feasible approach that introduces further structure and invariances in the dictionary, *e.g.* shift-invariance, and that makes each atom dependent on the whole signal instead of just individual patches. In this realm convolutional dictionaries have been introduced. Here atoms are given by convolution kernels and act on signal features via convolution, that is, $\mathbf{D}$ is a concatenation of Toeplitz matrices (a union of banded and circulant matrices). Set up like this, convolutional dictionaries render computationally feasible shift-invariant dictionaries, where atoms depend on the entire signal.

Convolutional sparse coding. Consider now the inverse problem of recovering $f_{\text{true}} \in X$ from (2.1) with the assumption that f_{true} is compressible with respect to convolution dictionary $\mathbb{D} := \{\phi_i\} \subset X$.

In convolutional sparse coding (CSC), this is done by performing a synthesis using convolutional dictionaries, that is, atoms act by convolutions. More precisely, the reconstruction operator $\mathcal{R}_\theta \colon Y \to X$ is given as

$$\mathcal{R}(g) := \sum_i \hat{\xi}_i * \phi_i, \tag{4.19}$$

where

$$\hat{\xi}_i \in \operatorname*{arg\,min}_{\xi_i \in X} \left\{ \mathcal{L}\left(\mathcal{A}\left(\sum_i \xi_i * \phi_i \right), g \right) + \lambda \|\xi_i\|_0 \right\}.$$

Computational methods for solving (4.19) for denoising use convex relaxation followed by the alternating direction method of multipliers (ADMM) in frequency space (Bristow, Eriksson and Lucey 2013) and its variants. See also Sreter and Giryes (2017) on using LISTA in this context. So far CSC has only been analysed in the context of denoising (Bristow *et al.* 2013, Wohlberg 2014, Gu *et al.* 2015, Garcia-Cardona and Wohlberg 2017) with theoretical properties given in Papyan, Sulam and Elad (2016*a*, 2016*b*).

Convolutional dictionary learning. Learning a dictionary in the context of CSC is called convolutional dictionary learning. Here, given unsupervised training data $f_1, \ldots, f_m \in X$ and a loss function $\ell_X \colon X \colon X \to X$, one solves

$$\operatorname*{arg\,min}_{\phi_i, \xi_{j,i} \in X} \left\{ \sum_{j=1}^m \ell_X\left(f_j, \sum_i \xi_{j,i} * \phi_i \right) + \lambda \sum_{j=1}^m \sum_i \|\xi_{j,i}\|_1 \right\}, \tag{4.20}$$

where $\|\phi_i\|_2 = 1$. For instance, Garcia-Cardona and Wohlberg (2017) solved (4.20) with a squared L_2-loss using an ADMM-type scheme. Extension of convolutional dictionary learning to a supervised data setting has been considered by Affara, Ghanem and Wonka (2018), for instance. Here, discriminative dictionaries instead of purely reconstructive ones are learned by introducing a supervised regularization term in the usual CSC objective that encourages the final dictionary elements to be discriminative.

Multi-layer convolutional sparse coding. A multi-layer extension of CSC, referred to as multi-layer convolutional sparse coding (ML-CSC), is proposed by Sulam, Papyan, Romano and Elad (2017). Given L convolutional dictionaries $\mathbb{D}_1, \ldots, \mathbb{D}_L \subset X$ with atoms $\mathbb{D}_k := \{\phi_j^k\}_j$, a model parameter $f \in X$ admits a representation in terms of the corresponding multi-layer convolutional sparse coding (ML-CSC) model if there are $s_1, \ldots, s_L \in \mathbb{R}$ such that

$$\begin{cases} f = \sum_j (\xi_j^1 * \phi_j^1), \\ \boldsymbol{\xi}^k = \sum_j (\xi_j^{k+1} * \phi_j^{k+1}) & \text{for } k = 1, \ldots, L-1, \end{cases}$$

and $\|\boldsymbol{\xi}^k\|_{0,\infty} \leq s_k$ for $k = 1, \ldots, L$. Hence, atoms $\phi_{k,i} \in \mathbb{D}_k$ in the kth convolution dictionary are compressible in the $(k+1)$th dictionary $\mathbb{D}_{k+1}$ for $k = 1, \ldots, L-1$.

The ML-CSC model is a special case of CSC where intermediate representations have a specific structure (Sulam *et al.* 2017, Lemma 1). Building on the theory for CSC, Sulam *et al.* (2017) provide a theoretical study of this novel model and its associated pursuits for dictionary learning and sparse coding in the context of denoising. Further, consequences for the theoretical analysis of CNNs can be extracted from ML-CSC using the fact that the resulting layered thresholding algorithm and the layered basis pursuit share many similarities with a forward pass of a deep CNN.

Indeed, Papyan, Romano and Elad (2017) show that ML-CSC yields a Bayesian model that is implicitly imposed on $\hat{f}$ when deploying a CNN, and that consequently characterizes signals belonging to the model behind a deep CNN. Among other properties, one can show that the CNN is guaranteed to recover an estimate of the underlying representations of an input signal, assuming these are sparse in a local sense (Papyan *et al.* 2017, Theorem 4) and the recovery is stable (Papyan *et al.* 2017, Theorems 8 and 10). Many of these results also hold for fully connected networks, and they can be used to formulate new algorithms for CNNs, for example to propose an alternative to the commonly used forward pass algorithm in CNN. This is related to both deconvolutional (Zeiler, Krishnan, Taylor and Fergus 2010, Pu *et al.* 2016) and recurrent networks (Bengio, Simard and Frasconi 1994).

An essential technique for proving the key results in the cited references for ML-CSC is based on unrolling, which establishes a link between sparsity-promoting regularization (compressed sensing) and deep neural networks. More precisely, one starts with a variational formulation like that in (4.20) and specifies a suitable iterative optimization scheme. In this setting one can prove several theoretical results, such as convergence, stability and error estimates. Next, one unrolls the truncated optimization iterates and identifies the updating between iterates as layers in a deep neural network (Section 4.9.1). The properties of this network, such as stability and convergence, can now be analysed using methods from compressed sensing.

Deep dictionary learning. Another recent approach in the context of dictionary learning is deep dictionary learning. Here, the two popular representation learning paradigms – dictionary learning and deep learning – come together. Conceptually, while dictionary learning focuses on learning a 'basis' and 'features' by matrix factorization, deep learning focuses on extracting features via learning 'weights' or 'filters' in a greedy layer-by-layer fashion. Deep dictionary learning, in turn, builds deeper architectures by using the layers of dictionary learning. See Tariyal, Majumdar, Singh and Vatsa (2016), who show that their approach is competitive against other deep learning approaches, such as stacked auto-encoders, deep belief networks and convolutional neural networks, regarding classification and clustering accuracies.

4.4.3. TV-type regularizer with learned sparsifying transforms.

The idea here is to learn the underlying sparsifying transform in a TV-type regularizer. One such approach is presented by Chun, Zheng, Long and Fessler (2017), who consider the case of a single sparsifying transform. This idea is further developed by Zheng, Ravishankar, Long and Fessler (2018), who consider a regularizer that is given by a union of sparsifying transforms (ULTRA), which act on image patches and quantify the sparsification error of each patch using its best-matched sparsifying transform: see Zheng *et al.* (2018, equation (3)). The resulting optimization problem is solved by an intertwined scheme that alternates between a CT image reconstruction step, calculated via a relaxed linearized augmented Lagrangian method with ordered subsets, and a sparse coding and clustering step, which simultaneously groups the training patches into a fixed number of clusters and learns a transform in each cluster along with the sparse coefficients (in the transform domain) of the patches.

A variant of Zheng *et al.* (2018) is considered by Ye, Ravishankar, Long and Fessler (2018*b*), who use the same pre-learned union of sparsifying transforms as in Zheng *et al.* (2018), but the alternating scheme also includes iteratively updating a quadratic surrogate functions for the data discrepancy. These variants of learned sparsifying transforms are all applied to low-dose CT image reconstruction. Finally, Chen *et al.* (2018) use essentially the method of Zheng *et al.* (2018) applied to cone-beam CT.

4.5. Scattering networks

Scattering networks refer to networks that share the hierarchical structure of deep neural networks but replace data-driven filters with wavelets (Mallat 2012, Bruna and Mallat 2013, Mallat 2016), thus providing an alternative method for parametrizing a regularizer (Dokmanić, Bruna, Mallat and de Hoop 2016). The networks can be made globally invariant to the translation group and locally invariant to small deformations, and they have many other desirable properties, such as stability and regularity. They also manifest better performance than regular CNNs for image classification problems and small-sample training data.

A key component in defining scattering networks is to have access to wavelet transforms at different scales, $\mathcal{W}_{s_k}\colon X \to X_{\mathbb{C}}$, where X is the set of real-valued functions on Ω and $X_{\mathbb{C}}$ is the set of complex-valued functions on Ω. We also have $\rho\colon X_{\mathbb{C}} \to X$, where $\rho(f) = |f|$. For N levels, we now define $\Gamma^N_{s_1,\dots,s_N}\colon X \to X$ by

$$\Gamma^N_{s_1,\dots,s_N}(f) = [\phi * (\rho \circ \mathcal{W}_{s_N}) \circ \dots \circ (\rho \circ \mathcal{W}_{s_1})](f),$$

and the scattering transform of order N is defined as

$$\Lambda^N(f) := (\phi * f, \Gamma^1_{s_1}(f), \dots, \Gamma^N_{s_1,\dots,s_N}(f))_{s_1,\dots,s_N \in \mathbb{Z}}.$$

The output of $\Lambda^N(f)$ is a hierarchically ordered sequence that forms the 'tree' of scattering coefficients up to order N.

Dokmanić *et al.* (2016) combine the Central Limit Theorem with some additional assumptions to conclude that $\Lambda^N(\mathbb{f})$ with $\mathbb{f} \sim \Pi_{\text{prior}}$ is approximately Gaussian with mean $\bar{f}$ and covariance Σ, where

$$\bar{f} := \mathbb{E}_{\mathbb{f} \sim \widehat{\Pi}_{\text{prior}}}[\Lambda^N(\mathbb{f})] \quad \text{and} \quad \Sigma = \mathrm{Cov}_{\mathbb{f} \sim \widehat{\Pi}_{\text{prior}}}[\Lambda^N(\mathbb{f})].$$

The regularizer $\mathcal{S}$ in (2.7) is now defined as

$$\mathcal{S} := \frac{1}{2}\|\bar{f} - \Lambda^N(f)\|^2_{\Sigma^{-1/2}}.$$

4.6. Black-box denoisers

Several approaches for solving (2.7) explicitly decouple the data discrepancy and regularization terms so that the latter can be treated with stand-alone methods. This is especially useful for cases where the data discrepancy term is differentiable but the regularizer is not.

The Plug-and-Play Prior (P^3) method. This approach, which was introduced in Venkatakrishnan, Bouman and Wohlberg (2013), is based on the observation that a split operator method for minimizing the objective in (2.7) can be posed as an equality constrained problem, that is,

$$\hat{f} := \arg\min_{f}\{\mathcal{L}(\mathcal{A}(f), g) + \lambda \mathcal{S}(f)\}$$

is equivalent to

$$(\hat{f}, \hat{h}) := \arg\min_{f,h}\{\mathcal{L}(\mathcal{A}(f), g) + \lambda \mathcal{S}(h)\} \quad \text{subject to } f = h.$$

The latter can be solved using ADMM (Section 8.2.7), where the update in h is computed using a proximal operator,

$$h^{k+1)} = \mathrm{prox}_{\tau \lambda \mathcal{S}}(h^{(k)} - f + u), \tag{4.21}$$

where u is a Lagrange (dual) variable.

The idea is now to replace the proximal operator with a generic denoising operator, which implies that *the regularization functional $\mathcal{S}$ is not necessarily explicitly defined.* This opens the door to switching in any demonstrably successful denoising algorithms without redesigning a reconstruction algorithm – hence the name 'Plug-and-Play'. However, the method comes with some disadvantages. The lack of an explicit representation of the regularizer detracts from a strict Bayesian interpretation of the regularizer as a prior probability distribution in MAP estimation and prevents explicit monitoring of the change in posterior probability of the iterative estimates of the solution. Next, the method is by design tied to the ADMM iterative scheme, which may not be optimal and which requires a non-trivial tuning of parameters of the ADMM algorithm itself (*e.g.* the Lagrangian penalty weighting term). Finally, it is not provably convergent for arbitrary denoising 'engines'.

Regularization by denoising (RED). The RED method (Romano, Elad and Milanfar 2017*a*) is motivated by the P^3 method. It is a variational method where the reconstruction operator is given as in (2.7), where the regularization functional is *explicitly* given as

$$\mathcal{S}(f) := \langle f, f - \Lambda(f) \rangle \quad \text{for some } \Lambda \colon X \to X. \tag{4.22}$$

The Λ operator above is a general (non-linear) *denoising operator*: it can for example be a trained deep neural network. It does, however, need to satisfy two two key properties, which are justifiable in terms of the desirable features of an image denoiser.

Local homogeneity. The denoising operator should locally commute with scaling, that is,

$$\Lambda(cf) = c\Lambda(f)$$

for all $f \in X$ and $|c - 1| \leq \epsilon$ with ϵ small.[5]

Strong passivity. The derivative of Λ should have a spectral radius less than unity:

$$\rho(\partial\Lambda(f)) \leq 1.$$

This is justified by imposing a condition that the effect of the denoiser should not increase the norm of the model parameter:

$$\|\Lambda(f)\| = \|\partial\Lambda(f)f\| \leq \|\rho(\partial\Lambda(f))\| \, \|f\| \leq \|f\|.$$

The key implication from local homogeneity is that the directional derivative of Λ along f is just the application of the denoiser to the f itself:

$$\partial\Lambda(f)f = \Lambda(f).$$

---

[5] This is a less restrictive condition than requiring equality for all $c \geq 0$.

The key implication from strong passivity is that it allows for convergence of the proposed RED methods by ensuring convexity of their associated regularization functionals.

Defining $W\colon X \to \mathbb{R}$ implicitly through the relation $W(f)f = \Lambda(f)$ and assuming local homogeneity and strong passivity yields the following computationally feasible expression for the gradient of the regularizer:

$$\nabla\, \mathcal{S}(f) = f - \Lambda(f) = (\mathrm{id} - W(f))(f). \tag{4.23}$$

Here, the operator $(\mathrm{id} - W(f))$ is further interpreted as an 'image adaptive Laplacian-based' regularizer. The above allows one to implement the RED framework in any optimization such as gradient-descent, fixed-point or ADMM in contrast to the P^3 approach, which is coupled to the ADMM scheme. See also Reehorst and Schniter (2018) for further clarifications and new interpretations of the regularizing properties of the RED method.

4.7. Deep neural networks as regularization functionals

In this section we review two recent approaches (Lunz, Öktem and Schönlieb 2018, Li, Schwab, Antholzer and Haltmeier 2018*b*) to using deep learning to train a regularizer $\mathcal{S}_\theta\colon X \to \mathbb{R}$ in (2.7).

4.7.1. Adversarial regularizer

A recent proposal by Lunz *et al.* (2018) for the construction of data-driven regularizers is inspired by how discriminative networks are trained using modern generative adversarial network (GAN) architectures. Our aim is to learn a regularizer $\mathcal{S}_\theta$, which in some cases (Section 3.3.2) can be interpreted as being proportional to the negative log-likelihood of the prior Π_{prior} in a MAP estimator.

Consider the statistical setting in (3.4) and let $f_i \in X$ be samples of the X-valued random variable $\mathrm{f} \sim \Pi_{\mathrm{prior}}$. Likewise, let $g_i \in X$ be samples of the Y-valued random variable $\mathrm{g} \sim \sigma$ that are independent of the samples f_i, that is, we have unmatched training data. We also assume there exists a (potentially regularizing) pseudo-inverse $\mathcal{A}^\dagger\colon Y \to X$ to the forward operator $\mathcal{A}$ in (3.2) and define the measure $\rho \in \mathscr{P}_X$ as $\rho := \mathcal{A}^\dagger_{\#}(\sigma)$ for $\sigma \in \mathscr{P}(Y)$. Note that both Π_{prior} and σ are replaced by their empirical counterparts given by the training data $f_i \in X$ and $g_i \in Y$, respectively.

The idea is to train the regularizer $\mathcal{S}_\theta$ parametrized by a neural network (see Section 7.4 for an exemplar architecture and application of this approach) in order to discriminate between the distributions Π_{prior} and ρ, the latter representing the distribution of imperfect solutions $\mathcal{A}^\dagger(g_i)$. More specifically, we compute

$$\mathcal{S}_{\widehat{\theta}}\colon X \to \mathbb{R}, \tag{4.24}$$

where $\widehat{\theta} \in \arg\min_\theta L(\theta)$, with the loss function $\theta \mapsto L(\theta)$ defined as

$$L(\theta) := \mathbb{E}_{\mathbb{f} \sim \Pi_{\mathrm{prior}}}[G_1(\mathcal{S}_\theta(\mathbb{f}))] - \mathbb{E}_{\mathbb{f} \sim \rho}[G_2(\mathcal{S}_\theta(\mathbb{f}))]. \tag{4.25}$$

Here, $G_1, G_2 \colon \mathbb{R} \to \mathbb{R}$ are monotone functions that have to be chosen. Popular choices for G_i are logarithms, as in the original GAN paper (Goodfellow *et al.* 2014), and the G_i associated with the Wasserstein loss, as in Arjovsky, Chintala and Bottou (2017) and Gulrajani *et al.* (2017). The heuristic behind this choice for the loss function for training a regularizer is that a network trained with (4.25) will penalize noise and artefacts generated by the pseudo-inverse (and contained in ρ). When used as a regularizer, it will hence prevent these undesirable features from occurring. Note also that in practical applications, the measures $\Pi_{\mathrm{prior}}, \rho \in \mathscr{P}_X$ are replaced with their empirical counterparts $\widehat{\Pi}_{\mathrm{prior}}$ and $\hat{\rho}$, given by training data f_i and $\mathcal{A}^\dagger(g_i)$, respectively. The training problem in (4.24) for computing $\widehat{\theta}$ then reads as

$$\widehat{\theta} \in \arg\min_\theta \left\{ \frac{1}{m} \sum_{i=1}^{m} G_1(\mathcal{S}_\theta(f_i)) - \frac{1}{n} \sum_{i=1}^{n} G_2(\mathcal{S}_\theta(\mathcal{A}^\dagger(g_i))) \right\}. \tag{4.26}$$

We also point out that, unlike other data-driven approaches for inverse problems, the above method can be adapted to work with only unsupervised training data. A special case is to have $g_i \approx \mathcal{A}(f_i)$, which gives a unsupervised formulation of (4.26).

Lunz *et al.* (2018) chose a Wasserstein-flavoured loss functional (Gulrajani *et al.* 2017) to train the regularizer, that is, one solves (4.24) with the loss function

$$L(\theta) := \mathbb{E}_{\mathbb{f} \sim \Pi_{\mathrm{prior}}}[\mathcal{S}_\theta(\mathbb{f})] - \mathbb{E}_{\mathbb{f} \sim \rho}[\mathcal{S}_\theta(\mathbb{f})] + \lambda \mathbb{E}[(\|\nabla \mathcal{S}_\theta(\mathbb{f})\| - 1)_+^2]. \tag{4.27}$$

The last term in the loss function serves to enforce the trained regularizer $\mathcal{S}_\theta$ to be Lipschitz-continuous with constant one (Gulrajani *et al.* 2017). Under appropriate assumptions on ρ and π (see Assumptions 4.4 and 4.5) and for the asymptotic case of $\mathcal{S}_{\widehat{\theta}}$ having been trained to perfection, the loss (4.27) coincides with the 1-Wasserstein distance defined in (B.2).

A list of qualitative properties of $\mathcal{S}_{\widehat{\theta}}$ can be proved: for example, Theorem 1 of Lunz *et al.* (2018) shows that under appropriate regularity assumptions on the Wasserstein distance between ρ and π, starting from elements in ρ and taking a gradient descent step of $\mathcal{S}_{\widehat{\theta}}$ (which results in a new distribution ρ_η) strictly decreases the Wasserstein distance between the new distribution ρ_η and π. This is a good indicator that using $\mathcal{S}_{\widehat{\theta}}$ as a variational regularization term, and consequently penalizing it, indeed introduces the highly desirable incentive to align the distribution of regularized solutions with the distribution of ground truth samples Π_{prior}. Another characterization of such a trained regularizer $\mathcal{S}_{\widehat{\theta}}$ using the Wasserstein loss in (B.2) is in terms of a distance function to a manifold of desirable solutions $\mathcal{M}$. To do so, we make the following assumptions.

Assumption 4.4 (weak data manifold assumption). Assume that the measure ρ is supported on a weakly compact set $\mathcal{M}$, *i.e.* $\rho(\mathcal{M}^c) = 0$

This assumption captures the intuition that real data lie in a lower-dimensional submanifold of X, which is a common assumption for analysing adversarial networks (Goodfellow *et al.* 2014). Moreover, it is assumed that the distributions π can be recovered from the distribution π through an appropriate projection onto the data manifold $\mathcal{M}$.

Assumption 4.5. Assume that π and ρ satisfy $(P_{\mathcal{M}})_{\#}(\rho) = \pi$, where $P_{\mathcal{M}}\colon D \to \mathcal{M}$ is the mapping $x \mapsto \arg\min_{y \in \mathcal{M}} \|x - y\|$. Here D denotes the set of points for which such a projection exists (which under weak assumptions on $\mathcal{M}$ and ρ can be assumed to cover all of ρ, *i.e.* $\rho(D) = 1$)

Note that Assumption 4.5 is weaker than assuming that any given f can be recovered by projecting the pseudo-inverse of the corresponding g back onto the data manifold. Assumption 4.5 can instead be considered as a low-noise assumption. These assumptions yield the following theorem.

Theorem 4.6 (Lunz, Öktem and Schönlieb 2018). The distance function to the data manifold $d_{\mathcal{M}}(x) := \min_{y \in \mathcal{M}} \|x - y\|$ is a maximizer to (B.2) under Assumptions 4.4 and 4.5.

Theorem 4.6 shows that, if $\mathcal{S}_\theta$ were trained to perfection, *i.e.* trained so that $\mathcal{S}_\theta$ solves (B.2), then $\mathcal{S}_\theta$ would be given by the L^2-distance function to $\mathcal{M}$. This is implicitly also done in the RED approach described in Section 4.6. Similarly, Wu, Kim, Fakhri and Li (2017) learn a regularizer in a variational model given as in Wu *et al.* (2017, equation (3)) from unsupervised data by means of a K-sparse auto-encoder. This yields a regularizer that minimizes the distance of the image to the data manifold.

Finally, a weak stability result for $\mathcal{S}_\theta$ is proved in the spirit of the classical theory provided in Engl *et al.* (2000). Since $\mathcal{S}_\theta$ is not necessarily bounded from below, the 1-Lipschitz property of $\mathcal{S}_\theta$ is used instead to prove this stability result.

Theorem 4.7 (Lunz, Öktem and Schönlieb 2018). Under appropriate assumptions on $\mathcal{A}\colon X \to Y$ given in Lunz *et al.* (2018, Appendix A), the following holds. Consider a sequence $\{g_n\}_n \subset Y$ with $g_n \to g$ in the norm topology in Y and let $\{f_n\} \subset X$ denote a sequence of corresponding minimizers of (2.7), that is,

$$f_n \in \arg\min_{f \in X}\{\|\mathcal{A}(f) - g_n\|^2 + \lambda\,\mathcal{S}_\theta(f)\}.$$

Then f_n has a subsequence that converges weakly to a minimizer of $f \mapsto \|\mathcal{A}(f) - g\|^2 + \lambda \mathcal{S}_\theta(f)$.

Theorem 4.7 constitutes a starting point for further investigation into the regularizing properties of adversarial regularizers $\mathcal{S}_\theta$. Section 7.4 shows the application of the adversarial regularizer to CT reconstruction.

4.7.2. The neural network Tikhonov (NETT) approach

Another proposal for learning a regularizer in (2.7) by a neural network is given in Li *et al.* (2018*b*) and called the NETT approach. NETT is based on composing a pre-trained network $\Psi_\theta \colon X \to \Xi$ with a regularization functional $\mathcal{S} \colon \Xi \to [0, \infty]$, such that $\mathcal{S} \circ \Psi_\theta \colon X \to [0, \infty]$ takes small values for desired model parameters and penalizes (larger values) model parameters with artefacts or other unwanted structures.

Here, $\Psi_\theta \colon X \to \Xi$ and $\mathcal{S} \colon \Xi \to [0, \infty]$ and the deep neural network Ψ_θ is allowed to be a rather general network model, a typical choice for Ψ_θ being an auto-encoder network. Once trained, the reconstruction operator is given by

$$\mathcal{R}_\theta := \arg\min_{f} \mathcal{J}_\theta(f), \tag{4.28}$$

where $\mathcal{J}_\theta(f) := \mathcal{L}(\mathcal{A}(f), g) + \lambda \mathcal{S}(\Psi_\theta(f))$.

The main focus of Li *et al.* (2018*b*) is a discussion on analytic conditions, which guarantees that the NETT approach is indeed a regularization method in the sense of functional analytic regularization. In particular, Li *et al.* discuss assumptions on $\mathcal{S}$ and Ψ_θ such that the functional analytic regularization theory of Grasmair *et al.* (2008) can be applied. This theory requires a weakly lower semicontinuous and coercive regularization term, which obviously holds for many deep neural networks Ψ_θ with coercive activation functions. Accordingly, Li *et al.* (2018*b*) discuss replacing the usual ReLU activation function with leaky ReLU, defined with a small $\tau > 0$ as

$$\ell\mathrm{ReLU}_\tau(s) := \max(\tau s, s).$$

For $s \to -\infty$, leaky ReLU also tends to $-\infty$, which in combination with the affine linear maps $\mathcal{W}$ in Ψ_θ yields a coercive and weakly lower semicontinuous regularization function $\mathcal{S} \circ \Psi_\theta$ for standard choices of $\mathcal{S}$, such as weighted ℓ_p-norms $\mathcal{S}(\xi) = \sum_i v_i |\xi_i|^p$ with uniformly positive weights v_i and $p \geq 1$. They even go beyond these classical results and, by introducing the novel concept of absolute Bregman distances (Li *et al.* 2018*b*), they obtain convergence results and convergence rates in the underlying function space norm.

Li *et al.* (2018*b*) discuss the application of NETT to PAT reconstruction from limited data. The neural network Ψ_θ is trained against supervised data $(f_i, h_i) \in X \times X$, where f_i serves as ground truth model parameter and $h_i = (\mathcal{A}^\dagger \circ \mathcal{A})(f_i)$, where $\mathcal{A}^\dagger \colon Y \to X$ is some (regularized) pseudo-inverse of the forward operator $\mathcal{A}$. In tomographic applications it is taken as FBP, so h_i will typically contain sampling artefacts. The network Ψ_θ is then modelled as an auto-encoder, more specifically as the encoder part of an encoder–decoder neural network. The parameters θ are trained by minimizing a loss function that evaluates the capability of the encoder–decoder pair to reproduce desirable f as well as $h = (\mathcal{A}^\dagger \circ \mathcal{A})(f)$ with artefacts using an appropriate distance function. After training Ψ_θ as described above, the NETT functional (4.28) is minimized by a generalized gradient descent method. The numerical results obtained with artificial data confirm the theoretical findings but lack comparison or seem to be slightly less favourable when compared with results obtained with other neural network approaches: see *e.g.* Hauptmann *et al.* (2018).

4.8. Learning optimal data acquisition schemes

Learning has also been used to determine data sampling patterns. An example is that of Baldassarre *et al.* (2016), who use training signals and develop a combinatorial training procedure which efficiently and effectively learns the structure inherent in the data. Thereby it is possible to design measurement matrices that directly acquire only the relevant information during acquisition. The resulting data sampling schemes not only outperform the existing state-of-the-art compressive sensing techniques on real-world datasets (including neural signal acquisition and magnetic resonance imaging): they also come with strong theoretical guarantees. In particular, Baldassarre *et al.* (2016) describe how to optimize the samples for the standard linear acquisition model along with the use of a simple linear decoder, and build towards optimizing the samples for non-linear reconstruction algorithms.

Gözcü *et al.* (2018) apply these techniques to MRI in order to learn optimal subsampling patterns for a specific reconstruction operator and anatomy, considering both the noiseless and noisy settings. Examples are for reconstruction operators given by sparsity-promoting variational regularization. The idea is to parametrize the data sampling (data acquisition and its digitization), then learn over these parameters in a supervised setting.

4.9. Data-driven optimization

Reconstruction methods given by variational methods or by the MAP estimator give rise to an optimization problem with an objective parametrized by $g \in Y$:

$$\mathcal{R}(g) := \arg\min_{f \in X} \mathcal{J}(f, g) \tag{4.29}$$

for $g \in Y$ and $\mathcal{J} \colon X \times Y \to \mathbb{R}$. A typical example is that of reconstruction methods in (2.7), where $\mathcal{J}(f, g) := \mathcal{L}(\mathcal{A}(f), g) + \mathcal{S}(f)$.

The *objective here is to use data-driven methods for faster evaluation of* $\mathcal{R}$, which is often computationally demanding. Given a parametrized family (an architecture) $\{\mathcal{R}_\theta\}_\theta$ of candidate solution operators, the 'best' approximation to $\mathcal{R}$ is given by $\mathcal{R}_{\widehat{\theta}} \colon Y \to X$, where $\widehat{\theta}$ solves the unsupervised learning problem:

$$\widehat{\theta} \in \arg\min_\theta \mathbb{E}_{\mathrm{g} \sim \sigma}[\mathcal{J}(\mathcal{R}_\theta(\mathrm{g}), \mathrm{g})]. \tag{4.30}$$

The probability distribution σ of the random variable g generating data is unknown. In (4.30) it is therefore replaced by its empirical counterpart derived from unsupervised training data $g_1, \ldots, g_m \in Y$ that are samples of $\mathrm{g} \sim \sigma$. In such a case, the unsupervised learning problem in (4.30) reads as

$$\widehat{\theta} \in \arg\min_\theta \left\{ \frac{1}{m} \sum_{i=1}^{m} \mathcal{J}(\mathcal{R}_\theta(g_i), g_i) \right\}, \tag{4.31}$$

where $g_1, \ldots, g_m \in Y$ are samples of $\mathrm{g} \sim \sigma$.

Note that the loss in (4.30) involves evaluations of the objective $f \mapsto \mathcal{J}(f, g)$, so training can be computationally quite demanding. However, the training is an off-line batch operation, so the computational performance requirements on $\mathcal{R}$ are much more relaxed during training than when $\mathcal{R}$ is used for reconstruction. Besides having access to a sufficient amount of unsupervised training data, a central part in successfully realizing the above scheme is to select an appropriate architecture for $\mathcal{R}_\theta$. It should be computationally feasible, yet one should be able to approximate $\mathcal{R}$ with reasonable accuracy by solving (4.31).

An early example in the context of tomographic reconstruction is that of Pelt and Batenburg (2013) and Plantagie and Batenburg (2015). Here, $\mathcal{R}$ is the TV-regularized reconstruction operator, which in many imaging applications is considered computationally unfeasible. Each $\mathcal{R}_\theta$ is given as a non-linear combination of FBP reconstruction operators that are fast to compute. These non-linear combinations of the reconstruction filters in the FBP reconstruction operators are all learned by training against the outcome of the TV regularization as in (4.31). The architecture used in the above-cited publications uses FBP operators, which only makes sense for inverse problems involving inversion of the ray transform. Another line of development initiated by Gregor and LeCun (2010) considers deep neural network architectures for approximating reconstruction operators $\mathcal{R}$ given by (2.20) (sparse coding). The idea is to incorporate selected components of an iterative scheme, in this case ISTA (Section 8.2.7), to solve the optimization problem in (2.20). This is done by 'unrolling' the iterative scheme and replacing its explicit updates with learned ones, which essentially amounts to optimizing over optimization solvers; see also Andrychowicz *et al.* (2016) for similar work. This has been used for more efficient solution of variational problems arising in a large-scale inverse problem. As an example, Hammernik, Knoll, Sodickson and Pock (2016) and Lee, Yoo and Ye (2017) consider unrolling for compressed sensing MRI reconstruction and Schlemper *et al.* (2018) for dynamic two-dimensional cardiac MRI, the latter aiming to solve the variational problem (Schlemper *et al.* 2018, equation (4)). Similarly, Meinhardt, Moeller, Hazirbas and Cremers (2017) consider unrolling for deconvolution and demosaicking.

Remark 4.8. The idea of optimizing over optimization solvers also appears in *reinforcement learning*. Furthermore, a similar problem is treated in Drori and Teboulle (2014), which considers the worst-case performance

$$\sup_{\theta, g}\{\mathcal{J}(\mathcal{R}_\theta(g), g) - \mathcal{J}(\mathcal{R}(g), g)\}$$

with $\mathcal{R}_\theta$ given by a gradient-based scheme that is stopped after N steps. It is assumed here that $f \mapsto \mathcal{J}(f, g)$ is continuously differentiable with Lipschitz-continuous gradients, and with a uniform upper bound on the Lipschitz constants. Subsequent work along the same lines can be found in Kim and Fessler (2016) and Taylor, Hendrickx and Glineur (2017).

4.9.1. General principle of unrolling

The aim of unrolling is to find a deep neural network architecture $\mathcal{R}_\theta \colon Y \to X$ that is especially suited to approximating an operator $\mathcal{R} \colon Y \to X$ that is implicitly defined via an iterative scheme. A typical example is when $\mathcal{R}(g)$ is implicitly defined via an iterative scheme that is designed to converge to a (local) minimum of $f \mapsto \mathcal{J}(f, g)$.

We start by giving an illustrative example of how to unroll an iterative gradient descent scheme. This is followed by an outline of the general principle of unrolling. Two specific cases are considered in Sections 4.9.2 and 4.9.3.

Unrolled gradient descent as a neural network. Consider the case when $\mathcal{J}(\,\cdot\,, g) \colon X \to \mathbb{R}$ is smooth for any $g \in Y$ and assume $\mathcal{R}(g)$ is given by a standard gradient descent algorithm, that is,

$$\mathcal{R}(g) = \lim_{k \to \infty} f^k \quad \text{where} \quad \begin{cases} f^0 = f_0 & \text{is given,} \\ f^k = f^{k-1} - \omega_k \nabla_f \mathcal{J}(f^k, g) & \text{for } k = 1, 2, \ldots. \end{cases}$$

Each parameter ω_k is a step length for the kth iteration, and normally these are chosen via the Goldstein rule or backtracking line search (Armijo rule) (Bertsekas 1999), which under suitable conditions ensures convergence to a minimum.

Computational feasibility of the above scheme is directly tied to the number of iterates necessary to get sufficiently close to the desired local minima. Time limitations imposed by many applications limit the maximum number of evaluations of $f \mapsto \mathcal{J}(f, g)$, that is, we have an *a priori* bound N on the number of iterates. Unrolling the gradient descent scheme defining $\mathcal{R}$ and stopping the iterates after N steps allows us to express the Nth iterate as

$$\mathcal{R}_\theta(g) := (\Lambda_{\omega_N} \circ \ldots \circ \Lambda_{\omega_1})(f_0) \quad \text{with } \theta := (\omega_1, \ldots, \omega_N), \tag{4.32}$$

where $\Lambda_{\omega_k} \colon X \to X$ are updating operators given by

$$\Lambda_{\omega_k} := \mathrm{id} - \omega_k \nabla \mathcal{J}(\,\cdot\,, g) \quad \text{for } k = 1, \ldots, N. \tag{4.33}$$

Now, note that $\mathcal{R}_\theta$ can be seen as a *feed-forward neural network* where each layer in the network evaluates Λ_{ω_k} and the parameters of the network are $\theta = (\omega_1, \ldots, \omega_N)$. Moreover, if the step length is fixed, *i.e.* $\omega_1 = \cdots = \omega_N = \omega$ for some ω, the gradient descent algorithm can in fact be interpreted as a *recurrent neural network*. For both cases, the best choice of step lengths $\omega_1, \ldots, \omega_N$ for approximating $\mathcal{R} \colon Y \to X$ with N gradient descent iterates can be obtained by unsupervised learning: simply solve (4.31) with $\mathcal{R}_\theta$ as in (4.32).

A further option is to replace the explicit updating in (4.33), used to define $\mathcal{R}_\theta$ in (4.32), with generic deep neural networks:

$$\Lambda_{\theta_k} := \mathrm{id} + \Gamma_{\theta_k}(\,\cdot\,, \nabla \mathcal{J}(\,\cdot\,, g)) \quad \text{where } \Gamma_{\theta_k} \colon X \times X \to X. \tag{4.34}$$

Each Γ_{θ_k} is now a deep neural network, for example a CNN, that is trained against unsupervised data by solving (4.31) with $\mathcal{R}_\theta$ as in (4.34).

Abstract unrolled schemes. The above example of unrolling a gradient descent and replacing its explicit updates with a neural network trained by unsupervised learning applies to many other iterative schemes. It essentially amounts to optimizing over optimization solvers. Since we seek $g \mapsto \mathcal{R}(g)$ given by (4.29), we are not only interested in optimizing a single objective, but rather an infinite family $\{\mathcal{J}(\,\cdot\,, g)\}_g$ of objectives parametrized by data $g \in Y$.

More precisely, if $f \mapsto \mathcal{J}(f, g)$ in (4.29) is smooth, then it is natural to consider an iterative scheme that makes use of the gradient of the objective. Given an initial model parameter $f^0 \in X$ (usually set to zero), such schemes can be written abstractly as

$$\begin{cases} f^0 = f_0 \in X \text{ chosen,} \\ f^{k+1} := \Gamma_{\theta_k}(f^k, \boldsymbol{f}_m^k, \nabla_f \mathcal{J}(f^k, g)), \end{cases}$$

for some updating operator $\Gamma_{\theta_k} \colon X \times X^m \times X \to X$. The above formulation includes accelerated schemes, like fast gradient methods (Nesterov 2004) and quasi-Newton methods (Nocedal and Wright 2006). Now, stopping and unrolling the above scheme after N iterates amounts to defining $\mathcal{R}_\theta\colon Y \to X$ with $\theta = (\theta_1, \ldots, \theta_N)$ as

$$\mathcal{R}_\theta(g) = (\Lambda_{\theta_N} \circ \ldots \circ \Lambda_{\theta_1})(f^0),$$

where $\Lambda_{\theta_k}\colon X \times X^m \to X \times X^m$ is defined as

$$\Lambda_{\theta_k}(f, \boldsymbol{h}) := (\Gamma_{\theta_k}(f, \boldsymbol{h}, \nabla_f \mathcal{J}(f, g)), \boldsymbol{h}') \quad \text{with } \boldsymbol{h}' := (h_2, \ldots, h_m) \in X^{m-1}$$

and $\mathcal{P}_X\colon X \times X^m \to X$ is the usual projection. The final step is to replace the hand-crafted updating operators $\Gamma_{\theta_k}\colon X \times X^m \times X \to X$ with deep neural networks that then train the resulting $\mathcal{R}_\theta\colon Y \to X$ against unsupervised data as in (4.31).

In inverse problem applications, the objective in (4.29) often has further structure that can be utilized. A typical example is (2.7), where the objective is of the form

$$\mathcal{J}(f, g) := \mathcal{L}(\mathcal{A}(f), g) + \mathcal{S}_\lambda(f). \tag{4.35}$$

A wide range of iterative schemes that better utilize such a structure include updating in both X (primal) and Y (dual) spaces. These can be written abstractly as

$$\begin{cases} (f^0, g^0) = (f_0, g) \in X \times Y \text{ with } f_0 \text{ chosen,} \\ (f^{k+1}, g^{k+1}) := \Gamma_{\theta_k}(f^k, \boldsymbol{f}^k_{m_1}, g^k, \boldsymbol{g}^k_{m_2}, [\partial\mathcal{A}(f^k)]^*(g^k), \mathcal{A}(f^k), \nabla\mathcal{S}_\lambda(f^k),) \end{cases} \tag{4.36}$$

for some updating operator

$$\Gamma_{\theta_k}\colon X \times X^{m_1} \times Y \times Y^{m_2} \times X \times Y \times X \to X \times Y. \tag{4.37}$$

Here, m_1, m_2 is the memory in the X- and Y-iterates, so

$$\boldsymbol{f}^k_{m_1} = (f^{k-1}, \ldots, f^{k-m_1}) \in X^{m_1} \quad \text{and} \quad \boldsymbol{g}^k_{m_2} = (g^{k-1}, \ldots, g^{k-m_2}) \in Y^{m_2},$$

with the convention that $f^{k-l} = f^0$ and $g^{k-l} = g$ whenever $k - l < 0$. Stopping and unrolling such a scheme after N iterates amounts to defining $\mathcal{R}_\theta\colon Y \to X$ with $\theta = (\theta_1, \ldots, \theta_N)$ as

$$\mathcal{R}_\theta(g) := (\mathcal{P}_X \circ \Lambda_{\theta_N} \circ \ldots \circ \Lambda_{\theta_1})(f^0, \boldsymbol{f}^0_{m_1}, g, \boldsymbol{g}^0_{m_2}), \tag{4.38}$$

where $\Lambda_{\theta_k}\colon X \times X^{m_1} \times Y \times Y^{m_2} \to X \times X^{m_1} \times Y \times Y^{m_2}$ for $k = 1, \ldots, N$ is defined as

$$\Lambda_{\theta_k}(f, \boldsymbol{h}, v, \boldsymbol{v}) := (\hat{f}_{\theta_k}, \boldsymbol{h}', \hat{v}_{\theta_k}, \boldsymbol{v}'),$$

with $\boldsymbol{h}' := (h_2, \ldots, h_{m_1}) \in X^{m_1-1}$, $\boldsymbol{v}' := (v_2, \ldots, v_{m_2}) \in X^{m_2-1}$, and

$$(\hat{f}_{\theta_k}, \hat{v}_{\theta_k}) := \Gamma_{\theta_k}(f, \boldsymbol{f}, v, \boldsymbol{v}, [\partial\mathcal{A}(f)]^*(v), \mathcal{A}(f), \nabla\mathcal{S}_\lambda(f)) \in X \times Y.$$

In the above, $\mathcal{P}_X\colon X \times X^{m_1} \times Y \times Y^{m_2} \to X$ is the usual projection and, just as before, these hand-crafted updating operators are replaced with deep neural networks and the resulting deep neural network $\mathcal{R}_\theta\colon Y \to X$ is trained against training data, for example as in (4.31) in the case with unsupervised data.

Likewise, in the setting where $\mathcal{S}_\lambda\colon X \to \mathbb{R}$ in (4.35) is non-smooth, the prototype proximal-gradient scheme without memory reads as

$$\begin{cases} f^{k+1/2} := \Lambda_{\theta_k^1}(f^k, \nabla_f \mathcal{L}(\mathcal{A}(f^k), g)), \\ f^{k+1} := \Lambda_{\theta_k^2}(f^k, f^{k+1/2}, \operatorname{prox}_{\gamma \mathcal{S}_\lambda}(f^{k+1/2})), \end{cases} \tag{4.39}$$

for updating operators $\Lambda_{\theta_k^1}\colon X \times X \to X$ and $\Lambda_{\theta_k^2}\colon X \times X \times X \to X$. Now, just as before, one can terminate and unroll the above scheme after N iterates and replace the updating operators with deep neural networks. The resulting deep neural network $\mathcal{R}_\theta\colon Y \to X$, where $\theta = (\theta_1^1, \ldots, \theta_N^1, \theta_1^2, \ldots, \theta_N^2)$, is then trained against unsupervised data as in (4.31). Just as in the smooth case, it is furthermore possible to add memory and/or 'break up' the objective even further, the latter in order to better account for the structure in the problem. For example, variational methods for linear inverse problems often result in minimizing an objective of the form

$$\mathcal{J}(f, g) := \mathcal{S}_\lambda(f) + \sum_{i=1}^m \mathcal{L}_i(\mathcal{A}_i(f), g_i)$$

where $\mathcal{A}_i\colon X \to Y_i$ are linear and $\mathcal{S}_\lambda\colon X \to [-\infty, \infty]$ and $\mathcal{L}_i\colon Y_i \times Y_i \to [-\infty, \infty]$ are proper, convex and lower semicontinuous. One can then consider unrolling iterative schemes that utilize this structure, such as operator splitting techniques: see Eckstein and Bertsekas (1992), Beck and Teboulle (2009), Chambolle and Pock (2011), Boyd *et al.* (2011), Combettes and Pesquet (2011, 2012), He and Yuan (2012), Boţ and Hendrich (2013), Boţ and Csetnek (2015), Ko, Yu and Won (2017), Latafat and Patrinos (2017) and Bauschke and Combettes (2017).

A natural question is to investigate the error in using $\mathcal{R}_\theta$ to approximate $\mathcal{R}$ as a function of N and properties of the unsupervised training data. Such estimates, which are referred to as time–accuracy trade-offs, have been proved for LISTA (Section 4.9.2) by Giryes, Eldar, Bronstein and Sapiro (2017); see also Oymak and Soltanolkotabi (2017) for further development along these lines. Another related type of investigation has been pursued by Banert *et al.* (2018), who derive conditions on the unrolled iterative scheme that is terminated after N iterates, so that training this yields a scheme that is convergent in the limit. It turns out that imposing such convergence constraints has only a minor impact on the performance at N iterates. Furthermore, it improves the generalization properties, that is, one can use the same trained solver even when the objective is slightly changed, for example by considering a different forward operator.

To summarize, the iterative update in most schemes for solving optimization problems has the structure of a composition of (possibly multiple) affine operations followed by a non-linear operation. If each iterate is identified with a layer and the non-linear operation plays the role of an activation function, *e.g.* $\mathcal{S}_\lambda(f) = \lambda\|f\|$, then $\mathrm{prox}_{\lambda\mathcal{S}}$ corresponds to soft thresholding, which in turn is very close to the well-known ReLU activation function. In this way the iterative scheme stopped after N steps can be represented via a neural network with an architecture that is adapted to that iterative scheme.

4.9.2. Learned Iterative Soft-Thresholding Algorithm (LISTA)

LISTA is the earliest example of unrolling an optimization scheme and it was first introduced in Gregor and LeCun (2010). It is the abstract unrolling scheme in Section 4.9.1 adapted to the specific case of ISTA iterates (Section 8.2.7). This results in a fully connected network with N-internal layers of identical size that is adapted to evaluating the solution operator for the following convex non-smooth optimization problem:

$$\mathcal{R}(g) := \mathcal{E}^*(\hat{\xi}), \tag{4.40}$$

where

$$\hat{\xi} \in \underset{\xi \in \Xi}{\arg\min}\{\|g - \mathcal{A}\circ\mathcal{E}^*(\xi)\|_2^2 + \lambda\|\xi\|_1\}.$$

Such optimization problems arise as the convex relaxation of (2.20), which is sparsity-promoting regularization of an ill-posed linear inverse problem (Section 2.7).

The ISTA iterative scheme for evaluating $\mathcal{R}$ reads as

$$\xi^{n+1} = \mathrm{prox}_{\lambda\tau\|\cdot\|_1} = S_{\lambda\tau}[\tau\,\mathcal{A}_{\mathcal{E}^*}^*\,g + (\mathrm{id} - \tau\,\mathcal{A}_{\mathcal{E}^*}^*\,\mathcal{A}_{\mathcal{E}^*})\xi^n], \tag{4.41}$$

where $0 < \tau < 2/L$ is the step length approximated by the reciprocal of the Lipschitz constant of $\mathcal{A}_{\mathcal{E}^*}$, and S is the shrinkage operator (see Section 8.2.7). Now the insight is to recognize $b_k(g) := \tau_k\,\mathcal{A}_{\mathcal{E}^*}^*\,g$ as a *bias*, $W_k := (\mathrm{id} - \tau_k\,\mathcal{A}_{\mathcal{E}^*}^*\,\mathcal{A}_{\mathcal{E}^*})$ as a fixed linear operator and $\psi_k = S_{\lambda\tau_k}$ as a *pointwise non-linearity*. Note that W_k is symmetric positive definite by construction. Assuming a fixed iteration number N, we may write the approximation to (4.40) as

$$\begin{cases} \xi^{(0)} \in \Xi \text{ given} \\ \xi^{(k)} = \psi(W_k\xi^{(k-1)} + b_k(g)) \end{cases} \qquad \text{for } k = 1,\dots,N.$$

In unrolled form, the above defines $\mathcal{R}_\theta\colon Y \to \Xi$ with $\theta_k := (\psi, \tau_k, W_k, b_k)$ as

$$\mathcal{R}_\theta := (\Lambda_{\theta_N} \circ \dots \circ \Lambda_{\theta_1})(\xi^{(0)}) \tag{4.42}$$

where $\Lambda_{\theta_k}\colon \Xi \to \Xi$ is $\Lambda_{\theta_k} := \psi_k \circ (W_k + b_k(g)\,\mathrm{id})$. This is precisely the form of a feed-forward network with N layers, each with identical fixed affine map $\mathcal{W}_k(\xi) := W_k\xi + b_k$; the activation function $\psi_k := S_{\lambda\tau_k}$ is given by the shrinkage operator (see Section 8.2.7).

Remark 4.9. The expression given in (4.42) has the form of an *encoder* $\Psi_W : Y \to \Xi$ for a fixed dictionary given by W.

The affine maps between the layers are assumed to be identical. LISTA then trains W and potentially ψ on some given unsupervised training data by solving (4.31). The derivation of Gregor and LeCun (2010) is framed in terms of learning a sparse encoder. Here we rephrase the idea in terms of a reconstruction problem and a learned *decoder* as follows.

Lemma 4.10. Let $\Psi_W^\dagger : \Xi \to X$ denote a fully connected network with input ξ^0 and N-internal layers. Furthermore, assume that the activation function is identical to a proximal mapping for a convex functional $\tau\lambda\,\mathcal{S} : \Xi \to \mathbb{R}$. Also, let W be restricted so that $\mathrm{id} - W$ is positive definite, *i.e.* there exists a matrix B such that

$$\mathrm{id} - W = \tau B^* B.$$

Finally, fix the bias term as $b = \tau B^* g$. Then $\Psi_W^\dagger(\xi)$ is the Nth iterate of an ISTA scheme with starting value $\xi^{(0)}$ for minimizing

$$\mathcal{J}_B(\xi) = \frac{1}{2}\|B\xi - g\|^2 + \lambda\,\mathcal{S}(\xi). \tag{4.43}$$

Note that the connection to ISTA is only given when the weights are the same across layers. The conclusion in the lemma follows directly from (A.10). In this sense, training a decoder network $\Psi_W^\dagger$ by minimizing

$$\frac{1}{2}\|\mathcal{A}\,\Psi_W^\dagger(\xi) - g\|^2$$

with respect to W, and computing $\hat{f} = \Psi_W^\dagger(\xi)$ is equivalent to computing $\hat{B}$ by minimizing a Tikhonov functional of the form (4.43) with respect to B, then computing $\hat{f} := \Psi_W^\dagger(\hat{B})$ as a solution to the inverse problem. Following these arguments, one can rephrase LISTA as a concept for learning the discrepancy term in a classical Tikhonov functional.

Remark 4.11. The restriction of activation functions to proximal mappings is not as severe as it might look at first glance. For example, as already mentioned in Section 4.9.1, ReLU is the proximal mapping for the indicator function of positive real numbers and soft shrinkage is the proximal mapping for the modulus function.

4.9.3. Learned proximal operators
Just as with LISTA, the *learned proximal operator* approach is the abstract unrolling scheme in Section 4.9.1 adapted to the specific case of unrolling

operator-splitting approaches, such as proximal gradient (PG), primal–dual hybrid gradient (PDHG) or ADMM; see also Section 8.2.7 and Chambolle and Pock (2016).

More precisely, consider the variational regularization that minimizes the functional in (4.35) under the Gaussian noise log-likelihood model with covariance Σ and a non-smooth regularizer, that is,

$$\mathcal{R}(g) := \hat{f}, \tag{4.44}$$

where

$$\hat{f} := \arg\min\{\lambda\,\mathcal{S}(f) + \|\mathcal{A}(f) - g\|_{\Sigma^{-1}}^2\}$$

and $\mathcal{S}\colon X \to \mathbb{R}$ is non-smooth. Our aim is to approximate $\mathcal{R}(g)$ in (4.44) by a deep neural network that is obtained by unrolling a suitable scheme.

Unrolling iterates of a PG method (Section 8.2.7) and replacing the proximal operator with a learned operator yields

$$\mathcal{R}_\theta(g) := f^N, \tag{4.45}$$

where

$$f^{k+1} := \Lambda_\theta(f^k + \tau\,\mathcal{A}^*\,\Sigma^{-1}(\mathcal{A}\,f^k - g)).$$

In the above, $\Lambda_\theta\colon X \to X$ is a deep neural network that replaces the proximal of $\mathcal{S}$ in (4.44). Similarly, the ADMM framework (Section 8.2.7) has the augmented Lagrangian

$$\mathcal{L}(f, h, u) = \frac{1}{2}\|\mathcal{A}(f) - g\|_{\Sigma^{-1}}^2 + \lambda\,\mathcal{S}(h) + \frac{\beta}{2}\left\|f - h + \frac{1}{\beta}u\right\|_2^2 - \frac{1}{2\beta}\|u\|_2^2, \tag{4.46}$$

which results in $\mathcal{R}_\theta(g) := f^N$, where

$$f^{k+1} := (\mathcal{A}^*\,\Sigma^{-1}\,\mathcal{A} + \beta\,\mathrm{id})^{-1}(\mathcal{A}^*\,\Sigma^{-1}(\mathcal{A}(f^k)) - g) + \beta(f^k - h^k) + u^k), \tag{4.47}$$

$$h^{k+1} := \Lambda_\theta\left(f^{k+1} + \frac{1}{\beta}u^k\right), \tag{4.48}$$

$$u^{k+1} := u^k + \beta(f^{k+1} - h^{k+1}), \tag{4.49}$$

where the general expression for the proximal operator of the log-likelihood from (A.12) has the direct form (4.47), and the general expression for the proximal operator of $\mathcal{S}$ from (A.13) is replaced with a network $\Lambda_\theta\colon X \to X$ in (4.48).

This compares to other P^3 approaches (Section 4.6) using a denoising method such as BM3D in place of the proximal operator. An important point noted in Meinhardt *et al.* (2017) is that by choosing $\beta = 1$ in (4.46), the networks do not need to be retrained on different noise levels if they can be considered as a scaling of the likelihood term. Related work includes 'deep image demosaicing' using a cascade of convolutional residual denoising networks (Kokkinos and Lefkimmiatis 2018, Lefkimmiatis 2017), which includes learning of the activation function.

A related method is that of Chang *et al.* (2017), who propose a general framework that implicitly learns a signal prior and a projection operator from a large image dataset and is predicated on the ADMM framework (Section 8.2.7). In place of an 'ideal' prior defined as the indicator function $\mathbb{I}_{X_0}\colon X \to \mathbb{R}$ on the set of natural images $X_0 \subset X$ and its proximal operator, they make use of a trained classifier $\mathcal{D}\colon X \to [0, 1]$ and a learned projection function $\mathcal{P}$ that maps an estimated $\hat{f} \in X$ to the set defined by the classifier; the learned $\mathcal{P}$ then replaces the proximal operator within the ADMM updates. They identify sufficient conditions for the convergence of the non-convex ADMM with the proposed projection operator, and use these conditions as guidelines to design the proposed projection network. They show that it is inefficient at solving generic linear inverse problems with state-of-the-art methods using specially trained networks. Experimental results also show that these are prone to being affected by changes in the linear operators and noise in the linear measurements. In contrast, the proposed method is more robust to these factors. Results are shown for the trained network applied to several problems including compressive sensing, denoising, in-painting (random pixels and block-wise) and super-resolution, and to different databases including MNIST, MS-Celeb-1M dataset and the ImageNet dataset.

4.9.4. Summary and concluding remarks on unrolling

The idea of unrolling can be seen as constructing a deep neural network architecture. Here we used it to approximate solution operators to optimization problems, which are defined implicitly through an iterative optimization solver. This principle can, however, be applied in a much wider context and it also establishes a link between numerical analysis and deep learning.

For example, the learned iterative scheme (Section 5.1.4) uses unrolling to construct a deep learning architecture for the solution of an inverse problem that incorporates a forward operator and the adjoint of its derivative. Another example of using unrolling is that of Gilton, Ongie and Willett (2019), who construct a deep neural network by unrolling a truncated Neumann series for the inverse of a linear operator. This is used to successfully solve inverse problems with a linear forward operator. Gilton *et al.* claim that the resulting Neumann network architecture outperforms functional analytic approaches (Section 2), model-free deep learning approaches (Section 5.1.3) and state-of-the-art learned iterative schemes (Section 5.1.4) on standard datasets.

One can also unroll iterative schemes for solving PDEs, as shown by Hsieh *et al.* (2019), who unroll an iterative solver tailored to a PDE and modify the updates using a deep neural network. A key part of their approach is ensuring that the learned scheme has convergence guarantees. See also Rizzuti, Siahkoohi and Herrmann (2019) for a similar approach to solving the Helmholtz equation.

Finally, we also mention Ingraham, Riesselman, Sander and Marks (2019), who unroll a Monte Carlo simulation as a model for protein folding. They compose a neural energy function with a novel and efficient simulator based on Langevin dynamics to build an end-to-end-differentiable model of atomic protein structure given amino acid sequence information.

4.10. Deep inverse priors

Deep inverse priors (DIPs) generalize deep image priors that were recently introduced in Ulyanov, Vedaldi and Lempitsky (2018) for some image processing tasks. We emphasize that deep inverse priors (DIPs) do not use a learned prior in the sense used so far in this paper. Deep inverse priors (DIPs), as we will see, more closely resemble non-linear Landweber schemes, but with a partly learned likelihood given by a trained neural network.

The choice of network design is crucial: it is assumed to provide a structural prior for the parameters or images to be reconstructed. More precisely, one assumes that the structure of a generator network is sufficient to capture most of the low-level image statistics prior to any learning. During the training of the network, more and more detailed information is added and an appropriate stopping criterion is essential to avoiding overfitting. In particular, a randomly initialized neural network can be used with excellent results in standard inverse problems such as denoising, super-resolution and in-painting.

For linear operators in a finite-dimensional setting, the task is to train a decoder network $\Psi_\theta^\dagger \colon \Xi \to X$ with fixed input $\xi_0 \in \Xi$. The reconstruction operator $\mathcal{R} \colon Y \to X$ is now given as the output of this decoder, that is,

$$\mathcal{R}(g) := \Psi_{\widehat{\theta}(g)}^\dagger(\xi_0), \qquad (4.50)$$

where

$$\widehat{\theta}(g) \in \arg\min_\theta \| \mathcal{A} \circ \Psi_\theta^\dagger(\xi_0) - g \|^2.$$

At the core of the DIP approach is the assumption that one can construct a (decoder) network $\Psi_\theta^\dagger \colon \Xi \to X$ which outputs elements in X, which are close to or have a high probability of belonging to the set of feasible parameters.

We emphasize that the training is with respect to θ: the input ξ_0 is kept fixed. Furthermore, machine learning approaches generally use large sets of training data, so it is somewhat surprising that deep image priors are trained on a single data set g. In summary, the main ingredients of deep inverse priors (DIPs) are a single data set, a well-chosen network architecture and a stopping criterion for terminating the training process (Ulyanov *et al.* 2018).

One might assume that the network architecture $\Psi_\theta^\dagger$ would need to incorporate rather specific details about the forward operator $\mathcal{A}$, or even more importantly about the prior distribution Π_{prior} of feasible model parameters. This seems not to be the case: empirical evidence suggests that rather generic network architectures work for different inverse problems, and the obtained numerical results demonstrate the potential of DIP approaches for large-scale inverse problems such as MPI.

So far, investigations related to DIP have been predominantly experimental and mostly restricted to problems that do not fall within the class of ill-posed inverse problems. However, some work has been done in the context of inverse problems: for example, Van Veen *et al.* (2018) consider the DIP approach to solve an inverse problem with a linear forward operator. They introduce a novel learned regularization technique which further reduces the number of measurements required to achieve a given reconstruction error. An approach similar to DIP is considered by Gupta *et al.* (2018) for CT reconstruction. Here one regularizes by projection onto a convex set (see Gupta *et al.* 2018, equation (3)) and the projector is constructed by training a U-Net against unsupervised data. Gupta *et al.* (2018, Theorem 3) also provide guarantees for convergence to a local minimum.

We now briefly summarize the known theoretical foundations of DIP for inverse problems based on the recent paper by Dittmer, Kluth, Maass and Baguer (2018), who analyse and prove that certain network architectures in combination with suitable stopping rules do indeed lead to regularization schemes, which lead to the notion of 'regularization by architecture'. We also include numerical results for the integration operator; more complex results for MPI are presented in Section 7.5.

4.10.1. DIP with a trivial generator

To better understand the regularizing properties of DIP, we begin by considering a trivial generator that simply takes a scalar value of unity on a single node (*i.e.* $\xi_0 = 1$) and outputs an element $h \in X$, that is, $\Psi_\theta^\dagger(\xi_0) = h$ independently of ξ_0. This implies simply that the network is a single layer with the value $\theta = h$, without any bias term or non-linear activation function. Then, the reconstruction operator in (4.50) reads as $\mathcal{R}(g) = \widehat{\theta}$, where

$$\widehat{\theta} \in \arg\min_\theta \| \mathcal{A}(\theta) - g \|^2. \tag{4.51}$$

In this setting θ can be directly identified with an element in X (and $\Xi \equiv X$), so training this network by gradient descent that seeks to minimize the objective in (4.51) is equivalent to the classical Landweber iteration. Despite its obvious trivialization of the neural network approach, this shows that there is potential for training networks with a single data point. Also, Landweber iterations converge from rather arbitrary starting points, indicating that the choice of ξ_0 in the general case is indeed of minor importance.

4.10.2. Analytic deep priors

The idea is based on the perspective that training a DIP network is, for certain network architectures, equivalent to minimizing a Tikhonov functional as in (2.10), that is,

$$\min_{f \in X}\left\{\frac{1}{2}\|\mathcal{B}(f) - g\|^2 + \lambda \mathcal{S}(f)\right\}. \tag{4.52}$$

This can be seen by considering networks similar to the unrolled schemes (see Section 4.9.2); that is, we consider a fully connected feed-forward network with N layers. We impose the further restriction that (i) the non-linearity (activation function) is identical to a proximal mapping $\mathrm{prox}_{\lambda \mathcal{S}}$ with respect to a convex functional $\mathcal{S}\colon X \to \mathbb{R}$, (ii) the affine linear mapping between layers allows the decomposition $\mathrm{id} - \mathcal{W} = \lambda \mathcal{B}^* \circ \mathcal{B}$ for some linear operator $\mathcal{B}\colon X \to Y$, and (iii) the bias term is fixed as $b = \lambda \mathcal{B}^* g$.

As described in Section 4.9.2, the output of a network using this architecture is equivalent to the Nth iterate of an ISTA scheme for approximating a minimizer of (4.52). With $\mathcal{W} = \mathrm{id} - \lambda \mathcal{B}^* \circ \mathcal{B}$, the network $\Psi_{\mathcal{W}}^\dagger(\xi_0)$ is given by the unrolled ISTA scheme (Section 4.9.1)

$$\begin{cases} f^0 = \xi_0, \\ f^{k+1} = \mathrm{prox}_{\lambda \mathcal{S}}(\mathcal{W}(f^k) + b), \\ \Psi_{\mathcal{W}}^\dagger(\xi_0) = f^N. \end{cases} \tag{4.53}$$

Such ISTA schemes converge as $N \to \infty$ for rather arbitrary starting points; hence, as pointed out above, the particular choice of ξ_0 in (4.50) is indeed of minor importance.

The starting point for a more in-depth mathematical analysis is the assumption that the above unrolled ISTA scheme has fully converged, that is,

$$\Psi_{\mathcal{W}}^\dagger(\xi_0) = \hat{f} := \arg\min_f \mathcal{J}_\mathcal{B}(f). \tag{4.54}$$

Using this characterization of $\hat{f}$, we define the *analytic deep prior* as the network, which is obtained by a gradient descent method with respect to $\mathcal{B}$ for

$$\mathcal{L}(\mathcal{B}, g) = \frac{1}{2}\|\mathcal{A}(\hat{f}) - g\|^2. \tag{4.55}$$

The resulting deep prior network has $\mathrm{prox}_{\lambda \mathcal{S}}$ as activation function, and the linear map $\mathcal{W}$ and its bias b are as described in (ii) and (iii). This allows us to obtain an explicit description of the gradient descent for $\mathcal{B}$, which in turn leads to an iteration of functionals $\mathcal{J}_\mathcal{B}$.

Below, we provide this derivation for a toy example that nevertheless highlights the differences between a classical Tikhonov minimizer and the solution of the DIP approach.

Simple example. We here examine analytic deep priors for linear inverse problems, *i.e.* $\mathcal{A}\colon X \to Y$ is linear, and compare them to classical Tikhonov regularization with $\mathcal{S}(f) = \frac{1}{2}\|f\|^2$. Let $f_\lambda \in X$ denote the solution obtained with the classical Tikhonov regularization, which by (2.11) can be expressed as

$$f_\lambda = (\mathcal{A}^* \circ \mathcal{A} + \lambda\,\mathrm{id})^{-1} \circ \mathcal{A}^*(g).$$

This is equivalent to the solution obtained by the analytic deep prior approach, with $\mathcal{B} = \mathcal{A}$ without any iteration. Now, take $\mathcal{B} = \mathcal{A}$ as a starting point for computing a gradient descent with respect to $\mathcal{B}$ using the DIP approach, and compare the resulting $\hat{f}$ with f_λ.

The proximal mapping for the functional R above is given by

$$\mathrm{prox}_{\lambda\mathcal{S}}(z) = \frac{1}{1+\lambda}z.$$

A rather lengthy calculation (see Dittmer, Kluth, Maass and Baguer 2018) yields an explicit formula for the derivative of F with respect to $\mathcal{B}$ in the iteration

$$\mathcal{B}^{k+1} = \mathcal{B}^k - \eta\partial F(\mathcal{B}^k).$$

The expression stated there can be made explicit for special settings. For illustration we assume the rather unrealistic case that $f^+ = h$, where $h \in X$ is a singular function for $\mathcal{A}$ with singular value σ. The dual singular function is denoted by $v \in Y$, *i.e.* $\mathcal{A}h = \sigma v$ and $\mathcal{A}^* v = \sigma h$, and we further assume that the measurement noise in g is in the direction of this singular function, *i.e.* $g = (\sigma + \delta)v$. In this case, the problem is indeed restricted to the span of h and the span of v, respectively. The iterates $\mathcal{B}^k$ only change the singular value β_k of h, that is,

$$\mathcal{B}^{k+1} = \mathcal{B}^k - c_k\langle\,\cdot\,, h\rangle v,$$

with a suitable $c_k = c(\lambda, \delta, \sigma, \eta)$.

Deep inverse priors for the integration operator. We now illustrate the use of deep inverse prior approaches for solving an inverse problem with the integration operator $\mathcal{A}\colon L^2([0,1]) \to L^2([0,1])$, defined by

$$\mathcal{A}(f)(t) = \int_0^t f(s)\,\mathrm{d}s. \tag{4.56}$$

Here $\mathcal{A}$ is linear and compact, hence the task of evaluating its inverse is an ill-posed inverse problem.

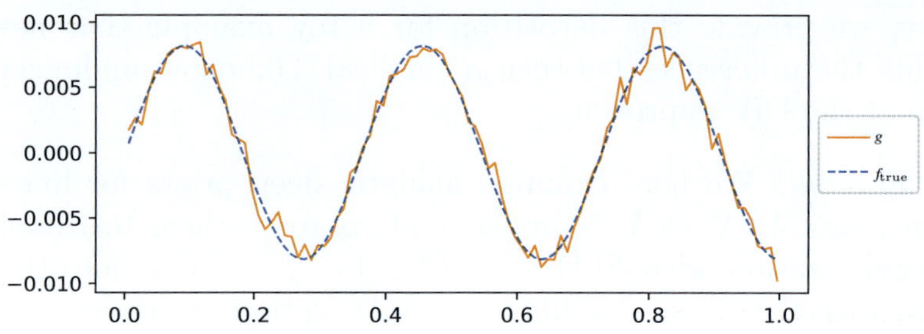

Figure 4.2. Example of g for $f_{\mathrm{true}} = u_5$ and 10% of noise.

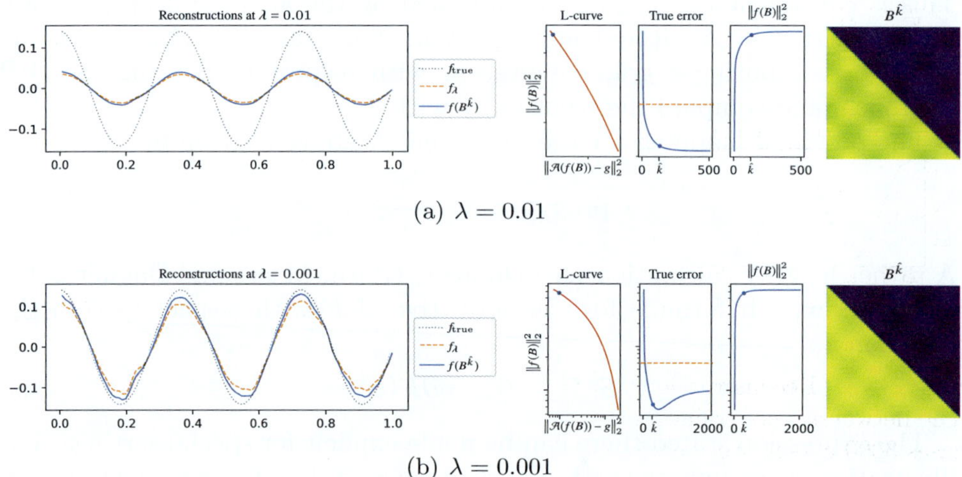

(a) $\lambda = 0.01$

(b) $\lambda = 0.001$

Figure 4.3. Comparison of Tikhonov reconstructions and results obtained with DIP. Reconstructions are shown for different fixed values of λ. The network was trained with the standard gradient descent method and a learning rate of 0.05. In (a) 500 epochs were used whereas in (b) 2000 were used.

Discretizing this operator with $n = 100$ yields a matrix $\mathbf{A}_n \in \mathbb{R}^{n \times n}$, which has $h/2$ on the main diagonal, h everywhere under the main diagonal and 0 above (here $h = 1/n$). We choose f_{true} to be one of the singular vectors of $\mathbf{A}$ and determine noisy data $g = \mathbf{A}_n f_{\mathrm{true}} + e$ with $e \sim \mathcal{N}(0, \sigma^2)$, where σ^2 is chosen as 10% of the largest coefficient of g; see Figure 4.2.

Figure 4.3 shows some reconstruction results with $N = 10$ layers. The first plot contains the true solution f_{true}, the standard Tikhonov solution f_λ, and the reconstruction obtained with the analytic deep inverse approach $f(\mathbf{B}_{\mathrm{opt}})$ after 2000 iterations for updating $\mathbf{B}$. For both choices of λ the training of $\mathbf{B}$ converges to a matrix $\mathbf{B}_{\mathrm{opt}}$, such that $f(\mathbf{B}_{\mathrm{opt}})$ has a smaller true error than f_λ. As can be observed in the last plot, the resulting matrix $\mathbf{B}_{\mathrm{opt}}$ contains some patterns that reflect what was predicted by the analytic deep prior.

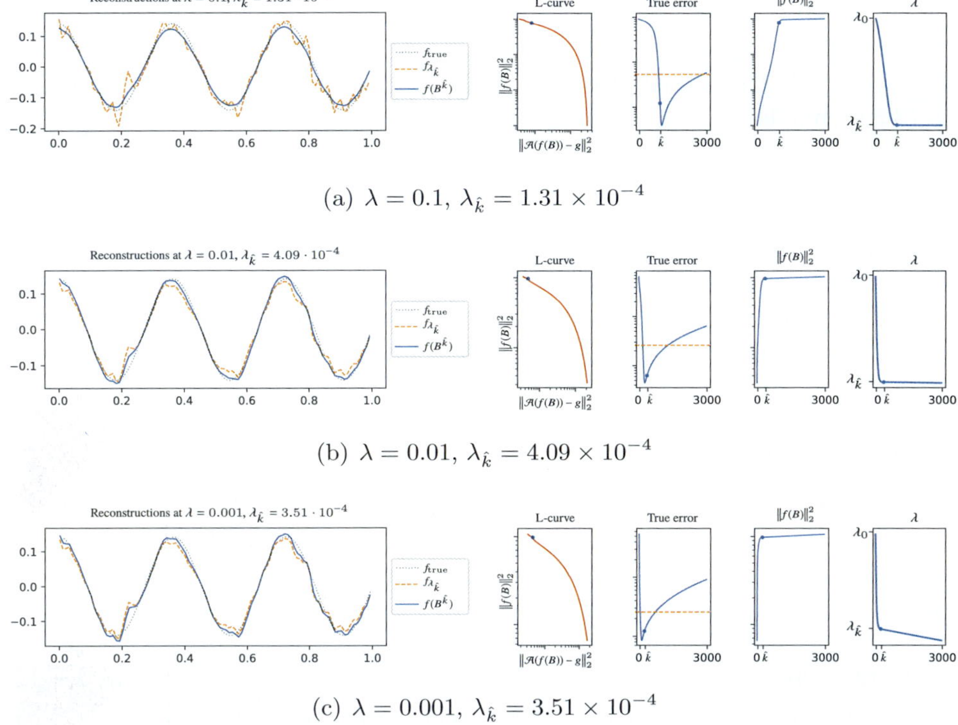

(a) $\lambda = 0.1$, $\lambda_{\hat{k}} = 1.31 \times 10^{-4}$

(b) $\lambda = 0.01$, $\lambda_{\hat{k}} = 4.09 \times 10^{-4}$

(c) $\lambda = 0.001$, $\lambda_{\hat{k}} = 3.51 \times 10^{-4}$

Figure 4.4. Reconstructions with an adaptive λ for different starting values λ_0. The networks were trained with gradient descent using 0.1 as learning rate. In all cases 3000 epochs were used.

So far, the regularization parameter λ has been assumed to be fixed. In a real application one needs to choose it via a technique such as the L-curve (Hansen 1992) or the discrepancy principle (4.1) in Section 4.1. However, they usually involve finding reconstructions for many different values of λ. In our case, that would mean retraining the network each time, which would lead to a really high computational cost. This motivates an adaptive choice of λ during the training, which could be achieved by letting λ also be a trainable weight of the network. The results for the same example and different starting values λ_0 are shown in Figure 4.4.

5. Learning in statistical regularization

The focus here is on various approaches for combining techniques from deep learning with Bayesian inversion and we begin by recapitulating the statistical setting in Section 3.1.2.

To recapitulate, we assume there exists a $(X \times Y)$-valued random variable $(\mathbb{f}, \mathbb{g}) \sim \mu$ that generates model parameters with associated data. The aim is to compute various estimators from the posterior Π^g_{post} where $g \in Y$ is single sample of $(\mathrm{g} \mid \mathrm{f} = f_{\mathrm{true}})$ with $f_{\mathrm{true}} \in X$ unknown. The data likelihood Π^f_{data}, which is the distribution of $(\mathrm{g} \mid \mathrm{f} = f)$, is here known for any $f \in X$ via (3.3), that is,

$$\mathrm{g} = \mathcal{A}(\mathrm{f}) + \mathrm{e},$$

with $\mathrm{e} \sim \Pi_{\mathrm{noise}}$ independent of f.

Ideally one would like to recover the entire posterior distribution $f \mapsto \Pi^g_{\mathrm{post}}$ for the measured data g. However, this is very challenging (Section 3.5), so many approaches settle for computing a selected estimator (Section 5.1). Alternatively, one may use deep neural nets to sample from the posterior, as surveyed in Section 5.2.

5.1. Learning an estimator

As outlined in Section 3.3, any reconstruction operator that can be represented by a deterministic measurable map $\mathcal{R}\colon Y \to X$ formally corresponds to a point estimator (also called a non-randomized decision rule). One can now use techniques from deep learning in computing such estimators.

5.1.1. Overview

There are various ways of combining techniques from deep learning with statistical regularization. The statistical characteristics of training data together with the choice of loss function determines the training problem one seeks to solve during learning. This in turn determines the type of estimator (reconstruction operator) one is approximating.

Supervised learning. The training data are given as samples $(f_i, g_i) \in X \times Y$ generated by $(\mathrm{f}, \mathrm{g}) \sim \mu$. One can then approximate the Bayes estimator, that is, we seek $\mathcal{R}_{\widehat{\theta}}\colon Y \to X$, where $\widehat{\theta}$ solves

$$\widehat{\theta} \in \arg\min_{\theta} \mathbb{E}_{(\mathrm{f},\mathrm{g})\sim\mu}[\ell_X(\mathcal{R}_\theta(\mathrm{g}), \mathrm{f})]. \tag{5.1}$$

The actual training involves replacing the joint law μ with its empirical counterpart induced by the supervised training data. Examples of methods that build on the above are surveyed in Section 5.1.2.

Learned prior. The training data $f_i \in X$ are samples generated by a μ_{f}-distributed random variable, where $\mu_{\mathrm{f}} \in \mathscr{P}_X$ is the f-marginal of μ. One can then learn the negative log prior density in a MAP estimator, that is, $\mathcal{R}_{\widehat{\theta}}\colon Y \to X$ is given by

$$\mathcal{R}_{\widehat{\theta}}(g) := \arg\min_{f \in X}\{-\log \pi_{\mathrm{data}}(g \mid f) + \mathcal{S}_{\widehat{\theta}}(f)\}.$$

Here $\pi_{\mathrm{data}}(\cdot \mid f)$ is the density for the data likelihood $\Pi^f_{\mathrm{data}} \in \mathscr{P}_Y$ and $\widehat{\theta}$

is learned such that $\mathcal{S}_{\widehat{\theta}}(f) \approx -\log(\pi_{\mathtt{f}}(f))$, with $\pi_{\mathtt{f}}$ denoting the density for $\mu_{\mathtt{f}} \in \mathscr{P}_X$, which is the $\mathtt{f}$-marginal of μ. The actual training involves replacing $\mu_{\mathtt{f}}$ with its empirical counterpart induced by the training data. Examples of methods that build on the above are surveyed in Section 4.7.

Unsupervised learning. The training data $\mathtt{g}_i \in Y$ are samples generated by a $\mu_{\mathtt{g}}$-distributed random variable where $\mu_{\mathtt{g}} \in \mathscr{P}_Y$ is the $\mathtt{g}$-marginal of μ. It is not possible to learn a prior in a MAP estimator from such training data, but one can improve upon the computational feasibility for evaluating a given MAP estimator. We do that by considering $\mathcal{R}_{\widehat{\theta}} \colon Y \to X$, where $\widehat{\theta}$ solves

$$\widehat{\theta} := \arg\min_{\theta} \mathbb{E}_{\mathtt{g} \sim \mu_{\mathtt{g}}}[-\log \pi_{\mathrm{data}}(\mathtt{g} \mid \mathcal{R}_\theta(\mathtt{g})) + \mathcal{S}_\lambda(\mathcal{R}_\theta(\mathtt{g}))]. \tag{5.2}$$

In the above, both the density $\pi_{\mathrm{data}}(\,\cdot\mid f)$ for the data likelihood $\Pi_{\mathrm{data}}^f \in \mathscr{P}_Y$ and the negative log density $\mathcal{S}_\lambda \colon X \to \mathbb{R}$ of the prior are handcrafted. The actual training involves replacing $\mu_{\mathtt{g}} \in \mathscr{P}_Y$ with its empirical counterpart induced by the training data. In the above, $\widehat{\mu}_{\mathtt{g}}$ is the empirical counterpart of $\widehat{\mu}_{\mathtt{g}}$ given by training data, $\mathcal{L} \colon Y \times Y \to \mathbb{R}$ is the negative data log-likelihood, and $\mathcal{S}_\lambda \colon X \to \mathbb{R}$ is the negative log-prior. The latter two are not learned. Examples of methods that build on the above are surveyed in Section 4.9.

Semi-supervised learning. The training data $\mathtt{f}_i \in X$ and $\mathtt{g}_i \in Y$ are semi-supervised, *i.e.* unpaired samples from the marginal distributions $\mu_{\mathtt{f}}$ and $\mu_{\mathtt{g}}$ of μ, respectively. One can then compute an estimator $\mathcal{R}_{\widehat{\theta}} \colon Y \to X$, where $\widehat{\theta}$ solves

$$\widehat{\theta} \in \arg\min_{\theta} \{ \mathbb{E}_{(\mathtt{f},\mathtt{g}) \sim \mu_{\mathtt{f}} \otimes \mu_{\mathtt{g}}}[\ell_Y(\mathcal{A}(\mathcal{R}_\theta(\mathtt{g})), \mathtt{g}) + \ell_X(\mathcal{R}_\theta(\mathtt{g}), \mathtt{f})]$$

$$+ \lambda \ell_{\mathscr{P}_X}((\mathcal{R}_\theta)_{\#}(\mu_{\mathtt{g}}), \mu_{\mathtt{f}}) \}. \tag{5.3}$$

In the above, $\ell_X \colon X \times X \to \mathbb{R}$ and $\ell_Y \colon Y \times Y \to \mathbb{R}$ are loss functions on X and Y, respectively. Next, $\ell_{\mathscr{P}_X} \colon \mathscr{P}_X \times \mathscr{P}_X \to \mathbb{R}$ is a distance notion between probability distributions on X and $(\mathcal{R}_\theta)_{\#}(\mu_{\mathtt{g}}) \in \mathscr{P}_X$ denotes the pushforward of the measure $\mu_{\mathtt{g}} \in \mathscr{P}_Y$ by $\mathcal{R}_\theta \colon Y \to X$. It is common to evaluate $\ell_{\mathscr{P}_X}$ using techniques from GANs, which introduce a separate deep neural network (discriminator/critic). Finally, the parameter λ controls the balance between the distributional consistency, noise suppression and data consistency. One can also consider further variants of the above, for example when there is access to a large sample of unpaired data combined with a small amount of paired data, or when parts of the probability distributions involved are known.

The choice of neural network architecture for the reconstruction operator $\mathcal{R}_{\widehat{\theta}} \colon Y \to X$ is formally independent of the choice of loss function and the set-up of the training problem. The choice does, however, impact the trainability of the learning, especially when there is little training data. In such cases, it is important to make use of all the information. In inverse problems one has explicit knowledge about how data are generated that comes in the form of a forward operator, or one might have an expression for the entire data likelihood. Architectures that embed such explicit knowledge, *e.g.* the forward operator and the adjoint of its derivative, perform better when there is little training data. They also have better generalization properties, and against adversarial attacks (Chakraborty *et al.* 2018, Akhtar and Mian 2018) they are more difficult to design since a successful attack needs to be consistent with how data are generated. Architectures that account for such information can be defined by unrolling (Section 4.9.4).

The remaining sections survey various approaches from the literature in computing the above estimators.

5.1.2. *Deep direct Bayes estimation*

The aim is to compute a Bayes estimator, which by the definition given in (3.14) amounts to finding a reconstruction operator $\mathcal{R}_\mu \colon Y \to X$ that solves

$$\mathcal{R}_\mu \in \underset{\mathcal{R} \colon Y \to X}{\arg\min} \, \mathbb{E}_{(\mathsf{f},\mathsf{g})\sim\mu}[\ell_X(\mathsf{f}, \mathcal{R}(\mathsf{g}))], \tag{5.4}$$

where $\ell_X \colon X \times X \to \mathbb{R}$ is a fixed loss function. The data likelihood is often known, and by the law of total probability we have

$$\mu(f, g) = \Pi_{\text{prior}}(f) \otimes \Pi_{\text{data}}^f, \tag{5.5}$$

so the joint law μ is known as soon as a prior has been selected.

As already mentioned (Section 3.5.1), selecting an appropriate prior that reflects the probability distribution of natural model parameters is very challenging, and current hand-crafted choices (Section 3.4) do not capture the full extent of the available *a priori* information about the true unknown model parameter $f_{\text{true}} \in X$. Next, the expression in (5.4) involves taking an expectation over $X \times Y$ as well as an optimization over all possible non-randomized decision rules. Both these operations easily become computationally overwhelming in large-scale inverse problem, such as those that arise in imaging.

These issues can be addressed by using techniques from supervised training. To start with, one can restrict the minimization in (5.4) to a family of reconstruction methods parametrized by a deep neural network architecture $\mathcal{R}_\theta \colon Y \to X$. Next, the unknown joint law μ can be replaced with its empirical counterpart given by supervised training data

$$\Sigma_m := \{(f_1, g_1), \ldots, (f_m, g_m)\} \subset X \times Y, \tag{5.6}$$

where (f_i, g_i) are generated by $(\mathsf{f}, \mathsf{g}) \sim \mu$. If there is a sufficient amount of such training data, then one can approximate the Bayes estimator in (5.4) by the neural network $\mathcal{R}_{\widehat{\theta}}\colon Y \to X$, where the finite-dimensional network parameter $\widehat{\theta} \in \Theta$ is learned from data by solving the following empirical risk minimization problem:

$$\widehat{\theta} \in \operatorname*{arg\,min}_{\theta \in \Theta} \frac{1}{m} \sum_{i=1}^{m} \ell_X(f_i, \mathcal{R}_\theta(g_i)), \tag{5.7}$$

where $(f_i, g_i) \in \Sigma_m$ as in (5.6). Note now that (5.7) does not explicitly require specifying a prior $f \mapsto \Pi_{\mathrm{prior}}(f)$ or a data likelihood $g \mapsto \Pi_{\mathrm{data}}^f$ that models how data are generated given a model parameter. Information about both of these is implicitly contained in supervised training data $\Sigma_m \subset X \times Y$.

Fully learned Bayes estimation (Section 5.1.3) refers to approaches where one assumes there is enough supervised training data to learn the joint law μ, that is, one disregards the explicit knowledge about the data likelihood. In contrast, *learned iterative schemes* (Section 5.1.4) include the information about the data likelihood by using an appropriate architecture of $\mathcal{R}_\theta\colon Y \to X$. *Learned post-processing* methods (Section 5.1.5) offer an alternative way to account for the data likelihood since these methods apply an initial reconstruction operator that maps data to a model parameter. This is actually an estimator different from the above Bayes estimator, but if the loss is the squared L^2-norm and the initial reconstruction operator is a linear sufficient statistic, then these estimators coincide in the 'large-sample' or 'small-noise' limit.

Regularizing the learning. The problem in (5.7) is ill-posed in itself, so one should not try to solve it in the formal sense. A wide range of techniques have been developed within supervised learning for implicitly or explicitly regularizing the empirical risk minimization problem in (5.7) as surveyed and categorized by Kukačka, Golkov and Cremers (2017). A key challenge is to handle the non-convexity, and the energy landscape for the objective in (5.7) typically has many local minima: for example, for binary classification there is an exponential number (in terms of network parameters) of distinct local minima (Auer, Herbster and Warmuth 1996).

Similar to Shai and Shai (2014, Section 2.1), we define a *training algorithm* for (5.7) as an operator mapping a probability measure on $X \times Y$ to a parameter in Θ that approximately solves (5.7):

$$\mathcal{T}\colon \mathscr{P}_{X \times Y} \to \Theta, \tag{5.8}$$

where $\mathcal{T}(\widehat{\mu}) \approx \widehat{\theta}$ with $\widehat{\theta} \in \Theta$ denoting a solution to (5.7). Thus, the training algorithm is a method for approximately solving (5.7) given a fixed neural network architecture. This also includes necessary regularization techniques: for example, a common strategy for solving (5.7) is to use some variant of stochastic gradient descent that is cleverly initialized (often at random from a specific distribution) along with techniques to ensure that the value of the objective (training error) decreases sufficiently rapidly. This, combined with early stopping (*i.e.* not fully solving (5.7)), warm-start, use of mini-batches, adding a regularization term to the objective, and so on, acts as a regularization (Ruder 2016, Kukačka *et al.* 2017, Bottou, Curtis and Nocedal 2018).

Concerning the model architecture, there is strong empirical evidence that the choice of neural network architecture has an influence on the ill-posedness of (5.7) (Li, Xu, Taylor and Goldstein 2018*c*, Draxler, Veschgini, Salmhofer and Hamprecht 2018). Many tasks that are successfully solved by supervised learning rely on (deep) neural networks, which can approximate a wide range of non-linear phenomena (large model capacity) without impairing computability. For this reason we consider deep neural networks to parametrize $\mathcal{R}_\theta\colon Y \to X$. Furthermore, empirical evidence indicates that deep neural network architectures yield a more favourable energy landscape for the objective of (5.7) than shallow ones: for example, most local minima are almost global (Choromanska *et al.* 2015, Becker, Zhang and Lee 2018). This intricate interplay between the choice of architecture and avoiding getting trapped in 'bad' local minima is poorly understood, and it is an active area of research within the machine learning community. Despite the lack of a theory, there is a consensus that an appropriate model architecture not only ensures computational feasibility but also acts as a kind of implicit regularization for (5.7).

To summarize, a training algorithm $\mathcal{T}$ as in (5.8) together with a specific model architecture regularizes (5.7), thereby resulting in the following approximation to the Bayes estimator (5.4):

$$\mathcal{R}_{\mathcal{T}(\widehat{\mu})}\colon Y \to X \tag{5.9}$$

for the empirical measure $\widehat{\mu}$ given by $\Sigma_m \subset X \times Y$ as in (5.6).

5.1.3. *Fully learned Bayes estimation*

Here the reconstruction operator $\mathcal{R}_\theta\colon Y \to X$ has a generic parametrization given by a deep neural network that does not explicitly account for the data likelihood.

An obvious difficulty with this approach is that the data space Y and model parameter space X are mathematically different. Without an explicit mapping from Y to X the action of convolutional layer operators cannot be properly defined. Therefore, fully learned approaches usually involve one or more 'fully connected layers' that represent a pseudo-inverse operator $\mathcal{B}_{\theta_1}\colon Y \to X$ mapping elements in Y to elements in X followed by a conventional neural network for $\mathcal{F}_{\theta_2}\colon X \to X$, that is, we get

$$\mathcal{R}_\theta := \mathcal{F}_{\theta_2} \circ \mathcal{B}_{\theta_1} \quad \text{with } \theta = (\theta_1, \theta_2). \tag{5.10}$$

In the discrete setting, where $Y = \mathbb{R}^n$ and $X = \mathbb{R}^m$, the simplest representation for $\mathcal{B}$ is a dense matrix $\mathbf{B} \in \mathbb{R}^{m \times n}$; the inclusion of an activation function makes this a non-linear mapping.

Initial examples of fully learned reconstruction in tomographic imaging include Paschalis *et al.* (2004) for SPECT imaging and Argyrou, Maintas, Tsoumpas and Stiliaris (2012) for transmission tomography. Both papers consider small-scale problems: for example, Paschalis *et al.* (2004) consider recovering 27×27 pixel SPECT images, and Argyrou *et al.* (2012) consider recovering 64×64 pixel images from tomographic data.

A more recent example is the automated transform by manifold approximation (AutoMap) method introduced in Zhu *et al.* (2018) as a tool for fully data-driven image reconstruction. Here, $\mathcal{R}_\theta \colon Y \to X$ is represented by a feed-forward deep neural network with fully connected layers followed by a sparse convolutional auto-encoder. This is in some sense similar to the Deep Cascade architecture in Schlemper *et al.* (2017), as it has one portion of the network for data consistency and the other for super-resolution/refinement of image quality. The encoder from data space to the model parameter space is implemented using three consecutive fully connected networks with sinh activation functions followed by two CNN layers with ReLU activation. We interpret this as a combination of a pseudo-inverse $\mathcal{R}_{\theta_3}^\dagger \colon Y \to X$ with a conventional convolutional auto-encoder:

$$\mathcal{R}_\theta = \underbrace{\Psi_{\theta_1}^\dagger \circ \Psi_{\theta_2}}_{\text{auto-encoder}} \circ \mathcal{R}_{\theta_3}^\dagger \quad \text{for } \theta = (\theta_1, \theta_2.\theta_3).$$

AutoMap was used to reconstruct 128×128 pixel images from MRI and PET imaging data. The dependence on fully connected layers results in a large number of neural network parameters that have to be trained. Primarily motivated by this difficulty, a further development of AutoMap is ETER-net (Oh *et al.* 2018), which uses a recurrent neural network architecture in place of the fully connected/convolutional auto-encoder architecture. Also addressing 128×128 pixel images from MRI, Oh *et al.* (2018) found a reduction in required parameters by over 80%. A method similar to AutoMap is used by Yoo *et al.* (2017) to solve the non-linear reconstruction problem in diffuse optical tomography. Here the forward problem is the Lippman–Schwinger equation but only a single fully connected layer is used in the backprojection step. Yoo *et al.* (2017) exploit the intrinsically ill-posed nature of the forward problem to argue that the mapping induced by the auto-encoder step is low-rank and therefore sets an upper bound on the dimension of the hidden convolution layers.

The advantage of fully learned Bayes estimation lies in its simplicity, since one avoids making use of an explicit forward operator (or data likelihood). On the other hand, any generic approach to reconstruction by deep neural networks requires having connected layers that represent the relation between model parameters and data. For this reason, generic fully learned Bayes estimation will always scale badly: for example, in three-dimensional tomographic reconstruction it is common to have an inverse problem which, when discretized, involves recovering a $(512 \times 512 \times 512 \approx 10^8)$-dimensional model parameter from data of the same order of magnitude. Hence, a fully learned generic approach would involve learning at least 10^{16} weights from supervised data! There have been several attempts to address the above issue by considering neural network architectures that are adapted to specific direct and inverse problems. One example is that of Khoo and Ying (2018), who provide a novel neural network architecture (SwitchNet) for solving inverse scattering problems involving the wave equation. By leveraging the inherent low-rank structure of the scattering problems and introducing a novel switching layer with sparse connections, the SwitchNet architecture uses far fewer parameters than a U-Net architecture for such problems. Another example is that of Ardizzone *et al.* (2018), who propose encoding the forward operator using a invertible neural network, also called a reversible residual network (Gomez, Ren, Urtasun and Grosse 2017). The reconstruction operator is then obtained as the inverse of the invertible neural network for the forward operator. However, it is unclear whether this is a clever approach to problems that are ill-posed, since an inverse of the forward operator is not stable. Another approach is that of Yoo *et al.* (2017), who apply an AutoMap-like architecture for non-linear reconstruction problems in diffuse optical tomography. Here the forward problem is the Lippman–Schwinger equation, but only a single fully connected layer is used in the backprojection step. Yoo *et al.* (2017) exploit the intrinsically ill-posed nature of the forward problem to argue that the mapping induced by the auto-encoder step is low-rank and therefore sets an upper bound on the dimension of the hidden convolution layers. The above approaches can also to some extent be seen as further refinements of methods in Section 4.2.

However, neither of the above efforts address the challenge of finding sufficient supervised training data necessary for the training. Furthermore, any changes to the acquisition protocol or instrumentation may require retraining, making the method impractical. In particular, due to the lack of training data, fully learned Bayes estimation is inapplicable to cases when data are acquired using novel instrumentation. A practical case would be spectral CT, where novel direct counting energy resolving detectors are being developed.

5.1.4. *Learned iterative schemes*

The idea here is to choose an architecture for $\mathcal{R}_\theta \colon Y \to X$ in (5.7) that contains an explicit expression for the data likelihood, which accounts for how a model parameter gives rise to data. This requires us to embed an explicit forward operator $\mathcal{A}\colon X \to Y$ into the architecture for $\mathcal{R}_\theta$, which is somewhat tricky since $\mathcal{R}_\theta$ and $\mathcal{A}$ are mappings that go in the reverse direction compared to each other.

One approach is presented by Mousavi and Baraniuk (2017), who suggest a CNN architecture (DeepInverse) adapted for solving a linear inverse problem. The architecture involves a fully connected layer (that is not learned) to represent the normal operator $\mathcal{A}^* \circ \mathcal{A}\colon X \to X$ followed by convolutional layers as in a regular CNN with ReLU activation, but here one dispenses with the downsampling (max-pooling operation) that is common in a CNN. The usefulness is limited, however, since the normal operator needs to have a certain structure for the sake of computational efficiency, for example when inverting the Fourier transform, which results in a block-circulant matrix.

Another class of methods is that of *learned iterative schemes*, which include a handcrafted forward operator and the adjoint of its derivative into the architecture by unrolling a suitable iterative scheme (Section 4.9.4). An early variant was presented by Yang, Sun, Li and Xu (2016), who define a learned iterative method based on unrolling an ADMM-type scheme. The network is trained against supervised data using a somewhat unusual asymmetric loss, namely

$$\ell_X(f, h) := \sqrt{\|f - h\|_2^2 / \|f\|_2^2}.$$

The trained network is used to invert the Fourier transform (MRI image reconstruction). However, the whole approach is unnecessarily complex, and it is now surpassed by learned iterative methods that have a more transparent logic. The survey will therefore focus on these latter variants.

Learned iterative in model parameter space. In the simplest setting, $\mathcal{R}_\theta$ in (5.7) is given as in (4.32) with an updating operator as in (4.34) where $\mathcal{J} := \mathcal{L}(\mathcal{A}(\,\cdot\,), g)$. Hence, given an initial model parameter $f^0 \in X$, we define $\mathcal{R}_\theta\colon Y \to X$ with $\theta = (\theta_1, \ldots, \theta_N)$ as

$$\mathcal{R}_\theta(g) := (\Lambda_{\theta_N} \circ \cdots \circ \Lambda_{\theta_1})(f^0), \tag{5.11}$$

where $\Lambda_{\theta_k} := \mathrm{id} + \Gamma_{\theta_k} \circ \nabla\mathcal{L}(\mathcal{A}(\,\cdot\,), g)$. In the above, $\Gamma_{\theta_k}\colon X \to X$ is learned from supervised data (5.6) by approximately solving (5.7) using a training algorithm as in (5.8). In contrast, $\nabla\mathcal{L}(\mathcal{A}(\,\cdot\,), g)\colon X \to X$ is not learned: it is derived from an explicit expression for the data likelihood. For example, a common choice for data where the observational noise is Gaussian is

$$\mathcal{L}(v, g) := \frac{1}{2}\|v - g\|_2^2 \implies \nabla\mathcal{L}(\mathcal{A}(f), g) = [\partial\mathcal{A}(f)]^*(\mathcal{A}(f) - g). \tag{5.12}$$

The reconstruction operator $\mathcal{R}_\theta\colon Y \to X$ in (5.11) can now be interpreted as a *residual neural network*, as popularized by He, Zhang, Ren and Sun (2016) for image classification. Furthermore, the operators $\Gamma_{\theta_k}\colon X \to X$ for

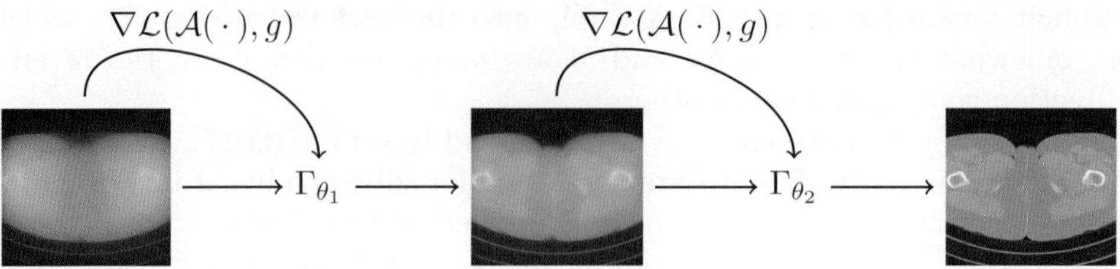

Figure 5.1. Learned iterative method in model parameter space. Illustration of the unrolled scheme in (5.11) for $N = 2$ in the context of CT image reconstruction (Section 7.3.1). Each $\Gamma_{\theta_1} \colon X \to X$ is a CNN, $g \in X$ is the measured data, and f^0 is an initial image, usually taken as zero.

$k = 1, \ldots, N$ are represented by deep neural networks with an architecture that is usually fixed: for example, in imaging problems one selects a suitable CNN architecture. See Figure 5.1 for an illustration of (5.11).

Following Section 4.9.1, the next level of complexity comes when the learned component is allowed to be less constrained by removing the explicit expression for $f \mapsto \nabla \mathcal{L}(\mathcal{A}(f),g)$ in (5.11) while keeping expressions for the forward operator and the adjoint of its derivative. This corresponds to defining $\mathcal{R}_\theta \colon Y \to X$ as in (4.38) but with no memory, that is,

$$\mathcal{R}_\theta(g) := (\mathcal{P}_X \circ \Lambda_{\theta_N} \circ \ldots \circ \Lambda_{\theta_1})(f^0, g), \qquad (5.13)$$

where $\mathcal{P}_X \colon X \times Y \to X$ is the usual projection and $\Lambda_{\theta_k} \colon X \times Y \to X \times Y$ for $k = 1, \ldots, N$ is

$$\Lambda_{\theta_k}(f, v) := \Gamma_{\theta_k}(f, v, [\partial \mathcal{A}(f)]^*(v), \mathcal{A}(f), \nabla \mathcal{S}_\lambda(f)) \quad \text{for } (f, v) \in X \times Y.$$

Here, $\Gamma_{\theta_k} \colon X \times Y \times X \times Y \times X \to X \times Y$ is the updating operator that is given by a deep neural network. The resulting deep neural network $\mathcal{R}_\theta \colon Y \to X$ is learned from supervised data (5.6) by approximately solving (5.7) using a training algorithm as in (5.8).

The constraints on the learning can be further decreased, at the expense of increased memory footprint and computational complexity, by allowing for some memory $l > 0$, that is, each of the learned updating operators account for more than the previous iterate. This leads to an architecture for $\mathcal{R}_\theta \colon Y \to X$ of the form (4.38) with updating operators as in (4.37). A special case of this formulation is the learned gradient method of Adler and Öktem (2017), who in turn present the recurrent inference machines of Putzky and Welling (2017) as special case.

Another special case is variational networks. These are defined by unrolling an iterative scheme for minimizing an explicit objective that has a data discrepancy component and a regularizer. The idea was introduced by Hammernik *et al.* (2016) for two-dimensional Fourier inversion, where the objective has a regularizer based on a reaction–diffusion model. Hammernik *et al.* (2018) develop it further: they consider a variational network (outlined in their Figure 1) obtained by unrolling the iterations of their equation (6). This corresponds to (5.11) with $\Lambda_{\theta_k} : X \to X$ given as

$$\Lambda_{\theta_k} := \mathrm{id} + \omega_k \nabla[\mathcal{L}(\mathcal{A}(\,\cdot\,), g) + \mathcal{S}_{\phi_k}(\,\cdot\,)] \quad \text{for } \theta_k = (\omega_k, \phi_k).$$

The regularizer $\mathcal{S}_{\phi_k} : X \to \mathbb{R}$ is chosen as the FoE model (see also Section 4.3.2):

$$\mathcal{S}_{\phi_k}(f) := \sum_j \Phi_{k,j}(f * K_{k,j}) \quad \text{for } \phi_k = (\Phi_{k,j}, K_{k,j}),$$

where the (potential) functionals $\Phi_{k,j} : X \to \mathbb{R}$ and the convolution kernels $K_{k,j} \in X$ are all parametrized by finite-dimensional parameters, so ϕ_k is a finite-dimensional parameter. See also Chen *et al.* (2019), who essentially apply the approach of Hammernik *et al.* (2018) to CT reconstruction. Another variant of variational networks is that of Bostan, Kamilov and Waller (2018), who unroll a proximal algorithm for an objective with a TV regularizer and replace the scalar soft-thresholding function with a parametrized variant (see Bostan, Kamilov and Waller 2018, equation (8)). This yields a proximal algorithm that uses a sequence of adjustable shrinkage functions in addition to self-tuning the step-size. In particular, unlike Mousavi and Baraniuk (2017), the method presented here does not rely on having a structured normal operator. A further variant of a variational network is given by Aggarwal, Mani and Jacob (2019), who unroll a gradient descent scheme for minimizing an objective whose regularizer is given by a CNN (see Aggarwal, Mani and Jacob 2019, equation (7)). A similar approach is also considered by Zhao, Zhang, Wang and Gao (2018a), who unroll an ADMM scheme and stop iterates according to a Morozov-type stopping criterion (not a fixed number of iterates), so the number of layers depends on the noise level in data.

Remark 5.1. An interesting aspect of the variational network of Hammernik *et al.* (2016) is that Λ_{θ_k} can be interpreted as a gradient descent step in an optimization scheme that minimizes an objective functional $f \mapsto \mathcal{L}(\mathcal{A}(f), g) + \mathcal{S}_{\phi_k}(f)$. In particular, if ϕ_k is the same for all k, then increasing the number of layers by $N \to \infty$ will in the limit yield a MAP estimator instead of a Bayes estimator.

Other applications of the learned gradient method include those of Gong *et al.* (2018) for deconvolution, Qin *et al.* (2019) for dynamic cardiac two-dimensional MRI (here one needs to exploit the temporal dependence), Hauptmann *et al.* (2018) for three-dimensional PAT, and Wu, Kim and Li (2018a) for three-dimensional CT reconstruction. The challenge in three-dimensional applications is to manage the high computational and memory cost for the training of the unrolled network. One approach is to replace the end-to-end training of the entire neural network and instead break down each iteration and train the sub-networks sequentially (gradient boosting). This is the approach taken by Hauptmann *et al.* (2018) and Wu *et al.* (2018*a*), but as shown by Wu *et al.* (2018*a*), the output quality has minor improvements over learned post-processing (Section 5.1.5), which also scales to the three-dimensional setting. A better alternative could be to use a reversible residual network architecture (Gomez *et al.* 2017, Ardizzone *et al.* 2018) for a learned iterative method, since these are much better at managing memory consumption in training (mainly when calculating gradients using backpropagation) as networks grow deeper and wider. However, this is yet to be done.

Learned iterative in both model parameter and data spaces. The final enhancement to the learned iterative schemes is to introduce an explicit learned updating in the data space as well. To see how this can be achieved, one can unroll a primal–dual-type scheme of the form

$$
\begin{cases}
v^0 = g \text{ and } f^0 \in X \text{ given} \\
v^{k+1} = \Gamma^{\mathrm{d}}_{\theta^{\mathrm{d}}_k}(v^k, \mathcal{A}(f^k), g)) \\
f^{k+1} = \Gamma^{\mathrm{m}}_{\theta^{\mathrm{m}}_k}(f^k, [\partial\,\mathcal{A}(f)]^*(v^{k+1}))
\end{cases}
\qquad \text{for } k = 0, \ldots, N-1. \qquad (5.14)
$$

Here,

$$
\Gamma^{\mathrm{m}}_{\theta^{\mathrm{m}}_k} : X \times X \to X \quad \text{and} \quad \Gamma^{\mathrm{d}}_{\theta^{\mathrm{d}}_k} : Y \times Y \times Y \to Y
$$

are the updating operators. This corresponds to defining $\mathcal{R}_\theta : Y \to X$ with $\theta_k = (\theta^{\mathrm{m}}_k, \theta^{\mathrm{d}}_k)$ as in (5.13), where $\Lambda_{\theta_k} : X \times Y \to X \times Y$ is given by

$$
\Lambda_{\theta_k}(f, v) := (\Gamma^{\mathrm{m}}_{\theta^{\mathrm{m}}_k}(f, [\partial\,\mathcal{A}(f)]^*(\Gamma^{\mathrm{d}}_{\theta^{\mathrm{d}}_k}(v, \mathcal{A}(f), g))), \Gamma^{\mathrm{data}}_{\theta^{\mathrm{d}}_k}(v, \mathcal{A}(f), g))). \qquad (5.15)
$$

This is illustrated in Figure 5.2, and similar networks are also suggested by Vogel and Pock (2017) and Kobler *et al.* (2017), who extend the approach of Hammernik *et al.* (2018) by parametrizing and learning the data discrepancy $\mathcal{L}$. Applications are for inverting the two-dimensional Fourier transform (two-dimensional MRI image reconstruction). See also He *et al.* (2019), who unroll an ADMM scheme with updates in both reconstruction and data spaces and apply that to two-dimensional CT reconstruction.

Finally, allowing for some memory in both model parameter and data spaces leads to the learned primal–dual scheme of Adler and Öktem (2018*b*), which is used for low-dose two-dimensional CT reconstruction. The robustness of this approach against uncertainties in the image and uncertainties in system settings is empirically studied by Boink, van Gils, Manohar and Brune (2018). The conclusion is that learning improves pure knowledge-

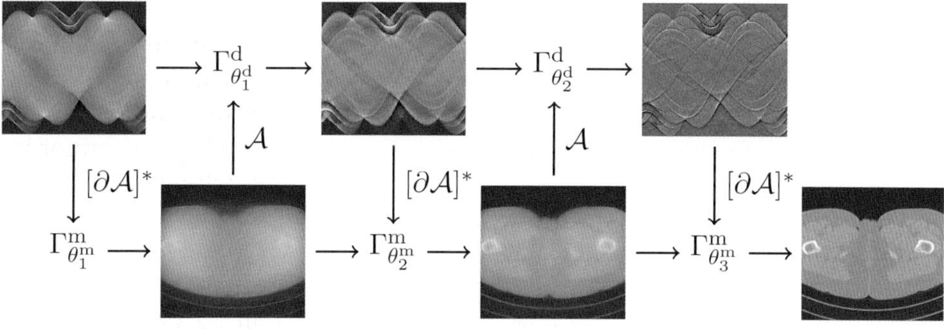

Figure 5.2. Learned iterative method in both model parameter and data spaces. Illustration of the operator obtained by unrolling the scheme in (5.14) for $N = 3$ in the context of CT image reconstruction (Section 7.3.1).

based reconstruction in terms of noise removal and background identification, and more variety in the training set increases the robustness against image uncertainty. Robustness against model uncertainty, however, is not readily obtained. An application of a learned primal dual scheme to breast tomosynthesis is given by Moriakov *et al.* (2018). To outperform existing reconstruction methods, one needs to encode the information about breast thickness into the learned updates for both primal and dual variables.

Further remarks. Successively reducing the constraints on the neural network architecture in the learned iterative scheme allows for larger model capacity, but training such a model also requires more training data. The well-known universal approximation property of deep neural networks (see Cybenko 1989, Hornik, Stinchcombe and White 1989, Hornik 1991, Barron 1994) implies that the learned iterative schemes in their most unconstrained formulation can approximate a Bayes estimator arbitrarily well given enough supervised data and neural network parameters (model capacity). This does not necessarily hold for the more constrained network architectures, such as those used by the variational networks of Hammernik *et al.* (2018). See also Section 8.2.1 for remarks on approximation properties of certain deep neural networks.

Unrolling an iterative scheme is a central theme in learned iterative methods, and it allows one to construct a deep neural network from an iterative scheme that is stopped after N steps. Here, a larger N simply means adding more layers, *i.e.* increasing the model capacity. The same idea is used to solve an optimization problem more rapidly (Section 4.9), but here the training is unsupervised and the loss is given by the objective that we seek to minimize. Such an architecture can be used for computing a Bayes estimator when trained using a supervised data or a MAP estimator when trained using unsupervised data. The same holds also for the GAN approach taken in

Mardani *et al.* (2017*b*) (Section 5.1.6), which uses the same architecture as learned iterative methods but a different loss, that is, it computes a different estimator.

5.1.5. Learned post- and/or pre-processing

One of the earliest applications of deep learning in inverse problems in imaging was as a post-processing tool. Here, an established non-learned reconstruction method is followed by a deep neural network that is trained to 'clean up' noise and artefacts introduced by the first inversion step. The first inversion step is typically performed using a pseudo-inverse of the forward operator, *e.g.* backprojection or FBP in CT or a zero-filling solution in MRI. This initial reconstruction method can be seen as a way to account for how data are generated, whereas the learning part acts only on the model parameter instead of data.

As one might expect, a vast number of papers apply deep learning to images obtained from some kind of image reconstruction. In principle, all of these qualify as a learned post-processing scheme. We will not attempt to prove a (near-complete) survey of these since the learning part is not directly related to the inverse problem. Instead we mention some key publications from imaging that have image reconstruction as their main theme, followed by a characterization of the estimator one seeks to approximate when using learned post-processing.

Selected publications focusing on image reconstruction. We start with surveying work related to learned post-processing for CT image reconstruction. Early approaches used a CNN to map a sparse-view CT reconstruction to a full-view one, as in Zhao, Chen, Zhang and Jin (2016). A similar approach to mapping low-dose CT images to normal-dose images (denoising) is that of Chen *et al.* (2017*b*), who train a CNN on image patches. See also Chen *et al.* (2017*a*) for an approach that uses a residual encoder–decoder CNN (RED-CNN) trained on image patches for the same purpose.

Denoising low-dose CT images can also be done using a U-Net, as in Jin, McCann, Froustey and Unser (2017). Another variant is to use the U-Net on directional wavelets (AAPM-Net), as in Kang, Min and Ye (2017). This method came second in the 2016 AAPM Low Dose CT Grand Challenge.[6] It has since been further developed and refined in a series of publications: for example, Kang, Chang, Yoo and Ye (2018) and Kang and Ye (2018) modify the AAPM-Net architecture by using a wavelet residual network (WavResNet), which is a deep CNN reinterpreted as cascaded convolution framelet signal representation. Another drawback of AAPM-Net is that

---

[6] The method that won was a variational method with a non-local regularizer (Kim, Fakhri and Li 2017), but that approach has a run-time that scales very poorly with problem size.

it does not satisfy the frame condition and it overly emphasizes the low-frequency component of the signal, which leads to blurring artefacts in the post-processed CT images. To address this, Han and Ye (2018) suggest a U-Net-based network architecture with directional wavelets that satisfy the frame condition. Finally, Ye *et al.* (2018) develop a mathematical framework to understand deep learning approaches for inverse problems based on these deep convolutional framelets. Such architectures represent a signal decomposition similar to using wavelets or framelets, but here the basis is learned from the training data. This idea of using techniques from applied harmonic analysis and sparse signal processing to analyse approximation properties of certain classes of deep neural networks bears similarities to Bölcskei, Grohs, Kutyniok and Petersen (2019) (see Section 8.2.1) and the scattering networks discussed in Section 4.5 as well as work related to multi-layer convolutional sparse coding outlined in Section 4.4.2.

Yet another CNN architecture (Mixed-Scale Dense CNN) is proposed in Pelt, Batenburg and Sethian (2018) for denoising and removing streak artefacts from limited angle CT reconstructions. Empirical evidence shows that this architecture comes with some advantages over encoder–decoder networks. It can be trained on relatively small training sets and the same hyper-parameters in training can often be re-used across a wide variety of problems. This removes the need to perform a time-consuming trial-and-error search for hyper-parameter values.

Besides architectures, one may also consider the choice of loss function, as in Zhang *et al.* (2018), who consider CNN-based denoising of CT images using a loss function that is a linear combination of squared L^2 and multi-scale structural similarity index (SSIM). A closely related work is that of Zhang and Yu (2018), which uses a CNN trained on image patches with a loss function that is a sum of squared L^2 losses over the patches. The aim here is to reduce streak artefacts from highly scattering media, such as metal implants. A number of papers use techniques from GANs to post-process CT images. Shan *et al.* (2018) use a conveying path-based convolutional encoder–decoder network. A novel feature of this approach is that an initial three-dimensional denoising model can be directly obtained by extending a trained two-dimensional CNN, which is then fine-tuned to incorporate three-dimensional spatial information from adjacent slices (transfer learning from two to three dimensions). The paper also contains a summary of deep learning network architectures for CT post-processing listing the loss function (squared L^2, adversarial or perpetual loss). A similar approach is taken by Yang *et al.* (2018c), who denoise CT images via a GAN with Wasserstein distance and perceptual similarity. The perceptual loss suppresses noise by comparing the perceptual features of a denoised output against those of the ground truth in an established feature space, while the generator focuses more on migrating the data noise distribution. Another approach is that of You *et al.* (2018*a*), who use a generator from a GAN to generate high-resolution CT images from low-resolution counterparts. The model is trained on semi-supervised training data and the training is regularized by enforcing a cycle-consistency expressed in terms of the Wasserstein distance. See also You *et al.* (2018*b*) for similar work, along with an investigation of the impact of different loss functions for training the GAN.

For PET image reconstruction, da Luis and Reader (2017) use a CNN to denoise PET reconstructions obtained by ML-EM. A more involved approach is presented in Yang, Ying and Tang (2018*a*), which learns a post-processing step for enhancing PET reconstructions obtained by MAP with a Green smoothness prior (see Yang, Ying and Tang 2018*a*, equation (6)). More precisely, this is supervised training on tuples consisting of ground truth and a number of MAP solutions where one varies parameters defining the prior. The training seeks to learn a mapping that takes a set of small image patches at the same location from the MAP solutions to the corresponding patch in the ground truth, thereby resulting in a learned patch-based image denoising scheme. A related approach is that of Kim *et al.* (2018), who train a CNN to map low-dose PET images to a full-dose one. Both low-dose and full-dose reconstructions are obtained using ordered subsets ML-EM. Since the resulting trained CNN denoiser produces additional bias induced by the disparity of noise levels, one considers learning a regularizer that includes the CNN denoiser (see Kim *et al.* 2018, equation (8)). The resulting variational problem is solved using the ADMM method.

Concerning MRI, most learned post-processing applications seek to train a mapping that takes a zero-filling reconstruction[7] obtained from under-sampled MRI data to the MRI reconstruction that corresponds to fully sampled data. For example, Wang *et al.* (2016) use a CNN for this purpose, and the deep learning output is either used as an initialization or as a regularization term in classical compressed sensing approaches to MRI image reconstruction. Another example is that of Hyun *et al.* (2018), who use a CNN to process a zero-filling reconstruction followed by a particular k-space correction. This outperforms plain zero-filling reconstruction as well as learned post-processing where Fourier inversion is combined with a trained denoiser based on a plain U-Net architecture.

Similar to CT image processing, there has been some work on using GANs to post-process MRI reconstructions. One example is that of Quan, Member, Nguyen-Duc and Jeong (2018), who use a generator within a GAN setting to learn a post-processing operator that maps a zero-filling reconstruction image to a full reconstruction image. Training is regularized using

---

[7] Zero-filling reconstruction is computed by setting to zero all Fourier coefficients that are not measured in the MRI data, and then applying a normal inverse Fourier transform.

a loss that includes a cyclic loss consistency term that promotes accurate interpolation of the given under-sampled k-space data. The generator consists of multiple end-to-end networks chained together, where the first network translates a zero-filling reconstruction image to a full reconstruction image, and the following networks improve accuracy of the full reconstruction image (refinement step). Another approach using a generator from a trained GAN is given by Yang *et al.* (2018*b*), who use a U-Net architecture with skip connections for the generator. The loss consists of an adversarial loss term, a novel content loss term considering both squared L^2 and a perceptual loss term defined by pre-trained deep convolutional networks. There is also a squared L^2 in both model parameter and data spaces, and the latter involves applying the forward operator to the training data to evaluate the squared L^2 in data space. See Yang *et al.* (2018*b*, equation (13)) for the full expression.

We conclude by mentioning some approaches that involve pre-processing. For CT imaging, deep-learning-based pre-processing targets sinogram inpainting, which is the task of mapping observed, sparsely sampled, CT projection data onto corresponding densely sampled CT projection data. Lee *et al.* (2019) achieve this via a plain CNN whereas Ghani and Karl (2018) use a generator from a trained conditional GAN. A CNN is also used by Hong *et al.* (2018) to pre-process PET data. Here one uses a deep residual CNN for PET super-resolution, that is, to map PET sinogram data from a scanner with large pixellated crystals to one with small pixellated crystals. The CNN-based method was designed and applied as an intermediate step between the projection data acquisition and the image reconstruction. Results are validated using both analytically simulated data, Monte Carlo simulated data and experimental pre-clinical data. In a similar manner, Allman, Reiter and Bell (2018) use a CNN to pre-process photoacoustic data to identify and remove noise artefacts. Finally, we cite Huizhuo, Jinzhu and Zhanxing (2018), who use a CNN to jointly pre- and post-process CT data and images. The pre-processing amounts to sinogram in-painting and the post-processing is image denoising, and the middle reconstruction step is performed using FBP. The loss function for this *joint pre- and post-processing* scheme is given in Huizhuo *et al.* (2018, equation (3)).

Characterizing the estimator. Here we consider post-processing; one can make analogous arguments for pre-processing. To understand what learned post-processing computes, consider the supervised learning setting. Learned post-processing seeks to approximate the Bayes estimator for the model parameter conditioned on the initial reconstruction, which is a different estimator from the one considered in deep direct Bayes estimation (Sections 5.1.2–5.1.4).

Stated formally, let $\mathbb{h}$ be a X-valued random variable defined as $\mathbb{h} := \mathcal{A}^{\dagger}(\mathbb{g})$, where $\mathbb{g}$ is the Y-valued random variable generating data. Also, let $\mathcal{A}^{\dagger} \colon Y \to X$ denote the fixed initial reconstruction operator that is not learned. We now seek the Bayes estimator for the conditional random variable $(\mathbb{f} \mid \mathbb{h} = h)$, where $h = \mathcal{A}^{\dagger}(g)$ with data $g \in Y$ being a single sample of $\mathbb{g}$. This yields a reconstruction operator $\mathcal{R} \colon Y \to X$ given as $\mathcal{R} := \mathcal{B}_{\sigma} \circ \mathcal{A}^{\dagger}$ with $\mathcal{B}_{\sigma} \colon X \to X$ solving

$$\mathcal{B}_{\sigma} \in \operatorname*{arg\,min}_{\mathcal{B}\colon X \to X} \mathbb{E}_{(\mathbb{f},\mathbb{h})\sim\sigma}[\ell_X(\mathbb{f}, \mathcal{B}(\mathbb{h}))], \tag{5.16}$$

where $\ell_X \colon X \times X \to \mathbb{R}$ is a fixed loss function. In the above, σ denotes the joint law for $(\mathbb{f}, \mathbb{h})$, which is clearly unknown. It can, however, be replaced by its empirical counterpart given from supervised training data,

$$\Sigma_m := \{(f_1, h_1), \ldots, (f_m, h_m)\} \subset X \times X, \tag{5.17}$$

where (f_i, h_i) are generated by $(\mathbb{f}, \mathbb{h}) \sim \sigma$. Furthermore, considering all possible estimators (non-randomized decision rules) $\mathcal{B} \colon X \to X$ in the minimization in (5.16) is computationally unfeasible. To address this, we consider a family $\{\mathcal{B}_{\theta}\}_{\theta\in\Theta}$ of estimators that is parametrized by a finite-dimensional parameter in Θ. Restricting attention to such a parametrized family of estimators yields the following empirical risk minimization problem:

$$\widehat{\theta} \in \operatorname*{arg\,min}_{\theta\in\Theta} \frac{1}{m} \sum_{i=1}^{m} \ell_X(f_i, \mathcal{B}_{\theta}(h_i)), \tag{5.18}$$

with $(f_i, h_i) \in \Sigma_m$ as in (5.17).

Now, consider the case when the initial reconstruction method $\mathcal{A}^{\dagger} \colon Y \to X$ is a linear sufficient statistic, that is,

$$\mathbb{E}[\mathbb{f} \mid \mathbb{g} = g] = \mathbb{E}[\mathbb{f} \mid \mathcal{A}^{\dagger}(\mathbb{g}) = \mathcal{A}^{\dagger}(g)].$$

For example, the operators given by the FBP and the backprojection are both linear sufficient statistics. If in addition the loss function is the squared L^2-norm, then $\mathcal{R} := \mathcal{B}_{\sigma} \circ \mathcal{A}^{\dagger}$ is also a Bayes estimator for $(\mathbb{f} \mid \mathbb{g} = g)$, that is, the reconstruction obtained from learned post-processing coincides with the one from deep direct Bayes estimation. Note, however, that this holds in the limit of infinite amount of training data and infinite model capacity. In fact, as we shall see in Section 7.3, when applied to finite number of training data and finite model capacity, learned post-processing differs from deep direct Bayes estimation.

5.1.6. *Other estimators*

Learning using a supervised GAN. Some recent work uses a GAN in a supervised learning setting, which leads to a training problem of the type (5.3). For example, Mardani *et al.* (2017*b*) defines the variant of (5.3) where ℓ_X

is the L^1-norm, $\ell_{\mathscr{P}_X}$ is the Pearson χ^2-divergence, which can be evaluated using a least-squares GAN (Mao *et al.* 2016), and $\mathcal{R}_\theta\colon Y \to X$ is given by an architecture adapted to MRI. Use of a 1-norm in the generator loss motivates the reference made to 'compressed sensing' made by the authors. See also Mardani *et al.* (2017*a*) and Schwab, Antholzer and Haltmeier (2018) for further work along these lines: for example, Schwab *et al.* (2018) more formally treat the manifold projection step that in Mardani *et al.* (2017*b*) is specially tailored for MRI imaging.

It is not easy to identify what estimator the above really corresponds to, but clearly the generator (after training) resembles a MAP estimator where the prior is given implicitly by the supervised training data. To some extent, one may view the above as a supervised variant of Section 4.7 that uses a GAN to learn a regularizer (prior) from unsupervised data.

Deep direct estimation of higher-order moments. As described in Adler and Öktem (2018*a*), one can train a deep neural network to directly approximate an estimator involving higher-order moments, such as pointwise variance and correlation. The starting point is the well-known result

$$\mathbb{E}_{\mathsf{w}}[\mathsf{w} \mid \mathsf{g} = \cdot] = \min_{\mathsf{h}\colon Y \to W} \mathbb{E}_{(\mathsf{g},\mathsf{w})}[\|\,\mathsf{h}(\mathsf{g}) - \mathsf{w}\|_W^2]. \tag{5.19}$$

In the above, w is *any* random variable taking values in some measurable Banach space W and the minimization is over all W-valued measurable maps on Y. This is useful since many estimators relevant for uncertainty quantification are expressible using terms of this form for appropriate choices of w.

Specifically, Adler and Öktem (2018*a*) consider two (deep) neural networks $\mathcal{R}_{\theta^*}\colon Y \to X$ and $\mathsf{h}_{\phi^*}\colon Y \to X$ with appropriate architectures that are trained according to

$$\theta^* \in \arg\min_\theta\{\mathbb{E}_{(\mathsf{f},\mathsf{g})\sim\mu}[\|\mathsf{f} - \mathcal{R}_\theta(\mathsf{g})\|_X^2]\},$$

$$\phi^* \in \arg\min_\phi\{\mathbb{E}_{(\mathsf{f},\mathsf{g})\sim\mu}[\|\,\mathsf{h}_\phi(\mathsf{g}) - (\mathsf{f} - \mathcal{R}_{\theta^*}(\mathsf{g}))^2\|_X^2]\}.$$

The joint law μ above can be replaced by its empirical counterpart given from supervised training data (f_i, g_i), so the μ-expectation is replaced by an averaging over training data. The resulting networks will then approximate the conditional mean and the conditional pointwise variance, respectively.

As already shown, by using (5.19) it is possible to rewrite many estimators as minimizers of an expectation. Such estimators can then be approximated using the direct estimation approach outlined here. This should coincide with computing the same estimator by posterior sampling (Section 5.2.1). Direct estimation is significantly faster, but not as flexible as posterior sampling since each estimator requires a new neural network that specifically trained for that estimator. Section 7.7 compares the outcomes of the two approaches.

5.2. Deep posterior sampling

The idea here is to use techniques from deep learning to sample from the posterior. This can then be used to perform various statistical computations relevant to solving the inverse problem.

Approaches to sampling from high-dimensional distributions that do not use neural networks are very briefly surveyed in Section 3.5.2. A drawback of these approaches is that they require access to an explicit prior, so they do not apply to cases where no explicit prior is available. Furthermore, despite significant algorithmic advances, these methods do not offer computationally feasible means for sampling from the posterior in large-scale inverse problems, such as those arising in three-dimensional imaging. Here we survey an alternative method that uses conditional GAN for the same purpose (Section 5.2.1). This approach has very desirable properties, so it does not require access to an explicit prior, and it is computationally very efficient.

Remark 5.2. Deep Gaussian mixture models (Viroli and McLachlan 2017) are multi-layered networks where the variables at each layer follow a mixture of Gaussian distributions. Hence, the resulting deep mixture model consists of a set of nested (non-linear) mixtures of linear models. Such models can be shown to be universal approximators of probability densities and they can be trained using ML-EM techniques. This is an interesting approach for sampling from the posterior in Bayesian inversion, but it is yet to be used in this context.

Recent work using conditional GAN for the same purpose (Section 5.2.1) gives very promising results. We do not cover deep Gaussian mixture models (Viroli and McLachlan 2017), which are multi-layered networks where, at each layer, the variables follow a mixture of Gaussian distributions. Thus, the deep mixture model consists of a set of nested mixtures of linear models, which globally provide a non-linear model able to describe the data in a very flexible way. These are universal approximators of probability densities that are trainable using ML-EM techniques. This is an interesting approach that is yet to be used in the context of inverse problems.

5.2.1. Conditional GAN

The approach taken was first introduced by Adler and Öktem (2018*a*), and it is a special case of variational Bayes inference (Section 3.5.2) where the variational family is parametrized via GAN. More precisely, the idea is to explore the posterior by sampling from a generator that has been trained using a conditional Wasserstein GAN discriminator.

To describe how a Wasserstein GAN can be used for this purpose, let data $g \in Y$ be fixed and assume that Π^g_{post}, the posterior of $\mathbf{f}$ at $\mathbf{g} = g$, can be approximated by elements in a parametrized family $\{\mathcal{G}_\theta(g)\}_{\theta \in \Theta}$ of probability measures on X. The best such approximation is defined as $\mathcal{G}_{\theta^*}(g)$, where $\theta^* \in \Theta$ solves

$$\theta^* \in \underset{\theta \in \Theta}{\arg\min}\ \ell_{\mathscr{P}_X}\big(\mathcal{G}_\theta(g), \Pi^g_{\text{post}}\big). \tag{5.20}$$

Here, $\ell_{\mathscr{P}_X} \colon \mathscr{P}_X \times \mathscr{P}_X \to \mathbb{R}$ quantifies the 'distance' between two probability measures on X. We are, however, interested in the best approximation for 'all data', so we extend (5.20) by including an averaging over all possible data. The next step is to choose a distance notion ℓ that is desirable from both a theoretical and a computational point of view. For example, the distance should be finite, and computational feasibility requires it to be differentiable almost everywhere, since this opens up using stochastic gradient descent (SGD)-type schemes. The Wasserstein 1-distance $\mathcal{W}$ (Section 8.2.7) has these properties (Arjovsky *et al.* 2017), and sampling from the posterior Π^g_{post} can then be replaced by sampling from the probability distribution $\mathcal{G}_{\theta^*}(g)$, where θ^* solves

$$\theta^* \in \underset{\theta \in \Theta}{\arg\min}\ \mathbb{E}_{\mathbf{g} \sim \sigma}\big[\mathcal{W}(\mathcal{G}_\theta(\mathbf{g}), \Pi^{\mathbf{g}}_{\text{post}})\big]. \tag{5.21}$$

In the above, σ is the probability distribution for data and the random variable $\mathbf{g} \sim \sigma$ generates data.

Observe now that evaluating the objective in (5.21) requires access to the very posterior that we seek to approximate. Furthermore, the distribution σ of data is often unknown, so an approach based on (5.21) is essentially useless if the purpose is to sample from an unknown posterior. Finally, evaluating the Wasserstein 1-distance directly from its definition is not computationally feasible.

On the other hand, as we shall see, *all* of these drawbacks can be circumvented by rewriting (5.21) as an expectation over the joint law $(\mathbf{f}, \mathbf{g}) \sim \mu$. This makes use of the Kantorovich–Rubinstein duality for the Wasserstein 1-distance (see (B.2)), and one obtains the following approximate version of (5.21):

$$\theta^* \in \underset{\theta \in \Theta}{\arg\min}\left\{ \sup_{\phi \in \Phi} \mathbb{E}_{(\mathbf{f},\mathbf{g}) \sim \mu}\big[D_\phi(\mathbf{f}, \mathbf{g}) - \mathbb{E}_{\mathbf{z} \sim \eta}[D_\phi(G_\theta(\mathbf{z}, \mathbf{g}), \mathbf{g})]\big] \right\}. \tag{5.22}$$

Here, $G_\theta \colon Z \times Y \to X$ (generator) is a deterministic mapping such that $G_\theta(\mathbf{z}, g) \sim \mathcal{G}_\theta(g)$, where $\mathbf{z} \sim \eta$ is a 'simple' Z-valued random variable in the sense that it can be sampled in a computationally feasible manner. Next, the mapping $D_\phi \colon X \times Y \to \mathbb{R}$ (discriminator) is a measurable mapping that is 1-Lipschitz in the X-variable.

At first sight, it might be unclear why (5.22) is better suited than (5.21) to sampling from the posterior, especially since the joint law μ in (5.22) is unknown. The advantage becomes clear when one has access to supervised training data for the inverse problem, *i.e.* i.i.d. samples $(f_1, g_1), \ldots, (f_m, g_m)$ generated by the random variable $(\mathsf{f}, \mathsf{g}) \sim \mu$. The μ-expectation in (5.22) can then be replaced by an averaging over training data.

To summarize, solving (5.22) given supervised training data in $X \times Y$ amounts to learning a generator $G_{\theta^*}(\mathsf{z}, \cdot) \colon Y \to X$ such that $G_{\theta^*}(\mathsf{z}, g)$ with $\mathsf{z} \sim \eta$ is approximately distributed as the posterior Π_{post}^g. In particular, for given $g \in Y$ we can sample from Π_{post}^g by generating values of $z \mapsto G_{\theta^*}(z, g) \in X$ in which $z \in Z$ is generated by sampling from $\mathsf{z} \sim \eta$.

An important part of the implementation is the concrete parametrizations of the generator and discriminator:

$$G_\theta \colon Z \times Y \to X \quad \text{and} \quad D_\phi \colon X \times Y \to \mathbb{R}.$$

We use deep neural networks for this purpose, and following Gulrajani *et al.* (2017), we softly enforce the 1-Lipschitz condition on the discriminator by including a gradient penalty term in the training objective function in (5.22). Furthermore, if (5.22) is implemented as is, then in practice z is not used by the generator (so called mode-collapse). To solve this problem, we introduce a novel conditional Wasserstein GAN discriminator that can be used with conditional WGAN without impairing its analytical properties: see Adler and Öktem (2018*a*) for more details.

We conclude by referring to Section 7.7 for an example of how the conditional Wasserstein GAN can be used in clinical image-guided decision making.

6. Special topics

In this section we address several topics of machine learning that do not strictly fall within the previously covered contexts of functional analytic or statistical regularization. In Section 6.1 we discuss regularization methods that go beyond pure reconstructions. These reconstructions include – at least partially – the decision process, which typically follows the solution of inverse problems, for example examination of a CT reconstruction by a medical expert. Then Section 6.2.1 aims at investigating the connections between neural networks and differential equations, and Section 6.2 discusses the case where the forward operator is incorrectly known. Finally, Section 6.2.2 discusses total least-squares approaches, which are classical tools for updating the operator as well as the reconstruction based on measured data. We are well aware that this is still very much an incomplete list of topics not covered in the previous sections. As already mentioned in Section 1, we apologize for our ignorance with respect to the missing material.

6.1. Task-adapted reconstruction

Estimating a model parameter in an inverse problem is often only one of many steps in a procedure where the reconstructed model parameter is used in a task. Consider a setting where the task is given by an operator $\mathcal{T}\colon X \to D$ (*task operator*) which maps a model parameter f to an element in some set D (decision space). Such tasks were introduced in Louis (2011) within the functional analytic framework (Section 2), and image segmentation served as the prime example.

A wider range of tasks can be accounted for if one adopts the statistical view as in Adler *et al.* (2018). Here we introduce a D-valued random variable $\mathbb{d}$ and interpret $\mathcal{T}$ as a non-randomized decision rule that is given as a Bayes estimator. The risk is given through the *task loss* $\ell_D\colon D \times D \to \mathbb{R}$ and the optimal task operator is the Bayes estimator with respect to the task loss, that is,

$$\mathcal{T}_{\mathrm{opt}} \in \underset{\mathcal{T}\colon X \to D}{\arg\min}\, \mathbb{E}_{(\mathbb{f},\mathbb{d})}[\ell_D(\mathcal{T}(\mathbb{f}), \mathbb{d})]. \tag{6.1}$$

In practice we restrict ourselves to a parametrized family of tasks $\mathcal{T}_\phi\colon X \to D$ with $\phi \in \Phi$, typically given by deep neural networks. There are essentially three ways to combine a neural network $\mathcal{R}_\theta\colon Y \to X$ for reconstruction with a neural network $\mathcal{T}_\phi\colon X \to D$ for the task that is given by (6.1). The approaches differ in the choice of loss used for learning $(\widehat{\theta}, \widehat{\phi}) \in \Theta \times \Phi$ in

$$\mathcal{T}_{\widehat{\phi}} \circ \mathcal{R}_{\widehat{\theta}}\colon Y \to D. \tag{6.2}$$

Sequential training. The optimal parameter $(\widehat{\theta}, \widehat{\phi}) \in \Theta \times \Phi$ in (6.2) is given as

$$\begin{cases} \widehat{\theta} \in \underset{\theta \in \Theta}{\arg\min}\, \mathbb{E}_{\mathbb{f},\mathbb{g}}[\ell_X(\mathcal{R}_\theta(\mathbb{g}), \mathbb{f})], \\[2mm] \widehat{\phi} \in \underset{\phi \in \Phi}{\arg\min}\, \mathbb{E}_{\mathbb{g},\mathbb{d}}[\ell_D(\mathcal{T}_\phi \circ \mathcal{R}_{\widehat{\theta}}(\mathbb{g}), \mathbb{d})]. \end{cases}$$

The training data are samples (f_i, g_i) generated by $(\mathbb{f}, \mathbb{g})$ for computing $\widehat{\theta}$ and (g_i, d_i) generated by $(\mathbb{g}, \mathbb{d})$ for computing $\widehat{\phi}$.

End-to-end training. The optimal parameter $(\widehat{\theta}, \widehat{\phi}) \in \Theta \times \Phi$ in (6.2) is given by directly minimizing the loss for the task, that is,

$$(\widehat{\phi}, \widehat{\theta}) \in \underset{(\phi,\theta) \in \Theta \times \Phi}{\arg\min}\, \mathbb{E}_{\mathbb{g},\mathbb{d}}[\ell_D(\mathcal{T}_\phi \circ \mathcal{R}_\theta(\mathbb{g}), \mathbb{d})].$$

The training data are samples (g_i, d_i) generated by $(\mathbb{g}, \mathbb{d})$.

Task-adapted training. This refers to anything in between sequential and end-to-end training. More precisely, $(\widehat{\theta}, \widehat{\phi}) \in \Theta \times \Phi$ in (6.2) is given by minimizing the following *joint expected loss (risk)*:

$$(\widehat{\theta}, \widehat{\phi}) \in \underset{(\theta,\phi)\in\Theta\times\Phi}{\arg\min} \; \mathbb{E}_{(\mathsf{f},\mathsf{g},\mathsf{d})}\left[(1-C)\ell_X(\mathcal{R}_\theta(\mathsf{g}),\mathsf{f}) + C\ell_D(\mathcal{T}_\phi \circ \mathcal{R}_\theta(\mathsf{g}),\mathsf{d})\right]. \quad (6.3)$$

The parameter $C \in [0,1)$ above is a tuning parameter where $C \approx 0$ corresponds to sequential training and $C \to 1$ to end-to-end training. The training data are samples (f_i, g_i, d_i) generated by $(\mathsf{f.g}, \mathsf{d})$.

Task-adapted training is a generic approach to adapting the reconstruction to a task with a plug-and-play structure for adapting to a specific inverse problem and a specific task. The former can be achieved by using a suitable neural network architecture, such as one given by a learned iterative method (Section 5.1.4). For the latter, note that the framework can handle *any task* that is given by a trainable neural network. This includes a wide range of tasks, such as semantic segmentation (Thoma 2016, Guo, Liu, Georgiou and Lew 2018), caption generation (Karpathy and Fei-Fei 2017, Li, Liang, Hu and Xing 2018*a*), in-painting (Xie, Xu and Chen 2012), depixelization/super-resolution (Romano, Isidoro and Milanfar 2017*b*), demosaicing (Syu, Chen and Chuang 2018), image translation (Wolterink *et al.* 2017), object recognition (Sermanet *et al.* 2013, He *et al.* 2016, Farabet, Couprie, Najman and LeCun 2013) and non-rigid image registration (Yang, Kwitt, Styner and Niethammer 2017, Ghosal and Ray 2017, Dalca, Balakrishnan, Guttag and Sabuncu 2018, Balakrishnan *et al.* 2019). Section 7.6 shows the performance of task-adapted reconstruction for joint tomographic image reconstruction and segmentation of white brain matter.

The importance of task-adapted reconstruction is also emphasized in the editorial of Wang, Ye, Mueller and Fessler (2018), which explicitly points out the potential in integrating reconstruction in an end-to-end workflow for medical imaging. They even coin the notion of 'rawdiomics', which is task-adapted reconstruction with the task corresponding to radiomics.[8] Most approaches to radiomics include some kind of classification that is performed by deep learning, and repeatability and reproducibility are among the main challenges (Rizzo *et al.* 2018, Traverso, Wee, Dekker and Gillies 2018). Typically, trained classifiers fail when confronted with images that are acquired using an acquisition protocol that is not represented in training data. This becomes especially problematic in multicentre studies where images are acquired using varying acquisition protocols and/or equipment. Clearly, the natural option is to include the information on how the images are generated, which naturally leads to task-adapted reconstruction.

---

[8] Radiomics seeks to identify distinctive imaging features between disease forms that may be useful for predicting prognosis and therapeutic response for various conditions.

6.2. Non-perfect forward operators

As already discussed in Section 1, classical inverse problems are based on a mathematical formulation of the forward operator $\mathcal{A}$. Those models are typically derived from physical first principles or other well-established laws and expert descriptions. These models are never complete. In most cases these models are regarded as sufficiently accurate to capture the main properties and a more detailed model would not help the reconstruction process in the presence of noisy data. However, there are certain cases, for example emerging new technologies, where models are still underdeveloped. Here one can aim to obtain an at least partially updated operator based on sets of test data.

Secondly, such a data-driven approach to model updates might also be necessary if one has a complete but very complex forward operator. The complexity of the model might lead to numerically very costly computations, which, for example in the case of optoacoustic tomography, are beyond any limits required for routine clinical applications. In this case one might resort to a much simpler analytical model, which is then updated using data-driven approaches. A third line of motivation for using partially learned operators refers to models that use so-called measured system matrices. These system matrices determine the linear forward operator experimentally, and hence their accuracy is limited by measurement accuracy.

6.2.1. Learning physics

Several recent papers have discussed the application of deep learning in forward problems: see Khoo, Lu and Ying (2017), Raissi and Karniadakis (2017), Sirignano and Spiliopoulos (2017), Tompson, Schlachter, Sprechmann and Perlin (2017), E, Han and Jentzen (2017) and Wu, Zhang, Shen and Zhai (2018).

Several authors have drawn the comparison between neural networks and PDEs. For example 'PDE-Net' (Long, Lu, Ma and Dong 2018) proposes designing a feed-forward neural network with convolution filters representing spatial derivatives up to a certain order, and multiplied by spatially varying weights. Training this system on dynamic data obtained with accurate numerical models allowed the discovery of appropriate PDEs for different physical problems. Other examples include learning coefficients of a PDE via optimal control (Liu, Lin, Zhang and Su 2010), as well as deriving CNN architectures motivated by diffusion processes (Chen *et al.* 2015) (compare also Section 4.3.2 and in particular (4.7) for learned reaction–diffusion equations), deriving stable architectures by drawing connections to ordinary differential equations (Haber and Ruthotto 2017) and constraining CNNs (Ruthotto and Haber 2018) by the interpretation as a partial differential equation. Another fascinating approach to learning first-principles physical models from data is that of Lam, Horesh, Avron and Willcox (2017).

6.2.2. Total least-squares

A widely used approach to integrating operator updates into regularization schemes for inverse problems can be formulated by generalized Tikhonov functionals. This is motivated by the total least-squares (TLS) approach (Golub and Van Loan 1980) (also known as 'errors-in-variable regression' in the statistical literature). One extension includes a regularization resulting in an approach called regularized total least-squares (R-TLS) (Golub, Hansen and O'Leary 1999, Markovsky and Van Huffel 2007), which for linear operators $\mathcal{A} = \mathbf{A}$ aims to learn an operator correction $\delta\mathbf{A}$ from the data by

$$\underset{\delta\mathbf{A},f}{\arg\min}\ \frac{1}{2}\|(\mathbf{A}+\delta\mathbf{A})f - g\|^2 + \frac{\alpha}{2}\|\mathbf{L}f\|^2 + \frac{\beta}{2}\|\delta\mathbf{A}\|_{\mathrm{F}}^2, \qquad (6.4)$$

where the operator norm is the Frobenius norm (typical choice $\beta = 1$). The linear operator (matrix) $\mathbf{L}$ is included in order to allow for more general regularization terms; for simplicity one may choose the identity $\mathbf{L} = \mathbf{I}$.

In the TLS literature, the minimization in (6.4) is commonly formulated with respect to $\widetilde{\mathbf{A}}$, *i.e.* defined by $\widetilde{\mathbf{A}} := \mathbf{A}+\delta\mathbf{A}$. This formulation uses a single data point g for simultaneously computing an operator update and for computing an approximation to the inverse problem. This is a heavily under-determined problem, which, however, leads to good results, at least for some applications: see Gutta *et al.* (2019), Kluth and Maass (2017) and Hirakawa and Parks (2006). The regularized TLS approach has been analysed by Golub *et al.* (1999), for example, who prove an equivalence result to classical Tikhonov regularization (Golub *et al.* 1999, Theorem 2.1), which we restate here.

Theorem 6.1. The solution $\hat{f}_\gamma$ to the problem

$$\underset{\widetilde{\mathbf{A}},f}{\min}\{\|\widetilde{\mathbf{A}}f - g\|^2 + \|\widetilde{\mathbf{A}} - \mathbf{A}\|_{\mathrm{F}}^2\}, \text{ subject to } \|\mathbf{L}f\| = \gamma$$

is a solution $(\mathbf{A}^T\mathbf{A}+\lambda_I I_n + \lambda_L \mathbf{L}^T\mathbf{L})f = \mathbf{A}^T g$ where

$$\lambda_I = -\frac{\|g - \mathbf{A}f\|^2}{1 + \|f\|^2} \quad \text{and} \quad \lambda_L = \mu(1 + \|f\|^2).$$

In the above, μ is the Lagrange multiplier in

$$\mathcal{L}(\widetilde{\mathbf{A}}, f, \mu) = \|\widetilde{\mathbf{A}}f - g\|^2 + \|\widetilde{\mathbf{A}} - \mathbf{A}\|_{\mathrm{F}}^2 + \mu(\|\mathbf{L}f\|^2 - \gamma^2).$$

The two parameters are related by

$$\lambda_L \gamma^2 = y^{\delta,T}(g - \mathbf{A}f) + \lambda_I$$

and the residual fulfils

$$\|\widetilde{\mathbf{A}}f - g\|^2 + \|\widetilde{\mathbf{A}} - \mathbf{A}\|_{\mathrm{F}}^2 = -\lambda_I.$$

Golub *et al.* (1999) conclude that if $\|\,\mathbf{L}\,f_\gamma\| < \gamma$ solves the R-TLS problem, then it also solves the TLS problem without regularization. Moreover, this approach has been extended to include sparsity constrained optimization (Zhu, Leus and Giannakis 2011), which equivalent formulation then reads

$$\arg\min_{\delta\mathbf{A},f}\left\{\frac{1}{2}\|(\mathbf{A}+\delta\mathbf{A})f - g\|^2 + \alpha\|f\|_1 + \frac{\beta}{2}\|\partial\,\mathbf{A}\|_{\mathrm{F}}^2\right\}.$$

The previous sparsity-promoting approach, as well as the original R-TLS approach, can be easily extended if sets of training data (f_i, g_i) are available. One either aims for a two-stage approach to first update the operator and then solve the inverse problem with some new data point g, or one can integrate both steps at once leading to

$$\arg\min_{\delta\mathbf{A},f}\left\{\sum_i \frac{1}{2}\|(\mathbf{A}+\delta\mathbf{A})f_i - g_i\|^2 + \frac{1}{2}\|(\mathbf{A}+\delta\mathbf{A})f - g\|^2 + \frac{\alpha}{2}\|\,\mathbf{L}\,f\|^2 + \frac{\beta}{2}\|\delta\mathbf{A}\|_{\mathrm{F}}^2\right\}.$$

Total least-squares is still an active field of research: see *e.g.* Markovsky and Van Huffel (2007) and Beck, Sabach and Teboulle (2016). Alternative problem formulations of the R-TLS problem in terms of given error bounds $\|y - g\| \le \delta$ and $\|A - \tilde{A}\| \le \epsilon$ (instead of $\|\,\mathbf{L}\,f\| \le \gamma$ as choosing an appropriate γ can be challenging) were further investigated by Lu, Pereverzev and Tautenhahn (2009) and Tautenhahn (2008). One further extension of the R-TLS is to include a regularization with respect to parameters determining the operator (operator deviation). This was considered in a general Hilbert space setting (Bleyer and Ramlau 2013) for image deblurring (see Buccini, Donatelli and Ramlau 2018, who call it 'semi-blind').

For the purposes of the present review article we highlight the properties of TLS for applications in MPI: see Section 7.5, Knopp, Gdaniec and Möddel (2017) and Kluth (2018) for further information on MPI. The following is a brief summary of the results in Kluth and Maass (2017). We seek to reconstruct a five-point phantom consisting of a glass capillary with a diameter of 1.1 mm filled with tracer with a concentration of 0.5 mol/l provided by the GitHub project page of Knopp *et al.* (2016). The data-driven reconstructions (obtained by using a measured noisy forward operator) are shown in Figure 6.1(a–d). We obtain smoothed reconstructions of the five points, which is typical for Tikhonov regularization: see Figure 6.1(a). In contrast, the minimization with sparsity constraints is able to obtain a better localization of the tracer. Signal energy from regions filled with tracer which are not included in the system matrix used may cause a larger concentration value than the expected 0.5 mol/l. Using the total least-squares

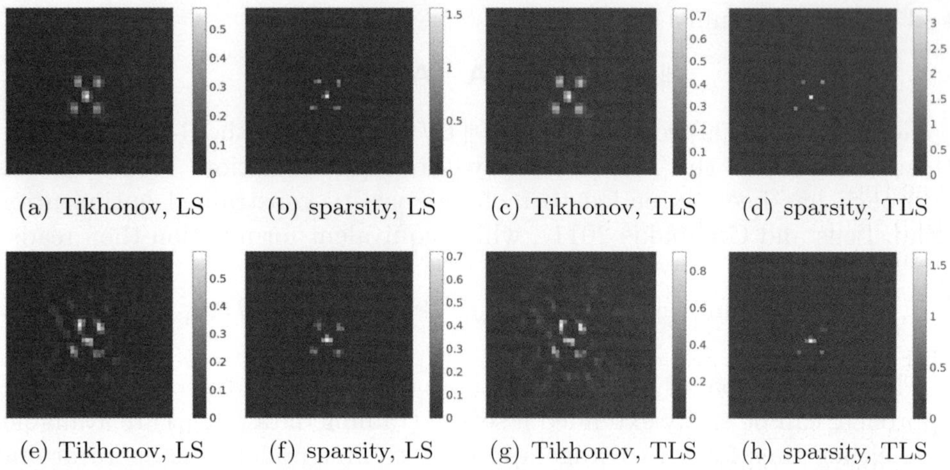

(a) Tikhonov, LS   (b) sparsity, LS   (c) Tikhonov, TLS   (d) sparsity, TLS

(e) Tikhonov, LS   (f) sparsity, LS   (g) Tikhonov, TLS   (h) sparsity, TLS

Figure 6.1. Reconstructions of a five-point phantom (pixel size 1 mm) provided by Knopp *et al.* (2016) obtained using Tikhonov (with $\alpha = 0.1 \times 10^{-6}$) and sparsity-promoting (with $\alpha = 0.1$) regularization with and without TLS. (a–d) Results from using a measured noisy forward operator. (e–h) Results from a knowledge-driven forward operator. Figure adapted from Kluth and Maass (2017).

approach further improves the localization in the sparse reconstruction for the data-based system matrices: see Figure 6.1(d).

A simplified model was fitted to measured data to obtain a knowledge-driven forward operator. As can be seen in Figure 6.1(e), using Tikhonov regularization results in a reconstruction of the five dots with additional background artefacts. Using the total least-squares approach in this set-up increases the contrast in concentration values but background artefacts are not significantly reduced. In contrast to the data-based reconstruction with Tikhonov regularization, the sparse knowledge-driven reconstruction in Figure 6.1(f) has a similar quality in terms of localization of the dots. By using the total least-squares approach, in Figure 6.1(h), the localization of the dots can be further improved such that the localization is similar in quality compared to the data-based sparse reconstruction.

6.2.3. Learned Landweber

Another approach to including a learning component into a forward operator is presented by Aspri, Banert, Öktem and Scherzer (2018). The starting point is to consider the iteratively regularized Landweber iteration (Scherzer 1998) (see also Kaltenbacher, Neubauer and Scherzer 2008), which amounts to computing the following iterative updates:

$$f^{k+1} := f^k - [\partial \mathcal{A}(f^k)]^*(\mathcal{A}(f^k) - g) - \lambda_k(f^k - f^0), \qquad (6.5)$$

where $f^0 \in X$ is an initial guess that incorporates *a priori* knowledge about the unknown f_{true} we seek. Next, one can introduce a data-driven damping factor in the above Landweber iteration:

$$f^{k+1} := f^k - [\partial\,\mathcal{A}(f^k)]^*(\mathcal{A}(f^k) - g) - \lambda_k[\partial\,\mathcal{B}(f^k)]^*(\mathcal{B}(f^k) - g). \qquad (6.6)$$

The (possibly non-linear) operator $\mathcal{B}\colon X \to Y$ can now be represented by a deep neural network that can be trained against supervised data by comparing the final iterate in (6.6) (iterates are stopped following the Morozov discrepancy principle (4.1)) against the ground truth for given data.

Convergence and stability for the scheme in (6.6) in infinite-dimensional Hilbert spaces is proved by Aspri *et al.* (2018). This theoretical results are complemented by several numerical experiments for solving linear inverse problems for the Radon transform and a non-linear inverse problem of Schlieren tomography. In these examples, however, Aspri *et al.* (2018) restrict attention to a linear operator $\mathcal{B}$.

6.3. Microlocal analysis

Microlocal analysis is a powerful mathematical theory for precisely describing how the singular part of a function, or more generally a distribution, is transformed under the action of an operator. Since its introduction to the mathematical community with the landmark publications by Sato (1971) and Hörmander (1971), it has proved itself useful in both pure and applied mathematical research. It is now a well-developed theory that can be used to study how singularities propagate under certain classes of operators, most notably Fourier integral operators, which include most differential and pseudo-differential operators as well as many integral operators frequently encountered in analysis, scientific computing and the physical sciences (Hörmander 1971, Candès, Demanet and Ying 2007).

The crucial underlying observation in microlocal analysis is that the information about the location of the singularities (singular support) needs to be complemented by specifying those 'directions' along which singularities may propagate. Making this precise leads to the notion of the wavefront set of the function (or distribution).

Role in inverse problems. Microlocal analysis is in particularly useful in inverse problems for a variety of reasons.

First, in many applications it is sufficient to recover the wavefront set of the model parameter from noisy data. For example, in imaging this would correspond to recovering the edges of the image from data. Such applications frequently arise when using imaging/sensing technologies where the transform is a pseudo-differential or Fourier integral operator (Krishnan and Quinto 2015). It turns out that one can use microlocal analysis to precisely describe how the wavefront set in data relates to the wavefront set for the model parameter, and this explicit relation is referred to as the (microlocal) canonical relation. Using the canonical relation one can recover the wavefront set from data without solving the inverse problem, a process that can be highly non-trivial.

Second, the canonical relation also describes which part of the wavefront set one can recover from data. This was done by Quinto (1993) for the case when the two- or three-dimensional ray transform is restricted to parallel lines, and by Quinto and Öktem (2008) for an analysis in the region-of-interest limited angle setting. Faber, Katsevich and Ramm (1995) derived a related principle for the three-dimensional ray transform restricted to lines given by helical acquisition, which is common in medical imaging. Similar principles hold for transforms integrating along other types of curves, *e.g.* ellipses with foci on the x-axis and geodesics (Uhlmann and Vasy 2012).

Finally, recovering the wavefront set of the model parameter from data is a less ill-posed procedure than attempting to recover the model parameter itself. This was demonstrated in Davison (1983), where the severely ill-posed reconstruction problem in limited angle CT becomes mildly ill-posed if one settles for recovering the wavefront. See also Quinto and Öktem (2008) for an application of this principle to cryo-electron tomography.

Data-driven extraction of the wavefront set. The above motivates the inverse problems community to work with the wavefront set. One difficulty that has limited use of the wavefront set is that it is virtually impossible to extract it numerically from a digitized signal. This is due to its definition, which depends on the asymptotic behaviour of the Fourier transform after a localization procedure. An alternative possibility is to identify the wavefront set after transforming the signal using a suitable representation, *e.g.* a curvelet or shearlet transform (Candès and Donoho 2005, Kutyniok and Labate 2009). This requires analysing the rate of decay of transformed signal, which again is unfeasible in large-scale imaging applications.

A recent paper (Andrade-Loarca, Kutyniok, Öktem and Petersen 2019) uses a data-driven approach to training a wavefront set extractor applicable to noisy digitized signals. The idea is to construct a deep neural network classifier that predicts the wavefront set from the shearlet coefficients of a signal. The approach is successfully demonstrated on two-dimensional imaging examples where it outperforms all conventional edge-orientation estimators as well as alternative data-driven methods including the current state of the art. This learned wavefront set extractor can now be combined with a learned iterative method using the framework in Section 6.1.

Using the canonical relation to guide data-driven recovery. In a recent paper Bubba *et al.* (2018) consider using the aforementioned microlocal canonical relation to steer a data-driven component in limited angle CT reconstruction, which is a severely ill-posed inverse problem.

Bubba *et al.* develop a hybrid reconstruction framework that fuses a knowledge-driven sparse regularization approach with a data-driven deep learning approach. The learning part is only applied to those parts that are not possible to recover (invisible part), which in turn can be characterized *a priori* through the canonical relation. The theoretically controllable sparse regularization is thus applied to the remaining parts that can be recovered (visible part).

This decomposition into visible and invisible parts is achieved numerically via the shearlet transform, which allows us to resolve wavefront sets in phase space. The neural network is then used to infer unknown shearlet coefficients associated with the invisible part.

7. Applications

In this section we revisit some of the machine learning methods for inverse problems discussed in the previous sections, and demonstrate their applicability to prototypical examples of inverse problems.

7.1. A simple example

We start the applications part of the paper by considering the exemplar inverse problem of ill-conditioned matrix inversion. This example should highlight the particular difficulties of applying learning to solve an ill-posed inverse problem. Surprisingly, even small 2×2 examples cannot be solved reliably by straightforward neural networks! The results of this section are based on Maass (2019).

This small-scale setting allows a somewhat complete analysis of the neural network; in particular, we can prove the shortcomings of such neural nets if the condition number of the matrix and the noise level in the data are in a critical relation. To be precise, in our most basic example we set

$$\mathcal{A}_\varepsilon(f) = \mathbf{A}_\varepsilon \cdot f \quad \text{where} \quad \mathbf{A}_\varepsilon = \begin{pmatrix} a_{11} & a_{12} \\ a_{21} & a_{22} \end{pmatrix} = \begin{pmatrix} 1 & 1 \\ 1 & 1+\varepsilon \end{pmatrix}.$$

This matrix has eigenvalues $\lambda_1 = 2 + \varepsilon/2 + O(\varepsilon^2)$ and $\lambda_2 = \varepsilon/2 + O(\varepsilon^2)$, with corresponding orthogonal eigenvectors

$$u_1 = \begin{pmatrix} 1 \\ 1 \end{pmatrix} + O(\varepsilon^2) \quad \text{and} \quad u_2 = \begin{pmatrix} 1 \\ -1 \end{pmatrix} + O(\varepsilon^2).$$

The ill-posedness of the problem, or rather the condition number of $\mathbf{A}_\varepsilon$, is controlled by $1/\varepsilon$. Typical values we have in mind here are $\varepsilon = 10^{-k}$ for $k = 0, \ldots, 10$.

We now compare two methods for solving the inverse problem of recovering f from $g = \mathbf{A}_\varepsilon \cdot f + e$. The first is classical Tikhonov regularization, which only uses information about the operator $\mathcal{A}_\varepsilon$. Here, given data g we estimate f_{true} by $\mathcal{R}_\sigma^{\text{Tik}}(g)$, where

$$\mathcal{R}_\sigma^{\text{Tik}}(g) := (\mathcal{A}_\varepsilon^* \circ \mathcal{A}_\varepsilon + \sigma^2 \,\text{id})^{-1} \circ \mathcal{A}_\varepsilon^*(g) = (\mathbf{A}_\varepsilon^T \cdot \mathbf{A}_\varepsilon + \sigma^2 \,\mathbf{I})^{-1} \cdot \mathbf{A}_\varepsilon^T \cdot g. \quad (7.1)$$

The second inversion is based on a trained neural network, that is, given data g we estimate f_{true} by $\mathcal{R}_{\mathbf{W}^*}^{\text{NN}}(g)$ where $\mathcal{R}_{\mathbf{W}^*}^{\text{NN}} : Y \to X$ is a trained neural network with $\mathbf{W}^*$ given by

$$\mathbf{W}^* \in \arg\min_{\mathbf{W}} \frac{1}{m} \sum_{i=1}^{m} \| \mathcal{R}_{\mathbf{W}}^{\text{NN}}(g^{(i)}) - f^{(i)} \|^2. \quad (7.2)$$

In the above, $(f^{(i)}, g^{(i)}) \in X \times Y$ with $i = 1, \ldots, m$, where coefficients in $f^{(i)}$ are i.i.d. samples of a $N(0,1)$ distributed random variable, and $g^{(i)} := \mathcal{A}\, f^{(i)} + e^{(i)}$, where $e^{(i)} \in \mathbb{R}^2$ are i.i.d. samples of a $N(0, \sigma^2)$ distributed random variable. This approach is fully data-driven and does not use any explicit knowledge about the operator $\mathcal{A}$.

Both methods are evaluated using a different set of test data and results are compared by computing the mean error:

$$E^{\text{Tik}} := \frac{1}{n} \sum_{i=1}^{n} \| \mathcal{R}_\sigma^{\text{Tik}}(g^{(i)}) - f^{(i)} \|^2 \quad \text{and} \quad E^{\text{NN}} := \frac{1}{n} \sum_{i=1}^{n} \| \mathcal{R}_{\mathbf{W}}^{\text{NN}}(g^{(i)}) - f^{(i)} \|^2,$$

for n test pairs $(f^{(i)}, g^{(i)}) \in X \times Y$ with $g^{(i)} := \mathcal{A}\, f^{(i)} + e^{(i)}$ as in the training set above but clearly distinct from the training examples.

The design of the network is crucial. We use a minimal network which allows us to reproduce a matrix vector multiplication. Hence the network is – in principle – capable of recovering the Tikhonov regularization operator or even an improvement of it. We use a network with a single hidden layer with four nodes. We restrict the eight weights connecting the two input variables with the first layer by setting

$$w_1 = -w_3 = w_{11}, \quad w_2 = -w_4 = w_{12},$$
$$w_5 = -w_7 = w_{21}, \quad w_6 = -w_8 = w_{22},$$

as depicted in Figure 7.1. We obtain a neural network depending on four variables $w_{11}, w_{12}, w_{21}, w_{22}$ and the network acts as a multiplication of the matrix

$$\mathbf{W} = \begin{pmatrix} w_{11} & w_{12} \\ w_{21} & w_{22} \end{pmatrix}$$

with the input vector $z = (z_1, z_2)$. We denote the output of such a neural network by $\mathcal{R}_{\mathbf{W}}^{\text{NN}}(z) = W z$.

For later use we define $(2 \times m)$ matrices $\mathbf{X}, \mathbf{Y}$ and $\mathbf{E}$ that store the vectors $f^{(i)}, g^{(i)}$ and $e^{(i)}$ column-wise, so the training data can be summarized as

$$\mathbf{Y} = \mathbf{A}_\varepsilon \cdot \mathbf{X} + \mathbf{E}. \quad (7.3)$$

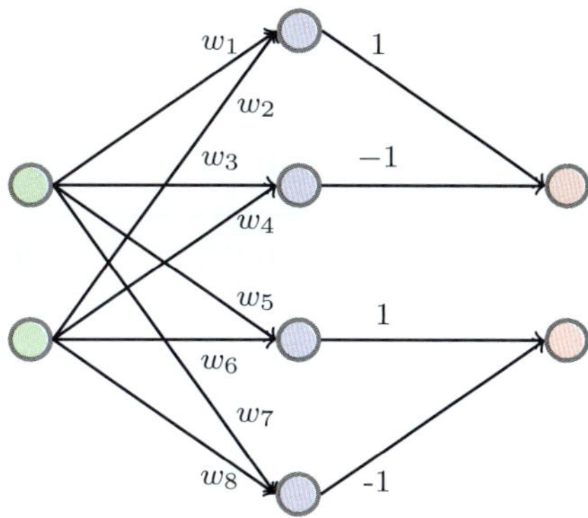

Figure 7.1. The network design with eight parameters, a setting that yields a matrix–vector multiplication of the input.

The training of such a network for modelling the forward problem is equivalent (using the Frobenius norm for matrices) to minimizing the expected mean square error

$$\min_{\mathbf{W}} \frac{1}{n} \sum_{i=1}^{n} \| \mathbf{W} f^{(i)} - g^{(i)} \|^2 = \min_{\mathbf{W}} \frac{1}{n} \| \mathbf{W}\mathbf{X} - \mathbf{Y} \|^2, \tag{7.4}$$

and the training model (7.2) for the inverse problem simplifies to

$$\min_{\mathbf{W}} \frac{1}{n} \sum_{i=1}^{n} \| \mathbf{W} g^{(i)} - f^{(i)} \|^2 = \min_{\mathbf{W}} \frac{1}{n} \| \mathbf{W}\mathbf{Y} - \mathbf{X} \|^2. \tag{7.5}$$

In the next paragraph we report some numerical examples before analysing these networks.

Testing error convergence for various values of ε. We train these networks using a set of training data $(f^{(i)}, g^{(i)})_{i=1,\dots,m}$ with $m = 10\,000$, i.e. $g^{(i)} = \mathcal{A}_\varepsilon f^{(i)} + e^{(i)}$. The network design with restricted coefficients as described above has four degrees of freedom $w = (w_{11}, w_{12}, w_{21}, w_{22})$. The corresponding loss function is minimized by a gradient descent algorithm, that is, the gradient of the loss function with respect to w is computed by backpropagation (Rumelhart, Hinton and Williams 1986, Martens and Sutskever 2012, Byrd, Chin, Nocedal and Wu 2012). We used 3000 iterations (epochs) of this gradient descent to minimize the loss function of a network for the forward operator using (7.4) or, respectively, for training a network for solving the inverse problem using (7.2). The MSE errors on the training data were close to zero in both cases.

After training we tested the resulting networks by drawing $n = 10\,000$ new data vectors $f^{(i)}$ as well as error vectors $e^{(i)}$. The $g^{(i)}$ were computed as above. Table 7.1 lists the resulting values using this set of test data

Table 7.1. The errors of the inverse net with an ill-conditioned matrix $\mathbf{A}_\varepsilon$ ($i.e.$ $\varepsilon \ll 1$) are large and the computed reconstructions with the test data are meaningless.

Error/choice of ε	1	0.1	0.01	0.0001
NMSE_{fwd}	0.002	0.013	0.003	0.003
NMSE_{inv}	0.012	0.8	10	10

for the network trained for the forward problem and the inverse problem, respectively:

$$\text{NMSE}_{\text{fwd}} := \frac{1}{n} \sum_{i=1}^{n} \| \mathbf{W} f^{(i)} - g^{(i)} \|^2,$$

$$\text{NMSE}_{\text{inv}} := \frac{1}{n} \sum_{i=1}^{n} \| \mathbf{W} f^{(i)} - g^{(i)} \|^2.$$

We observe that training the forward operator produces reliable results, as does the network for the inverse problem with $\varepsilon \geq 0.1$. However, training a network for the inverse problem with an ill-conditioned matrix $\mathbf{A}_\varepsilon$ with $\varepsilon \leq 0.01$ fails. This is confirmed by analysing the values of w and of the resulting matrix $\mathbf{W}$ after training. We would expect that in training the forward problem will produce values for $\mathbf{W}$ such that $\mathbf{W} \sim \mathbf{A}_\varepsilon$ and that training the inverse problems leads to $\mathbf{W} \sim \mathcal{R}_\sigma^{\text{Tik}}(g)$ for some regularization parameter σ. For the forward problem, the difference between $\mathbf{W}$ and $\mathbf{A}_\varepsilon$ is of order 10^{-3} or below, but for $\varepsilon \leq 0.01$ the training of the inverse problem leads to a matrix that has no similarity to the Tikhonov regularized inverse. Using a network with a single internal layer but with more nodes and no restriction on the structure of the weights did not yield any significant improvements.

Analysis of trivial neural networks for inverse problems. In this section we analyse the case where the application of the neural network is strictly equivalent to a matrix–vector multiplication, that is, training of the network is given by (7.5). The optimal $\mathbf{W}$ in (7.5) is given by

$$\mathbf{W}^T = (\mathbf{Y}\mathbf{Y}^T)^{-1}\mathbf{Y}\mathbf{X}^T. \tag{7.6}$$

Standard arguments show that, together with that hypothesized in the numerical discussion above, the $\mathbf{W}$ in (7.6) coincides with the Tikhonov regularizers $\mathcal{R}_\sigma^{\text{Tik}}(g)$ from (7.1). This is not a surprising result since (7.6) coincides with the classical maximum *a posteriori* (MAP) estimator of statistical inverse problems. However, analysing the variance $\mathbb{E}[\| \mathbf{W} - \mathbf{T} \|^2]$, where $\mathbf{T} = (\mathbf{A}_\varepsilon^T \cdot \mathbf{A}_\varepsilon + \sigma^2 \mathbf{I})^{-1} \cdot \mathbf{A}_\varepsilon^T$ is the Tikhonov reconstruction matrix, reflects the ill-posedness of the problem. Indeed, as is demonstrated in a series of numerical tests in Maass (2019), the deviation of $\mathbf{W}$ from $\mathbf{T}$ will be arbitrarily large if ε and σ are both small. Of course, we can also give this a positive meaning: the noise level acts as a regularizer, and large σ yields more stable matrices $\mathbf{W}$. See Maass (2019) for details.

This small example clearly illustrates that one needs to have some insight into the nature of inverse problems for successfully applying deep learning techniques. Performance of practical examples of more targeted deep learning approaches to inverse problems will be discussed in the following Sections 7.3, 7.4, 7.6 and 7.7.

7.2. Bilevel learning for total variation image denoising

In (4.4) bilevel learning of TV-type regularizers was discussed as a way to make functional analytic regularization more data-driven. In what follows, we showcase some results of this learning approach for the case of image denoising, *i.e.* $\mathcal{A} = \mathrm{id}$.

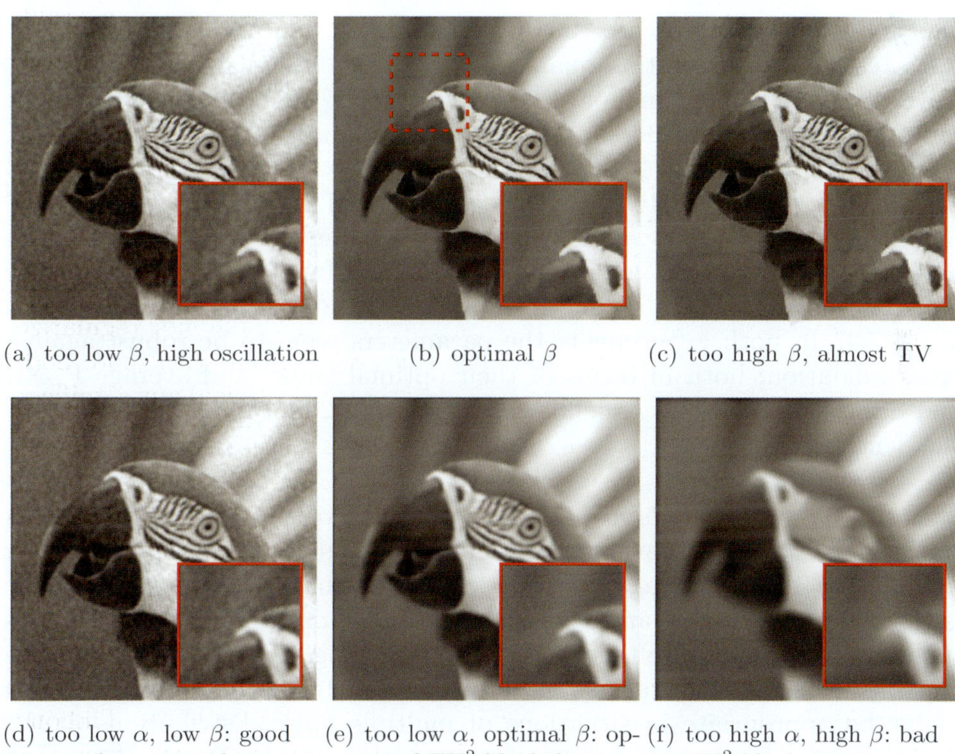

(a) too low β, high oscillation (b) optimal β (c) too high β, almost TV

(d) too low α, low β: good match to noisy data (e) too low α, optimal β: optimal TV^2-like behaviour (f) too high α, high β: bad TV^2-like behaviour

Figure 7.2. (a–c) Effect of choosing β on total generalized variation (TGV)^2 denoising with optimal α. (d–f) Effect of choosing α too large in TGV^2 denoising.

Optimal TV-regularizers for image denoising. The regularization effect of
TV and second-order TV approaches heavily depends on the choice of regularization parameters $\theta = \alpha$ (*i.e.* $= (\alpha, \beta)$ for second-order TV approaches).
In Figure 7.2 we show the effect of different choices of α and β in TGV^2
denoising. In what follows we show some results from De los Reyes *et al.*
(2017) applying the learning approach from (4.4) with the smoothed regularizer (4.6) to find optimal parameters in TV-type reconstruction models.
The regularization effect of TV and second-order TV approaches heavily
depends on the choice of regularization parameters $\theta = \alpha$ (*i.e.* $= (\alpha, \beta)$ for
second-order TV approaches).

The first example is TGV denoising of an image corrupted with white
Gaussian noise with PSNR of 24.72. The red dot in Figure 7.3 plots
the discovered regularization parameter $\widehat{\theta} = (\hat{\alpha}, \hat{\beta})$ reported in Figure 7.4.
Studying the location of the red dot, we may conclude that the Broyden–
Fletcher–Goldfarb–Shanno (BFGS) algorithm managed to find a nearly optimal parameter in very few iterations: see Table 7.2. Although the optimization problem for (α, β) is non-convex, in all of our experiments we
observed commendable convergence behaviour of the BFGS algorithm: see
De los Reyes *et al.* (2017) for further examples.

To test the generalization quality of the bilevel learning model in De los
Reyes *et al.* (2017), optimal parameters were cross-validated when tested for
image denoising on the Berkeley segmentation data set (BSDS300) (Martin,
Fowlkes, Tal and Malik 2001). A dataset of 200 images was selected and
split into two halves of 100 images each. Optimal parameters were learned
for each half individually, and then used to denoise the images of the other
half. The results for TV denoising with L^2-loss function and data fidelity
are reported in Table 7.3. The results for TGV denoising are reported in
Table 7.4. In both experiments the parameters seem to be robust against
cross-validation, both in terms of their optimal value, and average PSNR
and SSIM (Wang, Bovik, Sheikh and Simoncelli 2004) quality measures of
the denoised image.

Bilevel learning of the data discrepancy term in mixed noise scenarios. In
the examples in the previous paragraph we considered bilevel parameter
learning for TV-type image denoising, assuming that the noise in the image
is normally distributed and consequently an L^2-data discrepancy term is
the appropriate choice to take. The bilevel learning model (4.4), however, is
capable of linear combinations of data discrepancy terms as in (4.5) which
might be appropriate in situations of multiple noise distributions in the
data: see *e.g.* Lanza, Morigi, Sgallari and Wen (2014), De los Reyes and
Schönlieb (2013), Calatroni *et al.* (2017) and Calatroni (2015) and references
therein. Calatroni *et al.* (2017) also considered infimal convolutions of data
discrepancy functions.

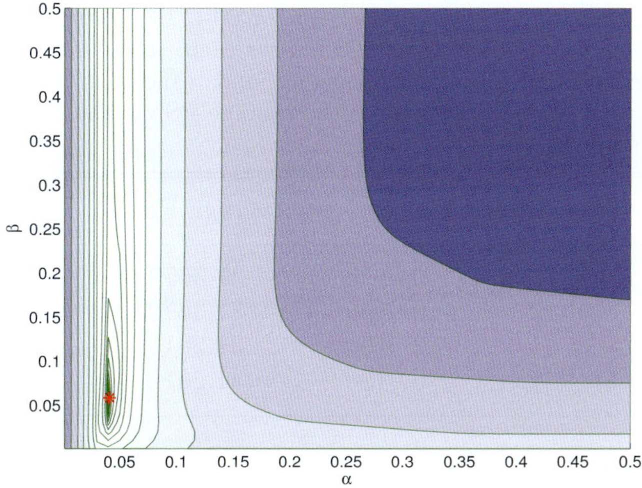

Figure 7.3. Contour plot of the objective functional in TGV^2 denoising in the (α, β)-plane.

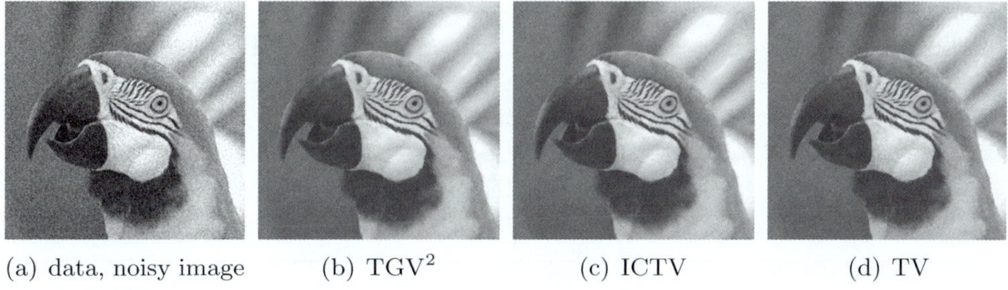

(a) data, noisy image   (b) TGV^2 (c) ICTV   (d) TV

Figure 7.4. Optimal denoising results for TGV^2, ICTV and TV, all with L_2^2 as data discrepancy.

Table 7.2. Quantified results for the parrot image ($s :=$ image width/height in pixels $= 256$), using L_2^2 discrepancy.

Denoise	Initial (α, β)	Result $(\hat{\alpha}, \hat{\beta})$	Objective	SSIM	PSNR	Its	Figure
TGV^2	$(\hat{\alpha}_{\mathrm{TV}}/s, \hat{\alpha}_{\mathrm{TV}})$	$(0.058/s^2, 0.041/s)$	6.412	0.890	31.992	11	7.4(b)
ICTV	$(\hat{\alpha}/s, \hat{\alpha}_{\mathrm{TV}})$	$(0.051/s^2, 0.041/s)$	6.439	0.887	31.954	7	7.4(c)
TV	$0.1/s$	$0.042/s$	6.623	0.879	31.710	12	7.4(d)

Table 7.3.   Cross-validated computations on the BSDS300 data set (Martin *et al.* 2001) split into two halves of 100 images each. TV regularization with L^2-discrepancy and fidelity function. 'Learning' and 'validation' indicate the halves used for learning α and for computing the average PSNR and SSIM, respectively. Noise variance $\sigma = 10$.

Validation	Learning	α	Average PSNR	Average SSIM
1	1	0.0190	31.3679	0.8885
1	2	0.0190	31.3672	0.8884
2	1	0.0190	31.2619	0.8851
2	2	0.0190	31.2612	0.8850

Table 7.4. Cross-validated computations on the BSDS300 data set (Martin *et al.* 2001) split into two halves of 100 images each. TGV^2 regularization with L^2-discrepancy. 'Learning' and 'validation' indicate the halves used for learning α and for computing the average PSNR and SSIM, respectively. Noise variance $\sigma = 10$.

Validation	Learning	$\vec{\alpha}$	Average PSNR	Average SSIM
1	1	(0.0187, 0.0198)	31.4325	0.8901
1	2	(0.0186, 0.0191)	31.4303	0.8899
2	1	(0.0186, 0.0191)	31.3281	0.8869
2	2	(0.0187, 0.0198)	31.3301	0.8870

Figures 7.5 and 7.6 present denoising results with optimally learned parameters for mixed Gaussian and impulse noise and for mixed Gaussian and Poisson noise, respectively. See Calatroni *et al.* (2017) for more details. The original image has been corrupted with Gaussian noise of zero mean and variance 0.005 and then a percentage of 5% of pixels has been corrupted with impulse noise. The parameters have been chosen to be $\gamma = 10^4$, $\mu = 10^{-15}$ and the mesh step size $h = 1/312$. The computed optimal weights are $\hat{\lambda}_1 = 734.25$ and $\hat{\lambda}_2 = 3401.2$. Together with an optimal denoised image, the results show the decomposition of the noise into its sparse and Gaussian components: see Calatroni *et al.* (2017) for more details.

Remark 7.1.   When optimizing only a handful of scalar parameters, as in the examples discussed above, bilevel optimization is by no means the most efficient approach for parameter learning. In fact, brute force line-search methods are in this context still computationally feasible as the dimensionality of the parameter space being explored is small. However, even in

(a)   (b)   (c)   (d)   (e)

Figure 7.5. Optimized impulse-Gaussian denoising: (a) original image, (b) noisy image with Gaussian noise of variance 0.005 and (c) with 5% of pixels corrupted with impulse noise, (d) impulse noise residuum, (e) Gaussian noise residuum. Optimal parameters $\hat{\lambda}_1 = 734.25$ and $\hat{\lambda}_2 = 3401.2$.

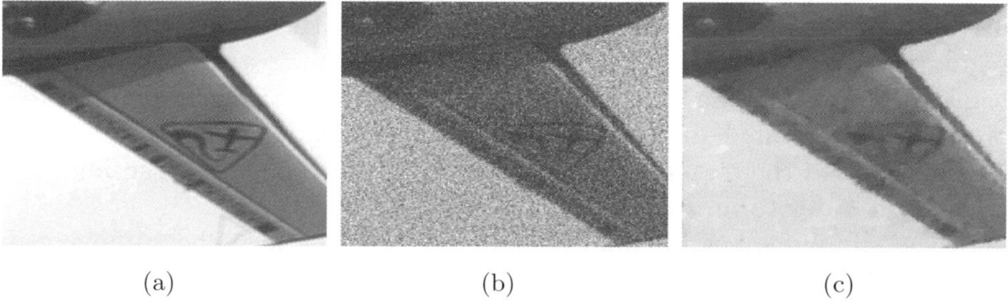

(a)   (b)   (c)

Figure 7.6. Optimized Poisson–Gauss denoising: (a) original image, (b) noisy image corrupted by Poisson noise and Gaussian noise with mean zero and variance 0.001, (c) denoised image. Optimal parameters $\hat{\lambda}_1 = 1847.75$ and $\hat{\lambda}_2 = 73.45$.

this small parameter example, investigating bilevel optimization methods is instructive, as it tells us something about the mathematical properties of parameter learning for the typically considered non-smooth variational regularization problems and the numerical approaches with which they can be tackled. Insight gained from this becomes particularly important when going to more advanced parametrizations, for instance when optimizing spatially varying regularization parameters (Van Chung *et al.* 2017) or different discrete parametrizations, as considered in Sections 4.3.2, 4.4 or 4.7.

7.3. Learned iterative reconstruction for computed tomography (CT) and photoacoustic tomography (PAT)

Learned iterative reconstruction schemes (Section 5.1.4) have been successfully applied to several large-scale inverse problems in imaging, such as image reconstruction in magnetic resonance imaging (MRI), CT and PAT. We will show examples from CT taken from Adler and Öktem (2017, 2018*b*) and PAT taken from Hauptmann *et al.* (2018).

7.3.1. *CT image reconstruction*

The learned primal–dual scheme in Adler and Öktem (2018*b*) is here tested on the two-dimensional CT image reconstruction problem, and its performance is evaluated in a simplified setting as well as a more realistic setting.

The forward operator for pre-log data is expressible in terms of the ray transform of (2.5), and for log data it is given by the ray transform of (2.6). The model parameter is a real-valued function defined on a domain $\Omega \subset \mathbb{R}^2$. This function represents the image we seek to recover and we assume $X \subset L^2(\Omega)$ is a suitable vector space of such functions.

This data set is used to train both the learned post-processing and learned iterative methods, where the former is filtered backprojection (FBP) reconstruction followed by a trained denoiser with a *U-Net* architecture and the latter is the learned primal–dual method in Adler and Öktem (2018*b*). Both networks were trained using the squared L^2-loss. The other two knowledge-driven reconstruction methods are the standard FBP and (isotropic) TV-regularized reconstruction. The FBP reconstructions is applied to log data using a Hann filter; the TV reconstruction was solved using 1000 iterations of the classical primal–dual hybrid gradient algorithm. The filter bandwidth in the FBP and the regularization parameter in the TV reconstruction were selected in order to maximize the PSNR.

In the simplified setting shown in Figure 7.8 (see also the summary in Table 7.5) the images are 128×128 pixel step functions, and we use supervised training data consisting of about 50 000 pairs of images and corresponding data as in Figure 7.7. Noise is 5% additive Gaussian. The images in the training data are randomly generated using a known probability distribution, and corresponding tomographic data (sinogram) are simulated with 5% additive Gaussian noise. This is a relatively small-scale problem, which allows us to also compute the conditional mean reconstruction using Markov chain Monte Carlo (MCMC) techniques (Section 3.5.2). The conditional mean reconstruction is useful since the learned iterative method approximates it. The same holds for learned post-processing since FBP is a linear sufficient statistic: see the discussion in Section 5.1.5. Hence, neither learned post-processing nor learned iterative will outperform the conditional mean, irrespective of the amount of training data and model capacity, that is, the conditional mean reconstruction serves as a theoretical limit for what one can recover.

In the realistic setting shown in Figure 7.9 (see also the summary in Table 7.5) the images are clinical CT scans. We use supervised training data consisting of about 2000 pairs of images from nine patients, and corresponding pre-log data are simulated with a Poisson noise corresponding to 10^4 incident photons per pixel before attenuation, which would correspond to a low-dose CT scan. Unfortunately we cannot compute the conditional mean as in the simplified setting, but we could compare against another

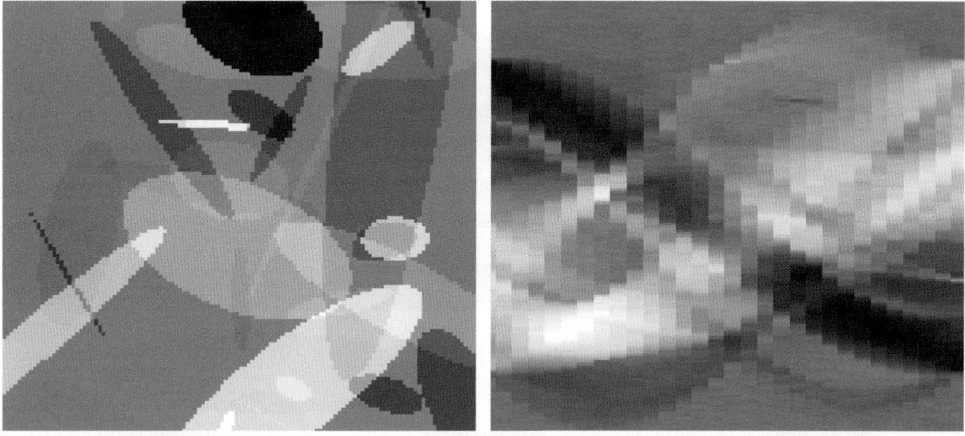

Figure 7.7. Example from supervised training data used to train the learned iterative and learned post-processing methods used in Figure 7.8.

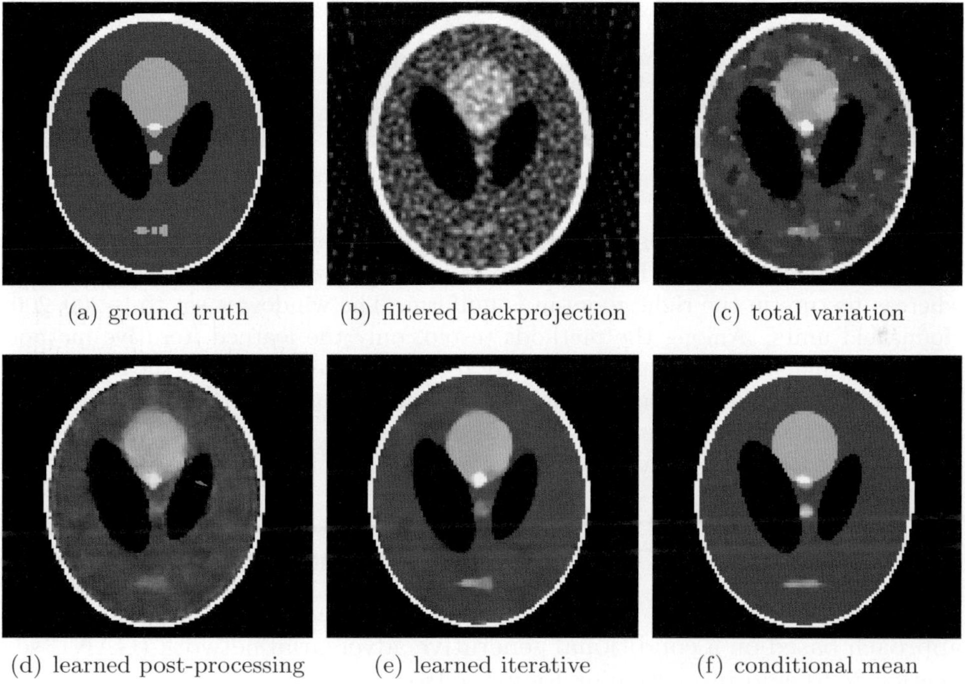

(a) ground truth · (b) filtered backprojection · (c) total variation

(d) learned post-processing · (e) learned iterative · (f) conditional mean

Figure 7.8. Reconstructions of the Shepp–Logan phantom using different methods. The window is set to $[0.1, 0.4]$, corresponding to the soft tissue of the modified Shepp–Logan phantom. We can see that the learned iterative method does indeed approximate the Bayes estimator, which here equals the conditional mean.

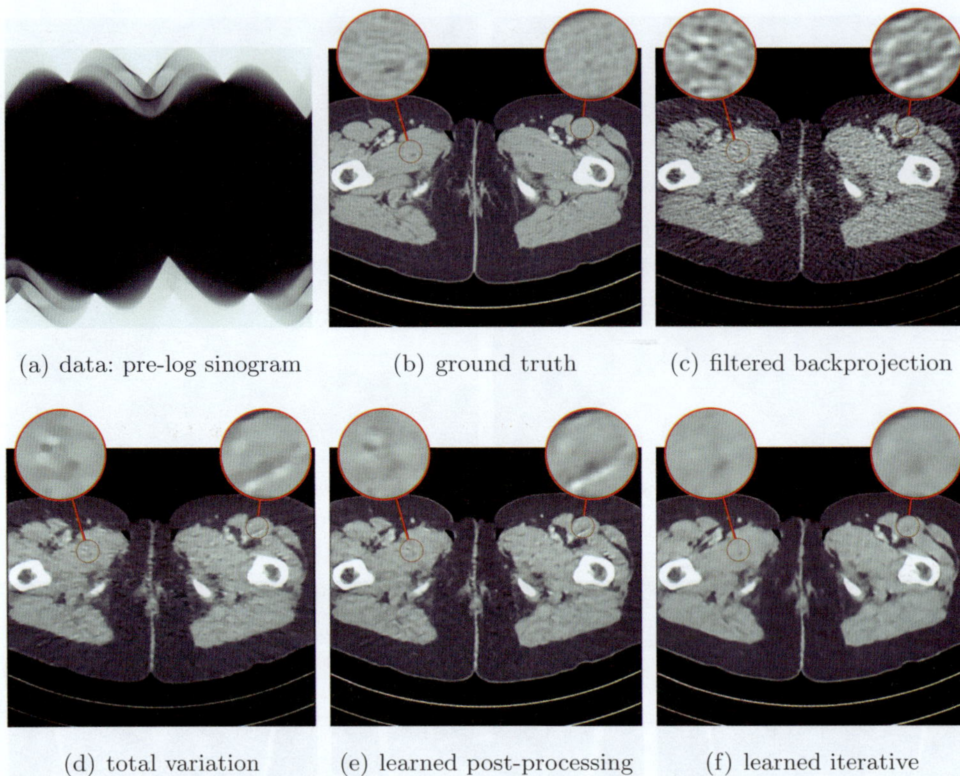

(a) data: pre-log sinogram　　　(b) ground truth　　　(c) filtered backprojection

(d) total variation　　　(e) learned post-processing　　　(f) learned iterative

Figure 7.9. Reconstructions of a 512×512 pixel human phantom along with two zoom-in regions indicated by small circles. The left zoom-in has a true feature whereas texture in the right zoom-in is uniform. The window is set to $[-200, 200]$ Hounsfield units. Among the methods tested, only the learned iterative method (learned primal–dual algorithm) correctly recovers these regions. In the others, the true feature in the left zoom-in is indistinguishable from other false features of the same size/contrast, and the right-zoom in has a streak artefact. The improvement that comes with using a learned iterative method thus translates into true clinical usefulness.

approach for computing the conditional mean, namely a sampling-based approach based on a conditional generative adversarial network (GAN) (see Section 5.2.1 and in particular Figure 7.16).

7.3.2. PAT reconstructions

Photoacoustic tomography (PAT) is a novel 'imaging from coupled physics' technique (Arridge and Scherzer 2012) that can obtain high-resolution three-dimensional *in vivo* images of absorbed optical energy by sensing laser-generated ultrasound (US) (Wang 2009, Beard 2011, Nie and Chen 2014,

Table 7.5. Summary of results shown in Figures 7.8 and 7.9 where an SSIM score of 1 corresponds to a perfect match. Note that the learned iterative method (learned primal–dual algorithm) significantly outperforms TV regularization even when reconstructing the Shepp–Logan phantom. With respect to run-time, the learned iterative method involves calls to the forward operator, and is therefore slower than learned post-processing by a factor of ≈ 6. Compared with TV-regularized reconstruction, all learned methods are at least two orders of magnitude faster.

Method	Results for Figure 7.8			Results for Figure 7.9		
	PSNR (dB)	SSIM	Parameters	PSNR (dB)	SSIM	Parameters
Filtered backprojection	19.75	0.597	1	33.65	0.823	1
Total variation	28.06	0.928	1	37.48	0.946	1
Learned post-processing	29.20	0.943	10^7	41.92	0.941	10^7
Learned iterative	38.28	0.988	2.4×10^5	44.11	0.969	2.4×10^5
Conditional expectation	45.46	0.993	0	–	–	–

Valluru, Wilson and Willmann 2016, Zhou, Yao and Wang 2016, Xia and Wang 2014). In the setting considered here, data are collected as a time series on a two-dimensional sensor $Y = [\Gamma \subset \mathbb{R}^2] \times [0, T]$ on the surface of a domain $X = \Omega \subset \mathbb{R}^3$. Several methods exist for reconstruction, including filtered backprojection-type inversions of the spherical Radon transform and numerical techniques such as time-reversal. As in problems such as CT (Section 2.2.4) and MRI (Section 2.2.5), data subsampling may be employed to accelerate image acquisition, which leads consequently to the need for regularization to prevent noise propagation and artefact generation. The long reconstruction times ensuing from conventional iterative reconstruction algorithms have motivated consideration of machine learning methods.

The deep gradient descent (DGD) method (Hauptmann *et al.* 2018) for PAT is an example of a learned iterative method (see Section 5.1.4). The main aspects can be summarized as follows.

- Each iteration adds an update by combining measurement information delivered via the gradient $\nabla \mathcal{L}(g, \mathcal{A} f_k) = \mathcal{A}^*(\mathcal{A} f_k - g)$ with an image processing step

$$f_{k+1} = \mathrm{G}_{\theta_k}(\nabla \mathcal{L}(g, \mathcal{A} f_k), f_k), \qquad (7.7)$$

  where the layer operators G_{θ_k} correspond to convolutional neural networks (CNNs) with different, learned parameters θ_k but with the same architecture. The initialization for the iterations was the backprojection of the data $f_0 = \mathcal{A}^* g$.

- The training data were taken from the publicly available data from the ELCAP Public Lung Image Database.[9] The data set consists of 50 whole-lung CT scans, from which about 1200 volumes of vessel structures were segmented, and scaled up to the final target size of $80 \times 240 \times 240$. Out of these volumes 1024 were chosen as the ground truth f_{true} for the training and simulated limited-view, subsampled data, using the same measurement set-up as in the *in vivo* data. Precomputing the gradient information for each CNN took about 10 hours.

- Initial results from training on synthetic data showed a failure to effectively threshold the noise-like artefacts in the low absorption regions (see Figure 7.10). This effect was ameliorated by simulating the effect of the low absorbing background as a Gaussian random field with short spatial correlation length. The synthetic CT volumes with the added background were then used for the data generation, *i.e.* $g_{\mathrm{back}}^i = \mathcal{A} f_{\mathrm{back}}^i + \varepsilon$, whereas the clean volumes f_{true} were used as reference for the training.

---

[9] http://www.via.cornell.edu/databases/lungdb.html

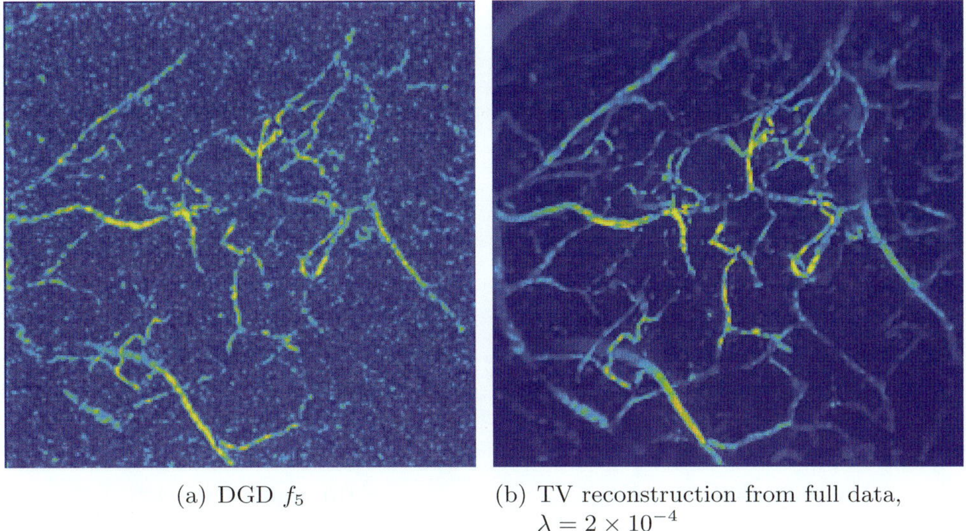

(a) DGD f_5 · (b) TV reconstruction from full data, $\lambda = 2 \times 10^{-4}$

Figure 7.10. Reconstruction from real measurement data of a human palm, without adjustments of the training data. The images shown are top-down maximum intensity projections. (a) Result of the DGD trained on images without added background. (b) TV reconstruction obtained from fully sampled data.

- The results were further improved using *transfer training* with a set of 20 (fully sampled) measurements of a human finger, wrist and palm from the same experimental system. To update the DGD an additional five epochs of training on the pairs $\{g_{\text{real}}, f_{\text{TV}}\}$ were performed with a reduced learning rate taking only 90 minutes. The effect of the updated DGD is shown in Figure 7.11.

7.4. Adversarial regularizer for CT

In Section 4.7 the concept of training a regularizer that is parametrized with a neural network in an adversarial manner has been presented. In what follows, we present numerical results as they are reported in Lunz *et al.* (2018). There, the performance of the adversarial regularizer for two-dimensional CT reconstruction is considered, that is, $\mathcal{A}$ is the ray transform as in (2.6). CT reconstruction is an application in which functional analytic inversion (see Section 2), and in particular the variational approach from Sections 2.5 and 2.6, is very widely used in practice. Here, it serves as a prototype inverse problem with non-trivial forward operator.

We compare the performance of TV-regularized reconstruction from Section 2.6 and (2.12), post-processing as in Section 5.1.5 (see in particular Gupta *et al.* 2018), regularization by denoising (RED) in Section 4.6 and

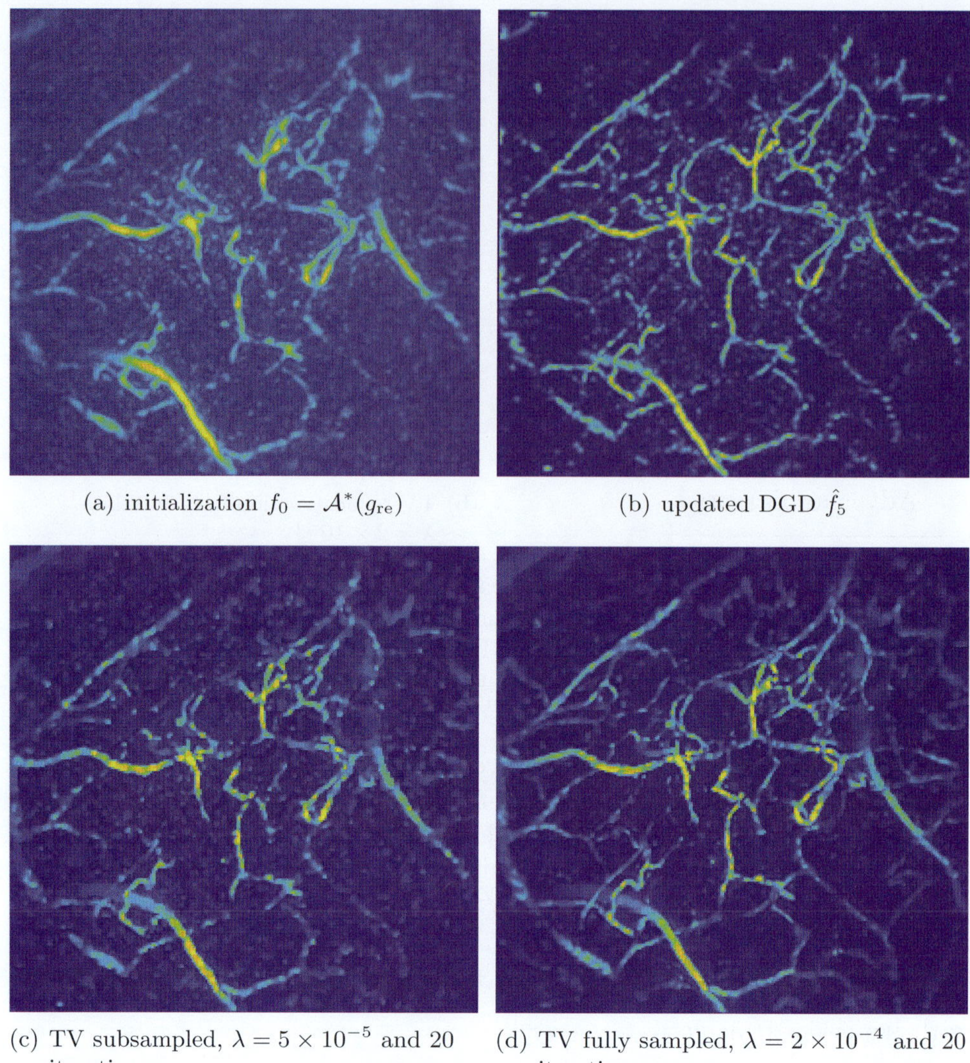

(a) initialization $f_0 = \mathcal{A}^*(g_{\mathrm{re}})$

(b) updated DGD $\hat{f}_5$

(c) TV subsampled, $\lambda = 5 \times 10^{-5}$ and 20 iterations

(d) TV fully sampled, $\lambda = 2 \times 10^{-4}$ and 20 iterations

Figure 7.11. Example of real measurement data of a human palm. Volumetric images are shown using top-down maximum intensity projections. (a) Initialization from subsampled data, and (b) the DGD $G_{\hat{\theta}_k}$ after five iterations. (c) TV reconstruction of subsampled data with an emphasis on the data fit. (d) Reference TV reconstruction from fully sampled limited-view data. All TV reconstructions were computed with 20 iterations.

Table 7.6. CT reconstruction on the LIDC dataset using various methods. Note that the learned post-processing and RED methods require training on supervised data, while the adversarial regularizer only requires training on unsupervised data.

| | High noise | | Low noise | |
Method	PSNR (dB)	SSIM	PSNR (dB)	SSIM
Knowledge-driven				
Filtered backprojection	14.9	0.227	23.3	0.604
Total variation	27.7	0.890	30.0	0.924
Supervised				
Learned post-processing	31.2	0.936	33.6	0.955
RED	29.9	0.904	32.8	0.947
Unsupervised				
Adversarial regularizer	30.5	0.927	32.5	0.946

the adversarial regularizer from Section 4.7 on the LIDC/IDRI database (Armato *et al.* 2011) of lung scans.

We used a simple eight-layer convolutional neural network with a total of four average pooling layers of window size 2×2, leaky ReLU ($\alpha = 0.1$) activations and two final dense layers for all experiments with the adversarial regularizer algorithm. Training and test measurements have been simulated by taking the ray transform of the two-dimensional CT slices, adding Gaussian white noise, and under-sampling the data by storing only 30 angles in the forward operator. Results are reported in Table 7.6 and Figure 7.12.

In Table 7.6 we see that TV is outperformed by the learned regularization techniques by a large margin. The reconstructions achieved by the adversarial regularizer are at least as good in visual quality as those obtained with supervised machine learning methods, despite having used unsupervised data only. The ability of the adversarial regularizer to be trained in an unsupervised fashion could be interesting for its application to practical inverse problems, where ground truth data are often scarce or unavailable. Further results of the adversarial regularizer and discussion can be found in Lunz *et al.* (2018).

7.5. Deep learning for magnetic particle imaging (MPI)

MPI is an imaging modality based on injecting ferromagnetic nanoparticles, which are then transported by the blood flow. Reconstructing the resulting spatial distribution $c(x)$ of those nanoparticles is based on exploiting the non-linear magnetization behaviour of ferromagnetic nanoparticles (Gleich and Weizenecker 2005).

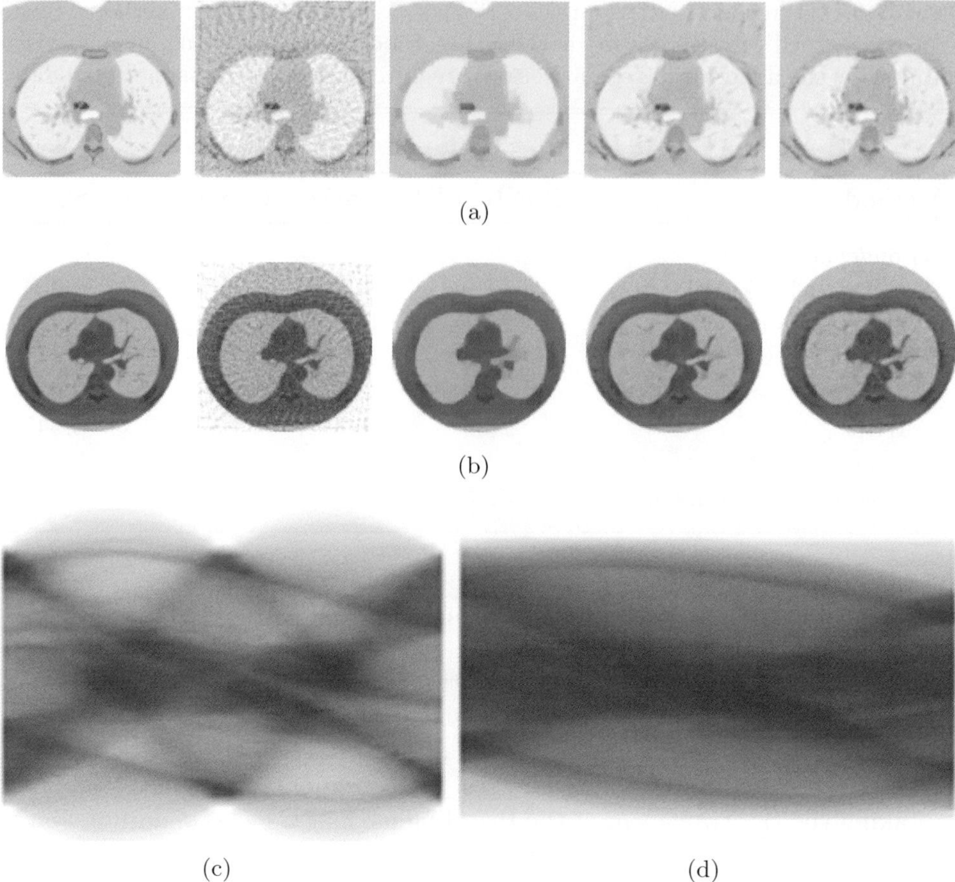

Figure 7.12. Exemplar CT reconstructions on the LIDC dataset under low-noise corruption. (a, b) Left to right: ground truth, FBP, TV, post-processing and adversarial regularization. (c,d) Data (CT sinograms): (c) data used for reconstructions in (a); (d) data used for reconstructions in (b).

More precisely, one applies a magnetic field, which is a superposition of a static gradient field, which generates a field-free point (FFP), and a highly dynamic spatially homogeneous field, which moves the FFP in space. The magnetic moment of the nanoparticles in the neighbourhood of the field-free point will oscillate, generating an electromagnetic field whose voltages can be measured by so-called receive coils. The time-dependent measurements $v_\ell(t)$ in the receive coils are the data for the inversion process, *i.e.* for reconstructing $c(x)$.

MPI benefits from a high temporal resolution and a potentially high spatial resolution which makes it suitable for several *in vivo* applications, such as imaging blood flow (Weizenecker *et al.* 2009, Khandhar *et al.* 2017), instrument tracking (Haegele *et al.* 2012) and guidance (Salamon *et al.* 2016), flow estimation (Franke *et al.* 2017), cancer detection (Yu *et al.* 2017) and treatment by hyperthermia (Murase *et al.* 2015). However, real-time applications are still far from being realized; also, the mathematical foundation of such dynamic inverse problems (see Schmitt and Louis 2002, Hahn 2015, Schuster, Hahn and Burger 2018) is just developing.

Due to the non-magnetic coating of the nanoparticles, which largely suppresses particle–particle interactions, MPI is usually modelled by a linear Fredholm integral equation of the first kind describing the relationship between particle concentration and the measured voltage. After subtracting the voltage induced by the applied magnetic field one obtains a measured signal in the ℓth receive coil as

$$y_\ell(t) = S_\ell c(t) := \int_\Omega c(x)\, s_\ell(x,t)\, \mathrm{d}t,$$

where s_ℓ denotes the kernel of the linear operator. Combining the measurements of all receive coils yields – after discretization – a linear system of equations $Sc = g$. Typically, the rows of S are normalized, resulting in the final form of the linearized inverse problem denoted by

$$\mathbf{A}\, c = g. \tag{7.8}$$

This is a coarse simplification of the physical set-up, which neglects non-linear magnetization effects of the nanoparticles as well as the non-homogeneity of the spatial sensitivity of the receive coils and also the small but non-negligible particle–particle interactions. Hence this is a perfect set-up for exploiting the potential of neural networks for matching complex and high-dimensional non-linear models.

We test the capability of the deep imaging prior approach to improving image reconstruction obtained by standard Tikhonov regularization. For the experiments we use datasets generated by the Bruker preclinical MPI system at the University Medical Center, Hamburg–Eppendorf.

We use the deep image prior network introduced by Ulyanov *et al.* (2018), specifically their U-Net architecture. Our implementation is based on TensorFlow (Abadi *et al.* 2015) and Keras (Chollet *et al.* 2015), and has the following specifications. Between the encoder and decoder part of the U-Net our skip connection has four channels. The convolutional encoder goes from the input to 32, 32, 64 and 128 channels, each with strides of 2×2 and filters of size 3×3. Then the convolutional decoder has the mirrored architecture plus first a resize-nearest-neighbour layer to reach the desired output shape and second an additional ReLU convolutional layer with filters of size 1. The number of channels of this last layers is three for data set 1 (DS1) to accommodate three slices (three two-dimensional scans, one above another) of a two-dimensional phantom centred at the central slice of the three. The

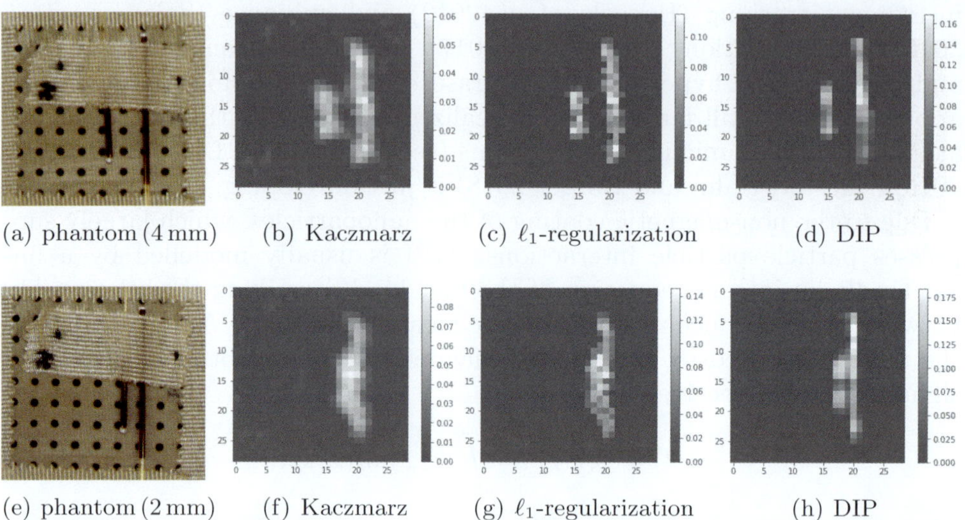

(a) phantom (4 mm) (b) Kaczmarz (c) ℓ_1-regularization (d) DIP

(e) phantom (2 mm) (f) Kaczmarz (g) ℓ_1-regularization (h) DIP

Figure 7.13. MPI reconstructions of two phantoms using different methods: (a)–(d) phantom with 4 mm distance between tubes containing ferromagnetic nano-particles; (e)–(h) phantom with 2 mm distance. The methods used are Kaczmarz with L^2-discrepancy ($\tilde{\lambda} = 5 \times 10^{-4}$), ℓ_1-regularization ($\tilde{\lambda} = 5 \times 10^{-3}$) and DIP ($\eta = 5 \times 10^{-5}$) for both cases. Photos of phantoms taken by T. Kluth at the University Medical Center, Hamburg–Eppendorf.

input of the network is given by a fixed Gaussian random input of size $1 \times 32 \times 32$.

For comparison with our deep inverse prior MPI reconstructions, we also compute sparse and classical Tikhonov reconstructions. We produce the Tikhonov reconstruction, usually associated with the minimization of the functional

$$\|\mathbf{A}\,c - g\|^2 + \lambda\|c\|^2, \tag{7.9}$$

via the algebraic reconstruction technique (Kaczmarz) as generalized to allow for the constraint $x \geq 0$ by Dax (1993). We produce the sparsity reconstruction, usually associated with the minimization of the functional

$$\|\mathbf{A}\,c - g\|^2 + \lambda\|c\|_1, \tag{7.10}$$

by simply implementing this functional in TensorFlow and minimizing it via gradient descent. In the end we set all negative entries to 0.

We start by presenting direct comparisons of the Kaczmarz, sparsity and DIP reconstructions in Figure 7.13. Beneath each image we state the parameters we used for the reconstruction $\tilde{\lambda} = \|\mathbf{A}\|_F^2\lambda$, where $\|\cdot\|_F$ denotes the

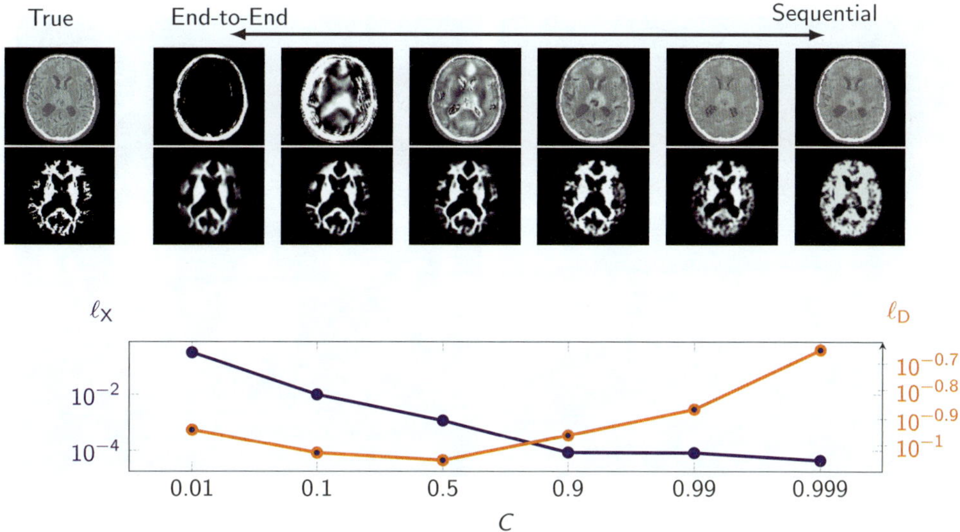

Figure 7.14. Joint tomographic reconstruction and segmentation of grey matter. Images shown using a $[-100, 100]$ HU window and segmentation using a $[0, 1]$ window. The choice $C = 0.9$ seems to be a good compromise for good reconstruction and segmentation, so clearly it helps to use a loss that includes the reconstruction and not only the task.

Frobenius norm and λ is the regularization parameter as used in (7.9) or (7.10) and η the learning rate used in training the network. For DIP we always used early stopping after 1000 optimization steps. The images started to deteriorate slowly for more iterations. For implementation details, as well as further numerical examples also showing the limitation of the DIP approach, see Dittmer *et al.* (2018).

7.6. Task-based reconstruction

We demonstrate the framework of Section 6.1 on joint tomographic image reconstruction and segmentation of white brain matter. $\mathcal{R}_\theta$ is given by a learned primal–dual method (Adler and Öktem 2018*b*), which incorporates a knowledge-based model for how data are generated into its architecture, and $\mathcal{T}_\phi$ is given by a U-Net (Ronneberger, Fischer and Brox 2015).

Some results are shown in Figure 7.14. Note in particular that (perhaps surprisingly) the 'best' segmentation is not obtained by a fully end-to-end approach: instead they are obtained when the reconstruction loss is included as a regularizer. Furthermore, it is clear that the reconstruction obtained for $C = 0.9$ over-emphasizes image features relevant for the task, for example white–grey matter contrast. This clearly 'helps' the task and also visually shows the image features used by the joint approach.

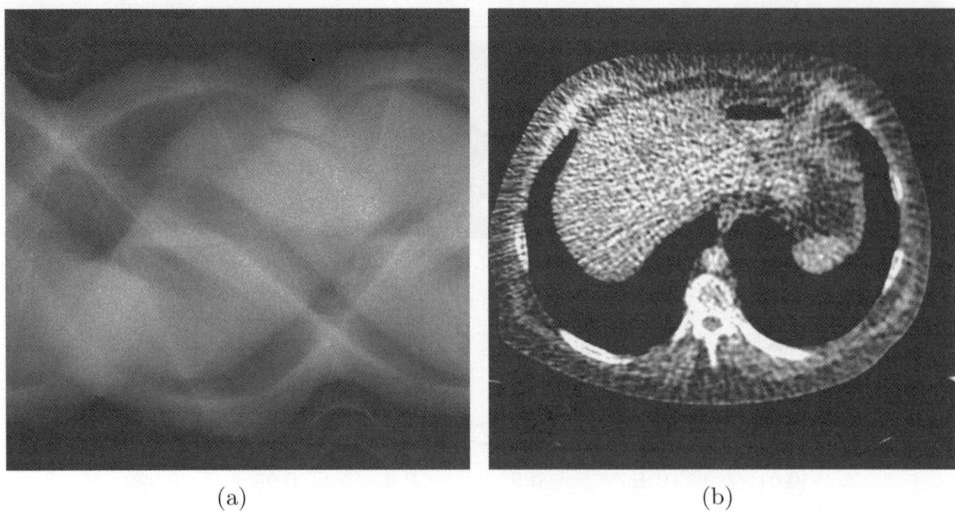

(a)   (b)

Figure 7.15. Test data: (a) subset of CT data from an ultra-low-dose three-dimensional helical scan and (b) the corresponding FBP reconstruction. Images are shown using a display window set to $[-150, 200]$ Hounsfield units.

7.7. Clinical image guided decision making

We show how to compute an estimator relevant for uncertainty quantification in the context of CT image reconstruction. As a practical example, we will compute a CT reconstruction from ultra-low-dose data (Figure 7.15(a)). The aim is to identify a feature (a potential tumour) and then seek to estimate the likelihood of its presence.

Formalizing the above, let Δ denote the difference in mean intensity in the reconstructed image between a region encircling the feature and the surrounding organ, which in our example is the liver. The feature is said to 'exist' whenever Δ is bigger than a certain threshold, say 10 Hounsfield units.

To use posterior sampling, start by computing the conditional mean image (top left in Figure 7.16) by sampling from the posterior using the conditional Wasserstein GAN approach in Section 5.2.1. There is a 'dark spot' in the liver (a possible tumour) and a natural clinical question is to statistically test for the presence of this feature. To do this, compute Δ for a number of samples generated by posterior sampling, which is the same 1000 samples used to compute the conditional mean. We estimate the probability p that $\Delta > 10$ Hounsfield units from the resulting histogram in Figure 7.17 and clearly $p > 0.95$, indicating that the 'dark spot' feature exists with at least 95% significance. This is confirmed by the ground truth image (Figure 7.17(a)). The conditional mean image also under-estimates Δ, whose

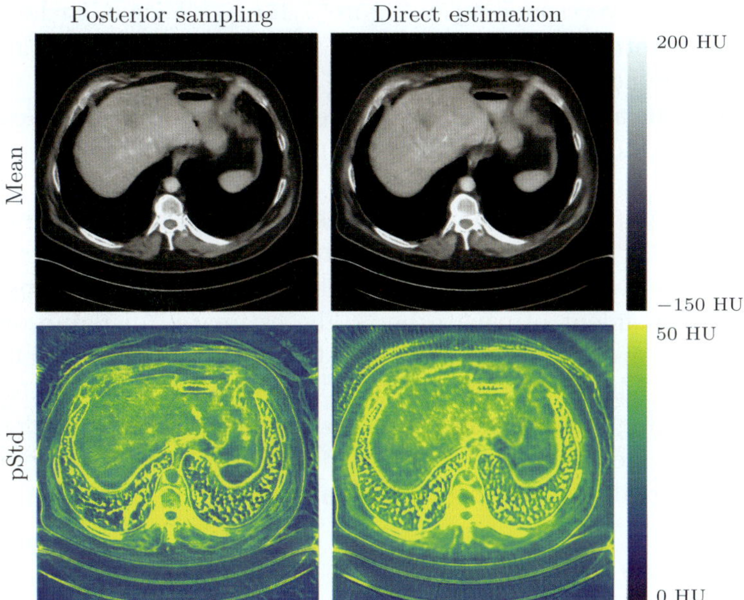

Figure 7.16. Conditional mean and pointwise standard deviation (pStd) computed from test data (Figure 7.15) using posterior sampling (Section 5.2.1) and direct estimation (Section 5.1.6).

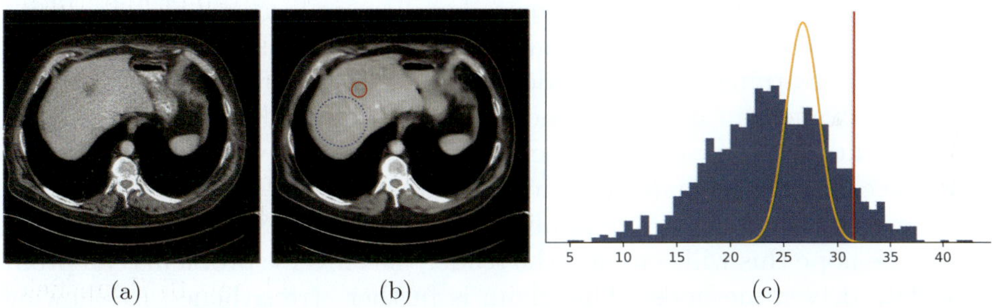

Figure 7.17. (b) Suspected tumour (red) and reference region (blue) shown in the sample posterior mean image. (c) Average contrast differences between the tumour and reference region. The histogram is computed by posterior sampling applied to test data (Figure 7.15); the yellow curve is from direct estimation (Section 5.1.6), and the true value is the red threshold. (a) The normal dose image that confirms the presence of the feature.

true value is the vertical line in Figure 7.17(c). This is to be expected since the prior introduces a bias towards homogeneous regions, a bias that decreases as the noise level decreases.

To perform the above analysis using direct estimation, start by computing the conditional mean image from the same ultra-low-dose data using direct estimation. As expected, the resulting image (top right in Figure 7.16) shows a 'dark spot' in the liver. Now, designing and training a neural network that directly estimates the distribution of Δ is unfeasible in a general setting. However, as shown in Section 5.1.6, this is possible if one assumes pixels are independent of each other. The estimated distribution of Δ is the curve in Figure 7.17 and we get $p > 0.95$, which is consistent with the result obtained using posterior sampling. The direct estimation approach is based on assuming independent pixels, so it will significantly underestimate the variance. In contrast, the approach based on posterior sampling seems to give a more realistic estimate of the variance.

8. Conclusions and outlook

8.1. Summary

In this survey we have tried to capture the state of the art in the still relatively young and fast-emerging field of machine learning approaches for solving inverse problems.

Our journey has taken us from more familiar applications of data-driven methods, such as dictionary learning (Section 4.4), bilevel learning (Section 4.3) and learning Markov random field-type regularizers (Section 4.3.2) to recent advances in using deep neural networks to solve inverse problems (Sections 4.6–4.10 and 5.1). These approaches are surveyed together with a brief account of the underlying mathematical setting (Sections 1–3), and their performance in some key applications is shown in Section 7. Taken together, we hope this will convince the reader that inverse problems can profit from data-driven methods. This claim is further strengthened by showing how data-driven methods can be used to compensate for inaccuracies in the forward model (Section 6.2) and how one can adapt the reconstruction to a specific task (Section 6.1).

The examples in Section 7 clearly show that some of these methods are very promising, regarding both output quality and computational feasibility. Strong empirical evidence suggests that using problem-specific deep neural networks that include knowledge-driven models outperform purely knowledge- or data-driven approaches. In contrast, there is little rigorous mathematical theory supporting these empirical observations, but the results clearly show that it is worth the effort to develop the necessary theory.

Below are some further key observations we believe are worth pointing out regarding the role of deep learning in solving inverse problems.

The functional analytic and Bayesian viewpoints. The way deep learning is used for solving an inverse problem depends on whether one adopts the functional analytic or the Bayesian viewpoint.

Within the functional analytic viewpoint, a deep neural network is simply a parametrized family of operators, and learning amounts to calibrating the parameters against example data by minimizing some appropriate loss function.

In Bayesian inversion, a deep neural network corresponds to a statistical decision rule, so methods from deep learning constitute a computational framework for statistical decision making in high dimensions. For example, many of the estimators that have previously been computationally unfeasible are now computable: for example, the conditional mean seems to be well approximated by learned iterative schemes (Section 5.1.4). Likewise, a trained generative network can be used to sample from the posterior (Section 5.2) in a computationally feasible manner, as shown in Section 5.2.

Computational feasibility. Essentially all methods from Bayesian inversion and many from functional analytic regularization are computationally very demanding. Those techniques based on unrolling an iterative scheme (Section 4.9.1) are for designing a deep neural network that approximates a computationally demanding operator, such as one that solves a large-scale optimization problem.

The training of such a deep neural network may take quite some time, but once it is trained, it is fast to evaluate. In some sense, this is a way to redistribute the computational burden from the execution to the training.

Handling lack of training data. Inverse problems in the sciences and engineering often have little training data compared to the dimensionality of the model parameter. Furthermore, it is impractical to have a method that requires retraining as soon as the measurement protocol changes. This becomes an issue is medical imaging where data in multi-centre studies is typically acquired using different CT or MRI scanners.

For these reasons, black-box machine learning algorithms (Section 7.1) are not suitable for solving such inverse problems. On the other hand, in these inverse problems there is often a knowledge-driven model for how data are generated and it is important to integrate this information into the data-driven method. Learned iterative schemes (Section 5.1.4) employ a deep neural network that embeds this model for data into its architecture.

Encoding a priori *information.* In functional analytic regularization, much of the theoretical research (Section 2) has focused on finding optimal convergence rates as the noise level tends to zero. Likewise, theoretical research in Bayesian inversion (Section 3) focuses on contraction rates for certain classes of priors and in deriving an asymptotic closed-form characterization of the posterior distribution as the noise level tends to zero. Here, the regularization functional (in functional analytic regularization) and the prior distribution (in Bayesian inversion) primarily act as a regularizers.

The above viewpoint does not acknowledge the potential that lies in encoding knowledge about the true model parameter into the regularization functional or prior. Furthermore, in applications data are fixed with some given noise level, and there is little, if any, guidance from the above theory on which regularizer or prior to select in such a setting. Empirical evidence suggests that instead of hand-crafting a regularization functional or a prior, one can learn it from example data. This allows us to pick up information related to the inverse problem that is difficult, if not impossible, to account for otherwise.

Unrolling. A key technique in many approaches for using deep learning to solve inverse problems is to construct problem-specific deep neural network architectures by unrolling an iterative scheme, as outlined in Section 4.9.1.

This technique allows us to use compressed sensing theory to derive properties for certain classes of deep neural networks, for example those resulting from multi-layer convolutional sparse coding (ML-CSC) (Section 4.4.2). Next, as shown in Section 4.9, the same technique is also useful in accelerating the evaluation of computationally demanding operators, such as those that solve large-scale optimization problems. Finally, unrolling is also used to embed a knowledge-driven forward operator and the adjoint of its derivative into a deep neural network that seeks to approximate an estimator (Section 5.1).

The above principle of unrolling can be used in a much wider context than solving an optimization problem or an inverse problem. It can be seen as constructing a deep neural network architecture for approximating an operator that is given implicitly through an iterative scheme. Hence, as pointed out in Section 4.9.4, unrolling establishes a link between numerical analysis and deep learning.

8.2. Outlook

We identify several interesting directions for future research in the context of inverse problems and machine learning.

8.2.1. Approximation and convergence properties of deep inversion

We believe a key component of future research is to analyse the mathematical–statistical approximation and convergence properties of inversion strategies that use deep neural networks, for example statistical recovery guarantees and generalization limits, as well as bounds on the number of training samples necessary for reaching prescribed accuracies, and estimates for uncertainty in terms of stability properties and statistical confidence statements for the algorithms used.

Convergence and stability properties of denoising by deep neural networks can be analysed using techniques from sparse signal processing, as in Papyan *et al.* (2017). Likewise, techniques from applied harmonic analysis can be used to analyse approximation properties of feed-forward deep neural networks, as in Bölcskei *et al.* (2019). This paper establishes a connection between the complexity of a function class in $L^2(\mathbb{R}^d)$ and the complexity (model capacity) of feed-forward deep neural networks approximating functions from this class to within a prescribed accuracy. A specific focus is on function classes in $L^2(\mathbb{R}^d)$ that are optimally approximated by general affine systems, which include a wide range of representation systems from applied harmonic analysis such as wavelets, ridgelets, curvelets, shearlets, α-shearlets and, more generally, α-molecules. The central result in Bölcskei *et al.* (2019) is that feed-forward deep neural networks achieve the optimum approximation properties of all affine systems combined with minimal connectivity and memory requirements.

None of these papers, however, consider deep neural networks in the context of inverse problems.

8.2.2. Robustness against adversarial attacks

Sensitivity towards adversarial attacks is a known issue that has mostly been studied in the context of classification (Szegedy *et al.* 2014); see also the surveys by Chakraborty *et al.* (2018) and Akhtar and Mian (2018). However, little work has been done regarding adversarial stability of reconstruction methods for solving inverse problems that are based on deep neural networks.

One recent work along these lines is that of Antun *et al.* (2019), who extend the approach in Szegedy *et al.* (2014) to the case of regression. The idea is to perturb the model parameter in a way that is hard to distinguish, yet the data from the perturbed model parameter have a large influence on the reconstruction. The perturbation (adversarial example) is computed by solving an optimization and, in contrast to classification, different optimization problems can be constructed to test for different types of instabilities. The adversarial stability test in Antun *et al.* (2019) is demonstrated on MRI image reconstruction. The paper tests two fully learned approaches given by Zhu *et al.* (2018) and Schlemper *et al.* (2018), two learned post-processing approaches given by Jin *et al.* (2017) and Yang *et al.* (2018*b*), and finally the learned iterative approach of Hammernik *et al.* (2018). All approaches show adversarial instability, which could come as a surprise for the learned iterative approach that seeks to approximate a conditional mean that is known to be stable (Section 3.2.2). One reason for this could be that the learned iterative approach of Hammernik *et al.* (2018), which is based on variational networks, has a model capacity too limited to properly approximate the conditional mean.

On a final note, a key element in Antun *et al.* (2019) is that data from the perturbation model parameter is noise-free. It is known that adding white noise to the input helps against adversarial attacks for classifiers (Cohen, Rosenfeld and Kolter 2019). Moreover, in inverse problems one always has noisy data, so it remains unclear whether the computed perturbation in Antun *et al.* (2019) actually acts as an adversarial example when noise is added.

Clearly, theory for robustness against adversarial attacks in the context of inverse problems is very much an emerging field.

8.2.3. *Theory for learned iterative reconstruction*

More specifically, the theory of statistical regularization applied to learned iterative methods in Section 5.1.4 is fairly incomplete, especially when interested in theoretical guarantees in the presence of empirical distributions for the data and model parameter. This will require studying estimates of the posterior in a non-asymptotic setting. Another key element is to estimate the generalization gap for learned iterative methods. Here one could consider theory for empirical Bayes methods, but current results focus on analysing Bayesian inversion methods where hyper-parameters defining a hierarchical prior are selected from data (Knapik, Szabó, van der Vaart and van Zanten 2016, Szabó, van der Vaart and van Zanten 2013).

8.2.4. *'Convergence' of training data*

For all approaches where supervised training data are used, there is a discrepancy between theoretical error estimates in the infinite-dimensional setting and the practical case of finite-dimensional training data used in presented data-driven inversion approaches. For instance, error estimates are needed between solutions of a neural network trained on finitely many samples (that describe a particular empirical distribution) and solutions trained with infinitely many samples from a joint distribution.

8.2.5. *Bespoke neural network architectures for inverse problems*

In many inverse problems the model parameter space X and the data space Y are not the same, and in particular the data g often live in non-Euclidean spaces. On the other hand, existing data-driven inversion models, as discussed in this survey, which make use of neural networks as data-driven parametrizations, usually employ off-the-shelf network architectures such as U-Net, for instance. For future research it would be interesting to investigate neural network architectures that are specifically designed as mappings between non-Euclidean spaces. Some developments along these lines can be found in the paper by Bronstein *et al.* (2017), who investigate how the notion of a CNN can be generalized to non-Euclidean structures such as graphs and manifolds.

The above is also closely related to work in developing neural network architectures that are equivariant and/or invariant to certain transformations given as group action. This is highly relevant when one seeks to solve inverse problems whose solutions enjoy such equivariance and/or invariance. Examples of work in this direction are those of Esteves, Allen-Blanchette, Makadia and Daniilidis (2017), Zhao *et al.* (2018), Weiler *et al.* (2018) and Veeling *et al.* (2018), but none of this is pursued in the context of inverse problems.

Another feature of inverse problems is that one can often prescribe how singularities in data are related to those in the model parameter (Section 6.3). Hence it is natural to seek network architectures that encode not only the forward operator but also such a relation. This is likely to further improve the robustness and generalization properties.

8.2.6. Continuous notion of neural network architectures

Some of the recent attempts to build a continuous framework for neural networks have been touched upon in Section 6.2.1. Continuous formulations to neural networks make them amenable to the rich toolkit of functional analysis and theoretical results as outlined in Section 2. Moreover, starting with a continuous model such as a partial differential equation, for instance, may give rise to new discretizations and new neural network architectures – a development that has clearly happened before in mathematical imaging (*e.g.* Perona and Malik 1990).

8.2.7. Theoretical guarantees for learning-to-optimize approaches

If a neural network is used to approximate and consequently computationally speed up a knowledge-driven approach (*e.g.* the learning-to-optimize methods in Section 4.9), it is important to understand the error committed by such an approximation. What is the correct notion of such an approximation error? How does it depend on the training set and the network architecture?

Acknowledgements

This article builds on lengthy discussions and long-standing collaborations with a large number of people. These include Jonas Adler, Sebastian Banert, Martin Benning, Marta Betcke, Luca Calatroni, Juan Carlos De Los Reyes, Andreas Hauptmann, Lior Horesh, Bangti Jin, Iasonas Kokkinos, Felix Lucka, Sebastian Lunz, Thomas Pock, Tuomo Valkonen and Olivier Verdier. The authors are moreover grateful to the following people for proofreading the manuscript and providing valuable feedback on its content and structure: Andrea Aspri, Martin Benning, Matthias Ehrhardt, Barbara Kaltenbacher, Yury Korolev, Mike McCann, Erkki Somersalo and Michael Unser.

SA acknowledges support from EPSRC grant EP/M020533/1. OÖ acknowledges support from the Swedish Foundation of Strategic Research grant AM13-004 and the Alan Turing Institute. CBS acknowledges support from the Leverhulme Trust project 'Breaking the non-convexity barrier', EPSRC grant EP/M00483X/1, EPSRC grant EP/N014588/1, the RISE projects CHiPS and NoMADS, the Cantab Capital Institute for the Mathematics of Information and the Alan Turing Institute.

Acronyms

ADMM alternating direction method of multipliers
AutoMap automated transform by manifold approximation
BFGS Broyden–Fletcher–Goldfarb–Shanno
CG conjugate gradient
CNN convolutional neural network
CSC convolutional sparse coding
CT computed tomography
DGD deep gradient descent
DIP deep inverse prior
FBP filtered backprojection
FFP field-free point
FoE Field of Experts
GAN generative adversarial network
ICA independent component analysis
ICTV infimal-convolution total variation
ISTA Iterative Soft-Thresholding Algorithm
KL Kullback–Leibler
LISTA Learned Iterative Soft-Thresholding Algorithm
MAP maximum *a posteriori*
MCMC Markov chain Monte Carlo
ML-CSC multi-layer convolutional sparse coding
ML-EM maximum likelihood expectationmaximization
MPI magnetic particle imaging
MRF Markov random field
MRI magnetic resonance imaging
NETT neural network Tikhonov
P^3 Plug-and-Play Prior
PAT photoacoustic tomography
PCA principal component analysis
PDE partial differential equation

PDHG primal–dual hybrid gradient
PET positron emission tomography
PG proximal gradient
PoE Product of Experts
PSNR peak signal-to-noise ratio
RED regularization by denoising
RIP restricted isometry property
R-TLS regularized total least-squares
SGD stochastic gradient descent
SPECT single photon emission computed tomography
SSIM structural similarity index
SVD singular value decomposition
TGV total generalized variation
TLS total least-squares
TV total variation

Appendices

A. Optimization of convex non-smooth functionals

Suppose in general that we want to optimize a problem defined as the sum of two parts,

$$\min_{f \in X} [\mathcal{J}(f) := \Phi(f) + \mathcal{S}(f)], \tag{A.1}$$

where $\Phi\colon X \to \mathbb{R}$ is a continuously differentiable convex function, and $\mathcal{S}\colon X \to \mathbb{R}$ is convex but possibly *non-differentiable*. We say that the combined function is *convex non-smooth*.

A.1. Proximal methods

First we define a *proximal operator* for a functional $\mathcal{S}\colon X \to \mathbb{R}$:

$$\operatorname{prox}_{\mathcal{S}}(h) := \arg\min_{f \in X} \left[\frac{1}{2}\|f - h\|^2 + \mathcal{S}(h) \right]. \tag{A.2}$$

Clearly, if $\mathcal{S}$ is differentiable, then $p = \operatorname{prox}_{\mathcal{S}}(h)$ satisfies

$$h - p = \nabla \mathcal{S}\,|_{f=p}. \tag{A.3}$$

When $\mathcal{S}$ is non-differentiable, we instead have

$$h - p \in \partial \mathcal{S}\,|_{f=p}, \tag{A.4}$$

where $\partial \mathcal{S}(f)$ is the *subdifferential* of $\mathcal{S}$. This allows us to write a formal expression for (A.2) as

$$\operatorname{prox}_{\mathcal{S}}(h) := (\operatorname{id} + \partial \mathcal{S})^{-1}(h). \tag{A.5}$$

A.2. Proximal gradient method for inverse problems

For inverse problems, $\Phi(f)$ corresponds to the data discrepancy $\mathcal{L}(\mathcal{A}(f), g)$ and $\mathcal{S}(f)$ to the regularization functional.

Now consider the minimization of $f \mapsto \mathcal{J}_\lambda(f)$ in (2.10). Defining $[\partial \mathcal{A}(f)]$ to be the Fréchet derivative of $\mathcal{A}$ at f, then, exploiting the first-order necessary conditions for such minima, we have

$$0 \in [\partial \mathcal{A}(f)]^* (\mathcal{A}(f) - g) + \lambda \partial \mathcal{S}(f), \tag{A.6}$$

which after multiplying both sides with τ, adding f on both sides and reordering terms yields the fixed-point condition for a minimizer:

$$f = \operatorname{prox}_{\tau \lambda \mathcal{S}}(f - \tau [\partial \mathcal{A}(f)]^* (\mathcal{A}(f) - g)). \tag{A.7}$$

The step length is given by $0 < \tau < 2/L$, where L is the Lipschitz constant of $\nabla \Phi$ (Combettes and Wajs 2005). For linear inverse problems, L can be approximated by the largest eigenvalue of $\mathcal{A}^* \mathcal{A}$, $i.e.$ the square of the largest singular value of the forward operator $\mathcal{A}$. For non-linear problems $L(f)$ is the square of the largest singular value of $[\partial \mathcal{A}(f)]$ and thus changes over iteration. We have the following examples.

- Multivariate Gaussian noise $e \sim \mathcal{N}(0, \Gamma_e)$: the likelihood is

$$\mathcal{L}(\mathcal{A}(f), g) = \|g - \mathcal{A} f\|_{\Gamma_e}^2,$$

  and

$$f^{(n+1)} \leftarrow f^{(n)} + \tau \lambda \mathcal{A}^* \Gamma_e^{-1}(g - \mathcal{A} f^{(n)}).$$

- Poisson noise $g = \operatorname{Poisson}(\mathcal{A} f)$: the likelihood is

$$\mathcal{L}(\mathcal{A}(f), g) = \int_\Omega g \ln \mathcal{A} f - \mathcal{A} f + g - \ln g,$$

  and

$$f^{(n+1)} \leftarrow f^{(n)} + \tau \mathcal{A}^* \left(1 - \frac{g}{\mathcal{A} f^{(n)}}\right).$$

The related iteration scheme to (A.7), which can also be derived by minimizing surrogate functionals (Daubechies $et\ al.$ 2004) or by the method of quadratic relaxation, yields the following algorithm.

Algorithm A.1 (generalized gradient projection method).

Choose f^0 and iterate for $k > 0$.

(1) Choose τ_k, $e.g.$ $\tau_k = \tau$ constant for all k.

(2) Determine $v^k = f^k - \tau_k [\partial \mathcal{A}(f)]^* (\mathcal{A}(f) - g)$.

(3) Determine $f^{k+1} = \operatorname{prox}_{\tau_k \lambda \mathcal{S}}(v^k)$.

This version, along with several accelerated variants incorporating a step size selection or primal–dual iterations, has been studied intensively (Bredies, Lorenz and Maass 2009, Nesterov 2007, Figueiredo *et al.* 2007).

The convergence proofs of such methods are typically based on rephrasing the first-order necessary condition.

Theorem A.1. Assume that $\mathcal{A} : X \to Y$ is Fréchet-differentiable and assume that $\mathcal{S} : X \to \mathbb{R}$ is proper and convex. Then, a (first-order) necessary condition for a minimizer $\hat{f}$ of $f \mapsto \mathcal{J}_\lambda(f)$ in (2.10) is given by

$$\langle [\partial \mathcal{A}(f)]^*(\mathcal{A}(f) - g), h - f \rangle_X \geq \mathcal{S}(f) - \mathcal{S}(h) \quad \text{for all } h \in X,$$

which is equivalent to

$$\langle [\partial \mathcal{A}(f)]^*(\mathcal{A}(f) - g), f \rangle + \mathcal{S}(f) = \min_{h \in X} \langle [\partial \mathcal{A}(f)]^*(\mathcal{A}(f) - g), h \rangle + \mathcal{S}(h).$$

This characterization motivates the definition of an auxiliary functional

$$D_\lambda(f^k) := \lambda(\mathcal{S}(f^k) - \mathcal{S}(f^{k+1})) + \langle [\partial \mathcal{A}(f^k)]^*(\mathcal{A}(f^k) - g), f^k - f^{k+1} \rangle, \quad \text{(A.8)}$$

which is decreased during the iteration and whose minimizing f allows an estimate in terms of the Bregman distance related to $\mathcal{S}$.

A.3. Iterative Soft-Thresholding Algorithm (ISTA)

The success of proximal methods usually depends on finding a fast 'trick' for performing the projection for a given functional $\mathcal{S}(f)$. One notable such method is for the L^1-functional $\mathcal{S}(f) = \lambda\|f\|_1$ whereby the proximal operator is the soft-thresholding (or shrinkage operator), denoted by

$$S_\alpha(z) := \begin{cases} z - \alpha & \text{if } z \geq \alpha, \\ 0 & \text{if } |z| \leq \alpha, \\ z + \alpha & \text{if } z \leq -\alpha. \end{cases} \quad \text{(A.9)}$$

We arrive at the following split method, known as the Iterative Soft-Thresholding Algorithm (ISTA) (Daubechies *et al.* 2004, Figueiredo *et al.* 2007):

$$\text{gradient descent} \quad f^{(n+1/2)} \leftarrow f^{(n)} - \tau \nabla \Phi(f),$$

$$\text{thresholding} \quad f^{(n+1)} \leftarrow S_{\tau\lambda}(f^{(n+1/2)}).$$

(Note that the threshold is the product of τ and λ.)

Now consider applying this principle to the problem of minimizing Tikhonov functionals of type $f \mapsto \mathcal{J}_\lambda(f)$ defined in (2.10). In this case $\Phi(f) := \frac{1}{2}\|\mathcal{A}(f) - g\|^2$ and the necessary first-order condition for a minimizer is given by

$$0 \in \mathcal{A}^*(\mathcal{A}f - g) + \lambda \partial \|f\|_1.$$

Multiplying with an arbitrary real positive real number τ and adding f plus rearranging yields

$$f - \tau \mathcal{A}^*(\mathcal{A}f - g) \in f + \tau\lambda\partial\|f\|_1.$$

Using (A.9) to invert the term on the right-hand side yields

$$S_{\tau\lambda}(f - \lambda\mathcal{A}^*(\mathcal{A}f - g)) = f.$$

Hence this is a fixed-point condition, which is a necessary condition for all minimizers of $f \mapsto \mathcal{J}_\lambda(f)$. Turning the fixed-point condition into an iteration scheme yields

$$\begin{aligned}
f^{k+1} &= S_{\tau\lambda}(f^k - \tau\mathcal{A}^*(\mathcal{A}f^k - g)) \\
&= S_{\tau\lambda}((\mathrm{id} - \tau\mathcal{A}^*\mathcal{A})f^k + \tau\mathcal{A}^*g).
\end{aligned} \tag{A.10}$$

A.4. Alternating direction method of multipliers (ADMM)
Consider solving (A.1) as a constrained problem,

$$\hat{f} = \arg\min_{f,v}[\Phi(f) + \mathcal{S}(v)] \quad \text{such that } f = v,$$

making use of the augmented Lagrangian with dual (adjoint) variable u,

$$\begin{aligned}
\mathcal{J}(f, v, u) &= \Phi(f) + \mathcal{S}(v) + \langle u, f - v\rangle + \frac{\beta}{2}\|f - v\|_2^2 \\
&= \Phi(f) + \mathcal{S}(v) + \frac{\beta}{2}\|f - v + \frac{1}{\beta}u\|_2^2 - \frac{1}{2\beta}\|u\|_2^2,
\end{aligned} \tag{A.11}$$

which results in the sequential update sequence

$$f^{(n+1)} \leftarrow \mathrm{prox}_{(1/\beta)\Phi}\left[v^{(n)} - \frac{1}{\beta}u^{(n)}\right], \tag{A.12}$$

$$v^{(n+1)} \leftarrow \mathrm{prox}_{(1/\beta)\mathcal{S}}\left[f^{(n+1)} + \frac{1}{\beta}u^{(n)}\right], \tag{A.13}$$

$$u^{(n+1)} \leftarrow u^{(n)} + \beta(f^{(n+1)} - v^{(n+1)}). \tag{A.14}$$

B. The Wasserstein 1-distance

Let X be a measurable separable Banach space and $\mathscr{P}_X$ the space of probability measures on X. The Wasserstein 1-distance $\mathcal{W}: \mathscr{P}_X \times \mathscr{P}_X \to \mathbb{R}$ is a metric on $\mathscr{P}_X$ that can be defined as (Villani 2009, Definition 6.1)

$$\mathcal{W}(p, q) := \inf_{\mu\in\Pi(p,q)} \mathbb{E}_{(\mathtt{f},\mathtt{h})\sim\mu}[\|\mathtt{f} - \mathtt{h}\|_X] \quad \text{for } p, q \in \mathscr{P}_X. \tag{B.1}$$

In the above, $\Pi(p, q) \subset \mathscr{P}_{X\times X}$ denotes the family of joint probability measures on $X \times X$ that has p and q as marginals. Note also that we assume

$\mathscr{P}_X$ only contains measures where the Wasserstein distance takes finite values (Wasserstein space): see Villani (2009, Definition 6.4) for the formal definition.

The Wasserstein 1-distance in (B.1) can be rewritten using the Kantorovich–Rubinstein dual characterization (Villani 2009, Remark 6.5, p. 95), resulting in

$$\mathcal{W}(p, q) = \sup_{\substack{D\colon X \to \mathbb{R} \\ D \in \mathrm{Lip}(X)}} \{\mathbb{E}_{f \sim q}[D(f)] - \mathbb{E}_{h \sim p}[D(h)]\} \quad \text{for } p, q \in \mathscr{P}_X. \tag{B.2}$$

Here, $\mathrm{Lip}(X)$ denotes real-valued 1-Lipschitz maps on X, that is,

$$D \in \mathrm{Lip}(X) \iff |D(f_1) - D(f_2)| \leq \|f_1 - f_2\|_X \quad \text{for all } f_1, f_2 \in X.$$

The above constraint can be hard to enforce in (B.2) as is, so following Gulrajani *et al.* (2017) and Adler and Lunz (2018) we prefer the *gradient characterization*

$$D \in \mathrm{Lip}(X) \iff \|\partial D(f)\|_{X^*} \leq 1 \quad \text{for all } f \in X,$$

where ∂ indicates the Fréchet derivative and X^* is the dual space of X. In our setting, X is an L_2-space, which is a Hilbert space so $X^* = X$, and the Fréchet derivative becomes the (Hilbert space) gradient of D.

REFERENCES[10]

M. Abadi *et al.* (2015), TensorFlow: Large-scale machine learning on heterogeneous systems. Software available from https://www.tensorflow.org.

B. Adcock and A. C. Hansen (2016), 'Generalized sampling and infinite-dimensional compressed sensing', *Found. Comput. Math.* **16**, 1263–1323.

J. Adler and S. Lunz (2018), Banach Wasserstein GAN. In *Advances in Neural Information Processing Systems 31 (NIPS 2018)* (S. Bengio *et al.*, eds), Curran Associates, pp. 6754–6763.

J. Adler and O. Öktem (2017), 'Solving ill-posed inverse problems using iterative deep neural networks', *Inverse Problems* **33**, 124007.

J. Adler and O. Öktem (2018*a*), Deep Bayesian inversion: Computational uncertainty quantification for large scale inverse problems. arXiv:1811.05910

J. Adler and O. Öktem (2018*b*), 'Learned primal–dual reconstruction', *IEEE Trans. Medical Imaging* **37**, 1322–1332.

J. Adler, S. Lunz, O. Verdier, C.-B. Schönlieb and O. Öktem (2018), Task adapted reconstruction for inverse problems. arXiv:1809.00948

L. Affara, B. Ghanem and P. Wonka (2018), Supervised convolutional sparse coding. arXiv:1804.02678

---

[10] The URLs cited in this work were correct at the time of going to press, but the publisher and the authors make no undertaking that the citations remain live or are accurate or appropriate.

S. Agapiou, S. Larsson and A. M. Stuart (2013), 'Posterior contraction rates for the Bayesian approach to linear ill-posed inverse problems', *Stoch. Process. Appl.* **123**, 3828–3860.

S. Agapiou, A. M. Stuart and Y. X. Zhang (2014), 'Bayesian posterior contraction rates for linear severely ill-posed inverse problems', *J. Inverse Ill-Posed Problems* **22**, 297–321.

H. K. Aggarwal, M. P. Mani and M. Jacob (2019), 'MoDL: Model-based deep learning architecture for inverse problems', *IEEE Trans. Medical Imaging* **38**, 394–405.

M. Aharon, M. Elad and A. M. Bruckstein (2006), 'K-SVD: An algorithm for designing of over-complete dictionaries for sparse representation', *IEEE Trans. Signal Process.* **54**, 4311–4322.

A. Ahmed, M. Aly, J. Gonzalez, S. Narayanamurthy and A. Smola (2012), Scalable inference in latent variable models. In *Proceedings of the Fifth ACM International Conference on Web Search and Data Mining (WSDM '12)*, pp. 123–132.

N. Akhtar and A. Mian (2018), Threat of adversarial attacks on deep learning in computer vision: A survey. arXiv:1801.00553

W. K. Allard, G. Chen and M. Maggioni (2012), 'Multi-scale geometric methods for data sets, II: Geometric multi-resolution analysis', *Appl. Comput. Harmon. Anal.* **32**, 435–462.

D. Allman, A. Reiter and M. A. L. Bell (2018), 'Photoacoustic source detection and reflection artifact removal enabled by deep learning', *IEEE Trans. Medical Imaging* **37**, 1464–1477.

L. Ambrosio, N. Fusco and D. Pallara (2000), *Functions of Bounded Variation and Free Discontinuity Problems*, Oxford University Press.

H. Andrade-Loarca, G. Kutyniok, O. Öktem and P. Petersen (2019), Extraction of digital wavefront sets using applied harmonic analysis and deep neural networks. arXiv:1901.01388

M. Andrychowicz, M. Denil, S. Gomez, M. Hoffman, D. Pfau, T. Schaul and N. de Freitas (2016), Learning to learn by gradient descent by gradient descent. In *Advances in Neural Information Processing Systems 29 (NIPS 2016)* (D. D. Lee *et al.*, eds), Curran Associates, pp. 3981–3989.

V. Antun, F. Renna, C. Poon, B. Adcock and A. C. Hansen (2019), On instabilities of deep learning in image reconstruction: Does AI come at a cost? arXiv:1902.05300v1

L. Ardizzone, J. Kruse, S. Wirkert, D. Rahner, E. W. Pellegrini, R. S. Klessen, L. Maier-Hein, C. Rother and U. Köthe (2018), Analyzing inverse problems with invertible neural networks. arXiv:1808.04730

M. Argyrou, D. Maintas, C. Tsoumpas and E. Stiliaris (2012), Tomographic image reconstruction based on artificial neural network (ANN) techniques. In *2012 IEEE Nuclear Science Symposium and Medical Imaging Conference (NSS/MIC)*, pp. 3324–3327.

M. Arjovsky, S. Chintala and L. Bottou (2017), Wasserstein generative adversarial networks. In *34th International Conference on Machine Learning (ICML '17)*, pp. 214–223.

S. Armato, G. McLennan, L. Bidaut, M. McNitt-Gray, C. Meyer, A. Reeves, B. Zhao, D. Aberle, C. Henschke, E. Hoffman *et al.* (2011), 'The lung image database consortium (LIDC) and image database resource initiative (IDRI): A completed reference database of lung nodules on CT scans', *Med. Phys.* **38**, 915–931.

S. R. Arridge and O. Scherzer (2012), 'Imaging from coupled physics', *Inverse Problems* **28**, 080201.

S. R. Arridge and J. C. Schotland (2009), 'Optical tomography: Forward and inverse problems', *Inverse Problems* **25**, 123010.

A. Aspri, S. Banert, O. Öktem and O. Scherzer (2018), A data-driven iteratively regularized Landweber iteration. arXiv:1812.00272

P. Auer, M. Herbster and M. K. Warmuth (1996), Exponentially many local minima for single neurons. In *8th International Conference on Neural Information Processing Systems (NIPS)*, MIT Press, pp. 316–322.

T. Bai, H. Yan, X. Jia, S. Jiang, G. Wang and X. Mou (2017), Volumetric computed tomography reconstruction with dictionary learning. In *14th International Meeting on Fully Three-Dimensional Image Reconstruction in Radiology and Nuclear Medicine (Fully3D 2017)*.

A. B. Bakushinskii (1984), 'Remarks on choosing a regularization parameter using the quasi-optimality and ratio criterion', *USSR Comput. Math. Math. Phys.* **24**, 181–182.

G. Bal, F. Chung and J. Schotland (2016), 'Ultrasound modulated bioluminescence tomography and controllability of the radiative transport equation', *SIAM J. Math. Anal.* **48**, 1332–1347.

G. Balakrishnan, A. Zhao, M. R. Sabuncu, J. Guttag and A. V. Dalca (2019), 'VoxelMorph: A learning framework for deformable medical image registration', *IEEE Trans. Imaging*, to appear. arXiv:1809.05231

L. Baldassarre, Y.-H. Li, J. Scarlett, B. Gözcü, I. Bogunovic and V. Cevher (2016), 'Learning-based compressive subsampling', *IEEE J. Selected Topics Signal Process.* **10**, 809–822.

A. Banerjee, X. Guo and H. Wang (2005), 'On the optimality of conditional expectation as a Bregman predictor', *IEEE Trans. Inform. Theory* **51**, 2664–2669.

S. Banert, A. Ringh, J. Adler, J. Karlsson and O. Öktem (2018), Data-driven nonsmooth optimization. arXiv:1808.00946

A. R. Barron (1994), 'Approximation and estimation bounds for artificial neural networks', *Machine Learning* **14**, 115–133.

F. Baus, M. Nikolova and G. Steidl (2014), 'Fully smoothed L1-TV models: Bounds for the minimizers and parameter choice', *J. Math. Imaging Vision* **48**, 295–307.

H. H. Bauschke and P. L. Combettes (2017), *Convex Analysis and Monotone Operator Theory in Hilbert Spaces*, second edition, CMS Books in Mathematics, Springer.

P. Beard (2011), 'Biomedical photoacoustic imaging', *Interface Focus* **1**, 602–631.

A. Beck and M. Teboulle (2009), 'A fast iterative shrinkage–thresholding algorithm for linear inverse problems', *SIAM J. Imaging Sci.* **2**, 183–202.

A. Beck, S. Sabach and M. Teboulle (2016), 'An alternating semiproximal method for nonconvex regularized structured total least squares problems', *SIAM J. Matrix Anal. Appl.* **37**, 1129–1150.

S. Becker, Y. Zhang and A. A. Lee (2018), Geometry of energy landscapes and the optimizability of deep neural networks. arXiv:1805.11572

Y. Bengio, P. Simard and P. Frasconi (1994), 'Learning long-term dependencies with gradient descent is difficult', *IEEE Trans. Neural Networks* **5**, 157–166.

M. Benning and M. Burger (2018), Modern regularization methods for inverse problems. In *Acta Numerica*, Vol. 27, Cambridge University Press, pp. 1–111.

M. Benning, C. Brune, M. Burger and J. Müller (2013), 'Higher-order TV methods: Enhancement via Bregman iteration', *J. Sci. Comput.* **54**, 269–310.

F. Benvenuto, A. L. Camera, C. Theys, A. Ferrari, H. Lantéri and M. Bertero (2008), 'The study of an iterative method for the reconstruction of images corrupted by Poisson and Gaussian noise', *Inverse Problems* **24**, 035016.

J. O. Berger (1985), *Statistical Decision Theory and Bayesian Analysis*, second edition, Springer.

R. Berinde, A. C. Gilbert, P. Indyk, H. Karloff and M. J. Strauss (2008), Combining geometry and combinatorics: A unified approach to sparse signal recovery. In *46th Annual Allerton Conference on Communication, Control, and Computing*, pp. 798–805.

M. Bertero and P. Boccacci (1998), *Introduction to Inverse Problems in Imaging*, Institute of Physics Publishing.

M. Bertero, H. Lantéri and L. Zanni (2008), Iterative image reconstruction: A point of view. In *Interdisciplinary Workshop on Mathematical Methods in Biomedical Imaging and Intensity-Modulated Radiation (IMRT)* (Y. Censor *et al.*, eds), pp. 37–63.

D. Bertsekas (1999), *Nonlinear Programming*, second edition, Athena Scientific.

J. Besag (1974), 'Spatial interaction and the statistical analysis of lattice systems', *J. Royal Statist. Soc. B* **36**, 192–236.

J. Besag and P. J. Green (1993), 'Spatial statistics and Bayesian computation', *J. Royal Statist. Soc. B* **55**, 25–37.

M. Betancourt (2017), A conceptual introduction to Hamiltonian Monte Carlo. arXiv:1701.02434

L. Biegler, G. Biros, O. Ghattas, M. Heinkenschloss, D. Keyes, B. Mallick, L. Tenorio, B. van Bloemen Waanders, K. Willcox and Y. Marzouk (2011), *Large-Scale Inverse Problems and Quantification of Uncertainty*, Vol. 712 of Wiley Series in Computational Statistics, Wiley.

N. Bissantz, T. Hohage, A. Munk and F. Ruymgaart (2007), 'Convergence rates of general regularization methods for statistical inverse problems and applications', *SIAM J. Numer. Anal.* **45**, 2610–2636.

A. Blake and A. Zisserman (1987), *Visual Reconstruction*, MIT Press.

D. M. Blei, A. Küçükelbir and J. D. McAuliffe (2017), 'Variational inference: A review for statisticians', *J. Amer. Statist. Assoc.* **112** (518), 859–877.

I. R. Bleyer and R. Ramlau (2013), 'A double regularization approach for inverse problems with noisy data and inexact operator', *Inverse Problems* **29**, 025004.

T. Blumensath (2013), 'Compressed sensing with nonlinear observations and related nonlinear optimization problems', *IEEE Trans. Inform. Theory* **59**, 3466–3474.

T. Blumensath and M. E. Davies (2008), 'Iterative thresholding for sparse approximations', *J. Fourier Anal. Appl.* **14**, 629–654.

N. Bochkina (2013), 'Consistency of the posterior distribution in generalized linear inverse problems', *Inverse Problems* **29**, 095010.

Y. E. Boink, S. A. van Gils, S. Manohar and C. Brune (2018), 'Sensitivity of a partially learned model-based reconstruction algorithm', *Proc. Appl. Math. Mech.* **18**, e201800222.

H. Bölcskei, P. Grohs, G. Kutyniok and P. Petersen (2019), 'Optimal approximation with sparsely connected deep neural networks', *SIAM J. Math. Data Sci.* **1**, 8–45.

J. F. Bonnans and D. Tiba (1991), 'Pontryagin's principle in the control of semilinear elliptic variational inequalities', *Appl. Math. Optim.* **23**, 299–312.

E. Bostan, U. S. Kamilov and L. Waller (2018), 'Learning-based image reconstruction via parallel proximal algorithm', *IEEE Signal Process. Lett.* **25**, 989–993.

R. Boţ and E. Csetnek (2015), 'On the convergence rate of a forward–backward type primal–dual splitting algorithm for convex optimization problems', *Optimization* **64**, 5–23.

R. Boţ and C. Hendrich (2013), 'A Douglas–Rachford type primal–dual method for solving inclusions with mixtures of composite and parallel-sum type monotone operators', *SIAM J. Optim.* **23**, 2541–2565.

L. Bottou, F. E. Curtis and J. Nocedal (2018), 'Optimization methods for large-scale machine learning', *SIAM Review* **60**, 223–311.

S. Boyd, N. Parikh, E. Chu, B. Peleato and J. Eckstein (2011), 'Distributed optimization and statistical learning via the alternating direction method of multipliers', *Found. Trends Mach. Learn.* **3**, 1–122.

K. Bredies and T. Valkonen (2011), Inverse problems with second-order total generalized variation constraints. In *9th International Conference on Sampling Theory and Applications (SampTA 2011)*.

K. Bredies, K. Kunisch and T. Pock (2011), 'Total generalized variation', *SIAM J. Imaging Sci.* **3**, 492–526.

K. Bredies, K. Kunisch and T. Valkonen (2013), 'Properties of l^1-TGV^2: The one-dimensional case', *J. Math. Anal. Appl.* **398**, 438–454.

K. Bredies, D. A. Lorenz and P. Maass (2009), 'A generalized conditional gradient method and its connection to an iterative shrinkage method', *Comput. Optim. Appl.* **42**, 173–193.

L. Breiman, L. Le Cam and L. Schwartz (1965), 'Consistent estimates and zero-one sets', *Ann. Math. Statist.* **35**, 157–161.

H. Bristow, A. Eriksson and S. Lucey (2013), Fast convolutional sparse coding. In *IEEE Conference on Computer Vision and Pattern Recognition (CVPR 2013)*, pp. 391–398.

M. M. Bronstein, J. Bruna, Y. LeCun, A. Szlam and P. Vandergheynst (2017), 'Geometric deep learning: Going beyond Euclidean data', *IEEE Signal Process. Mag.* **34**, 18–42.

A. M. Bruckstein, D. L. Donoho and M. Elad (2009), 'From sparse solutions of systems of equations to sparse modeling of signals and images', *SIAM Review* **51**, 34–18.

J. Bruna and S. Mallat (2013), 'Invariant scattering convolution networks', *IEEE Trans. Pattern Anal. Mach. Intel.* **35**, 1872–1886.

A. Buades, B. Coll and J.-M. Morel (2005), A non-local algorithm for image denoising. In *IEEE Computer Society Conference on Computer Vision and Pattern Recognition (CVPR 2005)*, Vol. 2, pp. 60–65.

T. A. Bubba, G. Kutyniok, M. Lassas, M. März, W. Samek, S. Siltanen and V. Srinivasan (2018), Learning the invisible: A hybrid deep learning–shearlet framework for limited angle computed tomography. arXiv:1811.04602

S. Bubeck (2015), 'Convex optimization: Algorithms and complexity', *Found. Trends Mach. Learn.* **8**, 231–357.

A. Buccini, M. Donatelli and R. Ramlau (2018), 'A semiblind regularization algorithm for inverse problems with application to image deblurring', *SIAM J. Sci. Comput.* **40**, A452–A483.

T. Bui-Thanh, K. Willcox and O. Ghattas (2008), 'Model reduction for large-scale systems with high-dimensional parametric input space', *SIAM J. Sci. Comput.* **30**, 3270–3288.

M. Burger and F. Lucka (2014), 'Maximum *a posteriori* estimates in linear inverse problems with log-concave priors are proper Bayes estimators', *Inverse Problems* **30**, 114004.

R. H. Byrd, G. M. Chin, J. Nocedal and Y. Wu (2012), 'Sample size selection in optimization methods for machine learning', *Math. Program.* **134**, 127–155.

C. L. Byrne (2008), *Applied Iterative Methods*, Peters/CRC Press.

L. Calatroni (2015), New PDE models for imaging problems and applications. PhD thesis, University of Cambridge.

L. Calatroni, C. Cao, J. C. De los Reyes, C.-B. Schönlieb and T. Valkonen (2016), 'Bilevel approaches for learning of variational imaging models', *Variational Methods* **18**, 252–290.

L. Calatroni, J. C. De los Reyes and C.-B. Schönlieb (2014), Dynamic sampling schemes for optimal noise learning under multiple nonsmooth constraints. In *26th IFIP Conference on System Modeling and Optimization (CSMO 2013)*, Springer, pp. 85–95.

L. Calatroni, J. C. De los Reyes and C.-B. Schönlieb (2017), 'Infimal convolution of data discrepancies for mixed noise removal', *SIAM J. Imaging Sci.* **10**, 1196–1233.

A. P. Calderón (1958), 'Uniqueness in the Cauchy problem for partial differential equations', *Amer. J. Math.* **80**, 16–36.

A. P. Calderón and A. Zygmund (1952), 'On the existence of certain singular integrals', *Acta Math.* **88**, 85.

A. P. Calderón and A. Zygmund (1956), 'On singular integrals', *Amer. J. Math.* **78**, 289–309.

D. Calvetti and E. Somersalo (2008), 'Hypermodels in the Bayesian imaging framework', *Inverse Problems* **24**, 034013.

D. Calvetti and E. Somersalo (2017), 'Inverse problems: From regularization to Bayesian inference', *WIREs Comput. Statist.* **10**, e1427.

D. Calvetti, B. Lewis and L. Reichel (2002), 'On the regularizing properties of the GMRES method', *Numer. Math.* **91**, 605–625.

D. Calvetti, E. Somersalo and A. Strang (2019), 'Hierarchical Bayesian models and sparsity: ℓ_2-magic', *Inverse Problems* **35**, 035003.

E. J. Candès and D. L. Donoho (2005), 'Continuous curvelet transform, I: resolution of the wavefront set', *Appl. Comput. Harmon. Anal.* **19**, 162–197.

E. J. Candès, L. Demanet and L. Ying (2007), 'Fast computation of Fourier integral operators', *SIAM J. Sci. Comput.* **29**, 2464–2493.

E. J. Candès, J. K. Romberg and T. Tao (2006), 'Robust uncertainty principles: Exact signal reconstruction from highly incomplete Fourier information', *IEEE Trans. Inform. Theory* **52**, 489–509.

M. Carriero, A. Leaci and F. Tomarelli (1996), A second order model in image segmentation: Blake & Zisserman functional. In *Variational Methods for Discontinuous Structures* (R. Serapioni and F. Tomarelli *et al.*), Springer, pp. 57–72.

I. Castillo and R. Nickl (2013), 'Nonparametric Bernstein–von Mises theorems in Gaussian white noise', *Ann. Statist.* **41**, 1999–2028.

I. Castillo and R. Nickl (2014), 'On the Bernstein–von Mises phenomenon for nonparametric Bayes procedures', *Ann. Statist.* **42**, 1941–1969.

A. Chakraborty, M. Alam, V. Dey, A. Chattopadhyay and D. Mukhopadhyay (2018), Adversarial attacks and defences: A survey. arXiv:1810.00069

A. Chambolle and P.-L. Lions (1997), 'Image recovery via total variation minimization and related problems', *Numer. Math.* **76**, 167–188.

A. Chambolle and T. Pock (2011), 'A first-order primal–dual algorithm for convex problems with applications to imaging', *J. Math. Imaging Vision* **40**, 120–145.

A. Chambolle and T. Pock (2016), An introduction to continuous optimization for imaging. In *Acta Numerica*, Vol. 25, Cambridge University Press, pp. 161–319.

A. Chambolle, M. Holler and T. Pock (2018), A convex variational model for learning convolutional image atoms from incomplete data. arXiv:1812.03077v1

T. F. Chan and J. Shen (2006), 'Image processing and analysis: Variational, PDE, wavelet, and stochastic methods', *BioMed. Engng OnLine* **5**, 38.

J. H. R. Chang, C.-L. Li, B. Póczos, B. V. K. V. Kumar and A. C. Sankaranarayanan (2017), One network to solve them all: Solving linear inverse problems using deep projection models. arXiv:1703.09912v1

B. Chen, K. Xiang, Z. Gong, J. Wang, and S. Tan (2018), 'Statistical iterative CBCT reconstruction based on neural network', *IEEE Trans. Medical Imaging* **37**, 1511–1521.

G. Chen and D. Needell (2016), 'Compressed sensing and dictionary learning', *Proc. Sympos. Appl. Math.* **73**, 201–241.

H. Chen, Y. Zhang, Y. Chen, J. Zhang, W. Zhang, H. Sun, Y. Lv, P. Liao, J. Zhou and G. Wang (2019), 'LEARN: Learned experts' assessment-based reconstruction network for sparse-data CT', *IEEE Trans. Medical Imaging* **37**, 1333–1347.

H. Chen, Y. Zhang, M. K. Kalra, F. Lin, Y. Chen, P. Liao, J. Zhou and G. Wang (2017*a*), 'Low-dose CT with a residual encoder–decoder convolutional neural network', *IEEE Trans. Medical Imaging* **36**, 2524–2535.

H. Chen, Y. Zhang, W. Zhang, P. Liao, K. Li, J. Zhou and G. Wang (2017*b*), Low-dose CT denoising with convolutional neural network. In *2017 IEEE 14th International Symposium on Biomedical Imaging (ISBI 2017)*, pp. 143–146.

Y. Chen, T. Pock and H. Bischof (2012), Learning ℓ_1-based analysis and synthesis sparsity priors using bi-level optimization. In *Workshop on Analysis Operator Learning vs. Dictionary Learning (NIPS 2012)*.

Y. Chen, T. Pock, R. Ranftl and H. Bischof (2013), Revisiting loss-specific training of filter-based MRFs for image restoration. In *German Conference on Pattern Recognition (GCPR 2013)*, Vol. 8142 of Lecture Notes in Computer Science, Springer, pp. 271–281.

Y. Chen, R. Ranftl and T. Pock (2014), 'Insights into analysis operator learning: From patch-based sparse models to higher order MRFs', *IEEE Trans. Image Process.* **23**, 1060–1072.

Y. Chen, W. Yu and T. Pock (2015), On learning optimized reaction diffusion processes for effective image restoration. In *IEEE Conference on Computer Vision and Pattern Recognition (CVPR 2015)*, pp. 5261–5269.

F. Chollet *et al.* (2015), Keras: The Python Deep Learning library. https://keras.io

A. Choromanska, M. Henaff, M. Mathieu, G. B. Arous and Y. LeCun (2015), The loss surfaces of multilayer networks. In *18th International Conference on Artificial Intelligence and Statistics (AISTATS 2015)*, pp. 192–204.

I. Y. Chun, X. Zheng, Y. Long and J. A. Fessler (2017), Sparse-view X-ray CT reconstruction using ℓ_1 regularization with learned sparsifying transform. In *14th International Meeting on Fully Three-Dimensional Image Reconstruction in Radiology and Nuclear Medicine (Fully3D 2017)*.

J. Chung and M. I. Espanol (2017), 'Learning regularization parameters for general-form Tikhonov', *Inverse Problems* **33**, 074004.

C. Clason, T. Helin, R. Kretschmann and P. Piiroinen (2018), Generalized modes in Bayesian inverse problems. arXiv:1806.00519

A. Cohen, W. Dahmen and R. DeVore (2009), 'Compressed sensing and best k-term approximation', *J. Amer. Math. Soc.* **22**, 211–231.

J. Cohen, E. Rosenfeld and J. Z. Kolter (2019), Certified adversarial robustness via randomized smoothing. arXiv:1902.02918v1

P. L. Combettes and J.-C. Pesquet (2011), Proximal splitting methods in signal processing. In *Fixed-Point Algorithms for Inverse Problems in Science and Engineering* (H. H. Bauschke *et al.*, eds), Vol. 49 of Springer Optimization and its Applications, Springer, pp. 185–212.

P. L. Combettes and J.-C. Pesquet (2012), 'Primal–dual splitting algorithm for solving inclusions with mixtures of composite, Lipschitzian, and parallel-sum type monotone operators', *Set-Valued Var. Anal.* **20**, 307–330.

P. L. Combettes and V. R. Wajs (2005), 'Signal recovery by proximal forward–backward splitting', *Multiscale Model. Simul.* **4**, 1168–1200.

R. Costantini and S. Susstrunk (2004), Virtual sensor design. In *Electronic Imaging 2004*, International Society for Optics and Photonics, pp. 408–419.

A. Courville, I. Goodfellow and Y. Bengio (2017), *Deep Learning*, MIT Press.

G. R. Cross and A. K. Jain (1983), 'Markov random field texture models', *IEEE Trans. Pattern Anal. Mach. Intel.* **5**, 25–39.

G. Cybenko (1989), 'Approximation by superpositions of a sigmoidal function', *Math. Control Signals Syst.* **2**, 303–314.

C. O. da Luis and A. J. Reader (2017), Deep learning for suppression of resolution-recovery artefacts in MLEM PET image reconstruction. In *2017 IEEE*

Nuclear Science Symposium and Medical Imaging Conference (NSS/MIC), pp. 1–3.

K. Dabov, A. Foi, V. Katkovnik and K. Egiazarian (2007), 'Image denoising by sparse 3-D transform-domain collaborative filtering', *IEEE Trans. Image Process.* **16**, 2080–2095.

A. V. Dalca, G. Balakrishnan, J. Guttag and M. R. Sabuncu (2018), Unsupervised learning for fast probabilistic diffeomorphic registration. In *21st International Conference on Medical Image Computing and Computer-Assisted Intervention (MICCAI 2018)* (A. F. Frangi *et al.*, eds), Vol. 11070 of Lecture Notes in Computer Science, Springer, pp. 729–738.

M. Dashti and A. M. Stuart (2017), The Bayesian approach to inverse problems. In *Handbook of Uncertainty Quantification* (R. Ghanem *et al.*, eds), Springer, chapter 10.

M. Dashti, K. J. H. Law, A. M. Stuart and J. Voss (2013), 'MAP estimators and their consistency in Bayesian nonparametric inverse problems', *Inverse Problems* **29**, 095017.

I. Daubechies, M. Defrise and C. De Mol (2004), 'An iterative thresholding algorithm for linear inverse problems with a sparsity constraint', *Commun. Pure Appl. Math.* **57**, 1413–1457.

I. Daubechies *et al.* (1991), *Ten Lectures on Wavelets*, Vol. 61 of CBMS-NSF Regional Conference Series in Applied Mathematics, SIAM.

M. E. Davison (1983), 'The ill-conditioned nature of the limited angle tomography problem', *SIAM J. Appl. Math.* **43**, 428–448.

E. Davoli and P. Liu (2018), 'One dimensional fractional order TGV: Gamma-convergence and bi-level training scheme', *Commun. Math. Sci.* **16**, 213–237.

A. Dax (1993), 'On row relaxation methods for large constrained least squares problems', *SIAM J. Sci. Comput.* **14**, 570–584.

J. C. De los Reyes (2011), 'Optimal control of a class of variational inequalities of the second kind', *SIAM J. Control Optim.* **49**, 1629–1658.

J. C. De los Reyes (2015), *Numerical PDE-Constrained Optimization*, Springer.

J. C. De los Reyes and C.-B. Schönlieb (2013), 'Image denoising: Learning the noise model via nonsmooth PDE-constrained optimization', *Inverse Problems* **7**, 1183–1214.

J. C. De los Reyes, C.-B. Schönlieb and T. Valkonen (2016), 'The structure of optimal parameters for image restoration problems', *J. Math. Anal. Appl.* **434**, 464–500.

J. C. De los Reyes, C.-B. Schönlieb and T. Valkonen (2017), 'Bilevel parameter learning for higher-order total variation regularisation models', *J. Math. Imaging Vision* **57**, 1–25.

A. P. Dempster, N. M. Laird, D. B. Rubin *et al.* (1977), 'Maximum likelihood from incomplete data via the EM algorithm', *J. Royal Statist. Soc. B* **39**, 1–38.

E. Y. Derevtsov, A. V. Efimov, A. K. Louis and T. Schuster (2011), 'Singular value decomposition and its application to numerical inversion for ray transforms in 2D vector tomography', *J. Inverse Ill-Posed Problems* **19**, 689–715.

A. Diaspro, M. Schneider, P. Bianchini, V. Caorsi, D. Mazza, M. Pesce, I. Testa, G. Vicidomini and C. Usai (2007), Two-photon excitation fluorescence microscopy. In *Science of Microscopy* (P. W. Hawkes and J. C. H. Spence, eds), Vol. 2, Springer, Chapter 11, pp. 751–789.

S. Dittmer, T. Kluth, P. Maass and D. O. Baguer (2018), Regularization by architecture: A deep prior approach for inverse problems. arXiv:1812.03889

I. Dokmanić, J. Bruna, S. Mallat and M. de Hoop (2016), Inverse problems with invariant multiscale statistics. arXiv:1609.05502

K. Doksum (1974), 'Tail-free and neutral random probabilities and their posterior distributions', *Ann. Probab.* **2**, 183–201.

W. Dong, G. Shi, Y. Ma and X. Li (2015), 'Image restoration via simultaneous sparse coding: Where structured sparsity meets Gaussian scale mixture', *Internat. J. Comput. Vision* **114**, 217–232.

W. Dong, L. Zhang, G. Shi and X. Li (2013), 'Nonlocally centralized sparse representation for image restoration', *IEEE Trans. Image Process.* **22**, 1620–1630.

W. Dong, L. Zhang, G. Shi and X. Wu (2011), 'Image deblurring and super-resolution by adaptive sparse domain selection and adaptive regularization', *IEEE Trans. Image Process.* **20**, 1838–1857.

D. L. Donoho, Y. Tsaig, I. Drori and J.-L. Starck (2012), 'Sparse solution of underdetermined systems of linear equations by stagewise orthogonal matching pursuit', *IEEE Trans. Inform. Theory* **58**, 1094–1121.

J. L. Doob (1948), Application of the theory of martingales. In *International Colloquium du CNRS: Probability Theory and its Application*, pp. 22–28.

F. Draxler, K. Veschgini, M. Salmhofer and F. Hamprecht (2018), 'Essentially no barriers in neural network energy landscape', *Proc. Mach. Learning Res.* **80**, 1309–1318.

Y. Drori and M. Teboulle (2014), 'Performance of first-order methods for smooth convex minimization: A novel approach', *Math. Program.* **145**, 451–482.

A. Durmus, E. Moulines and M. Pereyra (2018), 'Efficient Bayesian computation by proximal Markov chain Monte Carlo: When Langevin meets Moreau', *SIAM J. Imaging Sci.* **11**, 473–506.

W. E, J. Han and A. Jentzen (2017), Deep learning-based numerical methods for high-dimensional parabolic partial differential equations and backward stochastic differential equations. arXiv:1706.04702

J. Eckstein and D. Bertsekas (1992), 'On the Douglas–Rachford splitting method and the proximal point algorithm for maximal monotone operators', *Math. Program.* **55**, 293–318.

M. Elad (2010), *Sparse and Redundant Representations: From Theory to Applications in Signal and Image Processing*, Springer.

M. Elad and M. Aharon (2006), 'Image denoising via sparse and redundant representations over learned dictionaries', *IEEE Trans. Image Process.* **15**, 3736–3745.

Y. C. Eldar and G. Kutyniok (2012), *Compressed Sensing: Theory and Applications*, Cambridge University Press.

H. W. Engl, M. Hanke and A. Neubauer (2000), *Regularization of Inverse Problems*, Vol. 375 of Mathematics and its Applications, Springer.

H. W. Engl, K. Kunisch and A. Neubauer (1989), 'Convergence rates for Tikhonov regularisation of non-linear ill-posed problems', *Inverse Problems* **5**, 523.

C. Esteves, C. Allen-Blanchette, A. Makadia and K. Daniilidis (2017), Learning SO(3) equivariant representations with spherical CNNs. arXiv:1711.06721

S. N. Evans and P. B. Stark (2002), 'Inverse problems as statistics', *Inverse Problems* **18**, R1–R55.

V. Faber, A. I. Katsevich and A. G. Ramm (1995), 'Inversion of cone-beam data and helical tomography', *J. Inverse Ill-Posed Problems* **3**, 429–446.

C. Farabet, C. Couprie, L. Najman and Y. LeCun (2013), 'Learning hierarchical features for scene labeling', *IEEE Trans. Pattern Anal. Mach. Intel.* **35**, 1915–1929.

M. A. T. Figueiredo, R. D. Nowak and S. J. Wright (2007), 'Gradient projection for sparse reconstruction: Application to compressed sensing and other inverse problems', *IEEE J. Selected Topics Signal Process.* **1**, 586–598.

C. E. Floyd (1991), 'An artificial neural network for SPECT image reconstruction', *IEEE Trans. Medical Imaging* **10**, 485–487.

M. Fornasier and H. Rauhut (2008), 'Iterative thresholding algorithms', *Appl. Comput. Harmon. Anal.* **25**, 187–208.

S. Foucart (2016), 'Dictionary-sparse recovery via thresholding-based algorithms', *J. Funct. Anal. Appl.* **22**, 6–19.

S. Foucart and H. Rauhut (2013), *A Mathematical Introduction to Compressive Sensing*, Applied and Numerical Harmonic Analysis, Birkhäuser.

C. Fox and S. Roberts (2012), 'A tutorial on variational Bayes', *Artif. Intel. Rev.* **38**, 85–95.

J. Franke, R. Lacroix, H. Lehr, M. Heidenreich, U. Heinen and V. Schulz (2017), 'MPI flow analysis toolbox exploiting pulsed tracer information: An aneurysm phantom proof', *Internat. J. Magnetic Particle Imaging* **3**, 1703020.

D. Freedman (1963), 'On the asymptotic behavior of Bayes estimates in the discrete case, I', *Ann. Math. Statist.* **34**, 1386–1403.

D. Freedman (1965), 'On the asymptotic behavior of Bayes estimates in the discrete case, II', *Ann. Math. Statist.* **36**, 454–456.

A. Frommer and P. Maass (1999), 'Fast CG-based methods for Tikhonov–Phillips regularization', *SIAM J. Sci. Comput.* **20**, 1831–1850.

L. Fu, T.-C. Lee, S. M. Kim, A. M. Alessio, P. E. Kinahan, Z. Chang, K. Sauer, M. K. Kalra and B. D. Man (2017), 'Comparison between pre-log and post-log statistical models in ultra-low-dose CT reconstruction', *IEEE Trans. Medical Imaging* **36**, 707–720.

C. Garcia-Cardona and B. Wohlberg (2017), Convolutional dictionary learning. arXiv:1709.02893

M. U. Ghani and W. C. Karl (2018), Deep learning-based sinogram completion for low-dose CT. In *2018 IEEE 13th Image, Video, and Multidimensional Signal Processing Workshop (IVMSP)*.

S. Ghosal and N. Ray (2017), 'Deep deformable registration: Enhancing accuracy by fully convolutional neural net', *Pattern Recog. Lett.* **94**, 81–86.

S. Ghosal and A. W. van der Vaart (2017), *Fundamentals of Nonparametric Bayesian Inference*, Cambridge Series in Statistical and Probabilistic Mathematics, Cambridge University Press.

S. Ghosal, J. K. Ghosh and R. V. Ramamoorthi (1999), 'Posterior consistency of dirichlet mixtures in density estimation', *Ann. Statist.* **27**, 143–158.

S. Ghosal, J. K. Ghosh and A. W. van der Vaart (2000), 'Convergence rates of posterior distributions', *Ann. Statist.* **28**, 500–531.

G. Gilboa and S. Osher (2008), 'Nonlocal operators with applications to image processing', *Multiscale Model. Simul.* **7**, 1005–1028.

D. Gilton, G. Ongie and R. Willett (2019), Neumann networks for inverse problems in imaging. arXiv:1901.03707

M. Giordano and H. Kekkonen (2018), Bernstein–von Mises theorems and uncertainty quantification for linear inverse problems. arXiv:1811.04058v1

M. Girolami and B. Calderhead (2011), 'Riemann manifold Langevin and Hamiltonian Monte Carlo methods', *J. Royal Statist. Soc. B* **73**, 123–214.

R. Giryes, Y. C. Eldar, A. M. Bronstein and G. Sapiro (2017), Tradeoffs between convergence speed and reconstruction accuracy in inverse problems. arXiv:1605.09232v2

B. Gleich and J. Weizenecker (2005), 'Tomographic imaging using the nonlinear response of magnetic particles', *Nature* **435** (7046), 1214–1217.

G. H. Golub and C. F. Van Loan (1980), 'An analysis of the total least squares problem', *SIAM J. Numer. Anal.* **17**, 883–893.

G. H. Golub, P. C. Hansen and D. P. O'Leary (1999), 'Tikhonov regularization and total least squares', *SIAM J. Matrix Anal. Appl.* **21**, 185–194.

G. H. Golub, M. Heat and G. Wahba (1979), 'Generalized cross validation as a method for choosing a good ridge parameter', *Technometrics* **21**, 215–223.

A. N. Gomez, M. Ren, R. Urtasun and R. B. Grosse (2017), The reversible residual network: Backpropagation without storing activations. arXiv:1707.04585v1

D. Gong, Z. Zhang, Q. Shi, A. van den Hengel, C. Shen and Y. Zhang (2018), Learning an optimizer for image deconvolution. arXiv:1804.03368v1

I. Goodfellow, J. Pouget-Abadie, M. Mirza, B. Xu, D. Warde-Farley, S. Ozair, A. Courville and Y. Bengio (2014), Generative adversarial nets. In *Advances in Neural Information Processing Systems 27 (NIPS 2014)* (Z. Ghahramani *et al.*, eds), Curran Associates, pp. 2672–2680.

B. Gözcü, R. K. Mahabadi, Y.-H. Li, E. Ilcak, T. Çukur, J. Scarlett and V. Cevher (2018), 'Learning-based compressive MRI', *IEEE Trans. Medical Imaging* **37**, 1394–1406.

M. Grasmair, M. Haltmeier and O. Scherzer (2008), 'Sparse regularization with ℓ_q penalty term', *Inverse Problems* **24**, 055020.

P. J. Green, K. Łatuszysński, M. Pereyra and C. P. Robert (2015), 'Bayesian computation: A summary of the current state, and samples backwards and forwards', *Statist. Comput.* **25**, 835–862.

A. Greenleaf, Y. Kurylev, M. Lassas and G. Uhlmann (2007), 'Full-wave invisibility of active devices at all frequencies', *Comm. Math. Phys.* **275**, 749–789.

K. Gregor and Y. LeCun (2010), Learning fast approximations of sparse coding. In *27th International Conference on Machine Learning (ICML '10)*, pp. 399–406.

R. Gribonval and M. Nikolova (2018), On Bayesian estimation and proximity operators. arXiv:1807.04021

S. Gu, L. Zhang, W. Zuo and X. Feng (2014), Weighted nuclear norm minimization with application to image denoising. In *IEEE Conference on Computer Vision and Pattern Recognition (CVPR 2014)*, pp. 2862–2869.

S. Gu, W. Zuo, Q. Xie, D. Meng, X. Feng and L. Zhang (2015), Convolutional sparse coding for image super-resolution. In *IEEE International Conference on Computer Vision (ICCV 2015)*, pp. 1823–1831.

S. Gugushvili, A. van der Vaart and D. Yan (2018), Bayesian linear inverse problems in regularity scales. arXiv:1802.08992v1

I. Gulrajani, F. Ahmed, M. Arjovsky, V. Dumoulin and A. C. Courville (2017), Improved training of Wasserstein GANs. In *Advances in Neural Information Processing Systems 30 (NIPS 2017)* (I. Guyon *et al.*, eds), Curran Associates, pp. 5767–5777.

Y. Guo, Y. Liu, T. Georgiou and M. S. Lew (2018), 'A review of semantic segmentation using deep neural networks', *Internat. J. Multimedia Information Retrieval* **7**, 87–93.

H. Gupta, K. H. Jin, H. Q. Nguyen, M. T. McCann and M. Unser (2018), 'CNN-based projected gradient descent for consistent CT image reconstruction', *IEEE Trans. Medical Imaging* **37**, 1440–1453.

S. Gutta, M. Bhatt, S. K. Kalva, M. Pramanik and P. K. Yalavarthy (2019), 'Modeling errors compensation with total least squares for limited data photoacoustic tomography', *IEEE J. Selected Topics Quantum Electron.* **25**, 1–14.

E. Haber and L. Ruthotto (2017), 'Stable architectures for deep neural networks', *Inverse Problems* **34**, 014004.

E. Haber and L. Tenorio (2003), 'Learning regularization functionals', *Inverse Problems* **19**, 611–626.

E. Haber, L. Horesh and L. Tenorio (2010), 'Numerical methods for the design of large-scale nonlinear discrete ill-posed inverse problems', *Inverse Problems* **26**, 025002.

J. Hadamard (1902), 'Sur les problèmes aux dérivées partielles et leur signification physique', *Princeton University Bulletin*, pp. 49–52.

J. Hadamard (1923), *Lectures on Cauchy's Problem in Linear Partial Differential Equations*, Yale University Press.

J. Haegele, J. Rahmer, B. Gleich, J. Borgert, H. Wojtczyk, N. Panagiotopoulos, T. Buzug, J. Barkhausen and F. Vogt (2012), 'Magnetic particle imaging: Visualization of instruments for cardiovascular intervention', *Radiology* **265**, 933–938.

B. N. Hahn (2015), 'Dynamic linear inverse problems with moderate movements of the object: Ill-posedness and regularization', *Inverse Probl. Imaging* **9**, 395–413.

U. Hämarik, B. Kaltenbacher, U. Kangro and E. Resmerita (2016), 'Regularization by discretization in Banach spaces', *Inverse Problems* **32**, 035004.

K. Hammernik, T. Klatzer, E. Kobler, M. P. Recht, D. K. Sodickson, T. Pock and F. Knoll (2018), 'Learning a variational network for reconstruction of accelerated MRI data', *Magnetic Reson. Med.* **79**, 3055–3071.

K. Hammernik, F. Knoll, D. Sodickson and T. Pock (2016), Learning a variational model for compressed sensing MRI reconstruction. In *Proceedings of the International Society of Magnetic Resonance in Medicine (ISMRM)*.

Y. Han and J. C. Ye (2018), 'Framing U-Net via deep convolutional framelets: Application to sparse-view CT', *IEEE Trans. Medical Imaging* **37**, 1418–1429.

M. Hanke-Bourgeois (1995), *Conjugate Gradient Type Methods for Ill-Posed Problems*, Vol. 327 of Pitman Research Notes in Mathematics, Longman.

M. Hanke and P. C. Hansen (1993), 'Regularization methods for large-scale problems', *Surveys Math. Indust.* **3**, 253–315.

P. C. Hansen (1992), 'Analysis of discrete ill-posed problems by means of the L-curve', *SIAM Review* **34**, 561–580.

A. Hauptmann, F. Lucka, M. Betcke, N. Huynh, J. Adler, B. Cox, P. Beard, S. Ourselin and S. Arridge (2018), 'Model-based learning for accelerated limited-view 3-D photoacoustic tomography', *IEEE Trans. Medical Imaging* **37**, 1382–1393.

B. He and X. Yuan (2012), 'Convergence analysis of primal–dual algorithms for a saddle-point problem: From contraction perspective', *SIAM J. Imaging Sci.* **5**, 119–149.

J. He, Y. Yang, Y. Wang, D. Zeng, Z. Bian, H. Zhang, J. Sun, Z. Xu and J. Ma (2019), 'Optimizing a parameterized plug-and-play ADMM for iterative low-dose CT reconstruction', *IEEE Trans. Medical Imaging* **38**, 371–382.

K. He, X. Zhang, S. Ren and J. Sun (2016), Deep residual learning for image recognition. In *IEEE Conference on Computer Vision and Pattern Recognition (CVPR 2016)*, pp. 770–778.

T. Helin and M. Burger (2015), 'Maximum *a posteriori* probability estimates in infinite-dimensional Bayesian inverse problems', *Inverse Problems* **31**, 085009.

S. W. Hell, A. Schönle and A. Van den Bos (2007), Nanoscale resolution in far-field fluorescence microscopy. In *Science of Microscopy* (P. W. Hawkes and J. C. H. Spence, eds), Vol. 2, Springer, pp. 790–834.

C. F. Higham and D. J. Higham (2018), Deep learning: An introduction for applied mathematicians. arXiv:1801.05894

M. Hintermüller and C. N. Rautenberg (2017), 'Optimal selection of the regularization function in a weighted total variation model, I: Modelling and theory', *J. Math. Imaging Vision* **59**, 498–514.

M. Hintermüller and T. Wu (2015), 'Bilevel optimization for calibrating point spread functions in blind deconvolution', *Inverse Probl. Imaging* **9**, 1139–1169.

M. Hintermüller, A. Laurain, C. Löbhard, C. N. Rautenberg and T. M. Surowiec (2014), Elliptic mathematical programs with equilibrium constraints in function space: Optimality conditions and numerical realization. In *Trends in PDE Constrained Optimization* (G. Leugering *et al.*, eds), Springer, pp. 133–153.

M. Hintermüller, C. N. Rautenberg, T. Wu and A. Langer (2017), 'Optimal selection of the regularization function in a weighted total variation model, II: Algorithm, its analysis and numerical tests', *J. Math. Imaging Vision* **59**, 515–533.

K. Hirakawa and T. W. Parks (2006), 'Image denoising using total least squares', *IEEE Trans. Image Process.* **15**, 2730–2742.

B. Hofmann (1994), 'On the degree of ill-posedness for nonlinear problems', *J. Inverse Ill-Posed Problems* **2**, 61–76.

B. Hofmann, B. Kaltenbacher, C. Pöschl and O. Scherzer (2007), 'A convergence rates result for Tikhonov regularization in Banach spaces with non-smooth operators', *Inverse Problems* **23**, 987.

B. Hofmann, S. Kindermann *et al.* (2010), 'On the degree of ill-posedness for linear problems with noncompact operators', *Methods Appl. Anal.* **17**, 445–462.

T. Hohage and F. Weidling (2016), 'Characterizations of variational source conditions, converse results, and maxisets of spectral regularization methods', *SIAM J. Numer. Anal.* **55**, 598–620.

T. Hohage and F. Werner (2016), 'Inverse problems with Poisson data: Statistical regularization theory, applications and algorithms', *Inverse Problems* **32**, 093001.

X. Hong, Y. Zan, F. Weng, W. Tao, Q. Peng and Q. Huang (2018), 'Enhancing the image quality via transferred deep residual learning of coarse PET sinograms', *IEEE Trans. Medical Imaging* **37**, 2322–2332.

L. Hörmander (1971), 'Fourier integral operators, I', *Acta Math.* **127**, 79–183.

H. Hornik (1991), 'Approximation capabilities of multilayer feedforward networks', *Neural Networks* **4**, 251–257.

K. Hornik, M. Stinchcombe and H. White (1989), 'Multilayer feedforward networks are universal approximators', *Neural Networks* **2**, 359–366.

J.-T. Hsieh, S. Zhao, S. Eismann, L. Mirabella and S. Ermon (2019), Learning neural PDE solvers with convergence guarantees. In *Seventh International Conference on Learning Representations (ICLR 2019)*, to appear.

J. Huang and D. Mumford (1999), Statistics of natural images and models. In *IEEE Computer Society Conference on Computer Vision and Pattern Recognition (CVPR 1999)*, Vol. 1, pp. 541–547.

Y. Huizhuo, J. Jinzhu and Z. Zhanxing (2018), SIPID: A deep learning framework for sinogram interpolation and image denoising in low-dose CT reconstruction. In *2018 IEEE 15th International Symposium on Biomedical Imaging (ISBI 2018)*, pp. 1521–1524.

C. M. Hyun, H. P. Kim, S. M. Lee, S. Lee and J. K. Seo (2018), 'Deep learning for undersampled MRI reconstruction', *Phys. Med. Biol.* **63**, 135007.

M. Igami (2017), Artificial intelligence as structural estimation: Economic interpretations of Deep Blue, Bonanza, and AlphaGo. arXiv:1710.10967

N. Ikeda and S. Watanabe (1989), *Stochastic Differential Equations and Diffusion Processes*, second edition, North-Holland.

J. Ingraham, A. Riesselman, C. Sander and D. Marks (2019), Learning protein structure with a differentiable simulator. In *International Conference on Learning Representations (ICLR 2019)*.

E. Janssens, J. D. Beenhouwer, M. V. Dael, T. D. Schryver, L. V. Hoorebeke, P. Verboven, B. Nicolai and J. Sijbers (2018), 'Neural network Hilbert transform based filtered backprojection for fast inline X-ray inspection', *Measurement Sci. Tech.* **29**, 034012.

B. Jin and P. Maass (2012*a*), 'An analysis of electrical impedance tomography with applications to Tikhonov regularization', *ESAIM Control Optim. Calc. Var.* **18**, 1027–1048.

B. Jin and P. Maass (2012*b*), 'Sparsity regularization for parameter identification problems', *Inverse Problems* **28**, 123001.

K. H. Jin, M. T. McCann, E. Froustey and M. Unser (2017), 'Deep convolutional neural network for inverse problems in imaging', *IEEE Trans. Image Process.* **26**, 4509–4522.

F. John (1955*a*), 'A note on "improper" problems in partial differential equations', *Commun. Pure Appl. Math.* **8**, 591–594.

F. John (1955*b*), 'Numerical solution of the equation of heat conduction for preceding times', *Ann. Mat. Pura Appl.* (4) **40**, 129–142.

F. John (1959), Numerical solution of problems which are not well posed in the sense of Hadamard. In *Proc. Rome Symp. Prov. Int. Comp. Center*, pp. 103–116.

F. John (1960), 'Continuous dependence on data for solutions of partial differential equations with a prescribed bound', *Commun. Pure Appl. Math.* **13**, 551–585.

D. J. Kadrmas (2004), 'LOR-OSEM: Statistical PET reconstruction from raw line-of-response histograms', *Phys. Med. Biol.* **49**, 4731–4744.

J. P. Kaipio and E. Somersalo (2005), *Statistical and Computational Inverse Problems*, Vol. 160 of Applied Mathematical Sciences, Springer.

J. P. Kaipio and E. Somersalo (2007), 'Statistical inverse problems: Discretization, model reduction and inverse crimes', *J. Comput. Appl. Math.* **198**, 493–504.

O. Kallenberg (2002), *Foundations of Modern Probability*, second edition, Springer.

B. Kaltenbacher, A. Kirchner and B. Vexler (2011), 'Adaptive discretizations for the choice of a Tikhonov regularization parameter in nonlinear inverse problems', *Inverse Problems* **27**, 125008.

B. Kaltenbacher, A. Neubauer and O. Scherzer (2008), *Iterative Regularization Methods for Nonlinear Ill-posed Problems*, Vol. 6 of Radon Series on Computational and Applied Mathematics, De Gruyter.

E. Kang and J. C. Ye (2018), Framelet denoising for low-dose CT using deep learning. In *2018 IEEE 15th International Symposium on Biomedical Imaging (ISBI 2018)*, pp. 311–314.

E. Kang, W. Chang, J. Yoo and J. C. Ye (2018), 'Deep convolutional framelet denoising for low-dose CT via wavelet residual network', *IEEE Trans. Medical Imaging* **37**, 1358–1369.

E. Kang, J. Min and J. C. Ye (2017), 'A deep convolutional neural network using directional wavelets for low-dose X-ray CT reconstruction', *Med. Phys.* **44**, 360–375.

A. Karpathy and L. Fei-Fei (2017), 'Deep visual-semantic alignments for generating image descriptions', *IEEE Trans. Pattern Anal. Mach. Intel.* **39**, 664–676.

K. Kekkonen, M. Lassas and S. Siltanen (2016), 'Posterior consistency and convergence rates for Bayesian inversion with hypoelliptic operators', *Inverse Problems* **32**, 085005.

A. Khandhar, P. Keselman, S. Kemp, R. Ferguson, P. Goodwill, S. Conolly and K. Krishnan (2017), 'Evaluation of peg-coated iron oxide nanoparticles as blood pool tracers for preclinical magnetic particle imaging', *Nanoscale* **9**, 1299–1306.

Y. Khoo and L. Ying (2018), SwitchNet: A neural network model for forward and inverse scattering problems. arXiv:1810.09675v1

Y. Khoo, J. Lu and L. Ying (2017), Solving parametric PDE problems with artificial neural networks. arXiv:1707.03351v2

D. Kim and J. A. Fessler (2016), 'Optimized first-order methods for smooth convex minimization', *Math. Program.* **159**, 81–107.

K. Kim, G. E. Fakhri and Q. Li (2017), 'Low-dose CT reconstruction using spatially encoded nonlocal penalty', *Med. Phys.* **44**, 376–390.

K. Kim, D. Wu, K. Gong, J. Dutta, J. H. Kim, Y. D. Son, H. K. Kim, G. E. Fakhri and Q. Li (2018), 'Penalized PET reconstruction using deep learning prior and local linear fitting', *IEEE Trans. Medical Imaging* **37**, 1478–1487.

S.-J. Kim, K. Koh, M. Lustig, S. Boyd and D. Gorinevsky (2007), 'An interior-point method for large-scale ℓ_1-regularized least squares', *IEEE J. Selected Topics Signal Process.* **1**, 606–617.

S. Kindermann (2011), 'Convergence analysis of minimization-based noise level-free parameter choice rules for linear ill-posed problems', *Electron. Trans. Numer. Anal.* **38**, 233–257.

A. Kirsch (2011), *An Introduction to the Mathematical Theory of Inverse Problems*, second edition, Vol. 120 of Applied Mathematical Sciences, Springer.

T. Klatzer and T. Pock (2015), Continuous hyper-parameter learning for support vector machines. In *20th Computer Vision Winter Workshop (CVWW)*.

B. J. K. Kleijn and Y. Y. Zhao (2018), Criteria for posterior consistency. arXiv:1308.1263v5

T. Kluth (2018), 'Mathematical models for magnetic particle imaging', *Inverse Problems* **34**, 083001.

T. Kluth and P. Maass (2017), 'Model uncertainty in magnetic particle imaging: Nonlinear problem formulation and model-based sparse reconstruction', *Internat. J. Magnetic Particle Imaging* **3**, 1707004.

B. T. Knapik, B. T. Szabó, A. W. van der Vaart and J. H. van Zanten (2016), 'Bayes procedures for adaptive inference in inverse problems for the white noise model', *Probab. Theory Related Fields* **164**, 771–813.

B. T. Knapik, A. W. van der Vaart and J. H. van Zanten (2011), 'Bayesian inverse problems with Gaussian priors', *Ann. Statist.* **39**, 2626–2657.

B. T. Knapik, A. W. van der Vaart and J. H. van Zanten (2013), 'Bayesian recovery of the initial condition for the heat equation', *Commun. Statist. Theory Methods* **42**, 1294–1313.

T. Knopp, N. Gdaniec and M. Möddel (2017), 'Magnetic particle imaging: From proof of principle to preclinical applications', *Phys. Med. Biol.* **62**, R124.

T. Knopp, T. Viereck, G. Bringout, M. Ahlborg, J. Rahmer and M. Hofmann (2016), MDF: Magnetic particle imaging data format. arXiv:1602.06072

S. Ko, D. Yu and J.-H. Won (2017), On a class of first-order primal–dual algorithms for composite convex minimization problems. arXiv:1702.06234

E. Kobler, T. Klatzer, K. Hammernik and T. Pock (2017), Variational networks: connecting variational methods and deep learning. In *German Conference on Pattern Recognition (GCPR 2017)*, Vol. 10496 of Lecture Notes in Computer Science, Springer, pp. 281–293.

F. Kokkinos and S. Lefkimmiatis (2018), Deep image demosaicking using a cascade of convolutional residual denoising networks. arXiv:1803.05215

V. Kolehmainen, M. Lassas, K. Niinimäki and S. Siltanen (2012), 'Sparsity-promoting Bayesian inversion', *Inverse Problems* **28**, 025005.

V. P. Krishnan and E. T. Quinto (2015), *Microlocal Analysis in Tomography*, Handbook of Mathematical Methods in Imaging, Springer.

A. Küçükelbir, R. Ranganath, A. Gelman and D. Blei (2017), 'Automatic variational inference', *J. Mach. Learn. Res.* **18**, 430–474.

J. Kukačka, V. Golkov and D. Cremers (2017), Regularization for deep learning: A taxonomy. arXiv:1710.10686

K. Kunisch and T. Pock (2013), 'A bilevel optimization approach for parameter learning in variational models', *SIAM J. Imaging Sci.* **6**, 938–983.

H. Kushner and G. Yin (1997), *Stochastic Approximation Algorithms and Applications*, Springer.

G. Kutyniok and D. Labate (2009), 'Resolution of the wavefront set using continuous shearlets', *Trans. Amer. Math. Soc.* **361**, 2719–2754.

G. Kutyniok and D. Labate (2012), *Shearlets: Multiscale Analysis for Multivariate Data*, Springer.

R. R. Lam, L. Horesh, H. Avron and K. E. Willcox (2017), Should you derive, or let the data drive? An optimization framework for hybrid first-principles data-driven modeling. arXiv:1711.04374

F. Lanusse, J.-L. Starck, A. Woiselle and J. M. Fadili (2014), '3-D sparse representations', *Adv. Imaging Electron Phys.* **183**, 99–204.

A. Lanza, S. Morigi, F. Sgallari and Y.-W. Wen (2014), 'Image restoration with Poisson–Gaussian mixed noise', *Comput. Methods Biomech. Biomed. Engng Imaging Vis.* **2**, 12–24.

M. Lassas, E. Saksman and S. Siltanen (2009), 'Discretization invariant Bayesian inversion and Besov space priors', *Inverse Probl. Imaging* **3**, 87–122.

P. Latafat and P. Patrinos (2017), 'Asymmetric forward–backward–adjoint splitting for solving monotone inclusions involving three operators', *Comput. Optim. Appl.* **68**, 57–93.

R. Lattès and J.-L. Lions (1969), *The Method of Quasi-Reversibility: Applications to Partial Differential Equations*, Vol. 18 of Modern Analytic and Computational Methods in Science and Mathematics, American Elsevier.

L. Le Cam (1986), *Asymptotic Methods in Statistical Decision Theory*, Springer Series in Statistics, Springer.

Y. LeCun, Y. Bengio and G. Hinton (2015), 'Deep learning', *Nature* **521** (7553), 436–444.

D. Lee, J. Yoo and J. C. Ye (2017), Deep residual learning for compressed sensing MRI. In *2017 IEEE 14th International Symposium on Biomedical Imaging (ISBI 2017)*, pp. 15–18.

H. Lee, J. Lee, H. Kim, B. Cho and S. Cho (2019), 'Deep-neural-network based sinogram synthesis for sparse-view CT image reconstruction', *IEEE Trans. Radiation Plasma Med. Sci.* **3**, 109–119.

S. Lefkimmiatis (2017), Non-local color image denoising with convolutional neural networks. In *IEEE International Conference on Computer Vision and Pattern Recognition (CVPR 2017)*, pp. 3587–3596.

M. S. Lehtinen, L. Päivärinta and E. Somersalo (1989), 'Linear inverse problems for generalised random variables', *Inverse Problems* **5**, 599–612.

C. Y. Li, X. Liang, Z. Hu and E. P. Xing (2018a), Hybrid retrieval-generation reinforced agent for medical image report generation. arXiv:1805.08298

H. Li, J. Schwab, S. Antholzer and M. Haltmeier (2018*b*), NETT: Solving inverse problems with deep neural networks. arXiv:1803.00092

H. Li, Z. Xu, G. Taylor and T. Goldstein (2018*c*), Visualizing the loss landscape of neural nets. In *6th International Conference on Learning Representations (ICLR 2018)*.

F. Liese and K.-J. Miescke (2008), *Statistical Decision Theory: Estimation, Testing, and Selection*, Springer Series in Statistics, Springer.

R. Liu, Z. Lin, W. Zhang and Z. Su (2010), Learning PDEs for image restoration via optimal control. In *European Conference on Computer Vision (ECCV 2010)*, Vol. 6311 of Lecture Notes in Computer Science, Springer, pp. 115–128.

Z. Long, Y. Lu, X. Ma and B. Dong (2018), PDE-Net: Learning PDEs from data. arXiv:1710.09668v2

A. K. Louis (1989), *Inverse und schlecht gestellte Probleme*, Vieweg/Teubner.

A. K. Louis (1996), 'Approximate inverse for linear and some nonlinear problems', *Inverse Problems* **12**, 175–190.

A. K. Louis (2011), 'Feature reconstruction in inverse problems', *Inverse Problems* **27**, 065010.

A. K. Louis and P. Maass (1990), 'A mollifier method for linear operator equations of the first kind', *Inverse Problems* **6**, 427.

S. Lu, S. V. Pereverzev and U. Tautenhahn (2009), 'Regularized total least squares: computational aspects and error bounds', *SIAM J. Matrix Anal. Appl.* **31**, 918–941.

A. Lucas, M. Iliadis, R. Molina and A. K. Katsaggelos (2018), 'Using deep neural networks for inverse problems in imaging: Beyond analytical methods', *IEEE Signal Process. Mag.* **35**, 20–36.

S. Lunz, O. Öktem and C.-B. Schönlieb (2018), Adversarial regularizers in inverse problems. In *Advances in Neural Information Processing Systems 31 (NIPS 2018)* (S. Bengio *et al.*, eds), Curran Associates, pp. 8507–8516.

Z.-Q. Luo, J.-S. Pang and D. Ralph (1996), *Mathematical programs with equilibrium constraints*, Cambridge University Press.

P. Maass (2019), Deep learning for trivial inverse problems. In *Compressed Sensing and its Applications*, Birkhäuser.

J. Mairal and J. Ponce (2014), Sparse modeling for image and vision processing. arXiv:1411.3230v2

J. Mairal, F. Bach, J. Ponce and G. Sapiro (2010), 'Online learning for matrix factorization and sparse coding', *J. Mach. Learn. Res.* **11**, 19–60.

S. Mallat (2009), *A Wavelet Tour of Signal Processing: The Sparse Way*, third edition, Academic Press.

S. Mallat (2012), 'Group invariant scattering', *Commun. Pure Appl. Math.* **65**, 1331–1398.

S. Mallat (2016), 'Understanding deep convolutional networks', *Philos. Trans. Royal Soc. A* **374**, 20150203.

S. G. Mallat and Z. Zhang (1993), 'Matching pursuits with time-frequency dictionaries', *IEEE Trans. Signal Process.* **41**, 3397–3415.

A. Mandelbaum (1984), 'Linear estimators and measurable linear transformations on a Hilbert space', *Z. Wahrsch. verw. Gebiete* **65**, 385–397.

X. Mao, Q. Li, H. Xie, R. Y. Lau, Z. Wang and S. P. Smolley (2016), Least squares generative adversarial networks. arXiv:1611.04076

M. Mardani, E. Gong, J. Y. Cheng, J. Pauly and L. Xing (2017a), Recurrent generative adversarial neural networks for compressive imaging. In *IEEE 7th International Workshop on Computational Advances in Multi-Sensor Adaptive Processing (CAMSAP 2017)*.

M. Mardani, E. Gong, J. Y. Cheng, S. Vasanawala, G. Zaharchuk, M. Alley, N. Thakur, S. Han, W. Dally, J. M. Pauly and L. Xing (2017b), Deep generative adversarial networks for compressed sensing (GANCS) automates MRI. arXiv:1706.00051

M. Markkanen, L. Roininen, J. M. J. Huttunen and S. Lasanen (2019), 'Cauchy difference priors for edge-preserving Bayesian inversion', *J. Inverse Ill-Posed Problems* **27**, 225–240.

I. Markovsky and S. Van Huffel (2007), 'Overview of total least-squares methods', *Signal Processing* **87**, 2283–2302.

J. Martens and I. Sutskever (2012), Training deep and recurrent networks with Hessian-free optimization. In *Neural Networks: Tricks of the Trade*, Vol. 7700 of Lecture Notes in Computer Science, Springer, pp. 479–535.

D. Martin, C. Fowlkes, D. Tal and J. Malik (2001), A database of human segmented natural images and its application to evaluating segmentation algorithms and measuring ecological statistics. In *8th International Conference on Computer Vision (ICCV 2001)*, Vol. 2, pp. 416–423.

M. T. McCann and M. Unser (2019), Algorithms for biomedical image reconstruction. arXiv:1901.03565

M. T. McCann, K. H. Jin and M. Unser (2017), 'Convolutional neural networks for inverse problems in imaging: A review', *IEEE Signal Process. Mag.* **34**, 85–95.

T. Meinhardt, M. Moeller, C. Hazirbas and D. Cremers (2017), Learning proximal operators: Using denoising networks for regularizing inverse imaging problems. In *IEEE International Conference on Computer Vision (ICCV 2017)*, pp. 1799–1808.

K. Miller (1970), 'Least squares methods for ill-posed problems with a prescribed bound', *SIAM J. Math. Anal.* **1**, 52–74.

D. D. L. Minh and D. Le Minh (2015), 'Understanding the Hastings algorithm', *Commun. Statist. Simul. Comput.* **44**, 332–349.

T. Minka (2001), Expectation propagation for approximate Bayesian inference. In *17th Conference on Uncertainty in Artificial Intelligence (UAI '01)* (J. S. Breese and D. Koller, eds), Morgan Kaufmann, pp. 362–369.

F. Monard, R. Nickl and G. P. Paternain (2019), Efficient nonparametric Bayesian inference for X-ray transforms. *Ann. Statist.* **47**, 1113–1147.

N. Moriakov, K. Michielsen, J. Adler, R. Mann, I. Sechopoulos and J. Teuwen (2018), Deep learning framework for digital breast tomosynthesis reconstruction. arXiv:1808.04640

V. A. Morozov (1966), 'On the solution of functional equations by the method of regularization', *Soviet Math. Doklady* **7**, 414–417.

A. Mousavi and R. G. Baraniuk (2017), Learning to invert: Signal recovery via deep convolutional networks. In *2017 IEEE International Conference on Acoustics, Speech and Signal Processing (ICASSP)*, pp. 2272–2276.

J. L. Mueller and S. Siltanen (2012), *Linear and Nonlinear Inverse Problems with Practical Applications*, SIAM.

D. Mumford and J. Shah (1989), 'Optimal approximations by piecewise smooth functions and associated variational problems', *Commun. Pure Appl. Math.* **42**, 577–685.

K. Murase, M. Aoki, N. Banura, K. Nishimoto, A. Mimura, T. Kuboyabu and I. Yabata (2015), 'Usefulness of magnetic particle imaging for predicting the therapeutic effect of magnetic hyperthermia', *Open J. Medical Imaging* **5**, 85.

F. Natterer (1977), 'Regularisierung schlecht gestellter Probleme durch Projektionsverfahren', *Numer. Math.* **28**, 329–341.

F. Natterer (2001), *The Mathematics of Computerized Tomography*, Vol. 32 of Classics in Applied Mathematics, SIAM.

F. Natterer and F. Wübbeling (2001), *Mathematical Methods in Image Reconstruction*, SIAM.

R. M. Neal (2003), 'Slice sampling', *Ann. Statist.* **31**, 705–767.

D. Needell and J. A. Tropp (2009), 'CoSaMP: iterative signal recovery from incomplete and inaccurate samples', *Appl. Comput. Harmon. Anal.* **26**, 301–321.

D. Needell and R. Vershynin (2009), 'Uniform uncertainty principle and signal recovery via regularized orthogonal matching pursuit', *Found. Comput. Math.* **9**, 317–334.

Y. Nesterov (2004), *Introductory Lectures on Convex Optimization: A Basic Course*, Vol. 87 of Applied Optimization, Springer.

Y. Nesterov (2007), Gradient methods for minimizing composite objective function. CORE Discussion Papers no. 2007076, Center for Operations Research and Econometrics (CORE), Université Catholique de Louvain.

A. Neubauer and H. K. Pikkarainen (2008), 'Convergence results for the Bayesian inversion theory', *J. Inverse Ill-Posed Problems* **16**, 601–613.

R. Nickl (2013), *Statistical Theory*. Lecture notes, University of Cambridge. http://www.statslab.cam.ac.uk/~nickl/Site/_files/stat2013.pdf

R. Nickl (2017*a*), 'Bernstein–von Mises theorems for statistical inverse problems, I: Schrödinger equation,' *J. Eur. Math. Soc.*, to appear. arXiv:1707.01764

R. Nickl (2017*b*), 'On Bayesian inference for some statistical inverse problems with partial differential equations', *Bernoulli News* **24**, 5–9.

R. Nickl and J. Söhl (2017), 'Nonparametric Bayesian posterior contraction rates for discretely observed scalar diffusions', *Ann. Statist.* **45**, 1664–1693.

R. Nickl, S. van de Geer and S. Wang (2018), Convergence rates for penalised least squares estimators in PDE-constrained regression problems. arXiv:1809.08818

L. Nie and X. Chen (2014), 'Structural and functional photoacoustic molecular tomography aided by emerging contrast agents', *Chem. Soc. Review* **43**, 7132–70.

J. Nocedal and S. Wright (2006), *Numerical Optimization*, Springer Series in Operations Research and Financial Engineering, Springer.

C. Oh, D. Kim, J.-Y. Chung, Y. Han and H. W. Park (2018), ETER-net: End to end MR image reconstruction using recurrent neural network. In *International*

Workshop on Machine Learning for Medical Image Reconstruction (MLMIR 2018) (F. Knoll *et al.*, eds), Vol. 11074 of Lecture Notes in Computer Science, Springer.

B. A. Olshausen and D. J. Field (1997), 'Sparse coding with an overcomplete basis set: A strategy employed by V1?', *Vision Research* **37**, 3311–3325.

J. V. Outrata (2000), 'A generalized mathematical program with equilibrium constraints', *SIAM J. Control Optim.* **38**, 1623–1638.

S. Oymak and M. Soltanolkotabi (2017), 'Fast and reliable parameter estimation from nonlinear observations', *SIAM J. Optim.* **27**, 2276–2300.

V. Papyan, Y. Romano and M. Elad (2017), 'Convolutional neural networks analysed via convolutional sparse coding', *J. Mach. Learn. Res.* **18**, 1–52.

V. Papyan, J. Sulam and M. Elad (2016*a*), Working locally thinking globally, I: Theoretical guarantees for convolutional sparse coding. arXiv:1607.02005

V. Papyan, J. Sulam and M. Elad (2016*b*), Working locally thinking globally, II: Stability and algorithms for convolutional sparse coding. arXiv:1607.02009

P. Paschalis, N. D. Giokaris, A. Karabarbounis, G. K. Loudos, D. Maintas, C. N. Papanicolas, V. Spanoudaki, C. Tsoumpas and E. Stiliaris (2004), 'Tomographic image reconstruction using artificial neural networks', *Nucl. Instrum. Methods Phys. Res.* A **527**, 211–215.

D. M. Pelt and K. J. Batenburg (2013), 'Fast tomographic reconstruction from limited data using artificial neural networks', *IEEE Trans. Image Process.* **22**, 5238–5251.

D. M. Pelt, K. J. Batenburg and J. A. Sethian (2018), 'Improving tomographic reconstruction from limited data using mixed-scale dense convolutional neural networks', *J. Imaging* **4**, 128.

P. Perona and J. Malik (1990), 'Scale-space and edge detection using anisotropic diffusion', *IEEE Trans. Pattern Anal. Mach. Intel.* **12**, 629–639.

G. Peyré, S. Bougleux and L. D. Cohen (2011), 'Non-local regularization of inverse problems', *Inverse Probl. Imaging* **5**, 511–530.

D. L. Phillips (1962), 'A technique for the numerical solution of certain integral equations of the first kind', *J. Assoc. Comput. Mach.* **9**, 84–97.

L. Plantagie and J. K. Batenburg (2015), 'Algebraic filter approach for fast approximation of nonlinear tomographic reconstruction methods', *J. Electron. Imaging* **24**, 013026.

R. Plato and G. Vainikko (1990), 'On the regularization of projection methods for solving ill-posed problems', *Numer. Math.* **57**, 63–79.

Y. Pu, X. Yuan, A. Stevens, C. Li and L. Carin (2016), A deep generative deconvolutional image model. In *19th International Conference on Artificial Intelligence and Statistics (AISTATS 2016)*, pp. 741–750.

P. Putzky and M. Welling (2017), Recurrent inference machines for solving inverse problems. arXiv:1706.04008

C. Qin, J. Schlemper, J. Caballero, A. N. Price, J. V. Hajnal and D. Rueckert (2019), 'Convolutional recurrent neural networks for dynamic MR image reconstruction', *IEEE Trans. Medical Imaging* **38**, 280–290.

T. M. Quan, S. Member, T. Nguyen-Duc and W.-K. Jeong (2018), 'Compressed sensing MRI reconstruction using a generative adversarial network with a cyclic loss', *IEEE Trans. Medical Imaging* **37**, 1488–1497.

E. T. Quinto (1993), 'Singularities of the X-ray transform and limited data tomography in $\mathbb{R}^2$ and $\mathbb{R}^3$', *SIAM J. Math. Anal.* **24**, 1215–1225.

E. T. Quinto and O. Öktem (2008), 'Local tomography in electron microscopy', *SIAM J. Appl. Math.* **68**, 1282–1303.

J. Radon (1917), 'Über die Bestimmung von Funktionen durch ihre Integralwerte längs gewisser Mannigfaltigkeiten', *Ber. Verh. Sächs. Akad. Wiss. (Leipzig)* **69**, 262–277.

M. Raissi and G. E. Karniadakis (2017), Hidden physics models: Machine learning of nonlinear partial differential equations. arXiv:1708.00588v2

R. Ranftl and T. Pock (2014), A deep variational model for image segmentation. In *36th German Conference on Pattern Recognition (GCPR 2014)*, Vol. 8753 of Lecture Notes in Computer Science, Springer, pp. 107–118.

K. Ray (2013), 'Bayesian inverse problems with non-conjugate priors', *Electron. J. Statist.* **7**, 2516–2549.

E. T. Reehorst and P. Schniter (2018), Regularization by denoising: Clarifications and new interpretations. arXiv:1806.02296

A. Repetti, M. Pereyra and Y. Wiaux (2019), 'Scalable Bayesian uncertainty quantification in imaging inverse problems via convex optimization', *SIAM J. Imaging Sci.* **12**, 87–118.

W. Ring (2000), 'Structural properties of solutions to total variation regularization problems', *ESAIM Math. Model. Numer. Anal.* **34**, 799–810.

S. Rizzo, F. Botta, S. Raimondi, D. Origgi, C. Fanciullo, A. G. Morganti and M. Bellomi (2018), 'Radiomics: The facts and the challenges of image analysis', *European Radiol. Exp.* **2**, 36.

G. Rizzuti, A. Siahkoohi and F. J. Herrmann (2019), Learned iterative solvers for the Helmholtz equation. Submitted to *81st EAGE Conference and Exhibition 2019*. Available from https://www.slim.eos.ubc.ca/content/learned-iterative-solvers-helmholtz-equation.

H. Robbins and S. Monro (1951), 'A stochastic approximation method', *Ann. Math. Statist.* **22**, 400–407.

C. P. Robert and G. Casella (2004), *Monte Carlo Statistical Methods*, Springer Texts in Statistics, Springer.

R. T. Rockafellar and R. J.-B. Wets (1998), *Variational Analysis*, Springer.

Y. Romano and M. Elad (2015), Patch-disagreement as a way to improve K-SVD denoising. In *IEEE International Conference on Acoustics, Speech and Signal Processing (ICASSP)*, pp. 1280–1284.

Y. Romano, M. Elad and P. Milanfar (2017*a*), 'The little engine that could: Regularization by denoising (RED)', *SIAM J. Imaging Sci.* **10**, 1804–1844.

Y. Romano, J. Isidoro and P. Milanfar (2017*b*), 'RAISR: Rapid and accurate image super resolution', *IEEE Trans. Comput. Imaging* **3**, 110–125.

O. Ronneberger, P. Fischer and T. Brox (2015), U-Net: Convolutional networks for biomedical image segmentation. In *18th International Conference on Medical Image Computing and Computer-Assisted Intervention (MICCAI 2015)* (N. Navab *et al.*, eds), Vol. 9351 of Lecture Notes in Computer Science, Springer, pp. 234–241.

S. Roth and M. J. Black (2005), Fields of experts: A framework for learning image priors. In *IEEE Computer Society Conference on Computer Vision and Pattern Recognition (CVPR 2005)*, Vol. 2, pp. 860–867.

R. Rubinstein, A. M. Bruckstein and M. Elad. (2010), 'Dictionaries for sparse representation modeling', *Proc. IEEE* **98**, 1045–1057.

S. Ruder (2016), An overview of gradient descent optimization algorithms. arXiv:1609.04747

L. I. Rudin, S. Osher and E. Fatemi (1992), 'Nonlinear total variation based noise removal algorithms', *Phys. D* **60**, 259–268.

D. E. Rumelhart, G. E. Hinton and R. J. Williams (1986), Learning internal representation by error propagation. In *Parallel distributed processing: Explorations in the Microstructures of Cognition*, Vol. 1: *Foundations* (D. E. Rumelhart, J. L. McClelland and the PDP Research Group, eds), MIT Press, pp. 318–362.

L. Ruthotto and E. Haber (2018), Deep neural networks motivated by partial differential equations. arXiv:1804.04272

J. Salamon, M. Hofmann, C. Jung, M. G. Kaul, F. Werner, K. Them, R. Reimer, P. Nielsen, A. vom Scheidt, G. Adam, T. Knopp and H. Ittrich (2016), 'Magnetic particle/magnetic resonance imaging: *In-vitro* MPI-guided real time catheter tracking and 4D angioplasty using a road map and blood pool tracer approach', *PloS ONE* **11**, e0156899.

K. G. Samuel and M. F. Tappen (2009), Learning optimized map estimates in continuously-valued MRF models. In *IEEE Conference on Computer Vision and Pattern Recognition (CVPR 2009)*, pp. 477–484.

M. Sato (1971), Regularity of hyperfunctions solutions of partial differential equations. In *Actes du Congrès International des Mathématiciens*, Vol. 2, Gauthier-Villars, pp. 785–794.

A. Sawatzky, C. Brune, J. Müller and M. Burger (2009), Total variation processing of images with Poisson statistics. In *Computer Analysis of Images and Patterns* (X. Jiang and N. Petkov, eds), Vol. 5702 of Lecture Notes in Computer Science, Springer, pp. 533–540.

M. J. Schervish (1995), *Theory of Statistics*, Springer Series in Statistics, Springer.

O. Scherzer (1998), 'A modified Landweber iteration for solving parameter estimation problems', *Appl. Math. Optim.* **38**, 45–68.

O. Scherzer, M. Grasmair, H. Grossauer, M. Haltmeier and F. Lenzen (2009), *Variational Methods in Imaging*, Vol. 167 of Applied Mathematical Sciences, Springer.

J. Schlemper, J. Caballero, J. V. Hajnal, A. Price and D. Rueckert (2017), A deep cascade of convolutional neural networks for MR image reconstruction. In *25th International Conference on Information Processing in Medical Imaging (IPMI 2017)*, Vol. 10265 of Lecture Notes in Computer Science, Springer, pp. 647–658.

J. Schlemper, J. Caballero, J. V. Hajnal, A. N. Price and D. Rueckert (2018), 'A deep cascade of convolutional neural networks for dynamic MR image reconstruction', *IEEE Trans. Medical Imaging* **37**, 491–503.

U. Schmidt and S. Roth (2014), Shrinkage fields for effective image restoration. In *IEEE Conference on Computer Vision and Pattern Recognition (CVPR 2014)*, pp. 2774–2781.

U. Schmitt and A. K. Louis (2002), 'Efficient algorithms for the regularization of dynamic inverse problems, I: Theory', *Inverse Problems* **18**, 645–658.

T. Schuster (2007), *The Method of Approximate Inverse: Theory and Applications*, Vol. 1906 of Lecture Notes in Mathematics, Springer.

T. Schuster, B. Hahn and M. Burger (2018), 'Dynamic inverse problems: Modelling – regularization – numerics [Preface]', *Inverse Problems* **34**, 040301.

T. Schuster, B. Kaltenbacher, B. Hofmann and K. Kazimierski (2012), *Regularization Methods in Banach Spaces*, Radon Series on Computational and Applied Mathematics, De Gruyter.

J. Schwab, S. Antholzer and M. Haltmeier (2018), Deep null space learning for inverse problems: Convergence analysis and rates. arXiv:1806.06137

L. Schwartz (1965), 'On Bayes procedures', *Z. Wahrsch. verw. Gebiete* **4**, 10–26.

T. I. Seidman and C. R. Vogel (1989), 'Well-posedness and convergence of some regularisation methods for nonlinear ill posed problems', *Inverse Problems* **5**, 227–238.

P. Sermanet, D. Eigen, X. Zhang, M. Mathieu, R. Fergus and Y. LeCun (2013), OverFeat: Integrated recognition, localization and detection using convolutional networks. arXiv:1312.6229

S.-S. Shai and B.-D. Shai (2014), *Understanding Machine Learning: From Theory to Algorithms*, Cambridge University Press.

H. Shan, Y. Zhang, Q. Yang, U. Kruger, M. K. Kalra, L. Sun, W. Cong and G. Wang (2018), '3-D convolutional encoder–decoder network for low-dose CT via transfer learning from a 2-D trained network', *IEEE Trans. Medical Imaging* **37**, 1522–1534.

J. Sirignano and K. Spiliopoulos (2017), DGM: A deep learning algorithm for solving partial differential equations. arXiv:1708.07469v1

B. Sprungk (2017), Numerical methods for Bayesian inference in Hilbert spaces. PhD thesis, Technische Universität Chemnitz.

H. Sreter and R. Giryes (2017), Learned convolutional sparse coding. arXiv:1711.00328

R. L. Streit (2010), *Poisson Point Processes: Imaging, Tracking, and Sensing*, Springer.

D. M. Strong, T. F. Chan *et al.* (1996), Exact solutions to total variation regularization problems. In *UCLA CAM Report*, Citeseer.

A. M. Stuart (2010), Inverse problems: A Bayesian perspective. In *Acta Numerica*, Vol. 19, Cambridge University Press, pp. 451–559.

A. M. Stuart and A. L. Teckentrup (2018), 'Posterior consistency for Gaussian process approximations of Bayesian posterior distributions', *Math. Comp.* **87**, 721–753.

J. Sulam and M. Elad (2015), Expected patch log likelihood with a sparse prior. In *Energy Minimization Methods in Computer Vision and Pattern Recognition: Proceedings of the 10th International Conference (EMMCVPR 2015)*, pp. 99–111.

J. Sulam, V. Papyan, Y. Romano and M. Elad (2017), Multi-layer convolutional sparse modeling: Pursuit and dictionary learning. arXiv:1708.08705

N.-S. Syu, Y.-S. Chen and Y.-Y. Chuang (2018), Learning deep convolutional networks for demosaicing. arXiv:1802.03769

B. T. Szabó, A. W. van der Vaart and J. H. van Zanten (2013), 'Empirical Bayes scaling of Gaussian priors in the white noise model', *Electron. J. Statist.* **7**, 991–1018.

C. Szegedy, W. Zaremb, I. Sutskever, J. Bruna, D. Erhan, I. Goodfellow and R. Fergus (2014), Intriguing properties of neural networks. arXiv:1312.6199v4

M. F. Tappen (2007), Utilizing variational optimization to learn Markov random fields. In *IEEE Conference on Computer Vision and Pattern Recognition (CVPR 2007)*, pp. 1–8.

A. Tarantola (2005), *Inverse Problem Theory and Methods for Model Parameter Estimation*, second edition, SIAM.

A. Tarantola and B. Valette (1982), 'Inverse Problems = Quest for Information', *J. Geophys* **50**, 159–170.

S. Tariyal, A. Majumdar, R. Singh and M. Vatsa (2016), 'Deep dictionary learning', *IEEE Access* **4**, 10096–10109.

U. Tautenhahn (2008), 'Regularization of linear ill-posed problems with noisy right hand side and noisy operator', *J. Inverse Ill-Posed Problems* **16**, 507–523.

A. Taylor, J. Hendrickx and F. Glineur (2017), 'Smooth strongly convex interpolation and exact worst-case performance of first-order methods', *Math. Program.* **161**, 307–345.

M. Thoma (2016), A survey of semantic segmentation. arXiv:1602.06541

R. Tibshirani (1996), 'Regression shrinkage and selection via the Lasso', *J. Royal Statist. Soc. B* **58**, 267–288.

A. N. Tikhonov (1943), On the stability of inverse problems. *Dokl. Akad. Nauk SSSR* **39**, 195–198.

A. N. Tikhonov (1963), Solution of incorrectly formulated problems and the regularization method. *Dokl. Akad. Nauk.* **151**, 1035–1038.

A. N. Tikhonov and V. Y. Arsenin (1977), *Solutions of Ill-Posed Problems*, Winston.

J. Tompson, K. Schlachter, P. Sprechmann and K. Perlin (2017), Accelerating Eulerian fluid simulation with convolutional networks. arXiv:1607.03597v6

A. Traverso, L. Wee, A. Dekker and R. Gillies (2018), 'Repeatability and reproducibility of radiomic features: A systematic review', *Imaging Radiation Oncology* **102**, 1143–1158.

J. A. Tropp and A. C. Gilbert (2007), 'Signal recovery from random measurements via orthogonal matching pursuit', *IEEE Trans. Inform. Theory* **53**, 4655–4666.

G. Uhlmann and A. Vasy (2012), 'The inverse problem for the local geodesic X-ray transform', *Inventio. Math.* **205**, 83–120.

D. Ulyanov, A. Vedaldi and V. Lempitsky (2018), Deep image prior. In *IEEE Conference on Computer Vision and Pattern Recognition (CVPR 2018)*, pp. 9446–9454.

M. Unser and T. Blu (2000), 'Fractional splines and wavelets', *SIAM Review* **42**, 43–67.

K. Valluru, K. Wilson and J. Willmann (2016), 'Photoacoustic imaging in oncology: Translational preclinical and early clinical experience', *Radiology* **280**, 332–349.

C. Van Chung, J. De los Reyes and C.-B. Schönlieb (2017), 'Learning optimal spatially-dependent regularization parameters in total variation image denoising', *Inverse Problems* **33**, 074005.

D. Van Veen, A. Jalal, E. Price, S. Vishwanath and A. G. Dimakis (2018), Compressed sensing with deep image prior and learned regularization. arXiv:1806.06438

Y. Vardi, L. Shepp and L. Kaufman (1985), 'A statistical model for positron emission tomography', *J. Amer. Statist. Assoc.* **80** (389), 8–20.

B. S. Veeling, J. Linmans, J. Winkens, T. Cohen and M. Welling (2018), Rotation equivariant CNNs for digital pathology. arXiv:1806.03962

S. V. Venkatakrishnan, C. A. Bouman and B. Wohlberg (2013), Plug-and-play priors for model based reconstruction. In *IEEE Global Conference on Signal and Information Processing (GlobalSIP 2013)*, pp. 945–948.

R. Vidal, J. Bruna, R. Giryes and S. Soatto (2017), Mathematics of deep learning. arXiv:1712.04741

C. Villani (2009), *Optimal Transport: Old and New*, Vol. 338 of Grundlehren der mathematischen Wissenschaften, Springer.

C. Viroli and G. J. McLachlan (2017), Deep Gaussian mixture models. arXiv:1711.06929

C. Vogel and T. Pock (2017), A primal dual network for low-level vision problems. In *GCPR 2017: Pattern Recognition* (V. Roth and T. Vetter, eds), Vol. 10496 of Lecture Notes in Computer Science, Springer, pp. 189–202.

G. Wang, J. C. Ye, K. Mueller and J. A. Fessler (2018), 'Image reconstruction is a new frontier of machine learning', *IEEE Trans. Medical Imaging* **37**, 1289–1296.

L. V. Wang (2009), 'Multiscale photoacoustic microscopy and computed tomography', *Nature Photonics* **3**, 503–509.

S. Wang, Z. Su, L. Ying, X. Peng, S. Zhu, F. Liang, D. Feng and D. Liang (2016), Accelerating magnetic resonance imaging via deep learning. In *2016 IEEE 13th International Symposium on Biomedical Imaging (ISBI)*, pp. 514–517.

Y. Wang and D. M. Blei (2017), Frequentist consistency of variational Bayes. arXiv:1705.03439

Z. Wang, A. C. Bovik, H. R. Sheikh and E. P. Simoncelli (2004), 'Image quality assessment: From error visibility to structural similarity', *IEEE Trans. Image Process.* **13**, 600–612.

J. Weickert (1998), *Anisotropic Diffusion in Image Processing*, ECMI series, Teubner.

M. Weiler, M. Geiger, M. Welling, W. Boomsma and T. Cohen (2018), 3D steerable CNNs: Learning rotationally equivariant features in volumetric data. arXiv:1807.02547

J. Weizenecker, B. Gleich, J. Rahmer, H. Dahnke and J. Borgert (2009), 'Three-dimensional real-time *in vivo* magnetic particle imaging', *Phys. Med. Biol.* **54**, L1–L10.

M. Welling and Y. W. Teh (2011), Bayesian learning via stochastic gradient Langevin dynamics. In *28th International Conference on Machine Learning (ICML '11)*, pp. 681–688.

M. Welling, S. Osindero and G. E. Hinton (2003), Learning sparse topographic representations with products of Student-t distributions. In *15th International Conference on Neural Information Processing Systems (NIPS '02)*, MIT Press, pp. 1383–1390.

B. Wohlberg (2014), Efficient convolutional sparse coding. In *IEEE International Conference on Acoustics, Speech and Signal Processing (ICASSP)*, pp. 7173–7177.

J. M. Wolterink, A. M. Dinkla, M. H. F. Savenije, P. R. Seevinck, C. A. T. van den Berg and I. Išgum (2017), Deep MR to CT synthesis using unpaired data. In *Simulation and Synthesis in Medical Imaging SASHIMI 2017* (S. Tsaftaris *et al.*, eds), Vol. 10557 of Lecture Notes in Computer Science, Springer, pp. 14–23.

Y. Wu, P. Zhang, H. Shen and H. Zhai (2018), Visualizing neural network developing perturbation theory. arXiv:1802.03930v2

D. Wu, K. Kim and Q. Li (2018*a*), Computationally efficient cascaded training for deep unrolled network in CT imaging. arXiv:1810.03999v2

D. Wu, K. Kim, G. E. Fakhri and Q. Li (2017), 'Iterative low-dose CT reconstruction with priors trained by artificial neural network', *IEEE Trans. Medical Imaging* **36**, 2479–2486.

T. Würfl, M. Hoffmann, V. Christlein, K. Breininger, Y. Huang, M. Unberath and A. K. Maier (2018), 'Deep learning computed tomography: Learning projection-domain weights from image domain in limited angle problems', *IEEE Trans. Medical Imaging* **37**, 1454–1463.

J. Xia and L. V. Wang (2014), 'Small-animal whole-body photoacoustic tomography: A review', *IEEE Trans. Biomedical Engng* **61**, 1380–1389.

J. Xie, L. Xu and E. Chen (2012), Image denoising and inpainting with deep neural networks. In *Advances in Neural Information Processing Systems 25 (NIPS 2012)* (F. Pereira *et al.*, eds), Curran Associates, pp. 341–349.

Q. Xu, H. Yu, X. Mou, L. Zhang, J. Hsieh and G. Wang (2012), 'Low-dose X-ray CT reconstruction via dictionary learning', *IEEE Trans. Medical Imaging* **31**, 1682–1697.

B. Yang, L. Ying and J. Tang (2018*a*), 'Artificial neural network enhanced bayesian PET image reconstruction', *IEEE Trans. Medical Imaging* **37**, 1297–1309.

G. Yang, S. Yu, H. Dong, G. Slabaugh, P. L. Dragotti, X. Ye, F. Liu, S. Arridge, J. Keegan, Y. Guo and D. Firmin (2018*b*), 'DAGAN: Deep De-Aliasing Generative Adversarial Networks for fast compressed sensing MRI reconstruction', *IEEE Trans. Medical Imaging* **37**, 1310–1321.

Q. Yang, P. Yan, Y. Zhang, H. Yu, Y. Shi, X. Mou, M. K. Kalra, Y. Zhang, L. Sun and G. Wang (2018*c*), 'Low-dose CT image denoising using a generative adversarial network with Wasserstein distance and perceptual loss', *IEEE Trans. Medical Imaging* **37**, 1348–1357.

X. Yang, R. Kwitt, M. Styner and M. Niethammer (2017), 'Quicksilver: Fast predictive image registration: A deep learning approach', *NeuroImage* **158**, 378–396.

Y. Yang, J. Sun, H. Li and Z. Xu (2016), Deep ADMM-Net for compressive sensing MRI. In *Advances in Neural Information Processing Systems 29 (NIPS 2016)* (D. D. Lee *et al.*, eds), Curran Associates, pp. 10–18.

J. C. Ye, Y. S. Han and E. Cha (2018), 'Deep convolutional framelets: A general deep learning for inverse problems', *SIAM J. Imaging Sci.* **11**, 991–1048.

S. Ye, S. Ravishankar, Y. Long and J. A. Fessler (2018*b*), SPULTRA: Low-dose CT image reconstruction with joint statistical and learned image models. arXiv:1808.08791v2

J. Yoo, S. Sabir, D. Heo, K. H. Kim, A. Wahab, Y. Choi, S.-I. Lee, E. Y. Chae, H. H. Kim, Y. M. Bae, Y.-W. Choi, S. Cho and J. C. Ye (2017), Deep learning can reverse photon migration for diffuse optical tomography. arXiv:1712.00912

C. You, G. Li, Y. Zhang, X. Zhang, H. Shan, S. Ju, Z. Zhao, Z. Zhang, W. Cong, M. W. Vannier, P. K. Saha and G. Wang (2018*a*), CT super-resolution GAN constrained by the identical, residual, and cycle learning ensemble (GAN-CIRCLE). arXiv:1808.04256v3

C. You, Q. Yang, H. Shan, L. Gjesteby, G. Li, S. Ju, Z. Zhang, Z. Zhao, Y. Zhang, W. Cong and G. Wang (2018*b*), 'Structurally-sensitive multi-scale deep neural network for low-dose CT denoising', *IEEE Access* **6**, 41839–41855.

J. You and G. L. Zeng (2007), 'Hilbert transform based FBP algorithm for fan-beam CT full and partial scans', *IEEE Trans. Medical Imaging* **26**, 190–199.

E. Y. Yu, M. Bishop, B. Zheng, R. M. Ferguson, A. P. Khandhar, S. J. Kemp, K. M. Krishnan, P. W. Goodwill and S. M. Conolly (2017), 'Magnetic particle imaging: A novel *in vivo* imaging platform for cancer detection', *Nano Letters* **17**, 1648–1654.

M. D. Zeiler, D. Krishnan, G. W. Taylor and R. Fergus (2010), Deconvolutional networks. In *IEEE Conference on Computer Vision and Pattern Recognition (CVPR 2010)*, pp. 2528–2535.

C. Zhang, T. Zhang, M. Li, C. Peng, Z. Liu and J. Zheng (2016), 'Low-dose CT reconstruction via L1 dictionary learning regularization using iteratively reweighted least-squares', *BioMed. Engng OnLine* **15**, 66.

Y. Zhang and H. Yu (2018), 'Convolutional neural network based metal artifact reduction in X-ray computed tomography', *IEEE Trans. Medical Imaging* **37**, 1370–1381.

Z. Zhang, X. Liang, X. Dong, Y. Xie and G. Cao (2018), 'A sparse-view CT reconstruction method based on combination of densenet and deconvolution', *IEEE Trans. Medical Imaging* **37**, 1407–1417.

C. Zhao, J. Zhang, R. Wang and W. Gao (2018*a*), 'CREAM: CNN-REgularized ADMM framework for compressive-sensed image reconstruction', *IEEE Access* **6**, 76838–76853.

J. Zhao, Z. Chen, L. Zhang and X. Jin (2016), Few-view CT reconstruction method based on deep learning. In *2016 IEEE Nuclear Science Symposium, Medical Imaging Conference and Room-Temperature Semiconductor Detector Workshop (NSS/MIC/RTSD)*.

R. Zhao, Y. Hu, J. Dotzel, C. D. Sa and Z. Zhang (2018), Building efficient deep neural networks with unitary group convolutions. arXiv:1811.07755

X. Zheng, S. Ravishankar, Y. Long and J. A. Fessler (2018), 'PWLS-ULTRA: An efficient clustering and learning-based approach for low-dose 3D CT image reconstruction', *IEEE Trans. Medical Imaging* **37**, 1498–1510.

Y. Zhou, J. Yao and L. V. Wang (2016), 'Tutorial on photoacoustic tomography', *J. Biomedical Optics* **21**, 061007.

B. Zhu, J. Z. Liu, S. F. Cauley, B. R. Rosen and M. S. Rosen (2018), 'Image reconstruction by domain-transform manifold learning', *Nature* **555**, 487–492.

H. Zhu, G. Leus and G. B. Giannakis (2011), 'Sparsity-cognizant total least-squares for perturbed compressive sampling', *IEEE Trans. Signal Process.* **59**, 2002–2016.

S. C. Zhu, Y. Wu and D. Mumford (1998), 'Filters, random fields and maximum entropy (frame): Towards a unified theory for texture modeling', *Internat. J. Comput. Vision* **27**, 107–126.

D. Zoran and Y. Weiss (2011), From learning models of natural image patches to whole image restoration. In *IEEE International Conference on Computer Vision (ICCV 2011)*, pp. 479–486.

Acta Numerica (2019), pp. 175–286
doi:10.1017/S0962492919000023

Numerical analysis of hemivariational inequalities in contact mechanics

Weimin Han

Program in Applied Mathematical and Computational Sciences (AMCS),
and Department of Mathematics,
University of Iowa, Iowa City, IA 52242, USA
E-mail: weimin-han@uiowa.edu

Mircea Sofonea

Laboratoire de Mathématiques et Physique,
Université de Perpignan Via Domitia,
52 Avenue Paul Alduy, 66860 Perpignan, France
E-mail: sofonea@univ-perp.fr

Contact phenomena arise in a variety of industrial process and engineering applications. For this reason, contact mechanics has attracted substantial attention from research communities. Mathematical problems from contact mechanics have been studied extensively for over half a century. Effort was initially focused on variational inequality formulations, and in the past ten years considerable effort has been devoted to contact problems in the form of hemivariational inequalities. This article surveys recent development in studies of hemivariational inequalities arising in contact mechanics. We focus on contact problems with elastic and viscoelastic materials, in the framework of linearized strain theory, with a particular emphasis on their numerical analysis. We begin by introducing three representative mathematical models which describe the contact between a deformable body in contact with a foundation, in static, history-dependent and dynamic cases. In weak formulations, the models we consider lead to various forms of hemivariational inequalities in which the unknown is either the displacement or the velocity field. Based on these examples, we introduce and study three abstract hemivariational inequalities for which we present existence and uniqueness results, together with convergence analysis and error estimates for numerical solutions. The results on the abstract hemivariational inequalities are general and can be applied to the study of a variety of problems in contact mechanics; in particular, they are applied to the three representative mathematical models. We present numerical simulation results giving numerical evidence on the theoretically predicted optimal convergence order; we also provide mechanical interpretations of simulation results.

CONTENTS

1. Introduction

Processes of contact between deformable bodies abound in industry and everyday life. A few simple examples are brake pads in contact with wheels, tyres on roads, and pistons with skirts. Because of the importance of contact processes in structural and mechanical systems, considerable effort has been put into their modelling, analysis and numerical simulations. The literature on this field is extensive. The publications in the engineering literature are often concerned with specific settings, geometries or materials. Their aim is usually related to particular applied aspects of the problems. The publications on mathematical literature are concerned with the mathematical structures which underlie general contact problems with different constitutive laws, varied geometries and different contact conditions. They deal with the variational analysis of the corresponding models of contact. Once existence, uniqueness or non-uniqueness, and stability of solutions have been established, related important questions arise, such as numerical analysis of the solutions and how to construct reliable and efficient algorithms for their numerical approximations with guaranteed accuracy.

The first recognized publication on contact between deformable bodies was that of Hertz (1882). This was followed by Signorini (1933), who posed the problem in what is now termed a variational form. The Signorini problem was theoretically investigated by Fichera (1964, 1972). However, the general mathematical development for problems arising in contact mechanics began with the monograph by Duvaut and Lions (1976), who presented variational formulations of several contact problems and proved some basic existence and uniqueness results. A comprehensive treatment of unilateral contact problems, for linear and nonlinear elastic materials, and for both frictionless and frictional contact, was provided by Kikuchi and Oden (1988), who covered mathematical modelling of the contact phenomena, numerical analysis and implementation of numerical algorithms. A systematic coverage of numerical methods for solving unilateral contact problems, for both frictionless and frictional contact of linearly elastic materials, can be found in Hlaváček, Haslinger, Nečas and Lovíšek (1988), and an updated account of numerical methods for unilateral contact problems is given in Haslinger, Hlaváček and Nečas (1996).

For frictionless Signorini contact between two elastic bodies we refer to Haslinger and Hlaváček (1980, 1981a, 1981b), who proved the existence and uniqueness of a weak solution and provided numerical algorithms to solve the corresponding nonlinear boundary value problems. By introducing dual Lagrange multipliers for contact forces, a variational inequality in displacement can be reformulated as a saddle-point problem, which can be solved by semi-smooth Newton methods with a primal–dual active set strategy; see Wohlmuth and Krause (2003), Hüeber and Wohlmuth (2005a, 2005b), Wriggers and Fischer (2005) and the survey article by Wohlmuth (2011) for a summary account. Multigrid methods can be used to efficiently solve contact problems; *e.g.* Kornhuber and Krause (2001). For an optimal *a priori* error estimate for numerical solutions of the Signorini contact problem, we refer to the recent paper by Drouet and Hild (2015). A few steps in the mathematical analysis for models involving time-dependent unilateral contact between a deformable body and a rigid obstacle were made by Sofonea, Renon and Shillor (2004) and Renon, Montmitonnet and Laborde (2005). The quasistatic process of frictionless unilateral contact between a moving rigid obstacle and a viscoelastic body has been considered by Matei, Sitzmann, Willner and Wohlmuth (2017). Their model leads to a variational formulation with dual Lagrange multipliers. They obtained the existence of a solution by using a time discretization method combined with a saddle-point argument. Moreover, they used an efficient algorithm based on a primal–dual active set strategy, and presented three-dimensional numerical examples using the mortar method to discretize the contact constraints, without increasing the algebraic system size.

Monographs and books on mathematical problems in contact mechanics also include those by Panagiotopoulos (1985) for mechanical background, mathematical modelling and analysis, and engineering application, Han and Sofonea (2002) for mathematical modelling and analysis, as well as convergence analysis and optimal order error estimates of numerical methods for quasistatic contact problems of elastic, viscoelastic and viscoplastic materials, Shillor, Sofonea and Telega (2004) for mathematical modelling and analysis of contact problems, Eck, Jarušek and Krbec (2005) for variational analysis of unilateral contact problems in elasticity and viscoelasticity, and Capatina (2014) for a mathematical study of certain frictional contact problems. The books by Laursen (2002), Wriggers (2006) and Wriggers and

Laursen (2007) focus on numerical algorithms for solving contact problems and on engineering applications. The above references deal with variational inequality formulations of the problems in contact mechanics. In comparison, hemivariational inequality formulations are used in the study of contact problems with non-monotone mechanical relations in some more recent monographs, and we mention Panagiotopoulos (1993) for mathematical modelling, analysis and numerical simulation of contact problems, Migórski, Ochal and Sofonea (2013) for mathematical modelling and analysis of various contact problems, and Sofonea and Migórski (2018) for the modelling and analysis of static, history-dependent and evolutionary contact problems in the form of a special class of hemivariational inequalities called variational–hemivariational inequalities, in which both convex and non-convex functions are present.

Inequality problems in contact mechanics can be loosely classified into two main families: the family of variational inequalities, which is concerned with convex functionals (potentials), and the family of hemivariational inequalities, which is concerned with non-convex functionals (superpotentials). Some of the model problems considered in this paper are special kinds of inequalities, known as variational–hemivariational inequalities. In a variational–hemivariational inequality, we have the presence of both non-convex functionals and convex functionals. When the convex functionals are dropped from a general variational–hemivariational inequality, we have a 'pure' hemivariational inequality. Alternatively, when the non-convex functionals are dropped, we have a 'pure' variational inequality. Nevertheless, for simplicity, sometimes in this paper we use the term hemivariational inequality for both 'pure' hemivariational and variational–hemivariational inequalities. The theoretical results on the variational–hemivariational inequalities naturally lead to those for the corresponding variational and hemivariational inequalities.

Variational and hemivariational inequalities represent a powerful tool in the study of a large number of nonlinear boundary value problems. The theory of variational inequalities was first developed in the early 1960s, based on arguments of monotonicity and convexity, and properties of the subdifferential of a convex function. Representative references on mathematical studies of variational inequalities include Lions and Stampacchia (1967), Brézis (1972), Baiocchi and Capelo (1984) and Kinderlehrer and Stampacchia (2000), to name a few. Hemivariational inequalities were first introduced in the early 1980s by Panagiotopoulos in the context of applications in engineering problems. Studies of hemivariational inequalities can be found in several comprehensive references, for example Panagiotopoulos (1993), Naniewicz and Panagiotopoulos (1995) and Migórski, Ochal and Sofonea (2013), as well as in the volume edited by Han, Migórski and Sofonea (2015).

Since a closed-form solution formula can rarely be obtained for a general variational inequality or hemivariational inequality, numerical methods are essentially the only way to solve the inequality problems in practice. References on numerical analysis of general variational inequalities include the books by Glowinski, Lions and Trémolières (1981) and Glowinski (1984), and references on variational inequalities for contact problems include those of Kikuchi and Oden (1988), Hlaváček, Haslinger, Nečas and Lovíšek (1988), Haslinger, Hlaváček and Nečas (1996) and Han and Sofonea (2002). In comparison, the size of the literature on the numerical analysis of hemivariational inequalities is much smaller. The book by Haslinger, Miettinen and Panagiotopoulos (1999) is devoted to the finite element approximations of hemivariational inequalities, where convergence of numerical methods is discussed; however, no error estimates of the numerical solutions are derived. In recent years there have been efforts by various researchers to derive error estimates for numerical solutions of hemivariaional inequalities, and initially, only sub-optimal error estimates were reported. Han, Migórski and Sofonea (2014) were the first to give an optimal order error estimate for linear finite element solutions in solving hemivariational or variational–hemivariational inequalities. Then Barboteu, Bartosz, Han and Janiczko (2015) derived an optimal order error estimate for the numerical solution of a hyperbolic hemivariational inequality arising in dynamic contact when the linear finite element method is used for the spatial discretization and the backward Euler finite difference is used for the time derivative. With similar derivation techniques, various authors derived optimal order error estimates for the linear finite element method of a few individual hemivariational or variational–hemivariational inequalities, in several papers. More recently, general frameworks of convergence theory and error estimation for hemivariational or variational–hemivariational inequalities have been developed; see Han, Sofonea and Barboteu (2017) and Han, Sofonea and Danan (2018) for internal numerical approximations of general hemivariational and variational–hemivariational inequalities, and Han (2018) for both internal and external numerical approximations of general hemivariational and variational–hemivariational inequalities. In these recent papers, convergence is shown for numerical solutions by internal or external approximation schemes under minimal solution regularity condition, Céa-type inequalities are derived that serve as the starting point for error estimation for hemivariational and variational–hemivariational inequalities arising in contact mechanics, and optimal order error estimates for the linear finite element solutions are obtained.

The aim of this survey paper is to provide the state of the art on numerical analysis of some representative mathematical models which describe the contact of a deformable body with an obstacle, the so-called foundation, in the framework of the linearized strain theory. We present models for the processes, list the assumptions on the data and derive their weak formulation, which is in the form of a hemivariational inequality. We have tried to make this paper self-contained. Therefore, in addition to the numerical analysis of the contact models, we review the necessary background on the analysis of the related hemivariational inequalities, including existence and uniqueness results.

The paper is organized as follows. In Section 2 we introduce three representative models of contact and describe them in full detail. Then we list the assumptions on the data and state the weak formulations of the models, which are in the form of an elliptic, a history-dependent and an evolutionary hemivariational inequality, respectively. In Section 3 we present preliminary material on basic notions and results from non-smooth analysis that will be needed later in the well-posedness study and numerical analysis of the hemivariational inequalities. In addition, we also recall Banach's fixed-point theorem as well as Gronwall's inequalities in both the continuous version and discrete version. In Sections 4–6 we present well-posedness results and consider numerical approximations of three abstract hemivariational inequalities of the elliptic, history-dependent and evolutionary types. The results on the abstract hemivariational inequalities are applied in Sections 7–9 on the contact models, leading to statements of well-posedness of the contact problems, of convergence and optimal order error estimates of numerical methods. Numerical simulation results are shown to provide numerical evidence of the theoretically predicted first-order error estimate in the energy norm for linear finite element solutions. In Section 10, we comment on future research topics on the numerical solution of hemivariational inequalities, especially those arising in contact mechanics.

2. Three representative contact problems

Physical setting and mathematical models. A large number of processes of contact arising in various engineering applications can be cast in the following general physical setting: a deformable body is subjected to the action of body forces and surface tractions, is clamped on part of its surface and is in contact with a foundation on another part of its surface. We are interested in describing the evolution of the mechanical state of the body and, to this end, the first step is to construct a mathematical model which describes the physical setting above. Here and everywhere in this work, by a mathematical model we understand a system of partial differential equations with associated boundary conditions and with possibly initial conditions, for a specific contact process. Such models are constructed based on the general principles of solid mechanics, which can be found in Ciarlet (1988), Khludnev and Sokolowski (1997) and Temam and Miranville (2001), for instance.

To present a mathematical model in contact mechanics we need to combine several relations: the constitutive law, the balance equation, the boundary conditions, the interface laws, and for evolutionary problems, the initial conditions. Recall that a constitutive law represents a relation between the stress $\boldsymbol{\sigma}$ and the strain $\boldsymbol{\varepsilon}$, and the relation may involve derivatives and/or integrals of the the stress and/or strain. The constitutive law describes the mechanical reaction of the material with respect to the action of body forces and boundary tractions. Although the constitutive laws must satisfy some basic axioms and invariance principles, they originate mostly from experiments. We refer the reader to Han and Sofonea (2002) for a general description of several diagnostic experiments which provide information needed in constructing constitutive laws for specific materials. The balance equation for the stress field leads either to the equation of motion (used in the modelling of dynamic processes, *i.e.* processes in which the inertial terms are not neglected) or to the equation of equilibrium (used in the modelling of static and quasistatic processes, *i.e.* processes in which the inertial terms are neglected). The boundary conditions usually involve the displacement and the surface tractions. They express the fact that the body is held fixed on a part of the boundary and is acted upon by external forces on the other part. The interface laws are to be prescribed on the potential contact surface. These are divided naturally into conditions in the normal direction (called contact conditions) and those in the tangential directions (called friction laws). A comprehensive description of the interface laws used in the mathematical literature dedicated to modelling of contact problems can be found in Sofonea and Matei (2012) and Migórski, Ochal and Sofonea (2013).

Basic notation. In order to introduce the contact models, we let Ω be the reference configuration of the body, assumed to be an open, bounded, connected set in $\mathbb{R}^d$ with a Lipschitz boundary $\Gamma = \partial\Omega$. The dimension $d = 2$ or 3 for applications. The closure of Ω in $\mathbb{R}^d$ is denoted by $\overline{\Omega}$. To describe the boundary conditions, we split the boundary Γ into three disjoint, measurable parts Γ_1, Γ_2 and Γ_3. Here, $\mathrm{meas}\,(\Gamma_1) > 0$ and $\mathrm{meas}\,(\Gamma_3) > 0$, whereas Γ_2 is allowed to be empty. We use boldface letters for vectors and tensors, such as the outward unit normal $\boldsymbol{\nu}$ on Γ. A typical point in $\mathbb{R}^d$ is denoted by $\boldsymbol{x} = (x_i)$. The indices i, j, k, l run between 1 and d, and, unless stated otherwise, the summation convention over repeated indices is implied. An index that follows a comma indicates a partial derivative with respect to the corresponding component of the spatial variable $\boldsymbol{x}$. For integrals, we use dx for the infinitesimal volume element in Ω and da for the infinitesimal surface element on Γ. For a time-dependent contact problem, the time interval of interest will be denoted by I, which can be bounded or unbounded. A dot above a variable will represent the time derivative of the variable.

We are interested in mathematical models which describe the evolution or the equilibrium of the mechanical state of the body within the framework of the linearized strain theory. We use the symbols $\boldsymbol{u}$, $\boldsymbol{\sigma}$ and $\boldsymbol{\varepsilon} = \boldsymbol{\varepsilon}(\boldsymbol{u})$ for the displacement vector, the stress tensor and the linearized strain tensor, respectively. The components of the linearized strain tensor $\boldsymbol{\varepsilon}(\boldsymbol{u})$ are given by

$$\varepsilon_{ij}(\boldsymbol{u}) = (\boldsymbol{\varepsilon}(\boldsymbol{u}))_{ij} = \frac{1}{2}\left(u_{i,j} + u_{j,i}\right),$$

where $u_{i,j} = \partial u_i / \partial x_j$. These are functions of the spatial variable $\boldsymbol{x}$, and of the time variable t as well for time-dependent problems. Nevertheless, in what follows we do not indicate explicitly the dependence of these quantities on $\boldsymbol{x}$ and t: for example, we write $\boldsymbol{\sigma}$ instead of $\boldsymbol{\sigma}(\boldsymbol{x})$ or $\boldsymbol{\sigma}(\boldsymbol{x}, t)$. The displacement $\boldsymbol{u}$ and the stress $\boldsymbol{\sigma}$ play the roles of unknowns in the contact problems.

We let $\mathbb{S}^d$ denote the space of second-order symmetric tensors on $\mathbb{R}^d$. Equivalently, $\mathbb{S}^d$ can be viewed as the space of symmetric matrices of order d. The canonical inner products and the corresponding norms on $\mathbb{R}^d$ and $\mathbb{S}^d$ are given by

$$\boldsymbol{u} \cdot \boldsymbol{v} = u_i v_i, \quad \|\boldsymbol{v}\| = (\boldsymbol{v} \cdot \boldsymbol{v})^{1/2} \quad \text{for all } \boldsymbol{u} = (u_i), \boldsymbol{v} = (v_i) \in \mathbb{R}^d, \quad (2.1)$$

$$\boldsymbol{\sigma} \cdot \boldsymbol{\tau} = \sigma_{ij} \tau_{ij}, \quad \|\boldsymbol{\tau}\| = (\boldsymbol{\tau} \cdot \boldsymbol{\tau})^{1/2} \quad \text{for all } \boldsymbol{\sigma} = (\sigma_{ij}), \boldsymbol{\tau} = (\tau_{ij}) \in \mathbb{S}^d, \quad (2.2)$$

respectively. We use $\boldsymbol{0}$ for the zero element of the spaces $\mathbb{R}^d$ and $\mathbb{S}^d$.

Function spaces. We will use the standard notation for Sobolev and Lebesgue spaces over Ω or on Γ. In particular, we will use the spaces $L^2(\Omega; \mathbb{R}^d)$, $L^2(\Gamma_2; \mathbb{R}^d)$, $L^2(\Gamma_3; \mathbb{R}^d)$ and $H^1(\Omega; \mathbb{R}^d)$, endowed with their canonical inner products and associated norms. For an element $\boldsymbol{v} \in H^1(\Omega; \mathbb{R}^d)$ we write $\boldsymbol{v}$ for its trace $\gamma \boldsymbol{v} \in L^2(\Omega; \mathbb{R}^d)$ on Γ. Define the function spaces

$$V = \{\boldsymbol{v} \in H^1(\Omega; \mathbb{R}^d) \mid \boldsymbol{v} = \boldsymbol{0} \text{ on } \Gamma_1\}, \quad (2.3)$$

$$Q = \{\boldsymbol{\sigma} = (\sigma_{ij}) \mid \sigma_{ij} = \sigma_{ji} \in L^2(\Omega)\}, \quad (2.4)$$

for the displacement and the stress field, respectively. These are real Hilbert spaces endowed with the inner products

$$(\boldsymbol{u}, \boldsymbol{v})_V = \int_\Omega \boldsymbol{\varepsilon}(\boldsymbol{u}) \cdot \boldsymbol{\varepsilon}(\boldsymbol{v}) \, \mathrm{d}x, \quad (\boldsymbol{\sigma}, \boldsymbol{\tau})_Q = \int_\Omega \boldsymbol{\sigma} \cdot \boldsymbol{\tau} \, \mathrm{d}x.$$

The corresponding norms on the spaces are denoted by $\|\cdot\|_V$ and $\|\cdot\|_Q$, respectively. Since $\text{meas}\,(\Gamma_1) > 0$, Korn's inequality holds (see Nečas and Hlaváček 1981, p. 79):

$$\|\boldsymbol{v}\|_{H^1(\Omega; \mathbb{R}^d)} \leq c \, \|\boldsymbol{\varepsilon}(\boldsymbol{v})\|_Q \quad \text{for all } \boldsymbol{v} \in V,$$

for some constant $c > 0$. Thus, $\|\cdot\|_V$ defines a norm on V that is equivalent to the standard $H^1(\Omega;\mathbb{R}^d)$-norm. We will also use the function space

$$H = L^2(\Omega;\mathbb{R}^d) \tag{2.5}$$

with the canonical inner product and norm.

We let V^* denote the dual of the space V and $\langle\cdot,\cdot\rangle$ the corresponding duality pairing. For any element $\boldsymbol{v} \in V$, we let v_ν and $\boldsymbol{v}_\tau$ denote its normal and tangential components on Γ given by $v_\nu = \boldsymbol{v} \cdot \boldsymbol{\nu}$ and $\boldsymbol{v}_\tau = \boldsymbol{v} - v_\nu\boldsymbol{\nu}$, respectively. For a regular function $\boldsymbol{\sigma} : \overline{\Omega} \to \mathbb{S}^d$, we let σ_ν and $\boldsymbol{\sigma}_\tau$ denote its normal and tangential components on Γ, *i.e.* $\sigma_\nu = (\boldsymbol{\sigma}\boldsymbol{\nu}) \cdot \boldsymbol{\nu}$ and $\boldsymbol{\sigma}_\tau = \boldsymbol{\sigma}\boldsymbol{\nu} - \sigma_\nu\boldsymbol{\nu}$, and we recall that the following Green's formula holds:

$$\int_\Omega \boldsymbol{\sigma}\cdot\boldsymbol{\varepsilon}(\boldsymbol{v})\,\mathrm{d}x + \int_\Omega \mathrm{Div}\,\boldsymbol{\sigma}\cdot\boldsymbol{v}\,\mathrm{d}x = \int_\Gamma \boldsymbol{\sigma}\boldsymbol{\nu}\cdot\boldsymbol{v}\,\mathrm{d}a \quad \text{for all } \boldsymbol{v} \in H^1(\Omega,\mathbb{R}^d). \tag{2.6}$$

We also recall that

$$\|\boldsymbol{v}\|_{L^2(\Gamma;\mathbb{R}^d)} \leq \|\gamma\|\|\boldsymbol{v}\|_V \quad \text{for all } \boldsymbol{v} \in V, \tag{2.7}$$

where $\|\gamma\|$ represents the norm of the trace operator $\gamma : V \to L^2(\Gamma;\mathbb{R}^d)$. Inequality (2.7) represents a consequence of the Sobolev trace theorem.

We introduce a space of fourth-order tensors:

$$\mathbf{Q}_\infty = \{\mathcal{E} = (e_{ijkl}) \mid e_{ijkl} = e_{jikl} = e_{klij} \in L^\infty(\Omega),\ 1 \leq i,j,k,l \leq d\}. \tag{2.8}$$

It is a Banach space endowed with the norm

$$\|\mathcal{E}\|_{\mathbf{Q}_\infty} = \max_{1 \leq i,j,k,l \leq d} \|e_{ijkl}\|_{L^\infty(\Omega)}.$$

It is easy to see that

$$\|\mathcal{E}\boldsymbol{\tau}\|_Q \leq d\,\|\mathcal{E}\|_{\mathbf{Q}_\infty}\|\boldsymbol{\tau}\|_Q \quad \text{for all } \mathcal{E} \in \mathbf{Q}_\infty,\ \boldsymbol{\tau} \in Q. \tag{2.9}$$

Below, $\mathbb{N}$ represents the set of positive integers, I denotes either a bounded interval of the form $[0,T]$ with $T > 0$, or the unbounded interval $\mathbb{R}_+ = [0,+\infty)$ and X will be a Banach space. We let $C(I;X)$ denote the space of continuous functions on I with values in X. In addition, we let $C^1(I;X)$ denote the space of continuously differentiable functions on I with values in X. Therefore, $v \in C^1(I;X)$ if and only if $v \in C(I;X)$ and $\dot{v} \in C(I;X)$ where $\dot{v}$ represents the time derivative of the function v. In addition, for a subset $K \subset X$, we use the notation $C(I;K)$ for the set of functions defined on I with values in K.

In the case $I = [0,T]$ the space $C(I;X)$ is equipped with the norm

$$\|v\|_{C([0,T];X)} = \max_{t\in[0,T]} \|v(t)\|_X.$$

It is well known that $C(I;X)$ is a Banach space. In the case $I = \mathbb{R}_+$, $C(I;X)$ can be organized in a canonical way as a Fréchet space, *i.e.*, it is a complete metric space in which the corresponding topology is induced by a countable family of seminorms. The convergence of a sequence $\{v_k\}_k$ to an element v, in the space $C(\mathbb{R}_+; X)$, can be described as follows:

$$v_k \to v \text{ in } C(\mathbb{R}_+; X) \text{ as } k \to \infty \text{ if and only if}$$

$$\max_{r \in [0,n]} \|v_k(r) - v(r)\|_V \to 0 \text{ as } k \to \infty, \text{ for all } n \in \mathbb{N}.$$

In other words, the sequence $\{v_k\}_k$ converges to the element v in the space $C(\mathbb{R}_+; X)$ if and only if it converges to v in the space $C([0, n]; X)$ for all $n \in \mathbb{N}$.

Let $I = [0, T]$ and X be a Banach space. For $1 \le p \le \infty$, we define $L^p(I; X)$ to be the space of all measurable functions $v: I \to X$ such that $\|v\|_{L^p(I;X)} < \infty$, where the norm is defined by

$$\|v\|_{L^p(I;X)} = \begin{cases} \left(\int_I \|v(t)\|_X^p \, dt \right)^{1/p} & \text{if } 1 \le p < \infty, \\ \operatorname*{ess\,sup}_{t \in I} \|v(t)\|_X & \text{if } p = \infty. \end{cases}$$

The space $L^p(I; X)$ is a Banach space. When X is a Hilbert space with the inner product $(\cdot, \cdot)_X$ and $p = 2$, $L^2(I; X)$ is a Hilbert space with the inner product

$$(u, v) = \int_I (u(t), v(t))_X \, dt.$$

For $m \in \mathbb{N}$ and $1 \le p \le \infty$, we define the space

$$W^{m,p}(I; X) = \{v \in L^p(I; X) \mid \|v^{(i)}\|_{L^p(I;X)} < \infty, \ 0 \le i \le m\},$$

where $v^{(i)}(t)$ is the ith weak derivative of v with respect to t; the first two weak derivatives are usually also denoted as $\dot{v}, \ddot{v}$. For $1 \le p < \infty$, the norm in the space $W^{k,p}(I; X)$ is defined by

$$\|v\|_{W^{k,p}(I;X)} = \left(\int_I \sum_{0 \le i \le k} \|v^{(i)}(t)\|_X^p \, dt \right)^{1/p}.$$

For $p = \infty$, the norm is defined by

$$\|v\|_{W^{k,\infty}(I;X)} = \max_{0 \le i \le k} \operatorname*{ess\,sup}_{t \in I} \|v^{(i)}\|_X.$$

The space $W^{m,p}(I; X)$ is a Banach space. In the particular case where X is a Hilbert space with the inner product $(\cdot, \cdot)_X$ and $p = 2$, $W^{m,2}(I; X)$ is a Hilbert space, usually written as $H^m(I; X)$, with the inner product

$$(u, v)_{H^m(I;X)} = \sum_{0 \le i \le m} \int_I (u^{(i)}(t), v^{(i)}(t))_X \, dt.$$

With the time interval $I = [0, T]$ and the space V defined in (2.3), we will use the spaces

$$\mathcal{V} = L^2(I; V), \tag{2.10}$$
$$\mathcal{V}^* = L^2(I; V^*), \tag{2.11}$$
$$\mathcal{W} = \{w \in \mathcal{V} \mid \dot{w} \in \mathcal{V}^*\}. \tag{2.12}$$

Subdifferential boundary conditions. For the contact models, we will use contact laws expressed in terms of the subdifferential. Such conditions are of the form $\xi_\nu \in \partial j(u_\nu)$ in which ξ_ν represents an interface force, $u_\nu = \boldsymbol{u} \cdot \boldsymbol{\nu}$ denotes the normal displacement and ∂j represents the subdifferential in the sense of Clarke. The definition and some basic properties of this subdifferential for locally Lipschitz functions on Banach spaces will be provided in Section 3. Nevertheless, for the convenience of the reader, we introduce here the Clarke subdifferential for real-valued functions of a real variable, as we need it to describe the interface boundary conditions in this section.

Let $j \colon \mathbb{R} \to \mathbb{R}$ be a locally Lipschitz function. The generalized (Clarke) directional derivative of j at $x \in \mathbb{R}$ in the direction $v \in \mathbb{R}$ is defined by

$$j^0(x; v) := \limsup_{y \to x, \, \lambda \downarrow 0} \frac{j(y + \lambda v) - j(y)}{\lambda}.$$

The generalized subdifferential of j at x is a subset of $\mathbb{R}$ given by

$$\partial j(x) := \{\zeta \in \mathbb{R} \mid j^0(x; v) \geq \zeta v \text{ for all } v \in \mathbb{R}\}.$$

We are interested in functions j satisfying the following conditions.

$$\left.\begin{array}{l} \text{(a) } j \colon \mathbb{R} \to \mathbb{R} \text{ is locally Lipschitz.} \\[4pt] \text{(b) There exists a constant } c_j > 0 \text{ such that} \\[2pt] \quad |\partial j(r)| \leq c_j(1 + |r|) \text{ for all } r \in \mathbb{R}, \text{ all } \xi \in \partial j(r). \\[4pt] \text{(c) There exists a constant } \alpha_j \geq 0 \text{ such that} \\[2pt] \quad (\xi_1 - \xi_2) \cdot (r_1 - r_2) \geq -\alpha_j |r_1 - r_2|^2 \\[2pt] \quad \text{for all } r_1, r_2 \in \mathbb{R}, \ \xi_1, \xi_2 \in \mathbb{R}, \text{ with } \xi_i \in \partial j(r_i), \ i = 1, 2. \end{array}\right\} \tag{2.13}$$

Note that the inequality in (2.13(b)) means

$$|\xi| \leq c_j(1 + |r|) \quad \text{for all } r \in \mathbb{R}, \ \xi \in \partial j(r).$$

We shall use this kind of shorthand notation in various places in the paper. Moreover, note that condition (2.13(c)) is known as the relaxed monotonicity condition in the literature. It is equivalent to the condition

$$j^0(r_1; r_2 - r_1) + j^0(r_2; r_1 - r_2) \leq \alpha_j |r_1 - r_2|^2 \quad \text{for all } r_1, r_2 \in \mathbb{R}. \tag{2.14}$$

If $j \colon \mathbb{R} \to \mathbb{R}$ is convex, this condition is satisfied with $\alpha_j = 0$.

Below we present several examples of a function j satisfying the condition (2.13). These examples show that such a function is in general non-convex and its subdifferential can be multivalued with a non-monotone graph in $\mathbb{R}^2$.

Example 2.1. Let $\alpha > 0$ and $r_0 > 0$. Define $j : \mathbb{R} \to \mathbb{R}$ by

$$j(r) = \begin{cases} \frac{1}{2}\alpha r_0^2 - \alpha r_0(r + r_0) & \text{if } r < -r_0, \\ \frac{1}{2}\alpha r^2 & \text{if } |r| \le r_0, \\ \frac{1}{2}\alpha r_0^2 + \alpha r_0(r - r_0) & \text{if } r > r_0. \end{cases} \tag{2.15}$$

It is easy to see that j is a C^1 function and therefore condition (2.13(a)) is satisfied. From the formula

$$\partial j(r) = \begin{cases} -\alpha r_0 & \text{if } r < -r_0, \\ \alpha r & \text{if } |r| \le r_0, \\ \alpha r_0 & \text{if } r > r_0, \end{cases}$$

we see that $|\partial j(r)| \le \alpha r_0$ for all $r \in \mathbb{R}$. Hence, condition (2.13(b)) holds with $c_j = \alpha r_0$. Since ∂j is a monotone function, we deduce that condition (2.13(c)) holds with $\alpha_j = 0$.

Example 2.2. Let $\alpha \in [0, 1)$ and consider the function

$$j(r) = (\alpha - 1)\,\mathrm{e}^{-|r|} + \alpha|r| \quad \text{for all } r \in \mathbb{R}. \tag{2.16}$$

Then j satisfies condition (2.13(a)). It follows from the definition of the Clarke subdifferential that

$$\partial j(r) = \begin{cases} (\alpha - 1)\,\mathrm{e}^{r} - \alpha & \text{if } r < 0, \\ [-1, 1] & \text{if } r = 0, \\ (1 - \alpha)\,\mathrm{e}^{-r} + \alpha & \text{if } r > 0. \end{cases}$$

Thus, $|\xi| \le 1$ for all $\xi \in \partial j(r)$ and $r \in \mathbb{R}$, and condition (2.13(b)) holds with $c_j = 1$. A simple calculation shows that condition (2.13(c)) holds with $\alpha_j = 1$.

Example 2.3. Let $\alpha \ge 0$ and let $j : \mathbb{R} \to \mathbb{R}$ be defined by

$$j(r) = \begin{cases} 0 & \text{if } r < 0, \\ -\mathrm{e}^{-r} + \alpha r + 1 & \text{if } r \ge 0. \end{cases} \tag{2.17}$$

It is easy to see that j satisfies condition (2.13(a)). An elementary computation shows that

$$\partial j(r) = \begin{cases} 0 & \text{if } r < 0, \\ [0, 1 + \alpha] & \text{if } r = 0, \\ \mathrm{e}^{-r} + \alpha & \text{if } r > 0. \end{cases}$$

Thus, $|\xi| \le 1 + \alpha$ for all $\xi \in \partial j(r)$ and $r \in \mathbb{R}$. In other words, condition (2.13(b)) holds with $c_j = 1 + \alpha$. Condition (2.13(c)) holds with $\alpha_j = 1$.

Example 2.4. Let $p \colon \mathbb{R} \to \mathbb{R}$ be the function defined by

$$p(r) = \begin{cases} 0 & \text{if } r < 0, \\ r & \text{if } 0 \le r < 1, \\ 2 - r & \text{if } 1 \le r < 2, \\ \sqrt{r - 2} + r - 2 & \text{if } 2 \le r < 6, \\ r & \text{if } r \ge 6. \end{cases} \tag{2.18}$$

This function is continuous, yet it is neither monotone nor Lipschitz continuous. Define the function $j \colon \mathbb{R} \to \mathbb{R}$ by

$$j(r) = \int_0^r p(s)\, \mathrm{d}s \quad \text{for all } r \in \mathbb{R}. \tag{2.19}$$

Note that j is not convex. Since $j'(r) = p(r)$ for $r \in \mathbb{R}$, j is a C^1 function and thus is a locally Lipschitz function, that is, condition (2.13(a)) is satisfied. Since $|p(r)| \le |r|$ for $r \in \mathbb{R}$, (2.13(b)) is satisfied. Since the function $r \mapsto r + p(r) \in \mathbb{R}$ is non-decreasing,

$$(p(r_1) - p(r_2))(r_2 - r_1) \le (r_1 - r_2)^2 \quad \text{for all } r_1, r_2 \in \mathbb{R}.$$

We combine this inequality with the equality $j^0(r_1; r_2) = p(r_1)\, r_2$, valid for r_1, $r_2 \in \mathbb{R}$, to see that condition (2.14) is satisfied with $\alpha_j = 1$. Hence, j satisfies condition (2.13(c)).

2.1. A static frictional contact problem

For the first contact problem to be studied in this section we assume that the material is elastic, the process is static, the contact is with normal compliance and unilateral constraint (see Han, Sofonea and Barboteu 2017), associated with one version of Coulomb's law of dry friction (see Han, Migórski and Sofonea 2014). These mechanical assumptions lead to the following mathematical model.

Problem 2.5. Find a displacement field $\boldsymbol{u} \colon \Omega \to \mathbb{R}^d$, a stress field $\boldsymbol{\sigma} \colon \Omega \to \mathbb{S}^d$ and an interface function $\xi_\nu \colon \Gamma_3 \to \mathbb{R}$ such that

$$\boldsymbol{\sigma} = \mathcal{F}\varepsilon(\boldsymbol{u}) \quad \text{in } \Omega, \tag{2.20}$$

$$\operatorname{Div} \boldsymbol{\sigma} + \boldsymbol{f}_0 = \boldsymbol{0} \quad \text{in } \Omega, \tag{2.21}$$

$$\boldsymbol{u} = \boldsymbol{0} \quad \text{on } \Gamma_1, \tag{2.22}$$

$$\boldsymbol{\sigma}\boldsymbol{\nu} = \boldsymbol{f}_2 \quad \text{on } \Gamma_2, \tag{2.23}$$

$$u_\nu \leq g, \quad \sigma_\nu + \xi_\nu \leq 0, \quad (u_\nu - g)(\sigma_\nu + \xi_\nu) = 0, \\ \xi_\nu \in \partial j_\nu(u_\nu) \qquad \left.\right\} \quad \text{on } \Gamma_3, \qquad (2.24)$$

$$\|\boldsymbol{\sigma}_\tau\| \leq F_b(u_\nu), \quad -\boldsymbol{\sigma}_\tau = F_b(u_\nu)\frac{\boldsymbol{u}_\tau}{\|\boldsymbol{u}_\tau\|} \quad \text{if } \boldsymbol{u}_\tau \neq \boldsymbol{0} \quad \text{on } \Gamma_3. \qquad (2.25)$$

We now explain the equations and boundary conditions in Problem 2.5. Equation (2.20) represents the constitutive law of the material where $\mathcal{F} \colon \Omega \times \mathbb{S}^d \to \mathbb{S}^d$ is the elasticity operator, allowed to be nonlinear. For an isotropic linearly elastic material, the constitutive law (2.20) becomes

$$\boldsymbol{\sigma} = \lambda \operatorname{tr} \boldsymbol{\varepsilon}(\boldsymbol{u})\, \boldsymbol{I}_d + 2\,\mu\,\boldsymbol{\varepsilon}(\boldsymbol{u}),$$

where $\lambda > 0$ and $\mu > 0$ are the Lamé coefficients, and $\boldsymbol{I}_d \in \mathbb{S}^d$ is the identity tensor. Another example of elastic constitutive law of the form (2.20) is provided by

$$\boldsymbol{\sigma} = \mathcal{E}\boldsymbol{\varepsilon}(\boldsymbol{u}) + \beta\,(\boldsymbol{\varepsilon}(\boldsymbol{u}) - \mathcal{P}_K\boldsymbol{\varepsilon}(\boldsymbol{u})). \qquad (2.26)$$

Here $\mathcal{E}$ is a linear or nonlinear operator, $\beta > 0$, K is a closed convex subset of $\mathbb{S}^d$ such that $\boldsymbol{0} \in K$ and $\mathcal{P}_K : \mathbb{S}^d \to K$ denotes the projection operator. The corresponding elasticity operator is nonlinear and is given by

$$\mathcal{F}\boldsymbol{\varepsilon} = \mathcal{E}\boldsymbol{\varepsilon} + \beta\,(\boldsymbol{\varepsilon} - \mathcal{P}_K\boldsymbol{\varepsilon}). \qquad (2.27)$$

A common choice of the set K is

$$K = \{\boldsymbol{\varepsilon} \in \mathbb{S}^d \mid f(\boldsymbol{\varepsilon}) \leq 0\}, \qquad (2.28)$$

where $f : \mathbb{S}^d \to \mathbb{R}$ is a convex continuous function with $f(\boldsymbol{0}) < 0$.

Equation (2.21) is the equation of equilibrium, in which $\boldsymbol{f}_0$ represents the density of body forces. This equation is obtained from the equation of motion by neglecting the inertial term. Since the body is fixed on Γ_1, we impose the homogeneous displacement condition (2.22). If Γ_2 is non-empty, then the body is subject to a surface traction of density $\boldsymbol{f}_2$ and (2.23) represents the traction boundary condition.

Conditions (2.24) and (2.25) represent the frictional unilateral contact condition on the contact surface Γ_3. Condition (2.24) models the contact with a foundation made of a rigid body covered by a layer made of deformable material, say asperities, and is derived based on the following considerations.

First, the thickness of the deformable layer is described by the non-negative valued function g. Penetration of the elastic material to the foundation is allowed but, due to the presence of the rigid body, is limited by the unilateral constraint

$$u_\nu \leq g \quad \text{on } \Gamma_3. \qquad (2.29)$$

Next, we assume that the normal stress has an additive decomposition of the form

$$\sigma_\nu = \sigma_\nu^D + \sigma_\nu^R \quad \text{on } \Gamma_3, \tag{2.30}$$

in which σ_ν^D describes the reaction of the deformable material and σ_ν^R describes the reaction of the rigid body. We assume that σ_ν^D satisfies a normal compliance contact condition in a subdifferential form, that is,

$$-\sigma_\nu^D \in \partial j_\nu(u_\nu) \quad \text{on } \Gamma_3, \tag{2.31}$$

where ∂j_ν is the Clarke subdifferential of the potential functional j_ν, which is assumed to be locally Lipschitz. Examples, details and various mechanical interpretations of this condition can be found in Migórski, Ochal and Sofonea (2013). Here we merely mention that such a condition represents a generalization of the well-known normal compliance condition $-\sigma_\nu = p(u_\nu)$ introduced in Oden and Martins (1985). Note that in this condition $p : \mathbb{R} \to \mathbb{R}$ is usually assumed to be an increasing function, and therefore the contact condition is monotone and single-valued. In contrast, condition (2.31) describes a relation between the stress σ_ν^D and the normal displacement u_ν which can be non-monotone or multivalued.

The part σ_ν^R of the normal stress satisfies the Signorini condition in the form with the gap g, that is,

$$\sigma_\nu^R \leq 0, \quad \sigma_\nu^R(u_\nu - g) = 0 \quad \text{on } \Gamma_3. \tag{2.32}$$

The Signorini condition was introduced by Signorini (1933) to describe the contact with a rigid obstacle. It was used in a large number of papers as explained in Shillor, Sofonea and Telega (2004).

We denote $-\sigma_\nu^D = \xi_\nu$ and use (2.30) to see that

$$\sigma_\nu^R = \sigma_\nu + \xi_\nu \quad \text{on } \Gamma_3. \tag{2.33}$$

Then we substitute equality (2.33) in (2.32) and use (2.29), (2.31) to obtain the contact condition (2.24).

Finally, (2.25) represents a version of the static Coulomb law of dry friction. According to this law, the tangential traction $\boldsymbol{\sigma}_\tau$ is limited in size by F_b, the so-called friction bound, which is the maximal frictional resistance that the surface can generate, and once the friction bound is reached, a relative slip motion commences. When slip starts, the frictional resistance has magnitude F_b and acts in the direction opposite to the motion. In (2.25) we allow the friction bound to depend on the normal penetration, which appears to be reasonable from the physical point of view, as explained in Han, Migórski and Sofonea (2014). The original formulation by Coulomb was proposed for the description of contact between rigid bodies, and its use in the description of contact between deformable bodies in the pointwise sense is more recent. More comments on this matter can be found in Shillor, Sofonea and Telega (2004).

We now list the assumption on the problem data. For the elasticity operator $\mathcal{F}\colon \Omega \times \mathbb{S}^d \to \mathbb{S}^d$, assume:

(a) there exists a constant $L_{\mathcal{F}} > 0$ such that
$$\|\mathcal{F}(\boldsymbol{x}, \boldsymbol{\varepsilon}_1) - \mathcal{F}(\boldsymbol{x}, \boldsymbol{\varepsilon}_2)\| \le L_{\mathcal{F}}\|\boldsymbol{\varepsilon}_1 - \boldsymbol{\varepsilon}_2\|$$
for all $\boldsymbol{\varepsilon}_1, \boldsymbol{\varepsilon}_2 \in \mathbb{S}^d$, a.e. $\boldsymbol{x} \in \Omega$;

(b) there exists a constant $m_{\mathcal{F}} > 0$ such that
$$(\mathcal{F}(\boldsymbol{x}, \boldsymbol{\varepsilon}_1) - \mathcal{F}(\boldsymbol{x}, \boldsymbol{\varepsilon}_2)) \cdot (\boldsymbol{\varepsilon}_1 - \boldsymbol{\varepsilon}_2) \ge m_{\mathcal{F}}\|\boldsymbol{\varepsilon}_1 - \boldsymbol{\varepsilon}_2\|^2$$
for all $\boldsymbol{\varepsilon}_1, \boldsymbol{\varepsilon}_2 \in \mathbb{S}^d$, a.e. $\boldsymbol{x} \in \Omega$;

(c) $\mathcal{F}(\cdot, \boldsymbol{\varepsilon})$ is measurable on Ω for all $\boldsymbol{\varepsilon} \in \mathbb{S}^d$;

(d) $\mathcal{F}(\boldsymbol{x}, \boldsymbol{0}) = \boldsymbol{0}$ for a.e. $\boldsymbol{x} \in \Omega$.
$$\tag{2.34}$$

For the potential function $j_\nu \colon \Gamma_3 \times \mathbb{R} \to \mathbb{R}$, assume:

(a) $j_\nu(\cdot, r)$ is measurable on Γ_3 for all $r \in \mathbb{R}$ and there exists $\bar{e} \in L^2(\Gamma_3)$ such that $j_\nu(\cdot, \bar{e}(\cdot)) \in L^1(\Gamma_3)$;

(b) $j_\nu(\boldsymbol{x}, \cdot)$ is locally Lipschitz on $\mathbb{R}$ for a.e. $\boldsymbol{x} \in \Gamma_3$;

(c) $|\partial j_\nu(\boldsymbol{x}, r)| \le \bar{c}_0 + \bar{c}_1 |r|$ for a.e. $\boldsymbol{x} \in \Gamma_3$, for all $r \in \mathbb{R}$ with constants $\bar{c}_0, \bar{c}_1 \ge 0$;

(d) $j_\nu^0(\boldsymbol{x}, r_1; r_2 - r_1) + j_\nu^0(\boldsymbol{x}, r_2; r_1 - r_2) \le \alpha_{j_\nu} |r_1 - r_2|^2$ for a.e. $\boldsymbol{x} \in \Gamma_3$, all $r_1, r_2 \in \mathbb{R}$ with a constant $\alpha_{j_\nu} \ge 0$.
$$\tag{2.35}$$

For the friction bound $F_b \colon \Gamma_3 \times \mathbb{R} \to \mathbb{R}$, assume:

(a) there exists a constant $L_{F_b} > 0$ such that
$$|F_b(\boldsymbol{x}, r_1) - F_b(\boldsymbol{x}, r_2)| \le L_{F_b}|r_1 - r_2|$$
for all $r_1, r_2 \in \mathbb{R}$, a.e. $\boldsymbol{x} \in \Gamma_3$;

(b) $F_b(\cdot, r)$ is measurable on Γ_3 for all $r \in \mathbb{R}$;

(c) $F_b(\boldsymbol{x}, r) = 0$ for $r \le 0$, $F_b(\boldsymbol{x}, r) \ge 0$ for $r \ge 0$, a.e. $\boldsymbol{x} \in \Gamma_3$.
$$\tag{2.36}$$

For the densities of body forces and surface tractions, assume
$$\boldsymbol{f}_0 \in L^2(\Omega; \mathbb{R}^d), \quad \boldsymbol{f}_2 \in L^2(\Gamma_2; \mathbb{R}^d). \tag{2.37}$$

Finally, the bound g satisfies
$$g \in L^2(\Gamma_3), \quad g(\boldsymbol{x}) \ge 0 \quad \text{a.e. } \boldsymbol{x} \in \Gamma_3. \tag{2.38}$$

Problem 2.5 is the classical formulation of the problem, that is, the unknowns and the data are assumed to be smooth functions such that all the derivatives and all the relations are satisfied in the usual sense at each point. However, the frictional contact conditions introduce a mathematical difficulty since they are expressed in terms of non-differentiable functions and belong to the conditions dealt with in the part of mechanics called non-smooth mechanics. In general, the problem does not have a classical solution, that is, a solution which has all the necessary classical derivatives, and some of the conditions will be satisfied in a weak sense that has to be made precise. Moreover, the frictional contact conditions impose a ceiling on the regularity or smoothness of the solutions, even if all the problem data are smooth. This is in contrast with the usual boundary value problems of elliptic partial differential equations where higher regularity on the data leads to higher regularity for the solutions and represents a challenging feature in the analysis of contact problems.

Problem 2.5 is most naturally studied in a weak sense. The weak formulation of the problem is not only a mathematical necessity, but also very useful practically since it leads directly to efficient numerical methods to solve the problem.

To derive the weak formulation, we assume the classical formulation has a solution and all the functions involved are as smooth as is needed for the various mathematical operations to be justified, and so the derivation is formal. We shall return to this point once we obtain the weak formulation. We introduce the set of admissible displacements defined by

$$U = \{ \boldsymbol{v} \in V \mid v_\nu \le g \ \text{ a.e. on } \Gamma_3 \}, \tag{2.39}$$

where the space V is defined by (2.3).

Let $\boldsymbol{v} \in U$. We multiply the equation (2.21) by $(\boldsymbol{v} - \boldsymbol{u})$ and integrate over Ω:

$$- \int_\Omega \operatorname{Div} \boldsymbol{\sigma} \cdot (\boldsymbol{v} - \boldsymbol{u}) \, \mathrm{d}x = \int_\Omega \boldsymbol{f}_0 \cdot (\boldsymbol{v} - \boldsymbol{u}) \, \mathrm{d}x.$$

Apply Green's formula (2.6) to the integral on the left-hand side:

$$- \int_\Omega \operatorname{Div} \boldsymbol{\sigma} \cdot (\boldsymbol{v} - \boldsymbol{u}) \, \mathrm{d}x = - \int_\Gamma \boldsymbol{\sigma\nu} \cdot (\boldsymbol{v} - \boldsymbol{u}) \, \mathrm{d}a + \int_\Omega \boldsymbol{\sigma} \cdot (\boldsymbol{\varepsilon}(\boldsymbol{v}) - \boldsymbol{\varepsilon}(\boldsymbol{u})) \, \mathrm{d}x.$$

Then we split the surface integral into three sub-integrals on Γ_1, Γ_2 and Γ_3 and use the boundary condition (2.23) to deduce that

$$\int_\Omega \boldsymbol{\sigma} \cdot (\boldsymbol{\varepsilon}(\boldsymbol{v}) - \boldsymbol{\varepsilon}(\boldsymbol{u})) \, \mathrm{d}x = \int_\Omega \boldsymbol{f}_0 \cdot (\boldsymbol{v} - \boldsymbol{u}) \, \mathrm{d}x + \int_{\Gamma_1} \boldsymbol{\sigma\nu} \cdot (\boldsymbol{v} - \boldsymbol{u}) \, \mathrm{d}a$$

$$+ \int_{\Gamma_2} \boldsymbol{f}_2 \cdot (\boldsymbol{v} - \boldsymbol{u}) \, \mathrm{d}a + \int_{\Gamma_3} \boldsymbol{\sigma\nu} \cdot (\boldsymbol{v} - \boldsymbol{u}) \, \mathrm{d}a.$$

Next, using the identities

$$\boldsymbol{v} - \boldsymbol{u} = \boldsymbol{0} \quad \text{a.e. on } \Gamma_1,$$

$$\boldsymbol{\sigma}\boldsymbol{\nu} \cdot (\boldsymbol{v} - \boldsymbol{u}) = \sigma_\nu(v_\nu - u_\nu) + \boldsymbol{\sigma}_\tau \cdot (\boldsymbol{v}_\tau - \boldsymbol{u}_\tau) \quad \text{a.e. on } \Gamma_3,$$

we find that

$$\int_\Omega \boldsymbol{\sigma} \cdot (\boldsymbol{\varepsilon}(\boldsymbol{v}) - \boldsymbol{\varepsilon}(\boldsymbol{u}))\,\mathrm{d}x = \int_\Omega \boldsymbol{f}_0 \cdot (\boldsymbol{v} - \boldsymbol{u})\,\mathrm{d}x + \int_{\Gamma_2} \boldsymbol{f}_2 \cdot (\boldsymbol{v} - \boldsymbol{u})\,\mathrm{d}a$$

$$+ \int_{\Gamma_3} \sigma_\nu(v_\nu - u_\nu)\,\mathrm{d}a + \int_{\Gamma_3} \boldsymbol{\sigma}_\tau \cdot (\boldsymbol{v}_\tau - \boldsymbol{u}_\tau)\,\mathrm{d}a. \tag{2.40}$$

On the other hand, we use the identity

$$\sigma_\nu(v_\nu - u_\nu) = (\sigma_\nu + \xi_\nu)(v_\nu - g) + (\sigma_\nu + \xi_\nu)(g - u_\nu) - \xi_\nu(v_\nu - u_\nu)$$

combined with the contact boundary condition (2.24) and the definition of the subdifferential to see that

$$-\sigma_\nu(v_\nu - u_\nu) \leq j_\nu^0(u_\nu; v_\nu - u_\nu) \quad \text{a.e. on } \Gamma_3.$$

Therefore,

$$\int_{\Gamma_3} \sigma_\nu(v_\nu - u_\nu)\,\mathrm{d}a \geq - \int_{\Gamma_3} j_\nu^0(u_\nu; v_\nu - u_\nu)\,\mathrm{d}a. \tag{2.41}$$

From the friction law (2.25),

$$\boldsymbol{\sigma}_\tau \cdot (\boldsymbol{v}_\tau - \boldsymbol{u}_\tau) \geq F_b(u_\nu)(\|\boldsymbol{u}_\tau\| - \|\boldsymbol{v}_\tau\|) \quad \text{a.e. on } \Gamma_3.$$

Therefore

$$\int_{\Gamma_3} \boldsymbol{\sigma}_\tau \cdot (\boldsymbol{v}_\tau - \boldsymbol{u}_\tau)\,\mathrm{d}a \geq \int_{\Gamma_3} F_b(u_\nu)(\|\boldsymbol{u}_\tau\| - \|\boldsymbol{v}_\tau\|)\,\mathrm{d}a. \tag{2.42}$$

We now combine equality (2.40) with inequalities (2.41) and (2.42) to deduce that

$$\int_\Omega \boldsymbol{\sigma} \cdot (\boldsymbol{\varepsilon}(\boldsymbol{v}) - \boldsymbol{\varepsilon}(\boldsymbol{u}))\,\mathrm{d}x + \int_{\Gamma_3} F_b(u_\nu)(\|\boldsymbol{v}_\tau\| - \|\boldsymbol{u}_\tau\|)\,\mathrm{d}a$$

$$+ \int_{\Gamma_3} j_\nu^0(u_\nu; v_\nu - u_\nu)\,\mathrm{d}a \geq \int_\Omega \boldsymbol{f}_0 \cdot (\boldsymbol{v} - \boldsymbol{u})\,\mathrm{d}x + \int_{\Gamma_2} \boldsymbol{f}_2 \cdot (\boldsymbol{v} - \boldsymbol{u})\,\mathrm{d}a. \tag{2.43}$$

Finally, we substitute the constitutive law (2.20) in (2.43) to obtain the following weak formulation of Problem 2.5, in terms of the displacement.

Problem 2.6. Find a displacement field $\boldsymbol{u} \in U$ such that

$$
\int_\Omega \mathcal{F}\varepsilon(\boldsymbol{u}) \cdot (\varepsilon(\boldsymbol{v}) - \varepsilon(\boldsymbol{u})) \, \mathrm{d}x
$$
$$
+ \int_{\Gamma_3} F_b(u_\nu)(\|\boldsymbol{v}_\tau\| - \|\boldsymbol{u}_\tau\|) \, \mathrm{d}a + \int_{\Gamma_3} j_\nu^0(u_\nu; v_\nu - u_\nu) \, \mathrm{d}a
$$
$$
\geq \int_\Omega \boldsymbol{f}_0 \cdot (\boldsymbol{v} - \boldsymbol{u}) \, \mathrm{d}x + \int_{\Gamma_2} \boldsymbol{f}_2 \cdot (\boldsymbol{v} - \boldsymbol{u}) \, \mathrm{d}a \quad \text{for all } \boldsymbol{v} \in U. \quad (2.44)
$$

Thus, if Problem 2.5 has a sufficiently smooth solution for all of the operations above to be justified, it is also a solution of Problem 2.6. However, now we have a weak formulation, Problem 2.6, that may have solutions which do not have the necessary regularity or smoothness, and we call them weak solutions of the original problem. This shows why it is necessary to derive and study weak formulations. This also indicates that once the existence of a weak solution is established, there is considerable interest in establishing its regularity, since if the weak solution is sufficiently smooth, then it is also a classical solution.

We note in passing that even if a problem possesses smooth or classical solutions, the weak formulation is usually the first step in its analysis, since many of the modern mathematical tools are better suited for such a formulation. Moreover, the weak formulation can often be employed directly in the finite element method for numerical approximations of the problem.

The inequality (2.44) involves both convex and locally Lipschitz functions. Moreover, it is time-independent and the corresponding differential operator is elliptic. For this reason, we refer to this inequality as an elliptic variational–hemivariational inequality. The well-posedness and numerical analysis of Problem 2.6 will be provided in Section 7.

2.2. A history-dependent frictionless contact problem

The model presented in this subsection is time-dependent. We assume that the inertial term in the equation of motion can be neglected, that is, the process is quasistatic. We use a viscoelastic constitutive law with long memory and we assume that the contact is frictionless. Moreover, we use the time-dependent version of the contact condition with normal compliance and unilateral constraints in Problem 2.5. Let I be the time interval of interest, which can be either bounded (*i.e.* $I = [0, T]$ with $T > 0$) or unbounded (*i.e.* $I = [0, +\infty)$). The classical formulation of the problem is as follows.

Problem 2.7. Find a displacement field $\boldsymbol{u}\colon \Omega \times I \to \mathbb{R}^d$, a stress field $\boldsymbol{\sigma}\colon \Omega \times I \to \mathbb{S}^d$ and an interface function $\xi_\nu\colon \Gamma_3 \times I \to \mathbb{R}$ such that

$$\boldsymbol{\sigma}(t) = \mathcal{F}\boldsymbol{\varepsilon}(\boldsymbol{u}(t)) + \int_0^t \mathcal{B}(t-s)\boldsymbol{\varepsilon}(\boldsymbol{u}(s))\,\mathrm{d}s \quad \text{in } \Omega, \qquad (2.45)$$

$$\operatorname{Div}\boldsymbol{\sigma}(t) + \boldsymbol{f}_0(t) = \boldsymbol{0} \quad \text{in } \Omega, \qquad (2.46)$$

$$\boldsymbol{u}(t) = \boldsymbol{0} \quad \text{on } \Gamma_1, \qquad (2.47)$$

$$\boldsymbol{\sigma}(t)\boldsymbol{\nu} = \boldsymbol{f}_2(t) \quad \text{on } \Gamma_2, \qquad (2.48)$$

$$\left.\begin{array}{l} u_\nu(t) \le g, \quad \sigma_\nu(t) + \xi_\nu(t) \le 0, \\[2mm] (u_\nu(t) - g)(\sigma_\nu(t) + \xi_\nu(t)) = 0, \quad \xi_\nu(t) \in \partial j_\nu(u_\nu(t)) \end{array}\right\} \quad \text{on } \Gamma_3, \qquad (2.49)$$

$$\boldsymbol{\sigma}_\tau(t) = \boldsymbol{0} \quad \text{on } \Gamma_3, \qquad (2.50)$$

for all $t \in I$.

Equation (2.45) represents the viscoelastic constitutive law in which $\mathcal{F}$ is the elasticity operator, assumed to satisfy condition (2.34), and $\mathcal{B}$ is the relaxation tensor, assumed to have the regularity

$$\mathcal{B} \in C(I; \mathbf{Q}_\infty). \qquad (2.51)$$

Various results, examples and mechanical interpretations in the study of viscoelastic materials of the form (2.45) can be found in Drozdov (1996). Note that for such a constitutive law the current value of the stress may depend on all the strain values up to the current time; therefore, incorporating this constitutive law in the contact model makes the problem history-dependent.

Equation (2.46) is the equilibrium equation, since the process is assumed quasistatic. Conditions (2.47) and (2.48) represent the displacement and the traction boundary condition, respectively. Condition (2.49) is a time-dependent version of contact law used in Problem 2.5. Finally, condition (2.50) represents the frictionless condition. It shows that the friction force, $\boldsymbol{\sigma}_\tau$, vanishes during the contact process. This is an idealization of the process, since even completely lubricated surfaces generate shear resistance to tangential motion. However, the frictionless condition (2.50) is a sufficiently good approximation of the reality in some situations, especially when the contact surfaces are lubricated. For this reason it has been used in several publications: see Shillor, Sofonea and Telega (2004).

In the study of Problem 2.7 we assume that the potential function j_ν satisfies condition (2.35), the bound g satisfies condition (2.38), and the densities of body forces and surface tractions have the regularity

$$\boldsymbol{f}_0 \in C(I; L^2(\Omega; \mathbb{R}^d)), \quad \boldsymbol{f}_2 \in C(I; L^2(\Gamma_2; \mathbb{R}^d)). \qquad (2.52)$$

We use the space V, (2.3), and the set of admissible displacement fields U, (2.39). Then, the weak formulation of Problem 2.7, obtained by using arguments similar to those used in the previous subsection, is as follows.

Problem 2.8. Find a displacement field $\boldsymbol{u} \colon I \to U$ such that

$$\int_\Omega \mathcal{F}\varepsilon(\boldsymbol{u}(t)) \cdot (\varepsilon(\boldsymbol{v}) - \varepsilon(\boldsymbol{u}(t)))\, \mathrm{d}x$$

$$+ \int_\Omega \left(\int_0^t \mathcal{B}(t-s)\, \varepsilon(\boldsymbol{u}(s))\, \mathrm{d}s \right) \cdot (\varepsilon(\boldsymbol{v}) - \varepsilon(\boldsymbol{u}(t)))\, \mathrm{d}x$$

$$+ \int_{\Gamma_3} j_\nu^0(u_\nu(t); v_\nu - u_\nu(t))\, \mathrm{d}a$$

$$\geq \int_\Omega \boldsymbol{f}_0(t) \cdot (\boldsymbol{v} - \boldsymbol{u}(t))\, \mathrm{d}x + \int_{\Gamma_2} \boldsymbol{f}_2(t) \cdot (\boldsymbol{v} - \boldsymbol{u}(t))\, \mathrm{d}a \qquad (2.53)$$

for all $\boldsymbol{v} \in U$, $t \in I$.

Note that inequality (2.53) involves both convex and locally Lipschitz functions. Further, it is time-dependent and it includes a history-dependent term. For this reason, we refer to this inequality as a history-dependent variational–hemivariational inequality. The well-posedness and numerical analysis of Problem 2.8 will be provided in Section 8.

2.3. A dynamic frictional contact problem

For the last model, the process is assumed to be dynamic and is considered in a finite time interval $I = [0, T]$, $T > 0$. The constitutive law is viscoelastic with short memory and the contact is described by a non-monotone version of the normal damped response condition, associated with a subdifferential friction law. The classical formulation of the problem is as follows.

Problem 2.9. Find a displacement field $\boldsymbol{u} \colon \Omega \times [0, T] \to \mathbb{R}^d$ and a stress field $\boldsymbol{\sigma} \colon \Omega \times [0, T] \to \mathbb{S}^d$ such that

$$\boldsymbol{\sigma}(t) = \mathcal{A}\varepsilon(\dot{\boldsymbol{u}}(t)) + \mathcal{B}\varepsilon(\boldsymbol{u}(t)) \quad \text{in } \Omega, \qquad (2.54)$$

$$\rho\,\ddot{\boldsymbol{u}}(t) = \operatorname{Div} \boldsymbol{\sigma}(t) + \boldsymbol{f}_0(t) \quad \text{in } \Omega, \qquad (2.55)$$

$$\boldsymbol{u}(t) = \boldsymbol{0} \quad \text{on } \Gamma_1, \qquad (2.56)$$

$$\boldsymbol{\sigma}(t)\boldsymbol{\nu} = \boldsymbol{f}_2(t) \quad \text{on } \Gamma_2, \qquad (2.57)$$

$$-\sigma_\nu(t) \in \partial j_\nu(\dot{u}_\nu(t)) \quad \text{on } \Gamma_3, \qquad (2.58)$$

$$-\boldsymbol{\sigma}_\tau(t) \in \partial j_\tau(\dot{\boldsymbol{u}}_\tau(t)) \quad \text{on } \Gamma_3, \qquad (2.59)$$

for all $t \in [0, T]$, and

$$\boldsymbol{u}(0) = \boldsymbol{u}_0, \quad \dot{\boldsymbol{u}}(0) = \boldsymbol{w}_0 \quad \text{in } \Omega. \qquad (2.60)$$

A brief description of equations and boundary conditions in Problem 2.9 follows.

First, equation (2.54) is the constitutive law for viscoelastic materials with short memory in which $\mathcal{A}$ represents the viscosity operator and $\mathcal{B}$ represents the elasticity operator. In the linear case, the constitutive law (2.54) becomes the well-known Kelvin–Voigt law

$$\sigma_{ij} = a_{ijkl}\varepsilon_{kl}(\dot{\boldsymbol{u}}) + b_{ijkl}\varepsilon_{kl}(\boldsymbol{u}), \tag{2.61}$$

where a_{ijkl} represent the components of the viscosity tensor $\mathcal{A}$ and b_{ijkl} are the components of the elasticity tensor $\mathcal{B}$. Quasistatic contact problems for viscoelastic materials of the form (2.54) have been considered in Han and Sofonea (2002), Shillor, Sofonea and Telega (2004) and Sofonea and Matei (2012), in the context of variational inequalities. Numerical analysis of variational inequalities for such contact models can be found in Han and Sofonea (2002).

Equation (2.55) is the equation for motion in which ρ denotes the density of mass. Note that, for simplicity, we assume that ρ does not depend on $\boldsymbol{x} \in \Omega$. On Γ_1, we have the clamped boundary condition (2.56) and, on Γ_2, the surface traction boundary condition (2.57). Relation (2.58) represents the contact condition with normal damped response, in which ∂j_ν denotes the Clarke subdifferential of the given function j_ν. Such a contact condition was used in Han and Sofonea (2002), Shillor, Sofonea and Telega (2004) and Migórski, Ochal and Sofonea (2013) in order to model the setting when the foundation is covered with a thin lubricant layer such as oil. Condition (2.59) represents a friction law, written in a general subdifferential form. A particular example of the friction law is given by (2.59) with

$$j_\tau(\boldsymbol{\xi}) = \int_0^{\|\boldsymbol{\xi}\|} \mu(s)\,\mathrm{d}s \quad \text{for all } \boldsymbol{\xi} \in \mathbb{R}^d, \tag{2.62}$$

for a suitable friction bound $\mu\colon [0, +\infty) \to \mathbb{R}_+$. For this choice of $j_\tau(\cdot)$, (2.59) represents the subdifferential form of the friction law

$$\|\boldsymbol{\sigma}_\tau\| \leq \mu(\|\dot{\boldsymbol{u}}_\tau\|), \quad -\boldsymbol{\sigma}_\tau(t) = \mu(\|\dot{\boldsymbol{u}}_\tau\|)\frac{\dot{\boldsymbol{u}}_\tau(t)}{\|\dot{\boldsymbol{u}}_\tau(t)\|} \text{ if } \dot{\boldsymbol{u}}_\tau(t) \neq \boldsymbol{0} \quad \text{on } \Gamma_3, \tag{2.63}$$

which will be used in Section 9.3 of this paper. Additional examples of friction laws of the form (2.59) can be found in Migórski, Ochal and Sofonea (2013), for instance.

Finally, conditions (2.60) are the initial conditions in which $\boldsymbol{u}_0$ and $\boldsymbol{w}_0$ represent the initial displacement and the initial velocity, respectively.

We now introduce the assumptions of the data of Problem 2.9.

For the viscosity operator $\mathcal{A}\colon \Omega \times \mathbb{S}^d \to \mathbb{S}^d$, we assume:

$$
\left.
\begin{aligned}
&\text{(a) there exists a constant } L_{\mathcal{A}} > 0 \text{ such that}\\
&\qquad \|\mathcal{A}(\boldsymbol{x}, \varepsilon_1) - \mathcal{A}(\boldsymbol{x}, \varepsilon_2)\| \le L_{\mathcal{A}} \|\varepsilon_1 - \varepsilon_2\|\\
&\qquad \text{for all } \varepsilon_1, \varepsilon_2 \in \mathbb{S}^d, \text{ a.e. } \boldsymbol{x} \in \Omega;\\[4pt]
&\text{(b) there exists a constant } m_{\mathcal{A}} > 0 \text{ such that}\\
&\qquad (\mathcal{A}(\boldsymbol{x}, \varepsilon_1) - \mathcal{A}(\boldsymbol{x}, \varepsilon_2)) \cdot (\varepsilon_1 - \varepsilon_2) \ge m_{\mathcal{A}} \|\varepsilon_1 - \varepsilon_2\|^2\\
&\qquad \text{for all } \varepsilon_1, \varepsilon_2 \in \mathbb{S}^d, \text{ a.e. } \boldsymbol{x} \in \Omega;\\[4pt]
&\text{(c) } \mathcal{A}(\cdot, \varepsilon) \text{ is measurable on } \Omega \text{ for all } \varepsilon \in \mathbb{S}^d;\\[4pt]
&\text{(d) } \mathcal{A}(\boldsymbol{x}, \boldsymbol{0}) = \boldsymbol{0} \text{ a.e. } \boldsymbol{x} \in \Omega.
\end{aligned}
\right\} \tag{2.64}
$$

For the elasticity operator $\mathcal{B}\colon \Omega \times \mathbb{S}^d \to \mathbb{S}^d$, we assume:

$$
\left.
\begin{aligned}
&\text{(a) there exists a constant } L_{\mathcal{B}} > 0 \text{ such that}\\
&\qquad \|\mathcal{B}(\boldsymbol{x}, \varepsilon_1) - \mathcal{B}(\boldsymbol{x}, \varepsilon_2)\| \le L_{\mathcal{B}} \|\varepsilon_1 - \varepsilon_2\|\\
&\qquad \text{for all } \varepsilon_1, \varepsilon_2 \in \mathbb{S}^d, \text{ a.e. } \boldsymbol{x} \in \Omega;\\[4pt]
&\text{(b) } \mathcal{B}(\cdot, \varepsilon) \text{ is measurable on } \Omega \text{ for all } \varepsilon \in \mathbb{S}^d;\\[4pt]
&\text{(c) } \mathcal{B}(\boldsymbol{x}, \boldsymbol{0}) = \boldsymbol{0} \text{ a.e. } \boldsymbol{x} \in \Omega.
\end{aligned}
\right\} \tag{2.65}
$$

For the normal potential function $j_\nu\colon \Gamma_3 \times \mathbb{R} \to \mathbb{R}$, we assume:

$$
\left.
\begin{aligned}
&\text{(a) } j_\nu(\cdot, r) \text{ is measurable on } \Gamma_3 \text{ for all } r \in \mathbb{R} \text{ and there}\\
&\qquad \text{exists } \bar{e}_\nu \in L^2(\Gamma_3) \text{ such that } j_\nu(\cdot, \bar{e}_\nu(\cdot)) \in L^1(\Gamma_3);\\[4pt]
&\text{(b) } j_\nu(\boldsymbol{x}, \cdot) \text{ is locally Lipschitz on } \mathbb{R} \text{ for a.e. } \boldsymbol{x} \in \Gamma_3;\\[4pt]
&\text{(c) } |\partial j_\nu(\boldsymbol{x}, r)| \le \bar{c}_{0\nu} \text{ for a.e. } \boldsymbol{x} \in \Gamma_3, \text{ for } r \in \mathbb{R} \text{ with } \bar{c}_{0\nu} \ge 0;\\[4pt]
&\text{(d) } j_\nu^0(\boldsymbol{x}, r_1; r_2 - r_1) + j_\nu^0(\boldsymbol{x}, r_2; r_1 - r_2) \le \alpha_{j_\nu} |r_1 - r_2|^2\\
&\qquad \text{for a.e. } \boldsymbol{x} \in \Gamma_3, \text{ all } r_1, r_2 \in \mathbb{R} \text{ with } \alpha_{j_\nu} \ge 0.
\end{aligned}
\right\} \tag{2.66}
$$

For the tangential potential function $j_\tau\colon \Gamma_3 \times \mathbb{R}^d \to \mathbb{R}$, we assume:

$$
\left.
\begin{aligned}
&\text{(a) } j_\tau(\cdot, \boldsymbol{\xi}) \text{ is measurable on } \Gamma_3 \text{ for all } \boldsymbol{\xi} \in \mathbb{R}^d \text{ and there}\\
&\qquad \text{exists } \bar{\boldsymbol{e}}_\tau \in L^2(\Gamma_3)^d \text{ such that } j_\tau(\cdot, \bar{\boldsymbol{e}}_\tau(\cdot)) \in L^1(\Gamma_3);\\[4pt]
&\text{(b) } j_\tau(\boldsymbol{x}, \cdot) \text{ is locally Lipschitz on } \mathbb{R}^d \text{ for a.e. } \boldsymbol{x} \in \Gamma_3;\\[4pt]
&\text{(c) } |\partial j_\tau(\boldsymbol{x}, \boldsymbol{\xi})| \le \bar{c}_{0\tau} \text{ for a.e. } \boldsymbol{x} \in \Gamma_3, \text{ for } \boldsymbol{\xi} \in \mathbb{R}^d \text{ with } \bar{c}_{0\tau} \ge 0;\\[4pt]
&\text{(d) } j_\tau^0(\boldsymbol{x}, \boldsymbol{\xi}_1; \boldsymbol{\xi}_2 - \boldsymbol{\xi}_1) + j_\tau^0(\boldsymbol{x}, \boldsymbol{\xi}_2; \boldsymbol{\xi}_1 - \boldsymbol{\xi}_2) \le \alpha_{j_\tau} \|\boldsymbol{\xi}_1 - \boldsymbol{\xi}_2\|^2\\
&\qquad \text{for a.e. } \boldsymbol{x} \in \Gamma_3, \text{ all } \boldsymbol{\xi}_1, \boldsymbol{\xi}_2 \in \mathbb{R}^d \text{ with } \alpha_{j_\tau} \ge 0.
\end{aligned}
\right\} \tag{2.67}
$$

For the mass density assume it is a positive constant,

$$\rho > 0. \tag{2.68}$$

For the densities of body forces and surface tractions, assume

$$\boldsymbol{f}_0 \in L^2(0, T; L^2(\Omega; \mathbb{R}^d)), \quad \boldsymbol{f}_2 \in L^2(0, T; L^2(\Gamma_2; \mathbb{R}^d)), \tag{2.69}$$

and for the initial data,

$$\boldsymbol{u}_0 \in V, \quad \boldsymbol{w}_0 \in H. \tag{2.70}$$

Recall that the spaces V and H are defined by (2.3) and (2.5), respectively. Then, the weak formulation of Problem 2.9, obtained by using arguments similar to those used in Section 2.1, is as follows.

Problem 2.10. Find a displacement field $\boldsymbol{u} \colon [0, T] \to V$ such that

$$\int_\Omega \rho\, \ddot{\boldsymbol{u}}(t) \cdot \boldsymbol{v}\, \mathrm{d}x + \int_\Omega \mathcal{A}\varepsilon(\dot{\boldsymbol{u}}(t)) \cdot \varepsilon(\boldsymbol{v})\, \mathrm{d}x + \int_\Omega \mathcal{B}\varepsilon(\boldsymbol{u}(t)) \cdot \varepsilon(\boldsymbol{v})\, \mathrm{d}x$$

$$+ \int_{\Gamma_3} j_\nu^0(\dot{u}_\nu(t); v_\nu)\, \mathrm{d}a + \int_{\Gamma_3} j_\tau^0(\dot{\boldsymbol{u}}_\tau(t); \boldsymbol{v}_\tau)\, \mathrm{d}a$$

$$\geq \int_\Omega \boldsymbol{f}_0(t) \cdot \boldsymbol{v}\, \mathrm{d}x + \int_{\Gamma_2} \boldsymbol{f}_2(t) \cdot \boldsymbol{v}\, \mathrm{d}a \tag{2.71}$$

for all $\boldsymbol{v} \in V$, $t \in [0, T]$, and

$$\boldsymbol{u}(0) = \boldsymbol{u}_0, \quad \dot{\boldsymbol{u}}(0) = \boldsymbol{w}_0. \tag{2.72}$$

Note that inequality (2.71) involves the second time derivative of the unknown function $\boldsymbol{u}$. Consequently, the initial conditions (2.72) are introduced. We refer to such an inequality as an evolutionary hemivariational inequality. The well-posedness analysis and numerical approximations of Problem 2.6 will be provided in Section 9.

3. Preliminaries

As shown in the previous section, the weak formulations of contact problems lead to hemivariational or variational–hemivariational inequalities in which the unknown is the displacement field. In this section we present preliminary material needed to obtain an abstract existence and uniqueness result in the study of such inequalities. In this way we lay the background necessary in the well-posedness analysis and numerical approximation of the corresponding contact models. The preliminaries presented here include basic notions and results on non-smooth analysis, convex subdifferentials for convex functions, Clarke subdifferentials for locally Lipschitz functions, surjectivity for pseudomonotone multivalued operators, fixed-point theorems, and Gronwall inequalities. The material can be found in many textbooks and monographs. For this reason we present the theorems and propositions below without proofs and refer the reader to appropriate references.

Unless stated otherwise, in the rest of this section we assume X is a real reflexive Banach space. We let $\|\cdot\|_X$ denote its norm, X^* its topological dual, and $\langle\cdot,\cdot\rangle$ the canonical duality pairing between X^* and X. Moreover, 2^{X^*} denotes the set of all subsets of X^* and X^*_{w*} represents the space X^* equipped with the weak* topology. Finally, we let $\mathcal{L}(X,Y)$ denote the space of linear continuous operators defined on X with values in the normed space Y, equipped with the canonical norm $\|\cdot\|_{\mathcal{L}(X,Y)}$.

3.1. Basics on non-smooth analysis

Nonlinear operators of monotone type. We start with some definitions concerning single-valued operators.

Definition 3.1. An operator $A\colon X \to X^*$ is called:

(a) *bounded* if A maps bounded sets of X into bounded sets of X^*;

(b) *hemicontinuous* if, for all $u, v, w \in X$, the function $t \mapsto \langle A(u + t\, v), w\rangle$ is continuous for $t \in [0, 1]$;

(c) *monotone* if $\langle Au - Av, u - v\rangle \geq 0$ for all $u, v \in X$;

(d) *maximal monotone* if it is monotone and $\langle w - Av, u - v\rangle \geq 0$ for any $v \in X$ implies that $w = Au$;

(e) *u_0-coercive* if there exists a function $\alpha\colon \mathbb{R}_+ \to \mathbb{R}$ with $\lim_{r\to+\infty}\alpha(r) = +\infty$ such that $\langle Au, u - u_0\rangle \geq \alpha(\|u\|_X)\,\|u\|_X$ for all $u \in X$, where u_0 is a given element in X;

(f) *pseudomonotone* if it is bounded and $u_n \to u$ weakly in X with $\limsup \langle Au_n, u_n - u\rangle \leq 0$ imply $\liminf \langle Au_n, u_n - v\rangle \geq \langle Au, u - v\rangle$ for all $v \in X$;

(g) *radially continuous* if, for any $u, v \in X$, the function $t \mapsto \langle A(u+t\,v), v\rangle$ is continuous on $[0, 1]$.

Remark 3.2. It can be proved that an operator $A\colon X \to X^*$ is pseudomonotone if and only if it is bounded and $u_n \to u$ weakly in X together with $\limsup \langle Au_n, u_n - u\rangle \leq 0$ imply $\lim \langle Au_n, u_n - u\rangle = 0$ and $Au_n \to Au$ weakly in X^*. $\qquad\square$

We note that if $A\colon X \to X^*$ is continuous, then it is hemicontinuous. The following result can be deduced from Zeidler (1990, Proposition 27.6).

Proposition 3.3. If $A\colon X \to X^*$ is bounded, hemicontinuous and monotone, then it is pseudomonotone.

Next, we move to multivalued operators defined on the space X. Given a multivalued operator $T\colon X \to 2^{X^*}$, its domain $\mathcal{D}(T)$, range $\mathcal{R}(T)$ and graph $\mathcal{G}(T)$ are the sets defined by

$$\mathcal{D}(T) = \{x \in X \mid Tx \neq \emptyset\},$$
$$\mathcal{R}(T) = \bigcup \{Tx \mid x \in X\},$$
$$\mathcal{G}(T) = \{(x, x^*) \in X \times X^* \mid x^* \in Tx\},$$

respectively. We proceed with the following definitions.

Definition 3.4. An operator $T\colon X \to 2^{X^*}$ is called:

(a) *bounded* if the range of any bounded set in X is a bounded set in X^*;

(b) *monotone* if $\langle u^* - v^*, u - v \rangle \geq 0$ for all (u, u^*), $(v, v^*) \in \mathcal{G}(T)$;

(c) *maximal monotone* if it is monotone and maximal in the sense of inclusion of graphs in the family of monotone operators from X to 2^{X^*};

(d) u_0-*coercive* if there exists a function $\alpha\colon \mathbb{R}_+ \to \mathbb{R}$ with $\lim_{r \to +\infty} \alpha(r) = +\infty$ such that for all $(u, u^*) \in \mathcal{G}(T)$ we have

$$\langle u^*, u - u_0 \rangle \geq \alpha(\|u\|_X)\, \|u\|_X,$$

where u_0 is a given element in X;

(e) *generalized pseudomonotone* if, for any sequences $\{u_n\} \subset X$ and $\{u_n^*\} \subset X^*$ such that $u_n \to u$ weakly in X, $u_n^* \in Tu_n$ for all $n \in \mathbb{N}$, $u_n^* \to u^*$ weakly in X^* and $\limsup \langle u_n^*, u_n - u \rangle \leq 0$, we have $u^* \in Tu$ and

$$\lim \langle u_n^*, u_n \rangle = \langle u^*, u \rangle.$$

Given $u_0 \in X$ and $T\colon X \to 2^{X^*}$, we define a multivalued operator T_{u_0} by $T_{u_0}(v) = T(v + u_0)$ for $v \in X$. The following surjectivity result concerns the sum of a generalized pseudomonotone operator and a maximal monotone one. Its proof can be found in Naniewicz and Panagiotopoulos (1995).

Theorem 3.5. Let X be a reflexive Banach space, $T_1\colon X \to 2^{X^*}$, $T_2\colon X \to 2^{X^*}$ and $u_0 \in X$. Assume

(a) T_1 is bounded, generalized pseudomonotone and u_0-coercive;

(b) $T_1 u$ is a non-empty, closed and convex subset of X^*, for all $u \in X$;

(c) T_2 is a maximal monotone operator, and $u_0 \in \mathcal{D}(T_2)$.

Then $T_1 + T_2\colon X \to 2^{X^*}$ is surjective, *i.e.* $\mathcal{R}(T_1 + T_2) = X^*$.

Theorem 3.5 will be employed in the proof of the solvability of elliptic variational–hemivariational inequalities on the space X.

Convex functions. Let $\varphi\colon X \to \mathbb{R} \cup \{+\infty\}$. The effective domain of φ is the set $\operatorname{dom}\varphi = \{x \in X \mid \varphi(x) < +\infty\}$. We say that φ is convex if, for all $x,\,y \in \operatorname{dom}\varphi$ and all $\lambda \in (0,1)$, we have

$$\varphi(\lambda x + (1 - \lambda)y) \le \lambda\varphi(x) + (1 - \lambda)\varphi(y).$$

The function φ is said to be *proper* if $\operatorname{dom}\varphi \ne \emptyset$. The function $\varphi\colon X \to \mathbb{R} \cup \{+\infty\}$ is lower semicontinuous (l.s.c.) if, for any $u \in X$ and for any sequence $\{u_n\} \subset X$ such that $u_n \to u$ in X, we have

$$\liminf \varphi(u_n) \ge \varphi(u).$$

Next, we recall the notion of the subdifferential of a convex function.

Definition 3.6.   Let $\varphi\colon X \to \mathbb{R} \cup \{+\infty\}$ be a proper and convex function. The mapping $\partial_c\varphi\colon X \to 2^{X^*}$ defined by

$$\partial_c\varphi(x) = \big\{x^* \in X^* \mid \langle x^*, v - x\rangle \le \varphi(v) - \varphi(x) \text{ for all } v \in X\big\}$$

is called the *convex subdifferential* of φ. If $\partial_c\varphi(x)$ is non-empty, any element $x^* \in \partial_c\varphi(x)$ is called a *subgradient* of φ at x.

It is easy to see that $\partial_c\varphi\colon X \to 2^{X^*}$ is a monotone operator. Moreover, we have the following well-known result.

Theorem 3.7.   Let X be a Banach space and $\varphi\colon X \to \mathbb{R} \cup \{+\infty\}$ a proper, convex, l.s.c. function. Then $\partial_c\varphi\colon X \to 2^{X^*}$ is maximal monotone.

A proof of Theorem 3.7 can be found in Denkowski, Migórski and Papageorgiou (2003b).

In many situations we deal with functions $\varphi\colon K \to \mathbb{R}$ where K is a non-empty subset of X. In such cases it is convenient to extend the function φ to the space X by considering the function $\widetilde{\varphi}\colon X \to \mathbb{R} \cup \{+\infty\}$ defined by

$$\widetilde{\varphi}(v) = \begin{cases} \varphi(v) & \text{if } v \in K, \\ +\infty & \text{otherwise.} \end{cases}$$

We say that the function $\varphi\colon K \to \mathbb{R}$ is convex if its extension $\widetilde{\varphi}\colon X \to \mathbb{R} \cup \{+\infty\}$ is convex. It is lower semicontinuous (on K) if $\widetilde{\varphi}$ is lower semicontinuous. The special case $\varphi = 0$ deserves a more detailed treatment.

Example 3.8.   Given a non-empty subset K of X, the function I_K on X, defined by

$$I_K(x) = \begin{cases} 0 & \text{if } x \in K, \\ +\infty & \text{if } x \notin K, \end{cases}$$

is called the indicator function of K. It can be proved that the subset K of X is convex if and only if I_K is convex. Moreover, K is closed if and only if I_K is l.s.c. The subdifferential of I_K is the multivalued operator $\partial_c I_K : X \to 2^{X^*}$ given by

$$\partial_c I_K(x) = \begin{cases} \{x^* \in X^* \mid \langle x^*, x - v \rangle \geq 0 \text{ for all } v \in K\} & \text{if } x \in K, \\ \emptyset & \text{if } x \notin K. \end{cases} \tag{3.1}$$

It follows from (3.1) that

$$x^* \in \partial_c I_K(x) \quad \Longleftrightarrow \quad x \in K, \quad \langle x^*, x - v \rangle \geq 0 \quad \text{for all } v \in K. \tag{3.2}$$

For detailed discussion of the properties of the convex functions, including the convex subdifferential, we refer the reader to Ekeland and Temam (1976) and Kurdila and Zabarankin (2005), for instance.

Clarke subdifferential. We now review the notion of the Clarke subdifferential for a locally Lipschitz function. First, we recall that a function $j \colon X \to \mathbb{R}$ is said to be locally Lipschitz if, for every $x \in X$, there exists a neighbourhood of x, denoted by U_x, and a U_x-dependent constant $L_x > 0$ such that

$$|j(y) - j(z)| \leq L_x \|y - z\|_X \quad \text{for all } y, z \in U_x.$$

We also recall that a convex continuous function $j \colon X \to \mathbb{R}$ is locally Lipschitz.

Definition 3.9. Let $j \colon X \to \mathbb{R}$ be a locally Lipschitz function. The *generalized (Clarke) directional derivative* of j at a point $x \in X$ in a direction $v \in X$ is defined by

$$j^0(x; v) = \limsup_{y \to x, \lambda \downarrow 0} \frac{j(y + \lambda v) - j(y)}{\lambda}. \tag{3.3}$$

The *generalized gradient (subdifferential)* of j at x is a subset of the dual space X^* given by

$$\partial j(x) = \{\xi \in X^* \mid j^0(x; v) \geq \langle \xi, v \rangle \text{ for all } v \in X\}. \tag{3.4}$$

A locally Lipschitz function j is said to be *regular* (in the sense of Clarke) at the point $x \in X$ if, for all $v \in X$, the one-sided directional derivative $j'(x; v)$ exists and $j^0(x; v) = j'(x; v)$.

We now follow Clarke (1983) and recall the following properties of the generalized directional derivative and the generalized gradient.

Proposition 3.10. Assume that $j \colon X \to \mathbb{R}$ is a locally Lipschitz function. Then the following statements are valid.

(i) For every $x \in X$, the function $X \ni v \mapsto j^0(x; v) \in \mathbb{R}$ is positively homogeneous and subadditive, that is, $j^0(x; \lambda v) = \lambda j^0(x; v)$ for all $\lambda \geq 0$, $v \in X$ and $j^0(x; v_1 + v_2) \leq j^0(x; v_1) + j^0(x; v_2)$ for all v_1, $v_2 \in X$, respectively.

(ii) For every $v \in X$, we have $j^0(x; v) = \max\{\langle \xi, v \rangle \mid \xi \in \partial j(x)\}$.

(iii) The function $X \times X \ni (x, v) \mapsto j^0(x; v) \in \mathbb{R}$ is upper semicontinuous, that is, for all x, $v \in X$, $\{x_n\}$, $\{v_n\} \subset X$ such that $x_n \to x$ and $v_n \to v$ in X, we have $\limsup j^0(x_n; v_n) \leq j^0(x; v)$.

(iv) For every $x \in X$, the generalized gradient $\partial j(x)$ is a non-empty, convex, and weakly* compact subset of X^*.

(v) The graph of the generalized gradient ∂j is closed in $X \times X^*_{w*}$ topology, that is, if $\{x_n\} \subset X$ and $\{\xi_n\} \subset X^*$ are sequences such that $\xi_n \in \partial j(x_n)$ and $x_n \to x$ in X, $\xi_n \to \xi$ weakly* in X^*, then $\xi \in \partial j(x)$.

(vi) If $j \colon X \to \mathbb{R}$ is convex, then the subdifferential in the sense of Clarke $\partial j(x)$ at any $x \in X$ coincides with the convex subdifferential $\partial_c j(x)$.

Proposition 3.11. Let $j, j_1, j_2 \colon X \to \mathbb{R}$ be locally Lipschitz functions. Then we have the following.

(i) *Scalar multiples.* The equality $\partial(\lambda j)(x) = \lambda \partial j(x)$ holds for all $\lambda \in \mathbb{R}$ and all $x \in X$.

(ii) *Sum rules.* The inclusion

$$\partial(j_1 + j_2)(x) \subseteq \partial j_1(x) + \partial j_2(x) \tag{3.5}$$

holds for all $x \in X$ or, equivalently,

$$(j_1 + j_2)^0(x; v) \leq j_1^0(x; v) + j_2^0(x; v) \tag{3.6}$$

for all $x \in X$, $v \in X$.

(iii) If j_1, j_2 are regular at $x \in X$, then $j_1 + j_2$ is regular at $x \in X$ and we have equalities in (3.5) and (3.6).

A proof of the following result can be found in Lemma 4.2 of Migórski, Ochal and Sofonea (2010) and follows from the chain rule for the generalized gradient.

Proposition 3.12. Let X and Y be Banach spaces, let $\varphi \colon Y \to \mathbb{R}$ be locally Lipschitz and let $T \colon X \to Y$ be given by $Tx = Ax + y$ for $x \in X$, where $A \in \mathcal{L}(X, Y)$ and $y \in Y$ is fixed. Then the function $j \colon X \to \mathbb{R}$ defined by $j(x) = \varphi(Tx)$ is locally Lipschitz and

(i) $j^0(x; v) \leq \varphi^0(Tx; Av)$ for all $x, v \in X$,

(ii) $\partial j(x) \subseteq A^* \partial \varphi(Tx)$ for all $x \in X$,

where $A^* \in \mathcal{L}(Y^*, X^*)$ is the adjoint operator of A. Moreover, if φ (or $-\varphi$) is regular, then j (or $-j$) is regular and in (i) and (ii) we have equalities. These equalities are also true if instead of the regularity condition, we assume that A is surjective.

For detailed discussion of the properties of the Clarke subdifferential, we refer the reader to Clarke (1975, 1983) and Denkowski, Migórski and Papageorgiou (2003a, 2003b).

3.2. History-dependent operators

In the study of Problems 2.8 and 2.10, we will use the notion of a history-dependent operator. In contact mechanics, a history-dependent operator appears either in the constitutive law or in the contact boundary conditions. Such operators describe various memory effects. Abstract classes of quasi-variational inequalities with history-dependent operators were considered in Sofonea and Matei (2011) and Sofonea and Xiao (2016), where existence, uniqueness and regularity results were proved. A survey of some recent results for history-dependent variational–hemivariational inequalities can be found in Sofonea and Migórski (2018). There, the abstract well-posedness results are applied to various examples arising in contact mechanics. In this subsection we assume that X and Y are normed spaces endowed with the norms $\| \cdot \|_X$ and $\| \cdot \|_Y$, respectively. Let $I = [0, T]$ for some $T > 0$ or $I = \mathbb{R}_+$ be the time interval of interest.

Definition 3.13. An operator $\mathcal{S} \colon C(I; X) \to C(I; Y)$ is called a *history-dependent* operator if, for any compact subset $I_0 \subset I$, there exists a constant $L_{I_0} > 0$ such that

$$\|\mathcal{S}u_1(t) - \mathcal{S}u_2(t)\|_Y \leq L_{I_0} \int_0^t \|u_1(s) - u_2(s)\|_X \, \mathrm{d}s$$

$$\text{for all } u_1, u_2 \in C(I; X), \ t \in I_0. \qquad (3.7)$$

Similarly, an operator $\mathcal{S} \colon L^2(I; X) \to L^2(I; Y)$ is called a *history-dependent* operator if, for any compact subset $I_0 \subset I$, there exists a constant $L_{I_0} > 0$ such that

$$\|\mathcal{S}u_1(t) - \mathcal{S}u_2(t)\|_Y \leq L_{I_0} \int_0^t \|u_1(s) - u_2(s)\|_X \, \mathrm{d}s$$

$$\text{for all } u_1, u_2 \in L^2(I; X), \ \text{a.e. } t \in I_0. \qquad (3.8)$$

Note that here and below, when no confusion arises, we use the shorthand notation $\mathcal{S}u(t)$ to represent the value of the function $\mathcal{S}u$ at a point $t \in I$, i.e. $\mathcal{S}u(t) = (\mathcal{S}u)(t)$. We make some comments regarding Definition 3.13 for $\mathcal{S}\colon C(I;X) \to C(I;Y)$; similar comments can be stated for $\mathcal{S}\colon L^2(I;X) \to L^2(I;Y)$.

Remark 3.14.

(1) An operator $\mathcal{S}\colon C([0,T];X) \to C([0,T];Y)$ is a history-dependent operator if and only if there exists a constant $L > 0$ such that

$$\|\mathcal{S}u_1(t) - \mathcal{S}u_2(t)\|_Y \leq L \int_0^t \|u_1(s) - u_2(s)\|_X \, \mathrm{d}s$$

$$\text{for all } u_1, u_2 \in C([0,T];X), \ t \in [0,T]. \quad (3.9)$$

(2) An operator $\mathcal{S}\colon C(\mathbb{R}_+;X) \to C(\mathbb{R}_+;Y)$ is a history-dependent operator if and only if, for any $n \in \mathbb{N}$, there exists a constant $L_n > 0$ that varies with n, such that

$$\|\mathcal{S}u_1(t) - \mathcal{S}u_2(t)\|_Y \leq L_n \int_0^t \|u_1(s) - u_2(s)\|_X \, \mathrm{d}s$$

$$\text{for all } u_1, u_2 \in C(\mathbb{R}_+;X), \ t \in [0,n]. \quad (3.10)$$

Examples of operators satisfying condition (3.7) are given next.

Example 3.15. Let $u_0 \in X$ and $G\colon X \to Y$ be a Lipschitz continuous operator, that is, an operator which satisfies the inequality

$$\|Gu_1 - Gu_2\|_Y \leq L_G \|u_1 - u_2\|_X \quad \text{for all } u_1, u_2 \in X$$

with some constant $L_G > 0$. Define an operator $\mathcal{S}\colon C(I;X) \to C(I;Y)$ by

$$\mathcal{S}u(t) = G\left(\int_0^t u(s) \, \mathrm{d}s + u_0 \right) \quad \text{for all } u \in C(I;X), \ t \in I. \quad (3.11)$$

Then, for $u_1, u_2 \in C(I;X)$, we have

$$\|\mathcal{S}u_1(t) - \mathcal{S}u_2(t)\|_Y \leq L_G \int_0^t \|u_1(s) - u_2(s)\|_X \, \mathrm{d}s.$$

Thus, condition (3.7) holds with $L_{I_0} = L_G$ for any compact subset $I_0 \subset I$ and $\mathcal{S}$ is a history-dependent operator. In particular when $X = Y$ and G is the identity operator on X, (3.11) reduces to

$$\mathcal{S}u(t) = \int_0^t u(s) \, \mathrm{d}s + u_0 \quad \text{for all } u \in C(I;X), \ t \in I. \quad (3.12)$$

The operator $\mathcal{S}\colon C(I;X) \to C(I;X)$ defined by (3.12) is a history-dependent operator.

Example 3.16. Let $G \in C(I;\mathcal{L}(X,Y))$ and $\mathcal{S}\colon C(I;X) \to C(I;Y)$ be a *Volterra operator* given by

$$\mathcal{S}u(t) = \int_0^t G(t-s)\,u(s)\,\mathrm{d}s, \quad u \in C(I;X),\ t \in I. \tag{3.13}$$

Then for any compact set $I_0 \subset I$, inequality (3.7) holds with

$$L_{I_0} = \|G\|_{C(I_0;\mathcal{L}(X,Y))} = \max_{s \in I_0} \|G(s)\|_{\mathcal{L}(X,Y)}.$$

This shows that the operator $\mathcal{S}$ defined by (3.13) is a history-dependent operator.

3.3. A class of evolutionary inclusions

We now introduce an abstract result which will be used in the study of the dynamic Problem 2.10. The functional framework is as follows. Let $V \subset H \subset V^*$ be an evolution triple of spaces, that is, V is a separable, reflexive Banach space, H is a separable Hilbert space, the embedding $V \subset H$ is continuous and V is dense in H. We use $\langle \cdot, \cdot \rangle$ for the duality pairing between V^* and V. Given $0 < T < +\infty$, we introduce the function spaces

$$\mathcal{V} = L^2(0,T;V) \quad \text{and} \quad \mathcal{W} = \{w \in \mathcal{V} \mid \dot{w} \in \mathcal{V}^*\}.$$

We identify $\mathcal{H} = L^2(0,T;H)$ with its dual to obtain the following continuous embeddings $\mathcal{W} \subset \mathcal{V} \subset \mathcal{H} \subset \mathcal{V}^*$. The duality pairing between $\mathcal{V}^*$ and $\mathcal{V}$ is denoted by

$$\langle w, v \rangle_{\mathcal{V}^* \times \mathcal{V}} = \int_0^T \langle w(t), v(t) \rangle\,\mathrm{d}t \quad \text{for } w \in \mathcal{V}^*,\ v \in \mathcal{V}.$$

Now consider an operator $A\colon (0,T) \times V \to V^*$ and a function $\psi\colon (0,T) \times V \to \mathbb{R}$, assumed to be locally Lipschitz with respect to its second argument. We let $\partial\psi$ denote the Clarke generalized gradient of ψ with respect to its second argument. Given $f\colon (0,T) \to V^*$ and $w_0 \in V$, we consider the following evolutionary inclusion.

Problem 3.17. Find $w \in \mathcal{W}$ such that

$$\dot{w}(t) + A(t, w(t)) + \partial\psi(t, w(t)) \ni f(t) \text{ for a.e. } t \in (0,T),$$
$$w(0) = w_0.$$

In the study of this inclusion problem, we impose the following hypotheses on the data.

$A\colon (0,T) \times V \to V^*$ is such that:

(a) $A(\cdot, v)$ is measurable on $(0,T)$ for all $v \in V$;

(b) $A(t, \cdot)$ is demicontinuous on V for a.e. $t \in (0,T)$,

    i.e. $u_n \to u$ in $V \Longrightarrow Au_n \rightharpoonup Au$ in V^*;

(c) $\|A(t,v)\|_{V^*} \le a_0(t) + a_1\|v\|_V$ for all $v \in V$,

    a.e. $t \in (0,T)$ with $a_0 \in L^2(0,T), a_0 \ge 0$ and $a_1 \ge 0$;

(d) there is a constant $m_A > 0$ such that

$$\langle A(t,v_1) - A(t,v_2), v_1 - v_2 \rangle \ge m_A\|v_1 - v_2\|_V^2$$
    for all $v_1, v_2 \in V$, a.e. $t \in (0,T)$.
$$\tag{3.14}$$

$\psi\colon (0,T) \times V \to \mathbb{R}$ is such that:

(a) $\psi(\cdot, v)$ is measurable on $(0,T)$ for all $v \in V$;

(b) $\psi(t, \cdot)$ is locally Lipschitz on V for a.e. $t \in (0,T)$;

(c) $\|\partial\psi(t,v)\|_{V^*} \le c_0(t) + c_1\|v\|_V$ for all $v \in V$,

    a.e. $t \in (0,T)$ with $c_0 \in L^2(0,T), c_0 \ge 0, c_1 \ge 0$;

(d) there exists a constant $m_\psi \ge 0$ such that

$$\langle z_1 - z_2, v_1 - v_2 \rangle \ge -m_\psi\|v_1 - v_2\|_V^2$$
    for all $z_i \in \partial\psi(t,v_i), z_i \in V^*, v_i \in V, i = 1,2$, a.e. $t \in (0,T)$.
$$\tag{3.15}$$

$$f \in \mathcal{V}^*, \ w_0 \in V. \tag{3.16}$$

$$\max\{m_\psi, 2\sqrt{2}\,c_1\} < m_A. \tag{3.17}$$

We say a function $w \in \mathcal{W}$ is a solution of Problem 3.17 if there exists $w^* \in \mathcal{V}^*$ such that

$$\dot{w}(t) + A(t, w(t)) + w^*(t) = f(t) \text{ for a.e. } t \in (0,T),$$
$$w^*(t) \in \partial\psi(t, w(t)) \text{ for a.e. } t \in (0,T),$$
$$w(0) = w_0.$$

We have the following existence and uniqueness result.

Theorem 3.18. Assume that (3.14)–(3.17) hold. Then Problem 3.17 has a unique solution $w \in \mathcal{W}$.

A proof of Theorem 3.18 can be found in Sofonea and Migórski (2018, p. 183). The existence part is based on a surjectivity result with maximal monotone operators in reflexive Banach spaces. The uniqueness part follows from standard arguments and is based on the smallness assumption (3.17).

3.4. Fixed-point theorems

We will apply Banach's fixed-point theorem in solution existence proofs of various problems: see Zeidler (1985, Section 1.1) or Atkinson and Han (2009, Section 5.1).

Theorem 3.19. Let K be a non-empty closed set in a Banach space X, and let $\Lambda\colon K \to K$ be a contractive mapping, that is, for some constant $\alpha \in [0,1)$,

$$\|\Lambda u - \Lambda v\|_X \le \alpha\,\|u - v\|_X \quad \text{for all } u, v \in K.$$

Then $\Lambda\colon K \to K$ has a unique fixed point, that is, there exists a unique $u^* \in K$ such that

$$\Lambda u^* = u^*.$$

History-dependent operators have important fixed-point properties which are useful in proving the solvability of various classes of nonlinear equations and inequalities. The following result is proved in Sofonea, Avramescu and Matei (2008).

Theorem 3.20. Assume that X is a Banach space and $\Lambda\colon C(I;X) \to C(I;X)$ is a history-dependent operator. Then Λ has a unique fixed point, that is, there exists a unique element $\eta^* \in C(I;X)$ such that $\Lambda\eta^* = \eta^*$.

In case of a history-dependent operator on $L^2(0,T;X)$, we have a similar result from Sofonea and Migórski (2018, Theorem 67).

Theorem 3.21. Assume that X is a Banach space and $\Lambda\colon L^2(0,T;X) \to L^2(0,T;X)$ is a history-dependent operator. Then Λ has a unique fixed point, that is, there exists a unique element $\eta^* \in L^2(0,T;X)$ such that $\Lambda\eta^* = \eta^*$.

3.5. Some inequalities

We will use the modified Cauchy–Schwarz inequality

$$a\,b \le \epsilon\,a^2 + c\,b^2 \quad \text{for all } a, b \in \mathbb{R}, \tag{3.18}$$

where $\epsilon > 0$ is an arbitrary positive number and the constant $c > 0$ depends on ϵ; indeed, we may simply take $c = 1/(4\,\epsilon)$.

In the well-posedness analysis of contact problems, we will need the Gronwall inequality: see Han and Sofonea (2002, Section 7.4).

Lemma 3.22. Let $a,\,b \in \mathbb{R}$ with $a < b$ and assume $f, g : [a,b] \to \mathbb{R}$ are continuous functions satisfying

$$f(t) \le g(t) + c \int_a^t f(s)\,\mathrm{d}s \quad \text{for all } t \in [a,b], \tag{3.19}$$

where $c > 0$ is a constant. Then

$$f(t) \le g(t) + c \int_a^t g(s)\, e^{c\,(t-s)}\, ds \quad \text{for all } t \in [a, b]. \tag{3.20}$$

Moreover, if g is non-decreasing, then

$$f(t) \le g(t)\, e^{c\,(t-a)} \quad \text{for all } t \in [a, b]. \tag{3.21}$$

In error analysis for numerical solutions of the contact problems, we will need discrete versions of Gronwall's inequality. For a fixed T and a positive integer N, let $k = T/N$.

Lemma 3.23. Assume $\{g_n\}_{n=1}^N$ and $\{e_n\}_{n=1}^N$ are two sequences of non-negative numbers satisfying

$$e_n \le c\, g_n + c\, k \sum_{i=1}^{n-1} e_i, \quad n = 1, \ldots, N,$$

for a constant $c > 0$. Then, for a possibly different constant $c > 0$,

$$\max_{1 \le n \le N} e_n \le c \max_{1 \le n \le N} g_n. \tag{3.22}$$

Lemma 3.24. Assume $\{g_n\}_{n=1}^N$ and $\{e_n\}_{n=1}^N$ are two sequences of non-negative numbers satisfying

$$e_n \le c\, g_n + c\, k \sum_{i=1}^{n} e_i, \quad n = 1, \ldots, N,$$

for a constant $c > 0$. Then, if k is sufficiently small,

$$\max_{1 \le n \le N} e_n \le c \max_{1 \le n \le N} g_n. \tag{3.23}$$

More general versions of these results can be found in Han and Sofonea (2002, Section 7.4).

4. An elliptic variational–hemivariational inequality

In this section we consider an abstract elliptic variational–hemivariational inequality, present a well-posedness result, develop numerical methods for its solution, prove convergence of the numerical solutions, and provide a Céa-type inequality for error estimation of the numerical solutions. The results in this section will be applied in Section 7 in the analysis of Problem 2.6.

Let X be a reflexive Banach space. Given a set $K \subset X$, an operator $A \colon X \to X^*$, functions $\varphi \colon K \times K \to \mathbb{R}$, $j \colon X \to \mathbb{R}$ and $f \in X^*$, we consider the following problem.

Problem 4.1. Find an element $u \in K$ such that

$$\langle Au, v - u \rangle + \varphi(u, v) - \varphi(u, u) + j^0(u; v - u) \ge \langle f, v - u \rangle \quad \text{for all } v \in K.$$

Note that Problem 4.1 contains a function φ assumed to be convex with respect to its second argument and a function j assumed to be locally Lipschitz. Therefore, Problem 4.1 represents a variational–hemivariational inequality.

For the analysis of Problem 4.1, we consider the following hypotheses on the data.

$$K \text{ is a non-empty, closed, convex subset of } X. \tag{4.1}$$

$$\left.\begin{array}{l} A \colon X \to X^* \text{ is:} \\[4pt] \text{(a) bounded and hemicontinuous;} \\[4pt] \text{(b) strongly monotone, } i.e. \text{ for some constant } m_A > 0, \\[4pt] \quad \langle Av_1 - Av_2, v_1 - v_2 \rangle \geq m_A \|v_1 - v_2\|_X^2 \text{ for all } v_1, v_2 \in X. \end{array}\right\} \tag{4.2}$$

$$\left.\begin{array}{l} \varphi \colon K \times K \to \mathbb{R} \text{ is such that:} \\[4pt] \text{(a) } \varphi(\eta, \cdot) \colon K \to \mathbb{R} \text{ is convex and l.s.c. on } K \text{ for all } \eta \in K; \\[4pt] \text{(b) there exists a constant } \alpha_\varphi > 0 \text{ such that} \\[4pt] \quad \varphi(\eta_1, v_2) - \varphi(\eta_1, v_1) + \varphi(\eta_2, v_1) - \varphi(\eta_2, v_2) \\[4pt] \quad \leq \alpha_\varphi \|\eta_1 - \eta_2\|_X \|v_1 - v_2\|_X \text{ for all } \eta_1, \eta_2, v_1, v_2 \in K. \end{array}\right\} \tag{4.3}$$

$$\left.\begin{array}{l} j \colon X \to \mathbb{R} \text{ is such that:} \\[4pt] \text{(a) } j \text{ is locally Lipschitz;} \\[4pt] \text{(b) } \|\partial j(v)\|_{X^*} \leq c_0 + c_1 \|v\|_X \text{ for all } v \in X \\[4pt] \quad \text{with constants } c_0, c_1 \geq 0; \\[4pt] \text{(c) there exists a constant } \alpha_j > 0 \text{ such that} \\[4pt] \quad j^0(v_1; v_2 - v_1) + j^0(v_2; v_1 - v_2) \leq \alpha_j \|v_1 - v_2\|_X^2 \\[4pt] \quad \text{for all } v_1, v_2 \in X. \end{array}\right\} \tag{4.4}$$

$$f \in X^*. \tag{4.5}$$

Recall that, as usual, inequality in (4.4(b)) means

$$\|\xi\|_{X^*} \leq c_0 + c_1 \|v\|_X \quad \text{for all } v \in X, \ \xi \in \partial j(v).$$

Moreover, we recall that (4.4(c)) is equivalent to the following condition:

$$\langle \partial j(v_1) - \partial j(v_2), v_1 - v_2 \rangle \geq -\alpha_j \|v_1 - v_2\|_X^2 \quad \text{for all } v_1, v_2 \in X. \tag{4.6}$$

A proof of this statement can be found in Sofonea and Migórski (2018, p. 124).

4.1. Solution existence and uniqueness

The unique solvability of Problem 4.1 is provided by the following result.

Theorem 4.2.   Assume (4.1)–(4.5) and the smallness condition

$$\alpha_\varphi + \alpha_j < m_A. \tag{4.7}$$

Then, Problem 4.1 has a unique solution $u \in K$.

A version of Theorem 4.2 was proved in Migórski, Ochal and Sofonea (2017), under additional assumptions. The proof of the current version of the result is based on the same arguments used in the above-mentioned paper. Nevertheless, we present it below for the convenience of the reader.

Proof.   The proof consists of five steps.

(i) We show the coercivity of the operator A. We first prove that for all $u_0 \in K$ there exist $\beta, \gamma \in \mathbb{R}$ (which depend on u_0) such that

$$\langle Av, v - u_0 \rangle \geq m_A \|v\|_X^2 - \beta \|v\|_X - \gamma \quad \text{for all } v \in X. \tag{4.8}$$

Indeed, let $u_0 \in K$ and $v \in V$. We write

$$\langle Av, v - u_0 \rangle = \langle Av - Au_0, v - u_0 \rangle + \langle Au_0, v - u_0 \rangle$$
$$\geq m_A \|v - u_0\|_X^2 - \|Au_0\|_{X^*} \|v - u_0\|_X.$$

Then we use the inequalities

$$\big| \|v\|_X - \|u_0\|_X \big| \leq \|v - u_0\|_X,$$
$$\|Au_0\|_{X^*} \|v - u_0\|_X \leq \|Au_0\|_{X^*} \|v\|_X + \|Au_0\|_{X^*} \|u_0\|_X$$

to obtain

$$\langle Av, v - u_0 \rangle \geq m_A (\|v\|_X - \|u_0\|_X)^2 - \|Au_0\|_{X^*} \|v\|_X - \|Au_0\|_{X^*} \|u_0\|_X$$
$$= m_A \|v\|_X^2 - (2m_A \|u_0\|_X + \|Au_0\|_{X^*}) \|v\|_X$$
$$+ m_A \|u_0\|_X^2 - \|Au_0\|_{X^*} \|u_0\|_X,$$

which proves inequality (4.8). This inequality shows the u_0-coercivity of the operator A in the sense of Definition 3.1(d).

(ii) We introduce an auxiliary inclusion problem and show its unique solvability. For an arbitrarily fixed element $\eta \in K$, define a function $\widetilde{\varphi}_\eta \colon X \to \mathbb{R} \cup \{+\infty\}$ by

$$\widetilde{\varphi}_\eta(v) = \begin{cases} \varphi(\eta, v) & \text{if } v \in K, \\ +\infty & \text{otherwise.} \end{cases} \tag{4.9}$$

Then consider the following problem: find $u_\eta \in X$ such that

$$Au_\eta + \partial j(u_\eta) + \partial_c \widetilde{\varphi}_\eta(u_\eta) \ni f. \tag{4.10}$$

Let us apply Theorem 3.5 to prove that this inclusion has a solution. We fix an element $u_0 \in K$ and introduce multivalued operators $T_1, T_2 \colon X \to 2^{X^*}$

defined by

$$T_1 v = Av + \partial j(v), \quad T_2 v = \partial_c \widetilde{\varphi}_\eta(v) \quad \text{for } v \in X.$$

We check that the operators T_1 and T_2 satisfy conditions (a)–(c) in Theorem 3.5.

First, we note that by Proposition 3.3, under the assumptions (4.2), the operator A is pseudomonotone. Therefore, the boundedness of the operator T_1 follows easily from the boundedness of A and the growth condition (4.4(b)) on ∂j.

Next, we show that T_1 is generalized pseudomonotone. We use hypotheses (4.2(b)), (4.4(c)), (4.7) and inequality (4.6) to see that the operator T_1 is strongly monotone, that is,

$$\langle T_1 v_1 - T_1 v_2, v_1 - v_2 \rangle \geq (m_A - \alpha_j)\, \|v_1 - v_2\|_X^2 \quad \text{for all } v_1, v_2 \in X.$$

Assume now that $u_n \in X$, $u_n \to u$ weakly in X, $u_n^* \in T_1 u_n$, $u_n^* \to u^*$ weakly in X^* and $\limsup \langle u_n^*, u_n - u \rangle \leq 0$. We shall prove that $u^* \in T_1 u$ and $\langle u_n^*, u_n \rangle \to \langle u^*, u \rangle$. Using the strong monotonicity of T_1, from the relation

$$(m_A - \alpha_j)\|u_n - u\|_X^2 \leq \langle u_n^*, u_n - u \rangle - \langle T_1 u, u_n - u \rangle,$$

we deduce that $u_n \to u$ in X. From $u_n^* \in T_1 u_n$, we have $u_n^* = w_n + z_n$ with $w_n = Au_n$ and $z_n \in \partial j(u_n)$. Since A and ∂j are bounded operators, by passing to a subsequence, if necessary, we may assume that $w_n \to w$ and $z_n \to z$ both weakly in X^* with some $w, z \in X^*$. Therefore, from $u_n^* = w_n + z_n$, we find that $u^* = w + z$. Exploiting the equivalent condition for the pseudomonotonicity of A in Remark 3.2, we have $Au_n \to Au$ weakly in X^*, which gives $w = Au$. On the other hand, by Proposition 3.10(v) it follows that $X \ni v \mapsto \partial j(v) \in 2^{X^*}$ has a closed graph with respect to the strong topology in X and weak topology in X^*. Exploiting this property we infer that $z \in \partial j(u)$. Hence, $u^* = w + z \in Au + \partial j(u) = T_1 u$. Since $u_n^* \to u^*$ weakly in X^* and $u_n \to u$ in X, it is clear that $\langle u_n^*, u_n \rangle \to \langle u^*, u \rangle$. This shows that T_1 is generalized pseudomonotone.

In order to establish the u_0-coercivity of T_1, we use inequality (4.8), hypothesis (4.4(c)), inequality (4.6) and the following inequality, which is a consequence of (4.4(b)):

$$|\langle \partial j(u_0), v - u_0 \rangle| \leq (c_0 + c_1\|u_0\|_X)\|v - u_0\|_X.$$

We have

$$\langle T_1 v, v - u_0 \rangle = \langle Av, v - u_0 \rangle + \langle \partial j(v) - \partial j(u_0), v - u_0 \rangle + \langle \partial j(u_0), v - u_0 \rangle$$
$$\geq m_A\|v\|_X^2 - \beta\,\|v\|_X - \gamma - \alpha_j\|v - u_0\|_X^2$$
$$- (c_0 + c_1\|u_0\|_X)\|v - u_0\|_X$$

$$\geq (m_A - \alpha_j)\|v\|_X^2 - \|v\|_X(\beta + 2\alpha_j\|u_0\|_X + c_0 + c_1\|u_0\|_X)$$
$$- \gamma - \alpha_j\|u_0\|_X^2 - (c_0 + c_1\|u_0\|_X)\|u_0\|_X.$$

The u_0-coercivity of T_1 follows now from hypothesis (4.7).

We conclude from the above that T_1 is bounded, generalized pseudomonotone and u_0-coercive and therefore it satisfies condition (a) of Theorem 3.5.

Next, using Proposition 3.10(iv) we see that for all $v \in X$ the set $Av + \partial j(v)$ is non-empty, closed and convex in X^*. Therefore, condition (b) of Theorem 3.5 holds.

From hypothesis (4.3(a)) and the definition of $\widetilde{\varphi}_\eta$, we know that the function $\widetilde{\varphi}_\eta$ is proper, convex and lower semicontinuous with $\operatorname{dom}\widetilde{\varphi}_\eta = K$. By Theorem 3.7, the operator $T_2 = \partial_c\widetilde{\varphi}_\eta\colon X \to 2^{X^*}$ is maximal monotone with $\mathcal{D}(\partial_c\widetilde{\varphi}_\eta) = K$. This shows that condition (c) of Theorem 3.5 is satisfied.

Summarizing, we can apply Theorem 3.5 to deduce that there exists a solution $u_\eta \in X$ to the inclusion problem (4.10).

(iii) We introduce an auxiliary variational–hemivariational inequality and prove that it has a unique solution. Fix an element $\eta \in K$ and consider the auxiliary problem of finding an element $u_\eta \in K$ for which

$$\langle Au_\eta, v - u_\eta \rangle + \varphi(\eta, v) - \varphi(\eta, u_\eta) + j^0(u_\eta; v - u_\eta)$$
$$\geq \langle f, v - u_\eta \rangle \quad \text{for all } v \in K. \tag{4.11}$$

By making use of definition (4.9), we see that (4.11) is equivalent to the problem of finding $u_\eta \in X$ such that

$$\langle Au_\eta, v - u_\eta \rangle + \widetilde{\varphi}_\eta(v) - \widetilde{\varphi}_\eta(u_\eta) + j^0(u_\eta; v - u_\eta) \geq \langle f, v - u_\eta \rangle \quad \text{for all } v \in X. \tag{4.12}$$

We claim that every solution of inclusion (4.10) is a solution to problem (4.12). Indeed, let $u_\eta \in X$ be such that

$$Au_\eta + \xi_\eta + \theta_\eta = f \tag{4.13}$$

with $\xi_\eta \in \partial_c\widetilde{\varphi}_\eta(u_\eta)$ and $\theta_\eta \in \partial j(u_\eta)$. We have

$$\langle \xi_\eta, v - u_\eta \rangle \leq \widetilde{\varphi}_\eta(v) - \widetilde{\varphi}_\eta(u_\eta) \quad \text{for all } v \in X,$$
$$\langle \theta_\eta, v \rangle \leq j^0(u_\eta; v) \quad \text{for all } v \in X.$$

Combining (4.13) with the last two inequalities, we obtain

$$\langle Au_\eta, v - u_\eta \rangle + \widetilde{\varphi}_\eta(v) - \widetilde{\varphi}_\eta(u_\eta) + j^0(u_\eta; v - u_\eta) \geq \langle f, v - u_\eta \rangle \quad \text{for all } v \in X.$$

This implies that $u_\eta \in X$ solves problem (4.12) which concludes the proof of the claim.

By step (ii), inequality (4.12) has a solution. From the equivalence of inequalities (4.12) and (4.11), inequality (4.11) also has a solution $u_\eta \in K$. To prove its uniqueness assume that $u_1, u_2 \in K$ are two solutions of (4.11), that is,

$$\langle Au_1, v - u_1 \rangle + \varphi(\eta, v) - \varphi(\eta, u_1) + j^0(u_1; v - u_1) \geq \langle f, v - u_1 \rangle,$$
$$\langle Au_2, v - u_2 \rangle + \varphi(\eta, v) - \varphi(\eta, u_2) + j^0(u_2; v - u_2) \geq \langle f, v - u_2 \rangle,$$

for all $v \in K$. Taking $v = u_2$ in the first inequality, $v = u_1$ in the second one, and adding the resulting inequalities, we obtain

$$\langle Au_1 - Au_2, u_2 - u_1 \rangle + j^0(u_1; u_2 - u_1) + j^0(u_2; u_1 - u_2) \geq 0.$$

From the strong monotonicity of A and hypothesis (4.4(c)), we have

$$(m_A - \alpha_j)\|u_1 - u_2\|_X^2 \leq 0,$$

which, due to the smallness condition (4.7), implies $u_1 = u_2$. Thus, (4.11) has a unique solution.

(iv) We introduce an operator and apply the Banach fixed-point argument. Define an operator $\Lambda \colon K \to K$ by

$$\Lambda\eta = u_\eta \quad \text{for } \eta \in K, \tag{4.14}$$

where $u_\eta \in K$ denotes the unique solution of inequality (4.11). We prove that the operator Λ has a unique fixed point. For this purpose, let $\eta_1, \eta_2 \in K$ and $u_1 = u_{\eta_1}$, $u_2 = u_{\eta_2} \in K$ be the unique solutions of (4.11) corresponding to η_1, η_2, respectively. From the inequalities

$$\langle Au_1, v - u_1 \rangle + \varphi(\eta_1, v) - \varphi(\eta_1, u_1) + j^0(u_1; v - u_1) \geq \langle f, v - u_1 \rangle,$$
$$\langle Au_2, v - u_2 \rangle + \varphi(\eta_2, v) - \varphi(\eta_2, u_2) + j^0(u_2; v - u_2) \geq \langle f, v - u_2 \rangle,$$

valid for all $v \in K$, we have

$$\langle Au_1 - Au_2, u_1 - u_2 \rangle \leq \varphi(\eta_1, u_2) - \varphi(\eta_1, u_1) + \varphi(\eta_2, u_1) - \varphi(\eta_2, u_2)$$
$$+ j^0(u_1; u_2 - u_1) + j^0(u_2; u_1 - u_2).$$

Use the strong monotonicity of A and hypotheses (4.3(b)), (4.4(c)) to obtain

$$m_A\|u_1 - u_2\|_X^2 \leq \alpha_\varphi \|\eta_1 - \eta_2\|_X \|u_1 - u_2\|_X + \alpha_j \|u_1 - u_2\|_X^2.$$

Consequently,

$$\|\Lambda\eta_1 - \Lambda\eta_2\|_X = \|u_1 - u_2\|_X \leq \frac{\alpha_\varphi}{m_A - \alpha_j}\|\eta_1 - \eta_2\|_X.$$

From condition (4.7), by applying the Banach contraction principle, we deduce that there exists a unique $\eta^* \in K$ such that $\eta^* = \Lambda\eta^*$.

(v) To prove the existence part of Theorem 4.2, we write inequality (4.11) for $\eta = \eta^*$ and observe that $u_{\eta^*} = \Lambda\eta^* = \eta^*$. So the function $\eta^* \in K$ is a solution to Problem 4.1.

The uniqueness of a solution to Problem 4.1 is proved directly. Let $u_1, u_2 \in K$ be solutions, that is,

$$\langle Au_1, v - u_1 \rangle + \varphi(u_1, v) - \varphi(u_1, u_1) + j^0(u_1; v - u_1) \geq \langle f, v - u_1 \rangle,$$
$$\langle Au_2, v - u_2 \rangle + \varphi(u_2, v) - \varphi(u_2, u_2) + j^0(u_2; v - u_2) \geq \langle f, v - u_2 \rangle,$$

for all $v \in K$. From these inequalities, we obtain

$$\langle Au_1 - Au_2, u_1 - u_2 \rangle \leq \varphi(u_1, u_2) - \varphi(u_1, u_1) + \varphi(u_2, u_1) - \varphi(u_2, u_2)$$
$$+ j^0(u_1; u_2 - u_1) + j^0(u_2; u_1 - u_2).$$

Conditions (4.2(b)), (4.3(b)) and (4.4(c)) imply

$$m_A \|u_1 - u_2\|_X^2 \leq \alpha_\varphi \|u_1 - u_2\|_X^2 + \alpha_j \|u_1 - u_2\|_X^2$$

from which, due to the smallness assumption (4.7), it follows that $u_1 = u_2$. This completes the proof of the theorem. $\square$

We now follow Han (2018) and introduce a variant of Problem 4.1, which is more convenient for the numerical analysis as well as for applications in contact mechanics. Besides the reflexive Banach space X, we need two real Banach spaces X_φ and X_j, and two operators $\gamma_\varphi : X \to X_\varphi$, $\gamma_j : X \to X_j$. Moreover, we assume that $\varphi \colon X_\varphi \times X_\varphi \to \mathbb{R}$, $j \colon X_j \to \mathbb{R}$. The variational–hemivariational inequality we consider is stated as follows.

Problem 4.3. Find an element $u \in K$ such that

$$\langle Au, v - u \rangle + \varphi(\gamma_\varphi u, \gamma_\varphi v) - \varphi(\gamma_\varphi u, \gamma_\varphi u)$$
$$+ j^0(\gamma_j u; \gamma_j v - \gamma_j u) \geq \langle f, v - u \rangle \quad \text{for all } v \in K. \qquad (4.15)$$

For applications in contact mechanics, the functionals $\varphi(\cdot, \cdot)$ and $j(\cdot)$ are integrals over the contact boundary Γ_3. In such a situation, X_φ and X_j can be chosen to be $L^2(\Gamma_3)^d$ and/or $L^2(\Gamma_3)$.

For the analysis of Problem 4.3, we consider the following hypotheses on the data, with constants c_φ, c_j, α_φ, α_j, c_0 and c_1.

$$\gamma_\varphi \in \mathcal{L}(X, X_\varphi), \quad \|\gamma_\varphi v\|_{X_\varphi} \leq c_\varphi \|v\|_X \quad \text{for all } v \in X. \qquad (4.16)$$

$$\gamma_j \in \mathcal{L}(X, X_j), \quad \|\gamma_j v\|_{X_j} \leq c_j \|v\|_X \quad \text{for all } v \in X. \qquad (4.17)$$

$\varphi \colon X_\varphi \times X_\varphi \to \mathbb{R}$ is such that:

(a) $\varphi(\eta, \cdot) \colon X_\varphi \to \mathbb{R}$ is convex and l.s.c. for all $\eta \in X_\varphi$;

(b) there exists $\alpha_\varphi > 0$ such that
$$\varphi(z_1, z_4) - \varphi(z_1, z_3) + \varphi(z_2, z_3) - \varphi(z_2, z_4)$$
$$\leq \alpha_\varphi \|z_1 - z_2\|_{X_\varphi} \|z_3 - z_4\|_{X_\varphi} \text{ for all } z_1, z_2, z_3, z_4 \in X_\varphi.$$

$$\left. \vphantom{\begin{array}{c} a \\ a \\ a \\ a \\ a \\ a \end{array}} \right\} \quad (4.18)$$

$$
\left.\begin{array}{l}
j \colon X_j \to \mathbb{R} \text{ is such that:} \\[4pt]
\text{(a) } j \text{ is locally Lipschitz;} \\[4pt]
\text{(b) } \|\partial j(z)\|_{X_j^*} \le c_0 + c_1 \|z\|_{X_j} \text{ for all } z \in X_j \text{ with } c_0, c_1 \ge 0; \\[4pt]
\text{(c) there exists } \alpha_j > 0 \text{ such that} \\[4pt]
\quad j^0(z_1; z_2 - z_1) + j^0(z_2; z_1 - z_2) \le \alpha_j \|z_1 - z_2\|_{X_j}^2 \\[4pt]
\quad \text{for all } z_1, z_2 \in X_j.
\end{array}\right\} \qquad (4.19)
$$

The unique solvability of Problem 4.3 is given by the following result.

Theorem 4.4. Assume (4.1), (4.2), (4.5), (4.16)–(4.19) and

$$
\alpha_\varphi c_\varphi^2 + \alpha_j c_j^2 < m_A. \qquad (4.20)
$$

Then, Problem 4.3 has a unique solution $u \in K$.

Proof. We prove this result by applying Theorem 4.2. Define functions $\widetilde{\varphi} : X \times X \to \mathbb{R}$ and $\widetilde{j} : X \to \mathbb{R}$ by

$$
\widetilde{\varphi}(u, v) = \varphi(\gamma_\varphi u, \gamma_\varphi v) \quad \text{for all } u, v \in X, \qquad (4.21)
$$
$$
\widetilde{j}(v) = j(\gamma_j v) \quad \text{for all } v \in X. \qquad (4.22)
$$

Then, using assumption (4.18) on the function φ and inequality (4.16) it is easy to see that the function $\widetilde{\varphi}$ satisfies assumption (4.3) on the space X with constant $\alpha_{\widetilde{\varphi}} = \alpha_\varphi c_\varphi^2$. On the other hand, from arguments similar to those used in the proof of Lemma 6 in Sofonea and Migórski (2018) it follows that the function $\widetilde{j}$ satisfies condition (4.4) with constant $\alpha_{\widetilde{j}} = \alpha_j c_j^2$. This statement is based on the chain rule for the Clarke subgradient and assumption (4.17) which guarantee that

$$
\widetilde{j}^0(u; v) \le j^0(\gamma_j u; \gamma_j v) \quad \text{for all } u, v \in X. \qquad (4.23)
$$

Assume now that (4.20) holds. Then $\alpha_{\widetilde{\varphi}} + \alpha_{\widetilde{j}} \le m_A$. Therefore, we can deduce the existence of a unique element $u \in K$ such that

$$
\langle Au, v-u \rangle + \widetilde{\varphi}(u, v) - \widetilde{\varphi}(u, u) + \widetilde{j}^0(u; v-u) \ge \langle f, v-u \rangle \quad \text{for all } v \in K. \qquad (4.24)
$$

We now use equality (4.21) and inequalities (4.23), (4.24) to deduce that u is a solution of inequality (4.15). This proves the existence of the solution to Problem 4.3. The uniqueness of the solution follows from the same argument used at the end of the proof of Theorem 4.2. $\qquad \square$

Denote

$$
K_\varphi = \gamma_\varphi(K).
$$

We comment that X_φ can be replaced by K_φ in the assumption (4.18) and the statement of Theorem 4.4 is still valid. Moreover, Problem 4.1 may be viewed as a special case of Problem 4.3.

We have an equivalent formulation of Problem 4.3, similar to Minty's lemma for variational inequalities (see Atkinson and Han 2009, p. 435).

Theorem 4.5. Assume $K \subset X$ is convex, $A \colon X \to X^*$ is monotone and radially continuous, and for all $z \in K_\varphi$, $\varphi(z, \cdot)$ is convex on K_φ. Then $u \in K$ is a solution of Problem 4.3 if and only if it satisfies

$$\langle Av, v - u \rangle + \varphi(\gamma_\varphi u, \gamma_\varphi v) - \varphi(\gamma_\varphi u, \gamma_\varphi u)$$
$$+ j^0(\gamma_j u; \gamma_j v - \gamma_j u) \geq \langle f, v - u \rangle \quad \text{for all } v \in K. \tag{4.25}$$

Proof. By the monotonicity of A, we have

$$\langle Av, v - u \rangle \geq \langle Au, v - u \rangle \quad \text{for all } u, v \in X. \tag{4.26}$$

Then it is obvious that a solution of Problem 4.3 satisfies the inequality (4.25).

Conversely, assume $u \in K$ satisfies (4.25). Since K is convex, for any $v \in K$ and any $t \in [0, 1]$, $u + t(v - u)$ belongs to K. We replace v with $u + t(v - u)$ in (4.25):

$$t \langle A(u + t(v - u)), v - u \rangle + \varphi(\gamma_\varphi u, \gamma_\varphi u + t(\gamma_\varphi v - \gamma_\varphi u))$$
$$- \varphi(\gamma_\varphi u, \gamma_\varphi u) + t\, j^0(\gamma_j u; \gamma_j v - \gamma_j u) \geq t \langle f, v - u \rangle. \tag{4.27}$$

Note that

$$\varphi(\gamma_\varphi u, \gamma_\varphi u + t(\gamma_\varphi v - \gamma_\varphi u)) \leq t\, \varphi(\gamma_\varphi u, \gamma_\varphi v) + (1 - t)\varphi(\gamma_\varphi u, \gamma_\varphi u).$$

We deduce from (4.27) that for $t \in (0, 1)$,

$$\langle A(u + t(v - u)), v - u \rangle + \varphi(\gamma_\varphi u, \gamma_\varphi v) - \varphi(\gamma_\varphi u, \gamma_\varphi u) + j^0(\gamma_j u; \gamma_j v - \gamma_j u)$$
$$\geq \langle f, v - u \rangle.$$

We take the limit $t \to 0+$ in the above inequality to recover the inequality (4.15). $\qquad\square$

4.2. Numerical approximations

In the rest of this section, we assume (4.1), (4.2), (4.5), (4.16)–(4.20) so that Problem 4.3 has a unique solution.

Let $X^h \subset X$ be a finite-dimensional subspace with $h > 0$ being a spatial discretization parameter. Let K^h be a non-empty, closed and convex subset of X^h. Then, a Galerkin approximation of Problem 4.3 is as follows.

Problem 4.6. Find an element $u^h \in K^h$ such that

$$\langle Au^h, v^h - u^h \rangle + \varphi(\gamma_\varphi u^h, \gamma_\varphi v^h) - \varphi(\gamma_\varphi u^h, \gamma_\varphi u^h)$$
$$+ j^0(\gamma_j u^h; \gamma_j v^h - \gamma_j u^h) \geq \langle f, v^h - u^h \rangle \quad \text{for all } v^h \in K^h. \tag{4.28}$$

The approximation is external if $K^h \not\subset K$, and is internal if $K^h \subset K$. The internal approximation with the choice $K^h = X^h \cap K$ is considered in Han, Sofonea and Danan (2018).

Remark 4.7. We comment that for the numerical analysis of Problem 2.6 in Section 7, a discrete problem in a form of the type given by Problem 4.6 serves as an intermediate step. For Problem 2.6, the functional j in the inequality (4.15) is defined by the formula

$$j(\gamma_j \boldsymbol{u}) = \int_{\Gamma_3} j_\nu(\gamma_j \boldsymbol{u}) \, \mathrm{d}a, \quad \gamma_j \boldsymbol{u} = u_\nu.$$

In the numerical scheme Problem 7.2 for solving Problem 2.6, the term $\int_{\Gamma_3} j_\nu^0(u_\nu; v_\nu - u_\nu) \, \mathrm{d}a$ in (2.44) is approximated by $\int_{\Gamma_3} j_\nu^0(u_\nu^h; v_\nu^h - u_\nu^h) \, \mathrm{d}a$ (see (7.15)). We have

$$j^0(\gamma_j \boldsymbol{u}; \gamma_j \boldsymbol{v}) \leq \int_{\Gamma_3} j_\nu^0(\gamma_j \boldsymbol{u}; \gamma_j \boldsymbol{v}) \, \mathrm{d}a \tag{4.29}$$

(*e.g.* Migórski, Ochal and Sofonea 2013, Theorem 3.47), and under the additional assumption that j_ν is regular,

$$j^0(\gamma_j \boldsymbol{u}; \gamma_j \boldsymbol{v}) = \int_{\Gamma_3} j_\nu^0(\gamma_j \boldsymbol{u}; \gamma_j \boldsymbol{v}) \, \mathrm{d}a.$$

In the latter case, the two numerical schemes are equivalent. In this paper, we do not assume the regularity of j_ν and we get around this assumption by means of the following consideration.

For definiteness in the discussion here, let the functional j in Problem 4.3 be of the form

$$j(\gamma_j u) = \int_D j_0(\gamma_j u) \tag{4.30}$$

where the integrand j_0 is locally Lipschitz, and the integration region D can be a subset of the domain Ω or a part of the boundary $\partial\Omega$. For applications in the contact problems considered in this paper, $D = \Gamma_3$. Then, the numerical method for implementation in solving Problem 4.3 is to find an element $u^h \in K^h$ such that

$$\langle Au^h, v^h - u^h \rangle + \varphi(\gamma_\varphi u^h, \gamma_\varphi v^h) - \varphi(\gamma_\varphi u^h, \gamma_\varphi u^h)$$
$$+ \int_D j_0^0(\gamma_j u^h; \gamma_j v^h - \gamma_j u^h) \geq \langle f, v^h - u^h \rangle \quad \text{for all } v^h \in K^h. \tag{4.31}$$

Moreover, we introduce a further intermediate discrete problem of finding an element $u^h \in K^h$ such that

$$\langle Au^h, v^h - u^h \rangle + \varphi(\gamma_\varphi u^h, \gamma_\varphi v^h) - \varphi(\gamma_\varphi u^h, \gamma_\varphi u^h)$$
$$+ j^{h,0}(\gamma_j u^h; \gamma_j v^h - \gamma_j u^h) \geq \langle f, v^h - u^h \rangle \quad \text{for all } v^h \in K^h. \tag{4.32}$$

Here, $j^{h,0}$ is a finite-dimensional version of the Clarke generalized directional derivative (see (3.3))

$$j^{h,0}(z_1^h; z_2^h) = \limsup_{z_3^h \to z_1^h, \lambda \downarrow 0} \frac{j(z_3^h + \lambda z_2^h) - j(z_3^h)}{\lambda},$$

$$\text{for all } z_1^h, z_2^h \in X_j^h := \gamma_j(X^h) \tag{4.33}$$

where the limit is taken for $z_3^h \in X_j^h$ and $\lambda \in \mathbb{R}$. Easily, we see the inequality

$$j^{h,0}(z_1^h; z_2^h) \le j^0(z_1^h; z_2^h) \quad \text{for all } z_1^h, z_2^h \in X_j^h, \tag{4.34}$$

and from this,

$$j^{h,0}(z_1^h; z_2^h - z_1^h) + j^{h,0}(z_2^h; z_1^h - z_2^h) \le j^0(z_1^h; z_2^h - z_1^h) + j^0(z_2^h; z_1^h - z_2^h)$$

$$\text{for all } z_1^h, z_2^h \in X_j^h.$$

Then, we can apply the arguments of the proof of Theorem 4.4 in the setting of the finite-dimensional space X^h to conclude that the discretized hemivariational inequality (4.32) has a unique solution $u^h \in K^h$. Using the relation (4.34), we know that the solution $u^h \in K^h$ of (4.32) is a solution of Problem 4.6 which is also unique. Finally, in the case of (4.30), by (4.29),

$$j^0(z_1^h; z_2^h) \le \int_D j_0^0(z_1^h; z_2^h),$$

and we can verify that the solution $u^h \in K^h$ of (4.32) or Problem 4.6 is a solution of (4.31) which is also unique.

We choose to do numerical analysis of the abstract Problem 4.3 with Problem 4.6 in this section since the main ideas and techniques for convergence analysis and error estimation can be explained more concisely and in general forms, and the analysis of the numerical method such as (4.31) is conducted very similarly. $\qquad\square$

4.3. Convergence under basic solution regularity

In this subsection, we provide a general discussion of convergence for the numerical solution of Problem 4.6. The key point is that the convergence is shown under the minimal solution regularity $u \in K$ that is available from Theorem 4.4. For convergence analysis, we will need $\{K^h\}_h$ to approximate K in the following sense (see Glowinski, Lions and Trémolières 1981):

$$v^h \in K^h \text{ and } v^h \rightharpoonup v \text{ in } X \text{ imply } v \in K, \tag{4.35}$$

$$\text{for all } v \in K, \ \exists v^h \in K^h \text{ such that } v^h \to v \text{ in } X \text{ as } h \to 0. \tag{4.36}$$

Note that by (4.18), for any $z \in X_\varphi$, $\varphi(z, \cdot) \colon X_\varphi \to \mathbb{R}$ is convex and lower semicontinuous. Thus, $\varphi(z, \cdot) \colon X_\varphi \to \mathbb{R}$ is continuous.

The following uniform boundedness property will be useful for convergence analysis of the numerical solutions.

Proposition 4.8. The discrete solution u^h is uniformly bounded with respect to h: $\|u^h\|_X \le M$ for some constant $M > 0$ independent of h.

Proof. Since K is non-empty, there is an element $u_0 \in K$. We fix one such element. Then by (4.36), there exists $u_0^h \in K^h$ such that

$$u_0^h \to u_0 \text{ in } X \text{ as } h \to 0.$$

We let $v^h = u_0^h$ in (4.28) to get

$$\langle Au^h, u_0^h - u^h \rangle + \varphi(\gamma_\varphi u^h, \gamma_\varphi u_0^h) - \varphi(\gamma_\varphi u^h, \gamma_\varphi u^h)$$
$$+ j^0(\gamma_j u^h; \gamma_\varphi u_0^h - \gamma_j u^h) \ge \langle f, u_0^h - u^h \rangle.$$

Then from

$$m_A \|u^h - u_0^h\|_X^2 \le \langle Au^h, u^h - u_0^h \rangle - \langle Au_0^h, u^h - u_0^h \rangle,$$

we have

$$m_A \|u^h - u_0^h\|_X^2 \le \varphi(\gamma_\varphi u^h, \gamma_\varphi u_0^h) - \varphi(\gamma_\varphi u^h, \gamma_\varphi u^h) + j^0(\gamma_j u^h; \gamma_j u_0^h - \gamma_j u^h)$$
$$+ \langle f, u^h - u_0^h \rangle - \langle Au_0^h, u^h - u_0^h \rangle. \tag{4.37}$$

In (4.18), take $z_1 = z_3 = \gamma_\varphi u^h$ and $z_2 = z_4 = \gamma_\varphi u_0^h$,

$$\varphi(\gamma_\varphi u^h, \gamma_\varphi u_0^h) - \varphi(\gamma_\varphi u^h, \gamma_\varphi u^h) \le \varphi(\gamma_\varphi u_0^h, \gamma_\varphi u_0^h) - \varphi(\gamma_\varphi u_0^h, \gamma_\varphi u^h)$$
$$+ \alpha_\varphi \|\gamma_\varphi(u^h - u_0^h)\|_{X_\varphi}^2. \tag{4.38}$$

Use (4.18) again, this time taking $z_1 = z_4 = \gamma_\varphi u_0^h$, $z_2 = \gamma_\varphi u_0$ and $z_3 = \gamma_\varphi u^h$,

$$\varphi(\gamma_\varphi u_0^h, \gamma_\varphi u_0^h) - \varphi(\gamma_\varphi u_0^h, \gamma_\varphi u^h) \le \varphi(\gamma_\varphi u_0, \gamma_\varphi u_0^h)$$
$$- \varphi(\gamma_\varphi u_0, \gamma_\varphi u^h) + \alpha_\varphi \|\gamma_\varphi(u_0^h - u_0)\|_{X_\varphi} \|\gamma_\varphi(u^h - u_0)\|_{X_\varphi}. \tag{4.39}$$

Use the lower bound (see Atkinson and Han 2009, p. 433)

$$\varphi(\gamma_\varphi u_0, z) \ge c_3 + c_4 \|z\|_{X_\varphi} \quad \text{for all } z \in X_\varphi$$

for some constants c_3 and c_4, not necessarily positive. Then

$$-\varphi(\gamma_\varphi u_0, \gamma_\varphi u^h) \le -c_3 - c_4 \|\gamma_\varphi u^h\|_{X_\varphi}. \tag{4.40}$$

Take $z_1 = \gamma_j u^h$ and $z_2 = \gamma_j u_0^h$ in (4.19(c)) to obtain

$$j^0(\gamma_j u^h; \gamma_j u_0^h - \gamma_j u^h) \le \alpha_j \|\gamma_j(u_0^h - u^h)\|_{X_j}^2 - j^0(\gamma_j u_0^h; \gamma_j u^h - \gamma_j u_0^h).$$

By (4.19(b)),

$$-j^0(\gamma_j u_0^h; \gamma_j u^h - \gamma_j u_0^h) \le (c_0 + c_1 \|\gamma_j u_0^h\|_{X_j}) \|\gamma_j(u^h - u_0^h)\|_{X_j}.$$

Thus,

$$j^0(\gamma_j u^h; \gamma_j u_0^h - \gamma_j u^h) \le \alpha_j c_j^2 \|u_0^h - u^h\|_X^2$$
$$+ c_j(c_0 + c_1\|\gamma_j u_0^h\|_{X_j})\|u_0^h - u^h\|_X. \qquad (4.41)$$

Use (4.38), (4.39), (4.40), and (4.41) in (4.37) to obtain

$$m_A\|u^h - u_0^h\|_X^2 \le \varphi(\gamma_\varphi u_0, \gamma_\varphi u_0^h) - c_3 - c_4\|\gamma_\varphi u^h\|_{X_\varphi} + \alpha_\varphi\|\gamma_\varphi(u^h - u_0^h)\|_{X_\varphi}^2$$
$$+ \alpha_\varphi\|\gamma_\varphi(u_0^h - u_0)\|_{X_\varphi}\|\gamma_\varphi(u^h - u_0)\|_{X_\varphi}$$
$$+ (c_0 + c_1\|\gamma_j u_0^h\|_{X_j})\|\gamma_j(u^h - u_0^h)\|_{X_j}$$
$$+ \alpha_j\|\gamma_j(u_0^h - u^h)\|_{X_j}^2 + \langle f - Au_0^h, u^h - u_0^h\rangle.$$

Since $u_0^h \to u_0$ in X, we know that $\|u_0^h\|_X$, and then also $\|\gamma_j u_0^h\|_{X_j}$ and $\|Au_0^h\|_{X^*}$ are uniformly bounded with respect to h. Finally, by the smallness condition (4.20), we conclude that $\|u^h - u_0^h\|_X$, and then also $\|u^h\|_X$ is uniformly bounded in h. $\qquad\square$

We now prove the convergence of the numerical solutions under the minimal solution regularity $u \in K$. In applications to contact mechanics, γ_φ and γ_j are trace operators from an $H^1(\Omega)$-based space to $L^2(\Gamma_3)$-based spaces, and are thus compact operators.

Theorem 4.9. Assume (4.1), (4.2), (4.5), (4.16)–(4.20), (4.35) and (4.36). Assume further that $\gamma_\varphi\colon X \to X_\varphi$ and $\gamma_j\colon X \to X_j$ are compact operators, and $A\colon X \to X^*$ is continuous. Then,

$$u^h \to u \quad \text{in } X \text{ as } h \to 0. \qquad (4.42)$$

Proof. The proof consists of two steps. First we show the weak convergence of the numerical solutions. According to Theorem 4.5, the solution $u^h \in K^h$ of Problem 4.6 is characterized by the inequality

$$\langle Av^h, v^h - u^h\rangle + \varphi(\gamma_\varphi u^h, \gamma_\varphi v^h) - \varphi(\gamma_\varphi u^h, \gamma_\varphi u^h)$$
$$+ j^0(\gamma_j u^h; \gamma_j v^h - \gamma_j u^h) \ge \langle f, v^h - u^h\rangle \quad \text{for all } v^h \in K^h. \qquad (4.43)$$

Note that $\{u^h\}$ is bounded in X, by Proposition 4.8. Since X is reflexive and the operators $\gamma_\varphi\colon X \to X_\varphi$ and $\gamma_j\colon X \to X_j$ are compact, there exists a subsequence $\{u^{h'}\} \subset \{u^h\}$ and an element $w \in X$ such that

$$u^{h'} \rightharpoonup w \text{ in } X, \quad \gamma_\varphi u^{h'} \to \gamma_\varphi w \text{ in } X_\varphi, \quad \gamma_j u^{h'} \to \gamma_j w \text{ in } X_j.$$

By the assumption (4.35), we know that $w \in K$.

Fix an arbitrary element $v \in K$. By the assumption (4.36), we can find a sequence $v^{h'} \in K^{h'}$ such that $v^{h'} \to v$ in X as $h' \to 0$. Then, as $h' \to 0$, the following hold:

$$Av^{h'} \rightarrow Av, \quad \langle Av^{h'}, v^{h'} - u^{h'} \rangle \rightarrow \langle Av, v - w \rangle,$$

$$j^0(\gamma_j w; \gamma_j v - \gamma_j w) \geq \limsup j^0(\gamma_j u^{h'}; \gamma_j v^{h'} - \gamma_j u^{h'}),$$

$$\langle f, v^{h'} - u^{h'} \rangle \rightarrow \langle f, v - w \rangle.$$

From (4.18) with $z_1 = \gamma_\varphi w$, $z_2 = z_4 = \gamma_\varphi u^{h'}$, and $z_3 = \gamma_\varphi v^{h'}$, we have

$$\varphi(\gamma_\varphi u^{h'}, \gamma_\varphi v^{h'}) - \varphi(\gamma_\varphi u^{h'}, \gamma_\varphi u^{h'})$$
$$\leq \varphi(\gamma_\varphi w, \gamma_\varphi v^{h'}) - \varphi(\gamma_\varphi w, \gamma_\varphi u^{h'})$$
$$+ \alpha_\varphi \|\gamma_\varphi(w - u^{h'})\|_{X_\varphi} \|\gamma_\varphi(v^{h'} - u^{h'})\|_{X_\varphi}. \tag{4.44}$$

Use this inequality in (4.43) with $h = h'$,

$$\langle Av^{h'}, v^{h'} - u^{h'} \rangle + \varphi(\gamma_\varphi w, \gamma_\varphi v^{h'}) - \varphi(\gamma_\varphi w, \gamma_\varphi u^{h'})$$
$$+ \alpha_\varphi \|\gamma_\varphi(w - u^{h'})\|_{X_\varphi} \|\gamma_\varphi(v^{h'} - u^{h'})\|_{X_\varphi}$$
$$+ j^0(\gamma_j u^{h'}; \gamma_j v^{h'} - \gamma_j u^{h'}) \geq \langle f, v^{h'} - u^{h'} \rangle. \tag{4.45}$$

Note that $\|\gamma_\varphi(v^{h'} - u^{h'})\|_{X_\varphi}$ is bounded whereas $\|\gamma_\varphi(w - u^{h'})\|_{X_\varphi} \rightarrow 0$. Thus, taking the upper limit in (4.45) as $h' \rightarrow 0$, we find that

$$\langle Av, v - w \rangle + \varphi(\gamma_\varphi w, \gamma_\varphi v) - \varphi(\gamma_\varphi w, \gamma_\varphi w) + j^0(\gamma_j w; \gamma_j v - \gamma_j w) \geq \langle f, v - w \rangle.$$

This inequality holds for any $v \in K$. By Theorem 4.5, w is the solution u of Problem 4.3. So $u^{h'} \rightharpoonup u$ in X. Since the limit u does not depend on the subsequence $\{u^{h'}\}$, the entire family of numerical solutions converges weakly to u.

Next, we show the strong convergence $u^h \rightarrow u$ in X as $h \rightarrow 0$. By the assumption (4.36), there exists a sequence $\{\bar{u}^h\}$, $\bar{u}^h \in K^h$, such that $\bar{u}^h \rightarrow u$ in X as $h \rightarrow 0$. Applying (4.2(b)),

$$m_A \|u - u^h\|_X^2 \leq \langle Au - Au^h, u - u^h \rangle.$$

So

$$m_A \|u - u^h\|_X^2 \leq \langle Au, u - u^h \rangle - \langle Au^h, \bar{u}^h - u^h \rangle - \langle Au^h, u - \bar{u}^h \rangle.$$

From (4.28), we have

$$-\langle Au^h, \bar{u}^h - u^h \rangle \leq \varphi(\gamma_\varphi u^h, \gamma_\varphi \bar{u}^h) - \varphi(\gamma_\varphi u^h, \gamma_\varphi u^h)$$
$$+ j^0(\gamma_j u^h; \gamma_j \bar{u}^h - \gamma_j u^h) - \langle f, \bar{u}^h - u^h \rangle.$$

As in (4.44), we have

$$\varphi(\gamma_\varphi u^h, \gamma_\varphi \bar{u}^h) - \varphi(\gamma_\varphi u^h, \gamma_\varphi u^h) \leq \varphi(\gamma_\varphi u, \gamma_\varphi \bar{u}^h) - \varphi(\gamma_\varphi u, \gamma_\varphi u^h)$$
$$+ \alpha_\varphi \|\gamma_\varphi u - \gamma_\varphi u^h\|_{X_\varphi} \|\gamma_\varphi \bar{u}^h - \gamma_\varphi u^h\|_{X_\varphi}.$$

Combining the above three inequalities, we obtain that

$$m_A \|u - u^h\|_X^2 \leq \langle Au, u - u^h \rangle - \langle Au^h, u - \bar{u}^h \rangle$$
$$+ \varphi(\gamma_\varphi u, \gamma_\varphi \bar{u}^h) - \varphi(\gamma_\varphi u, \gamma_\varphi u^h)$$
$$+ \alpha_\varphi \|\gamma_\varphi u - \gamma_\varphi u^h\|_{X_\varphi} \|\gamma_\varphi \bar{u}^h - \gamma_\varphi u^h\|_{X_\varphi}$$
$$+ j^0(\gamma_j u^h; \gamma_j \bar{u}^h - \gamma_j u^h) - \langle f, \bar{u}^h - u^h \rangle. \tag{4.46}$$

Since $\varphi(\gamma_\varphi u, \cdot) \colon X_\varphi \to \mathbb{R}$ is continuous, as $h \to 0$,

$$\varphi(\gamma_\varphi u, \gamma_\varphi \bar{u}^h) \to \varphi(\gamma_\varphi u, \gamma_\varphi u),$$
$$\varphi(\gamma_\varphi u, \gamma_\varphi u^h) \to \varphi(\gamma_\varphi u, \gamma_\varphi u).$$

Also note that $\gamma_\varphi u^h \to \gamma_\varphi u$ in X_φ, $\gamma_j u^h \to \gamma_j u$ in X_j, $\bar{u}^h \to u$ in X and therefore $\|\gamma_\varphi(\bar{u}^h - u^h)\|_{X_\varphi} \to 0$, $\|\gamma_j(\bar{u}^h - u^h)\|_{X_j} \to 0$, and $\|\gamma_\varphi u - \gamma_\varphi u^h\|_{X_\varphi} \to 0$. Consequently, from (4.46),

$$\limsup_{h \to 0} \|u - u^h\|_X^2 \leq 0.$$

This implies the strong convergence $u^h \to u$ in X. $\square$

4.4. Error estimation

We now turn to the derivation of error estimates. In this subsection, we assume (4.1), (4.2), (4.5), (4.16)–(4.20). We will further assume that the operator $A \colon X \to X^*$ is Lipschitz continuous, that is, for a constant $L_A > 0$,

$$\|Au - Av\|_{X^*} \leq L_A \|u - v\|_X \quad \text{for all } u, v \in X. \tag{4.47}$$

We comment that this assumption implies (4.2(a)).

Let $v \in K$ and $v^h \in K^h$ be arbitrary. By (4.2(b)) with $v_1 = u$ and $v_2 = u^h$,

$$m_A \|u - u^h\|_X^2 \leq \langle Au - Au^h, u - u^h \rangle,$$

which is rewritten as

$$m_A \|u - u^h\|_X^2 \leq \langle Au - Au^h, u - v^h \rangle + \langle Au, v^h - u \rangle + \langle Au, v - u^h \rangle$$
$$+ \langle Au, u - v \rangle + \langle Au^h, u^h - v^h \rangle. \tag{4.48}$$

Applying (4.15),

$$\langle Au, u - v \rangle \leq \varphi(\gamma_\varphi u, \gamma_\varphi v) - \varphi(\gamma_\varphi u, \gamma_\varphi u)$$
$$+ j^0(\gamma_j u; \gamma_j v - \gamma_j u) - \langle f, v - u \rangle.$$

Applying (4.28),

$$\langle Au^h, u^h - v^h \rangle \leq \varphi(\gamma_\varphi u^h, \gamma_\varphi v^h) - \varphi(\gamma_\varphi u^h, \gamma_\varphi u^h)$$
$$+ j^0(\gamma_j u^h; \gamma_j v^h - \gamma_j u^h) - \langle f, v^h - u^h \rangle.$$

Using these inequalities in (4.48), after some rearrangement of the terms, we have

$$m_A \|u - u^h\|_X^2 \leq \langle Au - Au^h, u - v^h \rangle + R_u(v^h, u) + R_u(v, u^h)$$
$$+ I_\varphi(u^h, v^h) + I_j(v, v^h), \tag{4.49}$$

where

$$R_u(v, w) := \langle Au, v - w \rangle + \varphi(\gamma_\varphi u, \gamma_\varphi v) - \varphi(\gamma_\varphi u, \gamma_\varphi w)$$
$$+ j^0(\gamma_j u; \gamma_j v - \gamma_j w) - \langle f, v - w \rangle, \tag{4.50}$$

$$I_\varphi(u^h, v^h) := \varphi(\gamma_\varphi u, \gamma_\varphi u^h) + \varphi(\gamma_\varphi u^h, \gamma_\varphi v^h)$$
$$- \varphi(\gamma_\varphi u, \gamma_\varphi v^h) - \varphi(\gamma_\varphi u^h, \gamma_\varphi u^h), \tag{4.51}$$

$$I_j(v, v^h) := j^0(\gamma_j u; \gamma_j v - \gamma_j u) + j^0(\gamma_j u^h; \gamma_j v^h - \gamma_j u^h)$$
$$- j^0(\gamma_j u; \gamma_j v^h - \gamma_j u) - j^0(\gamma_j u; \gamma_j v - \gamma_j u^h). \tag{4.52}$$

Let us bound the first and the last two terms on the right-hand side of (4.49). First,

$$\langle Au - Au^h, u - v^h \rangle \leq L_A \|u - u^h\|_X \|u - v^h\|_X.$$

So for any $\varepsilon > 0$ arbitrarily small,

$$\langle Au - Au^h, u - v^h \rangle \leq \varepsilon \|u - u^h\|_X^2 + c \|u - v^h\|_X^2 \tag{4.53}$$

for some constant c depending on ε. By (4.19(c)), we have

$$I_\varphi(u^h, v^h) \leq \alpha_\varphi \|\gamma_\varphi u - \gamma_\varphi u^h\|_{X_\varphi} \|\gamma_\varphi u^h - \gamma_\varphi v^h\|_{X_\varphi}$$
$$\leq \alpha_\varphi c_\varphi^2 (\|u - u^h\|_X^2 + \|u - u^h\|_X \|u - v^h\|_X).$$

Thus,

$$I_\varphi(u^h, v^h) \leq (\alpha_\varphi c_\varphi^2 + \varepsilon) \|u - u^h\|_X^2 + c \|u - v^h\|_X^2 \tag{4.54}$$

for another constant c depending on $\varepsilon > 0$. Applying the subadditivity of the generalized directional derivative (see Proposition 3.10(i)),

$$j^0(z; z_1 + z_2) \leq j^0(z; z_1) + j^0(z; z_2) \quad \text{for all } z, z_1, z_2 \in X_j,$$

we have

$$j^0(\gamma_j u; \gamma_j v - \gamma_j u) \leq j^0(\gamma_j u; \gamma_j v - \gamma_j u^h) + j^0(\gamma_j u; \gamma_j u^h - \gamma_j u),$$
$$j^0(\gamma_j u^h; \gamma_j v^h - \gamma_j u^h) \leq j^0(\gamma_j u^h; \gamma_j v^h - \gamma_j u) + j^0(\gamma_j u^h; \gamma_j u - \gamma_j u^h).$$

Thus,

$$I_j(v, v^h) \leq j^0(\gamma_j u^h; \gamma_j v^h - \gamma_j u) - j^0(\gamma_j u; \gamma_j v^h - \gamma_j u)$$
$$+ j^0(\gamma_j u; \gamma_j u^h - \gamma_j u) + j^0(\gamma_j u^h; \gamma_j u - \gamma_j u^h).$$

By (4.19(c)),

$$j^0(\gamma_j u; \gamma_j u^h - \gamma_j u) + j^0(\gamma_j u^h; \gamma_j u - \gamma_j u^h) \leq \alpha_j \|\gamma_j u - \gamma_j u^h\|_{X_j}^2.$$

Moreover,

$$|j^0(\gamma_j u^h; \gamma_j v^h - \gamma_j u)| \leq (c_0 + c_1 \|\gamma_j u^h\|_{X_j}) \|\gamma_j v^h - \gamma_j u\|_{X_j},$$
$$|j^0(\gamma_j u; \gamma_j v^h - \gamma_j u)| \leq (c_0 + c_1 \|\gamma_j u\|_{X_j}) \|\gamma_j v^h - \gamma_j u\|_{X_j}.$$

Combining the above four inequalities and using the fact that $\|\gamma_j u^h\|_{X_j}$ is uniformly bounded (see Proposition 4.8), we find that

$$I_j(v, v^h) \leq \alpha_j \|\gamma_j u - \gamma_j u^h\|_{X_j}^2 + c \|\gamma_j u - \gamma_j v^h\|_{X_j} \qquad (4.55)$$

for some constant $c > 0$ independent of h. Using (4.53), (4.54) and (4.55) in (4.49), we have

$$(m_A - \alpha_\varphi c_\varphi^2 - \alpha_j c_j^2 - 2\,\varepsilon) \|u - u^h\|_X^2 \leq c \|u - v^h\|_X^2 + c \|\gamma_j u - \gamma_j v^h\|_{X_j}$$
$$+ R_u(v^h, u) + R_u(v, u^h).$$

Recall the smallness assumption, $\alpha_\varphi c_\varphi^2 + \alpha_j c_j^2 < m_A$. We then choose $\varepsilon = (m_A - \alpha_\varphi c_\varphi^2 - \alpha_j c_j^2)/4 > 0$ and get the inequality

$$\|u - u^h\|_X^2 \leq c \inf_{v^h \in K^h} \left[\|u - v^h\|_X^2 + \|\gamma_j(u - v^h)\|_{X_j} + R_u(v^h, u) \right]$$
$$+ c \inf_{v \in K} R_u(v, u^h).$$

We summarize the result in the form of a theorem.

Theorem 4.10. Assume (4.1), (4.2), (4.5), (4.16)–(4.20) and (4.47). Then for the solution u of Problem 4.3 and the solution u^h of Problem 4.6, we have the Céa-type inequality

$$\|u - u^h\|_X \leq c \inf_{v^h \in K^h} \left[\|u - v^h\|_X + \|\gamma_j(u - v^h)\|_{X_j}^{1/2} + |R_u(v^h, u)|^{1/2} \right]$$
$$+ c \inf_{v \in K} |R_u(v, u^h)|^{1/2}. \qquad (4.56)$$

We remark that in the literature on error analysis of numerical solutions of variational inequalities, it is standard that Céa-type inequalities involve the square root of the approximation error of the solution in certain norms, due to the inequality form of the problems; see Falk (1974), Kikuchi and Oden (1988) and Han and Sofonea (2002).

To proceed further, we need to bound the residual term (4.50) and this depends on the problem to be solved. We illustrate this point in Section 7 in the context of a contact problem.

In the special case where $K = X$, we have $K^h = X^h$. Then the approximation is always internal, and we have the following reduced Céa-type inequality:

$$\|u - u^h\|_X \le c \inf_{v^h \in X^h} \left[\|u - v^h\|_X + \|\gamma_\varphi(u - v^h)\|_{X_\varphi}^{1/2} + \|\gamma_j(u - v^h)\|_{X_j}^{1/2} \right]. \quad (4.57)$$

Note that in deriving error estimates, the application of (4.57) is straightforward, whereas the application of (4.56) is more involved as one has to bound the residual-type term R_u.

5. A history-dependent variational–hemivariational inequality

We now consider an abstract variational–hemivariational inequality with a history-dependent operator. Let X be a reflexive Banach space and Y be a normed space. Let K be a subset of X, and $A\colon X \to X^*$, $\mathcal{S}\colon C(I;X) \to C(I;Y)$ be given operators. Consider also a function $\varphi\colon Y \times K \times K \to \mathbb{R}$, a locally Lipschitz function $j\colon X \to \mathbb{R}$ and a function $f\colon I \to X^*$. We associate with these data the following problem.

Problem 5.1. Find a function $u \in C(I;K)$ such that for all $t \in I$, the following inequality holds:

$$\langle Au(t), v - u(t) \rangle + \varphi(\mathcal{S}u(t), u(t), v) - \varphi(\mathcal{S}u(t), u(t), u(t))$$
$$+ j^0(u(t); v - u(t)) \ge \langle f(t), v - u(t) \rangle \quad \text{for all } v \in K. \quad (5.1)$$

In the study of Problem 5.1, besides the assumptions on K, A and j already introduced, we consider the following hypotheses.

$$\mathcal{S}\colon C(I;X) \to C(I;Y) \text{ is a history-dependent operator.} \quad (5.2)$$

$\varphi\colon Y \times K \times K \to \mathbb{R}$ is a function such that:

(a) $\varphi(y, u, \cdot)\colon K \to \mathbb{R}$ is convex and l.s.c. on K,
   for all $y \in Y$, $u \in K$;

(b) there exist constants $\alpha_\varphi > 0$ and $\beta_\varphi > 0$ such that $\quad (5.3)$
$$\varphi(y_1, u_1, v_2) - \varphi(y_1, u_1, v_1) + \varphi(y_2, u_2, v_1) - \varphi(y_2, u_2, v_2)$$
$$\le (\alpha_\varphi \|u_1 - u_2\|_X + \beta_\varphi \|y_1 - y_2\|_Y) \|v_1 - v_2\|_X$$
   for all $y_1, y_2 \in Y$, $u_1, u_2, v_1, v_2 \in K$.

$$\alpha_\varphi + \alpha_j < m_A. \quad (5.4)$$
$$f \in C(I;X^*). \quad (5.5)$$

Note that the function φ is assumed to be convex with respect to its third argument, while the function j is locally Lipschitz and it can be non-convex. For this reason, inequality (5.1) represents a *variational–hemivariational inequality*. In addition, the function φ in (5.1) depends on the history-dependent operator $\mathcal{S}$. Therefore, we refer to Problem 5.1 as a *history-dependent variational–hemivariational inequality.*

5.1. *Solution existence and uniqueness*

We have the following existence and uniqueness result on Problem 5.1.

Theorem 5.2. Let X be a reflexive Banach space, Y a normed space, and assume that (4.1), (4.2), (4.4) and (5.2)–(5.5) hold. Then, Problem 5.1 has a unique solution $u \in C(I; K)$.

Proof. We use a fixed-point argument. Given an $\eta \in C(I; X)$, define

$$y_\eta(t) = \mathcal{S}\eta(t) \quad \text{for all } t \in I. \tag{5.6}$$

Then $y_\eta \in C(I; Y)$. Consider the auxiliary problem of finding a function $u_\eta : I \to K$ such that for all $t \in I$, the following inequality holds:

$$\langle Au_\eta(t), v - u_\eta(t) \rangle + \varphi(y_\eta(t), u_\eta(t), v) - \varphi(y_\eta(t), u_\eta(t), u_\eta(t))$$
$$+ j^0(u(t); v - u(t)) \geq \langle f(t), v - u(t) \rangle \quad \text{for all } v \in K. \tag{5.7}$$

Applying Theorem 4.2, we know that there exists a unique element $u_\eta(t) \in K$ which solves this inequality, for each $t \in I$. Moreover, it can be shown that the function $u_\eta : I \to X$ is continuous. Therefore, there exists a unique function $u_\eta \in C(I; K)$ such that (5.7) holds, for all $t \in I$. This allows us to define an operator $\Lambda \colon C(I; X) \to C(I; K) \subset C(I; X)$ via the relation

$$\Lambda\eta = u_\eta, \quad \eta \in C(I; X). \tag{5.8}$$

Let us prove that the operator Λ has a unique fixed point $\eta^* \in C(I; K)$. For two arbitrary functions $\eta_1, \eta_2 \in C(I; X)$, let u_i denote the solution of the variational–hemivariational inequality (5.7) for $\eta = \eta_i$, *i.e.* $u_i = u_{\eta_i}$, $i = 1, 2$. Let $I_0 \subset I$ be a compact set and let $t \in I_0$. We use definition (5.8), inequality (5.7) and the assumptions (5.2), (5.3), to derive the inequality

$$\|\Lambda\eta_1(t) - \Lambda\eta_2(t)\|_X \leq c\, L_{I_0} \int_0^t \|\eta_1(s) - \eta_2(s)\|_X \, \mathrm{d}s. \tag{5.9}$$

Here and below, c is a positive constant that depends on A, φ and j. This shows that the operator $\Lambda \colon C(I; X) \to C(I; K) \subset C(I; X)$ is a history-dependent operator. Applying Theorem 3.20, we know that the operator Λ has a unique fixed point $\eta^* \in C(I; X)$. Since Λ takes values in $C(I; K)$, we have $\eta^* \in C(I; K)$. Moreover, by the definition of Λ, η^* satisfies the inequality

$$\langle A\eta^*(t), v - \eta^*(t) \rangle + \varphi(\mathcal{S}\eta^*(t), \eta^*(t), v) - \varphi(\mathcal{S}\eta^*(t), \eta^*(t), \eta^*(t))$$
$$+ j^0(\eta^*(t); v - \eta^*(t)) \geq \langle f(t), v - \eta^*(t) \rangle \quad \text{for all } v \in K$$

for all $t \in I$, that is, $\eta^* \in C(I; K)$ is a solution of the variational–hemivariational inequality (5.1).

The uniqueness part is a consequence of the uniqueness of the fixed point of the operator Λ. A direct proof is also possible and goes as follows. Let u_1 and u_2 be two solutions of Problem 5.1. Let $I_0 \subset I$ be a compact subset and let $t \in I_0$. From (5.1),

$$\langle Au_1(t) - Au_2(t), u_1(t) - u_2(t) \rangle$$
$$\leq \varphi(\mathcal{S}u_1(t), u_1(t), u_2(t)) - \varphi(\mathcal{S}u_1(t), u_1(t), u_1(t))$$
$$+ \varphi(\mathcal{S}u_2(t), u_2(t), u_1(t)) - \varphi(\mathcal{S}u_2(t), u_2(t), u_2(t))$$
$$+ j^0(u_1(t); u_2(t) - u_1(t)) + j^0(u_2(t); u_1(t) - u_2(t)).$$

Using assumptions (4.2), (4.4) and (5.3), after some elementary manipulations, we deduce that

$$\|u_1(t) - u_2(t)\|_X \leq c \|\mathcal{S}u_1(t) - \mathcal{S}u_2(t)\|_Y.$$

We use this inequality and assumption (5.2) to find that

$$\|u_1(t) - u_2(t)\|_X \leq cL_{I_0} \int_0^t \|u_1(s) - u_2(s)\|_X \, ds.$$

Next, it follows from the Gronwall inequality (see Lemma 3.22) that $u_1(t) - u_2(t) = 0$ for all $t \in I_0$. This implies that $u_1(t) = u_2(t)$ for all $t \in I$, that is, the solution is unique. $\qquad\square$

We now follow Xu *et al.* (2019) and introduce a variant of Problem 5.1 which is more convenient for the numerical analysis as well as for the applications in contact mechanics. To this end, besides the reflexive Banach space X and the normed space Y, we consider a real Banach space X_j, as well as an operator $\gamma_j : X \to X_j$. Moreover, we assume that $j \colon X_j \to \mathbb{R}$. The history-dependent variational–hemivariational inequality we consider is stated as follows.

Problem 5.3. Find a function $u \in C(I; K)$ such that for all $t \in I$, the following inequality holds:

$$\langle Au(t), v - u(t) \rangle + \varphi(\mathcal{S}u(t), u(t), v) - \varphi(\mathcal{S}u(t), u(t), u(t))$$
$$+ j^0(\gamma_j u(t); \gamma_j v - \gamma_j u(t)) \geq \langle f(t), v - u(t) \rangle \quad \text{for all } v \in K. \qquad (5.10)$$

The unique solvability of Problem 5.3 is given by the following result.

Theorem 5.4. Assume (4.1), (4.2), (4.17), (4.19), (5.2), (5.3), (5.5) and

$$\alpha_\varphi + \alpha_j c_j^2 < m_A. \qquad (5.11)$$

Then, Problem 5.3 has a unique solution $u \in C(I; K)$.

Theorem 5.4 can be proved by applying Theorem 5.2, similar to the proof of Theorem 4.4 by applying Theorem 4.2.

5.2. Temporally semi-discrete approximation

For definiteness, in the discussion of numerical methods for solving Problem 5.3, we focus on the particular case where the operator $\mathcal{S}\colon C(I;X) \to C(I;Y)$ has the following form (see Examples 3.15, 3.16):

$$\mathcal{S}v(t) = G\left(\int_0^t q(t,s)\, v(s)\, \mathrm{d}s + a_{\mathcal{S}} \right) \quad \text{for all } v \in C(I;X),\ t \in I, \quad (5.12)$$

where $G \in \mathcal{L}(X;Y)$, $q \in C(I \times I; \mathcal{L}(X))$, $a_{\mathcal{S}} \in X$. It can be shown that the operator $\mathcal{S}$ given in (5.12) is a history-dependent operator.

In this subsection we study a temporally semi-discrete method for solving Problem 5.3 and derive an error bound. In the case of the unbounded time interval $\mathbb{R}_+$, we choose a $T \in \mathbb{R}_+$ and consider the numerical solution on the interval $[0,T]$. In other words, whether the time interval of Problem 5.3 is bounded or not, we will consider Problem 5.3 on $I = [0,T]$ for computation. For simplicity in exposition, we partition the interval I uniformly; however, the discussion can be directly extended to the case of general non-uniform partitions. For a positive integer N, let $k = T/N$ be the time step-size, and let $t_n = nk$, $0 \le n \le N$ denote the node points. For a continuous function $v(t)$ with values in a function space, we write $v_n = v(t_n)$, $0 \le n \le N$. For the operator $\mathcal{S}$ in the form (5.12), we use the trapezoidal rule to approximate the integral $\int_0^t q(t,s)\, v(s)\, \mathrm{d}s$ at $t = t_n$. Denote $\|G\| = \|G\|_{\mathcal{L}(X;Y)}$ and $\|q\| = \|q\|_{C(I \times I; \mathcal{L}(X))}$. The trapezoidal rule for a continuous function $w(t)$ is

$$\int_0^{t_n} w(s)\, \mathrm{d}s \approx k \sum_{i=0}^{n}{}' w_i, \quad 1 \le n \le N, \quad (5.13)$$

where a prime indicates that in the summation, the first and the last terms are to be halved. Then $\mathcal{S}_n := \mathcal{S}(t_n)$ is approximated by $\mathcal{S}_n^k$ defined by the relation

$$\mathcal{S}_n^k v := G\left(k \sum_{i=0}^{n}{}' q(t_n, t_i)v_i + a_{\mathcal{S}} \right), \quad 1 \le n \le N, \quad (5.14)$$

for a grid function $v = \{v_n\}_{n=0}^N$. Note that $\mathcal{S}_0^k v = G(a_{\mathcal{S}})$. For a continuous function $v \in C(I;X)$, we define $\mathcal{S}_n^k v$ by (5.14) with $v_i = v(t_i)$, $0 \le i \le n$.

Consider the following temporally semi-discrete scheme for Problem 5.3.

Problem 5.5. Find $u^k := \{u_n^k\}_{n=0}^N \subset K$ such that

$$\langle Au_n^k, v - u_n^k \rangle + \varphi(\mathcal{S}_n^k u^k, u_n^k, v) - \varphi(\mathcal{S}_n^k u^k, u_n^k, u_n^k)$$
$$+ j^0(\gamma_j u_n^k; \gamma_j v - \gamma_j u_n^k) \ge \langle f_n, v - u_n^k \rangle \quad \text{for all } v \in K. \quad (5.15)$$

Regarding the solution existence and uniqueness for Problem 5.5, we have the following result.

Theorem 5.6. Keep the assumptions stated in Theorem 5.4. If the time step-size is such that

$$k < \frac{2(m_A - \alpha_\varphi - \alpha_j c_j^2)}{\beta_\varphi \|G\| \|q\|}, \tag{5.16}$$

then Problem 5.5 has a unique solution.

Proof. We prove the result via induction. The following arguments also apply to show the existence and uniqueness of u_0^k. For $n \geq 1$, with $\{u_i^k\}_{i \leq n-1}$ known, let us show that the inequality (5.15) uniquely determines $u_n^k \in K$. The proof is completed via an application of the Banach fixed-point theorem.

For any $\eta \in X$, denote

$$y_\eta = S_n^k u^{k\eta}, \tag{5.17}$$

where $u^{k\eta} = \{u_0^k, \ldots, u_{n-1}^k, \eta\}$. According to Theorem 4.4, we know that there exists a unique $u_\eta \in K$ such that

$$\langle Au_\eta, v - u_\eta \rangle + \varphi(y_\eta, u_\eta, v) - \varphi(y_\eta, u_\eta, u_\eta)$$
$$+ j^0(\gamma_j u_\eta; \gamma_j v - \gamma_j u_\eta) \geq \langle f_n, v - u_\eta \rangle \quad \text{for all } v \in K \tag{5.18}$$

under the stated assumptions, where $f_n = f(t_n)$. This allows us to define an operator $\Lambda : X \to K$ by

$$\Lambda \eta = u_\eta. \tag{5.19}$$

Let us show that Λ is a contractive mapping. For any $\eta_1, \eta_2 \in X$, denote $y_i = y_{\eta_i}$, $u_i = u_{\eta_i}$, $u^{k\eta_i} = \{u_0^k, \ldots, u_{n-1}^k, \eta_i\}$, $i = 1, 2$. Then $u_1 \in K$ satisfies

$$\langle Au_1, v - u_1 \rangle + \varphi(y_1, u_1, v) - \varphi(y_1, u_1, u_1)$$
$$+ j^0(\gamma_j u_1; \gamma_j v - \gamma_j u_1) \geq \langle f_n, v - u_1 \rangle \quad \text{for all } v \in K, \tag{5.20}$$

and $u_2 \in K$ satisfies

$$\langle Au_2, v - u_2 \rangle + \varphi(y_2, u_2, v) - \varphi(y_2, u_2, u_2)$$
$$+ j^0(\gamma_j u_2; \gamma_j v - \gamma_j u_2) \geq \langle f_n, v - u_2 \rangle \quad \text{for all } v \in K. \tag{5.21}$$

Take $v = u_2$ in (5.20) and $v = u_1$ in (5.21), and add the two inequalities to obtain

$$\langle Au_1 - Au_2, u_1 - u_2 \rangle \leq \varphi(y_1, u_1, u_2) - \varphi(y_1, u_1, u_1)$$
$$+ \varphi(y_2, u_2, u_1) - \varphi(y_2, u_2, u_2)$$
$$+ j^0(\gamma_j u_1; \gamma_j u_2 - \gamma_j u_1) + j^0(\gamma_j u_2; \gamma_j u_1 - \gamma_j u_2).$$

We use assumptions (4.2(b)), (5.3(b)) and (4.19(c)) to get

$$(m_A - \alpha_\varphi - \alpha_j c_j^2) \|u_1 - u_2\|_X \leq \beta_\varphi \|y_1 - y_2\|_Y.$$

This inequality can be rewritten as

$$\|\Lambda\eta_1 - \Lambda\eta_2\|_X \leq \frac{\beta_\varphi}{m_A - \alpha_\varphi - \alpha_j c_j^2}\|y_1 - y_2\|_Y. \tag{5.22}$$

Now,

$$y_1 - y_2 = \mathcal{S}_n^k u^{k\eta_1} - \mathcal{S}_n^k u^{k\eta_2}$$

$$= G\left(k\sum_{i=0}^{n}{}' q(t_n, t_i)u_i^{k\eta_1} + a_\mathcal{S}\right) - G\left(k\sum_{i=0}^{n}{}' q(t_n, t_i)u_i^{k\eta_2} + a_\mathcal{S}\right).$$

Then,

$$\|y_1 - y_2\|_Y \leq \frac{k}{2}\|G\|\,\|q\|\,\|\eta_1 - \eta_2\|_X. \tag{5.23}$$

Use (5.23) in (5.22) to obtain

$$\|\Lambda\eta_1 - \Lambda\eta_2\|_X \leq \frac{k}{2}\frac{\beta_\varphi}{m_A - \alpha_\varphi - \alpha_j c_j^2}\|G\|\|q\|\|\eta_1 - \eta_2\|_X. \tag{5.24}$$

By the smallness condition (5.16),

$$\frac{k}{2}\frac{\beta_\varphi}{m_A - \alpha_\varphi - \alpha_j c_j^2}\|G\|\|q\| < 1.$$

Applying Theorem 3.19, we conclude that the operator Λ has a unique fixed point $\eta^* \in K$. By the definitions (5.17) and (5.19),

$$y_{\eta^*} = \mathcal{S}_n^k u^{k\eta^*}, \quad u_{\eta^*} = \eta^*. \tag{5.25}$$

Considering the inequality (5.18) with $\eta = \eta^*$, we know that $u_n^k = \eta^* \in K$ is the unique solution of (5.15). $\qquad\square$

Before deriving an error bound for the semi-discrete solution defined by Problem 5.5, we present a preliminary result.

Lemma 5.7. Let $\mathcal{S}\colon C(I;X) \to C(I;Y)$ be defined by (5.12) and let $\mathcal{S}_n^k$ be defined by (5.14), where $G \in \mathcal{L}(X;Y)$, $q \in C^2(I \times I; \mathcal{L}(X))$, $a_\mathcal{S} \in X$. Assume $u \in W^{2,\infty}(I;X)$. Then for a constant c depending on G and q,

$$\|\mathcal{S}_n u - \mathcal{S}_n^k u\|_Y \leq c\,k^2\|u\|_{W^{2,\infty}(I;X)}. \tag{5.26}$$

Proof. By definition,

$$\mathcal{S}_n u - \mathcal{S}_n^k u = G\left(\int_0^t q(t,s)\,u(s)\,\mathrm{d}s + a_\mathcal{S}\right) - G\left(k\sum_{i=0}^{n}{}' q(t_n, t_i)u_i + a_\mathcal{S}\right).$$

Then,

$$\|\mathcal{S}_n u - \mathcal{S}_n^k u\|_Y \leq \|G\|\left\|\int_0^{t_n} q(t_n, s)\,u(s)\,\mathrm{d}s - k\sum_{i=0}^{n}{}' q(t_n, t_i)\,u_i\right\|_X.$$

With the use of Taylor's expansion on each of the subintervals $[0, t_1]$, $[t_1, t_2]$, $\ldots$, $[t_{n-1}, t_n]$, we find that

$$\left\| \int_0^{t_n} q(t_n, s)\, u(s)\, \mathrm{d}s - k \sum_{i=0}^{n}{}' q(t_n, t_i)\, u_i \right\|_X$$

$$\leq c\, k^2 \left\| \left(\frac{\mathrm{d}}{\mathrm{d}s} \right)^2 [q(t_n, s)\, u(s)] \right\|_{L^\infty((0, t_n); X)}.$$

Thus, (5.26) holds. $\qquad\qquad\qquad\qquad\qquad\qquad\qquad\qquad\qquad\qquad\square$

Theorem 5.8. Keep the assumptions of Theorem 5.4 and Lemma 5.7. Then for the semi-discrete solution of Problem 5.5, we have the following error bound:

$$\max_{0 \leq n \leq N} \|u_n - u_n^k\|_X \leq c\, k^2. \tag{5.27}$$

Proof. We take $v = u_n^k$ in the inequality (5.10) at $t = t_n$ to get

$$\langle Au_n, u_n^k - u_n \rangle + \varphi(\mathcal{S}_n u, u_n, u_n^k) - \varphi(\mathcal{S}_n u, u_n, u_n)$$
$$+ j^0(\gamma_j u_n; \gamma_j u_n^k - \gamma_j u_n) \geq \langle f_n, u_n^k - u_n \rangle, \tag{5.28}$$

where $\mathcal{S}_n u = G(\int_0^{t_n} q(t_n, s)\, u(s)\, \mathrm{d}s + a_S)$. Take $v = u_n$ in (5.15),

$$\langle Au_n^k, u_n - u_n^k \rangle + \varphi(\mathcal{S}_n^k u^k, u_n^k, u_n) - \varphi(\mathcal{S}_n^k u^k, u_n^k, u_n^k)$$
$$+ j^0(\gamma_j u_n^k; \gamma_j u_n - \gamma_j u_n^k) \geq \langle f_n, u_n - u_n^k \rangle. \tag{5.29}$$

Add (5.28) and (5.29),

$$\langle Au_n - Au_n^k, u_n - u_n^k \rangle \leq \varphi(\mathcal{S}_n u, u_n, u_n^k) - \varphi(\mathcal{S}_n u, u_n, u_n)$$
$$+ \varphi(\mathcal{S}_n^k u^k, u_n^k, u_n) - \varphi(\mathcal{S}_n^k u^k, u_n^k, u_n^k)$$
$$+ j^0(\gamma_j u_n; \gamma_j u_n^k - \gamma_j u_n) + j^0(\gamma_j u_n^k; \gamma_j u_n - \gamma_j u_n^k).$$

By (4.2(b)),

$$m_A \|u_n - u_n^k\|_X^2 \leq \langle Au_n - Au_n^k, u_n - u_n^k \rangle.$$

By (5.3(b)),

$$\varphi(\mathcal{S}_n u, u_n, u_n^k) - \varphi(\mathcal{S}_n u, u_n, u_n) + \varphi(\mathcal{S}_n^k u^k, u_n^k, u_n) - \varphi(\mathcal{S}_n^k u^k, u_n^k, u_n^k)$$
$$\leq \alpha_\varphi \|u_n - u_n^k\|_X^2 + \beta_\varphi \|\mathcal{S}_n u - \mathcal{S}_n^k u^k\|_Y \|u_n - u_n^k\|_X.$$

By (4.19(c)) and (4.17),

$$j^0(\gamma_j u_n; \gamma_j u_n^k - \gamma_j u_n) + j^0(\gamma_j u_n^k; \gamma_j u_n - \gamma_j u_n^k) \leq \alpha_j c_j^2 \|u_n - u_n^k\|_X^2.$$

Thus,

$$m_A \|u_n - u_n^k\|_X^2 \leq \alpha_\varphi \|u_n - u_n^k\|_X^2 + \beta_\varphi \|\mathcal{S}_n u - \mathcal{S}_n^k u^k\|_Y \|u_n - u_n^k\|_X$$
$$+ \alpha_j c_j^2 \|u_n - u_n^k\|_X^2$$

or, equivalently,

$$(m_A - \alpha_\varphi - \alpha_j c_j^2)\|u_n - u_n^k\|_X \le \beta_\varphi \|\mathcal{S}_n u - \mathcal{S}_n^k u^k\|_Y. \tag{5.30}$$

Write

$$\|\mathcal{S}_n u - \mathcal{S}_n^k u^k\|_Y \le \|\mathcal{S}_n u - \mathcal{S}_n^k u\|_Y + \|\mathcal{S}_n^k u - \mathcal{S}_n^k u^k\|_Y.$$

By Lemma 5.7,

$$\|\mathcal{S}_n u - \mathcal{S}_n^k u\|_Y \le c\, k^2 \|u\|_{W^{2,\infty}(I;X)}.$$

It is easy to see that

$$\|\mathcal{S}_n^k u - \mathcal{S}_n^k u^k\|_Y \le k\,\|G\|\,\|q\| \sum_{i=0}^{n} \|u_i - u_i^k\|_X.$$

Hence,

$$\|\mathcal{S}_n u - \mathcal{S}_n^k u^k\|_Y \le c\,k^2 \|u\|_{W^{2,\infty}(I;X)} + k\|G\|\|q\| \sum_{i=0}^{n} \|u_i - u_i^k\|_X. \tag{5.31}$$

Using (5.31) in (5.30), we get

$$\|u_n - u_n^k\|_X \le k^2 \frac{c\,\beta_\varphi \|u\|_{W^{2,\infty}(I;X)}}{m_A - \alpha_\varphi - \alpha_j c_j^2} + k\frac{\beta_\varphi \|G\|\|q\|}{m_A - \alpha_\varphi - \alpha_j c_j^2} \sum_{i=0}^{n} \|u_i - u_i^k\|_X. \tag{5.32}$$

We then apply the discrete Gronwall's inequality (see Lemma 3.24) to derive the error bound (5.27) from (5.32). $\qquad\square$

5.3. Fully discrete approximation

Now we consider fully discrete approximations of Problem 5.3. In a fully discrete scheme, both the temporal and spatial variables are discretized. In addition to the notation and assumptions stated in Section 5.2 for the temporal discretization, we introduce a regular family of finite element partitions $\{\mathcal{T}^h\}$ for the spatial discretization. We use a finite element space $X^h \subset X$ that corresponds to the partition $\mathcal{T}^h$ and use a non-empty, convex and closed subset $K^h \subset X^h$ to approximate K. Focusing on internal approximations, we assume $K^h \subset K$. External approximations can be also considered, as in Section 4.

Apply the operator $\mathcal{S}_n^k$ defined in (5.14) on $u^{hk} = \{u_n^{hk}\}_{n=0}^{N} \subset X^h$:

$$\mathcal{S}_n^k u^{hk} := G\left(k \sum_{i=0}^{n}{}' q(t_n, t_i) u_i^{hk} + a_\mathcal{S}\right). \tag{5.33}$$

We then introduce a fully discrete approximation of Problem 5.3 as follows.

Problem 5.9. Find $u^{hk} := \{u_n^{hk}\}_{n=0}^N \subset K^h$ such that

$$\langle Au_n^{hk}, v^h - u_n^{hk} \rangle + \varphi(\mathcal{S}_n^k u^{hk}, u_n^{hk}, v^h) - \varphi(\mathcal{S}_n^k u^{hk}, u_n^{hk}, u_n^{hk})$$
$$+ j^0(\gamma_j u_n^{hk}; \gamma_j v^h - \gamma_j u_n^{hk}) \geq \langle f_n, v^h - u_n^{hk} \rangle \quad \text{for all } v^h \in K^h. \quad (5.34)$$

As in Theorem 5.6, Problem 5.9 has a unique solution under the assumptions stated in Theorem 5.6. Next we derive an error bound for the fully discrete scheme.

Theorem 5.10. Keep the assumptions stated in Theorem 5.4. Moreover, assume $A : X \to X^*$ is Lipschitz continuous, $q \in C^2(I \times I; \mathcal{L}(X))$, and $u \in W^{2,\infty}(I; X)$. Then for k sufficiently small, we have the error bound

$$\max_{0 \leq n \leq N} \|u_n - u_n^{hk}\|_X$$
$$\leq c \max_{0 \leq n \leq N} \inf_{v^h \in K^h} \left[\|u_n - v^h\|_X + \|\gamma_j(u_n - v^h)\|_{X_j}^{1/2} + |R_n(v^h, u_n)|^{1/2} \right] + c\,k^2,$$
$$(5.35)$$

where the residual-type term $R_n(v^h, u_n)$ is defined by

$$R_n(v^h, u_n) = \langle Au_n, v^h - u_n \rangle + \varphi(\mathcal{S}_n u, u_n, v^h) - \varphi(\mathcal{S}_n u, u_n, u_n)$$
$$+ j^0(\gamma_j u_n; \gamma_j v^h - \gamma_j u_n) - \langle f_n, v^h - u_n \rangle. \quad (5.36)$$

Proof. By (4.2(b)),

$$m_A \|u_n - u_n^{hk}\|_X^2 \leq \langle Au_n - Au_n^{hk}, u_n - u_n^{hk} \rangle. \quad (5.37)$$

Write

$$\langle Au_n - Au_n^{hk}, u_n - u_n^{hk} \rangle = \langle Au_n - Au_n^{hk}, u_n - v^h \rangle + \langle Au_n, v^h - u_n \rangle$$
$$+ \langle Au_n, u_n - u_n^{hk} \rangle + \langle Au_n^{hk}, u_n^{hk} - v^h \rangle. \quad (5.38)$$

Take $v = u_n^{hk}$ in (5.10) at $t = t_n$,

$$\langle Au_n, u_n^{hk} - u_n \rangle + \varphi(\mathcal{S}_n u, u_n, u_n^{hk}) - \varphi(\mathcal{S}_n u, u_n, u_n)$$
$$+ j^0(\gamma_j u_n; \gamma_j u_n^{hk} - \gamma_j u_n) \geq \langle f_n, u_n^{hk} - u_n \rangle,$$

that is,

$$\langle Au_n, u_n - u_n^{hk} \rangle \leq \varphi(\mathcal{S}_n u, u_n, u_n^{hk}) - \varphi(\mathcal{S}_n u, u_n, u_n)$$
$$+ j^0(\gamma_j u_n; \gamma_j u_n^{hk} - \gamma_j u_n) - \langle f_n, u_n - u_n^{hk} \rangle. \quad (5.39)$$

From (5.34),

$$\langle Au_n^{hk}, u_n^{hk} - v^h \rangle \leq \varphi(\mathcal{S}_n^k u^{hk}, u_n^{hk}, v^h) - \varphi(\mathcal{S}_n^k u^{hk}, u_n^{hk}, u_n^{hk})$$
$$+ j^0(\gamma_j u_n^{hk}; \gamma_j v^h - \gamma_j u_n^{hk}) - \langle f_n, v^h - u_n^{hk} \rangle. \quad (5.40)$$

Combining (5.37), (5.38), (5.39) and (5.40), we get

$$m_A\|u_n - u_n^{hk}\|_X^2 \le \langle Au_n - Au_n^{hk}, u_n - v^h\rangle + R_n(v^h, u_n)$$
$$+ E_{\varphi_1} + E_{\varphi_2} + E_j, \tag{5.41}$$

where $R_n(v^h, u_n)$ is defined in (5.36), and

$$E_{\varphi_1} = \varphi(\mathcal{S}_n u, u_n, u_n^{hk}) - \varphi(\mathcal{S}_n u, u_n, u_n)$$
$$+ \varphi(\mathcal{S}_n^k u^{hk}, u_n^{hk}, u_n) - \varphi(\mathcal{S}_n^k u^{hk}, u_n^{hk}, u_n^{hk}), \tag{5.42}$$
$$E_{\varphi_2} = \varphi(\mathcal{S}_n^k u^{hk}, u_n^{hk}, v^h) - \varphi(\mathcal{S}_n^k u^{hk}, u_n^{hk}, u_n)$$
$$+ \varphi(\mathcal{S}_n u, u_n, u_n) - \varphi(\mathcal{S}_n u, u_n, v^h), \tag{5.43}$$
$$E_j = j^0(\gamma_j u_n; \gamma_j u_n^{hk} - \gamma_j u_n) + j^0(\gamma_j u_n^{hk}; \gamma_j v^h - \gamma_j u_n^{hk})$$
$$- j^0(\gamma_j u_n; \gamma_j v^h - \gamma_j u_n). \tag{5.44}$$

Let us bound each of the terms on the right-hand side of (5.41). Let $L_A > 0$ denote the Lipschitz constant of the operator A. Then

$$\langle Au_n - Au_n^{hk}, u_n - v^h\rangle \le L_A\|u_n - u_n^{hk}\|_X\|u_n - v^h\|_X. \tag{5.45}$$

By (5.3(b)),

$$E_{\varphi_1} \le \alpha_\varphi\|u_n - u_n^{hk}\|_X^2 + \beta_\varphi\|\mathcal{S}_n u - \mathcal{S}_n^k u^{hk}\|_Y\|u_n - u_n^{hk}\|_X, \tag{5.46}$$
$$E_{\varphi_2} \le (\alpha_\varphi\|u_n - u_n^{hk}\|_X + \beta_\varphi\|\mathcal{S}_n u - \mathcal{S}_n^k u^{hk}\|_Y)\|u_n - v^h\|_X. \tag{5.47}$$

By the subadditivity of the generalized directional derivative (see Proposition 3.10),

$$j^0(\gamma_j u_n^{hk}; \gamma_j v^h - \gamma_j u_n^{hk}) \le j^0(\gamma_j u_n^{hk}; \gamma_j u_n - \gamma_j u_n^{hk})$$
$$+ j^0(\gamma_j u_n^{hk}; \gamma_j v^h - \gamma_j u_n).$$

Thus,

$$E_j \le j^0(\gamma_j u_n; \gamma_j u_n^{hk} - \gamma_j u_n) + j^0(\gamma_j u_n^{hk}; \gamma_j u_n - \gamma_j u_n^{hk})$$
$$+ j^0(\gamma_j u_n^{hk}; \gamma_j v^h - \gamma_j u_n) - j^0(\gamma_j u_n; \gamma_j v^h - \gamma_j u_n).$$

By (4.19(c)) and (4.17),

$$j^0(\gamma_j u_n; \gamma_j u_n^{hk} - \gamma_j u_n) + j^0(\gamma_j u_n^{hk}; \gamma_j u_n - \gamma_j u_n^{hk}) \le \alpha_j c_j^2\|u_n - u_n^{hk}\|_X^2.$$

By (4.19(b)) and (4.17), we have

$$|j^0(\gamma_j u_n; \gamma_j v^h - \gamma_j u_n)| \le (c_0 + c_1 c_j\|u_n\|_X)\|\gamma_j(u_n - v^h)\|_{X_j}$$

and

$$|j^0(\gamma_j u_n^{hk}; \gamma_j v^h - \gamma_j u_n)| \le (c_0 + c_1 c_j\|u_n^{hk}\|_X)\|\gamma_j(u_n - v^h)\|_{X_j}$$
$$\le c_1 c_j\|u_n - u_n^{hk}\|_X\|\gamma_j(u_n - v^h)\|_{X_j}$$
$$+ (c_0 + c_1 c_j\|u_n\|_X)\|\gamma_j(u_n - v^h)\|_{X_j}.$$

Hence,

$$E_j \leq (2c_0 + 2c_1 c_j \|u_n\|_X)\|\gamma_j(u_n - v^h)\|_{X_j} + \alpha_j c_j^2 \|u_n - u_n^{hk}\|_X^2$$
$$+ c_1 c_j^2 \|u_n - u_n^{hk}\|_X \|u_n - v^h\|_X. \tag{5.48}$$

From (5.41), (5.45)–(5.48), and using the condition (5.11), we deduce that

$$\|u_n - u_n^{hk}\|_X \leq c\big[\|u_n - v^h\|_X + \|\gamma_j(u_n - v^h)\|_{X_j}^{1/2} + |R_n(v^h, u_n)|^{1/2}\big]$$
$$+ c\|\mathcal{S}_n u - \mathcal{S}_n^k u^{hk}\|_Y. \tag{5.49}$$

By the triangle inequality,

$$\|\mathcal{S}_n u - \mathcal{S}_n^k u^{hk}\|_Y \leq \|\mathcal{S}_n u - \mathcal{S}_n^k u\|_Y + \|\mathcal{S}_n^k u - \mathcal{S}_n^k u^{hk}\|_Y.$$

The term $\|\mathcal{S}_n u - \mathcal{S}_n^k u\|_Y$ is bounded by Lemma 5.7. Since

$$\mathcal{S}_n^k u - \mathcal{S}_n^k u^{hk} = G\bigg(k\sum_{i=0}^n {}' q(t_n, t_i)u_i + a_{\mathcal{S}}\bigg) - G\bigg(k\sum_{i=0}^n {}' q(t_n, t_i)u_i^{hk} + a_{\mathcal{S}}\bigg),$$

we have

$$\|\mathcal{S}_n^k u - \mathcal{S}_n^k u^{hk}\|_Y \leq ck\sum_{i=0}^n {}' \|q(t_n, t_i)\|_{\mathcal{L}(X)}\|u_i - u_i^{hk}\|_X$$
$$\leq ck\sum_{i=0}^n \|u_i - u_i^{hk}\|_X.$$

Hence,

$$\|\mathcal{S}_n u - \mathcal{S}_n^k u^{hk}\|_Y \leq ck^2\|u\|_{W^{2,\infty}(I;X)} + ck\sum_{i=0}^n \|u_i - u_i^{hk}\|_X. \tag{5.50}$$

We combine (5.49) and (5.50) to get

$$\|u_n - u_n^{hk}\|_X \leq c\big[\|u_n - v^h\|_X + \|\gamma_j(u_n - v^h)\|_{X_j}^{1/2} + |R_n(v^h, u_n)|^{1/2}\big]$$
$$+ ck^2\|u\|_{W^{2,\infty}(I;X)} + ck\sum_{i=0}^n \|u_i - u_i^{hk}\|_X. \tag{5.51}$$

Applying the discrete Gronwall's inequality, Lemma 3.24, we derive the error bound (5.35) from (5.51). $\qquad \square$

Note that the square root involved in the error bound (5.35) results from the inequality feature of the problem. Proper bounding of the residual term $R_n(v^h, u_n)$ depends on the specific problem under consideration. In Section 8, we illustrate this point on the numerical solution of Problem 2.8.

6. An evolutionary hemivariational inequality

In this section we study a first-order hemivariational inequality involving a history-dependent operator. Let $V \subset H \subset V^*$ be an evolution triple of Banach spaces. Recall that $\langle \cdot, \cdot \rangle$ represents the duality pairing between V^* and V. We use $(\cdot, \cdot)$ and $\| \cdot \|_H$ for the inner product and norm of the space H. For $T > 0$, denote $\mathcal{V} = L^2(0, T; V)$, $\mathcal{V}^* = L^2(0, T; V^*)$, and $\mathcal{W} = \{v \in \mathcal{V} \mid \dot{v} \in \mathcal{V}^*\}$. Let V_1 and V_2 be two real Banach spaces, and let $\gamma_1 \in \mathcal{L}(V, V_1)$ and $\gamma_2 \in \mathcal{L}(V, V_2)$ be given and let c_1 and c_2 denote the operator norms of γ_1 and γ_2, respectively, that is,

$$\|\gamma_1 v\|_{V_1} \le c_1 \|v\|_V \quad \text{for all } v \in V,$$
$$\|\gamma_2 v\|_{V_2} \le c_2 \|v\|_V \quad \text{for all } v \in V.$$

For applications in the study of Problem 2.10 in Section 9, $V_1 = L^2(\Gamma_3)$, $V_2 = L^2(\Gamma_3)^d$, γ_1 is the normal trace operator on Γ_3, and γ_2 is the tangential trace operator on Γ_3.

6.1. Solution existence and uniqueness

The inequality under consideration reads as follows.

Problem 6.1.   Find $w \in \mathcal{W}$ such that

$$\langle \rho \dot{w}(t) + Aw(t) + \mathcal{S}w(t), v \rangle + j_1^0(\gamma_1 w(t); \gamma_1 v) + j_2^0(\gamma_2 w(t); \gamma_2 v)$$
$$\ge \langle f(t), v \rangle \quad \text{for all } v \in V, \text{ a.e. } t \in (0, T), \tag{6.1}$$
$$w(0) = w_0. \tag{6.2}$$

In the study of Problem 6.1, we will make assumptions on the data, commonly seen in literature in this areas. For $A \colon V \to V^*$, we assume:

(a) A is demicontinuous, *i.e.*
$$u_n \to u \text{ in } V \Longrightarrow Au_n \rightharpoonup Au \text{ in } V^*;$$

(b) $\|Av\|_{V^*} \le a_0 + a_1 \|v\|_V$ for all $v \in V$,
   with $a_0 \ge 0$ and $a_1 \ge 0$;

(c) A is strongly monotone, *i.e.* there is $m_A > 0$ such that
$$\langle Av_1 - Av_2, v_1 - v_2 \rangle \ge m_A \|v_1 - v_2\|_V^2$$
   for all $v_1, v_2 \in V$.

$$\left.\vphantom{\begin{array}{c} a \\ b \\ c \\ d \\ e \\ f \\ g \end{array}}\right\} \tag{6.3}$$

For $j_l \colon V_l \to \mathbb{R}$, $l = 1, 2$, we assume:

(a) $j_l(\cdot)$ is locally Lipschitz on V_l;

(b) $\|\partial j_l(z)\|_{V_l^*} \le c_{0l} + c_{1l} \|z\|_{V_l}$ for all $z \in V_l$, with $c_{0l}, c_{1l} \ge 0$;

(c) $j_l^0(z_1; z_2 - z_1) + j_l^0(z_2; z_1 - z_2) \le \alpha_l \|z_1 - z_2\|_{V_l}^2$
   for all $z_1, z_2 \in V_l$, with $\alpha_l \ge 0$,

$$\left.\vphantom{\begin{array}{c} a \\ b \\ c \\ d \\ e \\ f \end{array}}\right\} \tag{6.4}$$

and we impose a smallness assumption,

$$\max\{\alpha_1 c_1^2 + \alpha_2 c_2^2, 2\sqrt{2}(c_{11} + c_{12})\} < m_A. \tag{6.5}$$

For $\mathcal{S}\colon \mathcal{V} \to \mathcal{V}^*$, we assume it is a history-dependent operator, that is,

there is a constant $c_{\mathcal{S}} > 0$ such that $\tag{6.6}$

$$\|\mathcal{S}v_1(t) - \mathcal{S}v_2(t)\|_{\mathcal{V}^*} \le c_{\mathcal{S}} \int_0^t \|v_1(s) - v_2(s)\|_{\mathcal{V}}\, ds$$

for all $v_1, v_2 \in \mathcal{V}$, a.e. $t \in [0, T]$.

Finally, assume

$$f \in \mathcal{V}^*, \quad w_0 \in V, \quad \rho > 0. \tag{6.7}$$

We remark that it is possible to extend the following discussions to the case where ρ is a positive valued function. However, to simplify the presentation, we assume ρ is a positive constant. As usual, (6.4(b)) means

$$\|\xi\|_{V_l^*} \le c_{0l} + c_{1l}\|z\|_{V_l} \quad \text{for all } z \in V_l, \text{ for all } \xi \in \partial j_l(z).$$

The following existence and uniqueness result holds.

Theorem 6.2. Assume (6.3)–(6.7). Then Problem 6.1 has a unique solution.

Proof. We sketch the proof in four steps.

(i) *Existence of a solution to an intermediate problem.* Let $\xi \in \mathcal{V}^*$ and consider the following intermediate problem: find $w_\xi \in \mathcal{W}$ such that

$$\left. \begin{aligned} &\langle \rho \dot{w}_\xi(t) + A w_\xi(t) + \xi(t) - f(t), v - w_\xi(t)\rangle \\ &+ j_1^0(\gamma_1 w_\xi(t); \gamma_1 v - \gamma_1 w_\xi(t)) + j_2^0(\gamma_2 w_\xi(t); \gamma_2 v - \gamma_2 w_\xi(t)) \ge 0 \\ &\text{for all } v \in V, \text{ a.e. } t \in (0, T), \\[4pt] &w_\xi(0) = w_0. \end{aligned} \right\} \tag{6.8}$$

In order to find a solution to inequality (6.8), we define the functions $j\colon V \to \mathbb{R}$ and $\psi_\xi\colon (0, T) \times V \to \mathbb{R}$ by equalities

$$j(v) = j_1(\gamma_1 v) + j_2(\gamma_2 v), \tag{6.9}$$

$$\psi_\xi(t, v) = \langle \xi(t), v\rangle + j(v), \tag{6.10}$$

for all $v \in V$, a.e. $t \in (0, T)$. Then, we consider an additional intermediate problem, stated as follows: find $w_\xi \in \mathcal{W}$ such that

$$\left. \begin{aligned} &\rho\, \dot{w}_\xi(t) + A w_\xi(t) + \partial \psi_\xi(t, w_\xi(t)) \ni f(t) \text{ for a.e. } t \in (0, T), \\ &w_\xi(0) = w_0. \end{aligned} \right\} \tag{6.11}$$

The unique solvability of problem (6.11) follows from Theorem 3.18. Moreover, equality (6.10) implies that

$$\partial \psi_\xi(t, v) \subset \xi(t) + \partial j(v) \tag{6.12}$$

for all $v \in V$, a.e. $t \in (0, T)$. Therefore by (6.11) it is obvious that $w_\xi \in \mathcal{W}$ satisfies the following Cauchy problem:

$$\left. \begin{aligned} &\rho \dot{w}_\xi(t) + A w_\xi(t) + \partial j(w_\xi(t)) + \xi(t) \ni f(t) \quad \text{a.e. } t \in (0, T), \\ &w_\xi(0) = w_0. \end{aligned} \right\} \tag{6.13}$$

Furthermore, it is clear from the definitions of the Clarke subdifferential that every solution of problem (6.13) satisfies

$$\left. \begin{aligned} &\langle \rho \dot{w}_\xi(t) + A w_\xi(t) + \xi(t) - f(t), v - w_\xi(t) \rangle \\ &\quad + j^0(w_\xi(t); v - w_\xi(t)) \geq 0 \quad \text{for all } v \in V, \text{ a.e. } t \in (0, T), \\ &w_\xi(0) = w_0. \end{aligned} \right\} \tag{6.14}$$

Finally, we use equality (6.9) and Propositions 3.11, 3.12 to deduce that

$$j^0(w_\xi(t); v - w_\xi(t)) \leq j_1^0(\gamma_1 w_\xi(t); \gamma_1 v) + j_2^0(\gamma_2 w_\xi(t); \gamma_2 v) \tag{6.15}$$

for all $v \in V$, a.e. $t \in (0, T)$. We now combine (6.14) and (6.15) to see that $w_\xi \in \mathcal{W}$ is a solution of the intermediate problem (6.8).

(ii) *Uniqueness of a solution to the intermediate problem* (6.8). Let $w_1, w_2 \in \mathcal{W}$ be solutions to the problem (6.8). For simplicity in notation, in this part of the proof we skip the subscript ξ. We write the following two inequalities: the first one is for $w_1(t)$ with $w_2(t)$ as test function, the second one is for $w_2(t)$ with $w_1(t)$ as test function. We have $w_1(0) = w_2(0) = w_0$, and for a.e. $t \in (0, T)$,

$$\begin{aligned} &\langle \rho \dot{w}_1(t) + A w_1(t) - f(t) + \xi(t), w_2(t) - w_1(t) \rangle \\ &\quad + j_1^0(\gamma_1 w_1(t); \gamma_1 w_2(t) - \gamma_1 w_1(t)) + j_2^0(\gamma_2 w_1(t); \gamma_2 w_2(t) - \gamma_2 w_1(t)) \geq 0, \\ &\langle \rho \dot{w}_2(t) + A w_2(t) - f(t) + \xi(t), w_1(t) - w_2(t) \rangle \\ &\quad + j_1^0(\gamma_1 w_2(t); \gamma_1 w_1(t) - \gamma_1 w_2(t)) + j_2^0(\gamma_2 w_2(t); \gamma_2 w_1(t) - \gamma_2 w_2(t)) \geq 0. \end{aligned}$$

Adding these inequalities, we find that

$$\begin{aligned} &\rho \langle \dot{w}_1(t) - \dot{w}_2(t), w_1(t) - w_2(t) \rangle + \langle A w_1(t) - A w_2(t), w_1(t) - w_2(t) \rangle \\ &\quad \leq j_1^0(\gamma_1 w_1(t); \gamma_1 w_2(t) - \gamma_1 w_1(t)) + j_1^0(\gamma_1 w_2(t); \gamma_1 w_1(t) - \gamma_1 w_2(t)) \\ &\qquad + j_2^0(\gamma_2 w_1(t); \gamma_2 w_2(t) - \gamma_2 w_1(t)) + j_2^0(\gamma_2 w_2(t); \gamma_2 w_1(t) - \gamma_2 w_2(t)) \end{aligned}$$

for a.e. $t \in (0, T)$. Integrating the above inequality on the time interval

$(0, t)$, using conditions (6.3(c)) and (6.4(c)), we have

$$\frac{\rho}{2}\|w_1(t) - w_2(t)\|_H^2 - \frac{\rho}{2}\|w_1(0) - w_2(0)\|_H^2 + m_A \int_0^t \|w_1(s) - w_2(s)\|_V^2 \, ds$$

$$\leq (\alpha_1 c_1^2 + \alpha_2 c_2^2) \int_0^t \|w_1(s) - w_2(s)\|_V^2 \, ds$$

for all $t \in [0, T]$. Since $w_1(0) - w_2(0) = 0$ and $\alpha_1 c_1^2 + \alpha_2 c_2^2 < m_A$, by assumption (6.5) we obtain

$$\|w_1(t) - w_2(t)\|_H^2 = 0 \quad \text{for all } t \in [0, T].$$

This implies that $w_1(t) = w_2(t)$ for all $t \in [0, T]$, *i.e.* $w_1 = w_2$. In conclusion, a solution to the problem (6.8) is unique.

(iii) *A fixed-point property.* We now consider the operator $\Lambda \colon \mathcal{V}^* \to \mathcal{V}^*$ defined by

$$\Lambda \xi = \mathcal{S} w_\xi, \quad \xi \in \mathcal{V}^*,$$

where $w_\xi \in \mathcal{W}$ is the unique solution of the problem (6.8) corresponding to $\xi \in \mathcal{V}^*$. Using arguments similar to those used in the proof of the previous step, we prove that

$$\|\Lambda \xi_1(t) - \Lambda \xi_2(t)\|_{\mathcal{V}^*}^2 \leq c \int_0^t \|\xi_1(s) - \xi_2(s)\|_{\mathcal{V}^*}^2 \, ds \tag{6.16}$$

for all $\xi_i \in \mathcal{V}^*$, $i = 1, 2$ and a.e. $t \in (0, T)$ with $c > 0$. Then, using the arguments in the proof of Theorem 3.21, there exists a unique fixed point ξ^* of Λ, that is,

$$\xi^* \in \mathcal{V}^* \quad \text{and} \quad \Lambda \xi^* = \xi^*.$$

(iv) *Existence and uniqueness.* Let $\xi^* \in \mathcal{V}^*$ be the unique fixed point of the operator Λ. Let $w_{\xi^*} \in \mathcal{W}$ be the unique solution to the problem (6.8) corresponding to ξ^*. From the definition of operator Λ, we have

$$\xi^* = \mathcal{S} w_{\xi^*}.$$

Using this equality in problem (6.8), we conclude that w_{ξ^*} is the unique solution to Problem 6.1. $\qquad\square$

6.2. *Numerical analysis of a fully discrete scheme*

As in Sections 5.2 and 5.3, we introduce a uniform partition of the time interval into N subintervals. Then $k = T/N$ is the time step-size, and the temporal node points are $t_n = nk$, $0 \leq n \leq N$. For the spatial discretization, we use a finite-dimensional subspace V^h of V; in practice, V^h is usually a finite element space, h being the finite element mesh-size.

For simplicity, instead of $f \in V^*$ from (6.7), we assume

$$f \in C([0, T]; V^*). \tag{6.17}$$

Then the pointwise value $f_n = f(t_n) \in V^*$ is well-defined.

For definiteness, in the discussion of the numerical method for solving Problem 6.1, we consider the case where the operator $\mathcal{S} \colon V \to V^*$ is of the form (see (5.12))

$$\mathcal{S}v(t) = G\left(\int_0^t q(t, s)\, v(s)\, \mathrm{d}s + a_{\mathcal{S}}\right) \quad \text{for all } v \in V,\ t \in [0, T], \tag{6.18}$$

where $G \in \mathcal{L}(V; V^*)$, $q \in C([0, T]^2; \mathcal{L}(V))$, $a_{\mathcal{S}} \in V$. We use the backward divided difference to approximate the time derivative and use the left-point quadrature to define an approximation operator $\mathcal{S}^k$ for the history-dependent operator $\mathcal{S}$:

$$\mathcal{S}_n^k w^{hk} = G\left(k \sum_{i=0}^{n-1} q(t_n, t_i)\, w_i^{hk} + a_{\mathcal{S}}\right), \quad 1 \le n \le N, \tag{6.19}$$

for any $w^{hk} = \{w_i^{hk}\}_{i=0}^N \subset V^h$. Then there is a constant $c > 0$ such that

$$\|\mathcal{S}_n^k w^{hk}\|_{V^*} \le c\left(k \sum_{i=0}^{n-1} \|w_i^{hk}\|_V + 1\right). \tag{6.20}$$

The fully discrete scheme for Problem 6.1 is as follows.

Problem 6.3.   Find $w^{hk} = \{w_n^{hk}\}_{n=0}^N \subset V^h$ such that for $1 \le n \le N$,

$$\left\langle \rho\, \frac{w_n^{hk} - w_{n-1}^{hk}}{k} + A(w_n^{hk}) + \mathcal{S}_n^k w^{hk}, v^h \right\rangle + j_1^0(\gamma_1 w_n^{hk}; \gamma_1 v^h)$$

$$+ j_2^0(\gamma_2 w_n^{hk}; \gamma_2 v^h) \ge \langle f_n, v^h \rangle \quad \text{for all } v^h \in V^h, \tag{6.21}$$

and

$$w_0^{hk} = w_0^h, \tag{6.22}$$

where $w_0^h \in V^h$ is an approximation of w_0 with the property $w_0^h \to w_0$ in V as $h \to 0$.

Problem 6.3 has a unique solution. Let us explore the boundedness of the discrete solutions in Proposition 6.4 below.

Proposition 6.4.   Assume (6.3)–(6.7) and (6.17)–(6.19). There is a constant $c > 0$ such that

$$\max_{0 \le n \le N} \|w_n^{hk}\|_H^2 + \sum_{n=1}^N \|w_n^{hk} - w_{n-1}^{hk}\|_H^2 + k \sum_{n=1}^N \|w_n^{hk}\|_V^2 \le c. \tag{6.23}$$

Proof. First, from the conditions

$$\|Av\|_{V^*} \le a_0 + a_1 \|v\|_V,$$
$$\langle Au - Av, u - v \rangle \ge m_A \|u - v\|_V^2,$$

we obtain the inequality

$$\langle Av, v \rangle \ge m_A \|v\|_V^2 + \langle A0, v \rangle \ge m_A \|v\|_V^2 - a_0 \|v\|_V \quad \text{for all } v \in V.$$

Moreover, for $l = 1, 2$, from the conditions

$$j_l^0(z_1; z_2 - z_1) + j_l^0(z_2; z_1 - z_2) \le \alpha_l \|z_1 - z_2\|_{V_l}^2,$$
$$\|\partial j_l(z)\|_{V_l^*} \le c(1 + \|z\|_{V_l}),$$

we obtain the inequality

$$j_l^0(z; -z) \le \alpha_l \|z\|_{V_l}^2 + c \|z\|_{V_l} \quad \text{for all } z \in V_l. \tag{6.24}$$

We take $v^h = -w_n^{hk}$ in (6.21):

$$\left\langle \rho \frac{w_n^{hk} - w_{n-1}^{hk}}{k} + A(w_n^{hk}) + \mathcal{S}_n^k w^{hk}, w_n^{hk} \right\rangle$$
$$\le j_1^0(\gamma_1 w_n^{hk}; -\gamma_1 w_n^{hk}) + j_2^0(\gamma_2 w_n^{hk}; -\gamma_2 w_n^{hk}) + \langle f_n, w_n^{hk} \rangle. \tag{6.25}$$

Notice that

$$\langle \rho (w_n^{hk} - w_{n-1}^{hk}), w_n^{hk} \rangle = \frac{\rho}{2}(\|w_n^{hk}\|_H^2 - \|w_{n-1}^{hk}\|_H^2 + \|w_n^{hk} - w_{n-1}^{hk}\|_H^2).$$

Moreover, from (6.24),

$$j_l^0(\gamma_l w_n^{hk}; -\gamma_l w_n^{hk}) \le \alpha_l \|\gamma_l w_n^{hk}\|_{V_l}^2 + c \|\gamma_l w_n^{hk}\|_{V_l}, \quad l = 1, 2,$$

and then

$$j_l^0(\gamma_l w_n^{hk}; -\gamma_l w_n^{hk}) \le \alpha_l c_l^2 \|w_n^{hk}\|_V^2 + c \|w_n^{hk}\|_V, \quad l = 1, 2.$$

So, from (6.25) we deduce the following inequality:

$$\frac{\rho}{2k}(\|w_n^{hk}\|_H^2 - \|w_{n-1}^{hk}\|_H^2 + \|w_n^{hk} - w_{n-1}^{hk}\|_H^2) + m_A \|w_n^{hk}\|_V^2 - a_0 \|w_n^{hk}\|_V$$
$$\le (\alpha_1 c_1^2 + \alpha_2 c_2^2)\|w_n^{hk}\|_V^2 + c \|w_n^{hk}\|_V + \|f_n\|_{V^*}\|w_n^{hk}\|_V$$
$$- \langle \mathcal{S}_n^k w^{hk}, w_n^{hk} \rangle. \tag{6.26}$$

By (6.20), we have the bound

$$|\langle \mathcal{S}_n^k w^{hk}, w_n^{hk} \rangle| \le \|\mathcal{S}_n^k w^{hk}\|_{V^*}\|w_n^{hk}\|_V \le c\left(k \sum_{i=0}^{n-1} \|w_i^{hk}\|_V + 1 \right)\|w_n^{hk}\|_V.$$

Then from (6.26) we get

$$\frac{\rho}{2\,k}\left(\|w_n^{hk}\|_H^2 - \|w_{n-1}^{hk}\|_H^2 + \|w_n^{hk} - w_{n-1}^{hk}\|_H^2\right) + m_A\|w_n^{hk}\|_V^2$$

$$\leq (\alpha_1 c_1^2 + \alpha_2 c_2^2)\|w_n^{hk}\|_V^2 + c(\|w_n^{hk}\|_V + 1) + c\,\|w_n^{hk}\|_V\, k \sum_{i=0}^{n-1} \|w_i^{hk}\|_V. \tag{6.27}$$

Denote $c_0 = m_A - \alpha_1 c_1^2 - \alpha_2 c_2^2$, which is positive due to the smallness assumption (6.5). Applying the modified Cauchy–Schwarz inequality (3.18), for any $\epsilon > 0$, we have a positive constant c depending on ϵ such that

$$\frac{\rho}{2\,k}\left(\|w_n^{hk}\|_H^2 - \|w_{n-1}^{hk}\|_H^2 + \|w_n^{hk} - w_{n-1}^{hk}\|_H^2\right) + (c_0 - \epsilon)\|w_n^{hk}\|_V^2$$

$$\leq c + c\,k \sum_{i=1}^{n-1} \|w_i^{hk}\|_V^2.$$

Here, c depends on $\max_n \|f_n\|_{V^*}$ and an upper bound of $\|w_0^h\|_V$, and as an intermediate step of the derivation, we used

$$\left(k \sum_{i=0}^{n-1} \|w_i^{hk}\|_V\right)^2 \leq k^2 n \sum_{i=0}^{n-1} \|w_i^{hk}\|_V^2 \leq c\,k \sum_{i=1}^{n-1} \|w_i^{hk}\|_V^2 + c\,k\,\|w_0^h\|_V^2.$$

We choose $\epsilon = c_0/2$ to obtain

$$\rho\left(\|w_n^{hk}\|_H^2 - \|w_{n-1}^{hk}\|_H^2 + \|w_n^{hk} - w_{n-1}^{hk}\|_H^2\right) + c_0 k\,\|w_n^{hk}\|_V^2$$

$$\leq c\,k + c\,k^2 \sum_{i=1}^{n-1} \|w_i^{hk}\|_V^2.$$

We replace n with i in the above inequality and sum over i from 1 to n,

$$\rho\,\|w_n^{hk}\|_H^2 + \rho \sum_{i=1}^{n} \|w_i^{hk} - w_{i-1}^{hk}\|_H^2 + c_0 k \sum_{i=1}^{n} \|w_i^{hk}\|_V^2$$

$$\leq c + c\,k \sum_{i=1}^{n} k \sum_{l=1}^{i-1} \|w_l^{hk}\|_V^2$$

$$= c + c\,k \sum_{i=1}^{n-1} k \sum_{l=1}^{i} \|w_l^{hk}\|_V^2. \tag{6.28}$$

From (6.28), we have

$$k \sum_{i=1}^{n} \|w_i^{hk}\|_V^2 \leq c + c\,k \sum_{i=1}^{n-1} k \sum_{l=1}^{i} \|w_l^{hk}\|_V^2, \quad 1 \leq n \leq N.$$

Apply Lemma 3.23 to get

$$k \sum_{i=1}^{n} \|w_i^{hk}\|_V^2 \le c, \quad 1 \le n \le N.$$

By (6.28) again,

$$\rho \|w_n^{hk}\|_H^2 + \rho \sum_{i=1}^{n} \|w_i^{hk} - w_{i-1}^{hk}\|_H^2 \le c + c\, k \sum_{i=1}^{n-1} k \sum_{l=1}^{i} \|w_l^{hk}\|_V^2 \le c, \quad 1 \le n \le N.$$

Hence, (6.23) holds. $\qquad\qquad\qquad\qquad\qquad\qquad\qquad\qquad\qquad\qquad\square$

For error analysis, we additionally assume the Lipschitz continuity of the operator A,

$$\|Au - Av\|_{V^*} \le L_A \|u - v\|_V \quad \text{for all } u, v \in V, \tag{6.29}$$

and assume the smoothness

$$w \in H^1(0, T; V) \cap H^2(0, T; V^*), \quad q \in C^1([0, T]^2; \mathcal{L}(V)). \tag{6.30}$$

We start with an application of the strong monotonicity of A:

$$m_A \|w_n - w_n^{hk}\|_V^2 \le \langle A(w_n) - A(w_n^{hk}), w_n - w_n^{hk} \rangle,$$

which can be rewritten as, for any $v^h \in V^h$,

$$m_A \|w_n - w_n^{hk}\|_V^2 \le \langle A(w_n) - A(w_n^{hk}), w_n - v^h \rangle + \langle A(w_n), v^h - w_n \rangle$$
$$+ \langle A(w_n), w_n - w_n^{hk} \rangle + \langle A(w_n^{hk}), w_n^{hk} - v^h \rangle. \tag{6.31}$$

By (6.1),

$$\langle A(w_n), w_n - w_n^{hk} \rangle$$
$$\le \langle \rho \dot{w}_n + \mathcal{S}_n w, w_n^{hk} - w_n \rangle + j_1^0(\gamma_1 w_n; \gamma_1 w_n^{hk} - \gamma_1 w_n)$$
$$+ j_2^0(\gamma_2 w_n; \gamma_2 w_n^{hk} - \gamma_2 w_n) - \langle f_n, w_n^{hk} - w_n \rangle. \tag{6.32}$$

By (6.21),

$$\langle A(w_n^{hk}), w_n^{hk} - v^h \rangle \le \left\langle \rho\, \frac{w_n^{hk} - w_{n-1}^{hk}}{k} + \mathcal{S}_n^k w^{hk}, v^h - w_n^{hk} \right\rangle$$
$$+ j_1^0(\gamma_1 w_n^{hk}; \gamma_1 v^h - \gamma_1 w_n^{hk})$$
$$+ j_2^0(\gamma_2 w_n^{hk}; \gamma_2 v^h - \gamma_2 w_n^{hk}) - \langle f_n, v^h - w_n^{hk} \rangle. \tag{6.33}$$

Use (6.32) and (6.33) in (6.31) to obtain

$$m_A \|w_n - w_n^{hk}\|_V^2 \le \langle A(w_n) - A(w_n^{hk}), w_n - v^h \rangle$$
$$+ R_n(v^h - w_n) + \rho\, I_1 + I_2 + I_3, \tag{6.34}$$

where

$$R_n(v) = \langle \rho\,\ddot{w}_n + A(w_n) + \mathcal{S}_n w - f_n, v \rangle$$
$$+ j_1^0(\gamma_1 w_n; \gamma_1 v) + j_2^0(\gamma_2 w_n; \gamma_2 v), \quad v \in V \qquad (6.35)$$

is a residual-type term, and

$$I_1 = \langle \ddot{w}_n, w_n^{hk} - v^h \rangle + \left\langle \frac{w_n^{hk} - w_{n-1}^{hk}}{k}, v^h - w_n^{hk} \right\rangle, \qquad (6.36)$$

$$I_2 = \langle \mathcal{S}_n w - \mathcal{S}_n^k w^{hk}, w_n^{hk} - v^h \rangle, \qquad (6.37)$$

$$I_3 = j_1^0(\gamma_1 w_n; \gamma_1 w_n^{hk} - \gamma_1 w_n) + j_1^0(\gamma_1 w_n^{hk}; \gamma_1 v^h - \gamma_1 w_n^{hk})$$
$$- j_1^0(\gamma_1 w_n; \gamma_1 v^h - \gamma_1 w_n) + j_2^0(\gamma_2 w_n; \gamma_2 w_n^{hk} - \gamma_2 w_n)$$
$$+ j_2^0(\gamma_2 w_n^{hk}; \gamma_2 v^h - \gamma_2 w_n^{hk}) - j_2^0(\gamma_2 w_n; \gamma_2 v^h - \gamma_2 w_n). \qquad (6.38)$$

In the following, we let $\epsilon > 0$ be an arbitrarily fixed small positive number with its value to be chosen later. We will apply the modified Cauchy–Schwarz inequality (3.18) in several places, and the corresponding constant c may depend on ϵ.

By the Lipschitz continuity of A,

$$\langle A(w_n) - A(w_n^{hk}), w_n - v^h \rangle \leq L_A \|w_n - w_n^{hk}\|_V \|w_n - v^h\|_V.$$

Apply (3.18) to get

$$\langle A(w_n) - A(w_n^{hk}), w_n - v^h \rangle \leq \epsilon \|w_n - w_n^{hk}\|_V^2 + c\,\|w_n - v^h\|_V^2. \qquad (6.39)$$

To simplify the notation, we denote

$$E_n = \ddot{w}_n - \frac{w_n - w_{n-1}}{k}, \quad 1 \leq n \leq N. \qquad (6.40)$$

Rewrite I_1 as

$$I_1 = -\left\langle \frac{(w_n - w_n^{hk}) - (w_{n-1} - w_{n-1}^{hk})}{k}, w_n - w_n^{hk} \right\rangle - \langle E_n, w_n - w_n^{hk} \rangle$$
$$+ \left\langle \frac{(w_n - w_n^{hk}) - (w_{n-1} - w_{n-1}^{hk})}{k}, w_n - v^h \right\rangle + \langle E_n, w_n - v^h \rangle.$$

Note that

$$\left\langle \frac{(w_n - w_n^{hk}) - (w_{n-1} - w_{n-1}^{hk})}{k}, w_n - w_n^{hk} \right\rangle$$
$$\geq \frac{1}{2\,k}\left(\|w_n - w_n^{hk}\|_H^2 - \|w_{n-1} - w_{n-1}^{hk}\|_H^2 \right).$$

By (3.18),

$$|\langle E_n, w_n - w_n^{hk} \rangle| \leq \|E_n\|_{V^*}\|w_n - w_n^{hk}\|_V \leq \frac{\epsilon}{\rho}\|w_n - w_n^{hk}\|_V^2 + c\,\|E_n\|_{V^*}^2.$$

Also,

$$|\langle E_n, w_n - v^h\rangle| \le \|E_n\|_{V^*}\|w_n - v^h\|_V \le \frac{1}{2}(\|E_n\|_{V^*}^2 + \|w_n - v^h\|_V^2).$$

Thus, we can bound I_1 as follows:

$$I_1 \le -\frac{1}{2k}(\|w_n - w_n^{hk}\|_H^2 - \|w_{n-1} - w_{n-1}^{hk}\|_H^2) + \frac{\epsilon}{\rho}\|w_n - w_n^{hk}\|_V^2$$
$$+ \left\langle \frac{(w_n - w_n^{hk}) - (w_{n-1} - w_{n-1}^{hk})}{k}, w_n - v^h\right\rangle$$
$$+ c\|E_n\|_{V^*}^2 + c\|w_n - v^h\|_V^2. \tag{6.41}$$

To bound the term I_2, we write

$$|I_2| \le \|\mathcal{S}_n w - \mathcal{S}_n^k w^{hk}\|_{V^*}\|w_n^{hk} - v^h\|_V$$

and note that

$$\|\mathcal{S}_n w - \mathcal{S}_n^k w^{hk}\|_{V^*} \le \|\mathcal{S}_n w - \mathcal{S}_n^k w\|_{V^*} + \|\mathcal{S}_n^k w - \mathcal{S}_n^k w^{hk}\|_{V^*},$$
$$\|w_n^{hk} - v^h\|_V \le \|w_n - v^h\|_V + \|w_n - w_n^{hk}\|_V.$$

Applying (3.18), we have

$$|I_2| \le \epsilon\|w_n - w_n^{hk}\|_V^2 + c\|w_n - v^h\|_V^2$$
$$+ c\|\mathcal{S}_n^k w - \mathcal{S}_n^k w^{hk}\|_{V^*}^2 + c\|\mathcal{S}_n w - \mathcal{S}_n^k w\|_{V^*}^2. \tag{6.42}$$

By the definition of $\mathcal{S}^k$, and assumptions on G and q, it is easy to see that

$$\|\mathcal{S}_n^k w - \mathcal{S}_n^k w^{hk}\|_{V^*} \le ck\sum_{i=0}^{n-1}\|w_i - w_i^{hk}\|_V. \tag{6.43}$$

As in the derivation of (5.26), for $\mathcal{S}$ and $\mathcal{S}^k$ defined by (6.18) and (6.19), we have the bound

$$\|\mathcal{S}_n w - \mathcal{S}_n^k w\|_{V^*} \le ck\|w\|_{H^1(0,T;V)}. \tag{6.44}$$

Use (6.43) and (6.44) in (6.42) to obtain

$$|I_2| \le \epsilon\|w_n - w_n^{hk}\|_V^2 + c\|w_n - v^h\|_V^2$$
$$+ ck^2\|w\|_{H^1(0,T;V)}^2 + ck\sum_{i=0}^{n-1}\|w_i - w_i^{hk}\|_V^2. \tag{6.45}$$

To bound I_3, we first apply the subadditivity of the generalized directional derivative, with $l = 1, 2$,

$$j_l^0(\gamma_l w_n^{hk}; \gamma_l v^h - \gamma_l w_n^{hk}) \le j_l^0(\gamma_l w_n^{hk}; \gamma_l w_n - \gamma_l w_n^{hk})$$
$$+ j_l^0(\gamma_l w_n^{hk}; \gamma_l v^h - \gamma_l w_n),$$

and then apply the relaxed monotonicity,

$$j_l^0(\gamma_l w_n; \gamma_l w_n^{hk} - \gamma_l w_n) + j_l^0(\gamma_l w_n^{hk}; \gamma_l w_n - \gamma_l w_n^{hk}) \le \alpha_l c_j^2 \|w_n - w_n^{hk}\|_V^2.$$

As a result,

$$
\begin{aligned}
I_3 &\le (\alpha_1 c_1^2 + \alpha_2 c_2^2)\|w_n - w_n^{hk}\|_V^2 \\
&\quad + j_1^0(\gamma_1 w_n^{hk}; \gamma_1 v^h - \gamma_1 w_n) - j_1^0(\gamma_1 w_n; \gamma_1 v^h - \gamma_1 w_n) \\
&\quad + j_2^0(\gamma_2 w_n^{hk}; \gamma_2 v^h - \gamma_2 w_n) - j_2^0(\gamma_2 w_n; \gamma_2 v^h - \gamma_2 w_n).
\end{aligned}
$$

By the assumption on j_l,

$$
\begin{aligned}
|j_l^0(\gamma_l w_n^{hk}; \gamma_l v^h - \gamma_l w_n)| &\le (c_{0l} + c_{1l}\|\gamma_l w_n^{hk}\|_{V_l})\|\gamma_l(w_n - v^h)\|_{V_l}, \\
|j_l^0(\gamma_l w_n; \gamma_l v^h - \gamma_l w_n)| &\le (c_{0l} + c_{1l}\|\gamma_l w_n\|_{V_l})\|\gamma_l(w_n - v^h)\|_{V_l}.
\end{aligned}
$$

Since $\|\gamma_l w_n\|_{V_l}$ is uniformly bounded, we conclude that there is a constant c such that

$$
\begin{aligned}
I_3 &\le (\alpha_1 c_1^2 + \alpha_2 c_2^2)\|w_n - w_n^{hk}\|_V^2 \\
&\quad + c(1 + \|w_n^{hk}\|_V)(\|\gamma_1(w_n - v^h)\|_{V_1} + \|\gamma_2(w_n - v^h)\|_{V_2}). \tag{6.46}
\end{aligned}
$$

Denote the error

$$e_n = w_n - w_n^{hk}, \quad 0 \le n \le N. \tag{6.47}$$

Then, by applying (6.39), (6.41), (6.45) and (6.46) in (6.34), we have

$$
\begin{aligned}
&\frac{\rho}{2k}(\|e_n\|_H^2 - \|e_{n-1}\|_H^2) + (m_A - \alpha_1 c_1^2 - \alpha_2 c_2^2 - 3\epsilon)\|e_n\|_V^2 \\
&\quad \le c(\|w_n - v^h\|_V^2 + |R_n(v^h - w_n)| + \|E_n\|_{V^*}^2) + c k^2\|w\|_{H^1(0,T;V)}^2 \\
&\quad\quad + c(1 + \|w_n^{hk}\|_V)(\|\gamma_1(w_n - v^h)\|_{V_1} + \|\gamma_2(w_n - v^h)\|_{V_2}) \\
&\quad\quad + \left\langle \rho\frac{e_n - e_{n-1}}{k}, w_n - v^h \right\rangle + c k \sum_{i=0}^{n-1} \|e_i\|_V^2. \tag{6.48}
\end{aligned}
$$

Applying the smallness condition $\alpha_1 c_1^2 + \alpha_2 c_2^2 < m_A$ from (6.5) and choosing $\epsilon = (m_A - \alpha_1 c_1^2 - \alpha_2 c_2^2)/6$, we derive from (6.48) the following inequality, with $v^h \in V^h$ renamed as $v_n^h \in V^h$,

$$
\begin{aligned}
&\|e_n\|_H^2 - \|e_{n-1}\|_H^2 + k\|e_n\|_V^2 \\
&\quad \le ck(\|w_n - v_n^h\|_V^2 + |R_n(v_n^h - w_n)| + \|E_n\|_{V^*}^2) + c k^3\|w\|_{H^1(0,T;V)}^2 \\
&\quad\quad + c k(1 + \|w_n^{hk}\|_V)(\|\gamma_1(w_n - v_n^h)\|_{V_1} + \|\gamma_2(w_n - v_n^h)\|_{V_2}) \\
&\quad\quad + c\langle e_n - e_{n-1}, w_n - v_n^h \rangle + c k^2 \sum_{i=0}^{n-1} \|e_i\|_V^2. \tag{6.49}
\end{aligned}
$$

We replace n with i in (6.49) and make a summation over i from 1 to n,

$$\|e_n\|_H^2 - \|e_0\|_H^2 + k\sum_{i=1}^{n}\|e_i\|_V^2$$

$$\leq ck\sum_{i=1}^{n}(\|w_i - v_i^h\|_V^2 + |R_i(v_i^h - w_i)| + \|E_i\|_{V*}^2) + ck^2\|w\|_{H^1(0,T;V)}^2$$

$$+ ck\sum_{i=1}^{n}(1 + \|w_i^{hk}\|_V)(\|\gamma_1(w_i - v_i^h)\|_{V_1} + \|\gamma_2(w_i - v_i^h)\|_{V_2})$$

$$+ ck\sum_{i=1}^{n}\langle e_i - e_{i-1}, w_i - v_i^h\rangle + ck\sum_{i=0}^{n-1}k\sum_{l=0}^{i}\|e_l\|_V^2. \qquad (6.50)$$

Now

$$\sum_{i=1}^{n}(1 + \|w_i^{hk}\|_V)(\|\gamma_1(w_i - v_i^h)\|_{V_1} + \|\gamma_2(w_i - v_i^h)\|_{V_2})$$

$$\leq \left[\sum_{i=1}^{n}(1 + \|w_i^{hk}\|_V)^2\right]^{1/2}\left[\sum_{i=1}^{n}(\|\gamma_1(w_i - v_i^h)\|_{V_1} + \|\gamma_2(w_i - v_i^h)\|_{V_2})^2\right]^{1/2}.$$

By Proposition 6.4, $k\sum_{i=1}^{n}(1 + \|w_i^{hk}\|_V)^2$ is uniformly bounded. Thus,

$$ck\sum_{i=1}^{n}(1 + \|w_i^{hk}\|_V)(\|\gamma_1(w_i - v_i^h)\|_{V_1} + \|\gamma_2(w_i - v_i^h)\|_{V_2})$$

$$\leq c\left[k\sum_{i=1}^{n}(\|\gamma_1(w_i - v_i^h)\|_{V_1}^2 + \|\gamma_2(w_i - v_i^h)\|_{V_2}^2)\right]^{1/2}.$$

Write

$$\sum_{i=1}^{n}\langle e_i - e_{i-1}, w_i - v_i^h\rangle = \sum_{i=1}^{n}\langle e_i, w_i - v_i^h\rangle - \sum_{i=0}^{n-1}\langle e_i, w_{i+1} - v_{i+1}^h\rangle$$

$$= \langle e_n, w_n - v_n^h\rangle$$

$$+ \sum_{i=1}^{n-1}\langle e_i, (w_i - v_i^h) - (w_{i+1} - v_{i+1}^h)\rangle$$

$$- \langle e_0, w_1 - v_1^h\rangle$$

and we bound the terms on the right-hand side as follows. For the first term, with a small $\epsilon_0 > 0$,

$$|\langle e_n, w_n - v_n^h\rangle| \leq \|e_n\|_H\|w_n - v_n^h\|_H \leq \epsilon_0\|e_n\|_H^2 + c\|w_n - v_n^h\|_H^2.$$

For the second term,

$$|\langle e_i, (w_i - v_i^h) - (w_{i+1} - v_{i+1}^h)\rangle|$$
$$\leq \|e_i\|_H \|(w_i - v_i^h) - (w_{i+1} - v_{i+1}^h)\|_H$$
$$\leq \frac{k}{2}(\|e_i\|_H^2 + k^{-2}\|(w_i - v_i^h) - (w_{i+1} - v_{i+1}^h)\|_H^2).$$

For the last term,

$$|\langle e_0, w_1 - v_1^h\rangle| \leq \|e_0\|_H \|w_1 - v_1^h\|_H \leq \frac{1}{2}(\|e_0\|_H^2 + \|w_1 - v_1^h\|_H^2).$$

For the term E_n defined by (6.40), we can write

$$E_n = \frac{1}{k} \int_{t_{n-1}}^{t_n} (t - t_{n-1})\ddot{w}(t)\,dt.$$

Thus, we have the upper bound

$$\|E_n\|_{V^*} \leq \|\ddot{w}\|_{L^1(t_{n-1},t_n;V^*)}$$

and then

$$\sum_{i=1}^{n} \|E_i\|_{V^*}^2 \leq \sum_{i=1}^{n} \|\ddot{w}\|_{L^1(t_{i-1},t_i;V^*)}^2 \leq k\|\ddot{w}\|_{L^2(0,T;V^*)}^2.$$

Using the above inequalities with $\epsilon_0 > 0$ sufficiently small, we derive from (6.50) that

$$\|e_n\|_H^2 + k\sum_{i=1}^{n} \|e_i\|_V^2 \leq ck\sum_{i=1}^{n}(\|w_i - v_i^h\|^2 + |R_i(v_i^h - w_i)|)$$
$$+ ck^2(\|\ddot{w}\|_{L^2(0,T;V^*)}^2 + \|w\|_{H^1(0,T;V)}^2)$$
$$+ c\left[k\sum_{i=1}^{n}(\|\gamma_1(w_i - v_i^h)\|_{V_1}^2 + \|\gamma_2(w_i - v_i^h)\|_{V_2}^2)\right]^{1/2}$$
$$+ ck^{-1}\sum_{i=1}^{n-1}\|(w_i - v_i^h) - (w_{i+1} - v_{i+1}^h)\|_H^2$$
$$+ c(\|e_0\|_H^2 + k\|e_0\|_V^2 + \|w_1 - v_1^h\|_H^2 + \|w_n - v_n^h\|_H^2)$$
$$+ ck\sum_{i=0}^{n-1}\left(\|e_i\|_H^2 + k\sum_{l=1}^{i}\|e_l\|_V^2\right), \quad 1 \leq n \leq N.$$

We then apply Lemma 3.23 to find that

$$\max_{1\leq n\leq N}\|e_n\|_H^2 + k\sum_{n=1}^{N}\|e_n\|_V^2 \leq c\,k^2(\|\ddot{w}\|_{L^2(0,T;V^*)}^2 + \|w\|_{H^1(0,T;V)}^2)$$

$$+ c(\|e_0\|_H^2 + k\,\|e_0\|_V^2) + c\max_{1\leq n\leq N}\tilde{E}_n, \quad (6.51)$$

where

$$\tilde{E}_n = \inf_{v_i^h\in V^h,\, 1\leq i\leq n}\left\{ k\sum_{i=1}^{n}(\|w_i - v_i^h\|_V^2 + |R_i(v_i^h - w_i)|)\right.$$

$$+ \left[k\sum_{i=1}^{n}(\|\gamma_1(w_i - v_i^h)\|_{V_1}^2 + \|\gamma_2(w_i - v_i^h)\|_{V_2}^2)\right]^{1/2}$$

$$+ k^{-1}\sum_{i=1}^{n-1}\|(w_i - v_i^h) - (w_{i+1} - v_{i+1}^h)\|_{V^*}^2$$

$$\left. + \|w_1 - v_1^h\|_H^2 + \|w_n - v_n^h\|_H^2\right\}. \quad (6.52)$$

Summarizing, we state the result in the form of a theorem.

Theorem 6.5. Keep the assumptions of Theorem 6.2, and assume further that (6.17), (6.18), (6.19), (6.29) and (6.30) hold. Then for the numerical solution $w^{hk} = \{w_n^{hk}\}_{n=0}^{N} \subset V^h$ defined by Problem 6.3, we have the inequality

$$\max_{1\leq n\leq N}\|w_n - w_n^{hk}\|_H^2 + k\sum_{n=1}^{N}\|w_n - w_n^{hk}\|_V^2$$

$$\leq c\,k^2(\|\ddot{w}\|_{L^2(0,T;V^*)}^2 + \|w\|_{H^1(0,T;V)}^2)$$

$$+ c(\|w_0 - w_0^h\|_H^2 + k\,\|w_0 - w_0^h\|_V^2) + c\max_{1\leq n\leq N}\tilde{E}_n, \quad (6.53)$$

where $\tilde{E}_n$ is defined by (6.52).

The inequality (6.53) will be the starting point for concrete error estimation of numerical solutions for Problem 2.10 in Section 9.2.

We comment that for the approximate operator $\mathcal{S}^k$, instead of (6.19), we can also use (5.33) or define

$$\mathcal{S}_n^k w^{hk} = G\left(k\sum_{i=1}^{n}q(t_n,t_i)\,w_i^{hk} + a_{\mathcal{S}}\right), \quad 1\leq n\leq N, \quad (6.54)$$

for any $w^{hk} = \{w_i^{hk}\}_{i=0}^{N} \subset V^h$. For both these choices, the error bound (6.53) is still valid for k small.

7. Studies of the static contact problem

In this section we study the static contact problem, Problem 2.6. We first explore the solution existence and uniqueness, then introduce a linear finite element method to solve the problem and derive an optimal order error estimate under certain solution regularity assumptions. Finally, we present numerical simulation results.

7.1. Existence and uniqueness

We start with an existence and uniqueness result for Problem 2.6. In addition to the space V defined in (2.3), we introduce the spaces

$$V_\varphi = L^2(\Gamma_3)^d, \tag{7.1}$$

$$V_j = L^2(\Gamma_3). \tag{7.2}$$

Let $\lambda_1 > 0$ be the smallest eigenvalue of the eigenvalue problem

$$\boldsymbol{u} \in V, \quad \int_\Omega \boldsymbol{\varepsilon}(\boldsymbol{u}){\cdot}\boldsymbol{\varepsilon}(\boldsymbol{v})\,\mathrm{d}x = \lambda \int_{\Gamma_3} \boldsymbol{u}{\cdot}\boldsymbol{v}\,\mathrm{d}a \quad \text{for all } \boldsymbol{v} \in V \tag{7.3}$$

and let $\lambda_{1\nu} > 0$ be the smallest eigenvalue of the eigenvalue problem

$$\boldsymbol{u} \in V, \quad \int_\Omega \boldsymbol{\varepsilon}(\boldsymbol{u}){\cdot}\boldsymbol{\varepsilon}(\boldsymbol{v})\,\mathrm{d}x = \lambda \int_{\Gamma_3} u_\nu v_\nu\,\mathrm{d}a \quad \text{for all } \boldsymbol{v} \in V. \tag{7.4}$$

It is easy to check that

$$\lambda_1 = \inf\{\|\boldsymbol{\varepsilon}(\boldsymbol{v})\|^2_{L^2(\Omega;\mathbb{S}^d)} \mid \boldsymbol{v} \in V, \|\boldsymbol{v}\|_{L^2(\Gamma_3;\mathbb{R}^d)} = 1\},$$

$$\lambda_{1\nu} = \inf\{\|\boldsymbol{\varepsilon}(\boldsymbol{v})\|^2_{L^2(\Omega;\mathbb{S}^d)} \mid \boldsymbol{v} \in V, \|v_\nu\|_{L^2(\Gamma_3)} = 1\},$$

and both values exist and are positive.

Theorem 7.1. Assume (2.34)–(2.38) and

$$L_{F_b}\lambda_1^{-1} + \alpha_{j_\nu}\lambda_{1\nu}^{-1} < m_{\mathcal{F}}. \tag{7.5}$$

Then Problem 2.6 has a unique solution.

Proof. We apply Theorem 4.4 with $X = V$, $X_\varphi = V_\varphi$ of (7.1), $X_j = V_j$ of (7.2), $K = U$, $\gamma_\varphi \colon V \to V_\varphi$ being the trace operator, $\gamma_j \colon V \to V_j$ being the normal trace operator, and $A \colon V \to V^*$, $\varphi \colon V_\varphi \times V_\varphi \to \mathbb{R}$, $j \colon V_j \to \mathbb{R}$ and $\boldsymbol{f} \in V^*$ defined by

$$\langle A\boldsymbol{u}, \boldsymbol{v}\rangle = \int_\Omega \mathcal{F}\boldsymbol{\varepsilon}(\boldsymbol{u}) \cdot \boldsymbol{\varepsilon}(\boldsymbol{v})\,\mathrm{d}x, \quad \boldsymbol{u}, \boldsymbol{v} \in V, \tag{7.6}$$

$$\varphi(\boldsymbol{z}_1, \boldsymbol{z}_2) = \int_{\Gamma_3} F_b(z_{1,\nu})\,\|z_{2,\tau}\|\,\mathrm{d}a, \quad \boldsymbol{z}_1, \boldsymbol{z}_2 \in V_\varphi, \tag{7.7}$$

$$j(z) = \int_{\Gamma_3} j_\nu(z)\, da, \quad z \in V_j, \tag{7.8}$$

$$\langle \boldsymbol{f}, \boldsymbol{v} \rangle = \int_\Omega \boldsymbol{f}_0 \cdot \boldsymbol{v}\, dx + \int_{\Gamma_2} \boldsymbol{f}_2 \cdot \boldsymbol{v}\, da, \quad \boldsymbol{v} \in V. \tag{7.9}$$

Here and below, for any vector field $\boldsymbol{z}_i$ we use the notation $z_{i,\nu}$ and $\boldsymbol{z}_{i,\tau}$ to represent its normal and tangential components, respectively. Observe that (4.16) holds with $c_\varphi = \lambda_1^{-1/2}$, (4.17) holds with $c_j = \lambda_{1\nu}^{-1/2}$.

Then, (4.1) is obviously true: the set is non-empty since the zero function belongs to U. The operator A defined by (7.6) satisfies condition (4.2) with $m_A = m_{\mathcal{F}}$ and condition (4.47) with $L_A = L_{\mathcal{F}}$. Indeed, for $\boldsymbol{u}, \boldsymbol{v}, \boldsymbol{w} \in V$, by assumption (2.34(a)), we have

$$\langle A\boldsymbol{u} - A\boldsymbol{v}, \boldsymbol{w} \rangle \leq (\mathcal{F}\varepsilon(\boldsymbol{u}) - \mathcal{F}\varepsilon(\boldsymbol{v}), \varepsilon(\boldsymbol{w}))_Q \leq L_{\mathcal{F}} \|\boldsymbol{u} - \boldsymbol{v}\|_V \|\boldsymbol{w}\|_V.$$

Thus,

$$\|A\boldsymbol{u} - A\boldsymbol{v}\|_{V^*} \leq L_{\mathcal{F}} \|\boldsymbol{u} - \boldsymbol{v}\|_V \quad \text{for all } \boldsymbol{u}, \boldsymbol{v} \in V.$$

This shows that A is Lipschitz continuous. On the other hand,

$$\langle A\boldsymbol{u} - A\boldsymbol{v}, \boldsymbol{u} - \boldsymbol{v} \rangle = (\mathcal{F}\varepsilon(\boldsymbol{u}) - \mathcal{F}\varepsilon(\boldsymbol{v}), \varepsilon(\boldsymbol{u}) - \varepsilon(\boldsymbol{v}))_Q.$$

Then, assumption (2.34(b)) yields

$$\langle A\boldsymbol{u} - A\boldsymbol{v}, \boldsymbol{u} - \boldsymbol{v} \rangle \geq m_{\mathcal{F}} \|\boldsymbol{u} - \boldsymbol{v}\|_V^2 \quad \text{for all } \boldsymbol{u}, \boldsymbol{v} \in V. \tag{7.10}$$

This shows that condition (4.2(b)) is satisfied with $m_A = m_{\mathcal{F}}$. Since A is Lipschitz continuous it follows that A is bounded and hemicontinuous, that is, (4.2(a)) holds.

Next, for φ defined by (7.7), it is easy to check that assumption (2.36) implies (4.18) with $\alpha_\varphi = L_{F_b}$. On the other hand, hypothesis (2.35) allows us to apply a variant of Theorem 3.47 in Migórski, Ochal and Sofonea (2013). In this way we deduce that the function j given by (7.8) satisfies (4.19) with $c_0 = \sqrt{2\,\mathrm{meas}\,(\Gamma_3)}\,\bar{c}_0$, $c_1 = \sqrt{2}\,\bar{c}_1$ and $\alpha_j = \alpha_{j_\nu}$. Moreover,

$$j^0(\gamma_j \boldsymbol{u}; \gamma_j \boldsymbol{v}) \leq \int_{\Gamma_3} j_\nu^0(\boldsymbol{x}, u_\nu(\boldsymbol{x}); v_\nu(\boldsymbol{x}))\, da \quad \text{for all } \boldsymbol{u}, \boldsymbol{v} \in V. \tag{7.11}$$

This implies (4.19(c)).

For $\boldsymbol{f}$, assumption (2.37) implies (4.5). Finally, considering the above relationships among constants and noting that $c_\varphi = \lambda_1^{-1/2}$, $c_j = \lambda_{1\nu}^{-1/2}$, we see that assumption (7.5) implies the smallness condition (4.20).

Therefore, we apply Theorem 4.4 to conclude that there exists a unique element $\boldsymbol{u} \in U$ such that

$$\langle A\boldsymbol{u}, \boldsymbol{v} - \boldsymbol{u} \rangle + \varphi(\gamma_\varphi \boldsymbol{u}, \gamma_\varphi \boldsymbol{v}) - \varphi(\gamma_\varphi \boldsymbol{u}, \gamma_\varphi \boldsymbol{u})$$
$$+ j^0(\gamma_j \boldsymbol{u}; \gamma_j \boldsymbol{v} - \gamma_j \boldsymbol{u}) \geq \langle \boldsymbol{f}, \boldsymbol{v} - \boldsymbol{u} \rangle \quad \text{for all } \boldsymbol{v} \in U. \tag{7.12}$$

By the inequality (7.11), we see that $\boldsymbol{u} \in U$ satisfies (2.44), *i.e.*, it is a solution of Problem 2.6.

We now prove the uniqueness of the solution to Problem 2.6. Let $\boldsymbol{u}_1$, $\boldsymbol{u}_2 \in U$ be solutions to inequality (2.44). Then,

$$\int_\Omega \mathcal{F}\varepsilon(\boldsymbol{u}_1) \cdot (\varepsilon(\boldsymbol{v}) - \varepsilon(\boldsymbol{u}_1))\,\mathrm{d}x + \int_{\Gamma_3} F_b(u_{1,\nu})(\|\boldsymbol{v}_\tau\| - \|\boldsymbol{u}_{1,\tau}\|)\,\mathrm{d}a$$

$$+ \int_{\Gamma_3} j_\nu^0(u_{1,\nu}; v_\nu - u_{1,\nu})\,\mathrm{d}a \geq \int_\Omega \boldsymbol{f}_0 \cdot (\boldsymbol{v} - \boldsymbol{u}_1)\,\mathrm{d}x + \int_{\Gamma_2} \boldsymbol{f}_2 \cdot (\boldsymbol{v} - \boldsymbol{u}_1)\,\mathrm{d}a$$

and

$$\int_\Omega \mathcal{F}\varepsilon(\boldsymbol{u}_2) \cdot (\varepsilon(\boldsymbol{v}) - \varepsilon(\boldsymbol{u}_2))\,\mathrm{d}x + \int_{\Gamma_3} F_b(u_{2,\nu})(\|\boldsymbol{v}_\tau\| - \|\boldsymbol{u}_{2,\tau}\|)\,\mathrm{d}a$$

$$+ \int_{\Gamma_3} j_\nu^0(u_{2,\nu}; v_\nu - u_{2,\nu})\,\mathrm{d}a \geq \int_\Omega \boldsymbol{f}_0 \cdot (\boldsymbol{v} - \boldsymbol{u}_2)\,\mathrm{d}x + \int_{\Gamma_2} \boldsymbol{f}_2 \cdot (\boldsymbol{v} - \boldsymbol{u}_2)\,\mathrm{d}a$$

for all $\boldsymbol{v} \in U$. We take $\boldsymbol{v} = \boldsymbol{u}_2$ in the first inequality and $\boldsymbol{v} = \boldsymbol{u}_1$ in the second one, then we add the resulting inequalities. Next, we use the strong monotonicity of the operator $\mathcal{F}$, (2.34(b)), hypotheses (2.35), (2.36) and (2.7) to obtain

$$(m_\mathcal{F} - L_{F_b}\lambda_1^{-1} - \alpha_{j_\nu}\lambda_{1\nu}^{-1})\|\boldsymbol{u}_1 - \boldsymbol{u}_2\|_V^2 \leq 0.$$

Finally, we use the smallness condition (7.5) to deduce that $\boldsymbol{u}_1 = \boldsymbol{u}_2$, which concludes the proof. $\square$

Theorem 7.1 provides the unique weak solvability of Problem 2.5, in terms of displacement. Once the displacement field is obtained by solving Problem 2.5, then the stress field $\boldsymbol{\sigma}$ is uniquely determined by using the constitutive law (2.20). Nevertheless, the question of the uniqueness of the contact interface function ξ_ν is left open.

7.2. Finite element solution of the static contact problem

We now proceed with the discretization of Problem 2.6 using the finite element method. For simplicity, assume Ω is a polygonal/polyhedral domain and express the three parts of the boundary, Γ_k, $1 \leq k \leq 3$, as unions of closed flat components with disjoint interiors:

$$\overline{\Gamma_k} = \cup_{i=1}^{i_k}\Gamma_{k,i}, \quad 1 \leq k \leq 3.$$

Let $\{\mathcal{T}^h\}$ be a regular family of partitions of $\overline{\Omega}$ into triangles/tetrahedrons that are compatible with the partition of the boundary $\partial\Omega$ into $\Gamma_{k,i}$, $1 \leq i \leq i_k$, $1 \leq k \leq 3$, in the sense that if the intersection of one side/face of an element with one set $\Gamma_{k,i}$ has a positive measure with respect to $\Gamma_{k,i}$, then the side/face lies entirely in $\Gamma_{k,i}$. Then construct a linear element space corresponding to $\mathcal{T}^h$,

$$V^h = \{ \boldsymbol{v}^h \in C(\overline{\Omega})^d \mid \boldsymbol{v}^h|_T \in \mathbb{P}_1(T)^d \text{ for } T \in \mathcal{T}^h, \ \boldsymbol{v}^h = \boldsymbol{0} \text{ on } \Gamma_1 \}, \quad (7.13)$$

and a related finite element subset of the space V^h to approximate U:

$$U^h = \{ \boldsymbol{v}^h \in V^h \mid v_\nu^h \leq g \text{ at node points on } \Gamma_3 \}. \quad (7.14)$$

The finite element approximation of Problem 2.6 is as follows.

Problem 7.2. Find a displacement field $\boldsymbol{u}^h \in U^h$ such that

$$\int_\Omega \mathcal{F}\boldsymbol{\varepsilon}(\boldsymbol{u}^h) \cdot (\boldsymbol{\varepsilon}(\boldsymbol{v}^h) - \boldsymbol{\varepsilon}(\boldsymbol{u}^h)) \, \mathrm{d}x$$

$$+ \int_{\Gamma_3} F_b(u_\nu^h)(\|\boldsymbol{v}_\tau^h\| - \|\boldsymbol{u}_\tau^h\|) \, \mathrm{d}a + \int_{\Gamma_3} j_\nu^0(u_\nu^h; v_\nu^h - u_\nu^h) \, \mathrm{d}a$$

$$\geq \int_\Omega \boldsymbol{f}_0 \cdot (\boldsymbol{v}^h - \boldsymbol{u}^h) \, \mathrm{d}x + \int_{\Gamma_2} \boldsymbol{f}_2 \cdot (\boldsymbol{v}^h - \boldsymbol{u}^h) \, \mathrm{d}a \quad \text{for all } \boldsymbol{v}^h \in U^h.$$

$$(7.15)$$

As in Problem 2.6, we can use a discrete analogue of the arguments in Section 7.1 to conclude that Problem 7.2 admits a unique solution $\boldsymbol{u}^h \in U^h$. In the following, we assume g is a concave function so that $U^h \subset U$.

For an error analysis, we notice that the derivation of (4.56) in Section 4 can be easily adjusted by working directly on $j_\nu^0(\cdot; \cdot)$ rather than on $j^0(\cdot; \cdot)$, to yield the following Céa-type inequality for the solution $\boldsymbol{u}^h \in U^h$ of Problem 7.2:

$$\|\boldsymbol{u} - \boldsymbol{u}^h\|_V \leq c \inf_{\boldsymbol{v}^h \in U^h} \left[\|\boldsymbol{u} - \boldsymbol{v}^h\|_V + \|u_\nu - v_\nu^h\|_{L^2(\Gamma_3)}^{1/2} + |R_u(\boldsymbol{v}^h, \boldsymbol{u})|^{1/2} \right], \quad (7.16)$$

where the residual-type term from (4.50) is

$$R_u(\boldsymbol{v}^h, \boldsymbol{u}) = (\mathcal{F}(\boldsymbol{\varepsilon}(\boldsymbol{u})), \boldsymbol{\varepsilon}(\boldsymbol{v}^h - \boldsymbol{u}))_Q + \int_{\Gamma_3} F_b(u_\nu)(\|\boldsymbol{v}_\tau^h\| - \|\boldsymbol{u}_\tau\|) \, \mathrm{d}a$$

$$+ \int_{\Gamma_3} j_\nu^0(u_\nu; v_\nu^h - u_\nu) \, \mathrm{d}a - \langle \boldsymbol{f}, \boldsymbol{v}^h - \boldsymbol{u} \rangle. \quad (7.17)$$

To proceed further, we make the following solution regularity assumptions:

$$\boldsymbol{u} \in H^2(\Omega; \mathbb{R}^d), \quad \boldsymbol{\sigma} = \mathcal{F}(\boldsymbol{\varepsilon}(\boldsymbol{u})) \in H^1(\Omega; \mathbb{S}^d). \quad (7.18)$$

In many application problems, $\boldsymbol{\sigma} \in H^1(\Omega; \mathbb{S}^d)$ follows from $\boldsymbol{u} \in H^2(\Omega; \mathbb{R}^d)$, for example if the material is linearly elastic with suitably smooth coefficients, or if the elasticity operator $\mathcal{F}$ depends on $\boldsymbol{x}$ smoothly. In the latter case, we recall that $\mathcal{F}(\boldsymbol{x}, \boldsymbol{\varepsilon})$ is a Lipschitz function of $\boldsymbol{\varepsilon}$, and according to a general chain rule proved in Marcus and Mizel (1972), the composition of a Lipschitz continuous function and an $H^1(\Omega)$ function is an $H^1(\Omega)$ function.

Note that $\boldsymbol{\sigma} \in H^1(\Omega; \mathbb{S}^d)$ implies

$$\boldsymbol{\sigma\nu} \in L^2(\Gamma; \mathbb{R}^d). \tag{7.19}$$

For an appropriate upper bound on $R_{\boldsymbol{u}}(\boldsymbol{v}^h, \boldsymbol{u})$ defined in (7.17), we need to derive some pointwise relations for the weak solution $\boldsymbol{u}$ of Problem 2.6. We follow a procedure found in Han and Sofonea (2002, Section 8.2). Introduce a subset $\tilde{U}$ of U by

$$\tilde{U} := \{\boldsymbol{w} \in C^\infty(\overline{\Omega}; \mathbb{R}^d) \mid \boldsymbol{w} = \boldsymbol{0} \text{ on } \Gamma_1 \cup \Gamma_3\}. \tag{7.20}$$

We take $\boldsymbol{v} = \boldsymbol{u} + \boldsymbol{w}$ with $\boldsymbol{w} \in \tilde{U}$ in (2.44) to get

$$\int_\Omega \mathcal{F}(\boldsymbol{\varepsilon}(\boldsymbol{u})) \cdot \boldsymbol{\varepsilon}(\boldsymbol{w}) \, dx \geq \int_\Omega \boldsymbol{f}_0 \cdot \boldsymbol{w} \, dx + \int_{\Gamma_2} \boldsymbol{f}_2 \cdot \boldsymbol{w} \, da.$$

By replacing $\boldsymbol{w} \in \tilde{U}$ with $-\boldsymbol{w} \in \tilde{U}$ in the above inequality, we find the equality

$$\int_\Omega \mathcal{F}(\boldsymbol{\varepsilon}(\boldsymbol{u})) \cdot \boldsymbol{\varepsilon}(\boldsymbol{w}) \, dx = \int_\Omega \boldsymbol{f}_0 \cdot \boldsymbol{w} \, dx + \int_{\Gamma_2} \boldsymbol{f}_2 \cdot \boldsymbol{w} \, da \quad \text{for all } \boldsymbol{w} \in \tilde{U}. \tag{7.21}$$

Thus,

$$\int_\Omega \mathcal{F}(\boldsymbol{\varepsilon}(\boldsymbol{u})) \cdot \boldsymbol{\varepsilon}(\boldsymbol{w}) \, dx = \int_\Omega \boldsymbol{f}_0 \cdot \boldsymbol{w} \, dx \quad \text{for all } \boldsymbol{w} \in C_0^\infty(\Omega; \mathbb{R}^d),$$

and so in the distributional sense,

$$\operatorname{Div} \mathcal{F}(\boldsymbol{\varepsilon}(\boldsymbol{u})) + \boldsymbol{f}_0 = \boldsymbol{0}.$$

Since $\mathcal{F}(\boldsymbol{\varepsilon}(\boldsymbol{u})) \in H^1(\Omega; \mathbb{S}^d)$ and $\boldsymbol{f}_0 \in L^2(\Omega; \mathbb{R}^d)$, the above equality holds pointwise:

$$\operatorname{Div} \mathcal{F}(\boldsymbol{\varepsilon}(\boldsymbol{u})) + \boldsymbol{f}_0 = \boldsymbol{0} \quad \text{a.e. in } \Omega. \tag{7.22}$$

Performing integration by parts in (7.21) and using the relation (7.22), we have

$$\int_{\Gamma_2} \boldsymbol{\sigma\nu} \cdot \boldsymbol{w} \, da = \int_{\Gamma_2} \boldsymbol{f}_2 \cdot \boldsymbol{w} \, da \quad \text{for all } \boldsymbol{w} \in \tilde{U}.$$

Since $\boldsymbol{\sigma\nu} \in L^2(\Gamma; \mathbb{R}^d)$ (see (7.19)) and $\boldsymbol{w} \in \tilde{U}$ is arbitrary, we derive from the above equality that

$$\boldsymbol{\sigma\nu} = \boldsymbol{f}_2 \quad \text{a.e. on } \Gamma_2. \tag{7.23}$$

Now multiply (7.22) by $\boldsymbol{v} - \boldsymbol{u}$ with $\boldsymbol{v} \in U$, integrate over Ω, and integrate by parts to get

$$\int_\Gamma \boldsymbol{\sigma\nu} \cdot (\boldsymbol{v} - \boldsymbol{u}) \, da - \int_\Omega \mathcal{F}(\boldsymbol{\varepsilon}(\boldsymbol{u})) \cdot \boldsymbol{\varepsilon}(\boldsymbol{v} - \boldsymbol{u}) \, dx + \int_\Omega \boldsymbol{f}_0 \cdot (\boldsymbol{v} - \boldsymbol{u}) \, dx = 0,$$

that is,

$$\int_\Omega \mathcal{F}(\varepsilon(\boldsymbol{u}))\cdot\varepsilon(\boldsymbol{v}-\boldsymbol{u})\,\mathrm{d}x = \langle \boldsymbol{f}, \boldsymbol{v}-\boldsymbol{u}\rangle + \int_{\Gamma_3} \sigma\boldsymbol{\nu}\cdot(\boldsymbol{v}-\boldsymbol{u})\,\mathrm{d}a \quad \text{for all } \boldsymbol{v}\in U. \tag{7.24}$$

Thus,

$$R_{\boldsymbol{u}}(\boldsymbol{v}^h,\boldsymbol{u}) = \int_{\Gamma_3} \left[\sigma\boldsymbol{\nu}\cdot(\boldsymbol{v}^h-\boldsymbol{u}) + F_b(u_\nu)(\|\boldsymbol{v}_\tau^h\|-\|\boldsymbol{u}_\tau\|) + j_\nu^0(u_\nu; v_\nu^h - u_\nu)\right]\mathrm{d}a,$$

and then,

$$|R_{\boldsymbol{u}}(\boldsymbol{v}^h,\boldsymbol{u})| \le c\,\|\boldsymbol{u}-\boldsymbol{v}^h\|_{L^2(\Gamma_3)^d}. \tag{7.25}$$

Finally, from (7.16), we have the inequality

$$\|\boldsymbol{u}-\boldsymbol{u}^h\|_V \le c\,\inf_{\boldsymbol{v}^h\in U^h}\left[\|\boldsymbol{u}-\boldsymbol{v}^h\|_V + \|\boldsymbol{u}-\boldsymbol{v}^h\|_{L^2(\Gamma_3)^d}^{1/2}\right]. \tag{7.26}$$

Under additional solution regularity assumption

$$\boldsymbol{u}|_{\Gamma_{3,i}} \in H^2(\Gamma_{3,i};\mathbb{R}^d), \quad 1\le i\le i_3, \tag{7.27}$$

we apply standard finite element interpolation theory (*e.g.* Ciarlet 1978, Brenner and Scott 2008) and derive from (7.26) the following optimal order error bound:

$$\|\boldsymbol{u}-\boldsymbol{u}^h\|_V \le c\,h. \tag{7.28}$$

The constant c depends on $\|\boldsymbol{u}\|_{H^2(\Omega;\mathbb{R}^d)}$, $\|\sigma\boldsymbol{\nu}\|_{L^2(\Gamma_3;\mathbb{R}^d)}$ and $\|\boldsymbol{u}\|_{H^2(\Gamma_{3,i};\mathbb{R}^d)}$ for $1\le i\le i_3$.

We comment that similar results hold for the frictionless version of the model, that is, where the friction condition (2.25) is replaced with

$$\boldsymbol{\sigma}_\tau = \boldsymbol{0} \quad \text{on } \Gamma_3. \tag{7.29}$$

Then the problem is to solve the inequality (2.44) without the term

$$\int_{\Gamma_3} F_b(u_\nu)(\|\boldsymbol{v}_\tau\|-\|\boldsymbol{u}_\tau\|)\,\mathrm{d}a,$$

that is, to find a displacement field $\boldsymbol{u}\in U$ such that

$$\int_\Omega \mathcal{F}\varepsilon(\boldsymbol{u})\cdot(\varepsilon(\boldsymbol{v})-\varepsilon(\boldsymbol{u}))\,\mathrm{d}x + \int_{\Gamma_3} j_\nu^0(u_\nu; v_\nu - u_\nu)\,\mathrm{d}a$$

$$\ge \int_\Omega \boldsymbol{f}_0\cdot(\boldsymbol{v}-\boldsymbol{u})\,\mathrm{d}x + \int_{\Gamma_2} \boldsymbol{f}_2\cdot(\boldsymbol{v}-\boldsymbol{u})\,\mathrm{d}a \quad \text{for all } \boldsymbol{v}\in U. \tag{7.30}$$

The condition (7.5) reduces to

$$\alpha_{j_\nu}\lambda_{1\nu}^{-1} < m_{\mathcal{F}}. \tag{7.31}$$

The inequality (7.26) and the error bound (7.28) still hold for the linear finite element solution.

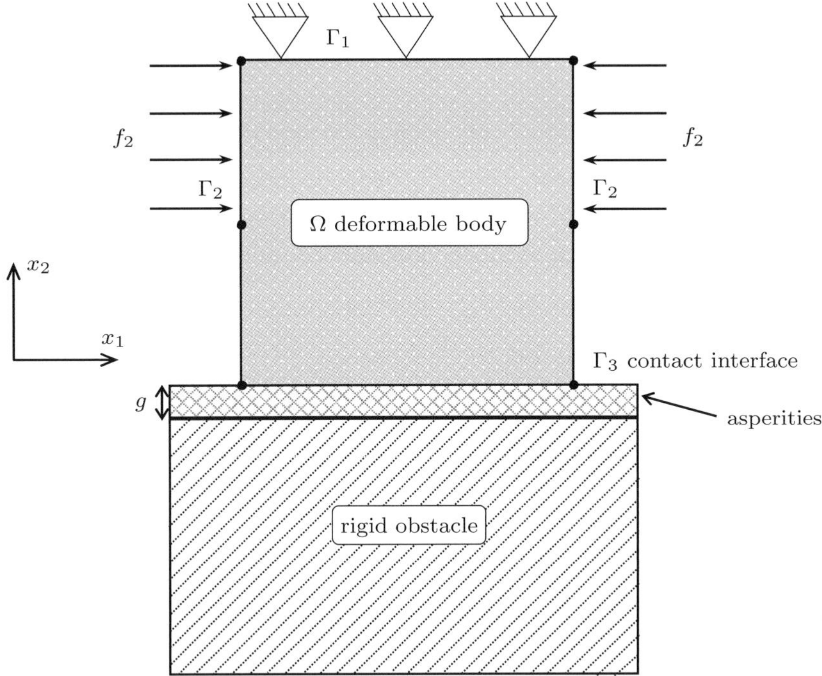

Figure 7.1. Reference configuration of the two-dimensional body.

7.3. Numerical examples

We now present numerical simulation results for the linear finite element solution of Problem 2.6 and its frictionless counterpart, *i.e.* the hemivariational inequality (7.30). The numerical results are adapted from Sofonea, Han and Barboteu (2017) and Han, Sofonea and Barboteu (2017), respectively. For both numerical examples, we use the physical setting shown in Figure 7.1. The domain Ω represents the cross-section of a three-dimensional linearly elastic body such that the plane stress hypothesis is valid. We take $\Omega = (0,1) \times (0,1) \subset \mathbb{R}^2$ to be the unit square and partition the boundary as follows:

$$\Gamma_1 = [0,1] \times \{1\}, \quad \Gamma_2 = (\{0\} \times (0,1)) \cup (\{1\} \times (0,1)), \quad \Gamma_3 = [0,1] \times \{0\}.$$

The body is clamped on Γ_1, and is subject to the actions of a vertical body force of constant density and of horizontally compressive forces on the part $(\{0\} \times [0.5,1)) \cup (\{1\} \times [0.5,1))$ of the boundary Γ_2. The part $(\{0\} \times (0,0.5)) \cup (\{1\} \times (0,0.5))$ is traction-free. The body is in contact with an obstacle on Γ_3. For numerical simulations, linear finite elements on uniform triangulations of the domain Ω are used. Each side of the boundary of the domain is divided into $1/h$ equal parts, and h is used as the discretization parameter.

The mechanical response of the material is described by a linear elastic constitutive law. In terms of the components, the elasticity tensor $\mathcal{F}$ is defined by the relations

$$(\mathcal{F}\boldsymbol{\tau})_{ij} = \frac{E\kappa}{1-\kappa^2}(\tau_{11} + \tau_{22})\delta_{ij} + \frac{E}{1+\kappa}\tau_{ij}, \quad 1 \le i,j \le 2, \text{ for all } \boldsymbol{\tau} \in \mathbb{S}^2,$$

where E and κ are the Young's modulus and Poisson ratio of the material, δ_{ij} is the Kronecker delta symbol. For both numerical examples, we use

$$E = 2000 \text{ N m}^{-2}, \quad \kappa = 0.4,$$
$$\boldsymbol{f}_0 = (0, -0.5 \times 10^{-3}) \text{ N m}^{-2},$$
$$\boldsymbol{f}_2 = \begin{cases} (8 \times 10^{-3}, 0) \text{ N m}^{-1} & \text{on } \{0\} \times [0.5, 1), \\ (-8 \times 10^{-3}, 0) \text{ N m}^{-1} & \text{on } \{1\} \times [0.5, 1). \end{cases}$$

To describe the contact condition on the subset $\Gamma_3 = [0,1] \times \{0\}$ of the boundary, we let $0 < r_{\nu 1} < r_{\nu 2}$ be given, and define two functions $p_\nu : \mathbb{R} \to \mathbb{R}$ and $j_\nu : \mathbb{R} \to \mathbb{R}$ by

$$p_\nu(r) = \begin{cases} 0 & \text{if } r \le 0, \\ c_{\nu 1}r & \text{if } r \in (0, r_{\nu 1}], \\ c_{\nu 1}r_{\nu 1} + c_{\nu 2}(r - r_{\nu 1}) & \text{if } r \in (r_{\nu 1}, r_{\nu 2}), \\ c_{\nu 1}r_{\nu 1} + c_{\nu 2}(r_{\nu 2} - r_{\nu 1}) + c_{\nu 3}(r - r_{\nu 2}) & \text{if } r \ge r_{\nu 2}, \end{cases} \tag{7.32}$$

and

$$j_\nu(r) = \int_0^r p_\nu(s)\,\mathrm{d}s, \quad r \in \mathbb{R}, \tag{7.33}$$

respectively. Here $c_{\nu 1} > 0$, $c_{\nu 2} < 0$ and $c_{\nu 3} > 0$ are given constants. Then

$$F_b(r) = \mu\, p_\nu(r), \quad r \in \mathbb{R}, \tag{7.34}$$

where $\mu \ge 0$ represents a given coefficient of friction. The function p_ν is continuous but not monotone, and therefore j_ν is a locally Lipschitz non-convex function. In both examples we take $c_{\nu 1} = 100 \text{ N m}^{-2}$, $c_{\nu 2} = -100 \text{ N m}^{-2}$, $c_{\nu 3} = 400 \text{ N m}^{-2}$, $r_{\nu 1} = 0.1 \text{ m}$, $r_{\nu 2} = 0.15 \text{ m}$ and $g = 0.15 \text{ m}$. Note that g represents the maximally allowed amount of penetration.

In the first numerical example, the frictional contact condition on Γ_3 takes the following form:

$$u_\nu \le g, \quad \sigma_\nu + \xi_\nu \le 0, \quad (u_\nu - g)(\sigma_\nu + \xi_\nu) = 0, \tag{7.35}$$
$$\xi_\nu = p_\nu(u_\nu), \tag{7.36}$$
$$\|\boldsymbol{\sigma}_\tau\| \le \mu\,\xi_\nu, \quad -\boldsymbol{\sigma}_\tau = \mu\,\xi_\nu \frac{\boldsymbol{u}_\tau}{\|\boldsymbol{u}_\tau\|} \quad \text{if } \boldsymbol{u}_\tau \ne \boldsymbol{0}. \tag{7.37}$$

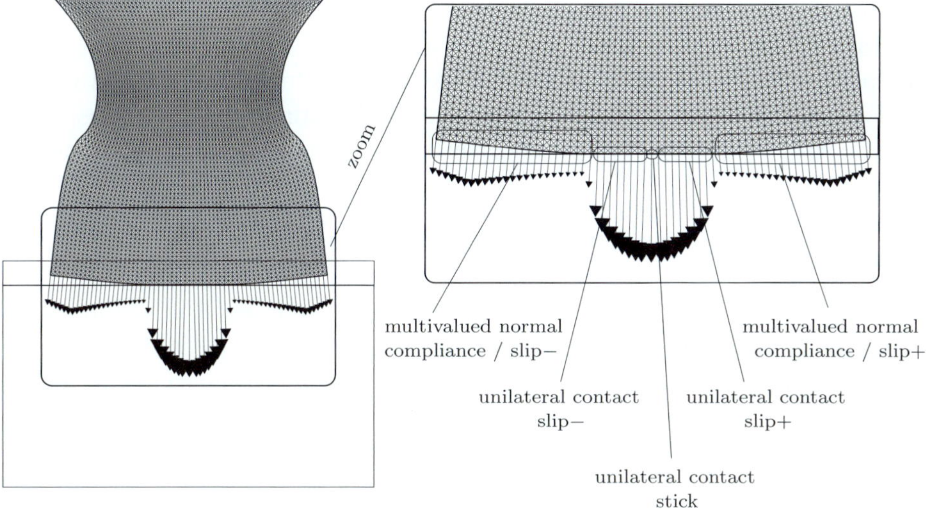

Figure 7.2. Deformed meshes and interface forces on Γ_3 for the first example.

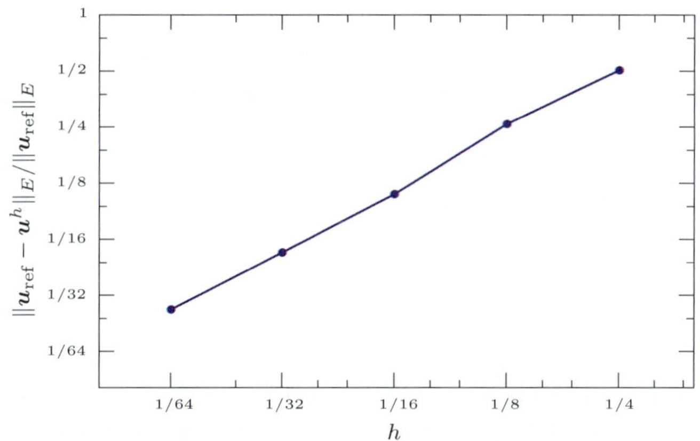

Figure 7.3. Relative errors in energy norm for the first example.

The deformed mesh and the distribution of the interface forces on Γ_3 are reported in Figure 7.2 corresponding to a mesh-size $h = 1/64$. We observe that the contact boundary Γ_3 can be split into two parts depending on whether the penetration bound is reached. More precisely, for some number $\delta_1 \in (0, 1/2)$, Γ_3 can be expressed as the union of three subsets $\Gamma_{31} = [0, 1/2-\delta_1) \times \{0\}$, $\Gamma_{32} = [1/2-\delta_1, 1/2+\delta_1] \times \{0\}$ and $\Gamma_{33} = (1/2+\delta_1, 1] \times \{0\}$ such that the contact nodes on Γ_{31} are in multivalued normal compliance status with backward slip (slip−), those on Γ_{33} are in multivalued normal compliance status with forward slip (slip+), and the nodes on Γ_{32} are in unilateral contact. Note that on $\Gamma_{31} \cup \Gamma_{33}$ the normal displacement u_ν

does not reach the penetration bound, *i.e.* $u_\nu < g$, whereas on Γ_{32} the penetration bound is reached, *i.e.* $u_\nu = g$. Most of the nodes on Γ_{32} are in slip status, except the node at the centre of Γ_{32} which is in stick status.

In Figure 7.3, we report relative errors of the numerical solutions in the energy norm, $\|\boldsymbol{u}_{\mathrm{ref}} - \boldsymbol{u}^h\|_E / \|\boldsymbol{u}_{\mathrm{ref}}\|_E$, where the energy norm is defined by the formula

$$\|\boldsymbol{v}\|_E := \frac{1}{\sqrt{2}} (\mathcal{F}(\boldsymbol{\varepsilon}(\boldsymbol{v})), \boldsymbol{\varepsilon}(\boldsymbol{v}))_Q^{1/2}.$$

Note that the energy norm $\|\boldsymbol{v}\|_E$ is equivalent to the norm $\|\boldsymbol{v}\|_V$, and the error bound (7.28) predicts an optimal first-order convergence of the numerical solutions measured in the energy norm, under the regularity assumptions (7.18) and (7.27), which take the form

$$\boldsymbol{u} \in H^2(\Omega; \mathbb{R}^2), \quad \boldsymbol{u}|_{\Gamma_3} \in H^2(\Gamma_3; \mathbb{R}^2). \tag{7.38}$$

Since the true solution $\boldsymbol{u}$ is not available, we use the numerical solution corresponding to a fine discretization of Ω as the 'reference' solution $\boldsymbol{u}_{\mathrm{ref}}$ in computing the solution errors. Here, the numerical solution with $h = 1/256$ is taken to be the 'reference' solution $\boldsymbol{u}_{\mathrm{ref}}$. We clearly observe the theoretically predicted optimal linear convergence of the numerical solutions.

In the second numerical example, the contact boundary conditions on Γ_3 are characterized by a frictionless multivalued normal compliance contact in which the penetration is restricted by the unilateral constraint. For simulations, on Γ_3, we use (7.35), (7.36), and replace (7.37) with

$$\boldsymbol{\sigma}_\tau = \mathbf{0}. \tag{7.39}$$

It follows from Han, Sofonea and Barboteu (2017) that for the linear element solution of the corresponding hemivariational inequality we again have the optimal order error bound (7.28) under the regularity assumptions (7.38).

The numerical results on the deformed mesh and the distribution of the interface forces on Γ_3 are shown in Figure 7.4 for a mesh-size $h = 1/64$. As in the first example, the contact boundary Γ_3 can be split into three subsets $\Gamma_{31} = [0, 1/2 - \delta_2) \times \{0\}$, $\Gamma_{32} = [1/2 - \delta_2, 1/2 + \delta_2] \times \{0\}$ and $\Gamma_{33} = (1/2 + \delta_2, 1] \times \{0\}$, for some number $\delta_2 \in (0, 1/2)$, such that the nodes on $\Gamma_{31} \cup \Gamma_{33}$ are in multivalued normal compliance status, whereas the nodes on Γ_{32} are in unilateral contact status where the penetration reaches the upper bound g.

In Figure 7.5 we report relative errors of the numerical solutions in the energy norm, $\|\boldsymbol{u}_{\mathrm{ref}} - \boldsymbol{u}^h\|_E / \|\boldsymbol{u}_{\mathrm{ref}}\|_E$. Again, we use the numerical solution corresponding to $h = 1/256$ as the 'reference' solution $\boldsymbol{u}_{\mathrm{ref}}$ in computing the solution errors. Once more, we clearly observe the theoretically predicted optimal linear convergence of the numerical solutions.

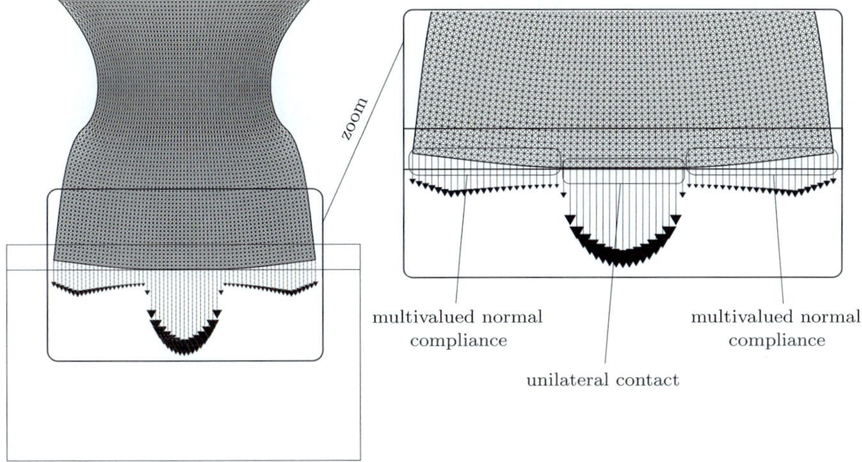

Figure 7.4. Deformed meshes and interface forces on Γ_3 for the second example.

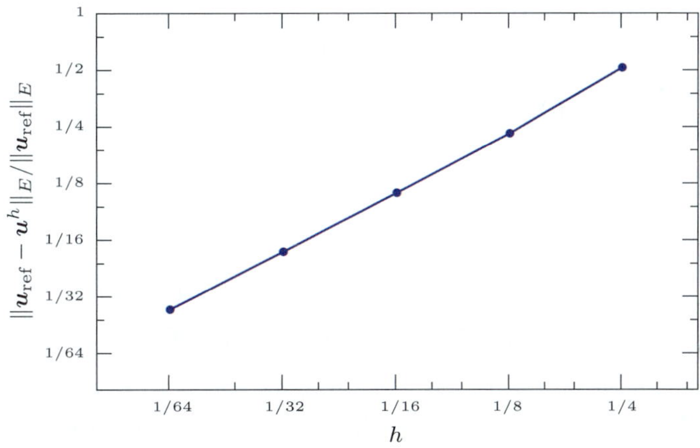

Figure 7.5. Relative errors in energy norm for the second example.

8. Studies of the history-dependent contact problem

In this section we study the history-dependent contact problem, Problem 2.8. We first explore the solution existence and uniqueness, then introduce a linear finite element method and derive an optimal order error estimate and, finally, present numerical simulation results.

8.1. Solution existence and uniqueness

Recall that $\lambda_{1\nu} > 0$ is the smallest eigenvalue of the eigenvalue problem (7.4). The unique solvability of Problem 2.8 is given by the following existence and uniqueness result.

Theorem 8.1. Assume (2.34), (2.35), (2.38), (2.51), (2.52) and

$$\alpha_{j_\nu} \lambda_{1\nu}^{-1} < m_{\mathcal{F}}. \tag{8.1}$$

Then Problem 2.8 has a unique solution $\boldsymbol{u} \in C(I; U)$.

Proof. Let $X = V$ as defined in (2.3), $Y = Q$ as defined in (2.4), $K = U$ as defined in (2.39), $X_j = V_j$ as defined in (7.2), and let $\gamma_j \colon V \to V_j$ be the normal trace operator. We introduce operators and functionals $A \colon V \to V^*$, $\mathcal{S} \colon C(I; V) \to C(I; Q)$, $\varphi \colon Q \times V \times V \to \mathbb{R}$, $j \colon V_j \to \mathbb{R}$ and $\boldsymbol{f} \colon I \to V^*$ as follows:

$$\langle A\boldsymbol{u}, \boldsymbol{v} \rangle = \int_\Omega \mathcal{F}\varepsilon(\boldsymbol{u}) \cdot \varepsilon(\boldsymbol{v}) \, \mathrm{d}x, \quad \boldsymbol{u}, \boldsymbol{v} \in V, \tag{8.2}$$

$$\mathcal{S}\boldsymbol{u}(t) = \int_0^t \mathcal{B}(t-s)\varepsilon(\boldsymbol{u}(s)) \, \mathrm{d}s, \quad \boldsymbol{u} \in C(I; V), \tag{8.3}$$

$$\varphi(\boldsymbol{y}, \boldsymbol{u}, \boldsymbol{v}) = \int_\Omega \boldsymbol{y} \cdot \varepsilon(\boldsymbol{v}) \, \mathrm{d}x, \quad \boldsymbol{y} \in Q, \, \boldsymbol{u}, \boldsymbol{v} \in V, \tag{8.4}$$

$$j(z) = \int_{\Gamma_3} j_\nu(z) \, \mathrm{d}a, \quad z \in V_j, \tag{8.5}$$

$$\langle \boldsymbol{f}(t), \boldsymbol{v} \rangle = \int_\Omega \boldsymbol{f}_0(t) \cdot \boldsymbol{v} \, \mathrm{d}x + \int_{\Gamma_2} \boldsymbol{f}_2(t) \cdot \boldsymbol{v} \, \mathrm{d}a, \quad \boldsymbol{v} \in V. \tag{8.6}$$

Consider the problem of finding a function $\boldsymbol{u} \colon I \to U$ such that for each $t \in I$, the following inequality holds:

$$\langle A\boldsymbol{u}(t), \boldsymbol{v} - \boldsymbol{u}(t) \rangle + \varphi(\mathcal{S}\boldsymbol{u}(t), \boldsymbol{u}(t), \boldsymbol{v}) - \varphi(\mathcal{S}\boldsymbol{u}(t), \boldsymbol{u}(t), \boldsymbol{u}(t))$$
$$+ j^0(\gamma_j \boldsymbol{u}(t); \gamma_j \boldsymbol{v} - \gamma_j \boldsymbol{u}(t)) \geq \langle \boldsymbol{f}(t), \boldsymbol{v} - \boldsymbol{u}(t) \rangle \quad \text{for all } \boldsymbol{v} \in U. \tag{8.7}$$

We can apply Theorem 5.4 to see that inequality (8.7) has a unique solution $\boldsymbol{u} \in C(I; U)$. The argument is similar to that used in the proof of Theorem 7.1, so we only give a sketch of the proof. Note that the function φ defined by (8.4) satisfies condition (5.3) with $\alpha_\varphi = 0$ and $\beta_\varphi = 1$. From assumption (2.51) and inequality (2.9) we deduce that the operator $\mathcal{S} \colon C(I; V) \to C(I; Q)$ defined by (8.3) is a history-dependent operator. The smallness condition (5.11) follows from (8.1). Use inequality (7.11) to see that $\boldsymbol{u}$ is a solution to Problem 2.8. This proves the existence part of Theorem 8.1. The uniqueness part follows from a standard argument, similar to that in the last part of the proof of Theorem 7.1, combined with the Gronwall argument. $\qquad\square$

8.2. *Numerical analysis of the problem*

Throughout this subsection, we keep the assumptions stated in Theorem 8.1 so that we are assured that Problem 2.8 has a unique solution $\boldsymbol{u} \in C(I; U)$. We now proceed with the discretization of Problem 2.8. We use the finite element space V^h and the finite element set U^h as in Section 7.2. Here and below for any vector field z_i we use the notation $z_{i,\nu}$ and $z_{i,\tau}$ to represent its normal and tangential components, respectively.

Assume g is a continuous concave function. Then, $U^h \subset U$. Thus the approximation is internal and the numerical method for Problem 2.8 is defined as follows.

Problem 8.2. Find a discrete displacement $\boldsymbol{u}^{hk} := \{\boldsymbol{u}_n^{hk}\}_{n=0}^N \subset U^h$ such that for $0 \leq n \leq N$ and for all $\boldsymbol{v}^h \in U^h$,

$$\int_\Omega \mathcal{F}\varepsilon(\boldsymbol{u}_n^{hk}) \cdot (\varepsilon(\boldsymbol{v}^h) - \varepsilon(\boldsymbol{u}_n^{hk})) \, \mathrm{d}x + \int_\Omega \mathcal{S}_n^k \boldsymbol{u}^{hk} \cdot (\varepsilon(\boldsymbol{v}^h) - \varepsilon(\boldsymbol{u}_n^{hk})) \, \mathrm{d}x$$

$$+ \int_{\Gamma_3} j_\nu^0(u_{n,\nu}^{hk}; v_\nu^h - u_{n,\nu}^{hk}) \, \mathrm{d}a$$

$$\geq \int_\Omega \boldsymbol{f}_0(t_n) \cdot (\boldsymbol{v}^h - \boldsymbol{u}_n^{hk}) \, \mathrm{d}x + \int_{\Gamma_2} \boldsymbol{f}_2(t_n) \cdot (\boldsymbol{v}^h - \boldsymbol{u}_n^{hk}) \, \mathrm{d}a, \qquad (8.8)$$

where

$$\mathcal{S}_n^k \boldsymbol{u}^{hk} = k \sum_{i=0}^n {}' \mathcal{B}(t_n - t_i) \, \varepsilon(\boldsymbol{u}_i^{hk}).$$

It is straightforward to check that the derivation of the error inequality (5.35) in Theorem 5.10 is valid when the term $j^0(\gamma_j u_n^{hk}; \gamma_j v^h - \gamma_j u_n^{hk})$ in (5.34) is replaced by $\int_{\Gamma_3} j_\nu^0(u_{n,\nu}^{hk}; v_\nu^h - u_{n,\nu}^{hk}) \, \mathrm{d}a$ as in (8.8). Recall that, here and below, $u_{n,\nu}^{hk}$ denotes the normal component of the vector field $\boldsymbol{u}_n^{hk}$.

To derive an error estimate for the numerical solution defined by Problem 8.2, we need to bound the residual term defined in (5.36):

$$R_n(\boldsymbol{v}^h, \boldsymbol{u}_n) = \langle A\boldsymbol{u}_n, \boldsymbol{v}^h - \boldsymbol{u}_n \rangle + \varphi(\mathcal{S}_n \boldsymbol{u}, \boldsymbol{u}_n, \boldsymbol{v}^h) - \varphi(\mathcal{S}_n \boldsymbol{u}, \boldsymbol{u}_n, \boldsymbol{u}_n)$$

$$+ \int_{\Gamma_3} j_\nu^0(u_{n,\nu}; v_\nu^h - u_{n,\nu}) \, \mathrm{d}a - \langle \boldsymbol{f}_n, \boldsymbol{v}^h - \boldsymbol{u}_n \rangle. \qquad (8.9)$$

For this purpose, we assume the following solution regularity property:

$$\boldsymbol{u} \in W^{2,\infty}(I; V). \qquad (8.10)$$

For all $t \in I$ define

$$\boldsymbol{\sigma}(t) = \mathcal{F}\varepsilon(\boldsymbol{u}(t)) + \int_0^t \mathcal{B}(t - s) \, \varepsilon(\boldsymbol{u}(s)) \, \mathrm{d}s \quad \text{in } \Omega.$$

Then

$$\boldsymbol{\sigma} \in C(I; H^1(\Omega; \mathbb{S}^d)), \quad \boldsymbol{\sigma}\boldsymbol{\nu} \in C(I; L^2(\Gamma; \mathbb{R}^d)). \qquad (8.11)$$

We now derive some pointwise relations for the weak solution $\boldsymbol{u}$, similar to what is done in Section 7.2. Define a subset of U:

$$\widetilde{U} = \{ \boldsymbol{v} \in C^\infty(\overline{\Omega})^d \mid \boldsymbol{v} = \boldsymbol{0} \text{ on } \Gamma_1, \ v_\nu = 0 \text{ on } \Gamma_3 \}.$$

We take $\boldsymbol{v} = \boldsymbol{u}(t) \pm \widetilde{\boldsymbol{v}}$ in (2.53), where $\widetilde{\boldsymbol{v}} \in \widetilde{U}$ is arbitrary; this leads to

$$(\boldsymbol{\sigma}(t), \boldsymbol{\varepsilon}(\widetilde{\boldsymbol{v}}))_{L^2(\Omega;\mathbb{S}^d)} = \langle \boldsymbol{f}(t), \widetilde{\boldsymbol{v}} \rangle_{V^* \times V} \quad \text{for all } \widetilde{\boldsymbol{v}} \in \widetilde{U}.$$

From this identity, we can deduce that

$$\operatorname{Div} \boldsymbol{\sigma}(t) + \boldsymbol{f}_0(t) = \boldsymbol{0} \quad \text{a.e. in } \Omega, \tag{8.12}$$

$$\boldsymbol{\sigma}(t)\boldsymbol{\nu} = \boldsymbol{f}_2(t) \quad \text{a.e. on } \Gamma_2, \quad \boldsymbol{\sigma}_\tau(t) = \boldsymbol{0} \quad \text{a.e. on } \Gamma_3. \tag{8.13}$$

Next, we multiply (8.12) by $\boldsymbol{v} - \boldsymbol{u}(t)$ with $\boldsymbol{v} \in U$, integrate over Ω, and perform an integration by parts:

$$- \int_\Omega \boldsymbol{\sigma}(t) \cdot (\boldsymbol{\varepsilon}(\boldsymbol{v}) - \boldsymbol{\varepsilon}(\boldsymbol{u}(t)))\, \mathrm{d}x$$

$$+ \int_\Gamma \boldsymbol{\sigma}(t)\boldsymbol{\nu} \cdot (\boldsymbol{v} - \boldsymbol{u}(t))\, \mathrm{d}a + \int_\Omega \boldsymbol{f}_0 \cdot (\boldsymbol{v} - \boldsymbol{u}(t))\, \mathrm{d}x = 0.$$

Thus, at any $t \in I$, for any $\boldsymbol{v} \in U$, we have

$$\int_\Omega \boldsymbol{\sigma}(t) \cdot (\boldsymbol{\varepsilon}(\boldsymbol{v}) - \boldsymbol{\varepsilon}(\boldsymbol{u}(t)))\, \mathrm{d}x = \langle \boldsymbol{f}(t), \boldsymbol{v} - \boldsymbol{u}(t) \rangle_{V^* \times V} + \int_{\Gamma_3} \sigma_\nu(t)(v_\nu - u_\nu(t))\, \mathrm{d}a. \tag{8.14}$$

Take $\boldsymbol{v} = \boldsymbol{v}^h \in U^h$ in (8.14) at $t = t_n$,

$$\int_\Omega \boldsymbol{\sigma}_n \cdot (\boldsymbol{\varepsilon}(\boldsymbol{v}^h) - \boldsymbol{\varepsilon}(\boldsymbol{u}_n))\, \mathrm{d}x = \langle \boldsymbol{f}_n, \boldsymbol{v}^h - \boldsymbol{u}_n \rangle_{V^* \times V} + \int_{\Gamma_3} \sigma_{n,\nu}(v^h_{n,\nu} - u_{n,\nu})\, \mathrm{d}a,$$

which can be rewritten as

$$\langle A\boldsymbol{u}_n, \boldsymbol{v}^h - \boldsymbol{u}_n \rangle + \varphi(\mathcal{S}_n \boldsymbol{u}, \boldsymbol{u}_n, \boldsymbol{v}^h) - \varphi(\mathcal{S}_n \boldsymbol{u}, \boldsymbol{u}_n, \boldsymbol{u}_n)$$

$$= \langle \boldsymbol{f}_n, \boldsymbol{v}^h - \boldsymbol{u}_n \rangle + \int_{\Gamma_3} \sigma_{n,\nu}(v^h_\nu - u_{n,\nu})\, \mathrm{d}a. \tag{8.15}$$

Thus, for the residual term of (8.9), we have

$$R_n(\boldsymbol{v}^h, \boldsymbol{u}_n) = \int_{\Gamma_3} \left[\sigma_{n,\nu}(v^h_\nu - u_{n,\nu}) + j^0_\nu(u_{n,\nu}; v^h_\nu - u_{n,\nu}) \right] \mathrm{d}a. \tag{8.16}$$

Using the solution regularity assumptions (8.10) and (8.11), we have

$$|R_n(\boldsymbol{v}^h, \boldsymbol{u}_n)| \leq c \|u_{n,\nu} - v^h_\nu\|_{L^2(\Gamma_3)}. \tag{8.17}$$

Applying (5.35), we get

$$\max_{0 \leq n \leq N} \|\boldsymbol{u}_n - \boldsymbol{u}^{hk}_n\|_V$$

$$\leq c \max_{0 \leq n \leq N} \inf_{\boldsymbol{v}^h \in V^h} \left[\|\boldsymbol{u}_n - \boldsymbol{v}^h\|_V + \|u_{n,\nu} - v^h_\nu\|^{1/2}_{L^2(\Gamma_3)} \right] + c\, k^2. \tag{8.18}$$

Recall that

$$\overline{\Gamma_3} = \bigcup_{1 \le i \le i_3} \Gamma_{3,i},$$

where $\Gamma_{3,i}$ $(1 \le i \le i_3)$ is a closed subset of an affine hyperplane. We further assume

$$\boldsymbol{u} \in C(I; H^2(\Omega; \mathbb{R}^d)), \quad u_\nu|_{\Gamma_{3,i}} \in C(I; H^2(\Gamma_{3,i})), \ 1 \le i \le i_3. \tag{8.19}$$

Then applying the finite element interpolation theory (*e.g.* Ciarlet 1978, Brenner and Scott 2008), we can derive the optimal order error bound

$$\max_{0 \le n \le N} \|\boldsymbol{u}_n - \boldsymbol{u}_n^{hk}\|_V \le c(h + k^2). \tag{8.20}$$

Hence, the method is first-order with respect to the spatial mesh-size and second-order with respect to the temporal step-size.

8.3. A numerical example

Here, we report a numerical example for Problem 2.8. The physical setting of the example is shown in Figure 8.1. The domain Ω represents the cross-section of a three-dimensional linearly viscoelastic body such that the plane stress hypothesis is valid. For the numerical example, we take $\Omega = (0, 2) \times (0, 1)$ with the metre as the length unit. The boundary $\Gamma = \partial\Omega$ is decomposed into three parts: Γ_1 where the body is fixed, Γ_2 where the body is subject to the action of surface traction, and Γ_3 where contact takes place. We take $\Gamma_1 = \{0\} \times [0, 1]$, $\Gamma_2 = \Gamma_{21} \cup \Gamma_{22}$ with $\Gamma_{21} = \{2\} \times (0, 1)$ and $\Gamma_{22} = (0, 2) \times \{1\}$, and $\Gamma_3 = [0, 2] \times \{0\}$.

The elasticity tensor $\mathcal{F}$ is determined by the relations

$$(\mathcal{F}\tau)_{ij} = \frac{E\kappa}{1 - \kappa^2}(\tau_{11} + \tau_{22})\delta_{ij} + \frac{E}{1 + \kappa}\tau_{ij}, \quad 1 \le i, j \le 2, \tag{8.21}$$

where E and κ are the Young's modulus and Poisson ratio of the material, and δ_{ij} denotes the Kronecker delta symbol. In the numerical simulation, we choose $E = 1 \ \text{N} \ \text{m}^{-2}$ and $\kappa = 0.3$.

The relaxation tensor is given by $\mathcal{B}(s) = (0.5 + s)^3 \mathcal{I}$, where $\mathcal{I}$ denotes the identity tensor. For a given value $S \ge 0$, the function $j_\nu(\cdot)$ is defined by

$$j_\nu(u_\nu) = S \int_0^{|u_\nu|} \mu(s) \, \mathrm{d}s$$

with

$$\mu(s) = (a - b) \, \mathrm{e}^{-\alpha s} + b$$

with $a \ge b > 0$ and $\alpha > 0$. We choose $S = 0.1 \ N$, $\alpha = 100$, $a = 0.04$,

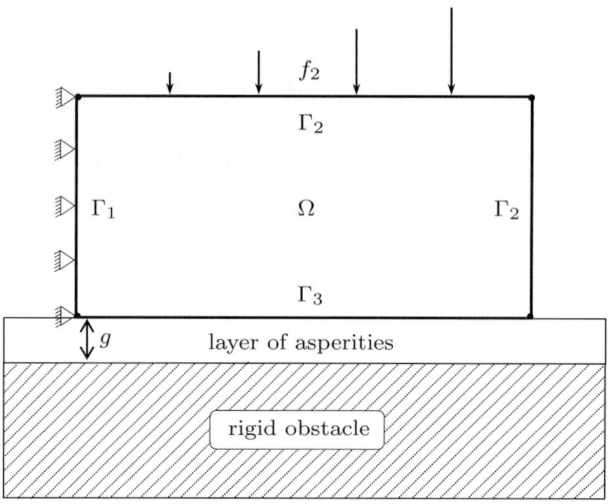

Figure 8.1. Reference configuration of the two-dimensional body.

$b = 0.02$. The body force is ignored, and for the surface traction

$$
\boldsymbol{f}_2(\boldsymbol{x}) = \begin{cases} (0,0) \text{ N m}^{-1} & \text{on } \Gamma_{21}, \\ (0, -0.1(1 - \mathrm{e}^{-2t})x_1) \text{ N m}^{-1} & \text{on } \Gamma_{22}. \end{cases}
$$

For the thickness function, we let $g = 0.2$ m.

For the numerical solution of the problem, we introduce a family of rectangular finite element partitions as follows: given a positive integer M, we divide the horizontal interval $[0, 2]$ and the vertical interval $[0, 1]$ into M equal parts, and form the corresponding rectangular mesh with M^2 rectangular elements. We then construct the bilinear element space V^h, with the finite element mesh parameter $h = 1/M$. Note that with the bilinear element replacing the linear element, the theoretical error estimate (8.20) stays the same.

We focus on the numerical convergence orders of the numerical solutions with respect to the mesh-size h and the time step-size k. Since the true solution is unknown, we use the numerical solution with $h = k = 1/256$ as the 'reference' solution to compute the numerical solution errors. In Figure 8.2, we report the numerical solution errors $\|\boldsymbol{u}_{\mathrm{ref}}(\cdot, 1) - \boldsymbol{u}_N^{hk}\|_1$ in $H^1(\Omega; \mathbb{R}^2)$-norm ($N = 1/k$) for $h = 1/8, 1/16, 1/32, 1/64$; a small fixed time step $k = 1/256$ is used. We observe the linear convergence of the error with respect to the mesh-size h. In Figure 8.3, we report the numerical solution errors $\|\boldsymbol{u}_{\mathrm{ref}}(\cdot, 1) - \boldsymbol{u}_N^{hk}\|_1$ ($N = 1/k$) for $k = 1/4, 1/8, 1/12, 1/16$; a small fixed mesh-size $h = 1/256$ is used. We observe the quadratic convergence of the error with respect to the time step k. These numerical results match the theoretical error bound (8.20) well.

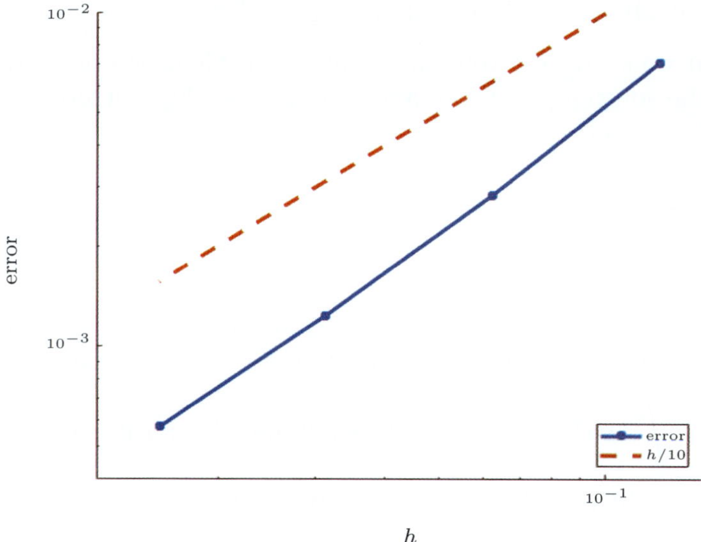

Figure 8.2. Numerical evidence of first-order convergence of $\|\boldsymbol{u}_{\mathrm{ref}}(\cdot, 1) - \boldsymbol{u}_N^{hk}\|_1$ ($N = 1/k$) with respect to h.

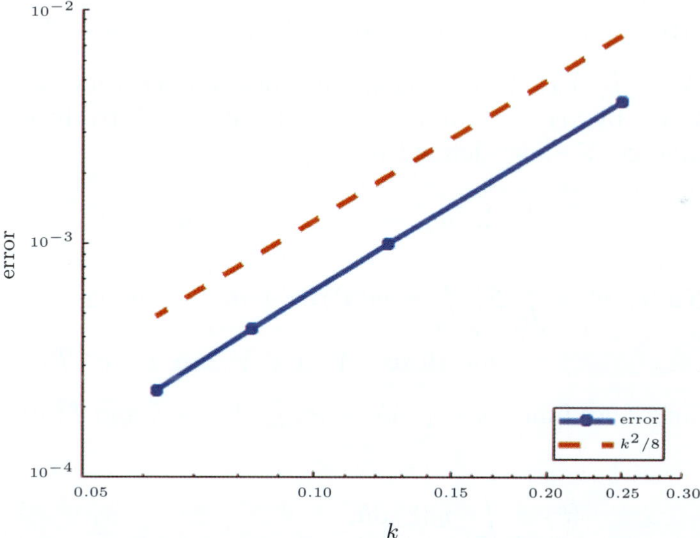

Figure 8.3. Numerical evidence of second-order convergence of $\|\boldsymbol{u}_{\mathrm{ref}}(\cdot, 1) - \boldsymbol{u}_N^{hk}\|_1$ ($N = 1/k$) with respect to k.

9. Studies of the dynamic contact problem

In this section we study the dynamic contact problem, Problem 2.10. We first explore the solution existence and uniqueness, then introduce a linear finite element method and derive an optimal order error estimate and, finally, present numerical simulation results. We will use the spaces V and H defined by (2.3) and (2.5), respectively.

9.1. Solution existence and uniqueness

Recall that $\lambda_{1\nu} > 0$ is the smallest eigenvalue of the eigenvalue problem (7.4). Let $\lambda_{1\tau} > 0$ be the smallest eigenvalue of the eigenvalue problem

$$\boldsymbol{u} \in V, \quad \int_{\Omega} \boldsymbol{\varepsilon}(\boldsymbol{u}){\cdot}\boldsymbol{\varepsilon}(\boldsymbol{v})\,\mathrm{d}x = \lambda \int_{\Gamma_3} \boldsymbol{u}_\tau{\cdot}\boldsymbol{v}_\tau \,\mathrm{d}a \quad \text{for all } \boldsymbol{v} \in V. \tag{9.1}$$

The unique solvability of Problem 2.10 is given by the following result.

Theorem 9.1. Assume (2.64)–(2.70) and

$$\alpha_{j_\nu}\lambda_{1\nu}^{-1} + \alpha_{j_\tau}\lambda_{1\tau}^{-1} < m_{\mathcal{A}}. \tag{9.2}$$

Then Problem 2.10 has a unique solution with regularity

$$\boldsymbol{u} \in H^1(0,T;V), \quad \dot{\boldsymbol{u}} \in \mathcal{W} \subset C([0,T];L^2(\Omega;\mathbb{R}^d)), \quad \ddot{\boldsymbol{u}} \in \mathcal{V}^*. \tag{9.3}$$

Proof. Let $V_1 = L^2(\Gamma_3)$, $V_2 = L^2(\Gamma_3)^d$ and define operators $\gamma_1 \in \mathcal{L}(V,V_1)$ and $\gamma_2 \in \mathcal{L}(V,V_2)$ by $\gamma_1\boldsymbol{v} = v_\nu$ and $\gamma_2\boldsymbol{v} = \boldsymbol{v}_\tau$ for $\boldsymbol{v} \in V$. Introduce operators $A\colon V \to V^*$ and $\mathcal{S}\colon \mathcal{V} \to \mathcal{V}^*$ defined by

$$\langle A\boldsymbol{u},\boldsymbol{v}\rangle = \int_{\Omega} \mathcal{A}\boldsymbol{\varepsilon}(\boldsymbol{u}) \cdot \boldsymbol{\varepsilon}(\boldsymbol{v})\,\mathrm{d}x \quad \text{for all } \boldsymbol{u},\boldsymbol{v} \in V, \tag{9.4}$$

$$\langle \mathcal{S}\boldsymbol{w}(t),\boldsymbol{v}\rangle = \int_{\Omega} \mathcal{B}\!\left(\int_0^t \boldsymbol{\varepsilon}(\boldsymbol{w}(s))\,\mathrm{d}s + \boldsymbol{u}_0\right) \cdot \boldsymbol{\varepsilon}(\boldsymbol{v})\,\mathrm{d}x$$

$$\text{for all } \boldsymbol{w} \in \mathcal{V}, \ \boldsymbol{v} \in V, \text{ a.e. } t \in (0,T). \tag{9.5}$$

In addition, introduce functions $j_1\colon V_1 \to \mathbb{R}$, $j_2\colon V_2 \to \mathbb{R}$ and $\boldsymbol{f}\colon (0,T) \to V^*$ given by

$$j_1(\xi) = \int_{\Gamma_3} j_\nu(\xi)\,\mathrm{d}a, \tag{9.6}$$

$$j_2(\boldsymbol{\xi}) = \int_{\Gamma_3} j_\tau(\boldsymbol{\xi})\,\mathrm{d}a, \tag{9.7}$$

$$\langle \boldsymbol{f}(t),\boldsymbol{v}\rangle = \int_{\Omega} \boldsymbol{f}_0(t) \cdot \boldsymbol{v}\,\mathrm{d}x + \int_{\Gamma_2} \boldsymbol{f}_2(t) \cdot \boldsymbol{v}\,\mathrm{d}a \tag{9.8}$$

for $\xi \in V_1$, $\boldsymbol{\xi} \in V_2$, $\boldsymbol{v} \in V$, a.e. $t \in (0,T)$. With the above notation we consider the following problem, in terms of the velocity.

Problem 9.2. Find $\boldsymbol{w} \in \mathcal{W}$ such that

$$\langle \rho\, \dot{\boldsymbol{w}}(t) + A\boldsymbol{w}(t) + \mathcal{S}\boldsymbol{w}(t) - \boldsymbol{f}(t), \boldsymbol{v} - \boldsymbol{w}(t)\rangle$$
$$+ j_1^0(w_\nu(t); v_\nu - w_\nu(t)) + j_2^0(\boldsymbol{w}_\tau(t); \boldsymbol{v}_\tau - \boldsymbol{w}_\tau(t)) \geq 0$$
$$\text{for all } \boldsymbol{v} \in V, \text{ a.e. } t \in (0, T),$$

$$\boldsymbol{w}(0) = \boldsymbol{w}_0.$$

For an analysis of Problem 9.2, we apply Theorem 6.2 in the functional framework described above. To this end, we check that the hypotheses (6.3)–(6.7) are satisfied.

First, we show that under hypothesis (2.64), the operator A defined by (9.4) satisfies hypothesis (6.3) with $m_A = m_{\mathcal{A}}$. It follows from (2.64(a)), (2.64(c)), and the Hölder inequality that

$$\langle A\boldsymbol{u}_1 - A\boldsymbol{u}_2, \boldsymbol{v}\rangle = \int_\Omega (\mathcal{A}\varepsilon(\boldsymbol{u}_1) - \mathcal{A}\varepsilon(\boldsymbol{u}_2)) \cdot \varepsilon(\boldsymbol{v})\, \mathrm{d}x$$
$$\leq \|\mathcal{A}\varepsilon(\boldsymbol{u}_1) - \mathcal{A}\varepsilon(\boldsymbol{u}_2)\|_{L^2(\Omega;\mathbb{S}^d)}\|\varepsilon(\boldsymbol{v})\|_{L^2(\Omega;\mathbb{S}^d)}$$
$$\leq L_{\mathcal{A}}\|\varepsilon(\boldsymbol{u}_1) - \varepsilon(\boldsymbol{u}_2)\|_{L^2(\Omega;\mathbb{S}^d)}\|\varepsilon(\boldsymbol{v})\|_{L^2(\Omega;\mathbb{S}^d)}$$

for all $\boldsymbol{u}_1, \boldsymbol{u}_2, \boldsymbol{v} \in V$. Hence, we infer that

$$\|A(t, \boldsymbol{u}_1) - A(t, \boldsymbol{u}_2)\|_{V^*} \leq L_{\mathcal{A}}\|\boldsymbol{u}_1 - \boldsymbol{u}_2\|_V \tag{9.9}$$

for all $\boldsymbol{u}_1, \boldsymbol{u}_2 \in V$. This inequality shows that A is Lipschitz continuous and, in particular, demicontinuous, which proves (6.3(a)).

From (9.9) and (2.64(d)), we also have $\|A\boldsymbol{u}\|_{V^*} \leq L_{\mathcal{A}}\|\boldsymbol{u}\|_V$ for all $\boldsymbol{u} \in V$, which gives (6.3(b)) with $a_0 = 0$ and $a_1 = L_{\mathcal{A}}$. Moreover, using (2.64(b)), we obtain

$$\langle A\boldsymbol{u}_1 - A\boldsymbol{u}_2, \boldsymbol{u}_1 - \boldsymbol{u}_2\rangle = \int_\Omega (\mathcal{A}\varepsilon(\boldsymbol{u}_1) - \mathcal{A}\varepsilon(\boldsymbol{u}_2)) \cdot (\varepsilon(\boldsymbol{u}_1) - \varepsilon(\boldsymbol{u}_2))\, \mathrm{d}x$$
$$\geq m_{\mathcal{A}} \int_\Omega \|\varepsilon(\boldsymbol{u}_1) - \varepsilon(\boldsymbol{u}_2)\|_{\mathbb{S}^d}^2\, \mathrm{d}x$$
$$= m_{\mathcal{A}}\|\boldsymbol{u}_1 - \boldsymbol{u}_2\|_V^2$$

for all $\boldsymbol{u}_1, \boldsymbol{u}_2 \in V$, which entails (6.3(c)) and completes the proof of (6.3).

Next, using assumption (2.66) and Theorem 3.47 in Migórski, Ochal and Sofonea (2013) it follows that the function j_1 defined by (9.6) satisfies condition (6.4) with $\alpha_1 = \alpha_{j_\nu}$, $c_{11} = 0$ and, moreover,

$$j_1^0(\xi; \eta) \leq \int_{\Gamma_3} j_\nu^0(\xi; \eta)\, \mathrm{d}a \quad \text{for all } \xi, \eta \in V_1. \tag{9.10}$$

Assumption (2.67) combined with the same argument shows that the function j_2 defined by (9.7) satisfies condition (6.4) with $\alpha_2 = \alpha_{j_\tau}$, $c_{12} = 0$ and, in addition,

$$j_2^0(\boldsymbol{\xi};\boldsymbol{\eta}) \le \int_{\Gamma_3} j_\tau^0(\boldsymbol{\xi};\boldsymbol{\eta})\,\mathrm{d}a \quad \text{for all } \boldsymbol{\xi},\boldsymbol{\eta} \in V_2. \tag{9.11}$$

It is easy to see that the smallness condition (6.5) follows from (9.2).

Moreover, the hypothesis (2.65) implies that the operator $\mathcal{S}$ defined by (9.5) satisfies conditions (6.6). Indeed, let $\boldsymbol{w}_1, \boldsymbol{w}_2 \in \mathcal{V}$, $\boldsymbol{v} \in V$ and $t \in (0,T)$. Using (2.65) and the Hölder inequality, we have

$$\left| \left(\mathcal{B}\left(\int_0^t \boldsymbol{\varepsilon}(\boldsymbol{w}_1(s))\,\mathrm{d}s + \boldsymbol{u}_0 \right) - \mathcal{B}\left(\int_0^t \boldsymbol{\varepsilon}(\boldsymbol{w}_2(s))\,\mathrm{d}s + \boldsymbol{u}_0 \right), \boldsymbol{\varepsilon}(\boldsymbol{v}) \right)_{L^2(\Omega;\mathbb{S}^d)} \right|$$

$$\le \left\| \mathcal{B}\left(\int_0^t \boldsymbol{\varepsilon}(\boldsymbol{w}_1(s))\,\mathrm{d}s + \boldsymbol{u}_0 \right) - \mathcal{B}\left(\int_0^t \boldsymbol{\varepsilon}(\boldsymbol{w}_2(s))\,\mathrm{d}s + \boldsymbol{u}_0 \right) \right\|_{L^2(\Omega;\mathbb{S}^d)} \|\boldsymbol{v}\|_V$$

$$\le L_\mathcal{B} \int_0^t \|\boldsymbol{\varepsilon}(\boldsymbol{w}_1(s) - \boldsymbol{w}_2(s))\|_{L^2(\Omega;\mathbb{S}^d)}\,\mathrm{d}s\,\|\boldsymbol{v}\|_V$$

$$= L_\mathcal{B} \left(\int_0^t \|\boldsymbol{w}_1(s) - \boldsymbol{w}_2(s)\|_V\,\mathrm{d}s \right) \|\boldsymbol{v}\|_V.$$

Therefore

$$|(\mathcal{S}\boldsymbol{w}_1(t) - \mathcal{S}\boldsymbol{w}_2(t), \boldsymbol{\varepsilon}(\boldsymbol{v}))_{L^2(\Omega;\mathbb{S}^d)}| \le L_\mathcal{B} \left(\int_0^t \|\boldsymbol{w}_1(s) - \boldsymbol{w}_2(s)\|_V\,\mathrm{d}s \right) \|\boldsymbol{v}\|_V,$$

which implies that the operator $\mathcal{S}$ satisfies (6.6) with $c_S = L_\mathcal{B}$.

Finally, condition (6.7) is a consequence of assumptions (2.69), (2.70) and definition (9.8).

Since all the hypotheses of Theorem 6.2 are satisfied, we conclude that Problem 9.2 has a unique solution $\boldsymbol{w} \in \mathcal{W}$. Then, we define the function $\boldsymbol{u}: [0,T] \to V$ by

$$\boldsymbol{u}(t) = \int_0^t \boldsymbol{w}(s)\,\mathrm{d}s + \boldsymbol{u}_0, \quad t \in [0,T]. \tag{9.12}$$

It follows from inequalities (9.10), (9.11) that $\boldsymbol{u}$ is a solution to Problem 2.10. This completes the proof of the existence part of the theorem. The regularity (9.3) is a consequence of the regularity $\boldsymbol{w} \in \mathcal{W}$, assumption $\boldsymbol{u}_0 \in V$, and equality (9.12).

Finally, the uniqueness part of Theorem 9.1 is a consequence of the smallness assumption (9.2) combined with a standard Gronwall argument. $\square$

9.2. Numerical analysis of the problem

We now proceed with the discretization of Problem 2.10. We use the symbol $\boldsymbol{w} := \dot{\boldsymbol{u}}$ to denote the velocity field and express the fully discrete scheme in terms of approximate velocities. As in Section 8.2, we use the linear finite element space V^h and a uniform partition of the time interval $[0, T]$ with step-size $k = T/N$ and partition points $t_n = n\,k$, $0 \leq n \leq N$. Corresponding to (6.17), we assume

$$\boldsymbol{f}_0 \in C([0, T]; L^2(\Omega; \mathbb{R}^d)), \quad \boldsymbol{f}_2 \in C([0, T]; L^2(\Gamma_2; \mathbb{R}^d)). \tag{9.13}$$

Denote $\boldsymbol{f}_{0,n} = \boldsymbol{f}_0(t_n)$, $\boldsymbol{f}_{2,n} = \boldsymbol{f}_2(t_n)$. Let $\boldsymbol{u}_0^h, \boldsymbol{w}_0^h \in V^h$ be appropriate approximations of the initial values $\boldsymbol{u}_0, \boldsymbol{w}_0$. Then the fully discrete numerical method is as follows.

Problem 9.3. Find $\boldsymbol{w}^{hk} = \{\boldsymbol{w}_n^{hk}\}_{n=0}^N \subset V^h$ such that for $1 \leq n \leq N$,

$$\int_\Omega \left[\rho \frac{\boldsymbol{w}_n^{hk} - \boldsymbol{w}_{n-1}^{hk}}{k} \cdot \boldsymbol{v}^h + \mathcal{A}\varepsilon(\boldsymbol{w}_n^{hk}) \cdot \varepsilon(\boldsymbol{v}^h) + \mathcal{B}\varepsilon(\boldsymbol{u}_n^{hk}) \cdot \varepsilon(\boldsymbol{v}^h) \right] dx$$

$$+ \int_{\Gamma_3} \left[j_\nu^0(w_{n,\nu}^{hk}; v_\nu^h) + j_\tau^0(\boldsymbol{w}_{n,\tau}^{hk}; \boldsymbol{v}_\tau^h) \right] da$$

$$\geq \int_\Omega \boldsymbol{f}_{0,n} \cdot \boldsymbol{v}^h \, dx + \int_{\Gamma_2} \boldsymbol{f}_{2,n} \cdot \boldsymbol{v}^h \, da \quad \text{for all } \boldsymbol{v}^h \in V^h, \tag{9.14}$$

and

$$\boldsymbol{w}_0^{hk} = \boldsymbol{w}_0^h. \tag{9.15}$$

Here

$$\boldsymbol{u}_n^{hk} = \boldsymbol{u}_0^h + k \sum_{i=0}^{n-1} \boldsymbol{w}_i^{hk} \tag{9.16}$$

and $\boldsymbol{u}_0^h \in V^h$ is an approximation of $\boldsymbol{u}_0$.

The main goal in this subsection is to derive an error estimate for the numerical solution defined by Problem 9.3. To this end, we will apply the error bound (6.53), which still holds when

$$j_1^0(\gamma_1 w_n^{hk}; \gamma_1 v^h), \quad j_2^0(\gamma_2 w_n^{hk}; \gamma_2 v^h)$$

in (6.21) are replaced by

$$\int_{\Gamma_3} j_\nu^0(w_{n,\nu}^{hk}; v_\nu^h) \, da, \quad \int_{\Gamma_3} j_\tau^0(\boldsymbol{w}_{n,\tau}^{hk}; \boldsymbol{v}_\tau^h) \, da$$

as in (9.14). We assume the solution regularity

$$\boldsymbol{u} \in C^1([0, T]; H^2(\Omega; \mathbb{R}^d)) \cap H^2(0, T; V) \cap H^3(0, T; V^*). \tag{9.17}$$

Note that (9.17) implies the following counterpart of (6.30):

$$\boldsymbol{w} \in C([0, T]; V) \cap H^2(0, T; V^*).$$

We will additionally assume

$$\boldsymbol{\sigma\nu} \in C([0,T]; L^2(\Gamma; \mathbb{R}^d)), \tag{9.18}$$

$$\dot{\boldsymbol{u}}|_{\Gamma_{3,i}} \in C([0,T]; H^2(\Gamma_{3,i}; \mathbb{R}^d)), \quad 1 \le i \le i_3. \tag{9.19}$$

Observe that if $\mathcal{A}$ and $\mathcal{B}$ are smooth, then (9.18) follows from the condition $\boldsymbol{u} \in C^1([0,T]; H^1(\Omega; \mathbb{R}^d))$ implied by (9.17).

We first consider the residual-type term defined by (6.35), which takes the following form for Problem 9.3:

$$R_n(\boldsymbol{v}) = \langle \rho\,\ddot{\boldsymbol{w}}_n, \boldsymbol{v}\rangle + \int_\Omega \boldsymbol{\sigma}_n{\cdot}\boldsymbol{\varepsilon}(\boldsymbol{v})\,\mathrm{d}x + \int_{\Gamma_3} [j_\nu^0(w_{n,\nu}; v_\nu) + j_\tau^0(\boldsymbol{w}_{n,\tau}; \boldsymbol{v}_\tau)]\,\mathrm{d}a$$

$$- \int_\Omega \boldsymbol{f}_{0,n} \cdot \boldsymbol{v}\,\mathrm{d}x - \int_{\Gamma_2} \boldsymbol{f}_{2,n} \cdot \boldsymbol{v}\,\mathrm{d}a, \tag{9.20}$$

where

$$\boldsymbol{\sigma}_n := \boldsymbol{\sigma}(t_n) = \mathcal{A}\boldsymbol{\varepsilon}(\boldsymbol{w}_n) + \mathcal{B}\boldsymbol{\varepsilon}(\boldsymbol{u}_n), \quad 1 \le n \le N. \tag{9.21}$$

Recall the space $\tilde{U}$ defined in (7.20). In the defining inequality (2.71), we take $\boldsymbol{v} \in \tilde{U}$ to obtain

$$\langle \rho\,\dot{\boldsymbol{w}}(t), \boldsymbol{v}\rangle + \int_\Omega \boldsymbol{\sigma}(t){\cdot}\boldsymbol{\varepsilon}(\boldsymbol{v})\,\mathrm{d}x = \int_\Omega \boldsymbol{f}_0(t){\cdot}\boldsymbol{v}\,\mathrm{d}x + \int_{\Gamma_2} \boldsymbol{f}_2(t){\cdot}\boldsymbol{v}\,\mathrm{d}a$$

$$\text{for all } \boldsymbol{v} \in \tilde{U}. \tag{9.22}$$

As in (7.22) and (7.23), we derive from (9.22) that for a.e. $t \in (0,T)$,

$$\rho\,\dot{\boldsymbol{w}}(t) - \mathrm{Div}\,\boldsymbol{\sigma}(t) = \boldsymbol{f}_0(t) \quad \text{a.e. in } \Omega \tag{9.23}$$

and

$$\boldsymbol{\sigma}(t)\boldsymbol{\nu} = \boldsymbol{f}_2(t) \quad \text{a.e. on } \Gamma_2. \tag{9.24}$$

Now we multiply the equation (9.23) by an arbitrary function $\boldsymbol{v} \in V$ and integrate over Ω:

$$\langle \rho\,\dot{\boldsymbol{w}}(t), \boldsymbol{v}\rangle - \int_\Omega \mathrm{Div}\,\boldsymbol{\sigma}(t){\cdot}\boldsymbol{v}\,\mathrm{d}x = \int_\Omega \boldsymbol{f}_0(t){\cdot}\boldsymbol{v}\,\mathrm{d}x.$$

Perform an integration by parts on the second integral and use (9.24) to get

$$\langle \rho\,\dot{\boldsymbol{w}}(t), \boldsymbol{v}\rangle + \int_\Omega \boldsymbol{\sigma}(t){\cdot}\boldsymbol{\varepsilon}(\boldsymbol{v})\,\mathrm{d}x - \int_{\Gamma_3} \boldsymbol{\sigma}(t)\boldsymbol{\nu}{\cdot}\boldsymbol{v}\,\mathrm{d}a$$

$$= \int_\Omega \boldsymbol{f}_0(t){\cdot}\boldsymbol{v}\,\mathrm{d}x + \int_{\Gamma_2} \boldsymbol{f}_2(t){\cdot}\boldsymbol{v}\,\mathrm{d}a \quad \text{for all } \boldsymbol{v} \in V,\ \text{a.e. } t \in (0,T). \tag{9.25}$$

Thus, the term $R_n(\cdot)$ given in (9.20) can be simplified to

$$R_n(\boldsymbol{v}) = \int_{\Gamma_3} [\boldsymbol{\sigma}_n\boldsymbol{\nu}{\cdot}\boldsymbol{v} + j_\nu^0(w_{n,\nu}; v_\nu) + j_\tau^0(\boldsymbol{w}_{n,\tau}; \boldsymbol{v}_\tau)]\,\mathrm{d}a, \quad \boldsymbol{v} \in V. \tag{9.26}$$

Therefore,

$$|R_n(\boldsymbol{v})| \le c \, \|\boldsymbol{v}\|_{L^2(\Gamma_3;\mathbb{R}^d)} \quad \text{for all } \boldsymbol{v} \in V. \tag{9.27}$$

Hence, from (6.53) and (6.52), we have

$$\max_{1 \le n \le N} \|\boldsymbol{w}_n - \boldsymbol{w}_n^{hk}\|_H^2 + k \sum_{n=1}^{N} \|\boldsymbol{w}_n - \boldsymbol{w}_n^{hk}\|_V^2$$

$$\le c \, k^2 (\|\ddot{\boldsymbol{w}}\|_{L^2(0,T;V^*)}^2 + \|\boldsymbol{w}\|_{H^1(0,T;V)}^2)$$

$$+ c(\|\boldsymbol{w}_0 - \boldsymbol{w}_0^h\|_H^2 + k \, \|\boldsymbol{w}_0 - \boldsymbol{w}_0^h\|_V^2) + c \max_{1 \le n \le N} \tilde{E}_n, \tag{9.28}$$

where

$$\tilde{E}_n = \inf_{\boldsymbol{v}_i^h \in V^h, \, 1 \le i \le n} \left\{ k \sum_{i=1}^{n} \|\boldsymbol{w}_i - \boldsymbol{v}_i^h\|_V^2 + \left[k \sum_{i=1}^{n} \|\boldsymbol{w}_i - \boldsymbol{v}_i^h\|_{L^2(\Gamma_3;\mathbb{R}^d)}^2 \right]^{1/2} \right.$$

$$+ k^{-1} \sum_{i=1}^{n-1} \|(\boldsymbol{w}_i - \boldsymbol{v}_i^h) - (\boldsymbol{w}_{i+1} - \boldsymbol{v}_{i+1}^h)\|_{V^*}^2$$

$$\left. + \|\boldsymbol{w}_1 - \boldsymbol{v}_1^h\|_H^2 + \|\boldsymbol{w}_n - \boldsymbol{v}_n^h\|_H^2 \right\}. \tag{9.29}$$

Note that (9.17) implies

$$\boldsymbol{w} \in C([0,T]; H^2(\Omega;\mathbb{R}^d)) \cap H^1(0,T;V) \cap H^2(0,T;V^*).$$

This in particular implies

$$\boldsymbol{w}_0 \in H^2(\Omega;\mathbb{R}^d).$$

Let $\boldsymbol{w}_0^h \in V^h$ be the finite element interpolant or projection of $\boldsymbol{w}_0$. Then

$$\|\boldsymbol{w}_0 - \boldsymbol{w}_0^h\|_H \le c \, h^2 \|\boldsymbol{w}_0\|_{H^2(\Omega)^d}. \tag{9.30}$$

Let $\boldsymbol{v}_i^h \in V^h$ be the finite element interpolant of $\boldsymbol{w}_i$, $1 \le i \le N$. Note that $(\boldsymbol{v}_i^h - \boldsymbol{v}_{i+1}^h)$ is the finite element interpolant of $(\boldsymbol{w}_i - \boldsymbol{w}_{i+1})$, $0 \le i \le N-1$. Moreover, $\boldsymbol{v}_i^h$ interpolates $\boldsymbol{w}_i$ on the boundary Γ_3. Then, from the finite element interpolation theory, under the stated solution regularities, we obtain from (9.28) and (9.29) the following optimal order error estimate:

$$\max_{1 \le n \le N} \|\boldsymbol{w}_n - \boldsymbol{w}_n^{hk}\|_H + \left[k \sum_{n=1}^{N} \|\boldsymbol{w}_n - \boldsymbol{w}_n^{hk}\|_V^2 \right]^{1/2} \le c(k+h), \tag{9.31}$$

for a constant $c > 0$ depending on $\boldsymbol{w}$, but not on the discretization parameters k and h.

We now turn to an error estimate for the displacement. We note that (9.17) implies

$$\ddot{\boldsymbol{u}} \in L^2(0, T; V), \quad \boldsymbol{u}_0 \in H^2(\Omega; \mathbb{R}^d).$$

From (9.12) and (9.16), we have

$$\boldsymbol{u}_n - \boldsymbol{u}_n^{hk} = \boldsymbol{u}_0 - \boldsymbol{u}_0^{hk} + k \sum_{i=0}^{n-1} (\boldsymbol{w}_i - \boldsymbol{w}_i^{hk}) + \sum_{i=0}^{n-1} \int_{t_i}^{t_{i+1}} (\boldsymbol{w}(t) - \boldsymbol{w}_i)\, \mathrm{d}t.$$

Then,

$$\|\boldsymbol{u}_n - \boldsymbol{u}_n^{hk}\|_V \le \|\boldsymbol{u}_0 - \boldsymbol{u}_0^{hk}\|_V + k \sum_{i=0}^{n-1} \|\boldsymbol{w}_i - \boldsymbol{w}_i^{hk}\|_V$$

$$+ \sum_{i=0}^{n-1} \int_{t_i}^{t_{i+1}} \|\boldsymbol{w}(t) - \boldsymbol{w}_i\|_V\, \mathrm{d}t. \tag{9.32}$$

From

$$\boldsymbol{w}(t) - \boldsymbol{w}_i = \int_{t_i}^{t} \dot{\boldsymbol{w}}(s)\, \mathrm{d}s,$$

we find

$$\|\boldsymbol{w}(t) - \boldsymbol{w}_i\|_V \le \int_{t_i}^{t_{i+1}} \|\dot{\boldsymbol{w}}(s)\|_V\, \mathrm{d}s, \quad t \in [t_i, t_{i+1}],$$

and then

$$\sum_{i=0}^{n-1} \int_{t_i}^{t_{i+1}} \|\boldsymbol{w}(t) - \boldsymbol{w}_i\|_V\, \mathrm{d}t \le k \|\dot{\boldsymbol{w}}\|_{L^1(0,T;V)}.$$

If $\boldsymbol{u}_0^h \in V^h$ is the finite element interpolant or projection of $\boldsymbol{u}_0$, then

$$\|\boldsymbol{u}_0 - \boldsymbol{u}_0^{hk}\|_V \le c\, h\, \|\boldsymbol{u}_0\|_{H^2(\Omega;\mathbb{R}^d)}.$$

Thus, from (9.32), we have

$$\|\boldsymbol{u}_n - \boldsymbol{u}_n^{hk}\|_V \le k \sum_{i=0}^{n-1} \|\boldsymbol{w}_i - \boldsymbol{w}_i^{hk}\|_V + c(h\, \|\boldsymbol{u}_0\|_{H^2(\Omega;\mathbb{R}^d)} + k\, \|\ddot{\boldsymbol{u}}\|_{L^1(0,T;V)}).$$

$$\tag{9.33}$$

Apply the Cauchy–Schwarz inequality and use (9.31) and (9.30):

$$k \sum_{i=1}^{n} \|\boldsymbol{w}_i - \boldsymbol{w}_i^{hk}\|_V \le (k\, n)^{1/2} \left[k \sum_{i=1}^{n} \|\boldsymbol{w}_i - \boldsymbol{w}_i^{hk}\|_V^2 \right]^{1/2}$$

$$\le c(k + h).$$

Therefore, from (9.33) we derive the optimal order error estimate

$$\max_{0\le n\le N} \|\boldsymbol{u}_n - \boldsymbol{u}_n^{hk}\|_V \le c(k+h). \tag{9.34}$$

9.3. A numerical example

In this subsection we report numerical results on a dynamic frictional contact problem following Barboteu, Bartosz, Han and Janiczko (2015). The contact problem represents a variant of Problem 2.9 and can be formulated as follows.

Problem 9.4. Find a displacement field $\boldsymbol{u}\colon \Omega \times [0,T] \to \mathbb{R}^d$ and a stress field $\boldsymbol{\sigma}\colon \Omega \times [0,T] \to \mathbb{S}^d$ such that

$$\boldsymbol{\sigma}(t) = \mathcal{A}\varepsilon(\dot{\boldsymbol{u}}(t)) + \mathcal{B}\varepsilon(\boldsymbol{u}(t)) \quad \text{in } \Omega, \tag{9.35}$$
$$\rho\,\ddot{\boldsymbol{u}}(t) = \operatorname{Div}\boldsymbol{\sigma}(t) + \boldsymbol{f}_0(t) \quad \text{in } \Omega, \tag{9.36}$$
$$\boldsymbol{u}(t) = \boldsymbol{0} \quad \text{on } \Gamma_1, \tag{9.37}$$
$$\boldsymbol{\sigma}(t)\boldsymbol{\nu} = \boldsymbol{f}_2(t) \quad \text{on } \Gamma_2, \tag{9.38}$$
$$u_\nu = 0 \quad \text{on } \Gamma_3, \tag{9.39}$$
$$|\boldsymbol{\sigma}_\tau| \le \mu(\|\dot{\boldsymbol{u}}_\tau\|), \quad -\boldsymbol{\sigma}_\tau = \mu(\|\dot{\boldsymbol{u}}_\tau\|)\frac{\dot{\boldsymbol{u}}_\tau}{\|\dot{\boldsymbol{u}}_\tau\|} \text{ if } \dot{\boldsymbol{u}}_\tau \neq \boldsymbol{0} \quad \text{on } \Gamma_3, \tag{9.40}$$

for all $t \in [0,T]$, and

$$\boldsymbol{u}(0) = \boldsymbol{u}_0, \quad \dot{\boldsymbol{u}}(0) = \boldsymbol{w}_0 \quad \text{in } \Omega. \tag{9.41}$$

The difference between Problems 2.9 and 9.4 lies in the contact boundary condition. The condition (2.58) is replaced by the bilateral contact condition (9.39). In the subdifferential friction law (2.59), we choose

$$j_\tau(\boldsymbol{\xi}) = \int_0^{\|\boldsymbol{\xi}\|} \mu(s)\,\mathrm{d}s, \quad \boldsymbol{\xi} \in \mathbb{R}^d. \tag{9.42}$$

Here, $\mu : [0,\infty) \to \mathbb{R}_+$ represents the friction bound and is assumed to satisfy the following conditions:

(a) μ is continuous;

(b) $|\mu(s)| \le c(1+s)$ for all $s \ge 0$, $c > 0$;

(c) $\mu(s_1) - \mu(s_2) \ge -\lambda(s_1 - s_2)$ for all $s_1 > s_2 \ge 0$ with $\lambda > 0$.

$$\left.\vphantom{\begin{array}{c}a\\a\\a\end{array}}\right\} \tag{9.43}$$

In the study of Problem 2.9, we use the function space

$$\widetilde{V} = \{\boldsymbol{v} \in V \mid v_\nu = 0 \text{ a.e. on } \Gamma_3\}. \tag{9.44}$$

Moreover, besides (9.43) we assume that (2.64), (2.65), (2.69) and (2.70) hold. Then, the weak formulation of Problem 2.9, obtained by using arguments similar to those used in Section 2.1, is as follows.

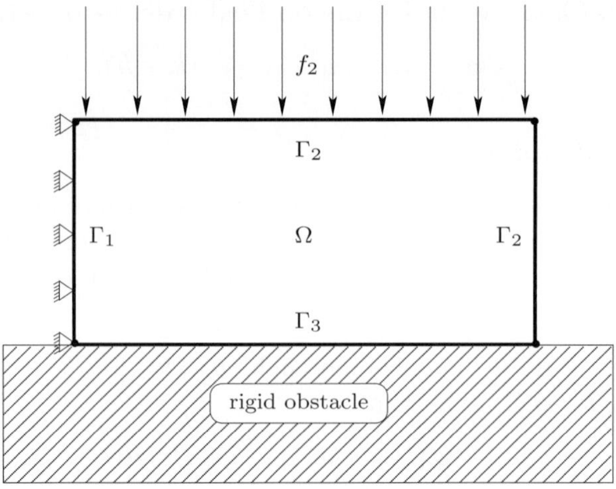

Figure 9.1. Physical setting of the contact problem.

Problem 9.5. Find a displacement field $\boldsymbol{u}\colon [0,T] \to \widetilde{V}$ such that

$$\int_{\Omega} \rho\,\ddot{\boldsymbol{u}}(t)\cdot\boldsymbol{v}\,\mathrm{d}x + \int_{\Omega} \mathcal{A}\varepsilon(\dot{\boldsymbol{u}}(t))\cdot\varepsilon(\boldsymbol{v})\,\mathrm{d}x + \int_{\Omega} \mathcal{B}\varepsilon(\boldsymbol{u}(t))\cdot\varepsilon(\boldsymbol{v})\,\mathrm{d}x$$

$$+ \int_{\Gamma_3} j_\tau^0(\dot{\boldsymbol{u}}_\tau(t);\boldsymbol{v}_\tau)\,\mathrm{d}a \geq \int_{\Omega} \boldsymbol{f}_0(t)\cdot\boldsymbol{v}\,\mathrm{d}x + \int_{\Gamma_2} \boldsymbol{f}_2(t)\cdot\boldsymbol{v}\,\mathrm{d}a \qquad (9.45)$$

for all $\boldsymbol{v} \in \widetilde{V}$, $t \in [0,T]$, and

$$\boldsymbol{u}(0) = \boldsymbol{u}_0, \quad \dot{\boldsymbol{u}}(0) = \boldsymbol{w}_0. \qquad (9.46)$$

The unique solvablity of Problem 9.5 can be obtained by using arguments similar to those used in Section 9.1, based on Theorem 6.2. Also, error estimates similar to those in Section 9.2 can be obtained.

For the numerical simulation, let $\Omega = (0,1) \times (0,0.5)$ with the metre as the length unit. The domain Ω represents the cross-section of a three-dimensional linearly viscoelastic body subjected to the action of tractions in such a way that a plane stress hypothesis is valid. The physical setting of the contact problem is shown in Figure 9.1. On $\Gamma_1 = \{0\} \times [0,0.5]$ the body is clamped, that is, the displacement field vanishes there. Let $\Gamma_2 = ((0,1) \times \{0.5\}) \cup (\{1\} \times (0,0.5))$; the part $(0,1) \times \{0.5\}$ is subject to vertical compressions and the part $\{1\} \times (0,0.5)$ is traction-free. No body forces are assumed to act on the elastic body during the process. On $\Gamma_3 = (0,1) \times \{0\}$, the body is in frictional bilateral contact with an obstacle. The friction follows the version (9.40) of Coulomb's law, in which the friction bound depends on the tangential velocity $\|\dot{\boldsymbol{u}}_\tau\|$. For the coefficient of friction, we

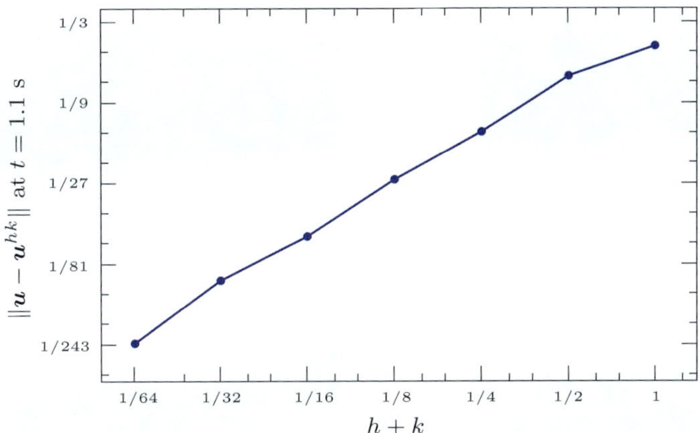

Figure 9.2. Numerical convergence orders.

choose a function $\mu : [0, \infty) \to \mathbb{R}$ of the form

$$\mu(r) = (a - b)\, e^{-\alpha r} + b, \qquad (9.47)$$

with $a \geq b > 0$ and $\alpha > 0$. We take $a = 1$, $b = 0.1$ and $\alpha = 200$ in the simulation. The friction law (9.40) with (9.47) describes the slip weakening phenomenon which appears in the study of geophysical problems; see Scholz (1990) for details. The coefficient of friction decreases with the slip rate from the value a to the limit value b.

The deformable material response is governed by a linearly viscoelastic constitutive law in which the viscosity tensor $\mathcal{A}$ and the elasticity tensor $\mathcal{B}$ are given by

$$(\mathcal{A}\tau)_{ij} = \mu_1(\tau_{11} + \tau_{22})\delta_{ij} + \mu_2\tau_{ij}, \quad 1 \leq i,j \leq 2,\ \tau \in \mathbb{S}^2,$$

$$(\mathcal{B}\tau)_{ij} = \frac{E\kappa}{(1+\kappa)(1-2\kappa)}(\tau_{11} + \tau_{22})\delta_{ij} + \frac{E}{1+\kappa}\tau_{ij}, \quad 1 \leq i,j \leq 2,\ \tau \in \mathbb{S}^2,$$

where μ_1 and μ_2 are viscosity constants, E and κ are the Young's modulus and Poisson ratio of the material and δ_{ij} denotes the Kronecker symbol. For the simulation, we use the values $\mu_1 = 50$ N m^{-2}, $\mu_2 = 100$ N m^{-2}, $E = 2000$ N m^{-2} and $\kappa = 0.3$. The mass density is chosen to be $\rho = 1000$ kg m^{-3}, and the force densities are

$$\boldsymbol{f}_0 = (0, -10^{-5})\ \text{N m}^{-2},$$

$$\boldsymbol{f}_2 = \begin{cases} (0,0)\ \text{N m}^{-1} & \text{on } \{1\} \times [0, 0.5], \\ (0, -600\,t)\ \text{N m}^{-1} & \text{on } [0,1] \times \{0.5\}. \end{cases}$$

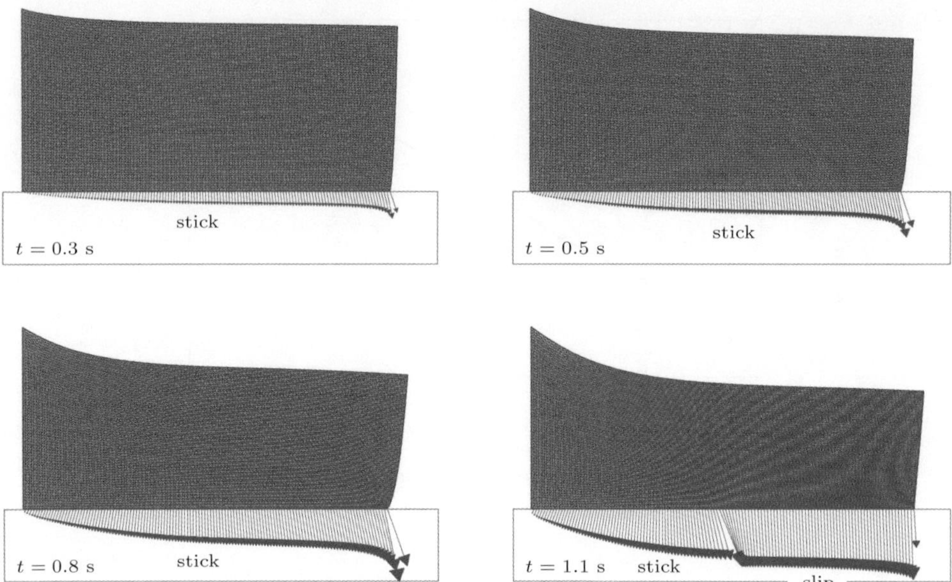

Figure 9.3. Evolution of deformed meshes and frictional contact forces during the dynamic compression process.

For the initial values, we choose

$$\boldsymbol{u}_0 = \boldsymbol{0} \text{ m}, \quad \boldsymbol{w}_0 = \boldsymbol{0} \text{ m s}^{-1}.$$

We compute a sequence of numerical solutions by using uniform triangulations of the domain Ω and uniform partitions of the time interval $[0, 1.1]$ (unit: second). We let h denote the leg of a right triangle in a uniform triangulation and let k be the time step-size. We start with $h = 1/2$ and $k = 1/2$ which are successively halved. The numerical solution corresponding to $h = 1/256$ and $k = 1/256$ is taken as the 'exact' solution and it is used to compute errors of the numerical solutions with larger h and k. The discrete problem corresponding to the fine triangulation with $h = 1/256$ has $133\,896$ degrees of freedom at each time level. The numerical errors $\|\boldsymbol{u} - \boldsymbol{u}^{hk}\|_V$ are reported in Figure 9.2. A first-order convergence of $\|\boldsymbol{u} - \boldsymbol{u}^{hk}\|_V$ with respect to $h + k$ is clearly observed.

Finally, we provide some graphical results on the mechanical behaviour of the solution. Figure 9.3 shows the deformed configuration as well as the interface forces on Γ_3 during the dynamic compression process at times $t = 0.3$ s, $t = 0.5$ s, $t = 0.8$ s and $t = 1.1$ s. At the beginning of the process, the contact nodes are all in stick status. As the compression force becomes stronger, the friction bound is reached at more contact nodes where the status switches from stick to slip: see the graph at $t = 1.1$ s.

10. Summary and outlook

Mathematical formulations of contact problems are naturally given in terms of inequalities. When the non-smooth and possibly multivalued constitutive relations and interface conditions are monotone, the contact problem has a convex structure and the mathematical formulation is in the form of a variational inequality. When the contact problem involves a non-monotone relation or condition, the mathematical formulation contains non-convex terms, and this leads to a hemivariational inequality which can be elliptic, history-dependent or evolutionary. The numerical solution of variational inequalities, in particular of those arising in contact mechanics, has been extensively studied in the literature. On the contrary, the numerical solution of hemivariational inequalities is still in an early stage, and many challenging issues remain to be addressed.

In this paper we have presented recent and new results on the numerical analysis of hemivariational inequalities arising in contact mechanics. We chose three representative contact problems to illustrate the main techniques used for the study of their numerical approximations and expected sample results. The three contact problems correspond to an elliptic, a history-dependent and an evolutionary hemivariational inequality, respectively. For the reader's convenience, we tried to make the paper self-contained. We provide a concise review of basic knowledge from non-smooth analysis and include main steps of solution existence and uniqueness proofs of the hemivariational inequalities. The discrete analogues of the proofs can be used to show existence and uniqueness of numerical solutions. The temporal derivative was approximated by the backward divided difference and the integral in a history-dependent operator was approximated with the trapezoidal rule; other temporal approximations can be introduced and can be analysed similarly. For spatial discretizations, we used the finite element method. We derived optimal order error estimates for the linear element solutions with the previously mentioned temporal approximations under appropriate solution regularity assumptions. For the elliptic hemivariational inequality, we discussed convergence of the numerical solutions under basic solution regularity proved in the solution existence and uniqueness result. For the history-dependent and evolutionary hemivariational inequalities, it is also possible to prove convergence of the numerical solutions without assuming additional solution regularity. Owing to space limitations we have not provided such a convergence discussion in this paper, and instead refer the reader to Han and Reddy (1999, 2000, 2013), where convergence of numerical solutions is proved for evolutionary variational inequalities arising in elasto-plasticity under basic solution regularities available from solution existence and uniqueness results. Hemivariational inequalities arise in many other applications, and it is worth studying their numerical approximations as well; for example, Han, Huang, Wang and Xu (2019) conduct numerical analysis on some hemivariational inequalities for applications related to semi-permeable media.

The results presented in this paper serve as a starting point in the development and analysis of more efficient numerical methods for solving hemivariational inequalities arising in contact mechanics. One promising research area is on *a posteriori* error analysis and adaptive mesh refinement for simulation of contact problems. Since the pioneering work of Babuška and Rheinboldt (1978*a*, 1978*b*), the field of *a posteriori* error analysis and adaptive algorithms for the numerical solution of differential equations has attracted many researchers, and a variety of different *a posteriori* error estimates have been proposed and analysed for mathematical models in various applications (see *e.g.* Ainsworth and Oden 2000, Babuška and Strouboulis 2001, Verfürth 2013). While a large percentage of the references in this area deal with boundary value problems of partial differential equations, one can also find papers on adaptive solution of variational inequalities arising in contact mechanics (*e.g.* Ben Belgacem, Bernardi, Blouza and Vohralík 2012, Han 2005, Bostan and Han 2009, Hild and Lleras 2009). A natural step is to introduce *a posteriori* error estimators for hemivariational inequalities, analyse their reliability and efficiency, and use them to solve related contact problems.

Another research direction is to develop discontinuous Galerkin (DG) methods to solve hemivariational inequalities in contact mechanics. In DG methods, finite element functions of lower global smoothness are employed. The methods enjoy several advantages, such as handling easily general meshes with hanging nodes and elements of different shapes, accommodating easier parallel implementation (Arnold, Brezzi, Cockburn and Marini 2002, Cockburn, Karniadakis and Shu 2000). DG methods have been successfully used to solve many different kind of mathematical problems. In particular, in the literature, one can find papers on DG methods for solving variational inequalities, including those arising in contact mechanics, for example Wang, Han and Cheng (2010, 2011, 2014) and Gudi and Porwal (2014, 2016). However, DG methods have not been used to solve hemivariational inequalities and it will be interesting to explore the potential of the methods in larger-scale computer simulation of the contact problems.

Since the pioneering work of Beirão da Veiga *et al.* (2013) and Alsaedi, Brezzi, Marini and Russo (2013), virtual element methods (VEMs) have been used to solve a wide variety of PDE problems arising in solid mechanics, fluid mechanics and other areas of science and engineering. VEMs offer great flexibility in handling complex geometries. One key point is that the methods do not require explicit evaluation of the shape functions. Recently, VEMs have been used to solve variational inequalities. Wriggers, Rust and Reddy (2016) simulated the problem of contact between two elastic bodies numerically using the VEM, though without theoretical analysis of the method. VEMs are used to solve an obstacle problem in Wang and Wei (2018a), and to solve a simplified friction problem in Wang and Wei (2018b); in both these papers, optimal order error estimates are derived. A general framework for VEMs is developed for solving elliptic variational inequalities of the second kind in Feng, Han and Huang (2019); the methods discussed are directly implementable and optimal order error estimates are derived. It looks promising to develop VEMs to solve the contact problems, especially those more complicated ones in the form of hemivariational inequalities.

Acknowledgements

The present paper is a result of cooperation between the authors during the last five years. Part of the material included in this survey is related to our joint work with several collaborators, to whom we express our gratitude: M. Barboteu, K. Bartosz, Z. Huang, S. Migórski, A. Ochal, C. Wang, W. Xu, *et al.* We thank J. Sullian for helping us with some of the figures. We extend our thanks to the Editorial Board of *Acta Numerica* for inviting us to contribute to this issue. The work was supported by NSF under the grant DMS-1521684 and the European Union's Horizon 2020 Research and Innovation Programme under the Marie Sklodowska-Curie Grant Agreement no. 823731 CONMECH.

REFERENCES[1]

M. Ainsworth and J. T. Oden (2000), *A Posteriori Error Estimation in Finite Element Analysis*, Wiley.

A. Alsaedi, F. Brezzi, L. Marini and A. Russo (2013), 'Equivalent projectors for virtual element methods', *Comput. Math. Appl.* **66**, 376–391.

D. N. Arnold, F. Brezzi, B. Cockburn and L. D. Marini (2002), 'Unified analysis of discontinuous Galerkin methods for elliptic problems', *SIAM J. Numer. Anal.* **39**, 1749–1779.

K. Atkinson and W. Han (2009), *Theoretical Numerical Analysis: A Functional Analysis Framework*, third edition, Springer.

I. Babuška and W. C. Rheinboldt (1978a), 'Error estimates for adaptive finite element computations', *SIAM J. Numer. Anal.* **15**, 736–754.

I. Babuška and W. C. Rheinboldt (1978b), '*A posteriori* error estimates for the finite element method', *Intern. J. Numer. Methods Engrg* **12**, 1597–1615.

I. Babuška and T. Strouboulis (2001), *The Finite Element Method and its Reliability*, Oxford University Press.

---

[1] The URLs cited in this work were correct at the time of going to press, but the publisher and the authors make no undertaking that the citations remain live or are accurate or appropriate.

C. Baiocchi and A. Capelo (1984), *Variational and Quasivariational Inequalities: Applications to Free-Boundary Problems*, Wiley.

M. Barboteu, K. Bartosz, W. Han and T. Janiczko (2015), 'Numerical analysis of a hyperbolic hemivariational inequality arising in dynamic contact', *SIAM J. Numer. Anal.* **53**, 527–550.

L. Beirão da Veiga, F. Brezzi, A. Cangiani, G. Manzini, L. D. Marini and A. Russo (2013), 'Basic principles of virtual element methods', *Math. Models Methods Appl. Sci.* **23**, 1–16.

F. Ben Belgacem, C. Bernardi, A. Blouza and M. Vohralík (2012), 'On the unilateral contact between membranes, 2: *a posteriori* analysis and numerical experiments', *IMA J. Numer. Anal.* **32**, 1147–1172.

V. Bostan and W. Han (2009), Adaptive finite element solution of variational inequalities with application in contact problems. In *Advances in Applied Mathematics and Global Optimization* (D. Y. Gao and H. D. Sherali, eds), Springer, pp. 25–106.

S. C. Brenner and L. R. Scott (2008), *The Mathematical Theory of Finite Element Methods*, third edition, Springer.

H. Brézis (1972), 'Problèmes unilatéraux', *J. Math. Pures Appl.* **51**, 1–168.

A. Capatina (2014), *Variational Inequalities and Frictional Contact Problems*, Vol. 31 of Advances in Mechanics and Mathematics, Springer.

P. G. Ciarlet (1978), *The Finite Element Method for Elliptic Problems*, North-Holland.

P. G. Ciarlet (1988), *Mathematical Elasticity, I: Three Dimensional Elasticity*, Vol. 20 of Studies in Mathematics and its Applications, North-Holland.

F. H. Clarke (1975), 'Generalized gradients and applications', *Trans. Amer. Math. Soc.* **205**, 247–262.

F. H. Clarke (1983), *Optimization and Nonsmooth Analysis*, Wiley-Interscience.

B. Cockburn, G. E. Karniadakis and C.-W. Shu, eds (2000), *Discontinuous Galerkin Methods. Theory, Computation and Applications*, Vol. 11 of Lecture Notes in Computational Science and Engineering, Springer.

Z. Denkowski, S. Migórski and N. S. Papageorgiou (2003*a*), *An Introduction to Nonlinear Analysis: Theory*, Kluwer Academic/Plenum.

Z. Denkowski, S. Migórski and N. S. Papageorgiou (2003*b*), *An Introduction to Nonlinear Analysis: Applications*, Kluwer Academic/Plenum.

G. Drouet and P. Hild (2015), 'Optimal convergence for discrete variational inequalities modelling Signorini contact in 2D and 3D without additional assumptions on the unknown contact set', *SIAM J. Numer. Anal.* **53**, 1488–1507.

A. D. Drozdov, *Finite Elasticity and Viscoelasticity: A Course in the Nonlinear Mechanics of Solids*, World Scientific.

G. Duvaut and J.-L. Lions (1976), *Inequalities in Mechanics and Physics*, Springer.

C. Eck, J. Jarušek and M. Krbec (2005), *Unilateral Contact Problems: Variational Methods and Existence Theorems*, Vol. 270 of Pure and Applied Mathematics, Chapman & Hall/CRC.

I. Ekeland and R. Temam (1976), *Convex Analysis and Variational Problems*, North-Holland.

R. S. Falk (1974), 'Error estimates for the approximation of a class of variational inequalities', *Math. Comp.* **28**, 963–971.

F. Feng, W. Han and J. Huang (2019), 'Virtual element methods for elliptic variational inequalities of the second kind', *J. Sci. Comput.* doi:10.1007/s10915-019-00929-y

G. Fichera (1964), Problemi elastostatici con vincoli unilaterali, II: Problema di Signorini con ambique condizioni al contorno, *Mem. Accas. Naz. Lincei*, Ser. VIII, Vol. VII, Sez. I, 5, 91–140.

G. Fichera (1972), Boundary value problems of elasticity with unilateral constraints. In *Linear Theories of Elasticity and Thermoelasticity* (C. Truesdell, ed.), Springer, pp. 391–424.

R. Glowinski (1984), *Numerical Methods for Nonlinear Variational Problems*, Springer.

R. Glowinski, J.-L. Lions and R. Trémolières (1981), *Numerical Analysis of Variational Inequalities*, North-Holland.

T. Gudi and K. Porwal (2014), '*A posteriori* error control of discontinuous Galerkin methods for elliptic obstacle problems', *Math. Comp.* **83**, 579–602.

T. Gudi and K. Porwal (2016), '*A posteriori* error estimates of discontinuous Galerkin methods for the Signorini problem', *J. Comput. Appl. Math.* **292**, 257–278.

W. Han (2005), *A Posteriori Error Analysis via Duality Theory, with Applications in Modeling and Numerical Approximations*, Springer.

W. Han (2018), 'Numerical analysis of stationary variational–hemivariational inequalities with applications in contact mechanics', *Math. Mech. Solids* **23**, 279–293.

W. Han, Z. Huang, C. Wang and W. Xu (2019), 'Numerical analysis of elliptic hemivariational inequalities for semipermeable media', *J. Comput. Math.* **37**, 543–560.

W. Han, S. Migórski and M. Sofonea (2014), 'A class of variational–hemivariational inequalities with applications to frictional contact problems', *SIAM J. Math. Anal.* **46**, 3891–3912.

W. Han, S. Migórski and M. Sofonea, eds (2015), *Advances in Variational and Hemivariational Inequalities: Theory, Numerical Analysis, and Applications*, Springer.

W. Han and B. D. Reddy (1999), 'Convergence analysis of discrete approximations of problems in hardening plasticity', *Comput. Methods Appl. Mech. Engrg* **171**, 327–340.

W. Han and B. D. Reddy (2000), 'Convergence of approximations to the primal problem in plasticity under conditions of minimal regularity', *Numer. Math.* **87**, 283–315.

W. Han and B. D. Reddy (2013), *Plasticity: Mathematical Theory and Numerical Analysis*, second edition, Springer.

W. Han and M. Sofonea (2002), *Quasistatic Contact Problems in Viscoelasticity and Viscoplasticity*, Vol. 30 of Studies in Advanced Mathematics, American Mathematical Society/International Press.

W. Han, M. Sofonea and M. Barboteu (2017), 'Numerical analysis of elliptic hemivariational inequalities', *SIAM J. Numer. Anal.* **55**, 640–663.

W. Han, M. Sofonea and D. Danan (2018), 'Numerical analysis of stationary variational–hemivariational inequalities', *Numer. Math.* **139**, 563–592.

J. Haslinger and I. Hlaváček (1980), 'Contact between two elastic bodies, I: Continuous problems', *Applikace Math.* **25**, 324–347.

J. Haslinger and I. Hlaváček (1981*a*), 'Contact between two elastic bodies, II: Finite element analysis', *Applikace Math.* **26**, 263–290.

J. Haslinger and I. Hlaváček (1981*b*), 'Contact between two elastic bodies, III: Dual finite element analysis', *Applikace Math.* **26**, 321–344.

J. Haslinger, I. Hlaváček and J. Nečas (1996), Numerical methods for unilateral problems in solid mechanics. In *Handbook of Numerical Analysis*, Vol. IV (P. G. Ciarlet and J.-L. Lions, eds), North-Holland, pp. 313–485.

J. Haslinger, M. Miettinen and P. D. Panagiotopoulos (1999), *Finite Element Method for Hemivariational Inequalities. Theory, Methods and Applications*, Kluwer Academic.

H. Hertz (1882), 'Über die Berührung fester Elastischer Körper', *J. Math.* (Crelle) **92**.

P. Hild and V. Lleras (2009), 'Residual error estimators for Coulomb friction', *SIAM J. Numer. Anal.* **47**, 3550–3583.

I. Hlaváček, J. Haslinger, J. Nečas and J. Lovíšek (1988), *Solution of Variational Inequalities in Mechanics*, Springer.

S. Hüeber and B. Wohlmuth (2005*a*), 'A primal–dual active set strategy for nonlinear multibody contact problems', *Comput. Meth. Appl. Mech. Engrg* **194**, 3147–3166.

S. Hüeber and B. Wohlmuth (2005*b*), 'An optimal *a priori* estimate for non-linear multibody contact problems', *SIAM J. Numer. Anal.* **43**, 157–173.

N. Kikuchi and J. T. Oden (1988), *Contact Problems in Elasticity: A Study of Variational Inequalities and Finite Element Methods*, SIAM.

D. Kinderlehrer and G. Stampacchia (2000), *An Introduction to Variational Inequalities and Their Applications*, Vol. 31 of Classics in Applied Mathematics, SIAM.

A. M. Khludnev and J. Sokolowski (1997), *Modelling and Control in Solid Mechanics*, Birkhäuser.

R. Kornhuber and R. Krause (2001), 'Adaptive multigrid methods for Signorini's problem in linear elasticity', *Comput Visual.* **4**, 9–20.

A. J. Kurdila and M. Zabarankin (2005), *Convex Functional Analysis*, Birkhäuser.

T. A. Laursen (2002), *Computational Contact and Impact Mechanics*, Springer.

J.-L. Lions and G. Stampacchia (1967), 'Variational inequalities', *Comm. Pure Appl. Math.* **20**, 493–519.

M. Marcus and V. Mizel (1972), 'Absolute continuity on tracks and mappings of Sobolev space', *Arch. Rat. Mech. Anal.* **45**, 294–302.

A. Matei, S. Sitzmann, K. Willner and B. I. Wohlmuth (2017), 'A mixed variational formulation for a class of contact problems in viscoelasticity', *Appl. Anal.* **97**, 1340–1356.

S. Migórski, A. Ochal and M. Sofonea (2010), 'Variational analysis of static frictional contact problems for electro-elastic materials', *Math. Nachr.* **283**, 1314–1335.

S. Migórski, A. Ochal and M. Sofonea (2013), *Nonlinear Inclusions and Hemivariational Inequalities: Models and Analysis of Contact Problems*, Vol. 26 of Advances in Mechanics and Mathematics, Springer.

S. Migórski, A. Ochal and M. Sofonea (2017), 'A class of variational–hemivariational inequalities in reflexive Banach spaces', *J. Elasticity* **127**, 151–178.

Z. Naniewicz and P. D. Panagiotopoulos (1995), *Mathematical Theory of Hemivariational Inequalities and Applications*, Dekker.

J. Nečas and I. Hlaváček (1981), *Mathematical Theory of Elastic and Elastico-Plastic Bodies: An Introduction*, Elsevier.

J. T. Oden and J. A. C. Martins (1985), 'Models and computational methods for dynamic friction phenomena', *Comput. Methods Appl. Mech. Engrg* **52**, 527–634.

P. D. Panagiotopoulos (1985), *Inequality Problems in Mechanics and Applications*, Birkhäuser.

P. D. Panagiotopoulos (1993), *Hemivariational Inequalities: Applications in Mechanics and Engineering*, Springer.

N. Renon, P. Montmitonnet and P. Laborde (2005), 'A 3D finite element model for soil/tool interaction in large deformation', *Eng. Comput.* **22**, 87–109.

C. H. Scholz (1990), *The Mechanics of Earthquakes and Faulting*, Cambridge University Press.

A. Signorini (1933), Sopra alcune questioni di elastostatica. *Atti della Società Italiana per il Progresso delle Scienze.*

M. Shillor, M. Sofonea and J. J. Telega (2004), *Models and Analysis of Quasistatic Contact*, Vol. 655 of Lecture Notes in Physics, Springer.

M. Sofonea, C. Avramescu and A. Matei (2008), 'A fixed point result with applications in the study of viscoplastic frictionless contact problems', *Comm. Pure Appl. Anal.* **7**, 645–658.

M. Sofonea, W. Han and M. Barboteu (2017), A variational–hemivariational inequality in contact mechanics. In *Mathematical Modelling in Solid Mechanics* (F. dell'Isola *et al.*, eds), Vol. 69 of Advanced Structured Materials, Springer, pp. 251–264.

M. Sofonea and A. Matei (2011), 'History-dependent quasivariational inequalities arising in contact mechanics', *Euro. J. Appl. Math.* **22**, 471–491.

M. Sofonea and A. Matei (2012), *Mathematical Models in Contact Mechanics*, Vol. 398 of London Mathematical Society Lecture Note Series, Cambridge University Press.

M. Sofonea and S. Migórski (2018), *Variational–hemivariational Inequalities with Applications*, Pure and Applied Mathematics, Chapman & Hall/CRC.

M. Sofonea, N. Renon and M. Shillor (2004), 'Stress formulation for frictionless contact of an elastic-perfectly-plastic body', *Appl. Anal.* **83**, 1157–1170.

M. Sofonea and Y. Xiao (2016), 'Fully history-dependent quasivariational inequalities in contact mechanics', *Appl. Anal.* **95**, 2464–2484.

R. Temam and A. Miranville (2001), *Mathematical Modeling in Continuum Mechanics*, Cambridge University Press.

R. Verfürth (2013), *A Posteriori Error Estimation Techniques for Finite Element Methods*, Oxford University Press.

F. Wang, W. Han and X.-L. Cheng (2010), 'Discontinuous Galerkin methods for solving elliptic variational inequalities', *SIAM J. Numer. Anal.* **48**, 708–733.

F. Wang, W. Han and X.-L. Cheng (2011), 'Discontinuous Galerkin methods for solving the Signorini problem', *IMA J. Numer. Anal.* **31**, 1754–1772.

F. Wang, W. Han and X.-L. Cheng (2014), 'Discontinuous Galerkin methods for solving a quasistatic contact problem', *Numer. Math.* **126**, 771–800.

F. Wang and H. Wei (2018*a*), 'Virtual element methods for the obstacle problem', *IMA J. Numer. Anal.* doi:10.1093/imanum/dry055

F. Wang and H. Wei (2018*b*), 'Virtual element method for simplified friction problem', *Appl. Math. Lett.* **85**, 125–131.

B. Wohlmuth (2011), Variationally consistent discretization schemes and numerical algorithms for contact problems. In *Acta Numerica*, Vol. 20, Cambridge University Press, pp. 569–734.

B. Wohlmuth and R. Krause (2003), 'Monotone methods on non-matching grids for nonlinear contact problems', *SIAM J. Sci. Comput.* **25**, 324–347.

P. Wriggers (2006), *Computational Contact Mechanics*, second edition, Springer.

P. Wriggers and K. Fischer (2005), 'Frictionless 2D contact formulations for finite deformations based on the mortar method', *Comput. Mech.* **36**, 226–244.

P. Wriggers and T. Laursen (2007), *Computational Contact Mechanics*, Vol. 298 of CISM Courses and Lectures, Springer.

P. Wriggers, W. T. Rust and B. D. Reddy (2016), 'A virtual element method for contact', *Comput. Mech.* **58**, 1039–1050.

W. Xu, Z. Huang, W. Han, W. Chen and C. Wang (2019), 'Numerical analysis of history-dependent variational–hemivariational inequalities with applications in contact mechanics', *J. Comput. Appl. Math.* **351**, 364–377.

E. Zeidler (1985), *Nonlinear Functional Analysis and its Applications, I: Fixed-point Theorems*, Springer.

E. Zeidler (1990), *Nonlinear Functional Analysis and its Applications, II/B: Nonlinear Monotone Operators*, Springer.

Acta Numerica (2019), pp. 287–404
doi:10.1017/S0962492919000060

Derivative-free optimization methods

Jeffrey Larson, Matt Menickelly and Stefan M. Wild
Mathematics and Computer Science Division,
Argonne National Laboratory, Lemont, IL 60439, USA
E-mail: jmlarson@anl.gov, mmenickelly@anl.gov, wild@anl.gov

Dedicated to the memory of Andrew R. Conn for his inspiring enthusiasm and his many contributions to the renaissance of derivative-free optimization methods.

In many optimization problems arising from scientific, engineering and artificial intelligence applications, objective and constraint functions are available only as the output of a black-box or simulation oracle that does not provide derivative information. Such settings necessitate the use of methods for derivative-free, or zeroth-order, optimization. We provide a review and perspectives on developments in these methods, with an emphasis on highlighting recent developments and on unifying treatment of such problems in the non-linear optimization and machine learning literature. We categorize methods based on assumed properties of the black-box functions, as well as features of the methods. We first overview the primary setting of deterministic methods applied to unconstrained, non-convex optimization problems where the objective function is defined by a deterministic black-box oracle. We then discuss developments in randomized methods, methods that assume some additional structure about the objective (including convexity, separability and general non-smooth compositions), methods for problems where the output of the black-box oracle is stochastic, and methods for handling different types of constraints.

CONTENTS

1. Introduction

The growth in computing for scientific, engineering and social applications has long been a driver of advances in methods for numerical optimization. The development of derivative-free optimization methods – those methods that do not require the availability of derivatives – has especially been driven by the need to optimize increasingly complex and diverse problems. One of the earliest calculations on MANIAC,[1] an early computer based on the von Neumann architecture, was the approximate solution of a six-dimensional non-linear least-squares problem using a derivative-free coordinate search (Fermi and Metropolis 1952). Today, derivative-free methods are used routinely, for example by Google (Golovin *et al.* 2017), for the automation and tuning needed in the artificial intelligence era.

In this paper we survey methods for derivative-free optimization and key results for their analysis. Since the field – also referred to as black-box optimization, gradient-free optimization, optimization without derivatives, simulation-based optimization and zeroth-order optimization – is now far too expansive for a single survey, we focus on methods for local optimization of continuous-valued, single-objective problems. Although Section 8 illustrates further connections, here we mark the following notable omissions.

- We focus on methods that seek a local minimizer. Despite users understandably desiring the best possible solution, the problem of global optimization raises innumerably more mathematical and computational challenges than do the methods presented here. We instead point to the survey by Neumaier (2004), which importantly addresses general

---

[1] Mathematical Analyzer, Integrator, And Computer. Other lessons learned from this application are discussed by Anderson (1986).

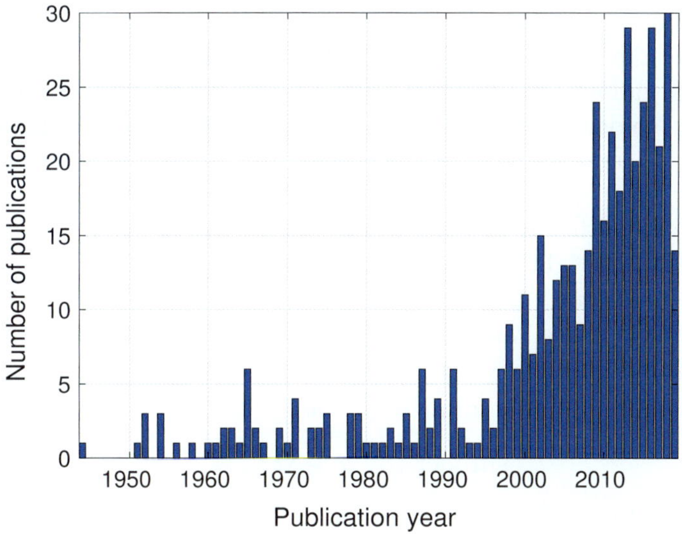

Figure 1.1. Histogram of the references cited in the bibliography.

constraints, and to the textbook by Forrester, Sobester and Keane (2008), which lays a foundation for global surrogate modelling.

- Multi-objective optimization and optimization in the presence of discrete variables are similarly popular tasks among users. Such problems possess fundamental challenges as well as differences from the methods presented here.

- In focusing on methods, we cannot do justice to the application problems that have driven the development of derivative-free methods and benefited from implementations of these methods. The recent textbook by Audet and Hare (2017) contains a number of examples and references to applications; Rios and Sahinidis (2013) and Auger *et al.* (2009) both reference a diverse set of implementations. At the persistent page

https://archive.org/services/purl/dfomethods

we intend to link all works that cite the entries in our bibliography and those that cite this survey; we hope this will provide a coarse, but dynamic, catalogue for the reader interested in potential uses of these methods.

Given these limitations, we particularly note the intersection with the foundational books by Kelley (1999*b*) and Conn, Scheinberg and Vicente (2009*b*). Our intent is to highlight recent developments in, and the evolution of, derivative-free optimization methods. Figure 1.1 summarizes our bias; over half of the references in this survey are from the past ten years.

Many of the fundamental inspirations for the methods discussed in this survey are detailed to a lesser extent. We note in particular the activity in the United Kingdom in the 1960s (see *e.g.* the works by Rosenbrock 1960, Powell 1964, Nelder and Mead 1965, Fletcher 1965 and Box 1966, and the later exposition and expansion by Brent 1973) and the Soviet Union (as evidenced by Rastrigin 1963, Matyas 1965, Karmanov 1974, Polyak 1987 and others). In addition to those mentioned later, we single out the work of Powell (1975), Wright (1995), Davis (2005) and Leyffer (2015) for insight into some of these early pioneers.

With our focus clear, we turn our attention to the deterministic optimization problem

$$\begin{aligned}\underset{\boldsymbol{x}}{\text{minimize}} \quad & f(\boldsymbol{x}) \\ \text{subject to} \quad & \boldsymbol{x} \in \boldsymbol{\Omega} \subseteq \mathbb{R}^n\end{aligned} \tag{DET}$$

and the stochastic optimization problem

$$\begin{aligned}\underset{\boldsymbol{x}}{\text{minimize}} \quad & f(\boldsymbol{x}) = \mathbb{E}_{\boldsymbol{\xi}}[\tilde{f}(\boldsymbol{x}; \boldsymbol{\xi})] \\ \text{subject to} \quad & \boldsymbol{x} \in \boldsymbol{\Omega}.\end{aligned} \tag{STOCH}$$

Although important exceptions are noted throughout this survey, the majority of the methods discussed assume that the objective function f in (DET) and (STOCH) is differentiable. This assumption may cause readers to pause (and some readers may never resume). The methods considered here do not necessarily address non-smooth optimization; instead they address problems where a (sub)gradient of the objective f or a constraint function defining $\boldsymbol{\Omega}$ is not available to the optimization method. Note that similar naming confusion has existed in non-smooth optimization, as evidenced by the introduction of Lemarechal and Mifflin (1978):

This workshop was held under the name Nondifferentiable Optimization, but it has been recognized that this is misleading, because it suggests 'optimization without derivatives'.

1.1. Alternatives to derivative-free optimization methods

Derivative-free optimization methods are sometimes employed for convenience rather than by necessity. Since the decision to use a derivative-free method typically limits the performance – in terms of accuracy, expense or problem size – relative to what one might expect from gradient-based optimization methods, we first mention alternatives to using derivative-free methods.

The design of derivative-free optimization methods is informed by the alternatives of algorithmic and numerical differentiation. For the former, the purpose seems clear: since the methods use only function values, they apply even in cases when one cannot produce a computer code for the function's derivative. Similarly, derivative-free optimization methods should be designed in order to outperform (typically measured in terms of the number of function evaluations) gradient-based optimization methods that employ numerical differentiation.

1.1.1. Algorithmic differentiation

Algorithmic differentiation[2] (AD) is a means of generating derivatives of mathematical functions that are expressed in computer code (Griewank 2003, Griewank and Walther 2008). The forward mode of AD may be viewed as performing differentiation of elementary mathematical operations in each line of source code by means of the chain rule, while the reverse mode may be seen as traversing the resulting computational graph in reverse order.

Algorithmic differentiation has the benefit of automatically exploiting function structure, such as partial separability or other sparsity, and the corresponding ability of producing a derivative code whose computational cost is comparable to the cost of evaluating the function code itself.

AD has seen significant adoption and advances in the past decade (Forth *et al.* 2012). Tools for algorithmic differentiation cover a growing set of compiled and interpreted languages, with an evolving list summarized on the community portal at

http://www.autodiff.org.

Progress has also been made on algorithmic differentiation of piecewise smooth functions, such as those with breakpoints resulting from absolute values or conditionals in a code; see, for example, Griewank, Walther, Fiege and Bosse (2016). The machine learning renaissance has also fuelled demand and interest in AD, driven in large part by the success of algorithmic differentiation in backpropagation (Baydin, Pearlmutter, Radul and Siskind 2018).

1.1.2. Numerical differentiation

Another alternative to derivative-free methods is to estimate the derivative of f by numerical differentiation and then to use the estimates in a derivative-based method. This approach has the benefit that only zeroth-order information (*i.e.* the function value) is needed; however, depending on the derivative-based method used, the quality of the derivative estimate may be a limiting factor. Here we remark that for the finite-precision (or even fixed-precision) functions encountered in scientific applications, finite-difference estimates of derivatives may be sufficient for many purposes; see Section 2.3.1.

---

[2] *Algorithmic differentiation* is sometimes referred to as *automatic differentiation*, but we follow the preferred convention of Griewank (2003).

When numerical derivative estimates are used, the optimization method must tolerate inexactness in the derivatives. Such methods have been classically studied for both non-linear equations and unconstrained optimization; see, for example, the works of Powell (1965), Brown and Dennis, Jr (1971) and Mifflin (1975) and the references therein. Numerical derivatives continue to be employed by recent methods (see *e.g.* the works of Cartis, Gould and Toint 2012 and Berahas, Byrd and Nocedal 2019). Use in practice is typically determined by whether the limit on the derivative accuracy and the expense in terms of function evaluations are acceptable.

1.2. *Organization of the paper*

This paper is organized principally by problem class: unconstrained domain (Sections 2 and 3), convex objective (Section 4), structured objective (Section 5), stochastic optimization (Section 6) and constrained domain (Section 7).

Section 2 presents deterministic methods for solving (DET) when $\Omega = \mathbb{R}^n$. The section is split between direct-search methods and model-based methods, although the lines between these are increasingly blurred; see, for example, Conn and Le Digabel (2013), Custódio, Rocha and Vicente (2009), Gramacy and Le Digabel (2015) and Gratton, Royer and Vicente (2016). Direct-search methods are summarized in far greater detail by Kolda, Lewis and Torczon (2003) and Kelley (1999*b*), and in the more recent survey by Audet (2014). Model-based methods that employ trust regions are given full treatment by Conn *et al.* (2009*b*), and those that employ stencils are detailed by Kelley (2011).

In Section 3 we review randomized methods for solving (DET) when $\Omega = \mathbb{R}^n$. These methods are often variants of the deterministic methods in Section 2 but require additional notation to capture the resulting stochasticity; the analysis of these methods can also deviate significantly from their deterministic counterparts.

In Section 4 we discuss derivative-free methods intended primarily for convex optimization. We make this delineation because such methods have distinct lines of analysis and can often solve considerably higher-dimensional problems than can general methods for non-convex derivative-free optimization.

In Section 5 we survey methods that address particular structure in the objective f in (DET). Examples of such structure include non-linear least-squares objectives, composite non-smooth objectives and partially separable objectives.

In Section 6 we address derivative-free stochastic optimization, that is, when methods have access only to a stochastic realization of a function in pursuit of solving (STOCH). This topic is increasingly intertwined with simulation optimization and Monte Carlo-based optimization; for these areas we refer to the surveys by Homem-de-Mello and Bayraksan (2014), Fu, Glover and April (2005), Amaran, Sahinidis, Sharda and Bury (2015) and Kim, Pasupathy and Henderson (2015).

Section 7 presents methods for deterministic optimization problems with constraints (*i.e.* $\Omega \subset \mathbb{R}^n$). Although many of these methods rely on the foundations laid in Sections 2 and 3, we highlight particular difficulties associated with constrained derivative-free optimization.

In Section 8 we briefly highlight related problem areas (including global and multi-objective derivative-free optimization), methods and other implementation considerations.

2. Deterministic methods for deterministic objectives

We now address deterministic methods for solving (DET). We discuss direct-search methods in Section 2.1, model-based methods in Section 2.2 and other methods in Section 2.3. At a coarse level, direct-search methods use comparisons of function values to directly determine candidate points, whereas model-based methods use a surrogate of f to determine candidate points. Naturally, some hybrid methods incorporate ideas from both model-based and direct-search methods and may not be so easily categorized. An early survey of direct-search and model-based methods is given in Powell (1998*a*).

2.1. Direct-search methods

Although Hooke and Jeeves (1961) are credited with originating the term 'direct search', there is no agreed-upon definition of what constitutes a direct-search method. We follow the convention of Wright (1995), wherein a direct-search method is a method that uses only function values and 'does not "in its heart" develop an approximate gradient'.

We first discuss simplex methods, including the Nelder–Mead method – perhaps the most widely used direct-search method. We follow this discussion with a presentation of directional direct-search methods; hybrid direct-search methods are discussed in Section 2.3. (The global direct-search method **DIRECT** is discussed in Section 8.3.)

2.1.1. Simplex methods

Simplex methods (not to be confused with Dantzig's simplex method for linear programming) move and manipulate a collection of $n + 1$ affinely independent points (*i.e.* the vertices of a simplex in $\mathbb{R}^n$) when solving (DET). The method of Spendley, Hext and Himsworth (1962) involves either taking the point in the simplex with the largest function value and reflecting it through the hyperplane defined by the remaining n points or moving the

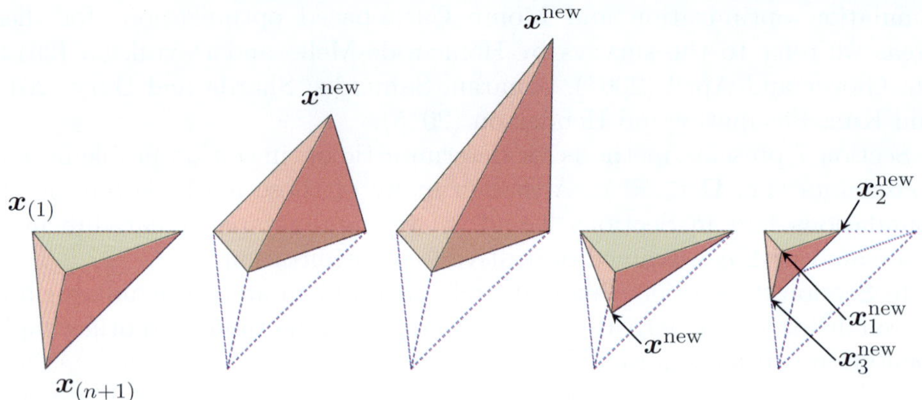

Figure 2.1. Primary Nelder–Mead simplex operations: original simplex, reflection, expansion, inner contraction, and shrink.

n worst points toward the best vertex of the simplex. In this manner, the geometry of all simplices remains the same as that of the starting simplex. (That is, all simplices are similar in the geometric sense.)

Nelder and Mead (1965) extend the possible simplex operations, as shown in Figure 2.1, by allowing the 'expansion' and 'contraction' operations in addition to the 'reflection' and 'shrink' operations of Spendley *et al.* (1962). These operations enable the Nelder–Mead simplex method to distort the simplex in order to account for possible curvature present in the objective function.

Nelder and Mead (1965) propose stopping further function evaluations when the standard error of the function values at the simplex vertices is small. Others, Woods (1985) for example, propose stopping when the size of the simplex's longest side incident to the best simplex vertex is small.

Nelder–Mead is an incredibly popular method, in no small part due to its inclusion in *Numerical Recipes* (Press, Teukolsky, Vetterling and Flannery 2007), which has been cited over 125 000 times and no doubt used many times more. The method (as implemented by Lagarias, Poonen and Wright 2012) is also the algorithm underlying fminsearch in MAT-LAB. Benchmarking studies highlight Nelder–Mead performance in practice (Moré and Wild 2009, Rios and Sahinidis 2013).

The method's popularity from its inception was not diminished by the lack of theoretical results proving its ability to identify stationary points. Woods (1985) presents a non-convex, two-dimensional function where Nelder–Mead converges to a non-stationary point (where the function's Hessian is singular). Furthermore, McKinnon (1998) presents a class of thrice-continuously differentiable, strictly convex functions on $\mathbb{R}^2$ where the Nelder–Mead simplex fails to converge to the lone stationary point. The only operation that Nelder–Mead performs on this relatively routine function is repeated 'inner contraction' of the initial simplex.

Researchers have continued to develop convergence results for modified or limited versions of Nelder–Mead. Kelley (1999a) addresses Nelder–Mead's theoretical deficiencies by restarting the method when the objective decrease on consecutive iterations is not larger than a multiple of the simplex gradient norm. Such restarts do not ensure that Nelder–Mead will converge: Kelley (1999a) shows an example of such behaviour. Price, Coope and Byatt (2002) embed Nelder–Mead in a different (convergent) algorithm using positive spanning sets. Nazareth and Tseng (2002) propose a clever, though perhaps superfluous, variant that connects Nelder–Mead to golden-section search.

Lagarias, Reeds, Wright and Wright (1998) show that Nelder–Mead (with appropriately chosen reflection and expansion coefficients) converges to the global minimizer of strictly convex functions when $n = 1$. Gao and Han (2012) show that the contraction and expansion steps of Nelder–Mead satisfy a descent condition on uniformly convex functions. Lagarias, Poonen and Wright (2012) show that a restricted version of the Nelder–Mead method – one that does not allow an expansion step – can converge to minimizers of any twice-continuously differentiable function with a positive-definite Hessian and bounded level sets. (Note that the class of functions from McKinnon (1998) have singular Hessians at only one point – their minimizers – and not at the point to which the simplex vertices are converging.)

The simplex method of Rykov (1980) includes ideas from model-based methods. Rykov varies the number of reflected vertices from iteration to iteration, following one of three rules that depend on the function value at the simplex centroid x_c. Rykov considers both evaluating f at the centroid and approximating f at the centroid using the values of f at the vertex. The non-reflected vertices are also moved in parallel with the reflected subset of vertices. In general, the number of reflected vertices is chosen so that x_c moves in a direction closest to $-\nabla f(x_c)$. This, along with a test of sufficient decrease in f, ensures convergence of the modified simplex method to a minimizer of convex, continuously differentiable functions with bounded level sets and Lipschitz-bounded gradients. (The sufficient-decrease condition is also shown to be efficient for the classical Nelder–Mead algorithm.)

Tseng (1999) proposes a modified simplex method that keeps the b_k best simplex vertices on a given iteration k and uses them to reflect the remaining vertices. Their method prescribes that 'the rays emanating from the reflected vertices toward the b_k best vertices should contain, in their convex hull, the rays emanating from a weighted centroid of the b_k best vertices toward the to-be-reflected vertices'. Their method also includes a *fortified descent* condition that is stronger than common sufficient-decrease conditions. If f is continuously differentiable and bounded below and b_k is fixed for all

---

Algorithm 1: $x^+ = \texttt{test\_descent}(f, x, P)$

---

1 Initialize $x^+ \leftarrow x$
2 **for** $p_i \in P$ **do**
3   Evaluate $f(p_i)$
4   **if** $f(p_i) - f(x)$ *acceptable* **then**
5          $x^+ \leftarrow p_i$
6   optional **break**

---

iterations, Tseng (1999) prove that every cluster point of the sequence of candidate points generated by their method is a stationary point.

Bűrmen, Puhan and Tuma (2006) propose a convergent version of a simplex method that does not require a sufficient descent condition to be satisfied. Instead, they ensure that evaluated points lie on a grid of points, and they show that this grid will be refined as the method proceeds.

2.1.2. Directional direct-search methods

Broadly speaking, each iteration of a directional direct-search (DDS) method generates a finite set of points near the current point x_k; these *poll points* are generated by taking x_k and adding terms of the form $\alpha_k d$, where α_k is a positive step size and d is an element from a finite set of directions D_k. Kolda *et al.* (2003) propose the term *generating set search methods* to encapsulate this class of methods.[3] The objective function f is then evaluated at all or some of the poll points, and x_{k+1} is selected to be some poll point that produces a (sufficient) decrease in the objective and the step size is possibly increased. If no poll point provides a sufficient decrease, x_{k+1} is set to x_k and the step size is decreased. In either case, the set of directions D_k can (but need not) be modified to obtain D_{k+1}.

A general DDS method is provided in Algorithm 2, which includes a *search step* where f is evaluated at any finite set of points Y_k, including $Y_k = \emptyset$. The search step allows one to (potentially) improve the performance of Algorithm 2. For example, points could be randomly sampled during the search step from the domain in the hope of finding a better local minimum, or a person running the algorithm may have problem-specific knowledge that can generate candidate points given the observed history of evaluated points and their function values. While the search step allows for this insertion of such heuristics, rigorous convergence results are driven by the more disciplined poll step. When testing for objective decrease in Algorithm 1, one can stop evaluating points in P (line 6) as soon as the first point is

---

[3] The term generating set arises from a need to generate a cone from the nearly active constraint normals when Ω is defined by linear constraints.

Algorithm 2: Directional direct-search method

---

1   Set parameters $0 < \gamma_{\text{dec}} < 1 \le \gamma_{\text{inc}}$
2   Choose initial point $\boldsymbol{x}_0$ and step size $\alpha_0 > 0$
3   **for** $k = 0, 1, 2, \ldots$ **do**
4   Choose and order a finite set $\boldsymbol{Y}_k \subset \mathbb{R}^n$ // (search step)
5      $\boldsymbol{x}_k^+ \leftarrow \texttt{test\_descent}(f, \boldsymbol{x}_k, \boldsymbol{Y}_k)$
6   **if** $\boldsymbol{x}_k^+ = \boldsymbol{x}_k$ **then**
7   Choose and order poll directions $\boldsymbol{D}_k \subset \mathbb{R}^n$ // (poll step)
8         $\boldsymbol{x}_k^+ \leftarrow \texttt{test\_descent}(f, \boldsymbol{x}_k, \{\boldsymbol{x}_k + \alpha_k \boldsymbol{d}_i : \boldsymbol{d}_i \in \boldsymbol{D}_k\})$
9   **if** $\boldsymbol{x}_k^+ = \boldsymbol{x}_k$ **then**
10         $\alpha_{k+1} \leftarrow \gamma_{\text{inc}} \alpha_k$
11   **else**
12         $\alpha_{k+1} \leftarrow \gamma_{\text{dec}} \alpha_k$
13      $\boldsymbol{x}_{k+1} \leftarrow \boldsymbol{x}_k^+$

---

identified where there is (sufficient) decrease in f. In this case, the polling (or search) step is considered *opportunistic*.

DDS methods are largely distinguished by how they generate the set of poll directions $\boldsymbol{D}_k$ at line 7 of Algorithm 2. Perhaps the first approach is *coordinate search*, in which the poll directions are defined as $\boldsymbol{D}_k = \{\pm \boldsymbol{e}_i : i = 1, 2, \ldots, n\}$, where $\boldsymbol{e}_i$ denotes the ith elementary basis vector (*i.e.* column i of the identity matrix in n dimensions). The first known description of coordinate search appears in the work of Fermi and Metropolis (1952) where the smallest positive integer l is sought such that $f(\boldsymbol{x}_k + l\alpha\boldsymbol{e}_1/2) > f(\boldsymbol{x}_k + (l-1)\alpha\boldsymbol{e}_1/2)$. If an increase in f is observed at $\boldsymbol{e}_1/2$ then $-\boldsymbol{e}_1/2$ is considered. After such an integer l is identified for the first coordinate direction, $\boldsymbol{x}_k$ is updated to $\boldsymbol{x}_k \pm l\boldsymbol{e}_1/2$ and the second coordinate direction is considered. If $\boldsymbol{x}_k$ is unchanged after cycling through all coordinate directions, then the method is repeated but with $\pm\boldsymbol{e}_i/2$ replaced with $\pm\boldsymbol{e}_i/16$, terminating when no improvement is observed for this smaller α. In terms of Algorithm 2 the search set $\boldsymbol{Y}_k = \emptyset$ at line 4, and the descent test at line 4 of Algorithm 1 merely tests for simple decrease, that is, $f(\boldsymbol{p}_i) - f(\boldsymbol{x}) < 0$. Other versions of acceptability in line 4 of Algorithm 1 are employed by methods discussed later.

Proofs that DDS methods converge first appeared in the works of Céa (1971) and Yu (1979), although both require the sequence of step-size parameters to be non-increasing. Lewis, Torczon and Trosset (2000) attribute the first global convergence proof for coordinate search to Polak (1971, p. 43). In turn, Polak cites the 'method of local variation' of Banichuk, Petrov and Chernous'ko (1966); although Banichuk *et al.* (1966) do develop parts of a convergence proof, they state in Remark 1 that 'the question of the strict formulation of the general sufficient conditions for convergence of the algorithm to a minimum remains open'.

Typical convergence results for DDS require that the set D_k is a *positive spanning set* (PSS) for the domain Ω; that is, any point $x \in \Omega$ can be written as

$$x = \sum_{i=1}^{|D_k|} \lambda_i d_i,$$

where $d_i \in D_k$ and $\lambda_i \geq 0$ for all i. Some of the first discussions of properties of positive spanning sets were presented by Davis (1954) and McKinney (1962), but recent treatments have also appeared in Regis (2016). In addition to requiring positive spanning sets during the poll step, earlier DDS convergence results depended on f being continuously differentiable. When f is non-smooth, no descent direction is guaranteed for these early DDS methods, even when the step size is arbitrarily small. See, for example, the modification of the Dennis–Woods (Dennis, Jr and Woods 1987) function by Kolda *et al.* (2003, Figure 6.2) and a discussion of why coordinate-search methods (for example) will not move when started at a point of non-differentiability; moreover, when started at differentiable points, coordinate-search methods tend to converge to a point that is not (Clarke) stationary.

The pattern-search method of Torczon (1991) revived interest in direct-search methods. The method therein contains ideas from both DDS and simplex methods. Given a simplex defined by $x_k, y_1, \ldots, y_n$ (where x_k is the simplex vertex with smallest function value), the polling directions are given by $D_k = \{y_i - x_k : i = 1, \ldots, n\}$. If a decrease is observed at the best poll point in $x_k + D_k$, the simplex is set to either $x_k \bigcup x_k + D_k$ or some expansion thereof. If no improvement is found during the poll step, the simplex is contracted. Torczon (1991) shows that if f is continuous on the level set of x_0 and this level set is compact, then a subsequence of $\{x_k\}$ converges to a stationary point of f, a point where f is non-differentiable, or a point where f is not continuously differentiable.

A generalization of pattern-search methods is the class of *generalized pattern-search* (GPS) *methods*. Early GPS methods did not allow for a search step; the search-poll paradigm was introduced by Audet (2004). GPS methods are characterized by fixing a positive spanning set D and selecting $D_k \subseteq D$ during the poll step at line 7 on each iteration of Algorithm 2. Torczon (1997) assumes that the test for decrease in line 4 in Algorithm 1 is simple decrease, that is, that $f(p_i) < f(x)$. Early analysis of GPS methods using simple decrease required the step size α_k to remain rational (Audet and Dennis, Jr 2002, Torczon 1997). Audet (2004) shows that such an assumption is necessary by constructing small-dimensional examples where

GPS methods do not converge if α_k is irrational. Works below show that if a sufficient (instead of simple) decrease is ensured, α_k can take irrational values.

A refinement of the analysis of GPS methods was made by Dolan, Lewis and Torczon (2003), which shows that when ∇f is Lipschitz-continuous, the step-size parameter α_k scales linearly with $\|\nabla f(\boldsymbol{x}_k)\|$. Therefore α_k can be considered a reliable measure of first-order stationarity and justifies the traditional approach of stopping a GPS method when α_k is small. Second-order convergence analyses of GPS methods have also been considered. Abramson (2005) shows that, when applied to a twice-continuously differentiable f, a GPS method that infinitely often has $\boldsymbol{D}_k$ include a fixed orthonormal basis and its negative will have a limit point satisfying a 'pseudo-second-order' stationarity condition. Building off the use of curvature information in Frimannslund and Steihaug (2007), Abramson, Frimannslund and Steihaug (2013) show that a modification of the GPS framework that constructs approximate Hessians of f will converge to points that are second-order stationary provided that certain conditions on the Hessian approximation hold (and a fixed orthonormal basis and its negative are in $\boldsymbol{D}_k$ infinitely often).

In general, first-order convergence results (there exists a limit point $\boldsymbol{x}_*$ of $\{\boldsymbol{x}_k\}$ generated by a GPS method such that $\nabla f(\boldsymbol{x}_*) = \boldsymbol{0}$) for GPS methods can be demonstrated when f is continuously differentiable. For general Lipschitz-continuous (but non-smooth) functions f, however, one can only demonstrate that on a particular subsequence $\mathcal{K}$, satisfying $\{\boldsymbol{x}_k\}_{k \in \mathcal{K}} \to \boldsymbol{x}_*$, for each $\boldsymbol{d}$ that appears infinitely many times in $\{\boldsymbol{D}_k\}_{k \in \mathcal{K}}$, it holds that $f'(\boldsymbol{x}_*; \boldsymbol{d}) \geq 0$; that is, the directional derivative at $\boldsymbol{x}_*$ in the direction $\boldsymbol{d}$ is non-negative.

The flexibility of GPS methods inspired various extensions. Abramson, Audet and Dennis, Jr (2004) consider adapting GPS to utilize derivative information when it is available in order to reduce the number of points evaluated during the poll step. Abramson, Audet, Dennis, Jr and Le Digabel (2009b) and Frimannslund and Steihaug (2011) re-use previous function evaluations in order to determine the next set of directions. Custódio and Vicente (2007) consider re-using previous function evaluations to compute simplex gradients; they show that the information obtained from simplex gradients can be used to reorder the poll points $\boldsymbol{P}$ in Algorithm 1. A similar use of simplex gradients in the non-smooth setting is considered by Custódio, Dennis, Jr and Vicente (2008). Hough, Kolda and Torczon (2001) discuss modifications to Algorithm 2 that allow for increased efficiency when concurrent, asynchronous evaluations of f are possible; an implementation of the method of Hough *et al.* (2001) is presented by Gray and Kolda (2006).

The early analysis of Torczon (1991, Section 7) of pattern-search methods when f is non-smooth carries over to GPS methods as well; such methods may converge to a non-stationary point. This motivated a further generalization of GPS methods, *mesh adaptive direct search* (MADS) methods (Audet and Dennis, Jr 2006, Abramson and Audet 2006). Inspired by Coope and Price (2000), MADS methods augment GPS methods by incorporating a mesh parametrized by a *mesh parameter* $\beta_k^m > 0$. In the kth iteration, given the fixed PSS $\boldsymbol{D}$ and the mesh parameter β_k^m, the MADS mesh around the current point $\boldsymbol{x}_k$ is

$$\mathcal{M}_k = \bigcup_{\boldsymbol{x} \in \boldsymbol{S}_k} \left\{ \boldsymbol{x} + \beta_k^m \sum_{j=1}^{|\boldsymbol{D}|} \lambda_j \boldsymbol{d}_j : \boldsymbol{d}_j \in \boldsymbol{D}, \lambda_j \in \mathbb{N} \bigcup \{0\} \right\},$$

where $\boldsymbol{S}_k$ is the set of points at which f has been evaluated prior to the kth iteration of the method.

MADS methods additionally define a frame

$$\mathcal{F}_k = \{\boldsymbol{x}_k + \beta_k^m \boldsymbol{d}^f : \boldsymbol{d}^f \in \boldsymbol{D}_k^f\},$$

where $\boldsymbol{D}_k^f$ is a finite set of directions, each of which is expressible as

$$\boldsymbol{d}^f = \sum_{j=1}^{|\boldsymbol{D}|} \lambda_j \boldsymbol{d}_j,$$

with each $\lambda_j \in \mathbb{N} \bigcup \{0\}$ and $\boldsymbol{d}_j \in \boldsymbol{D}$. Additionally, MADS methods define a *frame parameter* β_k^f and require that each $\boldsymbol{d}^f \in \boldsymbol{D}_k^f$ satisfies $\beta_k^m \|\boldsymbol{d}^f\| \le \beta_k^f \max\{\|\boldsymbol{d}\| : \boldsymbol{d} \in \boldsymbol{D}\}$. Observe that in each iteration, $\mathcal{F}_k \subsetneq \mathcal{M}_k$. Note that the mesh is never explicitly constructed nor stored over the domain. Rather, points are evaluated only at what *would be* nodes of some implicitly defined mesh via the frame.

In the poll step of Algorithm 2, the set of poll directions $\boldsymbol{D}_k$ is chosen as $\{\boldsymbol{y} - \boldsymbol{x}_k : \boldsymbol{y} \in \mathcal{F}_k\}$. The role of the step-size parameter α_k in Algorithm 2 is completely replaced by the behaviour of β_k^f, β_k^m. If there is no improvement at a candidate solution during the poll step, β_k^m is decreased, resulting in a finer mesh; likewise β_k^f is decreased, resulting in a finer local mesh around $\boldsymbol{x}_k$. MADS intentionally allows the parameters β_k^m and β_k^f to be decreased at different rates; roughly speaking, by driving β_k^m to zero faster than β_k^f is driven to zero, and by choosing the sequence $\{\boldsymbol{D}_k^f\}$ to satisfy certain conditions, the directions in $\mathcal{F}_k$ become asymptotically dense around limit points of $\boldsymbol{x}_k$. That is, it is possible to decrease β_k^m, β_k^f at rates such that poll directions will be arbitrarily close to any direction. This ensures that the Clarke directional derivative is non-negative in all directions around any limit point of the sequence of $\boldsymbol{x}_k$ generated by MADS; that is,

$$f_C'(\boldsymbol{x}_*; \boldsymbol{d}) \ge 0 \quad \text{for all directions } \boldsymbol{d}, \tag{2.1}$$

with an analogous result also holding for constrained problems, with (2.1) reduced to all *feasible* directions $\boldsymbol{d}$. (DDS methods for constrained optimization will be discussed in Section 7.) This powerful result highlights the ability of directional direct-search methods to address non-differentiable functions f.

MADS does not prescribe any one approach for adjusting β_k^m, β_k^f so that the poll directions are dense, but Audet and Dennis, Jr (2006) demonstrate an approach where randomized directions are completed to be a PSS and β_k^f either is $n\sqrt{\beta_k^m}$ or $\sqrt{\beta_k^m}$ results in a asymptotically dense poll directions for any convergent subsequence of $\{\boldsymbol{x}_k\}$. MADS does not require a sufficient-decrease condition.

Recent advances to MADS-based algorithms have focused on reducing the number of function evaluations required in practice by adaptively reducing the number of poll points queried; see, for example, Audet, Ianni, Le Digabel and Tribes (2014) and Alarie *et al.* (2018). Smoothing-based extensions to noisy deterministic problems include Audet, Ihaddadene, Le Digabel and Tribes (2018*b*). Vicente and Custódio (2012) show that MADS methods converge to local minima even for a limited class of *discontinuous* functions that satisfy some assumptions concerning the behaviour of the disconnected regions of the epigraph at limit points.

Worst-case complexity analysis. Throughout this survey, when discussing classes of methods, we will refer to their worst-case complexity (WCC). Generally speaking, WCC refers to an upper bound on the number of function evaluations N_ϵ required to attain an ϵ-accurate solution to a problem drawn from a problem class. Correspondingly, the definition of ϵ-accurate varies between different problem classes. For instance, and of particular immediate importance, if an objective function is assumed Lipschitz-continuously differentiable (which we denote by $f \in \mathcal{LC}^1$), then an appropriate notion of first-order ϵ-accuracy is

$$\|\nabla f(\boldsymbol{x}_k)\| \leq \epsilon. \tag{2.2}$$

That is, the WCC of a method applied to the class $\mathcal{LC}^1$ is characterized by N_ϵ, an upper bound on the number of function evaluations the method requires before (2.2) is satisfied for *any* $f \in \mathcal{LC}^1$. Similarly, we can define a notion of second-order ϵ-accuracy as

$$\max\{\|\nabla f(\boldsymbol{x}_k)\|, -\lambda_k\} \leq \epsilon, \tag{2.3}$$

where λ_k denotes the minimum eigenvalue of $\nabla^2 f(\boldsymbol{x}_k)$.

Note that WCCs can only be derived for methods for which convergence results have been established. Indeed, in the problem class $\mathcal{LC}^1$, first-order convergence results canonically have the form

$$\lim_{k \to \infty} \|\nabla f(\boldsymbol{x}_k)\| = 0. \tag{2.4}$$

The convergence in (2.4) automatically implies the weaker lim-inf-type result

$$\liminf_{k\to\infty} \|\nabla f(\boldsymbol{x}_k)\| = 0, \tag{2.5}$$

from which it is clear that for any $\epsilon > 0$, there must exist finite N_ϵ so that (2.2) holds. In fact, in many works, demonstrating a result of the form (2.5) is a stepping stone to proving a result of the form (2.4). Likewise, demonstrating a second-order WCC of the form (2.3) depends on showing

$$\lim_{k\to\infty} \max\{\|\nabla f(\boldsymbol{x}_k)\|, -\lambda_k\} = 0, \tag{2.6}$$

which guarantees the weaker lim-inf-type result

$$\liminf_{k\to\infty} \max\{\|\nabla f(\boldsymbol{x}_k)\|, -\lambda_k\} = 0. \tag{2.7}$$

Proofs of convergence for DDS methods applied to functions $f \in \mathcal{LC}^1$ often rely on a (sub)sequence of positive spanning sets $\{\boldsymbol{D}_k\}$ satisfying

$$\mathrm{cm}(\boldsymbol{D}_k) = \min_{\boldsymbol{v}\in\mathbb{R}^n\setminus\{\boldsymbol{0}\}} \max_{\boldsymbol{d}\in\boldsymbol{D}_k} \frac{\boldsymbol{d}^\top \boldsymbol{v}}{\|\boldsymbol{d}\|\|\boldsymbol{v}\|} \geq \kappa > 0, \tag{2.8}$$

where $\mathrm{cm}(\cdot)$ is the *cosine measure* of a set. Under Assumption (2.8), Vicente (2013) obtains a WCC of type (2.2) for a method in the Algorithm 2 framework. In that work, it is assumed that $\boldsymbol{Y}_k = \emptyset$ at every search step. Moreover, *sufficient* decrease is tested at line 4 of Algorithm 1; in particular, Vicente (2013) checks in this line whether $f(\boldsymbol{p}_i) < f(\boldsymbol{x}) - c\alpha_k^2$ for some $c > 0$, where α_k is the current step size in Algorithm 2. Under these assumptions, Vicente (2013) demonstrates a WCC in $O(\epsilon^{-2})$. Throughout this survey, we will refer to Table A.1 for more details concerning specific WCCs. In general, though, we will often summarize WCCs in terms of their ϵ-dependence, as this provides an asymptotic characterization of a method's complexity in terms of the accuracy to which one wishes to solve a problem.

When $f \in \mathcal{LC}^2$, work by Gratton *et al.* (2016) essentially augments the DDS method analysed by Vicente (2013), but forms an approximate Hessian via central differences from function evaluations obtained (for free) by using a particular choice of $\boldsymbol{D}_k$. Gratton *et al.* (2016) then demonstrate that this augmentation of Algorithm 2 has a subsequence that converges to a second-order stationary point. That is, they prove a convergence result of the form (2.7) and demonstrate a WCC result of type (2.3) in $O(\epsilon^{-3})$ (see Table A.1).

We are unaware of WCC results for MADS methods; this situation may be unsurprising since MADS methods are motivated by non-smooth problems, which depend on the generation of a countably infinite number of poll directions. However, WCC results are not necessarily impossible to obtain in *structured* non-smooth cases, which we discuss in Section 5. We will discuss a special case where *smoothing functions* of a non-smooth function are assumed to be available in Section 5.3.2.

2.2. Model-based methods

In the context of derivative-free optimization, model-based methods are methods whose updates are based primarily on the predictions of a model that serves as a surrogate of the objective function or of a related merit function. We begin with basic properties and construction of popular models; readers interested in algorithmic frameworks such as trust-region methods and implicit filtering can proceed to Section 2.2.4. Throughout this section, we assume that models are intended as a surrogate for the function f; in future sections, these models will be extended to capture functions arising, for example, as constraints or separable components. The methods in this section assume some smoothness in f and therefore operate with smooth models; in Section 5, we examine model-based methods that exploit knowledge of non-smoothness.

2.2.1. Quality of smooth model approximation

A natural first indicator of the quality of a model used for optimization is the degree to which the model locally approximates the function f and its derivatives. To say anything about the quality of such approximation, one must make an assumption about the smoothness of both the model and function. For the moment, we leave this assumption implicit, but it will be formalized in subsequent sections.

A function $m : \mathbb{R}^n \to \mathbb{R}$ is said to be a $\boldsymbol{\kappa}$-fully linear model of f on $\mathcal{B}(\boldsymbol{x}; \Delta) = \{\boldsymbol{y} : \|\boldsymbol{x} - \boldsymbol{y}\| \leq \Delta\}$ if

$$|f(\boldsymbol{x} + \boldsymbol{s}) - m(\boldsymbol{x} + \boldsymbol{s})| \leq \kappa_{\mathrm{ef}}\Delta^2, \quad \text{for all } \boldsymbol{s} \in \mathcal{B}(\boldsymbol{0}; \Delta), \tag{2.9a}$$

$$\|\nabla f(\boldsymbol{x} + \boldsymbol{s}) - \nabla m(\boldsymbol{x} + \boldsymbol{s})\| \leq \kappa_{\mathrm{eg}}\Delta, \quad \text{for all } \boldsymbol{s} \in \mathcal{B}(\boldsymbol{0}; \Delta), \tag{2.9b}$$

for $\boldsymbol{\kappa} = (\kappa_{\mathrm{ef}}, \kappa_{\mathrm{eg}})$. Similarly, for $\boldsymbol{\kappa} = (\kappa_{\mathrm{ef}}, \kappa_{\mathrm{eg}}, \kappa_{\mathrm{eH}})$, m is said to be a $\boldsymbol{\kappa}$-fully quadratic model of f on $\mathcal{B}(\boldsymbol{x}; \Delta)$ if

$$|f(\boldsymbol{x} + \boldsymbol{s}) - m(\boldsymbol{x} + \boldsymbol{s})| \leq \kappa_{\mathrm{ef}}\Delta^3, \quad \text{for all } \boldsymbol{s} \in \mathcal{B}(\boldsymbol{0}; \Delta), \tag{2.10a}$$

$$\|\nabla f(\boldsymbol{x} + \boldsymbol{s}) - \nabla m(\boldsymbol{x} + \boldsymbol{s})\| \leq \kappa_{\mathrm{eg}}\Delta^2, \quad \text{for all } \boldsymbol{s} \in \mathcal{B}(\boldsymbol{0}; \Delta), \tag{2.10b}$$

$$\|\nabla^2 f(\boldsymbol{x} + \boldsymbol{s}) - \nabla^2 m(\boldsymbol{x} + \boldsymbol{s})\| \leq \kappa_{\mathrm{eH}}\Delta, \quad \text{for all } \boldsymbol{s} \in \mathcal{B}(\boldsymbol{0}; \Delta). \tag{2.10c}$$

Extensions to higher-degree approximations follow a similar form, but the computational expense associated with achieving higher-order guarantees is not a strategy pursued by derivative-free methods that we are aware of.

Models satisfying (2.9) or (2.10) are called Taylor-like models. To understand why, consider the second-order Taylor model

$$m(\boldsymbol{x} + \boldsymbol{s}) = f(\boldsymbol{x}) + \nabla f(\boldsymbol{x})^{\mathrm{T}}\boldsymbol{s} + \frac{1}{2}\boldsymbol{s}^{\mathrm{T}}\nabla^2 f(\boldsymbol{x})\boldsymbol{s}. \tag{2.11}$$

This model is a κ-fully quadratic model of f, with

$$(\kappa_{\mathrm{ef}}, \kappa_{\mathrm{eg}}, \kappa_{\mathrm{eH}}) = (L_{\mathrm{H}}/6, L_{\mathrm{H}}/2, L_{\mathrm{H}}),$$

on any $\mathcal{B}(\boldsymbol{x}; \Delta)$, where f has a Lipschitz-continuous second derivative with Lipschitz constant L_{H}.

As illustrated in the next section, one also can guarantee that models that do not employ derivative information satisfy these approximation bounds in (2.9) or (2.10). This approximation quality is used by derivative-free algorithms to ensure that a sufficient reduction predicted by the model m yields an attainable reduction in the function f as Δ becomes smaller.

2.2.2. Polynomial models

Polynomial models are the most commonly used models for derivative-free local optimization. We let $\mathcal{P}^{d,n}$ denote the space of polynomials of n variables of degree d and $\boldsymbol{\phi} : \mathbb{R}^n \to \mathbb{R}^{\dim(\mathcal{P}^{d,n})}$ define a basis for this space. For example, quadratic models can be obtained by using the monomial basis

$$\boldsymbol{\phi}(\boldsymbol{x}) = [1, x_1, \ldots, x_n, x_1^2, \ldots x_n^2, x_1 x_2, \ldots, x_{n-1} x_n]^{\mathrm{T}}, \tag{2.12}$$

for which $\dim(\mathcal{P}^{2,n}) = (n+1)(n+2)/2$; linear models can be obtained by using the first $\dim(\mathcal{P}^{1,n}) = n+1$ components of (2.12); quadratic models with diagonal Hessians, which are considered by Powell (2003), can be obtained by using the first $2n+1$ components of (2.12).

Any polynomial model $m \in \mathcal{P}^{d,n}$ is defined by $\boldsymbol{\phi}$ and coefficients $\boldsymbol{a} \in \mathbb{R}^{\dim(\mathcal{P}^{d,n})}$ through

$$m(\boldsymbol{x}) = \sum_{i=1}^{\dim(\mathcal{P}^{d,n})} a_i \phi_i(\boldsymbol{x}). \tag{2.13}$$

Given a set of p points $\boldsymbol{Y} = \{\boldsymbol{y}_1, \ldots, \boldsymbol{y}_p\}$, a model that interpolates f on $\boldsymbol{Y}$ is defined by the solution $\boldsymbol{a}$ to

$$\boldsymbol{\Phi}(\boldsymbol{Y})\boldsymbol{a} = \begin{bmatrix} \boldsymbol{\phi}(\boldsymbol{y}_1) & \cdots & \boldsymbol{\phi}(\boldsymbol{y}_p) \end{bmatrix}^{\mathrm{T}} \boldsymbol{a} = \begin{bmatrix} f(\boldsymbol{y}_1) \\ \vdots \\ f(\boldsymbol{y}_p) \end{bmatrix}. \tag{2.14}$$

The existence, uniqueness and conditioning of a solution to (2.14) depend on the location of the sample points $\boldsymbol{Y}$ through the matrix $\boldsymbol{\Phi}(\boldsymbol{Y})$. We note that when $n > 1$, $|\boldsymbol{Y}| = \dim(\mathcal{P}^{d,n})$ is insufficient for guaranteeing that $\boldsymbol{\Phi}(\boldsymbol{Y})$ is non-singular (Wendland 2005). Instead, additional conditions, effectively on the geometry of the sample points $\boldsymbol{Y}$, must be satisfied.

Simplex gradients and linear interpolation models. The geometry conditions needed to uniquely define a linear model are relatively straightforward: the sample points Y must be affinely independent; that is, the columns of

$$Y_{-1} = \begin{bmatrix} y_2 - y_1 & \cdots & y_{n+1} - y_1 \end{bmatrix} \tag{2.15}$$

must be linearly independent. Such sample points define what is referred to as a simplex gradient g through $g = [a_2, \ldots, a_{n+1}]^{\mathrm{T}}$, when the monomial basis ϕ is used in (2.14).

Simplex gradients can be viewed as a generalization of first-order finite-difference estimates (*e.g.* the forward differences based on evaluations at the points $\{y_1, y_1 + \Delta e_1, \ldots, y_1 + \Delta e_n\}$); their use in optimization algorithms dates at least back to the work of Spendley *et al.* (1962) that inspired Nelder and Mead (1965). Other example usage includes pattern search (Custódio and Vicente 2007, Custódio *et al.* 2008) and noisy optimization (Kelley 1999*b*, Bortz and Kelley 1998); the study of simplex gradients continues with recent works such as those of Regis (2015) and Coope and Tappenden (2019).

Provided that (2.15) is non-singular, it is straightforward to show that linear interpolation models are κ-fully linear model of f in a neighbourhood of y_1. In particular, if $Y \subset \mathcal{B}(y_1; \Delta)$ and f has an L_g-Lipschitz-continuous first derivative on an open domain containing $\mathcal{B}(y_1; \Delta)$, then (2.9) holds on $\mathcal{B}(y_1; \Delta)$ with

$$\kappa_{\mathrm{eg}} = L_g(1 + \sqrt{n}\Delta\|Y_{-1}^{-1}\|/2) \quad \text{and} \quad \kappa_{\mathrm{ef}} = L_g/2 + \kappa_{\mathrm{eg}}. \tag{2.16}$$

The expressions in (2.16) also provide a recipe for obtaining a model with a potentially tighter error bound over $\mathcal{B}(y_1; \Delta)$: modify $Y \subset \mathcal{B}(y_1; \Delta)$ to decrease $\|Y_{-1}^{-1}\|$. We note that when Y_{-1} contains orthonormal directions scaled by Δ, one recovers $\kappa_{\mathrm{eg}} = L_g(1 + \sqrt{n}/2)$ and $\kappa_{\mathrm{ef}} = L_g(3 + \sqrt{n})/2$, which is the least value one can obtain from (2.16) given the restriction that $Y \subset \mathcal{B}(y_1; \Delta)$. Hence, by performing LU or QR factorization with pivoting, one can obtain directions (which are then scaled by Δ) in order to improve the conditioning of Y_{-1}^{-1} and hence the approximation bound. Such an approach is performed by Conn, Scheinberg and Vicente (2008*a*) for linear models and by Wild and Shoemaker (2011) for fully linear radial basis function models.

The geometric conditions on Y, induced by the approximation bounds in (2.9) or (2.10), can be viewed as playing a similar role to the geometric conditions (*e.g.* positive spanning) imposed on D in directional direct-search methods. Naturally, the choice of basis function used for any model affects the quantitative measure of that model's quality.

Note that many practical methods employ interpolation sets contained within a constant multiple of the trust-region radius (*i.e.* $Y \subset \mathcal{B}(y_1; c_1\Delta)$ for a constant $c_1 \in [1, \infty)$).

Quadratic interpolation models. Quadratic interpolation models have been used for derivative-free optimization for at least fifty years (Winfield 1969,

Winfield 1973) and were employed by a series of methods that revitalized interest in model-based methods; see, for example, Conn and Toint (1996), Conn, Scheinberg and Toint (1997b), Conn, Scheinberg and Toint (1997a) and Powell (1998b, 2002).

Of course, the quality of an interpolation model (quadratic or otherwise) in a region of interest is determined by the position of the underlying points being interpolated. For example, if a model m interpolates a function f at points far away from a certain region of interest, the model value may differ greatly from the value of f in that region. Λ-poisedness is a concept to measure how well a set of points is dispersed through a region of interest, and ultimately how well a model will estimate the function in that region.

The most commonly used metric for quantifying how well points are positioned in a region of interest is based on Lagrange polynomials. Given a set of p points $\boldsymbol{Y} = \{\boldsymbol{y}_1, \ldots, \boldsymbol{y}_p\}$, a basis of Lagrange polynomials satisfies

$$\ell_j(\boldsymbol{y}_i) = \begin{cases} 1 & \text{if } i = j, \\ 0 & \text{if } i \neq j. \end{cases} \qquad (2.17)$$

We now define Λ-poisedness. A set of points $\boldsymbol{Y}$ is said to be Λ-poised on a set $\boldsymbol{B}$ if $\boldsymbol{Y}$ is linearly independent and the Lagrange polynomials $\{\ell_1, \ldots, \ell_p\}$ associated with $\boldsymbol{Y}$ satisfy

$$\Lambda \geq \max_{1 \leq i \leq p} \max_{\boldsymbol{x} \in B} |\ell_i(\boldsymbol{x})|. \qquad (2.18)$$

(For an equivalent definition of Λ-poisedness, see Conn *et al.* (2009b, Definition 3.6).) Note that the definition of Λ-poisedness is independent of the function being modelled. Also, the points $\boldsymbol{Y}$ need not necessarily be elements of the set $\boldsymbol{B}$. Also, note that if a model is poised on a set $\boldsymbol{B}$, it is poised on any subset of $\boldsymbol{B}$. One is usually interested in the least value of Λ so that (2.18) holds.

Powell's unconstrained optimization by quadratic approximation method (UOBYQA) follows such an approach in maximizing the Lagrange polynomials. In Powell (1998b), Powell (2001) and Powell (2002), significant care is given to the linear algebra expense associated with this maximization and the associated change of basis as the methods change their interpolation sets. For example, in Powell (1998b), particular sparsity in the Hessian approximation is employed with the aim of capturing curvature while keeping linear algebraic expenses low.

Maintaining, and the question of to what extent it is necessary to maintain, this geometry for quadratic models has been intensely studied; see, for example, Fasano, Morales and Nocedal (2009), Marazzi and Nocedal (2002), D'Ambrosio, Nannicini and Sartor (2017) and Scheinberg and Toint (2010).

Underdetermined quadratic interpolation models. A fact not to be overlooked in the context of derivative-free optimization is that employing an interpolation set $\boldsymbol{Y}$ requires availability of the $|\boldsymbol{Y}|$ function values $\{f(\boldsymbol{y}_i) : \boldsymbol{y}_i \in \boldsymbol{Y}\}$. When the function f is computationally expensive to evaluate, the $(n+1)(n+2)/2$ points required by fully quadratic models can be a burden, potentially with little benefit, to obtain repeatedly in an optimization algorithm.

Beginning with Powell (2003), Powell investigated quadratic models constructed from fewer than $(n+1)(n+2)/2$ points. The most successful of these strategies was detailed in Powell (2004*a*) and Powell (2004*b*) and resolved the $(n+1)(n+2)/2 - |\boldsymbol{Y}|$ remaining degrees of freedom by solving problems of the form

$$\begin{aligned}
\underset{m \in \mathcal{P}^{2,n}}{\text{minimize}} \quad & \|\nabla^2 m(\check{\boldsymbol{x}}) - \boldsymbol{H}\|_F^2 \\
\text{subject to} \quad & m(\boldsymbol{y}_i) = f(\boldsymbol{y}_i), \quad \text{for all } \boldsymbol{y}_i \in \boldsymbol{Y}
\end{aligned} \tag{2.19}$$

to obtain a model m about a point of interest $\check{\boldsymbol{x}}$. Solutions to (2.19) are models with a Hessian closest in Frobenius norm to a specified $\boldsymbol{H} = \boldsymbol{H}^{\mathrm{T}}$ among all models that interpolate f on $\boldsymbol{Y}$. A popular implementation of this strategy is the NEWUOA solver (Powell 2006).

By using the basis

$$\begin{aligned}
\boldsymbol{\phi}(\check{\boldsymbol{x}} + \boldsymbol{x}) &= \left[\boldsymbol{\phi}_{\mathrm{fg}}(\check{\boldsymbol{x}} + \boldsymbol{x})^{\mathrm{T}} \quad \mid \boldsymbol{\phi}_{\mathrm{H}}(\check{\boldsymbol{x}} + \boldsymbol{x})^{\mathrm{T}}\right]^{\mathrm{T}} \tag{2.20} \\
&= \left[1, x_1, \ldots, x_n \mid \frac{1}{2}x_1^2, \ldots, \frac{1}{2}x_n^2, \frac{1}{\sqrt{2}}x_1 x_2, \ldots, \frac{1}{\sqrt{2}}x_{n-1}x_n\right]^{\mathrm{T}},
\end{aligned}$$

the problem (2.19) is equivalent to the problem

$$\underset{\boldsymbol{a}_{\mathrm{fg}}, \boldsymbol{a}_{\mathrm{H}}}{\text{minimize}} \quad \|\boldsymbol{a}_{\mathrm{H}}\|_2^2 \tag{2.21}$$

$$\text{subject to} \quad \boldsymbol{a}_{\mathrm{fg}}^{\mathrm{T}}\boldsymbol{\phi}_{\mathrm{fg}}(\boldsymbol{y}_i) + \boldsymbol{a}_{\mathrm{H}}^{\mathrm{T}}\boldsymbol{\phi}_{\mathrm{H}}(\boldsymbol{y}_i) = f(\boldsymbol{y}_i) - \frac{1}{2}\boldsymbol{y}_i^{\mathrm{T}}\boldsymbol{H}\boldsymbol{y}_i, \quad \text{for all } \boldsymbol{y}_i \in \boldsymbol{Y}.$$

Existence and uniqueness of solutions to (2.21) again depend on the positioning of the points in $\boldsymbol{Y}$. Notably, a necessary condition for there to be a unique minimizer of the seminorm is that at least $n+1$ of the points in $\boldsymbol{Y}$ be affinely independent. Lagrange polynomials can be defined for this case; Conn, Scheinberg and Vicente (2008*b*) establish conditions for Λ-poisedness (and hence a fully linear, or better, approximation quality) of such models.

Powell (2004*c*, 2007, 2008) develops efficient solution methodologies for (2.21) when $\boldsymbol{H}$ and m are constructed from interpolation sets that differ by at most one point, and employ these updates in NEWUOA and subsequent solvers. Wild (2008*b*) and Custódio *et al.* (2009) use $\boldsymbol{H} = \boldsymbol{0}$ in order to obtain tighter fully linear error bounds of models resulting from (2.21). A strategy of using even fewer interpolation points (including those in a proper subspace of $\mathbb{R}^n$) is developed by Powell (2013) and Zhang (2014).

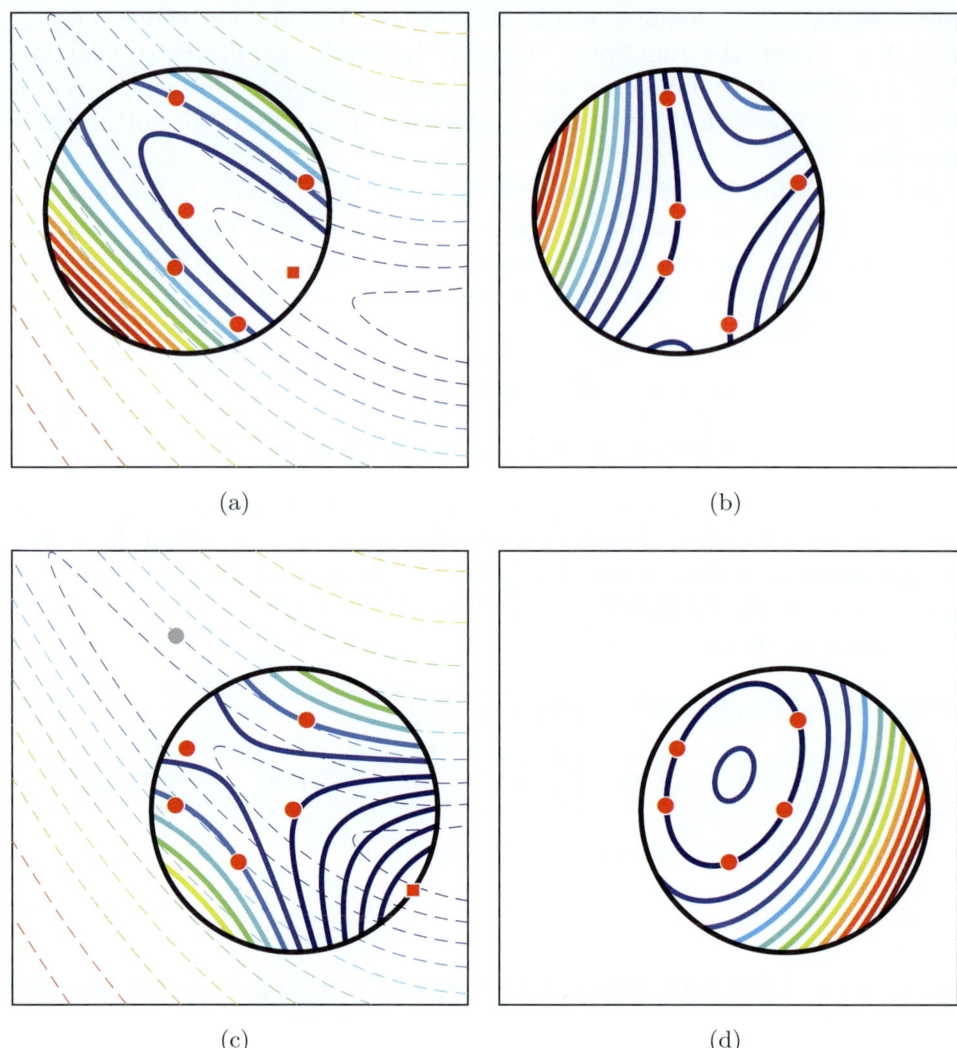

Figure 2.2. (a) Minimum-norm-Hessian model through five points in $\mathcal{B}(\boldsymbol{x}_k; \Delta_k)$ and its minimizer. (b) Absolute value of a sixth Lagrange polynomial for the five points. (c) Minimum-norm-Hessian model through five points in $\mathcal{B}(\boldsymbol{x}_{k+1}; \Delta_{k+1})$ and its minimizer. (d) Absolute value of a sixth Lagrange polynomial for the five points.

In Section 5.2, we summarize approaches that exploit knowledge of sparsity of the derivatives of f in building quadratic models that interpolate fewer than $(n+1)(n+2)/2$ points.

Figure 2.2 shows quadratic models in two dimensions that interpolate $(n+1)(n+2)/2 - 1 = 5$ points as well as the associated magnitude of the remaining Lagrange polynomial (note that this polynomial vanishes at the five interpolated points).

Regression models. Just as one can establish approximation bounds and geometry conditions when $\boldsymbol{Y}$ is linearly independent, the same can be done for overdetermined regression models (Conn *et al.* 2008*b*, Conn *et al.* 2009*b*). This can be accomplished by extending the definition of Lagrange polynomials from (2.17) to the regression case. That is, given a basis $\boldsymbol{\phi} : \mathbb{R}^n \to \mathbb{R}^{\dim(\mathcal{P}^{d,n})}$ and points $\boldsymbol{Y} = \{\boldsymbol{y}_1, \ldots, \boldsymbol{y}_p\} \subset \mathbb{R}^n$ with $p > \dim(\mathcal{P}^{d,n})$, the set of polynomials satisfies

$$\ell_j(\boldsymbol{y}_i) \stackrel{\text{l.s.}}{=} \begin{cases} 1 & \text{if } i = j, \\ 0 & \text{if } i \neq j, \end{cases} \tag{2.22}$$

where $\stackrel{\text{l.s.}}{=}$ denotes the least-squares solution. The regression model can be recovered finding the least-squares solution (now overdetermined) system from (2.14), and the definition of Λ-poisedness (in the regression sense) is equivalent to (2.18). Ultimately, given a linear regression model through a set of Λ-poised points $\boldsymbol{Y} \subset \mathcal{B}(\boldsymbol{y}_1; \Delta)$, and if f has an L_{g}-Lipschitz-continuous first derivative on an open domain containing $\mathcal{B}(\boldsymbol{y}_1; \Delta)$, then (2.9) holds on $\mathcal{B}(\boldsymbol{y}_1; \Delta)$ with

$$\kappa_{\mathrm{eg}} = \frac{5}{2}\sqrt{p}L_{\mathrm{g}}\Lambda \quad \text{and} \quad \kappa_{\mathrm{ef}} = \frac{1}{2}L_{\mathrm{g}} + \kappa_{\mathrm{eg}}. \tag{2.23}$$

Conn *et al.* (2008*b*) note the fact that the extension of Lagrange polynomials does not apply to the 1-norm or infinity-norm case. Billups, Larson and Graf (2013) show that the definition of Lagrange polynomials can be extended to the weighted regression case. Verdério, Karas, Pedroso and Scheinberg (2017) show that (2.9) can also be recovered for support vector regression models.

Efficiently minimizing the model (regardless of type) over a trust region is integral to the usefulness of such models within an optimization algorithm. In fact, this necessity is a primary reason for the use of low-degree polynomial models by the majority of derivative-free trust-region methods. For quadratic models, the resulting subproblem remains one of the most difficult non-convex optimization problems solvable in polynomial time, as illustrated by Moré and Sorensen (1983). As exemplified by Powell (1997), the implementation of subproblem solvers is a key concern in methods seeking to perform as few algebraic operations between function evaluations as possible.

2.2.3. Radial basis function interpolation models

An additional way to model non-linearity with potentially less restrictive geometric conditions is by using radial basis functions (RBFs). Such models take the form

$$m(\boldsymbol{x}) = \sum_{i=1}^{|\boldsymbol{Y}|} b_i \psi(\|\boldsymbol{x} - \boldsymbol{y}_i\|) + \boldsymbol{a}^{\mathrm{T}} \phi(\boldsymbol{x}), \tag{2.24}$$

where $\psi : \mathbb{R}_+ \to \mathbb{R}$ is a conditionally positive-definite univariate function and $\boldsymbol{a}^{\mathrm{T}}\phi(\boldsymbol{x})$ represents a (typically low-order) polynomial as before; see, for example, Buhmann (2000). Given a sample set $\boldsymbol{Y}$, RBF model coefficients $(\boldsymbol{a}, \boldsymbol{b})$ can be obtained by solving the augmented interpolation equations

$$\begin{bmatrix} \psi(\|\boldsymbol{y}_1 - \boldsymbol{y}_1\|) & \cdots & \psi(\|\boldsymbol{y}_1 - \boldsymbol{y}_{|\boldsymbol{Y}|}\|) & \phi(\boldsymbol{y}_1)^{\mathrm{T}} \\ \vdots & & \vdots & \vdots \\ \psi(\|\boldsymbol{y}_{|\boldsymbol{Y}|} - \boldsymbol{y}_1\|) & \cdots & \psi(\|\boldsymbol{y}_{|\boldsymbol{Y}|} - \boldsymbol{y}_{|\boldsymbol{Y}|}\|) & \phi(\boldsymbol{y}_{|\boldsymbol{Y}|})^{\mathrm{T}} \\ \phi(\boldsymbol{y}_1) & \cdots & \phi(\boldsymbol{y}_{|\boldsymbol{Y}|}) & \boldsymbol{0} \end{bmatrix} \begin{bmatrix} \boldsymbol{b} \\ \boldsymbol{a} \end{bmatrix} = \begin{bmatrix} f(\boldsymbol{y}_1) \\ \vdots \\ f(\boldsymbol{y}_{|\boldsymbol{Y}|}) \\ \boldsymbol{0} \end{bmatrix}. \tag{2.25}$$

That RBFs are conditionally positive-definite ensures that (2.25) is non-singular provided that the degree d of the polynomial ϕ is sufficiently large and that $\boldsymbol{Y}$ is poised for degree-d polynomial interpolation. For example, cubic $(\psi(r) = r^3)$ RBFs require a linear polynomial; multiquadric $(\psi(r) = -(\gamma^2 + r^2)^{1/2})$ RBFs require a constant polynomial; and inverse multiquadric $(\psi(r) = (\gamma^2 + r^2)^{-1/2})$ and Gaussian $(\psi(r) = \exp(-\gamma^{-2}r^2))$ RBFs do not require a polynomial. Consequently, RBFs have relatively unrestrictive geometric requirements on the interpolation points $\boldsymbol{Y}$ while allowing for modelling a wide range of non-linear behaviour.

This feature is typically exploited in global optimization (see *e.g.* Björkman and Holmström 2000, Gutmann 2001 and Regis and Shoemaker 2007), whereby an RBF surrogate model is employed to globally approximate f. However, works such as Oeuvray and Bierlaire (2009), Oeuvray (2005), Wild (2008a) and Wild and Shoemaker (2013) establish and use local approximation properties of these models. This approach is typically performed by relying on a linear polynomial $\boldsymbol{a}^{\mathrm{T}}\phi(\boldsymbol{x})$, which can be used to establish that the RBF model in (2.24) can be a fully linear local approximation of smooth f.

2.2.4. Trust-region methods

Having discussed issues of model construction, we are now ready to present a general statement of a model-based trust-region method in Algorithm 3.

A distinguishing characteristic of derivative-free model-based trust-region methods is how they manage $\boldsymbol{Y}_k$, the set of points used to construct the model m_k. Some methods ensure that $\boldsymbol{Y}_k$ contains a scaled stencil of points around $\boldsymbol{x}_k$; such an approach can be attractive since the objective at such points can be evaluated in parallel. A fixed stencil can also ensure that all models sufficiently approximate the objective. Other methods construct $\boldsymbol{Y}$ by using previously evaluated points near $\boldsymbol{x}_k$, for example, those points

---

Algorithm 3: Derivative-free model-based trust-region method

---

1 Set parameters $\epsilon > 0$, $0 < \gamma_{\mathrm{dec}} < 1 \leq \gamma_{\mathrm{inc}}$, $0 < \eta_0 \leq \eta_1 < 1$, $\Delta_{\max}$

2 Choose initial point $\boldsymbol{x}_0$, trust-region radius $0 < \Delta_0 \leq \Delta_{\max}$, and set of previously evaluated points $\boldsymbol{Y}_k$

3 **for** $k = 0, 1, 2 \ldots$ **do**

4 $\quad$ Select a subset of $\boldsymbol{Y}_k$ (or augment $\boldsymbol{Y}_k$ and evaluate f at new points) for model building

5 $\quad$ Build a model m_k using points in $\boldsymbol{Y}_k$ and their function values

6 $\quad$ **while** $\|\nabla m_k(\boldsymbol{x}_k)\| < \epsilon$ **do**

7 $\quad\quad$ **if** m_k *is accurate on* $\mathcal{B}(\boldsymbol{x}_k; \Delta_k)$ **then**

8 $\quad\quad\quad$ $\Delta_k \leftarrow \gamma_{\mathrm{dec}}\Delta_k$

9 $\quad\quad$ **else**

10 $\quad\quad\quad$ By updating $\boldsymbol{Y}_k$, make m_k accurate on $\mathcal{B}(\boldsymbol{x}_k; \Delta_k)$

11 $\quad$ Generate a direction $\boldsymbol{s}_k \in \mathcal{B}(\boldsymbol{0}; \Delta_k)$ so that $\boldsymbol{x}_k + \boldsymbol{s}_k$ approximately minimizes m_k on $\mathcal{B}(\boldsymbol{x}_k; \Delta_k)$

12 $\quad$ Evaluate $f(\boldsymbol{x}_k + \boldsymbol{s}_k)$ and $\rho_k \leftarrow \dfrac{f(\boldsymbol{x}_k) - f(\boldsymbol{x}_k + \boldsymbol{s}_k)}{m_k(\boldsymbol{x}_k) - m_k(\boldsymbol{x}_k + \boldsymbol{s}_k)}$

13 $\quad$ **if** $\rho_k < \eta_1$ *and* m_k *is inaccurate on* $\mathcal{B}(\boldsymbol{x}_k; \Delta_k)$ **then**

14 $\quad\quad$ Add model improving point(s) to $\boldsymbol{Y}_k$

15 $\quad$ **if** $\rho_k \geq \eta_1$ **then**

16 $\quad\quad$ $\Delta_{k+1} \leftarrow \min\{\gamma_{\mathrm{inc}}\Delta_k, \Delta_{\max}\}$

17 $\quad$ **else if** m_k *is accurate on* $\mathcal{B}(\boldsymbol{x}_k; \Delta_k)$ **then**

18 $\quad\quad$ $\Delta_{k+1} \leftarrow \gamma_{\mathrm{dec}}\Delta_k$

19 $\quad$ **else**

20 $\quad\quad$ $\Delta_{k+1} \leftarrow \Delta_k$

21 $\quad$ **if** $\rho_k \geq \eta_0$ **then** $\boldsymbol{x}_{k+1} \leftarrow \boldsymbol{x}_k + \boldsymbol{s}_k$ **else** $\boldsymbol{x}_{k+1} \leftarrow \boldsymbol{x}_k$

22 $\quad$ $\boldsymbol{Y}_{k+1} \leftarrow \boldsymbol{Y}_k$

---

within $\mathcal{B}(\boldsymbol{x}_k; c_1\Delta_k)$ for some constant $c_1 \in [1, \infty)$. Depending on the set of previously evaluated points, such methods may need to add points to $\boldsymbol{Y}_k$ that most improve the model quality. Determining which additional points to add to $\boldsymbol{Y}_k$ can be computationally expensive, but the method should be willing to do so in the hope of needing fewer evaluations of the objective function at new points in $\boldsymbol{Y}_k$. Most methods do not ensure that models are valid on every iteration but rather make a single step toward improving the model. Such an approach can ensure a high-quality model in a finite number of improvement steps. (Exceptional methods that ensure model quality before $\boldsymbol{s}_k$ is calculated are the methods of Powell and manifold sampling of Khan, Larson and Wild (2018).) The **ORBIT** method (Wild,

Regis and Shoemaker 2008) places a limit on the size of $\boldsymbol{Y}_k$ (*e.g.* in order to limit the amount of linear algebra or to prevent overfitting). In the end, such restrictions on $\boldsymbol{Y}_k$ may determine whether m_k is an interpolation or regression model.

Derivative-free trust-region methods share many similarities with traditional trust-region methods, for example, the use of a ρ-test to determine whether a step is taken or rejected. As in a traditional trust-region method, the ρ-test measures the ratio of actual decrease observed in the objective versus the decrease predicted by the model.

On the other hand, the management of the trust-region radius parameter Δ_k in Algorithm 3 differs remarkably from traditional trust-region methods. Derivative-free variants require an additional test of model quality, the failure of which results in shrinking Δ_k. When derivatives are available, Taylor's theorem ensures model accuracy for small Δ_k. In the derivative-free case, such a condition must be explicitly checked in order to ensure that Δ_k does not go to zero merely because the model is poor, hence the inclusion of tests of model quality. As a direct result of these considerations, $\Delta_k \to 0$ as Algorithm 3 converges; this is generally not the case in traditional trust-region methods.

As in derivative-based trust-region methods, the solution to the trust-region subproblem in line 11 of Algorithm 3 must satisfy a *Cauchy decrease condition*. Given the model m_k used in Algorithm 3, we define the optimal step length in the direction $-\nabla m_k(\boldsymbol{x}_k)$ by

$$t_k^C = \underset{t \geq 0 : \boldsymbol{x}_k - t\nabla m_k(\boldsymbol{x}_k) \in \mathcal{B}(\boldsymbol{x}_k;\Delta_k)}{\arg\min} m_k(\boldsymbol{x}_k - t\nabla m_k(\boldsymbol{x}_k)),$$

and the corresponding *Cauchy step*

$$\boldsymbol{s}_k^C = -t_k^C \nabla m_k(\boldsymbol{x}_k).$$

It is straightforward to show (see *e.g.* Conn, Scheinberg and Vicente 2009*b*, Theorem 10.1) that

$$m_k(\boldsymbol{x}_k) - m_k(\boldsymbol{x}_k + \boldsymbol{s}_k^C) \geq \frac{1}{2}\|\nabla m_k(\boldsymbol{x}_k)\| \min\left\{ \frac{\|\nabla m_k(\boldsymbol{x}_k)\|}{\|\nabla^2 m_k(\boldsymbol{x}_k)\|}, \Delta_k \right\}. \quad (2.26)$$

That is, (2.26) states that, provided that both $\Delta_k \approx \|\nabla m_k(\boldsymbol{x}_k)\|$ and a uniform bound exists on the norm of the model Hessian, the model decrease attained by the Cauchy step $\boldsymbol{s}_k^C$ is of the order of Δ_k^2. In order to prove convergence, it is desirable to ensure that each step $\boldsymbol{s}_k$ generated in line 11 of Algorithm 3 decreases the model m_k by no less than $\boldsymbol{s}_k^C$ does, or at least some fixed positive fraction of the decrease achieved by $\boldsymbol{s}_k^C$. Because successful iterations ensure that the actual decrease attained in an iteration is at least a constant fraction of the model decrease, the sequence of decreases of Algorithm 3 are square-summable, provided that $\Delta_k \to 0$. (This is indeed the case for derivative-free trust-region methods.) Hence, in most theoretical treatments of these methods, it is commonly stated as an assumption that the subproblem solution s_k obtained in line 11 of Algorithm 3 satisfies

$$m_k(\boldsymbol{x}_k) - m_k(\boldsymbol{x}_k + \boldsymbol{s}_k) \geq \kappa_{\text{fcd}}(m_k(\boldsymbol{x}_k) - m_k(\boldsymbol{x}_k + \boldsymbol{s}_k^C)), \tag{2.27}$$

where $\kappa_{\text{fcd}} \in (0, 1]$ is the fraction of the Cauchy decrease. In practice, when m_k is a quadratic model, subproblem solvers have been well studied and often come with guarantees concerning the satisfaction of (2.27) (Conn, Gould and Toint 2000). Wild *et al.* (2008, Figure 4.3) demonstrate the satisfaction of an assumption like (2.27) when the model m_k is a radial basis function.

Under reasonable smoothness assumptions, most importantly $f \in \mathcal{LC}^1$, algorithms in the Algorithm 3 framework have been shown to be first-order convergent (*i.e.* (2.4)) and second-order convergent (*i.e.* (2.6)), with the (arguably) most well-known proof given by Conn, Scheinberg and Vicente (2009*a*). In more recent work, Garmanjani, Jùdice and Vicente (2016) provide a WCC bound of the form (2.2) for Algorithm 3, recovering essentially the same upper bound on the number of function evaluations required by DDS methods found in Vicente (2013), that is, a WCC bound in $O(\epsilon^{-2})$ (see Table A.1). When $f \in \mathcal{LC}^2$, Gratton, Royer and Vicente (2019*a*) demonstrate a second-order WCC bound of the form (2.3) in $O(\epsilon^{-3})$; in order to achieve this result, fully quadratic models m_k are required. In Section 3.3, a similar result is achieved by using randomized variants that do not require a fully quadratic model in every iteration.

Early analysis of Powell's **UOBYQA** method shows that, with minor modifications, the algorithm can converge superlinearly in neighbourhoods of strict convexity (Han and Liu 2004). A key distinction between Powell's methods and other model-based trust-region methods is the use of separate neighbourhoods for model quality and trust-region steps, with each of these neighbourhoods changing dynamically. Convergence of such methods is addressed by Powell (2010, 2012).

The literature on derivative-free trust-region methods is extensive. We mention in passing several additional classes of trust-region methods that have not fallen neatly into our discussion thus far. *Wedge methods* (Marazzi and Nocedal 2002) explicitly enforce geometric properties (Λ-poisedness) of the sample set between iterations by adding additional constraints to the trust-region subproblem. Alexandrov, Dennis, Jr, Lewis and Torczon (1998) consider a trust-region method utilizing a hierarchy of model approximations. In particular, if derivatives can be obtained but are expensive, then the method of Alexandrov *et al.* (1998) uses a model that interpolates not only zeroth-order information but also first-order (gradient) information. For problems with deterministic noise, Elster and Neumaier (1995) propose a method that projects the solutions of a trust-region subproblem onto a dynamically refined grid, encouraging better practical behaviour. Similarly, for problems with deterministic noise, Maggiar, Wächter, Dolinskaya and Staum (2018) propose a model-based trust-region method that implicitly convolves the objective function with a Gaussian kernel, again yielding better practical behaviour.

2.3. Hybrid methods and miscellanea

While the majority of work in derivative-free methods for deterministic problems can be classified as direct-search or model-based methods, some work defies this simple classification. In fact, several works (Conn and Le Digabel 2013, Custódio *et al.* 2009, Dennis, Jr and Torczon 1997, Frimannslund and Steihaug 2011) propose methods that seem to hybridize these two classes, existing somewhere in the intersection. For example, Custódio and Vicente (2005) and Custódio *et al.* (2009) develop the SID-PSM method, which extends Algorithm 2 so that the search step consists of minimizing an approximate quadratic model of the objective (obtained either by minimum-Frobenius norm interpolation or by regression) over a trust region. Here, we highlight methods that do not neatly belong to the two aforementioned classes of methods.

2.3.1. Finite differences

As noted in Section 1.1.2, many of the earliest derivative-free methods employed finite-difference-based estimates of derivatives. The most popular first-order directional derivative estimates include the forward/reverse difference

$$\delta_{\mathrm{f}}(f; \boldsymbol{x}; \boldsymbol{d}; h) = \frac{f(\boldsymbol{x} + h\boldsymbol{d}) - f(\boldsymbol{x})}{h} \tag{2.28}$$

and central difference

$$\delta_{\mathrm{c}}(f; \boldsymbol{x}; \boldsymbol{d}; h) = \frac{f(\boldsymbol{x} + h\boldsymbol{d}) - f(\boldsymbol{x} - h\boldsymbol{d})}{2h}, \tag{2.29}$$

where $h \neq 0$ is the difference parameter and the non-trivial $\boldsymbol{d} \in \mathbb{R}^n$ defines the direction. Several recent methods, including the methods described in Sections 2.3.2, 2.3.3 and 3.1.2, use such estimates and employ difference parameters or directions that dynamically change.

As an example of a potentially dynamic choice of difference parameter, we consider the usual case of roundoff errors. We denote by $f'_{\infty}(\boldsymbol{x}; \boldsymbol{d})$ the directional derivative at $\boldsymbol{x}$ of the infinite-precision (*i.e.* based on real arithmetic) objective function f_{∞} in the unit direction $\boldsymbol{d}$ (*i.e.* $\|\boldsymbol{d}\| = 1$). We then have the following error for forward or reverse finite-difference estimates based on the function f available through computation:

$$|\delta_{\mathrm{f}}(f; \boldsymbol{x}; \boldsymbol{d}; h) - f'_{\infty}(\boldsymbol{x}; \boldsymbol{d})| \leq \frac{1}{2} L_{\mathrm{g}}(\boldsymbol{x})|h| + 2\frac{\epsilon_{\infty}(\boldsymbol{x})}{|h|}, \tag{2.30}$$

provided that $|f''_\infty(\cdot; \boldsymbol{d})| \le L_{\mathrm{g}}(\boldsymbol{x})$ and $|f_\infty(\cdot) - f(\cdot)| \le \epsilon_\infty(\boldsymbol{x})$ on the interval $[\boldsymbol{x}, \boldsymbol{x} + h\boldsymbol{d}]$. In Gill, Murray and Wright (1981) and Gill, Murray, Saunders and Wright (1983), the recommended difference parameter is $h = 2\sqrt{\epsilon_\infty(\boldsymbol{x})/L_{\mathrm{g}}(\boldsymbol{x})}$, which yields the minimum value $2\sqrt{\epsilon_\infty(\boldsymbol{x})L_{\mathrm{g}}(\boldsymbol{x})}$ of the upper bound in (2.30); when ϵ_∞ is a bound on the roundoff error and L_{g} is of order one, then the familiar $h \in O(\sqrt{\epsilon_\infty})$ is obtained.

Similarly, if one models the error between f_∞ and f as a stationary stochastic process (through the ansatz denoted by $f_{\boldsymbol{\xi}}$) with variance $\epsilon_{\mathrm{f}}(\boldsymbol{x})^2$, minimizing the upper bound on the mean-squared error,

$$\mathbb{E}_{\boldsymbol{\xi}}[(\delta_{\mathrm{f}}(f_{\boldsymbol{\xi}}; \boldsymbol{x}; \boldsymbol{d}; h) - f'_\infty(\boldsymbol{x}; \boldsymbol{d}))^2] \le \frac{1}{4}L_{\mathrm{g}}(\boldsymbol{x})^2 h^2 + 2\frac{\epsilon_{\mathrm{f}}(\boldsymbol{x})^2}{h^2}, \qquad (2.31)$$

yields the choice $h = (\sqrt{8}\epsilon_{\mathrm{f}}(\boldsymbol{x})/L_{\mathrm{g}}(\boldsymbol{x}))^{1/2}$ with an associated root-mean-squared error of $(\sqrt{2}\epsilon_\infty(\boldsymbol{x})L_{\mathrm{g}}(\boldsymbol{x}))^{1/2}$; see, for example, Moré and Wild (2012, 2014). A rough procedure for computing ϵ_{f} is provided in Moré and Wild (2011) and used in recent methods such as that of Berahas, Byrd and Nocedal (2019).

In both cases (2.30) and (2.31), the first-order error is $c\sqrt{\epsilon(\boldsymbol{x})L_{\mathrm{g}}(\boldsymbol{x})}$ (for a constant $c \le 2$), which can be used to guide the decision on whether the derivatives estimates are of sufficient accuracy.

2.3.2. Implicit filtering

Implicit filtering is a hybrid of a grid-search algorithm (evaluating all points on a lattice) and a Newton-like local optimization method. The gradient (and possible Hessian) estimates for local optimization are approximated by the central differences $\{\delta_{\mathrm{c}}(f; \boldsymbol{x}_k; \boldsymbol{e}_i; \Delta_k) : i = 1, \ldots, n\}$. The difference parameter Δ_k decreases when implicit filtering encounters a *stencil failure* at $\boldsymbol{x}_k$, that is,

$$f(\boldsymbol{x}_k) \le f(\boldsymbol{x}_k \pm \Delta_k \boldsymbol{e}_i), \qquad (2.32)$$

where $\boldsymbol{e}_i$ is the ith elementary basis vector. This is similar to direct-search methods, but notice that implicit filtering is not polling opportunistically: all polling points are evaluated on each iteration. The basic version of implicit filtering from Kelley (2011) is outlined in Algorithm 4. Note that most implementations of implicit filtering require a bound-constrained domain.

Considerable effort has been devoted to extensions of Algorithm 4 when f is 'noisy'. Gilmore and Kelley (1995) show that implicit filtering converges to local minima of (DET) when the objective f is the sum of a smooth function f_s and a high-frequency, low-amplitude function f_n, with $f_n \to 0$ quickly in a neighbourhood of all minimizers of f_s. Under similar assumptions, Choi and Kelley (2000) show that Algorithm 4 converges superlinearly if the step sizes Δ_k are defined as a power of the norm of the previous iteration's gradient approximation.

Algorithm 4: Implicit-filtering method

---

1   Set parameters `feval_max` > 0, $\Delta_{\min} > 0$, $\gamma_{\text{dec}} \in (0,1)$ and $\tau > 0$

2   Choose initial point $\boldsymbol{x}_0$ and step size $\Delta_0 \geq \Delta_{\min}$

3   $k \leftarrow 0$; evaluate $f(\boldsymbol{x}_0)$ and set `fevals` $\leftarrow 1$

4   **while** `fevals` $\leq$ `feval_max` *and* $\Delta_k \geq \Delta_{\min}$ **do**

5   Evaluate $f(\boldsymbol{x}_k \pm \Delta_k \boldsymbol{e}_i)$ for $i \in \{1, \ldots, n\}$ and approximate $\nabla f(\boldsymbol{x}_k)$
       via $\{\delta_{\text{c}}(f; \boldsymbol{x}_k; \boldsymbol{e}_i; \Delta_k) : i = 1, \ldots, n\}$

6   **if** *equation* (2.32) *is satisfied or* $\|\nabla f(\boldsymbol{x}_k)\| \leq \tau \Delta_k$ **then**

7          $\Delta_{k+1} \leftarrow \gamma_{\text{dec}} \Delta_k$

8          $\boldsymbol{x}_{k+1} \leftarrow \boldsymbol{x}_k$

9   **else**

10   Update Hessian estimate $\boldsymbol{H}_k$ (or set $\boldsymbol{H}_k \leftarrow \boldsymbol{I}$)

11         $\boldsymbol{s}_k \leftarrow -\boldsymbol{H}_k^{-1} \nabla f(\boldsymbol{x}_k)$

12   Perform a line search in the direction $\boldsymbol{s}_k$ to generate $\boldsymbol{x}_{k+1}$

13         $\Delta_{k+1} \leftarrow \Delta_k$

14      $k \leftarrow k + 1$

2.3.3. Adaptive regularized methods

Cartis, Gould and Toint (2012) perform an analysis of adaptive regularized cubic (ARC) methods and propose a derivative-free method, ARC-DFO. ARC-DFO is an extension of ARC whereby gradients are replaced with central finite differences of the form (2.29), with the difference parameter monotonically decreasing within a single iteration of the method. ARC-DFO is an intrinsically model-based method akin to Algorithm 3, but the objective within each subproblem regularizes third-order behaviour of the model. Thus, like a trust-region method, ARC-DFO employs trial steps and model gradients. During the main loop of ARC-DFO, if the difference parameter exceeds a constant factor of the minimum of the trial step norm or the model gradient norm, then the difference parameter is shrunk by a constant factor, and the iteration restarts to obtain a new trial step. This mechanism is structurally similar to a derivative-free trust-region method's checks on model quality. Cartis *et al.* (2012) show that ARC-DFO demonstrates a WCC result of type (2.2) in $O(\epsilon^{-3/2})$, the same asymptotic result (in terms of ϵ-dependence) that the authors demonstrate for derivative-based variants of ARC methods. In terms of dependence on ϵ, this result is a strict improvement over the WCC results of the same type demonstrated for DDS and trust-region methods, although this result is proved under the stronger assumption that $f \in \mathcal{LC}^2$.

In a different approach, Hare and Lucet (2013) show convergence of a derivative-free method that penalizes large steps via a proximal regularizer, thereby removing the necessity for a trust region. Lazar and Jarre (2016) regularize their line-search with a term seeking to minimize a weighted change of the model's third derivatives.

2.3.4. Line-search-based methods

Several line-search-based methods for derivative-free optimization have been developed. Grippo, Lampariello and Lucidi (1988) and De Leone, Gaudioso and Grippo (1984) (two of the few papers appearing in the 1980s concerning derivative-free optimization) both analyse conditions on the step sizes used in a derivative-free line-search algorithm, and provide methods for constructing such steps. Lucidi and Sciandrone (2002b) present methods that combine pattern-search and line-search approaches in a convergent framework. The VXQR method of Neumaier, Fendl, Schilly and Leitner (2011) performs a line search on a direction computed from a QR factorization of previously evaluated points. Neumaier et $al.$ (2011) apply VXQR to problems with $n = 1000$, a large problem dimension among the methods considered here.

Consideration has also been given to non-monotone line-search-based derivative-free methods. Since gradients are not available in derivative-free optimization, the search direction in a line-search method may not be a descent direction. Non-monotone methods allow one to still employ such directions in a globally convergent framework. Grippo and Sciandrone (2007) extend line-search strategies based on coordinate search and the method of Barzilai and Borwein (1988) to develop a globally convergent non-monotone derivative-free method. Grippo and Rinaldi (2014) extend such non-monotone strategies to broader classes of algorithms that employ simplex gradients, hence further unifying direct-search and model-based methods. Another non-monotone line-search method is proposed by Diniz-Ehrhardt, Martínez and Raydan (2008), who encapsulate early examples of randomized DDS methods (Section 3.2).

2.3.5. Methods for non-smooth optimization

In Section 2.1.2, we discuss how MADS handles non-differentiable objective functions by densely sampling directions on a mesh, thereby ensuring that all Clarke directional derivatives are non-negative ($i.e.$ (2.1)). Another early analysis of a DDS method on a class of non-smooth objectives was performed by García-Palomares and Rodríguez (2002).

Gradient sampling methods are a developing class of algorithms for general non-smooth non-convex optimization; see the recent survey by Burke et $al.$ (2018). These methods attempt to estimate the ϵ-$subdifferential$ at a point $\boldsymbol{x}$ by evaluating a random sample of gradients in the neighbourhood of $\boldsymbol{x}$ and constructing the convex hull of these gradients. In a derivative-free setting, the approximation of these gradients is not as immediately obvious in the presence of non-smoothness, but there exist gradient-sampling methods that use finite-difference estimates with specific smoothing techniques (Kiwiel 2010).

In another distinct line of research, Bagirov, Karasözen and Sezer (2007) analyse a derivative-free variant of subgradient descent, where subgradients are approximated via so-called discrete gradients. In Section 5.3, we will further discuss methods for minimizing *composite non-smooth objective functions* of the form $f = h \circ F$, where h is non-smooth but a closed-form expression is known and F is assumed smooth. These methods are characterized by their exploitation of the knowledge of h, making them less general than the methods for non-smooth optimization discussed so far.

3. Randomized methods for deterministic objectives

We now summarize randomized methods for solving (DET). Such methods often have promising theoretical properties, although some practitioners may dislike the non-deterministic behaviour of these methods. We discuss randomization within direct-search methods in Section 3.2 and within trust-region methods in Section 3.3, but we first begin with a discussion of random search as applied to deterministic objectives.

In any theoretical treatment of randomized methods, one must be careful to distinguish between random variables and their realizations. For the sake of terseness in this survey, we will intentionally conflate variables with realizations and refer to respective papers for more careful statements of theoretical results.

3.1. Random search

We highlight two randomized methods for minimizing a deterministic objective: pure random search and Nesterov random search.

3.1.1. Pure random search

Pure random search is a natural method to start with for randomized derivative-free optimization. Pure random search is popular for multiple reasons; in particular, it is easy to implement (with few or no user-defined tolerances), and (if the points generated are independent of one another) it exhibits perfect scaling in terms of evaluating f at many points simultaneously.

A pure random-search method is given in Algorithm 5, where points are generated randomly from Ω. For example, if $\Omega = \{x : c(x) \leq 0, l \leq x \leq u\}$, line 3 of Algorithm 5 may involve drawing points uniformly at random from $[l, u]$ and checking whether they satisfy $c(x) \leq 0$. If the procedure for generating points in line 3 of Algorithm 5 is independent of the function values observed, then the entire set of points used within pure random

---

Algorithm 5: Pure random search

---

1   Choose initial point $\hat{\boldsymbol{x}} \in \boldsymbol{\Omega}$, termination test, and point generation scheme

2   **while** *Termination test is not satisfied* **do**

3   Generate $\boldsymbol{x} \in \boldsymbol{\Omega}$

4   **if** $f(\boldsymbol{x}) < f(\hat{\boldsymbol{x}})$ **then**

5       $\hat{\boldsymbol{x}} \leftarrow \boldsymbol{x}$

---

search can be generated beforehand: we intentionally omit the index k in the statement of Algorithm 5.

Nevertheless, for the sake of analysis, it is useful to consider an ordering of the sequence of random points generated by Algorithm 5. With such a sequence $\{\boldsymbol{x}_k\}$, one can analyse the best points after N evaluations,

$$\hat{\boldsymbol{x}}_N \in \underset{k=1,\dots,N}{\arg\min} \, f(\boldsymbol{x}_k).$$

If f_* is the global minimum value, then

$$\mathbb{P}[f(\hat{\boldsymbol{x}}_N) \leq f_* + \epsilon] = 1 - \prod_{k=1}^{N}(1 - \mathbb{P}[\boldsymbol{x}_k \in \boldsymbol{\mathcal{L}}_{f_*+\epsilon}(f)]),$$

where $\epsilon \geq 0$ and $\boldsymbol{\mathcal{L}}_\alpha(f) = \{\boldsymbol{x} : f(\boldsymbol{x}) \leq \alpha\}$. Provided that the procedure used to generate points at line 3 of Algorithm 5 satisfies

$$\lim_{N\to\infty} \prod_{k=1}^{N}(1 - \mathbb{P}[\boldsymbol{x}_k \in \boldsymbol{\mathcal{L}}_{f_*+\epsilon}(f)]) = 0$$

for all $\epsilon > 0$, then $f(\hat{\boldsymbol{x}}_k)$ converges in probability to f_*. For example, if each $\boldsymbol{x}_k$ is drawn independently and uniformly over $\boldsymbol{\Omega}$, then one can calculate the number of evaluations required to ensure that the $\hat{\boldsymbol{x}}_k$ returned by Algorithm 5 satisfies $\hat{\boldsymbol{x}}_k \in \boldsymbol{\mathcal{L}}_{f_*+\epsilon}$ with probability $p \in (0,1)$, that is,

$$N \geq \frac{\log(p)}{\log\left(1 - \dfrac{\mu(\boldsymbol{\mathcal{L}}_{f_*+\epsilon} \bigcap \boldsymbol{\Omega})}{\mu(\boldsymbol{\Omega})}\right)},$$

provided $\mu(\boldsymbol{\mathcal{L}}_{f(\boldsymbol{x}_*)+\epsilon} \cap \boldsymbol{\Omega}) > 0$ and $\boldsymbol{\Omega}$ is measurable.

Random-search methods typically make few assumptions about f; see Zhigljavsky (1991) for further discussion about the convergence of pure random search. Naturally, a method that assumes only that f is measurable on $\boldsymbol{\Omega}$ is likely to produce function values that converge more slowly to f_* when applied to an $f \in \mathcal{C}^0$ than does a method that exploits the continuity of f. Heuristic modifications of random search have sought to improve empirical performance on certain classes of problems, while still maintaining random search's global optimization property; see, for example, the work of Zabinsky and Smith (1992) and Patel, Smith and Zabinsky (1989).

3.1.2. Nesterov random search

We refer to the method discussed in this section as Nesterov random search because of the seminal article by Nesterov and Spokoiny (2017), but the idea driving this method is much older. A similar method, for instance, is discussed in Polyak (1987, Chapter 3.4).

The method of Nesterov random search is largely motivated by Gaussian smoothing. In particular, given a covariance (*i.e.* symmetric positive-definite) matrix $\boldsymbol{B}$ and a smoothing parameter $\mu > 0$, consider the Gaussian smoothed function

$$f_\mu(\boldsymbol{x}) = \sqrt{\frac{\det(\boldsymbol{B})}{(2\pi)^n}} \int_{\mathbb{R}^n} f(\boldsymbol{x} + \mu\boldsymbol{u}) \exp\left(-\frac{1}{2}\boldsymbol{u}^\mathrm{T}\boldsymbol{B}\boldsymbol{u}\right) \mathrm{d}\boldsymbol{u}.$$

This smoothing has many desirable properties; for instance, if f is Lipschitz-continuous with constant L_f, then f_μ is Lipschitz-continuous with a constant no worse than L_f for all $\mu > 0$. Likewise, if f has Lipschitz-continuous gradients with constant L_g, then f_μ has Lipschitz-continuous gradients with a constant no worse than L_g for all $\mu > 0$. If f is convex, then f_μ is convex.

One can show that

$$\nabla f_\mu(\boldsymbol{x}) = \frac{1}{\mu}\sqrt{\frac{\det(\boldsymbol{B})}{(2\pi)^n}} \int_{\mathbb{R}^n} (f(\boldsymbol{x} + \mu\boldsymbol{u}) - f(\boldsymbol{x})) \exp\left(-\frac{1}{2}\boldsymbol{u}^\mathrm{T}\boldsymbol{B}\boldsymbol{u}\right) \boldsymbol{B}\boldsymbol{u} \,\mathrm{d}\boldsymbol{u}.$$

In other words, $\nabla f_\mu(\boldsymbol{x})$, which can be understood as an approximation of $\nabla f(\boldsymbol{x})$ in the smooth case, can be computed via an expectation over $\boldsymbol{u} \in \mathbb{R}^n$ weighted by the finite difference $f(\boldsymbol{x} + \mu\boldsymbol{u}) - f(\boldsymbol{x})$ and inversely weighted by a radial distance from $\boldsymbol{x}$. With this interpretation in mind, Nesterov and Spokoiny (2017) propose a collection of *random gradient-free oracles*, where one first generates a Gaussian random vector $\boldsymbol{u} \in \mathcal{N}(\boldsymbol{0}, \boldsymbol{B}^{-1})$ and then uses one of

$$\begin{aligned} \boldsymbol{g}_\mu(\boldsymbol{x}; \boldsymbol{u}) &= \delta_\mathrm{f}(f; \boldsymbol{x}; \boldsymbol{u}; \mu)\boldsymbol{B}\boldsymbol{u}, \quad \text{or} \\ \hat{\boldsymbol{g}}_\mu(\boldsymbol{x}; \boldsymbol{u}) &= \delta_\mathrm{c}(f; \boldsymbol{x}; \boldsymbol{u}; \mu)\boldsymbol{B}\boldsymbol{u}, \end{aligned} \tag{3.1}$$

for a difference parameter $\mu > 0$. Nesterov and Spokoiny also propose a third oracle, $\boldsymbol{g}_0(\boldsymbol{x}; \boldsymbol{u}) = f'(\boldsymbol{x}; \boldsymbol{u})\boldsymbol{B}\boldsymbol{u}$, intended for the optimization of non-smooth functions; this oracle assumes the ability to compute directional derivatives $f'(\boldsymbol{x}; \boldsymbol{u})$. For this reason, Nesterov and Spokoiny refer to all oracles as gradient-free instead of derivative-free. Given the scope of this survey, we focus on the derivative-free oracles $\boldsymbol{g}_\mu$ and $\hat{\boldsymbol{g}}_\mu$ displayed in (3.1).

With an oracle $\boldsymbol{g}$ chosen as either oracle in (3.1), Nesterov random-search methods are straightforward to define, and we do so in Algorithm 6. In Algorithm 6, $\mathrm{proj}(\cdot; \boldsymbol{\Omega})$ denotes projection onto a domain $\boldsymbol{\Omega}$.

Algorithm 6: Nesterov random search

---

1   Choose initial point $\boldsymbol{x}_0 \in \boldsymbol{\Omega}$, sequence of step sizes $\{\alpha_k\}_{k=0}^{\infty}$, oracle $\boldsymbol{g}$
    from (3.1), smoothing parameter $\mu > 0$ and covariance matrix $\boldsymbol{B}$

2   **for** $k = 0, 1, 2, \ldots$ **do**

3      $\hat{\boldsymbol{x}}_k \leftarrow \arg\min_{j \in \{0,1,\ldots,k\}} f(\boldsymbol{x}_j)$

4   Generate $\boldsymbol{u}_k \in \mathcal{N}(\boldsymbol{0}, \boldsymbol{B}^{-1})$; compute $\boldsymbol{g}(\boldsymbol{x}_k; \boldsymbol{u}_k)$

5      $\boldsymbol{x}_{k+1} \leftarrow \mathrm{proj}(\boldsymbol{x}_k - \alpha_k \boldsymbol{B}^{-1} \boldsymbol{g}(\boldsymbol{x}_k; \boldsymbol{u}_k), \boldsymbol{\Omega})$

A particularly striking result proved in Nesterov and Spokoiny (2017) was perhaps the first WCC result for an algorithm (Algorithm 6) in the case where $f \in \mathcal{LC}^0$ – that is, f may be both non-smooth and non-convex. Because of the randomized nature of iteratively sampling from a Gaussian distribution in Algorithm 6, complexity results are given as expectations. That is, letting $\boldsymbol{U}_k = \{\boldsymbol{u}_0, \boldsymbol{u}_1, \ldots, \boldsymbol{u}_k\}$ denote the random variables associated with the first k iterations of Algorithm 6, complexity results are stated in terms of expectations with respect to the filtration defined by these variables. A WCC is given as an upper bound on the number of f evaluations needed to attain the approximate ($\epsilon > 0$) optimality condition

$$\mathbb{E}_{\boldsymbol{U}_{k-1}}[\|\nabla f_{\check{\mu}}(\hat{\boldsymbol{x}}_k)\|] \leq \epsilon, \tag{3.2}$$

where $\hat{\boldsymbol{x}}_k = \arg\min_{j=0,1,\ldots,k-1} f(\boldsymbol{x}_j)$. By fixing a particular choice of $\check{\mu}$ (dependent on ϵ, n and Lipschitz constants), Nesterov and Spokoiny (2017) demonstrate that the number of f evaluations needed to attain (3.2) is in $O(\epsilon^{-3})$; see Table A.1. For $f \in \mathcal{LC}^1$ (but still non-convex), Nesterov and Spokoiny (2017) prove a WCC result of type (3.2) in $O(\epsilon^{-2})$ for the same method. WCC results of Algorithm 6 under a variety of stronger assumptions on the convexity and differentiability of f are also shown in Table A.1 and discussed in Section 4. We further note that some randomized methods of the form Algorithm 6 have also been developed for (STOCH), which we discuss in Section 6.

We remark on an undesirable feature of the convergence analysis for variants of Algorithm 6: the analysis of these methods supposes that the sequence $\{\alpha_k\}$ is chosen as a constant that depends on parameters, including L_{f}, that may not be available to the method. Similar assumptions concerning the preselection of $\{\alpha_k\}$ also appear in the convex cases discussed in Section 4, and we highlight these dependencies in Table A.1.

3.2. Randomized direct-search methods

Randomization has also been used in the DDS framework discussed in Section 2.1 in the hope of more efficiently using evaluations of f. Polling every point in a PSS requires at least $n + 1$ function evaluations; if f is expensive to evaluate and n is relatively large, this can be wasteful. A deterministic strategy for performing fewer evaluations on many iterations is opportunistic polling. Work in randomized direct-search methods attempts to address, formalize and analyse the situation where polling directions are *randomly sampled* from some distribution in each iteration. The ultimate goal is to replace the $O(n)$ per-iteration function evaluation cost with an $O(1)$ per-iteration cost,[4] while still guaranteeing some form of global convergence.

In Section 2.1, we mentioned MADS methods that consider the random generation of polling directions in each iteration (in order to satisfy the asymptotic density required of search directions for the minimization of non-smooth, but Lipschitz-continuous, f). Examples include Audet and Dennis, Jr (2006) and Van Dyke and Asaki (2013), which implement LTMADS and QRMADS, respectively. While this direction of research is within the scope of randomized methods, the purpose of randomization in MADS methods is to overcome particular difficulties encountered when optimizing general non-smooth objectives. This particular randomization does not fall within the scope of this section, where randomization is intended to decrease a method's dependence on n. In the remainder of this section, we focus on a body of work that seems to exist entirely for the unconstrained case where f is assumed sufficiently smooth.

Gratton, Royer, Vicente and Zhang (2015) extend the direct-search framework (Algorithm 2) by assuming that the set of polling directions $\boldsymbol{D}_k$ includes only a descent direction with probability p (as opposed to assuming $\boldsymbol{D}_k$ always includes a descent direction, which comes for free when $f \in \mathcal{LC}^1$ provided $\boldsymbol{D}_k$ is, for example, a PSS). To formalize, given $p \in (0, 1)$, a random sequence of polling directions $\{\boldsymbol{D}_k\}$ is said to be p-probabilistically κ_{d}-descent provided that, given a deterministic starting point $\boldsymbol{x}_0$,

$$\mathbb{P}[\mathrm{cm}([\boldsymbol{D}_0, -\nabla f(\boldsymbol{x}_0)]) \geq \kappa_{\mathrm{d}}] \geq p, \tag{3.3}$$

and for all $k \geq 1$,

$$\mathbb{P}[\mathrm{cm}([\boldsymbol{D}_k, -\nabla f(\boldsymbol{x}_k)]) \geq \kappa_{\mathrm{d}} \mid \boldsymbol{D}_0, \ldots, \boldsymbol{D}_{k-1}] \geq p, \tag{3.4}$$

where $\mathrm{cm}(\cdot)$ is the cosine measure in (2.8). A collection of polling directions $\boldsymbol{D}_k$ satisfying (3.3) and (3.4) can be obtained by drawing directions uniformly on the unit ball.

As with the other methods in this section, $\boldsymbol{x}_k$ in (3.4) is in fact a random variable due to the random sequence $\{\boldsymbol{D}_k\}$ generated by the algorithm, and hence it makes sense to view (3.4) as a probabilistic statement. In words, (3.4) states that with probability at least p, the set of polling directions

---

[4] We note that a $O(1)$ cost can naturally be achieved by deterministic DDS methods when derivatives are available (Abramson *et al.* 2004).

used in iteration k has a positive cosine measure with the steepest descent direction $-\nabla f(\boldsymbol{x}_k)$, regardless of the past history of the algorithm. Gratton *et al.* (2015) use Chernoff bounds in order to bound the worst-case complexity *with high probability*. Roughly, they show that if $f \in \mathcal{LC}^1$, and if (3.3) and (3.4) hold with $p > 1/2$ (this constant changing when $\gamma_{\mathrm{dec}} \neq 1/\gamma_{\mathrm{inc}}$), then $\|\nabla f(\boldsymbol{x}_k)\| \leq \epsilon$ holds within $O(\epsilon^{-2})$ function evaluations with a probability that increases exponentially to 1 as $\epsilon \to 0$. The WCC result of Gratton *et al.* (2015) demonstrates that as $p \to 1$ (*i.e.* $\boldsymbol{D}_k$ almost always includes a descent direction), the known WCC results for Algorithm 2 discussed in Section 2.1.2 are recovered. A more precise statement of this WCC result is included in Table A.1.

Bibi *et al.* (2019) propose a randomized direct-search method in which the two poll directions in each iteration are $\boldsymbol{D}_k = \{\boldsymbol{e}_i, -\boldsymbol{e}_i\}$, where $\boldsymbol{e}_i$ is the ith elementary basis vector. In the kth iteration, $\boldsymbol{e}_i$ is selected from $\{\boldsymbol{e}_1, \ldots, \boldsymbol{e}_n\}$ with a probability proportional to the Lipschitz constant of the ith partial derivative of f. Bibi *et al.* (2019) perform WCC analysis of this method assuming a known upper bound on Lipschitz constants of partial derivatives; this assumption leads to improved constant factors, but they essentially prove an upper bound on the number of iterations needed to attain $\mathbb{E}[\|\nabla f(\boldsymbol{x}_k)\|] \leq \epsilon$ in $O(\epsilon^{-2})$, where the expectation is with respect to the random sequence of $\boldsymbol{D}_k$. Bibi *et al.* (2019) prove additional WCC results in cases where f is convex or c-strongly convex.

An early randomized DDS derivative-free method that only occasionally employs a descent direction is developed by Diniz-Ehrhardt *et al.* (2008). There, a non-monotone line-search strategy is used to accommodate search directions along which descent may not be initially apparent. Belitz and Bewley (2013) develop a randomized DDS method that employs surrogates and an adaptive lattice.

3.3. Randomized trust-region methods

Whereas the theoretical convergence of a DDS method depends on the set of polling directions satisfying some spanning property (*e.g.* a cosine measure bounded away from zero), the theoretical convergence of a trust-region method (*e.g.* Algorithm 3) depends on the use of fully linear models. Analogous to how randomized DDS methods relax the requirement of the use of a positive spanning set in *every* iteration, (3.4), it is reasonable to ask whether one can relax the requirement of being fully linear in every iteration of a trust-region method. Practically speaking, in the unconstrained case it may not be necessary to ensure that every model is built by using a Λ-poised set of points (therefore ensuring that the model is fully linear) on every iteration, since ensuring Λ-poised sets entails additional function evaluations.

Bandeira, Scheinberg and Vicente (2014) consider a sequence of random models $\{m_k\}$ and a random sequence of trust-region centres and radii $\{x_k, \Delta_k\}$. They say that the sequence of random models is p-probabilistically κ-fully linear provided

$$\mathbb{P}[m_k \text{ is a } \kappa\text{-fully linear model of } f \text{ on } \mathcal{B}(x_k; \Delta_k) \mid \mathcal{H}_{k-1}] \geq p, \qquad (3.5)$$

where $\mathcal{H}_{k-1}$ is the filtration of the random process prior to the current iteration. That is, $\mathcal{H}_{k-1}$ is the σ-algebra generated by the algorithm's history. Under additional standard assumptions concerning Algorithm 3 (*e.g.* $\gamma_{\mathrm{dec}} = 1/\gamma_{\mathrm{inc}}$), the authors show that if (3.5) holds with $p > 1/2$, then $\lim_{k\to\infty} \|\nabla f(x_k)\| = 0$ almost surely (*i.e.* with probability one). Gratton, Royer, Vicente and Zhang (2018) build on this result; they demonstrate that, up to constants, the same (with high probability) WCC bound that was proved for DDS methods in Gratton *et al.* (2015) holds for the randomized trust-region method proposed by Bandeira *et al.* (2014). Higher-order versions of (3.5) also exist; in Section 5.2 we discuss settings for which Bandeira *et al.* (2014) obtain probabilistically κ-fully quadratic models by interpolating f on a set of fewer than $(n+1)(n+2)/2$ points.

4. Methods for convex objectives

As is true of derivative-based optimization, convexity in the objective of (DET) or (STOCH) can be exploited either when designing new methods or when analysing existing methods. Currently, this split falls neatly into two categories. The majority of work considering (DET) when f is convex sharpens the WCCs for frameworks already discussed in this survey. On the other hand, the influence of machine learning, particularly large-scale empirical risk minimization, has led to entirely new derivative-free methods for solving (STOCH) when f is convex.

4.1. Methods for deterministic convex optimization

We first turn our attention to the solution of (DET). In convex optimization, one can prove WCC bounds on the difference between a method's estimate of the global minimum of f and the value of f at a global minimizer $x_* \in \Omega$ (*i.e.* a point satisfying $f(x_*) \leq f(x)$ for all $x \in \Omega$). This differs from the local WCC bounds on the objective gradient, namely (2.4), that are commonly shown when f is not assumed to be convex. For convex f, under appropriate additional assumptions, one typically demonstrates that a method satisfies

$$\lim_{k\to\infty} f(x_k) - f(x_*) = 0, \qquad (4.1)$$

where x_k is the kth point of the method. Hence, an appropriate measure of ϵ-optimality when minimizing an unconstrained convex objective is the satisfaction of

$$f(\boldsymbol{x}_k) - f(\boldsymbol{x}_*) \le \epsilon. \tag{4.2}$$

For the DDS methods discussed in Section 2.1.2, Dodangeh and Vicente (2016) and Dodangeh, Vicente and Zhang (2016) analyse the worst-case complexity of Algorithm 2 when there is no search step and when f is smooth and convex, and has a bounded level set. By imposing an appropriate upper bound on the step sizes α_k, and (for $c > 0$) using a test of the form

$$f(\boldsymbol{p}_i) \le f(\boldsymbol{x}_k) - c\alpha_k^2, \tag{4.3}$$

in line 4 of Algorithm 2, Dodangeh and Vicente (2016) show that the worst-case number of f evaluations to achieve (2.2) is in $O(n^2 L_{\mathrm{g}}^2 \epsilon^{-1})$. Dodangeh *et al.* (2016) show that this n^2-dependence is optimal (in the sense that it cannot be improved) within the class of deterministic methods that employ positive spanning sets. Recall from Section 3.2 that randomized methods allow one to reduce this dependence to be linear in n. Under additional assumptions, which are satisfied, for example, when f is strongly convex, Dodangeh and Vicente (2016) prove R-linear convergence of Algorithm 2, yielding a WCC of type (4.2) with the dependence on ϵ reduced to $\log(\epsilon^{-1})$. We note that, in the convex setting, R-linear convergence had been previously established for a DDS method by Dolan *et al.* (2003). It is notable that the method analysed in Dolan *et al.* (2003) requires only strict decrease, whereas, to the authors' knowledge, the DDS methods for which WCC results have been established all require sufficient decrease.

Konečný and Richtárik (2014) propose a DDS method that does not allow for increases in the step size α_k and analyse the method on strongly convex, convex and non-convex objectives. Although Konečný and Richtárik (2014) demonstrate WCC bounds with the same dependence on n and ϵ as do Dodangeh and Vicente (2016), they additionally assume that one has explicit knowledge of a Lipschitz gradient constant L_{g} and can thus replace the test (4.3) explicitly with

$$f(\boldsymbol{p}_i) \le f(\boldsymbol{x}_k) - \frac{L_{\mathrm{g}}}{2}\alpha_k. \tag{4.4}$$

Exploiting this additional knowledge of L_{g}, the WCC result in Konečný and Richtárik (2014) exhibits a strictly better dependence on L_{g} than does the WCC result in Dodangeh and Vicente (2016), with a WCC of type (4.2) in $O(n^2 L_{\mathrm{g}} \epsilon^{-1})$. Additionally assuming f is c-strongly convex, Konečný and Richtárik (2014) provide a result showing a WCC of type (4.2) in $O(\log(\epsilon^{-1}))$.

In the non-convex case, Konečný and Richtárik (2014) recover the same WCC of type (2.2) from Vicente (2013); see the discussion in Section 2.1. Once again, however, the result by Konečný and Richtárik (2014) assumes knowledge of L_g and again recovers a strictly better dependence on L_g by using the test (4.4) in line 4 of Algorithm 2.

Recalling the Nesterov random search methods discussed in Section 3.1.2, we remark that Nesterov and Spokoiny (2017) explicitly give results for deterministic, convex f. In particular, Nesterov and Spokoiny prove WCCs of a specific type. Because of the randomized nature of Algorithm 6, WCCs are given as expectations of the form

$$\mathbb{E}_{\boldsymbol{U}_{k-1}}[f(\hat{\boldsymbol{x}}_k)] - f(\boldsymbol{x}_*) \le \epsilon, \tag{4.5}$$

where $\hat{\boldsymbol{x}}_k = \arg\min_{j \in \{0,1,\ldots,k-1\}} f(\boldsymbol{x}_j)$ and where $\boldsymbol{U}_{k-1} = \{\boldsymbol{u}_0, \boldsymbol{u}_1, \ldots, \boldsymbol{u}_{k-1}\}$ is the filtration of Gaussian samples. The form of ϵ-optimality represented by (4.5) can be interpreted as a probabilistic variant of (4.2). The WCC results of Nesterov and Spokoiny show that the worst-case number of f evaluations to achieve (4.5) is in $O(\epsilon^{-1})$ when $f \in \mathcal{LC}^1$. Additionally assuming that f is c-strongly convex yields an improved result; the WCC of type (4.5) is now in $O(\log(\epsilon^{-1}))$. Moreover, by mimicking the method of accelerated gradient descent (see *e.g.* Nesterov 2004, Chapter 2.2), Nesterov and Spokoiny present a variant of Algorithm 6 with a WCC of type (4.5) in $O(\epsilon^{-1/2})$. When $f \in \mathcal{LC}^0$, Nesterov and Spokoiny provide a WCC of type (4.5) in $O(\epsilon^{-2})$, but this result assumes that Algorithm 6 uses an oracle with access to exact directional derivatives of f. Thus, the method achieves the $O(\epsilon^{-2})$ result when $f \in \mathcal{LC}^0$ is not a derivative-free method.

As remarked in Section 3, these convergence results depend on preselecting a sequence of step sizes $\{\alpha_k\}$ for Algorithm 6; in the convex case, the necessary $\{\alpha_k\}$ depends not only on the Lipschitz constants but also on a bound $R_{\boldsymbol{x}}$ on the distance between the initial point and the global minimizer (*i.e.* $\|\boldsymbol{x}_0 - \boldsymbol{x}_*\| \le R_{\boldsymbol{x}}$). The aforementioned WCC results will hold only if one chooses $\{\alpha_k\}$ and μ (the difference parameter of the oracle used in Algorithm 6) that scale with L_g and $R_{\boldsymbol{x}}$ appropriately. When additionally assuming f is c-strongly convex, $\{\alpha_k\}$ and μ also depend on c. Stich, Müller and Gärtner (2013)[5] extend the framework of Algorithm 6 with an approximate line search that avoids the need for predetermined sequences of step sizes.

In general, the existing WCC results for derivative-free methods match the WCC results for their derivative-based counterparts in ϵ-dependence. The WCC results for derivative-free methods tend to involve an additional factor of n when compared with their derivative-based counterparts. This observation mirrors a common intuition in derivative-free optimization: since a number of f evaluations in $O(n)$ can guarantee a suitable approximation

---

[5] A careful reader may be caught off guard by the fact that Stich *et al.* (2013) was published before Nesterov and Spokoiny (2017). This is not a typo; Stich *et al.* (2013) build on the results from an early preprint of Nesterov and Spokoiny (2017).

of the gradient, then for any class of gradient-based method for which we re-place gradients with approximate gradients or model gradients, one should expect to recover the WCC of that method, but with an extra factor of n. WCC results such as those discussed thus far in this survey add credence to this intuition. As we will see in Section 4.2, however, this optimistic trend does not always hold: we will see problem classes for which derivative-free methods are provably worse than their derivative-based counterparts by a factor of ϵ^{-1}.

Bauschke, Hare and Moursi (2014) offer an alternative approach to Al-gorithm 6 for the solution of (DET) when f is assumed convex and, ad-ditionally, lower-$\mathcal{C}^2$. (Such an assumption on f is obviously stronger than plain convexity but contains, for example, functions that are defined as the pointwise maximum over a collection of convex functions.) Bauschke $et\ al.$ (2014) show that linear interpolation through function values is suffi-cient for obtaining approximate subgradients of convex, lower-$\mathcal{C}^2$ function; these approximate subgradients are used in lieu of subgradients in a mirror-descent algorithm (see $e.g.$ Srebro, Sridharan and Tewari 2011) similar to Algorithm 7 in Section 4.2.2. Bauschke $et\ al.$ (2014) establish convergence of their method in the sense of (4.1), and they demonstrate the performance of their method when applied to pointwise maxima of collections of convex quadratics.

4.2. Methods for convex stochastic optimization

We now turn our attention to the solution of the problem (STOCH) when $\tilde{f}(\boldsymbol{x}; \boldsymbol{\xi})$ is assumed convex in $\boldsymbol{x}$ for each realization $\boldsymbol{\xi}$. Up to this point in the survey, we have typically assumed that (STOCH) is unconstrained, that is, $\boldsymbol{\Omega} = \mathbb{R}^n$. In this section, however, it will become more frequent that $\boldsymbol{\Omega}$ is a compact set.

In the machine learning community, zeroth-order information ($i.e.$ eval-uations of $\tilde{f}$ only) is frequently referred to as $bandit\ feedback$,[6] due to the concept of multi-armed bandits from reinforcement learning. Multi-armed bandit problems are sequential allocation problems defined by a prescribed set of actions. Robbins (1952) formulates a multi-armed bandit problem as a gambler's desire to minimize the total losses accumulated from pulling discrete sequence (of length T) of $A < \infty$ slot machine arms. The gambler does not have to decide the full length-T sequence up front. Rather, at time $k \leq T$, the losses associated with the first $k-1$ pulls are known to the gam-bler when deciding which of the A arms to pull next. The gambler's decision of the kth arm to pull is represented by the scalar variable $x_k \in \boldsymbol{\Omega}$. Given

---

[6] A 'one-armed bandit' is an American colloquialism for a casino slot machine, the arm of which must be pulled to reveal a player's losses or rewards.

additional environmental variables outside the gambler's control, $\boldsymbol{\xi}_k \in \Xi$, which represents the stochastic nature of the slot machine,[7] the environment makes a decision $\boldsymbol{\xi}$ simultaneously with the gambler's decision x. The gambler's loss is then $\tilde{f}(x; \boldsymbol{\xi})$.

Within this multi-armed bandit setting, the typical metric of the gambler's performance is the *cumulative regret*. Provided that the expectation $\mathbb{E}_{\boldsymbol{\xi}}[\tilde{f}(x; \boldsymbol{\xi})] = f(x)$ exists for each $x \in \{1, \ldots, A\}$, then the gambler's best long-run strategy in terms of minimizing the expected total losses is to constantly play $x_* \in \arg\min_{x \in \{1, \ldots, A\}} f(x)$. If, over the course of T pulls, the gambler makes a sequence of decisions $x_1, \ldots, x_T$ and the environment makes a sequence of decisions $\boldsymbol{\xi}_1, \ldots, \boldsymbol{\xi}_T$ resulting in a sequence of realized losses $\tilde{f}(x_1; \boldsymbol{\xi}_1), \ldots, \tilde{f}(x_T; \boldsymbol{\xi}_T)$, then the cumulative regret r_T associated with the gambler's sequence of decisions is the difference between the cumulative loss incurred by the gambler's strategy $(x_1, \ldots, x_T)$ and the loss incurred by the best possible long-run strategy $(x_*, \ldots, x_*)$. Analysis of methods for bandit problems in this set-up is generally concerned with *expected cumulative regret*

$$\mathbb{E}_{\boldsymbol{\xi}}[r_T(x_1, \ldots, x_T)] = \mathbb{E}_{\boldsymbol{\xi}}\left[\sum_{k=1}^{T} \tilde{f}(x_k; \boldsymbol{\xi}_k)\right] - Tf(x_*), \qquad (4.6)$$

where the expectation is computed over $\boldsymbol{\xi} \in \Xi$, since we assume here that the sequence of $\boldsymbol{\xi}_k$ is independent and identically distributed.

This particular treatment of the bandit problem with a discrete space of actions $x \in \{1, \ldots, A\}$ was the one considered by Robbins (1952) and has been given extensive treatment (Auer, Cesa-Bianchi, Freund and Schapire 2003, Lai and Robbins 1985, Agrawal 1995, Auer, Cesa-Bianchi and Fischer 2002).

Extending multi-armed bandit methods to *infinite-armed* bandits makes the connections to derivative-free optimization – particularly derivative-free convex optimization – readily apparent. Auer (2002) extends the multi-armed bandit problem to allow for a compact (as opposed to discrete) set of actions for the gambler $\boldsymbol{x} \in \Omega \subset \mathbb{R}^n$ as well as a compact set of vectors for the environment, $\boldsymbol{\xi} \in \Xi \subset \mathbb{R}^n$. The vectors $\boldsymbol{\xi}$ in this set-up define linear functions; that is, if the gambler chooses $\boldsymbol{x}_k$ in their kth pull, and the environment chooses $\boldsymbol{\xi}_k \in \Xi$, then the gambler incurs loss $\tilde{f}(\boldsymbol{x}_k; \boldsymbol{\xi}_k) = \boldsymbol{\xi}_k^{\mathrm{T}} \boldsymbol{x}_k$. In this linear regime, expected regret takes a form remarkably similar to

---

[7] Depending on the problem set-up, the environment of a bandit problem may be either stochastic or adversarial. Because this section is discussing stochastic convex optimization, we will assume that losses are stochastic; that is, the $\boldsymbol{\xi}_k$ are i.i.d. and independent of the gambler's decisions x_k. See Bubeck and Cesa-Bianchi (2012) for a survey of bandit problems more general than those discussed in this survey.

(4.6), that is,

$$\mathbb{E}_{\xi}[r_T(\boldsymbol{x}_1,\ldots,\boldsymbol{x}_T)] = \mathbb{E}_{\xi}\left[\sum_{k=1}^{T}\tilde{f}(\boldsymbol{x}_k;\boldsymbol{\xi}_k)\right] - \min_{\boldsymbol{x}\in\Omega}\mathbb{E}_{\xi}\left[\sum_{k=1}^{T}\tilde{f}(\boldsymbol{x}_k;\boldsymbol{\xi}_k)\right]$$
$$= \left(\sum_{k=1}^{T}f(\boldsymbol{x}_k)\right) - Tf(\boldsymbol{x}_*), \tag{4.7}$$

with $\boldsymbol{x}_* \in \arg\min_{\boldsymbol{x}\in\Omega} f(\boldsymbol{x})$. As in (4.6), (4.7) defines the expected regret with respect to the best long-run strategy $\boldsymbol{x} \in \Omega$ that the gambler could have played for the T rounds (*i.e.* the strategy that would minimize the expected cumulative losses).[8]

By using a bandit method known as Thompson sampling, one can show (under appropriate additional assumptions) that if $\tilde{f}(\boldsymbol{x};\boldsymbol{\xi}) = \boldsymbol{\xi}^{\mathrm{T}}\boldsymbol{x}$, then (4.7) can be bounded as

$$\mathbb{E}_{\xi}[r_T(\boldsymbol{x}_1,\ldots,\boldsymbol{x}_T)] \in O(n\sqrt{T\log(T)})$$

(Russo and Van Roy 2016). Analysis of bounds on (4.7) in the linear case raises an interesting question: to what extent can similar analysis be performed for classes of functions $\tilde{f}(\boldsymbol{x};\boldsymbol{\xi})$ that are non-linear in $\boldsymbol{x}$?

In much of the bandit literature, the additional structure defining a class of non-linear functions is convexity; here, by convexity, we mean that $\tilde{f}(\boldsymbol{x};\boldsymbol{\xi})$ is convex in $\boldsymbol{x}$ for each realization $\boldsymbol{\xi} \in \Xi$.

Regret bounds on (4.7) automatically imply WCC results for stochastic convex optimization. To see this, define an *average point*

$$\bar{\boldsymbol{x}}_k = \frac{1}{k}\sum_{t=1}^{k}\boldsymbol{x}_t.$$

Because of the convexity of f,

$$f(\bar{\boldsymbol{x}}_T) - f(\boldsymbol{x}_*) \le \frac{1}{T}\sum_{k=1}^{T}f(\boldsymbol{x}_k) - f(\boldsymbol{x}_*) = \frac{\mathbb{E}_{\xi}[r_T(\boldsymbol{x}_1,\ldots,\boldsymbol{x}_T)]}{T}, \tag{4.8}$$

where the inequality follows from an application of Jensen's inequality. We see in (4.8) that, given an upper bound of $\bar{r}(T)$ on the expected regret $\mathbb{E}_{\xi}[r_T]$, we can automatically derive, provided $\bar{r}(T)/T \le \epsilon$, a WCC result for stochastic convex optimization of the form

$$\mathbb{E}_{\xi}[\tilde{f}(\bar{\boldsymbol{x}}_T;\boldsymbol{\xi}) - \tilde{f}(\boldsymbol{x}_*;\boldsymbol{\xi})] = f(\bar{\boldsymbol{x}}_T) - f(\boldsymbol{x}_*) \le \epsilon. \tag{4.9}$$

---

[8] We remark that many of the references we provide here may also refer to methods minimizing (4.7) as methods of *online optimization*; one reference in particular (Bach and Perchet 2016) even suggests a taxonomy identifying bandit learning as a restrictive case of online optimization.

Equation (4.9) corresponds to a stochastic version of a WCC result of type (4.2) where $N_\epsilon = T$.

The converse implication, however, is not generally true: a small optimization error does not imply a small regret. That is, WCC results of type (4.9) do not generally imply bounds on expected regret (4.7). This is particularly highlighted by Shamir (2013), who considers a class of strongly convex quadratic objectives f. For such objectives, Shamir (2013) establishes a strict gap between the upper bound on the optimization error (the left-hand side of (4.9)) and the lower bound on the expected regret (4.7) attainable by any method in the bandit setting. Such a gap, between expected regret and optimization error, has also been proved for bandit methods applied to problems of logistic regression (Hazan, Koren and Levy 2014).

Results such as those of Shamir (2013), Jamieson, Nowak and Recht (2012) and Hazan *et al.* (2014) have led researchers to consider methods of derivative-free (stochastic) optimization within the paradigm of convex bandit learning. In this survey, we group these methods into two categories of assumptions on the type of bandit feedback (*i.e.* the observed realizations of $\tilde{f}$) available to the method: one-point bandit feedback or two-point (multi-point) bandit feedback. Although many of the works we cite have also analysed regret bounds for these methods, we focus on optimization error.

4.2.1. One-point bandit feedback

In one-point bandit feedback, a method is assumed to have access to an oracle $\tilde{f}$ that returns unbiased estimates of f. In particular, given two points $x_1, x_2 \in \Omega$, two separate calls to the oracle will return $\tilde{f}(x_1; \xi_1)$ and $\tilde{f}(x_2; \xi_2)$; methods do not have control over the selection of the random variables ξ_1 and ξ_2. Many one-point bandit methods do not fall neatly into the frameworks discussed in this survey (Agarwal *et al.* 2011, Belloni, Liang, Narayanan and Rakhlin 2015, Bubeck, Lee and Eldan 2017). See Table 4.1 for a summary of best known WCC results of type (4.9) for one-point bandit feedback methods.

One example of a method using one-point bandit feedback, whose development falls naturally into our discussion thus far, is given by Flaxman, Kalai and McMahan (2005); they analyse a method resembling Algorithm 6, but the gradient-free oracle is chosen as

$$g_\mu(x; u; \xi) = \frac{\tilde{f}(x + \mu u; \xi)}{\mu} u, \tag{4.10}$$

where u is drawn uniformly from the unit n-dimensional sphere. That is, given a realization $\xi \in \Xi$, a stochastic gradient estimator based on a stochastic evaluation at the point $x + \mu u$ is computed via (4.10). For this method, and for general convex f, Flaxman *et al.* (2005) demonstrate a

Table 4.1. Best known WCCs for N_ϵ, the number of evaluations required to bound (4.9), for one-point bandit feedback. N_ϵ is given only in terms of n and ϵ. See text for the definition of β-smooth; here $\beta > 2$. Method types include random search (RS), mirror descent (MD) and ellipsoidal.

Assumption on f	N_ϵ	Method type (citation)
convex, $f \in \mathcal{LC}^0$	$n^2\epsilon^{-4}$	RS (Flaxman *et al.* 2005)
	$n^{13/2}\epsilon^{-2}$	ellipsoidal (Belloni *et al.* 2015)
c-strongly convex, $f \in \mathcal{LC}^0$	$n^2\epsilon^{-3}$	RS (Flaxman *et al.* 2005)
convex, $f \in \mathcal{LC}^1$	$n\epsilon^{-3}$	MD (Gasnikov *et al.* 2017)
	$n^{13/2}\epsilon^{-2}$	ellipsoidal (Belloni *et al.* 2015)
c-strongly convex, $f \in \mathcal{LC}^1$	$n^2\epsilon^{-2}$	MD (Gasnikov *et al.* 2017)
convex, β-smooth	$n^2\epsilon^{-2\beta/(\beta-1)}$	RS (Bach and Perchet 2016)
c-strongly convex, β-smooth	$n^2\epsilon^{-(\beta+1)/(\beta-1)}$	RS (Bach and Perchet 2016)

WCC bound of type (4.9) in $O(n^2\epsilon^{-4})$ for general convex f. For strongly convex f, Flaxman *et al.* (2005) demonstrate a WCC bound of type (4.9) in $O(n^2\epsilon^{-3})$. Various extensions of these results are given in Saha and Tewari (2011), Dekel, Eldan and Koren (2015) and Gasnikov *et al.* (2017). To the best of our knowledge, the best known upper bound on the WCC for smooth, strongly convex problems is in $O(n^2\epsilon^{-2})$ and the best known upper bound on the WCC for smooth, convex problems is in $O(n\epsilon^{-3})$ (Gasnikov *et al.* 2017).

For the solution of (STOCH) when $\Omega = \mathbb{R}^n$, Bach and Perchet (2016) analyse a method resembling Algorithm 6 wherein the gradient-free oracle is chosen as

$$g_\mu(\boldsymbol{x}; \boldsymbol{u}; \boldsymbol{\xi}_+, \boldsymbol{\xi}_-) = \frac{\tilde{f}(\boldsymbol{x} + \mu\boldsymbol{u}; \boldsymbol{\xi}_+) - \tilde{f}(\boldsymbol{x} - \mu\boldsymbol{u}; \boldsymbol{\xi}_-)}{\mu}\boldsymbol{u}, \tag{4.11}$$

where $\boldsymbol{u}$ is again drawn uniformly from the unit n-dimensional sphere. We remark that $\boldsymbol{\xi}_+, \boldsymbol{\xi}_- \in \Xi$ in (4.11) are *different* realizations; this distinction will become particularly relevant in Section 4.2.2. For the solution of (STOCH) when $\Omega \subsetneq \mathbb{R}^n$, Bach and Perchet (2016) also consider a gradient-free oracle like (4.10). Bach and Perchet (2016) are particularly interested in how the smoothness of f can be exploited to improve complexity bounds in the one-point bandit feedback paradigm; they define a parameter β and say that a function f is β-smooth provided f is almost everywhere $(\beta - 1)$-times Lipschitz-continuously differentiable (a strictly weaker condition than assuming $f \in \mathcal{LC}^{\beta-1}$). When $\beta = 2$, Bach and Perchet (2016) recover similar results to those seen in Table 4.1 for when $f \in \mathcal{LC}^1$. When $\beta > 2$, however, in both the constrained and unconstrained case, Bach and Perchet (2016) prove a WCC result of type (4.9) in $O(n^2 \epsilon^{2\beta/(1-\beta)})$ when f is convex and β-smooth. Further assuming that f is c-strongly convex, Bach and Perchet (2016) prove a WCC result of type (4.9) in $O(n^2 \epsilon^{(\beta+1)/(1-\beta)}/c)$. Notice that asymptotically, as $\beta \to \infty$, this bound is in $O(n^2 \epsilon^{-1}/c)$. This asymptotic result is particularly important because it attains the lower bound on optimization error demonstrated by Shamir (2013) for strongly convex ∞-smooth (quadratic) f.

We also note that Bubeck *et al.* (2017) conjecture that a particular kernel method can achieve a WCC of type (4.9) in $O(n^3 \epsilon^{-2})$ for general convex functions. In light of well-known results in deterministic convex optimization, the WCCs summarized in Table 4.1 may be surprising. In particular, for any c-strongly convex function $f \in \mathcal{LC}^1$, the best known WCC results are in $O(\epsilon^{-2})$. We place particular emphasis on this result because it illustrates a gap between derivative-free and derivative-based optimization that is not just a factor of n. In this particular one-point bandit feedback setting, there do not seem to exist methods that achieve the optimal[9] $O(\epsilon^{-1})$ convergence rate attainable by gradient-based methods for smooth strongly convex stochastic optimization. Hu, Prashanth, György and Szepesvári (2016) partially address this issue concerning one-point bandit feedback, which they refer to as 'uncontrolled noise'. These observations motivated the study of two-point (multi-point) bandit feedback, which we will discuss in the next section, Section 4.2.2.

We further remark that every WCC in Table 4.1 has a polynomial dependence on the dimension n, raising natural questions about the applicability of these methods in high-dimensional settings. Wang, Du, Balakrishnan and Singh (2018) consider a mirror descent method employing a special gradient-free oracle computed via a compressed sensing technique. They prove a WCC of type (4.9) in $O(\log(d)^{3/2} n_z^2 \epsilon^{-3})$, under additional assumptions on derivative sparsity, most importantly, that for all $x \in \Omega$, $\|\nabla f(x)\|_0 \leq n_z$ for some n_z. Thus, provided $n_z \ll n$, the polynomial dependence on n becomes a logarithmic dependence on n, at the expense of a WCC with a strictly worse dependence on ϵ than ϵ^{-2}. In Section 5.2, we discuss methods that similarly exploit known sparsity of objective function derivatives.

---

[9] Optimal here is meant in a minimax information-theoretic sense (Agarwal, Wainwright, Bartlett and Ravikumar 2009).

In an extension of Algorithm 6, Chen (2015, Chapter 3) dynamically updates the difference parameter by exploiting knowledge of the changing variance of $\boldsymbol{\xi}$.

4.2.2. Two-point (multi-point) bandit feedback

We now focus on the stochastic paradigm of two-point (or multi-point) bandit feedback. In this setting of bandit feedback, we do not encounter the same gaps in WCC results between derivative-free and derivative-based optimization that exist for one-point bandit feedback.

The majority of methods analysed in the two-point bandit feedback setting are essentially random-search methods of the form Algorithm 6. The gradient-free oracles from (3.1) in the two-point setting takes one of the two forms

$$\begin{aligned}
\boldsymbol{g}_{\mu_k}(\boldsymbol{x}; \boldsymbol{u}; \boldsymbol{\xi}) &= \delta_{\mathrm{f}}(\tilde{f}(\cdot, \boldsymbol{\xi}); \boldsymbol{x}; \boldsymbol{u}; \mu_k)\boldsymbol{B}\boldsymbol{u}, \quad \text{or} \\
\hat{\boldsymbol{g}}_{\mu_k}(\boldsymbol{x}; \boldsymbol{u}; \boldsymbol{\xi}) &= \delta_{\mathrm{c}}(\tilde{f}(\cdot, \boldsymbol{\xi}); \boldsymbol{x}; \boldsymbol{u}; \mu_k)\boldsymbol{B}\boldsymbol{u}.
\end{aligned} \tag{4.12}$$

The key observation in (4.12) is that $\boldsymbol{\xi}$ denotes a single realization used in the computation of both function values in the definitions of the oracles (see (2.28) and (2.29)). This assumption of 'controllable realizations' separates two-point bandit feedback from the more pessimistic one-point bandit feedback. This property of being able to recall a single realization $\boldsymbol{\xi}$ for two (or more) evaluations of $\tilde{f}$ is precisely why this setting of bandit feedback is called 'two-point' (or 'multi-point'). This property will also be exploited in Section 6.1.

The early work of Agarwal, Dekel and Xiao (2010) directly addresses the discussed gap in WCC results and demonstrates that a random-search method resembling Algorithm 6, but applied in the two-point (or multi-point) setting as opposed to the one-point setting, attains a WCC of type (4.9) in $O(\epsilon^{-1})$; this is a complexity result matching the optimal rate (in terms of ϵ-dependence) shown by Agarwal et $al.$ (2009). See Table 4.2 for a summary of best known WCC results of type (4.9) for two-point bandit feedback methods.

Nesterov and Spokoiny (2017) provide a WCC result for Algorithm 6 using the stochastic gradient-free oracles (4.12), but strictly better WCCs have since been established. We also note that, in contrast with the gradient-free oracles of (3.1) in Section 3.1.2, the difference parameter μ is written as μ_k in (4.12), indicating that a selection for μ_k must be made in the kth iteration. In the works that we discuss here, μ_k in (4.12) is either chosen as a constant sufficiently small or else $\mu_k \to 0$ at a rate typically of the order of $1/k$. We also remark that many of the results discussed in this section trivially hold for deterministic convex problems, and can be seen as an extension of results concerning methods of the form of Algorithm 6.

Algorithm 7: Mirror-descent method with two-point gradient estimate

---

1 Choose initial point $\boldsymbol{x}_0$, sequence of step sizes $\{\alpha_k\}$, sequence of difference parameters $\{\mu_k\}$, and distribution of $\boldsymbol{\xi}$
2 **for** $k = 1, 2, \ldots, T - 1$ **do**
3 $\quad$ Sample $\boldsymbol{u}_k$ uniformly from the unit sphere $\mathcal{B}(\boldsymbol{0}; 1)$
4 $\quad$ Sample a realization $\boldsymbol{\xi}_k$
5 $\quad$ $\boldsymbol{g}_k \leftarrow \boldsymbol{g}_{\mu_k}(\boldsymbol{x}_k; \boldsymbol{u}_k; \boldsymbol{\xi}_k)$ using an oracle from (4.12)
6 $\quad$ $\boldsymbol{x}_{k+1} \leftarrow \arg\min_{\boldsymbol{y} \in \Omega} \boldsymbol{g}_k^{\mathrm{T}} \boldsymbol{y} + \frac{1}{\alpha_k} D_\psi(\boldsymbol{y}, \boldsymbol{x}_k)$

Provided $f \in \mathcal{LC}^0$, the best known WCCs of type (4.9) (with variations in constants) can be found in Duchi, Jordan, Wainwright and Wibisono (2015), Gasnikov *et al.* (2017) and Shamir (2017). These works consider variants of mirror-descent methods with approximate gradients given by estimators of the form (4.12); see Algorithm 7 for a description of a basic mirror-descent method. Algorithm 7 depends on the concept of a *Bregman divergence* D_ψ, used in line 6 of Algorithm 7 to define a proximal-point subproblem. The Bregman divergence is defined by a function $\psi : \Omega \to \mathbb{R}$, which is assumed to be 1-strongly convex with respect to the norm $\|\cdot\|_p$. To summarize many findings in this area (Duchi *et al.* 2015, Gasnikov *et al.* 2017, Shamir 2017), if $\|\cdot\|_q$ is the dual norm to $\|\cdot\|_p$ (*i.e.* $p^{-1} + q^{-1} = 1$) where $p \in \{1, 2\}$ and R_p denotes the radius of the feasible set in the $\|\cdot\|_p$-norm, then a bound on WCC of type (4.9) in $O(n^{2/q} R_p^2 \epsilon^{-2})$ can be established for a method like Algorithm 7 in the two-point feedback setting where $f \in \mathcal{LC}^0$. These WCCs for methods like Algorithm 7 are responsible for the popularity of mirror-descent methods in machine learning. For many machine learning problems, solutions are typically sparse, and so, in some sense, $R_1 \le R_2$. Thus, using a function ψ that is 1-strongly convex with respect to the $\|\cdot\|_1$-norm (*e.g.* simply letting $\psi = \|\cdot\|_1$) may be preferable to $p = 2$, in both theory and practice.

Duchi *et al.* (2015) also provide an information-theoretic lower bound on convergence rates for any method in the (non-strongly) convex, $f \in \mathcal{LC}^0$, two-point feedback setting. This bound matches the best known WCCs up to constants, demonstrating that these results are tight. This lower bound is still of the order of ϵ^{-2}, matching the result of Agarwal *et al.* (2009) in ϵ-dependence in the case where $f \in \mathcal{LC}^0$. It is also remarkable that this result is only a factor of $\sqrt{n}$ worse than the bounds provided by Agarwal *et al.* (2009) for the derivative-based case, as opposed to the factor of n that one may expect.

Additionally assuming $f(\boldsymbol{x})$ is strongly convex (but still assuming $f \in \mathcal{LC}^0$), Agarwal *et al.* (2010) prove, for a method like Algorithm 6, a WCC

Table 4.2. Best known WCCs of type (4.9) for two-point bandit feedback. N_ϵ is given only in terms of n and ϵ. See text for the definition of p, q. R_p denotes the size of the feasible set in the $\|\cdot\|_p$-norm. If f is c_p-strongly convex, then f is strongly convex with respect to the $\|\cdot\|_p$-norm with constant c_p. σ is the standard deviation on the gradient estimator $\boldsymbol{g}_\mu(\boldsymbol{x}; \boldsymbol{u}; \boldsymbol{\xi})$ (i.e. $\mathbb{E}_{\boldsymbol{\xi}}[\|\boldsymbol{g}_\mu(\boldsymbol{x}; \boldsymbol{u}; \boldsymbol{\xi}) - \nabla f(\boldsymbol{x})\|^2] \leq \sigma^2$). The Lipschitz constant of the gradient L_g is defined by the $\|\cdot\|_2$-norm. $\star$ denotes the additional assumption that $\mathbb{E}_{\boldsymbol{\xi}}[\|\boldsymbol{g}_\mu(\boldsymbol{x}; \boldsymbol{u}; \boldsymbol{\xi})\|] < \infty$. Method types include random search (RS), mirror descent (MD) and accelerated mirror descent (AMD).

Assumption on f	N_ϵ	Method type (citation)
convex	$n^{2/q} R_p \epsilon^{-2}$	MD (Duchi *et al.* 2015, Gasnikov *et al.* 2017, Shamir 2017)
c_p-strongly convex	$n^{2/q} c_p^{-1} \epsilon^{-1}$	MD (Gasnikov *et al.* 2017)
convex, smooth	$\max\left\{ \dfrac{n L_g R_2}{\epsilon}, \dfrac{n\sigma^2}{\epsilon^2} \right\}$	RS (Ghadimi and Lan 2013)
	$\max\left\{ \dfrac{n^{2/q} L_g R_p^2}{\epsilon}, \dfrac{n^{2/q} \sigma^2 R_p^2}{\epsilon^2} \right\}$	MD (Dvurechensky, Gasnikov and Gorbunov 2018)
	$\max\left\{ n^{1/2+1/q} \sqrt{\dfrac{L_g R_p^2}{\epsilon}}, \dfrac{n^{2/q} \sigma^2 R_p^2}{\epsilon^2} \right\}$	AMD (Dvurechensky *et al.* 2018)
convex, smooth, $\star$	$n^{2/q} R_p \epsilon^{-2}$	MD (Duchi *et al.* 2015)
c_p-strongly conv., smooth, $\star$	$n^{2/q} c_p^{-1} \epsilon^{-1}$	MD (Gasnikov *et al.* 2017)

in $O(n^2\epsilon^{-1})$. Using a method like Algorithm 7, Gasnikov *et al.* (2017) improve the dependence on n in this WCC to $O(n^{2/q}c_p^{-1}\epsilon^{-1})$, provided f is c_p-strongly convex with respect to $\|\cdot\|_p$.

We now address the case where f is convex and $f \in \mathcal{LC}^1$. Given an assumption that $\|g_\mu(x;u;\xi)\|$ is uniformly bounded, Agarwal *et al.* (2010) demonstrate a WCC of type (4.9) in $O(n^2\epsilon^{-1})$. Dropping this somewhat restrictive assumption on the gradient-free oracle and assuming instead that the oracle used in a method like Algorithm 6 has bounded variance (*i.e.* the oracle satisfies $\mathbb{E}_\xi[\|g_\mu(x;u;\xi) - \nabla f(x)\|^2] \leq \sigma^2$), Ghadimi and Lan (2013) prove a WCC of a type similar to (4.9) (we avoid a discussion of randomized stopping) in $O(\max\{nL_g\|x_0 - x_*\|\epsilon^{-1}, n\sigma^2 L_g\|x_0 - x_*\|\epsilon^{-2}\})$. We mention that Nesterov and Spokoiny (2017) hinted at a similar WCC result, but with a strictly worse dependence on n, and different assumptions on ξ.

5. Methods for structured objectives

The methods discussed in Sections 2 and 3 assume relatively little about the structure of the objective function f beyond some differentiability required for analysis. Section 4 considered the case where f is convex, which resulted, for example, in improved worst-case complexity results. In this section, we consider a variety of assumptions about additional known structure in f (including non-linear least squares, sparse, composite and minimax-based functional forms) and methods designed to exploit this additional structure. Although problems in this section could be solved by the general-purpose methods discussed in Sections 2 and 3, practical gains should be expected by exploiting the additional structure.

5.1. Non-linear least squares

A frequently encountered objective in many applications of computational science, engineering and industry is

$$f(x) = \frac{1}{2}\|F(x)\|_2^2 = \frac{1}{2}\sum_{i=1}^{p} F_i(x)^2. \tag{5.1}$$

For example, data-fitting problems are commonly cast as (5.1); given data y_i collected at design sites θ_i, one may need to estimate the parameters x of a non-linear model or simulation output that best fit the data. In this scenario, F_i is represented by $F_i(x) = w_i(S_i(\theta;x) - y_i)$, which is a weighted residual between the simulation output S_i and target data y_i. In this way, objectives of the form (5.1) (and their correlated residual generalizations) encapsulate both the solution of non-linear equations and statistical estimation problems.

The methods of Zhang, Conn and Scheinberg (2010), Zhang and Conn (2012) and Wild (2017) use the techniques of Section 2.2 to construct models of the individual F_i (thereby obtaining a model of the Jacobian ∇F) in (5.1). These models are then used to generate search directions in a trust-region framework resembling the Levenberg–Marquardt method (Levenberg 1944, Marquardt 1963, Moré 1978). The analysis by Zhang *et al.* (2010) and Zhang and Conn (2012) demonstrates – under certain assumptions, such as $f(x_*) = 0$ at an optimal solution x_* (*i.e.* that the associated data-fitting problem has zero residual) – that the resulting methods achieve the same local quadratic convergence rate does as the Levenberg–Marquardt method. **POUNDERS** is a trust-region-based method for minimizing objectives of the form (5.1) that uses a full Newton approach for each residual F_i (Wild 2017, Dener *et al.* 2018). Another model-based method (implemented in **DFO-LS** (Cartis, Fiala, Marteau and Roberts 2018)) for minimizing functions of the form (5.1) – more closely resembling a Gauss–Newton method – is analysed by Cartis and Roberts (2017). Their method is shown to converge to stationary points of (5.1) even when $f(x_*) > 0$, at the expense of slightly weaker theoretical guarantees on the convergence rate.

Kelley (2003) proposes a hybridization of a Gauss–Newton method with implicit filtering (Algorithm 4 from Section 2.3.2) that estimates the Jacobian of F by building linear models of each component F_i using central differences (2.29) with an algorithmically updated difference parameter. This hybrid method is shown to demonstrate superlinear convergence for zero-residual (*i.e.* $f(x_*) = 0$) problems.

Earlier methods also used the vector F in order to more efficiently address objectives of the form (5.1). Spendley (1969) develops a simplex-based algorithm that employs quadratic approximations obtained by interpolating the vector F on the current simplex. Peckham (1970) proposes an iterative process that refines models of each component of F using between $n+1$ and $n+3+n/3$ points. Ralston and Jennrich (1978) also develop derivative-free Gauss–Newton methods and highlight their performance relative to methods that do not exploit the structure in (5.1). Brown and Dennis, Jr (1971) consider a variant of the Levenberg–Marquardt method that approximates gradients using appropriately selected difference parameters.

Li and Fukushima (2000) analyse the convergence of a derivative-free line-search method when assuming that the square of the Jacobian (*i.e.* $\nabla F(x)^{\mathrm{T}} \nabla F(x)$) of (5.1) is positive-definite everywhere. Grippo and Sciandrone (2007) and La Cruz, Martínez and Raydan (2006) augment a non-monotone line-search method to incorporate information about F in the case where $p = n$ in (5.1). Li and Li (2011) develop a line-search method that exploits a monotonicity property assumed about F. La Cruz (2014) and Morini, Porcelli and Toint (2018) address (5.1) in the case where simple convex constraints are present.

5.2. *Sparse objective derivatives*

In some applications, it is known that

$$\nabla^2 f(\boldsymbol{x})_{ij} = \nabla^2 f(\boldsymbol{x})_{ji} = 0 \quad \text{for all } (i,j) \in S, \tag{5.2}$$

for all $\boldsymbol{x} \in \Omega$, where the index set S defines the sparsity pattern of the Hessian. Similarly, one can consider *partially separable* objectives of the form

$$f(\boldsymbol{x}) = \sum_{i=1}^{p} F_i(\boldsymbol{x}) = \sum_{i=1}^{p} F_i(\{\boldsymbol{x}_j\}_{j \in S_i}), \tag{5.3}$$

where each F_i depends only on some subset of indices $S_i \subset \{1, 2, \ldots, n\}$. The extreme cases of (5.3) are totally separable functions, where $p = n$ and $S_i = \{i\}$ for $i \in \{1, \ldots, n\}$. In this special case, (DET) reduces to the minimization of n univariate functions.

Given the knowledge encoded in (5.2) and (5.3), derivative-free optimization methods need not consider interactions between certain components of $\boldsymbol{x}$ because they are known to be exactly zero. In the context of the model-based methods of Section 2.2, particularly when using quadratic models, using this knowledge amounts to dropping monomials in $\phi(\boldsymbol{x})$ in (2.12) corresponding to the non-interacting (i,j) pairs from (5.2). Intuitively, such an action reduces the degrees of freedom in (2.14) when building models, necessitating fewer function evaluations in the right-hand side of (2.14).

Colson and Toint (2005) propose a trust-region method for functions of the form (5.3) that builds and maintains separate fully linear models for the individual F_i in an effort to use fewer objective function evaluations. Similarly, Colson and Toint (2001) propose a method for the case when $\nabla^2 f$ has a band or block structure that exploits knowledge when building models of the objective; the work of Colson and Toint (2002) extends this work to general sparse objective Hessians. Bagirov and Ugon (2006) develop an algorithm that exploits the fact that efficient discrete gradient estimates can be obtained for f having the form (5.3).

In the context of pattern-search methods (discussed in Section 2.1.2), Price and Toint (2006) exploit knowledge of f having the form (5.3) to choose a particular stencil of search directions when forming a positive spanning set. In particular, the stencil is chosen as $\{\boldsymbol{e}_1, \ldots, \boldsymbol{e}_n, \boldsymbol{e}_{n+1}, \ldots, \boldsymbol{e}_{n+p}\}$, where $\{\boldsymbol{e}_1, \ldots, \boldsymbol{e}_n\}$ are the elementary basis vectors and

$$\boldsymbol{e}_{n+i} = \sum_{j \in S_i} -\boldsymbol{e}_j,$$

for $i \in \{1, \ldots, p\}$. Frimannslund and Steihaug (2010) also develop a DDS method for (5.3), with the search directions determined based on a smoothed quadratic formed from previously evaluated points.

Bandeira, Scheinberg and Vicente (2012) also assume that $\nabla^2 f(\boldsymbol{x})$ is sparse but do not assume knowledge of the sparsity structure (*i.e.* S in (5.2) is not known). They develop a quadratic model-based trust-region method where the models are selected by a minimum 1-norm solution to an underdetermined interpolation system (2.14). Under certain assumptions on f, if n_z is the number of non-zeros in the (unknown) sparsity pattern for $\nabla^2 f$, Bandeira *et al.* (2012) prove that $\boldsymbol{Y}$ in (2.14) must contain only $O((n_z + n)\log(n_z + n)\log(n))$ (as opposed to $O(n^2)$) randomly generated points in order to ensure that the constructed interpolation models are fully quadratic models of f with high probability. This work motivated the analysis of randomized trust-region methods discussed in Section 3.3 because the random underdetermined interpolation models of Bandeira *et al.* (2012) satisfy the assumptions made in Bandeira *et al.* (2014).

In Section 4.2.1, we noted the work of Wang *et al.* (2018), who used assumptions of gradient and Hessian sparsity (in particular, $\|\nabla f(\boldsymbol{x})\|_0 \leq n_z$) to improve the reduce the dependence on n in a WCC of type (4.9) from polynomial to logarithmic. Note that, similar to Bandeira *et al.* (2012), this is an assumption on knowing a universal bound (*i.e.* for all $\boldsymbol{x} \in \boldsymbol{\Omega}$) on the cardinality $\|\nabla f(\boldsymbol{x})\|_0$ rather than the actual non-zero components. Under a similar sparsity assumption, Balasubramanian and Ghadimi (2018) consider the two-point bandit feedback setting discussed in Section 4.2.2 and show that a truncated[10] version of the method proposed in Ghadimi and Lan (2013) has a WCC of type (4.9) in $O(n_z \log(n)^2/\epsilon^2)$. Like the result of Wang *et al.* (2018), this WCC result also exhibits a logarithmic dependence on n, provided $n_z \ll n$. As we will discuss in Section 6.4, Ghadimi and Lan (2013) analyse a method resembling Algorithm 6 to be applied to non-convex f in (STOCH). Balasubramanian and Ghadimi (2018) prove that the unaltered method of Ghadimi and Lan (2013) applied to problems in this sparse setting achieves a WCC of type

$$\mathbb{E}[\|\nabla f(\boldsymbol{x}_k)\|] \leq \epsilon, \tag{5.4}$$

in $O(n_z^2 \log(n)^2 \epsilon^{-4})$; this WCC result once again eliminates a polynomial dependence on n that would otherwise exist in a non-sparse setting. This WCC result, however, maintains the same ϵ-dependence as the method of Ghadimi and Lan (2013) in the non-sparse setting; in this sense, the method of Ghadimi and Lan (2013) is 'automatically' tuned for the sparse setting.

5.3. Composite non-smooth optimization

Sometimes, the objective f in (DET) is known to be non-smooth. Often, one has knowledge about the form of non-smoothness present in the objective,

---

[10] That is, all but the n_z largest values in $\boldsymbol{x}_{k+1}$ are set to 0 in line 5 of Algorithm 6.

and we discuss methods that exploit specific forms of non-smoothness in this section. For methods that do not access any structural information when optimizing non-smooth functions f, see Sections 2.1.2 and 2.3.

We define composite non-smooth functions as those of the form

$$f(\boldsymbol{x}) = h(\boldsymbol{F}(\boldsymbol{x})), \tag{5.5}$$

where $h : \mathbb{R}^p \to \mathbb{R}$ is a non-smooth function (in contrast to smooth h such as the sum of squares in (5.1)), and $\boldsymbol{F} : \mathbb{R}^n \to \mathbb{R}^p$ is continuously differentiable. In some of the works we cite, the definition of a composite non-smooth objective may include an additional smooth function g so that the objective function has the form $f(\boldsymbol{x}) + g(\boldsymbol{x})$, but we omit a discussion of this for the sake of focusing on the non-smooth aspect in (5.5).

5.3.1. Convex h

When h in (5.5) is convex (note that f may still be non-convex due to non-convexity in $\boldsymbol{F}$), one thrust of research extends the techniques in derivative-based composite non-smooth optimization; see the works of Yuan (1985) and Fletcher (1987, Chapter 14). For example, Yuan (1985) use derivatives to construct convex first-order approximations of f near $\boldsymbol{x}$,

$$\ell(\boldsymbol{x} + \boldsymbol{s}) = h(\boldsymbol{F}(\boldsymbol{x}) + \nabla\boldsymbol{F}(\boldsymbol{x})\boldsymbol{s}), \tag{5.6}$$

where $\nabla\boldsymbol{F}$ denotes the Jacobian of $\boldsymbol{F}$; see Hare (2017) for properties of such approximations in the derivative-free setting. By replacing $\nabla\boldsymbol{F}$ in (5.6) with the matrix $\boldsymbol{M}(\boldsymbol{x}_k)$ containing the gradients of a fully linear approximation to $\boldsymbol{F}$ at $\boldsymbol{x}_k$, Grapiglia, Yuan and Yuan (2016) and Garmanjani et al. (2016) independently analyse a model-based trust-region method similar to Algorithm 3 from Section 2.2.4 that uses the non-smooth trust-region subproblem

$$\operatorname*{minimize}_{\boldsymbol{s}:\|\boldsymbol{s}\|\leq\Delta_k} \ell(\boldsymbol{x}_k + \boldsymbol{s}) = h(\boldsymbol{F}(\boldsymbol{x}_k) + \boldsymbol{M}(\boldsymbol{x}_k)\boldsymbol{s}). \tag{5.7}$$

Note that only $\boldsymbol{F}$ is assumed to be a black-box; these methods exploit the fact that h is convex with a known form in order to appropriately solve (5.6). Both Grapiglia et al. (2016) and Garmanjani et al. (2016) use the stationarity measure of Yuan (1985) in their analysis,

$$\Psi(\boldsymbol{x}) = \ell(\boldsymbol{x}) - \min_{\|\boldsymbol{s}\|\leq 1} \ell(\boldsymbol{x} + \boldsymbol{s}), \tag{5.8}$$

for which it is known that $\Psi(\boldsymbol{x}_*) = 0$ if and only if $\boldsymbol{x}_*$ is a critical point of f in the sense that $\ell(\boldsymbol{x}_*) \leq \ell(\boldsymbol{x}_* + \boldsymbol{s})$ for all $\boldsymbol{s} \in \mathbb{R}^n$. Worst-case complexity results that bound the effort required to attain $\|\Psi(\boldsymbol{x}_k)\| \leq \epsilon$ are included in Table A.1.

Methods using the local approximation (5.6) require convexity of h in order for Ψ in (5.8) to be interpreted as a stationarity measure. From a practical perspective, the form of h directly affects the difficulty of solving the trust-region subproblem. Grapiglia *et al.* (2016) demonstrate this approach on a collection of problems of the form

$$f(\boldsymbol{x}) = \max_{i=1,\dots,p} F_i(\boldsymbol{x}), \tag{5.9}$$

where each F_i is assumed smooth. Garmanjani *et al.* (2016) test their method on a collection of problems of the form

$$f(\boldsymbol{x}) = \|\boldsymbol{F}(\boldsymbol{x})\|_1 = \sum_{i=1}^{p} |F_i(\boldsymbol{x})|. \tag{5.10}$$

For objectives of the form (5.9) or (5.10), the associated trust-region subproblems (5.7) can be cast as linear programs when the ∞-norm defines the trust region. (An early example of such an approach appears in Madsen (1975), where linear approximations to each F_i in (5.9) are constructed.) Although more general convex h could fit into this framework, one must be wary of the difficulty of the resulting subproblems.

Direct-search methods have also been adapted for composite non-smooth functions of specific forms. In these variants, knowledge of h in (5.5) informs the selection of search directions in a manner similar to that described in Section 5.2. Ma and Zhang (2009) and Bogani, Gasparo and Papini (2009) consider the cases of f of the form (5.9) and (5.10), respectively.

Objectives of the form (5.9) are also addressed by Hare, Planiden and Sagastizábal (2019), who develop an algorithm that decomposes such problems into orthogonal subspaces associated with directions of non-smoothness and directions of smoothness. The resulting derivative-free VU-algorithm employs model-based estimates of gradients to form and update this decomposition (Hare 2014). Liuzzi, Lucidi and Sciandrone (2006) address finite minimax problems by converting the original problem into a smooth problem using an exponential penalty function. Their DDS method adjusts the penalty parameter via a rule that depends on the current step size in order to guarantee convergence to a Clarke stationary point.

Approximate gradient-sampling methods are developed and analysed by Hare and Macklem (2013) and Hare and Nutini (2013) for the finite minimax problem (5.9). These methods effectively exploit the subdifferential structure of $h(\boldsymbol{y}) = \max_{i=1,\dots,p} y_i$ and employ derivative-free approximations of each $\nabla F_i(\boldsymbol{x})$. Larson, Menickelly and Wild (2016) propose a variant of gradient sampling, called *manifold sampling*, for objectives of the form (5.10). Unlike (approximate) gradient sampling, manifold sampling does not depend on a random sample of points to estimate the ϵ-subdifferential.

Algorithm 8: Smoothing method

---

1   Set initial smoothing parameter $\mu_1 > 0$, terminal smoothing parameter
    $\mu_* < \mu_1$, and decrease parameter $\gamma \in (0, 1)$
2   Choose initial point $\boldsymbol{x}_0$ and smooth optimization method $\mathfrak{M}$
3   $k \leftarrow 1$
4   **while** $\mu_k < \mu_*$ **do**
5   Apply $\mathfrak{M}$ to f_{μ_k}, supplying $\boldsymbol{x}_{k-1}$ as an initial point to $\mathfrak{M}$, until a
      termination criterion is satisfied and $\boldsymbol{x}_k$ is returned
6      $\mu_{k+1} \leftarrow \gamma\mu_k$
7      $k \leftarrow k + 1$

5.3.2. Non-convex h

When h is non-convex, minimization of (5.5) is considerably more challenging than when h is convex. Few methods exist that exploit the structure of non-convex h. One of the many challenges is that the model in (5.6) may no longer be an underestimator of h. Khan *et al.* (2018) propose a manifold sampling method for piecewise linear h; in contrast to the previously discussed methods, this method does not require that h be convex. Other methods applicable for non-convex h employ *smoothing functions*.

As mentioned in Section 2.1.2, the worst-case complexity of DDS methods applied to non-smooth (Lipschitz-continuous) objective functions is difficult to analyse. The reason that DDS methods generate an asymptotically dense set of polling directions is to ensure that no descent directions exist. An exception to this generality, however, is functions for which an appropriate smoothing function exists. Given a locally Lipschitz-continuous f, we say that $f_\mu : \mathbb{R}^n \to \mathbb{R}$ is a smoothing function for f provided that for any $\mu \in (0, \infty)$, f_μ is continuously differentiable and that

$$\lim_{\boldsymbol{z} \to \boldsymbol{x}, \mu \to 0^+} f_\mu(\boldsymbol{z}) = f(\boldsymbol{x}),$$

for all $\boldsymbol{x} \in \mathbb{R}^n$.

Thus, if a smoothing function f_μ exists for f, it is natural to iteratively apply a method for smooth unconstrained optimization to obtain approximate solutions $\boldsymbol{x}_k$ to $\min_{\boldsymbol{x}} f_{\mu_k}(\boldsymbol{x})$ while decreasing μ_k. We roughly prescribe such a smoothing method in Algorithm 8.

Garmanjani and Vicente (2012) consider the DDS framework analysed by Vicente (2013) as the method $\mathfrak{M}$ in Algorithm 8. They terminate $\mathfrak{M}$ when the step-size parameter α of Algorithm 2 is sufficiently small, where the notion of sufficiently small scales with μ_k in Algorithm 8. Garmanjani and Vicente (2012) prove a first-order stationarity result of the form

$$\liminf_{k \to \infty} \|\nabla f_{\mu_k}(\boldsymbol{x}_k)\| = 0. \tag{5.11}$$

Under certain assumptions (for instance, that h satisfies some regularity conditions at $\boldsymbol{F}(\boldsymbol{x}_*)$) this first-order stationarity result is equivalent to $0 \in \partial f(\boldsymbol{x}_*)$; that is, $\boldsymbol{x}_*$ is Clarke stationary.

Garmanjani and Vicente (2012) consider the decrease to be sufficient in line 4 of Algorithm 1 if $f(\boldsymbol{p}_i) < f(\boldsymbol{x}) - c_1 \alpha^{3/2}$ for some $c_1 > 0$. If, furthermore, $\mathfrak{M}$ terminates in each iteration of Algorithm 8 when $\alpha < c_2 \mu_k^2$ for some $c_2 > 0$, then an upper bound on the number of function evaluations needed to obtain

$$\|\nabla f_{\mu_*}(\boldsymbol{x}_k)\| \leq \epsilon, \tag{5.12}$$

for $\epsilon \in (0,1)$ and $\mu_* \in O(n^{-1/2}\epsilon)$, is in $O(\epsilon^{-3})$; see Table A.1. We note that while the sequence of smoothing parameters μ_k induces a type of limiting behaviour of the gradients (as seen in (5.12)) returned by the method $\mathfrak{M}$ used in Algorithm 8, this still does not necessarily recover elements of the Clarke subdifferential of f. The smoothing functions f_{μ_k} must satisfy an additional *gradient consistency property* in order for Algorithm 8 to produce a sequence of points $\boldsymbol{x}_k$ converging to Clarke stationary points (Rockafellar and Wets 2009, Theorem 9.67).

Garmanjani *et al.* (2016) consider the use of a model-based trust-region method $\mathfrak{M}$ in Algorithm 8. The authors demonstrate the first-order convergence result (5.11); they also prove the same WCC as is proved by Garmanjani and Vicente (2012).

5.4. Bilevel and general minimax problems

Bilevel optimization addresses problems where a *lower-level* objective is embedded within an *upper-level* problem. Bilevel problems take the form

$$\begin{aligned} \underset{\boldsymbol{x}\in\Omega}{\text{minimize}} \quad & f^u(\boldsymbol{x}, \boldsymbol{x}^l) \\ \text{subject to} \quad & \boldsymbol{x}^l \in \arg\min_{\boldsymbol{z}\in\Omega^l}\{f^l(\boldsymbol{x}, \boldsymbol{z})\}, \end{aligned} \tag{5.13}$$

where $f^u : \Omega \subseteq \mathbb{R}^n \to \mathbb{R}$ and $f^l : \Omega^l \subseteq \mathbb{R}^{n^l} \to \mathbb{R}$. Conn and Vicente (2012) propose a model-based trust-region method for solving (5.14) in the absence of derivative information. They show how to obtain approximations of the upper-level objective by solving the lower-level problem to sufficient accuracy. Mersha and Dempe (2011) and Zhang and Lin (2014) develop DDS-based algorithms for (5.13) under particular assumptions (*e.g.* strict convexity of the lower-level problem).

A special case of (5.13) is when $f^l = -f^u$, which results in the minimax problem (DET), where the objective is given by a maximization:

$$f(\boldsymbol{x}) = \max_{\boldsymbol{x}^l \in \Omega^l} f^l(\boldsymbol{x}, \boldsymbol{x}^l). \tag{5.14}$$

In contrast to the finite minimax problem (5.9), the objective in (5.14) involves a potentially infinite set Ω^l.

Bertsimas, Nohadani and Teo (2010) and Bertsimas and Nohadani (2010) consider (5.14) when exact gradients of f^l may not be available. The authors assume that approximate gradients of f^l are available and propose methods with convergence analysis restricted to functions f in (5.14) that are convex in $\boldsymbol{x}$. Ciccazzo *et al.* (2015) and Latorre, Habal, Graeb and Lucidi (2019) develop derivative-free methods that employ approximate solutions of the inner problem in (5.14). Menickelly and Wild (2019) also consider (5.14) and develop a derivative-free method of outer approximations for more general f. Their analysis shows that the resulting limit points are Clarke stationary for f.

6. Methods for stochastic optimization

We now turn our attention to methods for solving the stochastic optimization problem (STOCH). In Section 4.2, we considered the case where $f(\boldsymbol{x}) = \mathbb{E}_{\boldsymbol{\xi}}[\tilde{f}(\boldsymbol{x}; \boldsymbol{\xi})]$ is convex. In this section, we lift the assumption of convexity to consider a more general class of stochastic functions $\tilde{f}$.

In general, the analysis of methods for stochastic optimization requires assumptions on the random variable $\boldsymbol{\xi}$. In this section, we use the convention that $\boldsymbol{\xi} \sim \Xi$ denotes that the random variable $\boldsymbol{\xi}$ is from a distribution Ξ and that $\boldsymbol{\xi} \in \Xi$ refers to a random variable in the support of this distribution. Frequently, realizations $\boldsymbol{\xi} \sim \Xi$ are assumed to be independent and identically distributed (i.i.d.). Throughout this section, we assume that $\mathbb{E}_{\boldsymbol{\xi}}[\tilde{f}(\boldsymbol{x}; \boldsymbol{\xi})]$ exists for each $\boldsymbol{x} \in \Omega$ and $f(\boldsymbol{x}) = \mathbb{E}_{\boldsymbol{\xi}}[\tilde{f}(\boldsymbol{x}; \boldsymbol{\xi})]$; that is, the objective of (STOCH) is well-defined. Another common assumption in the stochastic optimization literature is that some bound on the variance of $\tilde{f}(\boldsymbol{x}; \boldsymbol{\xi})$ is assumed, that is,

$$\mathbb{E}_{\boldsymbol{\xi}}[(\tilde{f}(\boldsymbol{x}; \boldsymbol{\xi}) - f(\boldsymbol{x}))^2] < \sigma^2 < \infty \quad \text{for all } \boldsymbol{x} \in \Omega. \tag{6.1}$$

If, for a given $\boldsymbol{x}$, $\nabla_{\boldsymbol{x}} \tilde{f}(\boldsymbol{x}; \boldsymbol{\xi})$ exists for each $\boldsymbol{\xi} \in \Xi$, then under certain regularity conditions it follows that $\nabla f(\boldsymbol{x}) = \mathbb{E}_{\boldsymbol{\xi}}[\nabla_{\boldsymbol{x}} \tilde{f}(\boldsymbol{x}; \boldsymbol{\xi})]$; one such regularity condition is that

$$\tilde{f}(\cdot; \boldsymbol{\xi}) \text{ is } L_{\tilde{f}(\cdot; \boldsymbol{\xi})}\text{-Lipschitz-continuous and } \mathbb{E}_{\boldsymbol{\xi}}\big[L_{\tilde{f}(\cdot; \boldsymbol{\xi})}\big] < \infty.$$

We note that when first-order information is available, the assumption (6.1) is often replaced by an assumption on the variance of the expected gradient norm; see *e.g.* Bottou, Curtis and Nocedal (2018, Assumption 4.3). In this setting, a key class of methods for (STOCH) are *stochastic approximation* (SA) methods; see the paper proposing SA methods by Robbins and Monro (1951) and a survey of modern SA methods (often also referred to as 'stochastic gradient' methods when first-order information is available) by Bottou *et al.* (2018). Here we focus on situations where no objective derivative information is available; that is, stochastic gradient methods are not directly applicable. That said, some of the work we discuss attempts to approximate stochastic gradients, which are then used in an SA framework. As discussed in Section 1, we will not address global optimization methods, such as Bayesian optimization.[11]

Section 6.1 discusses stochastic approximation methods, and Section 6.2 presents direct-search methods for stochastic optimization. In Section 6.3 we highlight modifications to derivative-free model-based methods to address (STOCH), and in Section 6.4 we discuss bandit methods for (non-convex) stochastic optimization.

6.1. Stochastic and sample-average approximation

One of the first analysed approaches for solving (STOCH) is the method of Kiefer and Wolfowitz (1952), inspired by the SA method of Robbins and Monro (1951). We state the basic Kiefer–Wolfowitz framework in Algorithm 9. Since Kiefer and Wolfowitz (1952) consider only univariate problems, Algorithm 9 is in fact the multivariate extension first of Blum (1954b). In Algorithm 9, $\nabla f(\boldsymbol{x}_k)$ is approximated by observing realizations of $\tilde{f}$ using central differences. That is, $\nabla f(\boldsymbol{x}_k)$ is approximated by

$$
\boldsymbol{g}^{\mathrm{K}}(\boldsymbol{x}_k; \mu_k; \boldsymbol{\xi}_k) = \begin{bmatrix} \dfrac{\tilde{f}(\boldsymbol{x}_k + \mu_k \boldsymbol{e}_1; \boldsymbol{\xi}_1^+) - \tilde{f}(\boldsymbol{x}_k - \mu_k \boldsymbol{e}_1; \boldsymbol{\xi}_1^-)}{2\mu_k} \\ \vdots \\ \dfrac{\tilde{f}(\boldsymbol{x}_k + \mu_k \boldsymbol{e}_n; \boldsymbol{\xi}_n^+) - \tilde{f}(\boldsymbol{x}_k - \mu_k \boldsymbol{e}_n; \boldsymbol{\xi}_n^-)}{2\mu_k} \end{bmatrix}, \qquad (6.2)
$$

where $\mu_k > 0$ is a difference parameter, $\boldsymbol{e}_i$ is the ith elementary basis vector, and $2n$ realizations $\boldsymbol{\xi} \sim \Xi$ are employed. The next point $\boldsymbol{x}_{k+1}$ is then set to be $\boldsymbol{x}_k - \alpha_k \boldsymbol{g}^{\mathrm{K}}(\boldsymbol{x}_k; \mu_k; \boldsymbol{\xi}_k)$, where $\alpha_k > 0$ is a step-size parameter. As in Section 3, we note that $\boldsymbol{x}_{k+1}$ is a random variable that depends on the filtration generated by the method before $\boldsymbol{x}_{k+1}$ is realized; this will be the case throughout this section. In the SA literature, the sequences $\{\alpha_k\}$ and $\{\mu_k\}$ are often referred to as *gain* sequences.

Because evaluation of the function f requires computing an expectation (and in contrast to the primarily monotone algorithms in Section 2), stochastic optimization methods generally do not monotonically decrease f. This is exemplified by Algorithm 9, which updates $\boldsymbol{x}_{k+1}$ without considering the value of $\tilde{f}(\boldsymbol{x}_{k+1}; \boldsymbol{\xi})$ for *any* realization of $\boldsymbol{\xi}$.

---

[11] We recommend Shahriari *et al.* (2016) and Frazier (2018) to readers interested in recent surveys of Bayesian optimization.

Algorithm 9: Kiefer–Wolfowitz method

---

1   Choose initial point $\boldsymbol{x}_0$, sequence of step sizes $\{\alpha_k\}$ and sequence of difference parameters $\{\mu_k\}$

2   **for** $k = 0, 1, 2, \ldots$ **do**

3   Generate $\boldsymbol{\xi}_k = (\boldsymbol{\xi}_1^+, \boldsymbol{\xi}_1^-, \ldots, \boldsymbol{\xi}_n^+, \boldsymbol{\xi}_n^-)$

4   Compute gradient estimate $\boldsymbol{g}_k^{\mathrm{K}}(\boldsymbol{x}_k; \mu_k; \boldsymbol{\xi}_k)$ via (6.2)

5      $\boldsymbol{x}_{k+1} \leftarrow \boldsymbol{x}_k - \alpha_k \boldsymbol{g}_k^{\mathrm{K}}(\boldsymbol{x}_k; \mu_k; \boldsymbol{\xi}_k)$

Historically, Algorithm 9 has been analysed by a community more concerned with stochastic processes than with optimization. Hence, convergence results differ from those commonly found in the optimization literature. For example, many results in the SA literature consider a continuation of the dynamics of Algorithm 9 applied to the deterministic f as an ordinary differential equation (ODE) in terms of $\boldsymbol{x}(t) : \mathbb{R} \to \mathbb{R}^n$. That is, they consider

$$\frac{\mathrm{d}\boldsymbol{x}}{\mathrm{d}t} = -\nabla f(\boldsymbol{x}), \quad \boldsymbol{x} = \boldsymbol{x}(t).$$

and define the set of fixed points of the ODE, $\boldsymbol{S} = \{\boldsymbol{x} : \nabla f(\boldsymbol{x}) = 0\}$. Many convergence results then demonstrate that the continuation $\boldsymbol{x}(t)$ satisfies $\boldsymbol{x}(t) \to \boldsymbol{S}$ with probability one as the continuous iteration counter $t \to \infty$; see Kushner and Yin (2003) for a complete treatment of such ODE results.

In order to prove that the sequence of points $\boldsymbol{x}_k$ generated by Algorithm 9 converges almost surely (*i.e.* with probability one), conditions must be placed on the objective function, step sizes and difference parameters. In the SA literature there is no single consistent set of conditions, but there are nearly always conditions on the sequence of step sizes $\{\alpha_k\}$ requiring $\alpha_k \to 0$ and $\sum_k \alpha_k = \infty$. Intuitively, this divergence condition ensures that any point in the domain $\boldsymbol{\Omega}$ can be reached, independent of the history of iterations. As one example of convergence conditions, Bhatnagar, Prasad and Prashanth (2013) prove almost sure convergence of Algorithm 9 under the following assumptions (simplified for presentation).

(1) The sequences of step sizes and difference parameters satisfy $\alpha_k > 0$, $\mu_k > 0$, $\alpha_k \to 0$, $\mu_k \to 0$, $\sum_k \alpha_k = \infty$ and $\sum_k \alpha_k^2 \mu_k^{-2} < \infty$.

(2) The realizations $\boldsymbol{\xi} \sim \boldsymbol{\Xi}$ are i.i.d. and the distribution $\boldsymbol{\Xi}$ has a finite second moment.

(3) The function f is in $\mathcal{LC}^1$.

(4) $\sup_k\{\|\boldsymbol{x}_k\|\} < \infty$ with probability one.

Similar assumptions on algorithms of the form Algorithm 9 appear throughout the SA literature (Blum 1954a, Derman 1956, Sacks 1958, Fabian 1971,

Kushner and Huang 1979, Ruppert 1991, Spall 2005). Convergence of Algorithm 9 under similar assumptions to those above, but with the modification that μ_k is fixed in every iteration to a sufficiently small constant (that scales inversely with L_{g}), is additionally demonstrated by Bhatnagar *et al.* (2013).

In terms of WCCs, the convergence rates that have been historically derived for Algorithm 9 are also non-standard for optimization. In particular, results concerning convergence rates are typically shown as a convergence in distribution (Durrett 2010, Chapter 3.2): given a fixed $\boldsymbol{x}_* \in \boldsymbol{S}$,

$$\frac{1}{k^\gamma}(\boldsymbol{x}_k - \boldsymbol{x}_*) \to \mathcal{N}(\boldsymbol{0}, \boldsymbol{B}), \tag{6.3}$$

where $\gamma > 0$ and $\boldsymbol{B}$ is a covariance matrix, the entries of which depend on algorithmic parameters and $\nabla^2 f(\boldsymbol{x}_*)$ (provided it exists). With few assumptions on $\boldsymbol{\xi}$, it has been shown that (6.3) holds with $\gamma = 1/3$ (Spall 1992, L'Ecuyer and Yin 1998). Observe that these convergence rates are distinct from WCC results like those in (2.2).

Later, the use of *common random numbers* (CRNs) was considered. In contrast to (6.2), which employs a realization $\boldsymbol{\xi}_k = (\boldsymbol{\xi}_1^+, \boldsymbol{\xi}_1^-, \dots, \boldsymbol{\xi}_n^+, \boldsymbol{\xi}_n^-)$, a gradient estimator in the CRN regime uses a single realization $\boldsymbol{\xi}_k$ and has the form

$$\boldsymbol{g}^{\mathrm{K}}(\boldsymbol{x}_k; \mu_k; \boldsymbol{\xi}_k) = \begin{bmatrix} \delta_{\mathrm{c}}(\tilde{f}(\cdot; \boldsymbol{\xi}_k); \boldsymbol{x}_k; \boldsymbol{e}_1; \mu_k) \\ \vdots \\ \delta_{\mathrm{c}}(\tilde{f}(\cdot; \boldsymbol{\xi}_k); \boldsymbol{x}_k; \boldsymbol{e}_n; \mu_k) \end{bmatrix}, \tag{6.4}$$

where $\delta_{\mathrm{c}}(\cdot)$ is defined in (2.29)

The difference between (6.2) and (6.4) is analogous to the difference between one-point and two-point bandit feedback in the context of bandit problems (see Section 4.2). In the CRN regime, we can recall a single realization $\boldsymbol{\xi}_k$ to compute a finite-difference approximation in each coordinate direction. By using (6.4) as the gradient estimator in Algorithm 9, the rate (6.3) holds with $\gamma = 1/2$ (L'Ecuyer and Yin 1998, Kleinman, Spall and Naiman 1999). Thus, as in the analysis of bandit methods, the use of CRNs allows for strictly better convergence rate results.

Dai (2016*a*, 2016*b*) studies the complexity of Algorithm 9, as well as a method that uses the estimator (6.2) in Algorithm 7, under varying assumptions on $\boldsymbol{\Xi}$. Dai considers a gradient estimate of the form (6.4) with $\delta_{\mathrm{f}}(\tilde{f}(\cdot; \boldsymbol{\xi}_k); \boldsymbol{x}_k; \boldsymbol{e}_i; \mu_k)$ replacing each central difference; recall the definition of $\delta_{\mathrm{f}}(\cdot)$ in (2.28). Dai demonstrates that the best rate of the form (6.3) achievable by Algorithm 9 with forward differences has $\gamma = 1/3$, even when common random numbers are used. However, a rate of the form (6.3) with $\gamma = 1/2$ can be achieved using forward differences in Algorithm 7; Dai

(2016a, 2016b) draws parallels between this result and the WCC of Duchi *et al.* (2015), discussed in Section 4.2.2.

We remark that the gradient estimate (6.2) used in Algorithm 9 requires $2n$ evaluations of $\tilde{f}$ per iteration. Although replacing $\delta_{\mathrm{c}}(\tilde{f}(\cdot\,;\boldsymbol{\xi}_k);\boldsymbol{x}_k;\boldsymbol{e}_i;\mu_k)$ with $\delta_{\mathrm{f}}(\tilde{f}(\cdot\,;\boldsymbol{\xi}_k);\boldsymbol{x}_k;\boldsymbol{e}_i;\mu_k)$ could reduce this cost to $n+1$ evaluations of $\tilde{f}$ per iteration, it is still desirable to reduce this per-iteration cost from $O(n)$ to $O(1)$ evaluations. The SPSA method of Spall (1992) achieves this goal by using the gradient estimator

$$\boldsymbol{g}^{\mathrm{S}}(\boldsymbol{x}_k;\mu_k;\boldsymbol{\xi}_k;\boldsymbol{u}_k)=\delta_{\mathrm{c}}(\tilde{f}(\cdot\,;\boldsymbol{\xi}_k);\boldsymbol{x}_k;\boldsymbol{u}_k;\mu_k)\begin{bmatrix}\dfrac{1}{[\boldsymbol{u}_k]_1}\\[4pt]\vdots\\[4pt]\dfrac{1}{[\boldsymbol{u}_k]_n}\end{bmatrix},\tag{6.5}$$

where $\boldsymbol{u}_k\in\mathbb{R}^n$ is randomly generated from some distribution in each iteration.

The construction of (6.5) requires evaluations of $\tilde{f}(\cdot\,;\boldsymbol{\xi}_k)$ at exactly two points. Algorithm 9 is then modified by replacing the gradient estimator $\boldsymbol{g}_k^{\mathrm{K}}(\boldsymbol{x}_k;\mu_k;\boldsymbol{\xi}_k)$ with $\boldsymbol{g}_k^{\mathrm{S}}(\boldsymbol{x}_k;\mu_k;\boldsymbol{\xi}_k;\boldsymbol{u}_k)$. Informally, the conditions on the distribution governing $\boldsymbol{u}_k$ originally proposed by Spall (1992) cause each entry of $\boldsymbol{u}_k$ to be bounded away from 0 with high probability (intuitively, to avoid taking huge steps). A simple example distribution satisfying these properties is to let each entry of $\boldsymbol{u}_k$ independently follow a Bernoulli distribution with support $\{1,-1\}$, both events occurring with probability $1/2$. Under appropriate assumptions resembling those for Algorithm 9, the sequence $\{\boldsymbol{x}_k\}$ generated by SPSA can be shown to converge in the same sense as Algorithm 9. Convergence rates of the form (6.3) matching those obtained for Algorithm 9 have also been established (Gerencsér 1997, Kleinman *et al.* 1999).

The performance of SA methods is highly sensitive to the chosen sequence of step sizes $\{\alpha_k\}$ (Hutchison and Spall 2013). This mirrors the situation in gradient-based SA methods where the tuning of algorithmic parameters is an active area of research (Diaz, Fokoue-Nkoutche, Nannicini and Samulowitz 2017, Ilievski, Akhtar, Feng and Shoemaker 2017, Balaprakash *et al.* 2018).

The SA methods above consider only a single evaluation of the stochastic function $\tilde{f}$ at any point. Other methods more accurately estimate $f(\boldsymbol{x}_k)$ by querying $\tilde{f}(\boldsymbol{x}_k;\boldsymbol{\xi})$ for multiple, different realizations ('samples') of $\boldsymbol{\xi}$. These methods belong to the framework of sample average approximation, wherein the original problem (STOCH) is replaced with a (sequence of) deterministic *sample-path problem(s)*:

$$\underset{\boldsymbol{x}\in\Omega}{\text{minimize}}\ \frac{1}{p}\sum_{i=1}^{p}\tilde{f}(\boldsymbol{x};\boldsymbol{\xi}_i).\tag{6.6}$$

Retrospective approximation methods (Chen and Schmeiser 2001) vary the number of samples, p, in a predetermined sequence $\{p_0, p_1, \ldots\}$; the accuracy to which each instance of (6.6) subproblem is solved can also vary as a sample average approximation method progresses. Naturally, the performance of such a method depends critically on the sequence of sample sizes and accuracies used at each iteration; Pasupathy (2010) characterizes a class of sequences of predetermined sample sizes and accuracies for which derivative-free retrospective approximation methods can be shown to converge for smooth objectives.

Other approaches dynamically adjust the number of samples p from iteration to the next. For example, the method of Pasupathy, Glynn, Ghosh and Hashemi (2018) adjusts the number of samples p_k to balance the contributions from deterministic and stochastic errors in iteration k. The stochastic error at $\boldsymbol{x}_k$ is then

$$\left| f(\boldsymbol{x}_k) - \frac{1}{p_k} \sum_{i=1}^{p_k} \tilde{f}(\boldsymbol{x}_k; \boldsymbol{\xi}_{k,i}) \right|.$$

The deterministic error is the difference between the objective f and a specified approximation; for example, the deterministic error at $\boldsymbol{x}_k - \alpha_k \nabla f(\boldsymbol{x}_k)$ using a first-order Taylor approximation is

$$|f(\boldsymbol{x}_k - \alpha_k \nabla f(\boldsymbol{x}_k)) - (f(\boldsymbol{x}_k) - \alpha_k \|\nabla f(\boldsymbol{x}_k)\|^2)|.$$

Pasupathy *et al.* (2018) establish convergence rates for a variant of Algorithm 9 drawing independent samples $\{\boldsymbol{\xi}_{k,1}, \ldots, \boldsymbol{\xi}_{k,p_k}\}$ in each iteration.

6.2. Direct-search methods for stochastic optimization

Unsurprisingly, researchers have modified methods for deterministic objectives in order to produce methods appropriate for stochastic optimization. For example, in the paper inspiring Nelder and Mead (1965), Spendley *et al.* (1962) propose re-evaluating the point corresponding to the best simplex vertex if it hasn't changed in $n + 1$ iterations, saying that if the vertex is best 'only by reason of errors of observation, it is unlikely that the repeat observation will [be the best observed point], and the point will be eliminated in due course'. Barton and Ivey, Jr (1996) propose modifications to the Nelder–Mead method in order to avoid premature termination due to repeated shrinking. To alleviate this problem, they suggest reducing the amount the simplex is shrunk, re-evaluating the best point after each shrink operation, and re-evaluating each reflected point before performing a contraction. Chang (2012) proposes a Nelder–Mead variant that samples candidate points and all other points in the simplex an increasing number of times; this method ultimately ensures that stochasticity in the function evaluations will not affect the correct ranking of simplex vertices.

Sriver, Chrissis and Abramson (2009) augment a GPS method with a ranking and selection procedure and dynamically determine the number of samples performed for each polling point. The ranking and selection procedure allows the method to also address cases where $\boldsymbol{x}$ contains discrete variables. For the case of additive unbiased, Gaussian noise (*i.e.* $\tilde{f}(\boldsymbol{x};\boldsymbol{\xi}) = f(\boldsymbol{x}) + \sigma\boldsymbol{\xi}$ with $\boldsymbol{\xi}$ from a standard normal distribution and $\sigma > 0$ finite), they prove that the resulting method converges almost surely to a stationary point of f. For problems involving more general distributions, Kim and Zhang (2010) consider a DDS method that employs the sample mean

$$\frac{1}{p_k}\sum_{i=1}^{p_k}\tilde{f}(\boldsymbol{x};\boldsymbol{\xi}_{k,i}), \tag{6.7}$$

with a dynamically increasing sample size p_k. They establish a consistency result and appeal to the convergence properties of DDS methods. Sankaran, Audet and Marsden (2010) propose a surrogate-assisted method for stochastic optimization inspired by stochastic collocation techniques (see *e.g.* Gunzburger, Webster and Zhang 2014). Convergence for the method is established by appealing to the GPS and MADS mechanisms underlying the method.

Chen and Kelley (2016) consider an implicit-filtering method in which values of f are observable only through the sample average (6.7). Chen and Kelley (2016) demonstrate that the sequence of points generated by the method converges (*i.e.* $\{\nabla f(\boldsymbol{x}_k)\}$ admits a subsequence that converges to zero) with probability one if the sample size p_k increases to infinity. Algorithmically, p_k is adjusted to scale with the square of the inverse of the stencil step size (Δ_k in Algorithm 4).

Chen, Kelley, Xu and Zhang (2018*b*) consider the bound-constrained minimization of a composite non-smooth function of the form (5.5), where h is Lipschitz-continuous (but non-smooth) and $\boldsymbol{F}$ is continuously differentiable. However, they assume that values of $\boldsymbol{F}$ are observable only through sample averages and that a smoothing function h_μ of h (as discussed in Section 5.3.2) is available. They show that with probability one, the sequence of points from a smoothed implicit-filtering method converges to a first-order stationary point, where the stationarity measure is appropriate for non-smooth optimization.

6.3. Model-based methods for stochastic optimization

Analysis of the model-based trust-region methods in Section 2.2 generally depends on the construction of fully linear models of a deterministic function f; see (2.9). In particular, methods of the form of Algorithm 3 typically require that a model m_k satisfy

$$|f(\boldsymbol{x}_k + \boldsymbol{s}) - m_k(\boldsymbol{x}_k + \boldsymbol{s})| \leq \kappa_{\mathrm{ef}}\Delta_k^2 \quad \text{for all } \boldsymbol{s} \in \mathcal{B}(\boldsymbol{0};\Delta_k).$$

A natural model-based trust-region approach to stochastic optimization is to build a model m_k of the function f by fitting the model to observed values of the stochastic function $\tilde{f}$. Intuitively, if such models satisfy (2.9), then an extension of the analysis described in Section 2.2.4 should also apply to the minimization of f in (STOCH). The methods described here formalize the approximation properties of such models (which are stochastic because of their dependence on $\boldsymbol{\xi}$) and employ the models in a trust-region framework. For example, by employing an estimator $\bar{f}_p$ of f at each interpolation point $\boldsymbol{x}$ used in model construction, we can replace each function value $f(\boldsymbol{x})$ with $\bar{f}_p(\boldsymbol{x})$ in the interpolation system (2.14). One example of such an estimator $\bar{f}_p$ is the sample average (6.7).

Early work in applying derivative-free trust-region methods for stochastic optimization includes that of Deng and Ferris (2006), which modifies the UOBYQA method of Powell (2002). The kth iteration of the method of Deng and Ferris (2006) uses Bayesian techniques to dynamically update a budget of p_k new $\tilde{f}$ evaluations. This budget is then apportioned among the current set of interpolation points $\boldsymbol{y} \in \boldsymbol{Y}$ in order to reduce the variance in each value of $\bar{f}_{p_k}(\boldsymbol{y})$, with the authors using the sample mean for the estimator $\bar{f}_{p_k}$. Deng and Ferris (2009) show that, given assumptions on the sequence of evaluated $\boldsymbol{\xi}$ (*i.e.* the sample path), every limit point $\boldsymbol{x}_*$ produced by this method is stationary with probability 1.

Another method in this vein, STRONG, was proposed by Chang, Hong and Wan (2013) and combines response surface methodology (Box and Draper 1987) with a trust-region mechanism. In the analysis of STRONG, it is assumed that model gradients $\nabla m_k(\boldsymbol{x}_k)$ almost surely equal the true gradients $\nabla f(\boldsymbol{x}_k)$ as $k \to \infty$, which is algorithmically encouraged by monotonically increasing the sample size p_k in an inner loop. QNSTOP by Castle (2012) presents a similar approach using response surface models in a trust-region framework, but its convergence analysis and assumptions mirror those of stochastic approximation methods.

Both Larson and Billups (2016) and Chen, Menickelly and Scheinberg (2018*a*) build on the idea of probabilistically fully linear models in (3.5), which essentially says that the condition (2.9) needs to hold on a given iteration only with some probability (Bandeira *et al.* 2014). In contrast to the usage of such models in randomized methods for deterministic objectives (the subject of Section 3.3), in stochastic optimization the filtration in (3.5) also includes the realizations of the stochastic evaluations of $\tilde{f}$. This probabilistic notion of uniform local model quality is powerful. For example, although the connection is not made by Regier, Jordan and McAuliffe (2017), this notion of model quality implies probabilistic descent properties such as those required by Regier *et al.* (2017). This implication is an example of a setting in which stochastic gradient estimators can be replaced by gradients of probabilistically fully linear models.

One way to satisfy (3.5) is to build a regression model using randomly sampled points. For example, Menickelly (2017, Theorem 4.2.6) shows that evaluating $\tilde{f}$ on a sufficiently large set of points uniformly sampled from $\mathcal{B}(\boldsymbol{x}_k; \Delta_k)$ can be used to construct a probabilistically fully linear regression model.

Larson and Billups (2016) prove convergence of a probabilistic variant of Algorithm 3 in the sense that, for any $\epsilon > 0$,

$$\lim_{k \to \infty} \mathbb{P}[\|\nabla f(\boldsymbol{x}_k)\| > \epsilon] = 0.$$

Under similar assumptions, Chen *et al.* (2018*a*) prove almost sure convergence to a stationary point, that is,

$$\lim_{k \to \infty} \|\nabla f(\boldsymbol{x}_k)\| = 0 \quad \text{with probability one.} \tag{6.8}$$

Blanchet, Cartis, Menickelly and Scheinberg (2019) provide a WCC result for the variant of Algorithm 3 presented by Chen *et al.* (2018*a*). Blanchet *et al.* (2019) extend the analysis of Cartis and Scheinberg (2018) to study the stopping time of the stochastic process generated by the method of Chen *et al.* (2018*a*). In contrast to previous WCC results discussed in this survey, which bound the number of function evaluations N_ϵ needed to attain some form of *expected ϵ-optimality* (*e.g.* (3.2) or (5.4)), Blanchet *et al.* (2019) prove that the *expected number of iterations*, $\mathbb{E}[T_\epsilon]$, needed to achieve (2.2) is in $O(\epsilon^{-2})$. Paquette and Scheinberg (2018) apply similar analysis to a derivative-free stochastic line-search method, where they demonstrate that for non-convex f, $\mathbb{E}[T_\epsilon] \in O(\epsilon^{-2})$, while for convex and strongly convex f, $\mathbb{E}[T_\epsilon] \in O(\epsilon^{-1})$ and $\mathbb{E}[T_\epsilon] \in O(\log(\epsilon^{-1}))$, respectively. Since the number of function evaluations per iteration of the derivative-free methods of Blanchet *et al.* (2019) and Paquette and Scheinberg (2018) is highly variable across iterations, the total work (in terms of function evaluations) is not readily apparent from such WCC results.

Larson and Billups (2016) and Chen *et al.* (2018*a*) demonstrate that sampling $\tilde{f}$ on $\mathcal{B}(\boldsymbol{x}_k; \Delta_k)$ of the order of Δ_k^{-4} times will ensure that (2.9) holds (*i.e.* one can obtain a fully linear model) with high probability. Shashaani, Hunter and Pasupathy (2016) and Shashaani, Hashemi and Pasupathy (2018) take a related but distinct approach. As opposed to requiring that models be probabilistically fully linear, their derivative-free trust-region method performs adaptive Monte Carlo sampling both at current points $\boldsymbol{x}_k$ and interpolation points; the number of samples p_k is chosen to balance a measure of statistical error with the optimality gap at $\boldsymbol{x}_k$. Shashaani *et al.* (2018) prove that their method achieves almost sure convergence of the form (6.8).

A model-based trust-region method for constrained stochastic optimization, **SNOWPAC**, is developed by Augustin and Marzouk (2017). Their method addresses the stochasticity by employing Gaussian process-based models of robustness measures such as expectation and conditional value at risk. The approach used is an extension of the constrained deterministic method NOWPAC of Augustin and Marzouk (2014), which we discuss in Section 7.

6.4. Bandit feedback methods

While much of the literature on bandit methods for stochastic optimization focuses on convex objectives f (as discussed in Section 4.2), here we discuss treatment of non-convex objectives f. We recall our notation and discussion from Section 4.2, in particular the notion of regret minimization shown in (4.7).

In the absence of convexity, regret bounds do not translate into bounds on optimization error as easily as in (4.8). Some works address the case where each $\tilde{f}(\cdot; \boldsymbol{\xi}_k)$ in (4.7) is Lipschitz-continuous and employ a partitioning of the feasible region $\boldsymbol{\Omega}$ (Kleinberg, Slivkins and Upfal 2008, Bubeck, Stoltz and Yu 2011b, Bubeck, Munos, Stoltz and Szepesvári 2011a, Valko, Carpentier and Munos 2013, Zhang, Yang, Jin and Zhou 2015). These methods employ global optimization strategies that we do not discuss further here.

In another line of work, Ghadimi and Lan (2013) consider the application of an algorithm like Algorithm 6 with the choice of gradient estimator $\boldsymbol{g}_\mu(\boldsymbol{x}; \boldsymbol{u}; \boldsymbol{\xi})$ from (4.12). Under an assumption of bounded variance of the estimator (*i.e.* $\mathbb{E}_{\boldsymbol{\xi}}[\|\boldsymbol{g}_\mu(\boldsymbol{x}; \boldsymbol{u}; \boldsymbol{\xi}) - \nabla f(\boldsymbol{x})\|^2] \leq \sigma^2$), Ghadimi and Lan (2013) prove a WCC result similar to the one they obtained in the convex case; see Section 4.2.2. They show that an upper bound on the (randomized) number of iterations needed to attain

$$\mathbb{E}[\|\nabla f(\boldsymbol{x}_k)\|^2] \leq \epsilon \tag{6.9}$$

is in $O(\max\{nL_g R_{\boldsymbol{x}}\epsilon^{-1}, nL_g R_{\boldsymbol{x}}\sigma^2\epsilon^{-2}\})$. Notice that the stationarity condition given in (6.9) involves a square on the gradient norm, making it distinct from a result like (3.2) or (5.4). Thus, assuming σ^2 is sufficiently large, the result of Ghadimi and Lan (2013) translates to a WCC of type (5.4) in $O(n^2\epsilon^{-4})$.

Balasubramanian and Ghadimi (2018, 2019) propose a method that uses two-point bandit feedback (*i.e.* a gradient estimator from (4.12)) within a derivative-free conditional gradient method (Ghadimi 2019). The gradient estimator is used to define a linear model, which is minimized over $\boldsymbol{\Omega}$ to produce a trial step. If $\boldsymbol{\Omega}$ is bounded, they show a WCC of type (5.4) that again grows like ϵ^{-4}.

By replacing gradients with estimators of the form (4.12) in the stochastic variance-reduced gradient framework of machine learning (Reddi *et al.* 2016), Liu *et al.* (2018) prove a WCC of type (6.9) in $O(n\epsilon^{-1} + b^{-1})$, where b is the size of a minibatch drawn with replacement in each iteration. Gu, Huo and

Huang (2016) prove a similar WCC result in an asynchronous parallel computing environment for a distinct method using minibatches for variance reduction.

7. Methods for constrained optimization

In this section, we discuss derivative-free methods for problems where the feasible region Ω is a proper subset of $\mathbb{R}^n$. In the derivative-free setting, such constrained optimization problems can take many forms since an additional distinction is associated with the derivative-free nature of objective and constraint functions. For example, and in contrast to the preceding sections, a derivative-free constrained optimization problem may involve an objective function f for which a gradient is made available to the optimization method. The problem is still derivative-free if there is a constraint function defining the feasible region Ω for which a (sub)gradient is not available to the optimization method.

As is common in many application domains where derivative-free methods are applied, the feasible region Ω may also involve discrete choices. In particular, these choices can include categorical variables that are either ordinal (*e.g.* letter grades in $\{A, B, C, D, F\}$) or non-ordinal (*e.g.* compiler type in $\{\text{flang, gfortran, ifort}\}$). Although ordinal categorical variables can be mapped to a subset of the reals, the same cannot be done for non-ordinal variables. Therefore, we generalize the formulations of (DET) and (STOCH) to the problem

$$\begin{aligned}
\underset{x,y}{\text{minimize}} \quad & f(x,y) \\
\text{subject to} \quad & x \in \Omega \subset \mathbb{R}^n \\
& y \in \mathbf{N},
\end{aligned} \qquad \text{(CON)}$$

where y represents a vector of non-ordinal variables and $\mathbf{N}$ is a finite set of feasible values. Here we assume that discrete-valued ordinal variables are included in x. Furthermore, most of the methods we discuss do not explicitly treat non-ordinal variables y; hence, except where indicated, we will drop the use of y.

Similar to Section 5, here we distinguish methods based on the assumptions made about the problem structure. We organize these assumptions based on the black-box optimization constraint taxonomy of Le Digabel and Wild (2015), which characterizes the type of constraint functions that occur in a particular specification of a derivative-free optimization problem. When constraints are explicitly stated (*i.e.* 'known' to the method), this taxonomy takes the form of the tree in Figure 7.1.

The first distinction in Figure 7.1 is whether a constraint is algebraically available to the optimization method or whether it depends on a black-box

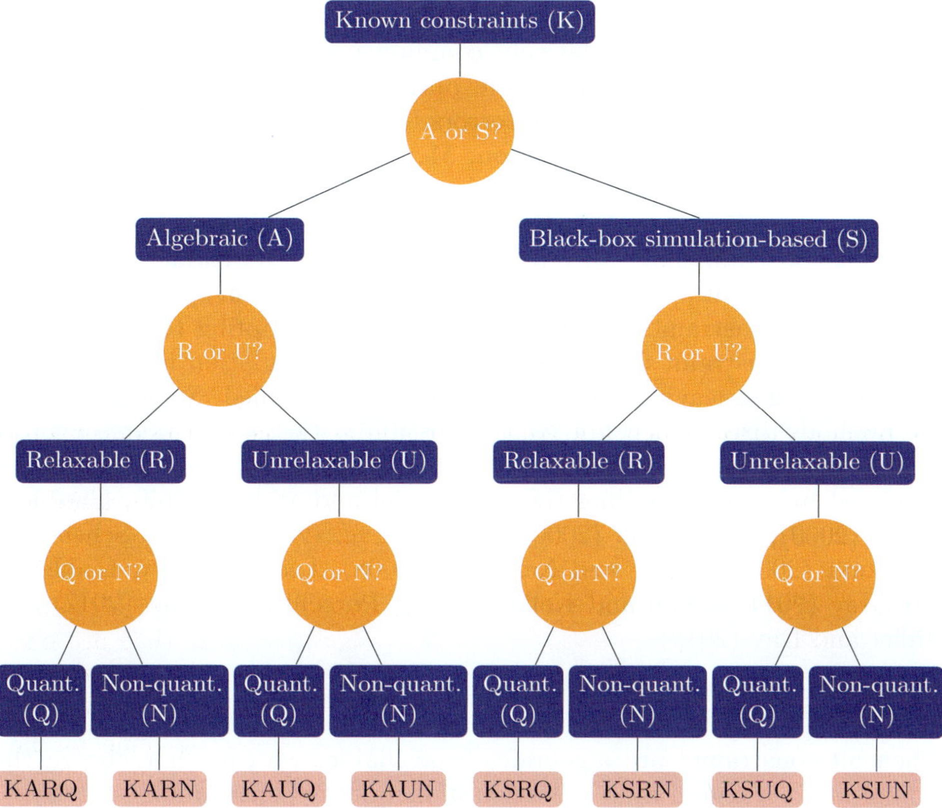

Figure 7.1. Tree-based taxonomy of known (*i.e.* non-hidden) constraints from Le Digabel and Wild (2015).

simulation. In the context of derivative-free optimization, we will assume that it is these latter constraint functions for which a (sub)gradient is not made available to the optimization method. Algebraic constraints are those for which a functional form or simple projection operator is provided to the optimization method. Section 7.1 discusses methods that exclusively handle algebraic constraints. Examples of such algebraic constraints have been discussed earlier in this paper (*e.g.* Sections 3.1 and 4), wherein it is assumed that satisfaction of the constraints (*e.g.* through a simple projection) is trivial relative to evaluation of the objective. This imbalance between the ease of the constraint and objective functions is also the subject of recent WCC analysis (Cartis, Gould and Toint 2018).

Section 7.2 discusses methods that target situations where one or more constraints do not have available derivatives.

The next distinction in Figure 7.1 is whether a constraint can be relaxed or whether the constraint must be satisfied in order to obtain meaningful information for the objective f and/or other constraint functions. Unrelaxable constraints are a relatively common occurrence in derivative-free optimization. In contrast to classic optimization, constraints are sometimes introduced solely to prevent errors in the evaluation of, for example, a simulation-based objective function. Methods for addressing relaxable algebraic constraints are discussed in Section 7.1.1, and unrelaxable algebraic constraints are the focus of Section 7.1.2.

Hidden constraints are not represented in Figure 7.1. Hidden constraints are constraints that are not explicitly stated in a problem specification. Violating these constraints is detected only when attempting to evaluate the objective or constraint functions; for example, a simulation may fail to return output, thus leaving one of these functions undefined. Some derivative-free methods directly account for the possibility that such failures may be present despite not being explicitly stated. Hidden constraints have been addressed in works including those of Avriel and Wilde (1967), Choi and Kelley (2000), Choi *et al.* (2000), Carter *et al.* (2001), Conn, Scheinberg and Toint (2001), Huyer and Neumaier (2008), Lee, Gramacy, Linkletter and Gray (2011), Chen and Kelley (2016), Porcelli and Toint (2017) and Müller and Day (2019).

7.1. Algebraic constraints

When all constraints are algebraically available, we can characterize the ordinal feasible region by a collection of inequality constraints:

$$\Omega = \{\boldsymbol{x} \in \mathbb{R}^n : c_i(\boldsymbol{x}) \leq 0, \text{ for all } i \in I\}, \tag{7.1}$$

where each $c_i : \mathbb{R}^n \to \mathbb{R} \cup \{\infty\}$ and the set I is finite for all of the methods discussed. Problems with semi-infinite constraints can be addressed by using structured approaches as in Section 5.4. In this setting, we define the constraint function $\boldsymbol{c} : \mathbb{R}^n \to (\mathbb{R} \cup \{\infty\})^{|I|}$, where the ith entry of the vector $\boldsymbol{c}(\boldsymbol{x})$ is given by $c_i(\boldsymbol{x})$. Equality constraints can be represented in (7.1) by including both $c_i(\boldsymbol{x})$ and $-c_i(\boldsymbol{x})$; however, this practice should be avoided since it can hamper both theoretical and empirical performance.

7.1.1. Relaxable algebraic constraints

Relaxable algebraic constraints are the constraints that are typically treated in derivative-based non-linear optimization. We will organize our discussion into three primary types of methods: penalty approaches, filter approaches, and approaches with subproblems that employ models of the constraint functions.

Penalty approaches. Given constraints defined by (7.1), it is natural in the setting of relaxable constraints to quantify the violation of the ith constraint via the value of $\max\{0, c_i(\boldsymbol{x})\}$. In fact, given a penalty parameter $\rho > 0$, a common approach in relaxable constrained optimization is to replace the minimization of $f(x)$ with the minimization of a merit function such as

$$f(x) + \frac{\rho}{2} \sum_{i \in I} \max\{0, c_i(x)\}. \tag{7.2}$$

The merit function in (7.2) is typically called an *exact penalty function*, because for a sufficiently large (but finite) value of $\rho > 0$, every local minimum x_* of (CON) is also a local minimum of the merit function in (7.2). We note that each summand $\max\{0, c_i(x)\}$ is generally non-smooth; the summand is still convex provided $c_i(x)$ is convex. Through the mapping

$$F(x) = \begin{bmatrix} f(x) \\ c(x) \end{bmatrix},$$

functions of the form (7.2) can be seen as cases of the composite non-smooth function (5.5) and are hence amenable to the methods discussed in Section 5.3. In contrast to this non-smooth approach, a more popular merit function historically has been the *quadratic penalty function*,

$$f(x) + \rho \sum_{i \in I} \max\{0, c_i(x)\}^2. \tag{7.3}$$

However, the merit function in (7.3) lacks the same exactness guarantees that come with (7.2); even as ρ grows arbitrarily large, local minima of (CON) need not correspond in any way with minima of (7.3).

Another popular means of maintaining the smoothness (and convexity, when applicable) of (7.3) but regaining the exactness of (7.2) is to consider Lagrangian-based merit functions. Associating multipliers λ_i with each of the constraints in (7.1), the Lagrangian of (CON) is

$$L(x; \lambda) = f(x) + \sum_{i \in I} \lambda_i \, c_i(x). \tag{7.4}$$

Combining (7.4) with (7.3) yields the *augmented Lagrangian* merit function

$$L_A(x; \lambda; \rho) = f(x) + \sum_{i \in I} \lambda_i \, c_i(x) + \frac{\rho}{2} \sum_{i \in I} \max\{0, c_i(x)\}^2 \tag{7.5}$$

with the desired properties; that is, for non-negative λ and ρ, $L_A(x; \lambda; \rho)$ is smooth (convex) provided that c is.

In all varieties of these methods, which we broadly refer to as penalty approaches, the parameter ρ is dynamically updated between iterations. Methods typically increase ρ in order to promote feasibility; penalty methods tend to approach solutions from outside of Ω and hence typically assume that the penalized constraints are relaxable. For a review of general penalty approaches, see Fletcher (1987, Chapter 12).

Lewis and Torczon (2002) adapt the augmented Lagrangian approach of Conn, Gould and Toint (1991) in one of the first proofs that DDS methods can be globally convergent for non-linear optimization. They utilize pattern search (see the discussion in Section 2.1.2) to approximately minimize the augmented Lagrangian function (7.5) in each iteration of their method. That is, each iteration of their method solves a subproblem

$$\underset{x}{\text{minimize}}\{L_A(x;\lambda;\rho) : l \le x \le u\}. \tag{7.6}$$

Lewis and Torczon (2002) prove global convergence of their method to first-order Karush–Kuhn–Tucker (KKT) points. We note that the algebraic availability of bound constraints is explicitly used in (7.6). Other constraints could be algebraic or simulation-based because the method used to approximately solve (7.6) does not require availability of the derivative $\nabla_x L_A(x;\lambda;\rho)$. The approach of Lewis and Torczon (2002) is expanded by Lewis and Torczon (2010), who demonstrate the benefits of treating linear constraints (including bound constraints) outside of the augmented Lagrangian merit function. That is, they consider subproblems of the form

$$\underset{x}{\text{minimize}}\{L_A(x;\lambda;\rho) : Ax \le b\}. \tag{7.7}$$

Bueno, Friedlander, Martínez and Sobral (2013) propose an inexact restoration method for problems (CON) where Ω is given by equality constraints. The inexact restoration method alternates between improving feasibility (measured through the constraint violation $\|c(x)\|_2$ in this equality-constrained case) and then approximately minimizing a $\|\cdot\|_2$-based exact penalty function before dynamically adjusting the penalty parameter ρ. Because of the separation of the feasibility and optimality phases of the inexact restoration method, the feasibility phase requires no evaluations of f. This feasibility phase is easier when constraint functions are available algebraically because (sub)derivative-based methods can be employed. Bueno *et al.* (2013) prove global convergence to first-order KKT points of this method under appropriate assumptions.

Amaioua, Audet, Conn and Le Digabel (2018) study the performance of a search step in **MADS** when solving (CON). One of their approaches uses the exact penalty (7.2), a second approach uses the augmented Lagrangian (7.5) and a third combines these two.

Audet, Le Digabel and Peyrega (2015) show that the convergence properties of **MADS** extend to problems with linear equality constraints. They explicitly address these algebraic constraints by reformulating the original problem into a new problem without equality constraints (and possibly fewer variables); other constraints are treated as will be discussed in Section 7.2.

Filter approaches. Whereas a penalty approach combines an objective function f and a measure of constraint violation into a single merit function to be minimized approximately, a *filter method* can be understood as a biobjective method minimizing the objective and the constraint violation simultaneously. For this general discussion, we will refer to the measure of constraint violation as $h(\boldsymbol{x})$. For example, in (7.2),

$$h(\boldsymbol{x}) = \sum_{i \in I} \max\{0, c_i(\boldsymbol{x})\}.$$

From the perspective of biobjective optimization, a *filter* can be understood as a subset of non-dominated points in the (f, h) space. A two-dimensional point $(f(\boldsymbol{x}_l), h(\boldsymbol{x}_l))$ is non-dominated, in the finite set of points $\{\boldsymbol{x}_j : j \in J\}$ evaluated by a method, provided there is no $j \in J \setminus \{l\}$ with

$$f(\boldsymbol{x}_j) \leq f(\boldsymbol{x}_l) \quad \text{and} \quad h(\boldsymbol{x}_j) \leq h(\boldsymbol{x}_l).$$

Unlike biobjective optimization, however, filter methods adaptively vary the subset of non-dominated points considered in order to identify feasible points (*i.e.* points where h vanishes). Different filter methods employ different mechanisms for managing the filter and generating new points.

Brekelmans, Driessen, Hamers and den Hertog (2005) employ a filter for handling relaxable algebraic constraints. Their model-based method attempts to have model-improving points satisfy the constraints. Ferreira, Karas, Sachine and Sobral (2017) extend the inexact-restoration method of Bueno *et al.* (2013) by replacing the penalty formulation with a filter mechanism and again prove global convergence to first-order KKT points.

Approaches with subproblems using modelled constraints. Another means of constraint handling is to construct local models m^{c_i} of each constraint c_i in (7.1). Given a local model m^f of the objective function f, such methods generally employ a sequence of subproblems of the form

$$\operatorname*{minimize}_{\boldsymbol{s}}\{m^f(\boldsymbol{s}) : c_i(\boldsymbol{x} + \boldsymbol{s}) \leq 0, \text{ for all } i \in I\}. \tag{7.8}$$

As an example approach, sequential quadratic programming (SQP) methods are popular derivative-based methods that employ a quadratic model of the objective function and linear models of the constraint functions. Several derivative-free approaches of this form exist, which we detail in this section. We mention that many of these approaches will generally impose an additional trust-region constraint (*i.e.* $\|\boldsymbol{s}\| \leq \Delta$) on (7.8). As in Section 2.2.4, this trust-region constraint often has the additional role of monitoring the quality of the model m^f. Furthermore, such a trust-region constraint ensures that whenever $\boldsymbol{s} = \boldsymbol{0}$ is feasible for (7.8), the feasible region of (7.8) is compact.

Conn, Scheinberg and Toint (1998) consider an adaptation of a model-based trust-region method to constrained problems with differentiable algebraic constraints treated via the trust-region subproblem (7.8). They target problems where they deem the algebraic constraints to be 'easy', meaning that the resulting trust-region subproblem is not too difficult to solve. This method is implemented in the solver DFO (Conn *et al.* 2001).

The CONDOR method of Vanden Berghen (2004) and Vanden Berghen and Bersini (2005) extends the unconstrained UOBYQA method of Powell (2002) to address algebraic constraints. The trust-region subproblem considered takes the form

$$\underset{\boldsymbol{x}}{\text{minimize}}\{m^f(\boldsymbol{x}) : c_i(\boldsymbol{s}) \leq 0, \text{ for all } i \in I'; \boldsymbol{A}\boldsymbol{x} \leq \boldsymbol{b}; \|\boldsymbol{s}\| \leq \Delta\}, \qquad (7.9)$$

where m^f is a quadratic model and $I' \subseteq I$ captures the non-linear constraints in (7.1). In solving (7.9), the linear constraints are enforced explicitly and the non-linear constraints are addressed via an SQP approach. As will be discussed in Section 7.1.2, this corresponds to the linear constraints being treated as unrelaxable.

The LINCOA model-based method of Powell (2015) addresses linear inequality constraints. The LINCOA trust-region subproblem, which can be seen as (7.9) with $I' = \emptyset$, enforces the linear constraints via an active set approach. The active set decreases the degrees of freedom in the variables by restricting $\boldsymbol{x}$ to an affine subspace. Numerically efficient conjugate gradient and Krylov methods are proposed for working in the resulting subspace. Although considerable care is taken to have most points satisfy the linear constraints $\boldsymbol{A}\boldsymbol{x} \leq \boldsymbol{b}$, these constraints are ultimately treated as relaxable, since the method does not enforce these constraints when attempting to improve the quality of the model m^f.

Conejo *et al.* (2013) propose a trust-region algorithm when Ω is closed and convex. They assume that it is easy to compute the projection onto Ω, which facilitates enforcement of the constraints via the trust-region subproblem (7.8). This approach is extended to include more general forms of Ω by Conejo, Karas and Pedroso (2015). As with LINCOA, although subproblem solutions are feasible, the constraints are treated as relaxable since they may be violated in the course of improving the model m^f.

Martínez and Sobral (2012) propose a feasibility restoration method intended for problems with inequality constraints where the feasible region is 'thin': for example, if Ω is defined by both $c_i(\boldsymbol{x}) \leq 0$ and $-c_i(\boldsymbol{x}) \leq 0$ for some i. Each iteration contains two steps: one that seeks to minimize the objective and one that seeks to decrease infeasibility using many evaluation of the constraint functions (without evaluating the objective). Similar to the progressive-barrier method discussed in Section 7.2, the method by Martínez and Sobral (2012) dynamically updates a tolerable level of infeasibility.

7.1.2. Unrelaxable algebraic constraints

We now address the case when all of the constraints are available algebraically but an unrelaxable constraint also exists. In this setting, such unrelaxable constraints are typically necessary to ensure meaningful output of a black-box objective function. Consequently, methods must always maintain feasibility (or at least establish feasibility and then maintain it) with respect to the unrelaxable constraints.

An early example of a method for unrelaxable constraints is the 'complex' method of Box (1965). This extension of the simplex method of Spendley *et al.* (1962) treats unrelaxable bound constraints by modifying the simplex operations to project into the interior of any potentially violated bound constraint. May (1974, 1979) extends the unconstrained derivative-free method of Mifflin (1975) to address unrelaxable linear constraints. The method of May (1979) uses finite-difference estimates, but care is taken to ensure that the perturbed points never violate the constraints.

As seen in Section 7.1.1, several approaches treat non-linear algebraic constraints via a merit function and enforce unrelaxable linear constraints via a constrained subproblem. These include the works of Lewis and Torczon (2002) for bound constraints in (7.6), Lewis and Torczon (2010) for inequality constraints in (7.7), and Vanden Berghen (2004) for inequality constraints in (7.9). Another merit function relevant for unrelaxable constraints is the extended-value merit function

$$h(\boldsymbol{x}) = f(\boldsymbol{x}) + \infty \, \delta_{\boldsymbol{\Omega}^C}(\boldsymbol{x}), \tag{7.10}$$

where $\delta_{\boldsymbol{\Omega}^C}$ is the indicator function of $\boldsymbol{\Omega}^C$. Such an *extreme-barrier* approach (see *e.g.* the discussion by Lewis and Torczon 1999) is particularly relevant for simulation-based constraints. Hence, with the exception of explicit treatment of unrelaxable algebraic constraints, we postpone significant discussion of extreme-barrier methods until Section 7.2.

DDS methods for unrelaxable algebraic constraints. Within DDS methods, an intuitive approach to handling unrelaxable constraints is to limit poll directions $\boldsymbol{D}_k$ so that $\boldsymbol{x}_k + \boldsymbol{d}_k$ is feasible with respect to the unrelaxable constraints. Lewis and Torczon (1999) and Lucidi and Sciandrone (2002a), respectively, develop pattern-search and coordinate-search methods for unrelaxable bound-constrained problems. By modifying the polling directions Lewis and Torczon (2000) show that pattern-search methods are also convergent in the presence of unrelaxable linear constraints. Chandramouli and Narayanan (2019) address unrelaxable bound constraints within a DDS method that employs a model-based method in the search step in addition to a bound-constrained line-search step. Kolda, Lewis and Torczon (2006) develop and analyse a new condition, related to the tangent cone of nearby active constraints, on the sets of directions used within a generating set search method when solving linearly constrained problems. The condition ensures that evaluated points are guaranteed to satisfy the linear constraints. Lucidi, Sciandrone and Tseng (2002) propose feasible descent methods that sample the objective over a finite set of search directions. Each iteration considers a set of ϵ-active constraints (*i.e.* those constraints for which $c_i(x_k) \geq -\epsilon$) for general algebraic inequality constraints. Poll steps are projected in order to ensure they are feasible with respect to these ϵ-active constraints. The analysis of Lucidi *et al.* (2002) extends that of Lewis and Torczon (2000) and establishes convergence to a first-order KKT point under standard assumptions.

As introduced in Section 7.1.1, Audet *et al.* (2015) reformulate optimization problems with unrelaxable linear equality constraints in the context of MADS.

Gratton, Royer, Vicente and Zhang (2019*b*) extend the randomized DDS method of Gratton *et al.* (2015) to linearly constrained problems; candidate points are accepted only if they are feasible. Gratton *et al.* (2019*b*) establish probabilistic convergence and complexity results using a stationary measure appropriate for linearly constrained problems.

Model-based methods for unrelaxable algebraic constraints. Model-based methods are more challenging to design in the presence of unrelaxable constraints because enforcing guarantees of model quality such as those in (2.9) can be difficult. For a fixed value of κ in (2.9), it may be impossible to obtain a κ-fully linear model using only feasible points. As an example, consider two linear constraints for which the angle between the constraints is too small to allow for κ-fully linear model construction; avoiding interpolation points drawn from such thin regions motivated development of the wedge-based method of Marazzi and Nocedal (2002) from Section 2.2.4.

Powell (2009) proposes **BOBYQA**, a model-based trust-region method for bound-constrained optimization without derivatives that extends the unconstrained method **NEWUOA** in Powell (2006). **BOBYQA** ensures that all points at which f is evaluated satisfy the bound constraints. Arouxét, Echebest and Pilotta (2011) modify **BOBYQA** to use an active-set strategy in solving the bound-constrained trust-region subproblems; an $\| \cdot \|_\infty$-trust region is employed so that these subproblems correspond to minimization of a quadratic over a compact, bound-constrained domain. Wild (2008*a*, Section 6.3) develops an RBF-model-based method for unrelaxable bound constraints by enforcing the bounds during both model improvement and $\| \cdot \|_\infty$-trust-region subproblems. Gumma, Hashim and Ali (2014) extend the **NEWUOA** method to address linearly constrained problems. The linear constraints are enforced both when solving the trust-region subproblem and when seeking to improve the geometry of the interpolation points. Gratton, Toint and Tröltzsch (2011) propose a model-based method for unrelaxable bound-constrained optimization, which restricts the construction of fully linear models to subspaces defined by nearly active constraints. Working in such a reduced space means that the machinery for unconstrained models in Section 2.2.4 again applies.

Methods for problems with unrelaxable discrete constraints. Constraints that certain variables take discrete values are often unrelaxable in derivative-free optimization. For example, a black-box simulation may be unable to assign meaningful output when input variables take non-integer values. That such integer constraints are unrelaxable presents challenges distinct from those typically arising in mixed-integer non-linear optimization (Belotti *et al.* 2013).

Naturally, researchers have modified derivative-free methods for continuous optimization to address integer constraints. Audet and Dennis, Jr (2000) and Abramson, Audet, Chrissis and Walston (2009*a*), respectively, propose integer-constrained pattern-search and MADS methods to ensure that evaluated points respect integer constraints. Abramson, Audet and Dennis, Jr (2007) develop a pattern-search method that employs a filter that handles general inequality constraints and ensures that integer-constrained variables are always integer.

Porcelli and Toint (2017) propose the 'brute-force optimizer' BFO, a DDS method for mixed-variable problems (including those with ordinal categorical variables) that aligns the poll points to respect the discrete constraints. A recursive call of the method reduces the number of discrete variables by fixing a subset of these variables.

Liuzzi, Lucidi and Rinaldi (2011) solve mixed-integer bound-constrained problems by using both a local discrete search (to address integer variables) and a line search (for continuous variables). This approach is extended by Liuzzi, Lucidi and Rinaldi (2015) to also address mixed-integer problems with general constraints using the SQP approach from Liuzzi, Lucidi and Sciandrone (2010). Liuzzi, Lucidi and Rinaldi (2018) solve constrained integer optimization problems by performing non-monotone line searches along feasible *primitive directions* D in a neighbourhood of the current point x_k. Feasible primitive directions are those $d \in \mathbb{Z}^n \cap \Omega$ satisfying $\mathrm{GCD}(d_1, \ldots, d_{|D|}) = 1$, that is, directions in a bounded neighbourhood that are not integer multiples of one another.

The method by Rashid, Ambani and Cetinkaya (2012) for mixed-integer problems builds multiquadric RBF models. Candidate points are produced by using gradient-based mixed-integer optimization techniques; the authors' relaxation-based approach employs a 'proxy model' that coincides with function values from points satisfying the unrelaxable integer constraints. The methods of Müller, Shoemaker and Piché (2013*a*, 2013*b*) and Müller (2016) similarly employ a global RBF model over the integer lattice, with various strategies for generating trial points based on this model. Newby and Ali (2015) build on BOBYQA to address bound-constrained mixed-integer problems. They outline an approach for building interpolation models of objectives using only points that are feasible. Their trust-region subproblems consist of minimizing a quadratic objective subject to bound and integer constraints.

Many of the discussed methods have been shown to converge to points that are mesh-isolated local solutions; see Newby and Ali (2015) for discussion of such 'local minimizers'. When an objective is convex, one can do better. Larson, Leyffer, Palkar and Wild (2019) propose a method for certifying a global minimum of a convex objective f subject to unrelaxable integer constraints. They form a piecewise linear underestimator by interpolating f through subsets of $n + 1$ affinely independent points. The resulting underestimator is then used to generate new candidate points until global optimality has been certified.

7.2. Simulation-based constraints

As opposed to the preceding section, methods in this section are not limited to constraints that have closed-form solutions but also address constraints that depend on the output from some calculated function. Many methods address such simulation-based constraints by using approaches similar to those used for algebraic constraints.

Filter approaches. Filter methods for simulation-based constraints, as with algebraic constraints, seek to simultaneously decrease the objective and constraint violation. For example, Audet and Dennis, Jr (2004) develop a pattern-search method for general constrained optimization that accepts steps that improve either the objective or some measure of violation of simulation-based constraints. Their hybrid approach applies an extreme barrier to points that violate linear or bound constraints. Audet (2004) provides examples where the method by Audet and Dennis, Jr (2004) does not converge to stationary points.

Pourmohamad (2016) models objective and constraint functions using Gaussian process models in a filter-based method. Because these models are stochastic, point acceptability is determined by criteria such as probability of filter acceptability or expected area of dominated region (in the filter space).

Echebest, Schuverdt and Vignau (2015) develop a derivative-free method in the inexact feasibility restoration filter method framework of Gonzaga, Karas and Vanti (2004). Echebest *et al.* (2015) employ fully linear models of the objective and constraint functions and show that the resulting limit points are first-order KKT points.

Penalty approaches. The original MADS method (Audet and Dennis, Jr 2006) converts constrained problems to unconstrained problems by using the extreme-barrier approach mentioned above; that is, the merit function (7.10) effectively assigns a value of infinity to points that violate any constraint. A similar approach to general constraints is used by the 'complex' method of Box (1965); the simplex (complex) is updated to maintain the feasibility of the vertices of the simplex. As a consequence of its generality, the extreme-barrier approach is applicable for algebraic constraints, simulation-based constraints and even hidden constraints. Furthermore, because (7.10) is independent of the degree of both constraint satisfaction and constraint violation, the extreme barrier is able to address non-quantifiable constraints.

In contrast, the progressive-barrier method by Audet and Dennis, Jr (2009) employs a quadratic constraint penalty similar to (7.3) for relaxable simulation-based constraints $\{c_i : i \in I_r\}$ and an extreme-barrier penalty for unrelaxable simulation-based constraints $\{c_i : i \in I_u\}$. Their progressive-barrier method maintains a non-increasing threshold value ϵ_k that quantifies the allowable relaxable constraint violation in each iteration. Their approach effectively uses the merit function

$$
h_k(\boldsymbol{x}) = \begin{cases} f(\boldsymbol{x}) & \text{if } \boldsymbol{x} \in \boldsymbol{\Omega}_u \text{ and } \sum_{i \in I_r} \max\{0, c_i(\boldsymbol{x})\}^2 < \varepsilon_k, \\ \infty & \text{otherwise,} \end{cases}
\tag{7.11}
$$

where $\boldsymbol{\Omega}_u = \{\boldsymbol{x} : c_i(\boldsymbol{x}) \le 0, \ \forall i \in I_u\}$ denotes the feasible domain with respect to the unrelaxable constraints. The progressive-barrier method maintains a set of feasible and infeasible incumbent points and seeks to decrease the threshold ϵ_k to 0 based on the value of (7.11) at infeasible incumbent points. Trial steps are accepted as incumbents based on criteria resembling, but distinct from, those used by filter methods. Convergence to Clarke stationary points is obtained for particular sequences of the incumbent points. The NOMAD (Le Digabel 2011) implementation of MADS allows users to choose to address inequality constraints handled via extreme-barrier, progressive-barrier or filter approaches.

Also within the DDS framework, Gratton and Vicente (2014) use an extreme-barrier approach to handle unrelaxable constraints and an exact penalty function to handle the relaxable constraints. That is, step acceptability is based on satisfaction of the unrelaxable constraints as well as sufficient decrease in the merit function (7.2), with the set I containing only those constraints that are relaxable. As the algorithm progresses, relaxable constraints are transferred to the set of constraints treated by the extreme barrier; this approach is similar to that underlying the progressive-barrier approach.

Liuzzi and Lucidi (2009) and Liuzzi *et al.* (2010) consider line-search methods that apply a penalty to simulation-based constraints; Liuzzi and Lucidi (2009) employ an exact penalty function (a smoothed version of

$\|\cdot\|_\infty$), whereas Liuzzi *et al.* (2010) employ a sequence of quadratic penalty functions of the form (7.3). Fasano, Liuzzi, Lucidi and Rinaldi (2014) propose a similar line-search approach to address constraint and objective functions that are not differentiable.

Primarily concerned with equality constraints, Sampaio and Toint (2015, 2016) propose a derivative-free variant of trust-funnel methods, a class of methods proposed by Gould and Toint (2010) that avoid the use of both merit functions and filters.

Diniz-Ehrhardt, Martínez and Pedroso (2011) propose a method that models objective and constraint functions in an augmented Lagrangian framework. Similarly, Picheny, Gramacy, Wild and Le Digabel (2016) use an augmented Lagrangian framework, wherein the merit function in (7.5) uses Gaussian process models of the objective and constraint functions in place of the actual objective and constraint functions.

Approaches with subproblems using modelled constraints. In early work, Glass and Cooper (1965) develop a coordinate-search method that also uses linear models of the objective and constraint functions. On each iteration, after the coordinate directions are polled, the models are used in a linear program to generate new points; points are accepted only if they are feasible. Extending this idea, Powell (1994) develops the constrained optimization by linear approximation (**COBYLA**) method, which builds linear interpolation models of the objective and constraint functions on a common set of $n + 1$ affinely independent points. Care is taken to maintain the non-degeneracy of this simplex. The method can handle both inequality and equality constraints, with candidate points obtained from a linearly constrained subproblem and then accepted based on a merit function of the form (7.2).

Bűrmen, Olenšek and Tuma (2015) propose a variant of **MADS** with a specialized model-based search step

$$\operatorname*{minimize}_{x}\{m^f(x) : \boldsymbol{A}\boldsymbol{x} \leq \boldsymbol{b}\}, \tag{7.12}$$

where m^f is a strongly convex quadratic model of f and $(\boldsymbol{A}, \boldsymbol{b})$ are determined from linear regression models of the constraint functions. Both the search and poll steps are accepted only if they are feasible; this corresponds to the method effectively treating the constraints with an extreme-barrier approach.

Gramacy and Le Digabel (2015) extend the MADS framework by using treed Gaussian processes to model both the objective and simulation-based constraint functions. The resulting models are used both within the search step and to order the poll points (within an opportunistic polling paradigm) using a filter-based approach.

A number of methods work with restrictions of the domain Ω in order to promote feasibility (typically with respect to the simulation-based constraints) of the generated points. Such strategies are often motivated by a desire to avoid the situation where feasibility is established only asymptotically. An example of such a restricted domain is the set

$$\Omega_{\mathrm{res}}(\boldsymbol{\epsilon}) = \{\boldsymbol{x} \in \mathbb{R}^n : c_i(\boldsymbol{x}) \leq 0 \;\forall i \in I_a, \; m^{c_i}(\boldsymbol{x}) + \epsilon_i(\boldsymbol{x}) \leq 0 \;\forall i \in I_s\}, \quad (7.13)$$

where algebraic constraints (corresponding to $i \in I_a$) are explicitly enforced and a parameter (or function of $\boldsymbol{x}$) $\boldsymbol{\epsilon}$ controls the degree of restriction for the modelled simulation-based constraints (corresponding to $i \in I_s$).

The methods of Regis (2013) utilize interpolating radial basis function surrogates of the objective and constraint functions. Acceptance of infeasible points is allowed and is followed by a constraint restoration phase that minimizes a quadratic penalty based on the modelled constraint violation. When the current point is feasible, a subproblem is solved with a feasible set defined by (7.13) in addition to a constraint that lower-bounds the distance between the trial point and the current point. Each parameter ϵ_i is adjusted based on the feasibility of constraint $i \in I_s$ in recent iterations.

Augustin and Marzouk (2014) develop a trust-region method employing fully linear models of both constraint and objective functions. They introduce a *path augmentation* scheme intended to locally convexify the simulation-based constraints. Their trust-region subproblem at the current point $\boldsymbol{x}_k$ minimizes the model of the objective function subject to a trust-region constraint and the restricted feasible set (7.13), where $\epsilon_i(\boldsymbol{x}) = \epsilon_0 \|\boldsymbol{x} - \boldsymbol{x}_k\|^{2/(1+p)}$ and where $\epsilon_0 > 0$ and $p \in (0, 1)$ are fixed constants. Augustin and Marzouk (2014) establish convergence of their method from feasible starting points; that is, they show a first-order criticality measure asymptotically tends to 0. Augustin and Marzouk (2014) produce a code, NOWPAC, that employs minimum-Frobenius norm quadratic models of both the objective and constraint functions. This work is extended by Augustin and Marzouk (2017) to the stochastic optimization problem (STOCH).

Whereas Augustin and Marzouk (2014) consider a local convexification of inequality constraints through the addition of a convex function to the constraint models, Regis and Wild (2017) consider a similar model-based approach but define an envelope around models of nearly active constraints. In particular, at the current point $\boldsymbol{x}_k$, the restricted feasible set (7.13) uses the parameter

$$\epsilon_i(\boldsymbol{x}) = \begin{cases} 0 & \text{if } c_i(\boldsymbol{x}_k) > -\xi_i, \\ \xi_i & \text{if } c_i(\boldsymbol{x}_k) \leq -\xi_i, \end{cases}$$

where $\{\xi_i : i \in I_s\}$ is fixed. This form of $\boldsymbol{\epsilon}$ ensures that trust-region subproblems remain non-empty and avoids applying a restriction when the algorithm is sufficiently close to the level set $\{\boldsymbol{x} : c_i(\boldsymbol{x}) = 0\}$.

Tröltzsch (2016) considers an SQP method in the style of Omojokun (1989), which applies a two-step process that first seeks to improve a measure of constraint violation and then solves a subproblem restricted to the null space of modelled constraint gradients. Tröltzsch (2016) uses linear models of the constraint functions and quadratic models of the objective function, with these models replacing c and f in the augmented Lagrangian merit function in (7.5). Step acceptance uses a merit function (an exact penalty function).

Müller and Woodbury (2017) develop a method for addressing computationally inexpensive objectives while satisfying computationally expensive constraints. Their two-phase method first seeks feasibility by solving a multi-objective optimization problem (a problem class that is the subject of Section 8.4) in which the constraint violations are minimized simultaneously; the second phase seeks to reduce the objective subject to constraints derived from cubic RBF models of the constraint functions.

Bajaj, Iyer and Hasan (2018) propose a two-phase method. In the feasibility phase, a trust-region method is applied to a quadratic penalty function that employs models of the simulation-based constraints. The trust-region subproblem at iteration k takes the form

$$\underset{\boldsymbol{x}}{\text{minimize}}\left\{\sum_{i\in I_s}\max\{0, m^{c_i}(\boldsymbol{x})\}^2 : c_i(\boldsymbol{x}) \le 0 \ \forall i \in I_a,\ \boldsymbol{x} \in \mathcal{B}(\boldsymbol{x}_k; \Delta_k)\right\}$$

$$(7.14)$$

and thus explicitly enforces the algebraic constraints ($i \in I_a$) and penalizes violation of the modelled simulation-based constraints ($i \in I_s$). In the optimality phase, a trust-region method is applied to a model of the objective function, and the modelled constraint violation is bounded by that achieved in the feasibility phase; that is, the trust-region subproblem is

$$\begin{aligned}
\underset{\boldsymbol{x}}{\text{minimize}} \quad & m^f(\boldsymbol{x}) \\
\text{subject to} \quad & c_i(\boldsymbol{x}) \le 0 \quad \text{for all } i \in I_a \\
& m^{c_i}(\boldsymbol{x}) \le c_i(\boldsymbol{x}_{\text{pen}}) \quad \text{for all } i \in I_s \\
& \boldsymbol{x} \in \mathcal{B}(\boldsymbol{x}_k; \Delta_k),
\end{aligned}$$

where $\boldsymbol{x}_{\text{pen}}$ is the point returned from the feasibility phase.

Hare and Lewis (2005) present an approach for approximating the normal and tangent cones; their approach is quite general and applies to the case when the domain is defined by non-quantifiable black-box constraints. Davis and Hare (2013) consider a simplex-gradient-based approach for approximating normal cones when the black-box constraints are quantifiable. Naturally, such approximate cones could be used to determine if a method's candidate solution approximately satisfies a stationarity condition.

8. Other extensions and practical considerations

We conclude with a cursory look at extensions of the methods presented, especially highlighting active areas of development.

8.1. Methods allowing for concurrent function evaluations

A number of the methods presented in this survey readily allow for the concurrent evaluation of the objective function at multiple points $x \in \mathbb{R}^n$. Performing function evaluations concurrently through the use of parallel computing resources should decrease the wall-clock time required by a given method. Depending on the method, there is a natural limit to the amount of concurrency that can be utilized efficiently. Below we summarize such methods and their limits for concurrency.

The simplex methods discussed in Section 2.1 benefit from performing n concurrent evaluations of the objective when a shrink operation is performed. Also, the points corresponding to the expansion and reflection operations could be evaluated in parallel. Non-opportunistic directional direct-search methods are especially amenable to parallelization (Dennis, Jr and Torczon 1991) because the $|D_k|$ poll directions can be evaluated concurrently.

Model-based methods from Section 2.2 can use concurrent evaluations during model building when, for example, evaluating up to $\dim(\mathcal{P}^{d,n})$ additional points for use in (2.14). In another example, CONDOR (Vanden Berghen and Bersini 2005, Vanden Berghen 2004) utilizes concurrent evaluations of the objective to replace points far away from the current trust-region centre by maximizing the associated Lagrange polynomial. The thesis by Olsson (2014) considers three ways of using concurrent resources within a model-based algorithm: using multiple starting points, evaluating different models in order to better predict a point's value, and generating multiple points with each model (*e.g.* solving with the trust-region subproblem with different radii). A similar approach of generating multiple trial points concurrently is employed in the parallel direct-search, trust-region method of Hough and Meza (2002).

Finite-difference-based approaches (*e.g.* Section 2.3) allow for n concurrent evaluations with forward differences (2.28) or $2n$ concurrent evaluations with central differences (2.29). Implicit filtering also performs such a central-difference calculation that can utilize $2n$ concurrent evaluations (line 5 of Algorithm 4). Line-search methods can evaluate multiple points concurrently during their line-search procedure. The methods of García-Palomares and Rodríguez (2002) and García-Palomares, García-Urrea and Rodríguez-Hernández (2013) also consider using parallel resources to concurrently evaluate points in a neighbourhood of interest.

When using a set of independently generated points, pure random search exhibits perfect scaling as the level of available concurrency increases. Otherwise, the randomized methods for deterministic objectives from Section 3 can utilize concurrent evaluations in a manner similar to that of their deterministic counterparts. Nesterov random search can use n or $2n$ concurrent objective evaluations when computing an approximate gradient in (3.1). Randomized DDS methods can concurrently evaluate $|\boldsymbol{D}_k|$ poll points, and randomized trust-region methods can concurrently evaluate points needed for building and improving models.

In addition to the above approaches for using parallel resources, methods from Section 5 for structured problems can use concurrent evaluations to calculate parts of the objective. For example, methods for optimizing separable objectives such as (5.1) or (5.3) can evaluate the p component functions F_i concurrently.

The various gradient approximations used by methods in Section 6 are amenable to parallelization in the same manner as previously discussed, but with the additional possibility of also evaluating at multiple $\boldsymbol{\xi}$ values. SA methods can use $2n$ concurrent evaluations of $\tilde{f}$ in calculating (6.2) or (6.4) and **SPSA** can use two concurrent evaluations when calculating (6.5). Methods employing the sample mean estimator (6.7) can utilize p_k evaluations concurrently.

8.2. Multistart methods

A natural approach for addressing non-convex objectives for which it is not known whether multiple local minima exist is to start a local optimization method from different points in the domain in the hope of identifying different local minima. Such multistart approaches also allow for the use of methods that are specialized for optimizing problems with known structure.

Multistart methods allow for the use of concurrent objective evaluations if two or more local optimization runs are being performed at the same time. Multistart methods also allow one to utilize additional computational resources; this ability is especially useful when an objective evaluation does not become faster with additional resources or when the local optimization method is inherently sequential.

Boender, Rinnooy Kan, Timmer and Stougie (1982) derive confidence intervals on the objective value of a global minimizer when starting a local optimization method at uniformly drawn points. Their analysis gives rise to the multilevel single linkage (**MLSL**) method (Rinnooy Kan and Timmer 1987a, Rinnooy Kan and Timmer 1987b). Iteration k of the method draws N points uniformly over the domain and starts a local optimization method from sampled points that do not have any other point within a specific distance, depending on k and N, with a smaller objective value. With this rule, and under assumptions on the distance between minimizers in Ω and properties of the local optimization method used, MLSL is shown to almost surely identify all local minima while starting the local optimization method from only finitely many points. Larson and Wild (2016, 2018) generalize MLSL by showing similar theoretical results when starting-point selection utilizes points both from the random sampling and from those generated by local optimization runs.

If a meaningful variance exists in the objective evaluation times, batched evaluation of points may result in an inefficient use of computational resources. Such concerns have motivated the development of a number of methods including the HOPSPACK framework (Plantenga 2009), which supports the sharing of information between different local optimization methods. Shoemaker and Regis (2003) also use information from multiple optimization methods to determine points at which to evaluate the objective function. Similarly, the SNOBFIT method by Huyer and Neumaier (2008) uses concurrent objective evaluations while combining local searches in a global framework. The software focuses on robustness in addressing many practical concerns including soft constraints, hidden constraints, and a problem domain that is modified by the user as the method progresses.

Instead of coordinating concurrent instances of a pattern-search method, Audet, Dennis, Jr and Le Digabel (2008a) propose an implementation of MADS that decomposes the domain into subspaces to be optimized over in parallel. Alarie *et al.* (2018) study different approaches for selecting subsets of variables to define subproblems in such an approach. Custódio and Madeira (2015) maintain concurrent instances of a pattern-search method, and merge those instances that become sufficiently close. Taddy, Lee, Gray and Griffin (2009) use a global treed-Gaussian process to guide a local pattern-search method to encourage the identification of better local minima.

8.3. Other global optimization methods

Guarantees of global optimality for general continuous functions rely on candidate points being generated densely in the domain (Törn and Žilinskas 1989, Theorem 1.3); such candidate points can be generated in either a deterministic or randomized fashion. When f is Lipschitz-continuous on Ω and the Lipschitz constant L_f is available to the optimization method, one need not generate points densely in the domain. In particular, if $\hat{x}$ is an approximate minimizer of f and x is a point satisfying $f(x) > f(\hat{x})$, no global minimizer can lie in – and therefore no point needs to be sampled from – $\mathcal{B}(x; (f(x) - f(\hat{x}))/L_f)$. Naturally, the benefit of exploiting this fact requires accurate knowledge of the Lipschitz constant. One can empirically observe a lower bound on L_f, but obtaining useful upper bounds on L_f may not be possible. Methods that exploit this Lipschitz knowledge may suffer a considerable performance decrease when overestimating L_f (Hansen, Jaumard and Lu 1991).

Motivated by situations where the Lipschitz constant of f is unavailable, Jones, Perttunen and Stuckman (1993) develop the DIRECT (DIviding RECTangles) method. DIRECT partitions a bound-constrained Ω into $2n + 1$ hyper-rectangles (hence the method's name) with an evaluated point at the centre of each. Each hyper-rectangle is scored via a combination of the length of its longest side and the function value at its centre. This scoring favours hyper-rectangles exhibiting both long sides and small function values; the best-scoring hyper-rectangles are further divided. (As such, DIRECT's performance can be significantly affected by adding a constant value to the objective (Finkel and Kelley 2006).) DIRECT generates centres that are dense in Ω and will therefore identify the global minimizers of f over Ω, even when f is non-smooth (Jones *et al.* 1993, Finkel and Kelley 2004, Finkel and Kelley 2009). Several versions of DIRECT that perform concurrent function evaluations take significant care to ensure the sequence of points generated is the same as that produced by DIRECT (He, Verstak, Sosonkina and Watson 2009*a*, He, Verstak, Watson and Sosonkina 2007, He, Verstak, Watson and Sosonkina 2009*b*, He, Watson and Sosonkina 2009*c*). Similar hyper-rectangle partitioning strategies are used by the methods of Munos (2011). The multilevel coordinate-search (MCS) method by Huyer and Neumaier (1999) is inspired by DIRECT in many ways. MCS maintains a partitioning of the domain and subdivides hyper-rectangles based on their size and value. MCS uses the function values at boundary points, rather than the centre points, to determine the value of a hyper-rectangle; such boundary points can be shared by more than one hyper-rectangle. Huyer and Neumaier (2008) show that a version of MCS needs to consider only finitely many hyper-rectangles before identifying a global minimizer.

Many randomized approaches for generating points densely in a domain Ω have been developed. These include Bayesian optimization methods and related variants (Mockus 1989, Jones, Schonlau and Welch 1998, Frazier 2018), some of which have established complexity rates (Bull 2011). Such randomized samplings of Ω can be used to produce a global surrogate; similar to other model-based methods, this global model can be minimized to produce points where the objective should be evaluated. Although minimizing such a global surrogate may be difficult, such a subproblem may be easier than the original problem, which typically entails a computationally expensive objective function for which derivatives are unavailable. Vu, D'Ambrosio, Hamadi and Liberti (2016) provide a recent survey of such surrogate-based methods for global optimization.

8.4. Methods for multi-objective optimization

Multi-objective optimization problems are typically stated as

$$
\begin{aligned}
\underset{\boldsymbol{x}}{\text{minimize}} \quad & \boldsymbol{F}(\boldsymbol{x}) \\
\text{subject to} \quad & \boldsymbol{x} \in \Omega \subset \mathbb{R}^n,
\end{aligned}
\tag{MOO}
$$

where $p > 1$ objective functions $f_i : \mathbb{R}^n \to \mathbb{R}$ for $i = 1, \ldots, p$ define the vector-valued mapping $\boldsymbol{F}$ via $\boldsymbol{F}(\boldsymbol{x}) = [f_1(\boldsymbol{x}), \ldots, f_p(\boldsymbol{x})]$. Given potentially conflicting objectives $f_1, \ldots, f_p$, the problem (MOO) is well-defined only when given an ordering on the vector of objective values $\boldsymbol{F}(\boldsymbol{x})$. Given distinct points $\boldsymbol{x}_1, \boldsymbol{x}_2 \in \mathbb{R}^n$, $\boldsymbol{x}_1$ *Pareto dominates* $\boldsymbol{x}_2$ provided

$$
f_i(\boldsymbol{x}_1) \le f_i(\boldsymbol{x}_2) \quad \text{for all } i = 1, \ldots, p \quad \text{and} \quad f_j(\boldsymbol{x}_1) < f_j(\boldsymbol{x}_2) \quad \text{for some } j.
$$

The set of all feasible points that are not Pareto-dominated by any other feasible point is referred to as the *Pareto(-optimal) set* of (MOO). An in-depth treatment of such problems is provided by Ehrgott (2005).

Ideally, a method designed for the solution of (MOO) should return an approximation of the Pareto set. If at least one objective $f_1, \ldots, f_p$ is non-convex, however, approximating the Pareto set can be challenging. Consequently, methods for multi-objective optimization typically pursue *Pareto stationarity*, which is a form of local optimality characterized by a first-order stationarity condition. If $\Omega = \mathbb{R}^n$, a point $\boldsymbol{x}_*$ is a Pareto stationary point of $\boldsymbol{F}$ provided that for each $\boldsymbol{d} \in \mathbb{R}^n$, there exists $j \in \{1, \ldots, p\}$ such that $f_j'(\boldsymbol{x}_*; \boldsymbol{d}) \ge 0$. This notion of stationarity is an extension of the one given for single-objective optimization in (2.1).

Typical methods for (MOO) return a collection of points that are not known to be Pareto-dominated and thus serve as an approximation to the set of Pareto points. From a theoretical point of view, most methods endeavour only to demonstrate that all accumulation points are Pareto stationary, and rarely prove the existence of more than one such point. From a practical point of view, comparing the approximate Pareto sets returned by a method for multi-objective optimization is not straightforward. For discussions of some comparators used in multi-objective optimization, see Knowles and Corne (2002) and Audet *et al.* (2018*a*).

Various derivative-free methods discussed in this survey have been extended to address (MOO). The method of Audet, Savard and Zghal (2008*b*) solves biobjective optimization problems by iteratively combining the two objectives into a single objective (for instance, by considering a weighted sum of the two objectives); MADS is then applied to this single-objective problem. Audet, Savard and Zghal (2010) extend the method of Audet *et al.* (2008*b*) to multi-objective problems with more than two objectives. Audet *et al.* (2008*b*, 2010) demonstrate that all refining points of the sequence of candidate points produced by these methods are Pareto stationary.

Custódio, Madeira, Vaz and Vicente (2011) propose *direct-multisearch* methods, a multi-objective analogue of direct-search methods. Like direct-search methods, direct-multisearch methods involve both a search step and a poll step. Direct-multisearch methods maintain a list of non-dominated points; at the start of an iteration, one non-dominated point must be selected to serve as the centre for a poll step. Custódio *et al.* (2011) demonstrate that at any accumulation point $\boldsymbol{x}_*$ of the maintained sequence of non-dominated points from a direct-multisearch method, it holds that for any direction $\boldsymbol{d}$ that appears in a poll step infinitely often, $f'_j(\boldsymbol{x}_*, \boldsymbol{d}) \geq 0$ for at least one j. In other words, accumulation points of the method are Pareto stationary when restricted to these directions $\boldsymbol{d}$. Custódio and Madeira (2016) incorporate these direct-multisearch methods within a multistart framework in an effort to find multiple Pareto stationary points and thus to better approximate the Pareto set.

For stochastic biobjective problems, Kim and Ryu (2011) employ sample average approximation to estimate $\boldsymbol{F}(\boldsymbol{x}) = \mathbb{E}_{\boldsymbol{\xi}}[\tilde{\boldsymbol{F}}(\boldsymbol{x}; \boldsymbol{\xi})]$ and propose a model-based trust-region method. Ryu and Kim (2014) adapt the approach of Kim and Ryu (2011) to the deterministic biobjective setting. At the start of each iteration, these methods construct fully linear models of both objectives around a (currently) non-dominated point. These methods solve three trust-region subproblems – one for each of the two objectives, and a third that weights the two objectives as in Audet *et al.* (2008*b*) – and accept all non-dominated trial points. If both objectives are in $\mathcal{LC}^1$, Ryu and Kim (2014) prove that one of the three objectives satisfies a lim-inf convergence result of the form (2.5), implying the existence of a Pareto-stationary accumulation point.

Liuzzi, Lucidi and Rinaldi (2016) propose a method for constrained multi-objective non-smooth optimization that separately handles each objective and constraint via an exact penalty (see (7.2)) in order to determine whether a point is non-dominated. Given the non-sequential nature of how non-dominated points are selected, Liuzzi *et al.* (2016) identify and link the subsequences implied by a lim-inf convergence result. They show that limit points of these linked sequences are Pareto stationary provided the search directions used in each linked sequence are asymptotically dense in the unit sphere.

Cocchi, Liuzzi, Papini and Sciandrone (2018) extend implicit filtering to the multi-objective case. They approximate each objective gradient separately using implicit-filtering techniques; they combine these approximate gradients in a disciplined way to generate search directions. Cocchi *et al.* (2018) demonstrate that their method generates at least one accumulation point and that every such accumulation point is Pareto stationary.

8.5. Methods for multifidelity optimization

Multifidelity optimization concerns the minimization of a high-fidelity objective function $f = f_0$ in situations where a lower-fidelity version f_ϵ (for $\epsilon > 0$) also exists. Evaluations of the lower-fidelity function f_ϵ are less computationally expensive than are evaluations of f_0; hence, a goal in multifidelity optimization is to exploit the existence of the lower-fidelity f_ϵ in order to perform as few evaluations of f_0 as possible. An example of such a setting occurs when there exist multiple grid resolutions defining discretizations for the numerical solution of partial differential equations that defines f_0 and f_ϵ.

Polak and Wetter (2006) develop a pattern-search method that exploits the existence of multiple levels of fidelity. The method begins at the coarsest available level and then monotonically refines the level of fidelity (*i.e.* decreases ϵ) after a sufficient number of consecutive unsuccessful iterations occur.

A method that both decreases ϵ and increases ϵ (akin to the V- and W-cycles of multigrid methods (Xu and Zikatanov 2017)), is the multilevel method of Frandi and Papini (2013). The method follows the MG/Opt framework of Nash (2000) and instantiates runs of a coordinate-search method at specified fidelity and solution accuracy levels. Another multigrid-inspired method is developed by Liu, Zeng and Yang (2015), wherein a hierarchy of DIRECT runs are performed at varying fidelity and budget levels.

Model-based methods have also been extended to the multifidelity setting. For example, March and Willcox (2012) employ a fully linear RBF model to interpolate the error between two different fidelity levels. Their method then employs this model within a trust-region framework, but uses f_0 to determine whether to accept a given step.

Another model-based approach for multifidelity optimization is co-kriging; see, for example, Xiong, Qian and Wu (2013) and Le Gratiet and Cannamela (2015). In such approaches, a statistical surrogate (typically a Gaussian process model) is constructed for each fidelity level with the aim of modelling the relationships among the fidelity levels in areas of the domain relevant to optimization.

Derivative-free methods for multifidelity, multi-objective and concurrent/parallel optimization remain an especially open avenue of future research.

Acknowledgements

We are grateful to referees and colleagues whose comments greatly improved the manuscript; these include Mark Abramson, Charles Audet, Ana Luísa Custódio, Sébastien Le Digabel, Warren Hare, Paul Hovland, Jorge Moré, Raghu Pasupathy, Margherita Porcelli, Francesco Rinaldi, Clément Royer, Katya Scheinberg, Sara Shashaani, Aekaansh Verma and Zaikun Zhang. We are especially indebted to Gail Pieper and Glennis Starling for their invaluable editing. This material is based upon work supported by the applied mathematics and SciDAC activities of the Office of Advanced Scientific Computing Research, Office of Science, US Department of Energy, under Contract DE-AC02-06CH11357.

Appendix: Collection of WCC results

Table A.1 contains select WCC bounds for methods appearing in the literature. Given $\epsilon > 0$, all WCC bounds in this appendix are given in terms of N_ϵ, an upper bound on the number of *function evaluations* of a method to guarantee that the specified condition is met. We present results in this form because function evaluation complexity of derivative-free methods is often of greater interest than is iteration complexity. We present N_ϵ in terms of four parameters:

- the accuracy ϵ;
- the dimension n;
- the Lipschitz constant of the function L_{f}, the Lipschitz constant of the function gradient L_{g} or the Lipschitz constant of the function Hessian L_{H} (provided these constants are well-defined); and
- a measure of how far the starting point $\boldsymbol{x}_0$ is from a stationary point $\boldsymbol{x}_*$. In this appendix, this measure is either $f(\boldsymbol{x}_0) - f(\boldsymbol{x}_*)$,

$$R_{\mathrm{level}} = \sup_{\boldsymbol{x} \in \mathbb{R}^n} \left\{ \|\boldsymbol{x} - \boldsymbol{x}_*\| : f(\boldsymbol{x}) \leq f(\boldsymbol{x}_0) \right\} \tag{8.1}$$

or

$$R_{\boldsymbol{x}} \geq \|\boldsymbol{x}_0 - \boldsymbol{x}_*\|. \tag{8.2}$$

We present additional constants in N_ϵ when particularly informative.

Naturally, each method in Table A.1 has additional algorithmic parameters that influence algorithmic behaviour. We have omitted the dependence of each method's WCC on the selection of algorithmic parameters to allow for an easier comparison of methods.

We recall that, with the exception of the methods from Nesterov and Spokoiny (2017) and Konečný and Richtárik (2014), the methods referenced in Table A.1 do not require knowledge of the value of the relevant Lipschitz constants.

Table A.1. Known WCC bounds on the number of function evaluations needed to achieve a given stationarity measure.

Rate type	Method type (citation)[NOTES]	N_ϵ
$f \in \mathcal{LC}^1$		
$\|\nabla f(\boldsymbol{x}_k)\| \leq \epsilon$	DDS (Konečný and Richtárik 2014)	$\dfrac{n^2 L_{\mathrm{g}}(f(\boldsymbol{x}_0) - f(\boldsymbol{x}_*))}{\epsilon^2}$
	TR (Garmanjani *et al.* 2016)	$\dfrac{n^2 L_{\mathrm{g}}^2(f(\boldsymbol{x}_0) - f(\boldsymbol{x}_*))}{\epsilon^2}$
	ARC-DFO (Cartis *et al.* 2012)[A]	$\dfrac{n^2 \max\{L_{\mathrm{H}}, L_{\mathrm{g}}\}^{3/2}(f(\boldsymbol{x}_0) - f(\boldsymbol{x}_*))}{\epsilon^{3/2}}$
$\mathbb{E}_{\boldsymbol{U}_{k-1}}[\|\nabla f(\hat{\boldsymbol{x}}_k)\|]\leq \epsilon$	RS (Nesterov and Spokoiny 2017)[B]	$\dfrac{n L_{\mathrm{g}}(f(\boldsymbol{x}_0) - f(\boldsymbol{x}_*))}{\epsilon^2}$
$\|\nabla f(\boldsymbol{x}_k)\| \leq \epsilon$ w.p. $1 - p_1$	DDS (Gratton *et al.* 2015)[C]	$\dfrac{mn L_{\mathrm{g}}^2(f(\boldsymbol{x}_0) - f(\boldsymbol{x}_*))}{\epsilon^2}$
$\|\nabla f(\boldsymbol{x}_k)\| \leq \epsilon$ w.p. $1 - p_2$	TR (Gratton *et al.* 2018)[C,D]	$\dfrac{m \max\{\kappa_{\mathrm{ef}}, \kappa_{\mathrm{eg}}\}^2(f(\boldsymbol{x}_0) - f(\boldsymbol{x}_*))}{\epsilon^2}$
$f \in \mathcal{LC}^2$		
$\max\{\|\nabla f(\boldsymbol{x}_k)\|, -\lambda_k\} \leq \epsilon$	DDS (Gratton *et al.* 2016)	$\dfrac{n^5 \max\{L_{\mathrm{H}}, L_{\mathrm{g}}\}^3(f(\boldsymbol{x}_0) - f(\boldsymbol{x}_*))}{\epsilon^3}$
	TR (Gratton *et al.* 2019*a*)	$\dfrac{n^5 \max\{L_{\mathrm{H}}^3, L_{\mathrm{g}}^2\}(f(\boldsymbol{x}_0) - f(\boldsymbol{x}_*))}{\epsilon^3}$
$\max\{\|\nabla f(\boldsymbol{x}_k)\|, -\lambda_k\} \leq \epsilon$ w.p. $1 - p_3$	TR (Gratton *et al.* 2018)[C,D]	$\dfrac{m \max\{\kappa_{\mathrm{eg}}, \kappa_{\mathrm{eH}}\}^3(f(\boldsymbol{x}_0) - f(\boldsymbol{x}_*))}{\epsilon^3}$

Table A.1 continued.

Rate type	Method type (citation)[NOTES]	N_ϵ
$f \in \mathcal{LC}^1$, f is λ-strongly convex		
$f(\boldsymbol{x}_k) - f(\boldsymbol{x}_*) \leq \epsilon$	DDS (Konečný and Richtárik 2014)	$\dfrac{n^2 L_{\mathrm{g}}}{\lambda} \log\left(\dfrac{1}{\epsilon}\right)$
$\mathbb{E}_{\boldsymbol{U}_{k-1}}[f(\hat{\boldsymbol{x}}_k)] - f(\boldsymbol{x}_*) \leq \epsilon$	RS (Nesterov and Spokoiny 2017)[B]	$\dfrac{n L_{\mathrm{g}}}{\lambda} \log\left(\dfrac{L_{\mathrm{g}} R_{\boldsymbol{x}}{}^2}{\epsilon}\right)$
$f \in \mathcal{LC}^1$, f is convex		
$f(\boldsymbol{x}_k) - f(\boldsymbol{x}_*) \leq \epsilon$	DDS (Konečný and Richtárik 2014)[E]	$\dfrac{n^2 L_{\mathrm{g}} R_{\mathrm{level}}}{\epsilon}$
$\mathbb{E}_{\boldsymbol{U}_{k-1}}[f(\hat{\boldsymbol{x}}_k)] - f(\boldsymbol{x}_*) \leq \epsilon$	RS (Nesterov and Spokoiny 2017)[B]	$\dfrac{n L_{\mathrm{g}} R_{\boldsymbol{x}}{}^2}{\epsilon}$
$f \in \mathcal{LC}^0$, f is convex		
$\mathbb{E}_{\boldsymbol{U}_{k-1}}[f(\hat{\boldsymbol{x}}_k)] - f(\boldsymbol{x}_*) \leq \epsilon$	RS (Nesterov and Spokoiny 2017)[B]	$\dfrac{n^2 L_{\mathrm{f}}^2 R_{\boldsymbol{x}}{}^2}{\epsilon^2}$
$f \in \mathcal{LC}^0$		
$\mathbb{E}_{\boldsymbol{U}_{k-1}}[\|\|\nabla f_{\bar{\mu}}(\hat{\boldsymbol{x}}_k)\|\|] \leq \epsilon, \bar{\mu} = \dfrac{\epsilon}{L_{\mathrm{f}}\sqrt{n}}$	RS (Nesterov and Spokoiny 2017)[B]	$\dfrac{n^3 L_{\mathrm{f}}^5 (f(\boldsymbol{x}_0) - f(\boldsymbol{x}_*))}{\epsilon^3}$

Table A.1 continued.

Rate type	Method type (citation)[NOTES]	N_ϵ
$f = h \circ \boldsymbol{F}$, convex $h \in \mathcal{LC}^0$, $\boldsymbol{F} \in \mathcal{LC}^1$		
$\Psi(\boldsymbol{x}_k) \leq \epsilon$	TR (Garmanjani $et\ al.$ 2016)[F]	$\dfrac{pn^2 L_{\mathrm{g}}(\boldsymbol{F})^2 L_{\mathrm{f}}(h)^2(f(\boldsymbol{x}_0) - f(\boldsymbol{x}_*))}{\epsilon^2}$
A smoothed $f_\mu(\boldsymbol{x})$ for f		
$\|\nabla f_{\mu_k}(\boldsymbol{x}_k)\| \leq \epsilon$ where $\mu_k \in O\left(\dfrac{\epsilon}{\sqrt{n}}\right)$	DDS (Garmanjani and Vicente 2012)[G]	$\dfrac{n^{5/2}\left[-\log(\epsilon) + \log(n)\right](f(\boldsymbol{x}_0) - f(\boldsymbol{x}_*))}{\epsilon^3}$
	TR (Garmanjani $et\ al.$ 2016)[G]	$\dfrac{n^{5/2}\left[\lvert\log(\epsilon)\rvert + \log(n)\right](f(\boldsymbol{x}_0) - f(\boldsymbol{x}_*))}{\epsilon^3}$

A We omit an additional $\lvert\log(\epsilon)\rvert$ dependence.

B $\hat{\boldsymbol{x}}_k = \arg\min_{j=1,\dots,k} f(\boldsymbol{x}_j)$.

C m is the number of function evaluations performed in each iteration, independent of n.

D Gratton $et\ al.$ (2018) prove results for an arbitrary model-building scheme that assumes the ability to yield p-probabilistically $\boldsymbol{\kappa}_Q$-fully quadratic models (where $\boldsymbol{\kappa}_Q = (\kappa_{\mathrm{ef}}, \kappa_{\mathrm{eg}}, \kappa_{\mathrm{eH}})$) when $f \in \mathcal{LC}^2$ and p-probabilistically $\boldsymbol{\kappa}_L$-fully linear models (where $\boldsymbol{\kappa}_L = (\kappa_{\mathrm{ef}}, \kappa_{\mathrm{eg}})$) when $f \in \mathcal{LC}^1$. The construction of probabilistically fully quadratic models or probabilistically fully linear models when $m \ll (n+1)(n+2)/2$ remains an open question. Note that when $p = 1$, it is known that by using $m \in O(n^2)$ points, one can guarantee $\boldsymbol{\kappa}_Q$-fully quadratic models with $\kappa_{\mathrm{ef}}, \kappa_{\mathrm{eg}}, \kappa_{\mathrm{eH}} \in O(nL_{\mathrm{H}})$ (Conn $et\ al.$ 2008a, Theorem 3). In this case, the result of Gratton $et\ al.$ (2018) yields a rate weaker than that obtained by Gratton $et\ al.$ (2016) by a factor of L_{g}. Similarly, when $p = 1$, it is known that by using $m \in O(n)$ points, one can guarantee $\boldsymbol{\kappa}_L$-fully linear models with $\kappa_{\mathrm{ef}}, \kappa_{\mathrm{eg}} \in O(n^{1/2}L_{\mathrm{g}})$ (Conn $et\ al.$ 2008a, Theorem 2). In this case, the result of Gratton $et\ al.$ (2018) yields a rate comparable to that obtained by Garmanjani $et\ al.$ (2016).

E Vicente (2013) derives the same bound but with L_{g}^2 instead of L_{g}; however, the method of Vicente (2013) does not require the value L_{g}.

F $L_{\mathrm{g}}(\boldsymbol{F})$ is the Lipschitz constant of the Jacobian $J(F)$, $L_{\mathrm{f}}(h)$ is the Lipschitz constant of h, and p is the dimension of the domain of h. A bound for a similar method with an additional $\lvert\log(\epsilon)\rvert$ dependence appears in Grapiglia $et\ al.$ (2016).

G Lipschitz constants do not appear because they are 'cancelled' by choosing the rate at which smoothing parameter $\mu_k \to 0$.

In Table A.1, we employ the constants

$$p_1 = \exp\left(-\frac{nL_g^2}{\epsilon^2}(f(\boldsymbol{x}_0) - f(\boldsymbol{x}_*))\right),$$

$$p_2 = \exp\left(-\frac{\max\{\kappa_{\mathrm{ef}}, \kappa_{\mathrm{eg}}\}^2}{\epsilon^2}(f(\boldsymbol{x}_0) - f(\boldsymbol{x}_*))\right),$$

$$p_3 = \exp\left(-\frac{\max\{\kappa_{\mathrm{eg}}, \kappa_{\mathrm{eH}}\}^3}{\epsilon^3}(f(\boldsymbol{x}_0) - f(\boldsymbol{x}_*))\right).$$

REFERENCES[12]

M. A. Abramson (2005), 'Second-order behavior of pattern search', *SIAM J. Optim.* **16**, 515–530.

M. A. Abramson and C. Audet (2006), 'Convergence of mesh adaptive direct search to second-order stationary points', *SIAM J. Optim.* **17**, 606–619.

M. A. Abramson, C. Audet and J. E. Dennis, Jr (2004), 'Generalized pattern searches with derivative information', *Math. Program.* **100**, 3–25.

M. A. Abramson, C. Audet, J. Chrissis and J. Walston (2009*a*), 'Mesh adaptive direct search algorithms for mixed variable optimization', *Optim. Lett.* **3**, 35–47.

M. A. Abramson, C. Audet, J. E. Dennis, Jr and S. Le Digabel (2009*b*), 'Ortho-MADS: A deterministic MADS instance with orthogonal directions', *SIAM J. Optim.* **20**, 948–966.

M. A. Abramson, L. Frimannslund and T. Steihaug (2013), 'A subclass of generating set search with convergence to second-order stationary points', *Optim. Methods Software* **29**, 900–918.

M. Abramson, C. Audet and J. E. Dennis, Jr (2007), 'Filter pattern search algorithms for mixed variable constrained optimization problems', *Pacific J. Optim.* **3**, 477–500.

A. Agarwal, O. Dekel and L. Xiao (2010), Optimal algorithms for online convex optimization with multi-point bandit feedback. In *23rd Conference on Learning Theory (COLT 2010)* (A. T. Kalai and M. Mohri, eds), pp. 28–40.

A. Agarwal, D. P. Foster, D. J. Hsu, S. M. Kakade and A. Rakhlin (2011), Stochastic convex optimization with bandit feedback. In *Advances in Neural Information Processing Systems 24* (J. Shawe-Taylor *et al.*, eds), Curran Associates, pp. 1035–1043.

A. Agarwal, M. J. Wainwright, P. L. Bartlett and P. K. Ravikumar (2009), Information-theoretic lower bounds on the oracle complexity of convex optimization. In *Advances in Neural Information Processing Systems 22* (Y. Bengio *et al.*, eds), Curran Associates, pp. 1–9.

---

[12] The URLs cited in this work were correct at the time of going to press, but the publisher and the authors make no undertaking that the citations remain live or are accurate or appropriate.

R. Agrawal (1995), 'Sample mean based index policies with $O(\log n)$ regret for the multi-armed bandit problem', *Adv. Appl. Probab.* **27**, 1054–1078.

S. Alarie, N. Amaioua, C. Audet, S. Le Digabel and L.-A. Leclaire (2018), Selection of variables in parallel space decomposition for the mesh adaptive direct search algorithm. Technical report, Cahier du GERAD G-2018-38, GERAD.

N. Alexandrov, J. E. Dennis, Jr, R. M. Lewis and V. Torczon (1998), 'A trust region framework for managing the use of approximation models in optimization', *Struct. Optim.* **15**, 16–23.

N. Amaioua, C. Audet, A. R. Conn and S. Le Digabel (2018), 'Efficient solution of quadratically constrained quadratic subproblems within the mesh adaptive direct search algorithm', *European J. Oper. Res.* **268**, 13–24.

S. Amaran, N. V. Sahinidis, B. Sharda and S. J. Bury (2015), 'Simulation optimization: A review of algorithms and applications', *Ann. Oper. Res.* **240**, 351–380.

H. L. Anderson (1986), 'Scientific uses of the MANIAC', *J. Statist. Phys.* **43**, 731–748.

M. B. Arouxét, N. Echebest and E. A. Pilotta (2011), 'Active-set strategy in Powell's method for optimization without derivatives', *Comput. Appl. Math.* **30**, 171–196.

C. Audet (2004), 'Convergence results for generalized pattern search algorithms are tight', *Optim. Engng* **5**, 101–122.

C. Audet (2014), A survey on direct search methods for blackbox optimization and their applications. In *Mathematics Without Boundaries: Surveys in Interdisciplinary Research* (P. M. Pardalos and T. M. Rassias, eds), Springer, pp. 31–56.

C. Audet and J. E. Dennis, Jr (2000), 'Pattern search algorithms for mixed variable programming', *SIAM J. Optim.* **11**, 573–594.

C. Audet and J. E. Dennis, Jr (2002), 'Analysis of generalized pattern searches', *SIAM J. Optim.* **13**, 889–903.

C. Audet and J. E. Dennis, Jr (2004), 'A pattern search filter method for nonlinear programming without derivatives', *SIAM J. Optim.* **14**, 980–1010.

C. Audet and J. E. Dennis, Jr (2006), 'Mesh adaptive direct search algorithms for constrained optimization', *SIAM J. Optim.* **17**, 188–217.

C. Audet and J. E. Dennis, Jr (2009), 'A progressive barrier for derivative-free nonlinear programming', *SIAM J. Optim.* **20**, 445–472.

C. Audet and W. L. Hare (2017), *Derivative-Free and Blackbox Optimization*, Springer.

C. Audet, J. Bigeon, D. Cartier, S. Le Digabel and L. Salomon (2018*a*), Performance indicators in multiobjective optimization. Technical report, Cahier du GERAD G-2018-90, GERAD.

C. Audet, J. E. Dennis, Jr and S. Le Digabel (2008*a*), 'Parallel space decomposition of the mesh adaptive direct search algorithm', *SIAM J. Optim.* **19**, 1150–1170.

C. Audet, A. Ianni, S. Le Digabel and C. Tribes (2014), 'Reducing the number of function evaluations in mesh adaptive direct search algorithms', *SIAM J. Optim.* **24**, 621–642.

C. Audet, A. Ihaddadene, S. Le Digabel and C. Tribes (2018*b*), 'Robust optimization of noisy blackbox problems using the mesh adaptive direct search algorithm', *Optim. Lett.* **12**, 675–689.

C. Audet, S. Le Digabel and M. Peyrega (2015), 'Linear equalities in blackbox optimization', *Comput. Optim. Appl.* **61**, 1–23.

C. Audet, G. Savard and W. Zghal (2008*b*), 'Multiobjective optimization through a series of single-objective formulations', *SIAM J. Optim.* **19**, 188–210.

C. Audet, G. Savard and W. Zghal (2010), 'A mesh adaptive direct search algorithm for multiobjective optimization', *European J. Oper. Res.* **204**, 545–556.

P. Auer (2002), 'Using confidence bounds for exploitation-exploration trade-offs', *J. Mach. Learn. Res.* **3**, 397–422.

P. Auer, N. Cesa-Bianchi and P. Fischer (2002), 'Finite-time analysis of the multi-armed bandit problem', *Machine Learning* **47**, 235–256.

P. Auer, N. Cesa-Bianchi, Y. Freund and R. E. Schapire (2003), 'The nonstochastic multiarmed bandit problem', *SIAM J. Comput.* **32**, 48–77.

A. Auger, N. Hansen, J. Perez Zerpa, R. Ros and M. Schoenauer (2009), Experimental comparisons of derivative free optimization algorithms. In *Experimental Algorithms (SEA 2009)* (J. Vahrenhold, ed.), Vol. 5526 of Lecture Notes in Computer Science, Springer, pp. 3–15.

F. Augustin and Y. M. Marzouk (2014), NOWPAC: A provably convergent derivative-free nonlinear optimizer with path-augmented constraints. arXiv:1403.1931

F. Augustin and Y. M. Marzouk (2017), A trust-region method for derivative-free nonlinear constrained stochastic optimization. arXiv:1703.04156

M. Avriel and D. J. Wilde (1967), 'Optimal condenser design by geometric programming', *Ind. Engng Chem. Process Des. Dev.* **6**, 256–263.

F. Bach and V. Perchet (2016), Highly-smooth zero-th order online optimization. In *29th Annual Conference on Learning Theory (COLT 2016)* (V. Feldman *et al.*, eds). *Proc. Mach. Learn. Res.* **49**, 257–283.

A. M. Bagirov and J. Ugon (2006), 'Piecewise partially separable functions and a derivative-free algorithm for large scale nonsmooth optimization', *J. Global Optim.* **35**, 163–195.

A. M. Bagirov, B. Karasözen and M. Sezer (2007), 'Discrete gradient method: Derivative-free method for nonsmooth optimization', *J. Optim. Theory Appl.* **137**, 317–334.

I. Bajaj, S. S. Iyer and M. M. F. Hasan (2018), 'A trust region-based two phase algorithm for constrained black-box and grey-box optimization with infeasible initial point', *Comput. Chem. Engng* **116**, 306–321.

P. Balaprakash, M. Salim, T. D. Uram, V. Vishwanath and S. M. Wild (2018), DeepHyper: Asynchronous hyperparameter search for deep neural networks. In *25th IEEE International Conference on High Performance Computing, Data, and Analytics (HiPC18)*. doi:10.1007/s10589-019-00063-3

K. Balasubramanian and S. Ghadimi (2018), Zeroth-order (non)-convex stochastic optimization via conditional gradient and gradient updates. In *Advances in Neural Information Processing Systems 31* (S. Bengio *et al.*, eds), Curran Associates, pp. 3459–3468.

K. Balasubramanian and S. Ghadimi (2019), Zeroth-order nonconvex stochastic optimization: Handling constraints, high-dimensionality, and saddle-points. arXiv:1809.06474

A. S. Bandeira, K. Scheinberg and L. N. Vicente (2012), 'Computation of sparse low degree interpolating polynomials and their application to derivative-free optimization', *Math. Program.* **134**, 223–257.

A. S. Bandeira, K. Scheinberg and L. N. Vicente (2014), 'Convergence of trust-region methods based on probabilistic models', *SIAM J. Optim.* **24**, 1238–1264.

N. V. Banichuk, V. M. Petrov and F. L. Chernous'ko (1966), 'The solution of variational and boundary value problems by the method of local variations', *USSR Comput. Math. Math. Phys.* **6**, 1–21.

R. R. Barton and J. S. Ivey, Jr (1996), 'Nelder–Mead simplex modifications for simulation optimization', *Manag. Sci.* **42**, 954–973.

J. Barzilai and J. M. Borwein (1988), 'Two-point step size gradient methods', *IMA J. Numer. Anal.* **8**, 141–148.

H. H. Bauschke, W. L. Hare and W. M. Moursi (2014), 'A derivative-free comirror algorithm for convex optimization', *Optim. Methods Software* **30**, 706–726.

A. G. Baydin, B. A. Pearlmutter, A. A. Radul and J. M. Siskind (2018), 'Automatic differentiation in machine learning: A survey', *J. Mach. Learn. Res.* **18** (153), 1–43.

P. Belitz and T. Bewley (2013), 'New horizons in sphere-packing theory, II: Lattice-based derivative-free optimization via global surrogates', *J. Global Optim.* **56**, 61–91.

A. Belloni, T. Liang, H. Narayanan and A. Rakhlin (2015), Escaping the local minima via simulated annealing: Optimization of approximately convex functions. In *28th Conference on Learning Theory (COLT 2015)*. *Proc. Mach. Learn. Res.* **40**, 240–265.

P. Belotti, C. Kirches, S. Leyffer, J. Linderoth, J. Luedtke and A. Mahajan (2013), Mixed-integer nonlinear optimization. In *Acta Numerica*, Vol. 22, Cambridge University Press, pp. 1–131.

A. S. Berahas, R. H. Byrd and J. Nocedal (2019), 'Derivative-free optimization of noisy functions via quasi-Newton methods', *SIAM J. Optim.*, to appear. arXiv:1803.10173

D. Bertsimas and O. Nohadani (2010), 'Robust optimization with simulated annealing', *J. Global Optim.* **48**, 323–334.

D. Bertsimas, O. Nohadani and K. M. Teo (2010), 'Robust optimization for unconstrained simulation-based problems', *Oper. Res.* **58**, 161–178.

S. Bhatnagar, H. Prasad and L. A. Prashanth (2013), Kiefer–Wolfowitz algorithm. In *Stochastic Recursive Algorithms for Optimization: Simultaneous Perturbation Methods*, Vol. 434 of Lecture Notes in Control and Information Sciences, Springer, pp. 31–39.

A. Bibi, E. H. Bergou, O. Sener, B. Ghanem and P. Richtárik (2019), A stochastic derivative-free optimization method with importance sampling. arXiv:1902.01272

S. C. Billups, J. Larson and P. Graf (2013), 'Derivative-free optimization of expensive functions with computational error using weighted regression', *SIAM J. Optim.* **55**, 27–53.

M. Björkman and K. Holmström (2000), 'Global optimization of costly nonconvex functions using radial basis functions', *Optim. Engng* **1**, 373–397.

J. Blanchet, C. Cartis, M. Menickelly and K. Scheinberg (2019), 'Convergence rate analysis of a stochastic trust region method via submartingales', *INFORMS J. Optim.*, to appear. arXiv:1609.07428

J. R. Blum (1954*a*), 'Approximation methods which converge with probability one', *Ann. Math. Statist.* **25**, 382–386.

J. R. Blum (1954*b*), 'Multidimensional stochastic approximation methods', *Ann. Math. Statist.* **25**, 737–744.

C. G. E. Boender, A. H. G. Rinnooy Kan, G. T. Timmer and L. Stougie (1982), 'A stochastic method for global optimization', *Math. Program.* **22**, 125–140.

C. Bogani, M. G. Gasparo and A. Papini (2009), 'Generalized pattern search methods for a class of nonsmooth optimization problems with structure', *J. Comput. Appl. Math.* **229**, 283–293.

D. M. Bortz and C. T. Kelley (1998), The simplex gradient and noisy optimization problems. In *Computational Methods for Optimal Design and Control* (J. Borggaard *et al.*, eds), Vol. 24 of Progress in Systems and Control Theory, Birkhäuser, pp. 77–90.

L. Bottou, F. E. Curtis and J. Nocedal (2018), 'Optimization methods for large-scale machine learning', *SIAM Review* **60**, 223–311.

G. E. P. Box and N. R. Draper (1987), *Empirical Model Building and Response Surfaces*, Wiley.

M. J. Box (1965), 'A new method of constrained optimization and a comparison with other methods', *Comput. J.* **8**, 42–52.

M. J. Box (1966), 'A comparison of several current optimization methods, and the use of transformations in constrained problems', *Comput. J.* **9**, 67–77.

R. Brekelmans, L. Driessen, H. Hamers and D. den Hertog (2005), 'Constrained optimization involving expensive function evaluations: A sequential approach', *European J. Oper. Res.* **160**, 121–138.

R. P. Brent (1973), *Algorithms for Minimization Without Derivatives*, Prentice Hall.

K. M. Brown and J. E. Dennis, Jr (1971), 'Derivative free analogues of the Levenberg–Marquardt and Gauss algorithms for nonlinear least squares approximation', *Numer. Math.* **18**, 289–297.

S. Bubeck and N. Cesa-Bianchi (2012), 'Regret analysis of stochastic and non-stochastic multi-armed bandit problems', *Found. Trends Mach. Learn.* **5**, 1–122.

S. Bubeck, Y. T. Lee and R. Eldan (2017), Kernel-based methods for bandit convex optimization. In *49th Annual ACM SIGACT Symposium on Theory of Computing (STOC 2017)*, ACM, pp. 72–85.

S. Bubeck, R. Munos, G. Stoltz and C. Szepesvári (2011*a*), '$\mathcal{X}$-armed bandits', *J. Mach. Learn. Res.* **12**, 1655–1695.

S. Bubeck, G. Stoltz and J. Y. Yu (2011*b*), Lipschitz bandits without the Lipschitz constant. In *Algorithmic Learning Theory (ALT 2011)* (J. Kivinen *et al.*, eds), Vol. 6925 of Lecture Notes in Computer Science, Springer, pp. 144–158.

L. F. Bueno, A. Friedlander, J. M. Martínez and F. N. C. Sobral (2013), 'Inexact restoration method for derivative-free optimization with smooth constraints', *SIAM J. Optim.* **23**, 1189–1213.

M. D. Buhmann (2000), Radial basis functions. In *Acta Numerica*, Vol. 9, Cambridge University Press, pp. 1–38.

A. D. Bull (2011), 'Convergence rates of efficient global optimization algorithms', *J. Mach. Learn. Res.* **12**, 2879–2904.

J. V. Burke, F. E. Curtis, A. S. Lewis, M. L. Overton and L. E. A. Simões (2018), Gradient sampling methods for nonsmooth optimization. arXiv:1804.11003

Á. Bűrmen, J. Olenšek and T. Tuma (2015), 'Mesh adaptive direct search with second directional derivative-based Hessian update', *Comput. Optim. Appl.* **62**, 693–715.

Á. Bűrmen, J. Puhan and T. Tuma (2006), 'Grid restrained Nelder–Mead algorithm', *Comput. Optim. Appl.* **34**, 359–375.

R. G. Carter, J. M. Gablonsky, A. Patrick, C. T. Kelley and O. J. Eslinger (2001), 'Algorithms for noisy problems in gas transmission pipeline optimization', *Optim. Engng* **2**, 139–157.

C. Cartis and L. Roberts (2017), A derivative-free Gauss–Newton method. arXiv:1710.11005

C. Cartis and K. Scheinberg (2018), 'Global convergence rate analysis of unconstrained optimization methods based on probabilistic models', *Math. Program.* **169**, 337–375.

C. Cartis and J. Fiala and B. Marteau and L. Roberts (2018), Improving the flexibility and robustness of model-based derivative-free optimization solvers. arXiv:1804.00154

C. Cartis, N. I. M. Gould and P. L. Toint (2012), 'On the oracle complexity of first-order and derivative-free algorithms for smooth nonconvex minimization', *SIAM J. Optim.* **22**, 66–86.

C. Cartis, N. I. M. Gould and P. L. Toint (2018), Sharp worst-case evaluation complexity bounds for arbitrary-order nonconvex optimization with inexpensive constraints. arXiv:1811.01220

B. Castle (2012), Quasi-Newton methods for stochastic optimization and proximity-based methods for disparate information fusion. PhD thesis, Indiana University.

J. Céa (1971), *Optimisation: Théorie et Algorithmes*, Méthodes Mathématiques de l'Informatique, Dunod.

G. Chandramouli and V. Narayanan (2019), A scaled conjugate gradient based direct search algorithm for high dimensional box constrained derivative free optimization. arXiv:1901.05215

K.-H. Chang (2012), 'Stochastic Nelder–Mead simplex method: A new globally convergent direct search method for simulation optimization', *European J. Oper. Res.* **220**, 684–694.

K.-H. Chang, L. J. Hong and H. Wan (2013), 'Stochastic trust-region response-surface method (STRONG): A new response-surface framework for simulation optimization', *INFORMS J. Comput.* **25**, 230–243.

H. Chen and B. W. Schmeiser (2001), 'Stochastic root finding via retrospective approximation', *IIE Trans.* **33**, 259–275.

R. Chen (2015), Stochastic derivative-free optimization of noisy functions. PhD thesis, Lehigh University.

R. Chen, M. Menickelly and K. Scheinberg (2018*a*), 'Stochastic optimization using a trust-region method and random models', *Math. Program.* **169**, 447–487.

X. Chen and C. T. Kelley (2016), 'Optimization with hidden constraints and embedded Monte Carlo computations', *Optim. Engng* **17**, 157–175.

X. Chen, C. T. Kelley, F. Xu and Z. Zhang (2018*b*), 'A smoothing direct search method for Monte Carlo-based bound constrained composite nonsmooth optimization', *SIAM J. Sci. Comput.* **40**, A2174–A2199.

T. D. Choi and C. T. Kelley (2000), 'Superlinear convergence and implicit filtering', *SIAM J. Optim.* **10**, 1149–1162.

T. D. Choi, O. J. Eslinger, C. T. Kelley, J. W. David and M. Etheridge (2000), 'Optimization of automotive valve train components with implicit filtering', *Optim. Engng* **1**, 9–27.

A. Ciccazzo, V. Latorre, G. Liuzzi, S. Lucidi and F. Rinaldi (2015), 'Derivative-free robust optimization for circuit design', *J. Optim. Theory Appl.* **164**, 842–861.

G. Cocchi, G. Liuzzi, A. Papini and M. Sciandrone (2018), 'An implicit filtering algorithm for derivative-free multiobjective optimization with box constraints', *Comput. Optim. Appl.* **69**, 267–296.

B. Colson and P. L. Toint (2001), 'Exploiting band structure in unconstrained optimization without derivatives', *Optim. Engng* **2**, 399–412.

B. Colson and P. L. Toint (2002), A derivative-free algorithm for sparse unconstrained optimization problems. In *Trends in Industrial and Applied Mathematics* (A. H. Siddiqi and M. Kočvara, eds), Vol. 72 of Applied Optimization, Springer, pp. 131–147.

B. Colson and P. L. Toint (2005), 'Optimizing partially separable functions without derivatives', *Optim. Methods Software* **20**, 493–508.

P. D. Conejo, E. W. Karas and L. G. Pedroso (2015), 'A trust-region derivative-free algorithm for constrained optimization', *Optim. Methods Software* **30**, 1126–1145.

P. D. Conejo, E. W. Karas, L. G. Pedroso, A. A. Ribeiro and M. Sachine (2013), 'Global convergence of trust-region algorithms for convex constrained minimization without derivatives', *Appl. Math. Comput.* **220**, 324–330.

A. R. Conn and S. Le Digabel (2013), 'Use of quadratic models with mesh-adaptive direct search for constrained black box optimization', *Optim. Methods Software* **28**, 139–158.

A. R. Conn and P. L. Toint (1996), An algorithm using quadratic interpolation for unconstrained derivative free optimization. In *Nonlinear Optimization and Applications* (G. Di Pillo and F. Giannessi, eds), Springer, pp. 27–47.

A. R. Conn and L. N. Vicente (2012), 'Bilevel derivative-free optimization and its application to robust optimization', *Optim. Methods Software* **27**, 561–577.

A. R. Conn, N. I. M. Gould and P. L. Toint (1991), 'A globally convergent augmented Lagrangian algorithm for optimization with general constraints and simple bounds', *SIAM J. Numer. Anal.* **28**, 545–572.

A. R. Conn, N. I. M. Gould and P. L. Toint (2000), *Trust-Region Methods*, SIAM.

A. R. Conn, K. Scheinberg and P. L. Toint (1997*a*), On the convergence of derivative-free methods for unconstrained optimization. In *Approximation Theory and Optimization: Tributes to M. J. D. Powell* (A. Iserles and M. Buhmann, eds), Cambridge University Press, pp. 83–108.

A. R. Conn, K. Scheinberg and P. L. Toint (1997*b*), 'Recent progress in unconstrained nonlinear optimization without derivatives', *Math. Program.* **79**, 397–414.

A. R. Conn, K. Scheinberg and P. L. Toint (1998), A derivative free optimization algorithm in practice. In *7th AIAA/USAF/NASA/ISSMO Symposium on Multidisciplinary Analysis and Optimization*, American Institute of Aeronautics and Astronautics.

A. R. Conn, K. Scheinberg and P. L. Toint (2001), DFO (derivative free optimization software). https://projects.coin-or.org/Dfo

A. R. Conn, K. Scheinberg and L. N. Vicente (2008*a*), 'Geometry of interpolation sets in derivative free optimization', *Math. Program.* **111**, 141–172.

A. R. Conn, K. Scheinberg and L. N. Vicente (2008*b*), 'Geometry of sample sets in derivative free optimization: Polynomial regression and underdetermined interpolation', *IMA J. Numer. Anal.* **28**, 721–748.

A. R. Conn, K. Scheinberg and L. N. Vicente (2009*a*), 'Global convergence of general derivative-free trust-region algorithms to first and second order critical points', *SIAM J. Optim.* **20**, 387–415.

A. R. Conn, K. Scheinberg and L. N. Vicente (2009*b*), *Introduction to Derivative-Free Optimization*, SIAM.

I. D. Coope and C. J. Price (2000), 'Frame based methods for unconstrained optimization', *J. Optim. Theory Appl.* **107**, 261–274.

I. D. Coope and R. Tappenden (2019), 'Efficient calculation of regular simplex gradients', *Comput. Optim. Appl.* **72**, 561–588.

A. L. Custódio and J. F. A. Madeira (2015), 'GLODS: Global and local optimization using direct search', *J. Global Optim.* **62**, 1–28.

A. L. Custódio and J. F. A. Madeira (2016), MultiGLODS: Global and local multiobjective optimization using direct search. Technical report, Universidade Nova de Lisboa.

A. L. Custódio and L. N. Vicente (2005), SID-PSM: A pattern search method guided by simplex derivatives for use in derivative-free optimization. http://www.mat.uc.pt/sid-psm

A. L. Custódio and L. N. Vicente (2007), 'Using sampling and simplex derivatives in pattern search methods', *SIAM J. Optim.* **18**, 537–555.

A. L. Custódio, J. E. Dennis, Jr and L. N. Vicente (2008), 'Using simplex gradients of nonsmooth functions in direct search methods', *IMA J. Numer. Anal.* **28**, 770–784.

A. L. Custódio, J. F. A. Madeira, A. I. F. Vaz and L. N. Vicente (2011), 'Direct multisearch for multiobjective optimization', *SIAM J. Optim.* **21**, 1109–1140.

A. L. Custódio, H. Rocha and L. N. Vicente (2009), 'Incorporating minimum Frobenius norm models in direct search', *Comput. Optim. Appl.* **46**, 265–278.

L. Dai (2016*a*), Convergence rates of finite difference stochastic approximation algorithms, I: General sampling. In *Sensing and Analysis Technologies for Biomedical and Cognitive Applications* (L. Dai *et al.*, eds), SPIE.

L. Dai (2016*b*), Convergence rates of finite difference stochastic approximation algorithms, II: Implementation via common random numbers. In *Sensing and Analysis Technologies for Biomedical and Cognitive Applications* (L. Dai *et al.*, eds), SPIE.

C. D'Ambrosio, G. Nannicini and G. Sartor (2017), 'MILP models for the selection of a small set of well-distributed points', *Oper. Res. Lett.* **45**, 46–52.

C. Davis (1954), 'Theory of positive linear dependence', *Amer. J. Math.* **76**, 733–746.

P. Davis (2005), Michael J. D. Powell: An oral history. In *History of Numerical Analysis and Scientific Computing Project*, SIAM.

C. Davis and W. L. Hare (2013), 'Exploiting known structures to approximate normal cones', *Math. Oper. Res.* **38**, 665–681.

R. De Leone, M. Gaudioso and L. Grippo (1984), 'Stopping criteria for linesearch methods without derivatives', *Math. Program.* **30**, 285–300.

O. Dekel, R. Eldan and T. Koren (2015), Bandit smooth convex optimization: Improving the bias-variance tradeoff. In *Advances in Neural Information Processing Systems 28* (C. Cortes *et al.*, eds), Curran Associates, pp. 2926–2934.

A. Dener, A. Denchfield, T. Munson, J. Sarich, S. M. Wild, S. Benson and L. Curfman McInnes (2018), TAO 3.10 users manual. Technical Memorandum ANL/MCS-TM-322, Argonne National Laboratory.

G. Deng and M. C. Ferris (2009), 'Variable-number sample-path optimization', *Math. Program.* **117**, 81–109.

G. Deng and M. C. Ferris (2006), Adaptation of the UOBYQA algorithm for noisy functions. In *Proceedings of the 2006 Winter Simulation Conference*, IEEE, pp. 312–319.

J. E. Dennis, Jr and V. Torczon (1991), 'Direct search methods on parallel machines', *SIAM J. Optim.* **1**, 448–474.

J. E. Dennis, Jr and V. Torczon (1997), Managing approximation models in optimization. In *Multidisciplinary Design Optimization: State-of-the-Art* (N. M. Alexandrov and M. Y. Hussaini, eds), SIAM, pp. 330–347.

J. E. Dennis, Jr and D. J. Woods (1987), Optimization on microcomputers: The Nelder–Mead simplex algorithm. In *New Computing Environments: Microcomputers in Large-Scale Computing* (A. Wouk, ed.), SIAM, pp. 116–122.

C. Derman (1956), 'An application of Chung's lemma to the Kiefer–Wolfowitz stochastic approximation procedure', *Ann. Math. Statist.* **27**, 532–536.

G. I. Diaz, A. Fokoue-Nkoutche, G. Nannicini and H. Samulowitz (2017), 'An effective algorithm for hyperparameter optimization of neural networks', *IBM J. Res. Develop.* **61**, 9:1–9:11.

M. A. Diniz-Ehrhardt, J. M. Martínez and L. G. Pedroso (2011), 'Derivative-free methods for nonlinear programming with general lower-level constraints', *Comput. Appl. Math.* **30**, 19–52.

M. A. Diniz-Ehrhardt, J. M. Martínez and M. Raydan (2008), 'A derivative-free nonmonotone line-search technique for unconstrained optimization', *J. Comput. Appl. Math.* **219**, 383–397.

M. Dodangeh and L. N. Vicente (2016), 'Worst case complexity of direct search under convexity', *Math. Program.* **155**, 307–332.

M. Dodangeh, L. N. Vicente and Z. Zhang (2016), 'On the optimal order of worst case complexity of direct search', *Optim. Lett.* **10**, 699–708.

E. D. Dolan, R. M. Lewis and V. Torczon (2003), 'On the local convergence of pattern search', *SIAM J. Optim.* **14**, 567–583.

J. C. Duchi, M. I. Jordan, M. J. Wainwright and A. Wibisono (2015), 'Optimal rates for zero-order convex optimization: The power of two function evaluations', *IEEE Trans. Inform. Theory* **61**, 2788–2806.

R. Durrett (2010), *Probability: Theory and Examples*, Cambridge University Press.

P. Dvurechensky, A. Gasnikov and E. Gorbunov (2018), An accelerated method for derivative-free smooth stochastic convex optimization. arXiv:1802.09022

N. Echebest, M. L. Schuverdt and R. P. Vignau (2015), 'An inexact restoration derivative-free filter method for nonlinear programming', *Comput. Appl. Math.* **36**, 693–718.

M. Ehrgott (2005), *Multicriteria Optimization*, second edition, Springer.

C. Elster and A. Neumaier (1995), 'A grid algorithm for bound constrained optimization of noisy functions', *IMA J. Numer. Anal.* **15**, 585–608.

V. Fabian (1971), Stochastic approximation. In *Optimizing Methods in Statistics*, Elsevier, pp. 439–470.

G. Fasano, G. Liuzzi, S. Lucidi and F. Rinaldi (2014), 'A linesearch-based derivative-free approach for nonsmooth constrained optimization', *SIAM J. Optim.* **24**, 959–992.

G. Fasano, J. L. Morales and J. Nocedal (2009), 'On the geometry phase in model-based algorithms for derivative-free optimization', *Optim. Methods Software* **24**, 145–154.

E. Fermi and N. Metropolis (1952), Numerical solution of a minimum problem. Technical report LA-1492, Los Alamos Scientific Laboratory of the University of California.

P. S. Ferreira, E. W. Karas, M. Sachine and F. N. C. Sobral (2017), 'Global convergence of a derivative-free inexact restoration filter algorithm for nonlinear programming', *Optimization* **66**, 271–292.

D. E. Finkel and C. T. Kelley (2004), Convergence analysis of the DIRECT algorithm. Technical report 04-28, Center for Research in Scientific Computation, North Carolina State University. https://projects.ncsu.edu/crsc/reports/ftp/pdf/crsc-tr04-28.pdf

D. E. Finkel and C. T. Kelley (2006), 'Additive scaling and the DIRECT algorithm', *J. Global Optim.* **36**, 597–608.

D. E. Finkel and C. T. Kelley (2009), 'Convergence analysis of sampling methods for perturbed Lipschitz functions', *Pacific J. Optim.* **5**, 339–350.

A. Flaxman, A. Kalai and B. McMahan (2005), Online convex optimization in the bandit setting: Gradient descent without a gradient. In *16th Annual ACM–SIAM Symposium on Discrete Algorithms (SODA '05)*, ACM, pp. 385–394.

R. Fletcher (1965), 'Function minimization without evaluating derivatives: A review', *Comput. J.* **8**, 33–41.

R. Fletcher (1987), *Practical Methods of Optimization*, second edition, Wiley.

A. Forrester, A. Sobester and A. Keane (2008), *Engineering Design via Surrogate Modelling: A Practical Guide*, Wiley.

S. Forth, P. Hovland, E. Phipps, J. Utke and A. Walther, eds (2012), *Recent Advances in Algorithmic Differentiation*, Springer.

E. Frandi and A. Papini (2013), 'Coordinate search algorithms in multilevel optimization', *Optim. Methods Software* **29**, 1020–1041.

P. I. Frazier (2018), A tutorial on Bayesian optimization. arXiv:1807.02811

L. Frimannslund and T. Steihaug (2007), 'A generating set search method using curvature information', *Comput. Optim. Appl.* **38**, 105–121.

L. Frimannslund and T. Steihaug (2010), A new generating set search algorithm for partially separable functions. In *4th International Conference on Advanced Engineering Computing and Applications in Sciences (ADVCOMP 2010)*, IARIA, pp. 65–70.

L. Frimannslund and T. Steihaug (2011), 'On a new method for derivative free optimization', *Internat. J. Adv. Software* **4**, 244–255.

M. C. Fu, F. W. Glover and J. April (2005), Simulation optimization: A review, new developments, and applications. In *Proceedings of the 2005 Winter Simulation Conference*, IEEE.

F. Gao and L. Han (2012), 'Implementing the Nelder–Mead simplex algorithm with adaptive parameters', *Comput. Optim. Appl.* **51**, 259–277.

U. M. García-Palomares and J. F. Rodríguez (2002), 'New sequential and parallel derivative-free algorithms for unconstrained minimization', *SIAM J. Optim.* **13**, 79–96.

U. M. García-Palomares, I. J. García-Urrea and P. S. Rodríguez-Hernández (2013), 'On sequential and parallel non-monotone derivative-free algorithms for box constrained optimization', *Optim. Methods Software* **28**, 1233–1261.

R. Garmanjani and L. N. Vicente (2012), 'Smoothing and worst-case complexity for direct-search methods in nonsmooth optimization', *IMA J. Numer. Anal.* **33**, 1008–1028.

R. Garmanjani, D. Jùdice and L. N. Vicente (2016), 'Trust-region methods without using derivatives: Worst case complexity and the nonsmooth case', *SIAM J. Optim.* **26**, 1987–2011.

A. V. Gasnikov, E. A. Krymova, A. A. Lagunovskaya, I. N. Usmanova and F. A. Fedorenko (2017), 'Stochastic online optimization: Single-point and multi-point non-linear multi-armed bandits; Convex and strongly-convex case', *Automat. Remote Control* **78**, 224–234.

L. Gerencsér (1997), Rate of convergence of moments of Spall's SPSA method. In *Stochastic Differential and Difference Equations*, Vol. 23 of Progress in Systems and Control Theory, Birkhäuser, pp. 67–75.

S. Ghadimi (2019), 'Conditional gradient type methods for composite nonlinear and stochastic optimization', *Math. Program.* **173**, 431–464.

S. Ghadimi and G. Lan (2013), 'Stochastic first- and zeroth-order methods for nonconvex stochastic programming', *SIAM J. Optim.* **23**, 2341–2368.

P. E. Gill, W. Murray and M. H. Wright (1981), *Practical Optimization*, Academic Press.

P. E. Gill, W. Murray, M. A. Saunders and M. H. Wright (1983), 'Computing forward-difference intervals for numerical optimization', *SIAM J. Sci. Statist. Comput.* **4**, 310–321.

P. Gilmore and C. T. Kelley (1995), 'An implicit filtering algorithm for optimization of functions with many local minima', *SIAM J. Optim.* **5**, 269–285.

H. Glass and L. Cooper (1965), 'Sequential search: A method for solving constrained optimization problems', *J. Assoc. Comput. Mach.* **12**, 71–82.

D. Golovin, B. Solnik, S. Moitra, G. Kochanski, J. Karro and D. Sculley (2017), Google Vizier: A service for black-box optimization. In *23rd ACM SIGKDD International Conference on Knowledge Discovery and Data Mining (KDD '17)*, ACM, pp. 1487–1495.

C. C. Gonzaga, E. Karas and M. Vanti (2004), 'A globally convergent filter method for nonlinear programming', *SIAM J. Optim.* **14**, 646–669.

N. I. M. Gould and P. L. Toint (2010), 'Nonlinear programming without a penalty function or a filter', *Math. Program.* **122**, 155–196.

R. B. Gramacy and S. Le Digabel (2015), 'The mesh adaptive direct search algorithm with treed Gaussian process surrogates', *Pacific J. Optim.* **11**, 419–447.

G. N. Grapiglia, J. Yuan and Y.-x. Yuan (2016), 'A derivative-free trust-region algorithm for composite nonsmooth optimization', *Comput. Appl. Math.* **35**, 475–499.

S. Gratton and L. N. Vicente (2014), 'A merit function approach for direct search', *SIAM J. Optim.* **24**, 1980–1998.

S. Gratton, C. W. Royer and L. N. Vicente (2016), 'A second-order globally convergent direct-search method and its worst-case complexity', *Optimization* **65**, 1105–1128.

S. Gratton, C. W. Royer and L. N. Vicente (2019*a*), 'A decoupled first/second-order steps technique for nonconvex nonlinear unconstrained optimization with improved complexity bounds', *Math. Program.* doi:10.1007/s10107-018-1328-7

S. Gratton, C. W. Royer, L. N. Vicente and Z. Zhang (2015), 'Direct search based on probabilistic descent', *SIAM J. Optim.* **25**, 1515–1541.

S. Gratton, C. W. Royer, L. N. Vicente and Z. Zhang (2018), 'Complexity and global rates of trust-region methods based on probabilistic models', *IMA J. Numer. Anal.* **38**, 1579–1597.

S. Gratton, C. W. Royer, L. N. Vicente and Z. Zhang (2019*b*), 'Direct search based on probabilistic feasible descent for bound and linearly constrained problems', *Comput. Optim. Appl.* **72**, 525–559.

S. Gratton, P. L. Toint and A. Tröltzsch (2011), 'An active-set trust-region method for derivative-free nonlinear bound-constrained optimization', *Optim. Methods Software* **26**, 873–894.

G. A. Gray and T. G. Kolda (2006), 'Algorithm 856: APPSPACK 4.0: Asynchronous parallel pattern search for derivative-free optimization', *ACM Trans. Math. Software* **32**, 485–507.

A. Griewank (2003), A mathematical view of automatic differentiation. In *Acta Numerica*, Vol. 12, Cambridge University Press, pp. 321–398.

A. Griewank and A. Walther (2008), *Evaluating Derivatives: Principles and Techniques of Algorithmic Differentiation*, SIAM.

A. Griewank, A. Walther, S. Fiege and T. Bosse (2016), 'On Lipschitz optimization based on gray-box piecewise linearization', *Math. Program.* **158**, 383–415.

L. Grippo and F. Rinaldi (2014), 'A class of derivative-free nonmonotone optimization algorithms employing coordinate rotations and gradient approximations', *Comput. Optim. Appl.* **60**, 1–33.

L. Grippo and M. Sciandrone (2007), 'Nonmonotone derivative-free methods for nonlinear equations', *Comput. Optim. Appl.* **37**, 297–328.

L. Grippo, F. Lampariello and S. Lucidi (1988), 'Global convergence and stabilization of unconstrained minimization methods without derivatives', *J. Optim. Theory Appl.* **56**, 385–406.

B. Gu, Z. Huo and H. Huang (2016), Zeroth-order asynchronous doubly stochastic algorithm with variance reduction. arXiv:1612.01425

E. A. E. Gumma, M. H. A. Hashim and M. M. Ali (2014), 'A derivative-free algorithm for linearly constrained optimization problems', *Comput. Optim. Appl.* **57**, 599–621.

M. D. Gunzburger, C. G. Webster and G. Zhang (2014), Stochastic finite element methods for partial differential equations with random input data. In *Acta Numerica*, Vol. 23, Cambridge University Press, pp. 521–650.

H.-M. Gutmann (2001), 'A radial basis function method for global optimization', *J. Global Optim.* **19**, 201–227.

L. Han and G. Liu (2004), 'On the convergence of the UOBYQA method', *J. Appl. Math. Comput.* **16**, 125–142.

P. Hansen, B. Jaumard and S.-H. Lu (1991), 'On the number of iterations of Piyavskii's global optimization algorithm', *Math. Oper. Res.* **16**, 334–350.

W. L. Hare (2014), 'Numerical analysis of $\mathcal{VU}$-decomposition, $\mathcal{U}$-gradient, and $\mathcal{U}$-Hessian approximations', *SIAM J. Optim.* **24**, 1890–1913.

W. L. Hare (2017), 'Compositions of convex functions and fully linear models', *Optim. Lett.* **11**, 1217–1227.

W. L. Hare and A. S. Lewis (2005), 'Estimating tangent and normal cones without calculus', *Math. Oper. Res.* **30**, 785–799.

W. L. Hare and Y. Lucet (2013), 'Derivative-free optimization via proximal point methods', *J. Optim. Theory Appl.* **160**, 204–220.

W. L. Hare and M. Macklem (2013), 'Derivative-free optimization methods for finite minimax problems', *Optim. Methods Software* **28**, 300–312.

W. L. Hare and J. Nutini (2013), 'A derivative-free approximate gradient sampling algorithm for finite minimax problems', *Comput. Optim. Appl.* **56**, 1–38.

W. L. Hare, C. Planiden and C. Sagastizábal (2019), A derivative-free $\mathcal{VU}$-algorithm for convex finite-max problems. arXiv:1903.11184

E. Hazan, T. Koren and K. Y. Levy (2014), Logistic regression: Tight bounds for stochastic and online optimization. In *27th Conference on Learning Theory (COLT 2014)* (M. F. Balcan et al., eds). *Proc. Mach. Learn. Res.* **35**, 197–209.

J. He, A. Verstak, M. Sosonkina and L. T. Watson (2009a), 'Performance modeling and analysis of a massively parallel DIRECT, part 2', *Internat. J. High Performance Comput. Appl.* **23**, 29–41.

J. He, A. Verstak, L. T. Watson and M. Sosonkina (2007), 'Design and implementation of a massively parallel version of DIRECT', *Comput. Optim. Appl.* **40**, 217–245.

J. He, A. Verstak, L. T. Watson and M. Sosonkina (2009b), 'Performance modeling and analysis of a massively parallel DIRECT, part 1', *Internat. J. High Performance Comput. Appl.* **23**, 14–28.

J. He, L. T. Watson and M. Sosonkina (2009*c*), 'Algorithm 897: VTDIRECT95: Serial and parallel codes for the global optimization algorithm DIRECT', *ACM Trans. Math. Software* **36**, 1–24.

T. Homem-de-Mello and G. Bayraksan (2014), 'Monte Carlo sampling-based methods for stochastic optimization', *Surv. Oper. Res. Manag. Sci.* **19**, 56–85.

R. Hooke and T. A. Jeeves (1961), '"Direct search" solution of numerical and statistical problems', *J. Assoc. Comput. Mach.* **8**, 212–229.

P. D. Hough and J. C. Meza (2002), 'A class of trust-region methods for parallel optimization', *SIAM J. Optim.* **13**, 264–282.

P. D. Hough, T. G. Kolda and V. J. Torczon (2001), 'Asynchronous parallel pattern search for nonlinear optimization', *SIAM J. Sci. Comput.* **23**, 134–156.

X. Hu, L. A. Prashanth, A. György and C. Szepesvári (2016), (Bandit) convex optimization with biased noisy gradient oracles. In *19th International Conference on Artificial Intelligence and Statistics. Proc. Mach. Learn. Res.* **51**, 819–828.

D. W. Hutchison and J. C. Spall (2013), Stochastic approximation. In *Encyclopedia of Operations Research and Management Science*, Springer, pp. 1470–1476.

W. Huyer and A. Neumaier (1999), 'Global optimization by multilevel coordinate search', *J. Global Optim.* **14**, 331–355.

W. Huyer and A. Neumaier (2008), 'SNOBFIT: Stable noisy optimization by branch and fit', *ACM Trans. Math. Software* **35**, 9:1–9:25.

I. Ilievski, T. Akhtar, J. Feng and C. Shoemaker (2017), Efficient hyperparameter optimization for deep learning algorithms using deterministic RBF surrogates. In *31st AAAI Conference on Artificial Intelligence (AAAI '17)*, pp. 822–829.

K. G. Jamieson, R. D. Nowak and B. Recht (2012), Query complexity of derivative-free optimization. In *Advances in Neural Information Processing Systems 25* (F. Pereira *et al.*, eds), Curran Associates, pp. 2672–2680.

D. R. Jones, C. D. Perttunen and B. E. Stuckman (1993), 'Lipschitzian optimization without the Lipschitz constant', *J. Optim. Theory Appl.* **79**, 157–181.

D. R. Jones, M. Schonlau and W. J. Welch (1998), 'Efficient global optimization of expensive black-box functions', *J. Global Optim.* **13**, 455–492.

V. G. Karmanov (1974), 'Convergence estimates for iterative minimization methods', *USSR Comput. Math. Math. Phys.* **14**, 1–13.

C. T. Kelley (1999*a*), 'Detection and remediation of stagnation in the Nelder–Mead algorithm using a sufficient decrease condition', *SIAM J. Optim.* **10**, 43–55.

C. T. Kelley (1999*b*), *Iterative Methods for Optimization*, SIAM.

C. T. Kelley (2003), Implicit filtering and nonlinear least squares problems. In *System Modeling and Optimization XX* (E. W. Sachs and R. Tichatschke, eds), Vol. 130 of IFIP: The International Federation for Information Processing, Springer, pp. 71–90.

C. T. Kelley (2011), *Implicit Filtering*, SIAM.

K. A. Khan, J. Larson and S. M. Wild (2018), 'Manifold sampling for optimization of nonconvex functions that are piecewise linear compositions of smooth components', *SIAM J. Optim.* **28**, 3001–3024.

J. Kiefer and J. Wolfowitz (1952), 'Stochastic estimation of the maximum of a regression function', *Ann. Math. Statist.* **22**, 462–466.

S. Kim and J.-h. Ryu (2011), 'A trust-region algorithm for bi-objective stochastic optimization', *Procedia Comput. Sci.* **4**, 1422–1430.

S. Kim and D. Zhang (2010), Convergence properties of direct search methods for stochastic optimization. In *Proceedings of the 2010 Winter Simulation Conference*, IEEE.

S. Kim, R. Pasupathy and S. G. Henderson (2015), A guide to sample average approximation. In *Handbook of Simulation Optimization* (M. Fu, ed.), Vol. 216 of International Series in Operations Research & Management Science, Springer, pp. 207–243.

K. C. Kiwiel (2010), 'A nonderivative version of the gradient sampling algorithm for nonsmooth nonconvex optimization', *SIAM J. Optim.* **20**, 1983–1994.

R. Kleinberg, A. Slivkins and E. Upfal (2008), Multi-armed bandits in metric spaces. In *40th Annual ACM Symposium on Theory of Computing (STOC 2008)*, ACM, pp. 681–690.

N. L. Kleinman, J. C. Spall and D. Q. Naiman (1999), 'Simulation-based optimization with stochastic approximation using common random numbers', *Manag. Sci.* **45**, 1570–1578.

J. Knowles and D. Corne (2002), On metrics for comparing nondominated sets. In *Proceedings of the 2002 Congress on Evolutionary Computation*, Vol. 1, IEEE, pp. 711–716.

T. G. Kolda, R. M. Lewis and V. Torczon (2006), 'Stationarity results for generating set search for linearly constrained optimization', *SIAM J. Optim.* **17**, 943–968.

T. G. Kolda, R. M. Lewis and V. J. Torczon (2003), 'Optimization by direct search: New perspectives on some classical and modern methods', *SIAM Review* **45**, 385–482.

J. Konečný and P. Richtárik (2014), Simple complexity analysis of simplified direct search. arXiv:1410.0390

H. Kushner and G. Yin (2003), *Stochastic Approximation and Recursive Algorithms and Applications*, Springer.

H. J. Kushner and H. Huang (1979), 'Rates of convergence for stochastic approximation type algorithms', *SIAM J. Control Optim.* **17**, 607–617.

W. La Cruz (2014), 'A projected derivative-free algorithm for nonlinear equations with convex constraints', *Optim. Methods Software* **29**, 24–41.

W. La Cruz, J. M. Martínez and M. Raydan (2006), 'Spectral residual method without gradient information for solving large-scale nonlinear systems of equations', *Math. Comput.* **75** (255), 1429–1449.

J. C. Lagarias, B. Poonen and M. H. Wright (2012), 'Convergence of the restricted Nelder–Mead algorithm in two dimensions', *SIAM J. Optim.* **22**, 501–532.

J. C. Lagarias, J. A. Reeds, M. H. Wright and P. E. Wright (1998), 'Convergence properties of the Nelder–Mead simplex algorithm in low dimensions', *SIAM J. Optim.* **9**, 112–147.

T. L. Lai and H. Robbins (1985), 'Asymptotically efficient adaptive allocation rules', *Adv. Appl. Math.* **6**, 4–22.

J. Larson and S. C. Billups (2016), 'Stochastic derivative-free optimization using a trust region framework', *Comput. Optim. Appl.* **64**, 619–645.

J. Larson and S. M. Wild (2016), 'A batch, derivative-free algorithm for finding multiple local minima', *Optim. Engng* **17**, 205–228.

J. Larson and S. M. Wild (2018), 'Asynchronously parallel optimization solver for finding multiple minima', *Math. Program. Comput.* **10**, 303–332.

J. Larson, S. Leyffer, P. Palkar and S. M. Wild (2019), A method for convex black-box integer global optimization. arXiv:1903.11366

J. Larson, M. Menickelly and S. M. Wild (2016), 'Manifold sampling for ℓ_1 nonconvex optimization', *SIAM J. Optim.* **26**, 2540–2563.

V. Latorre, H. Habal, H. Graeb and S. Lucidi (2019), 'Derivative free methodologies for circuit worst case analysis', *Optim. Lett.* doi:10.1007/s11590-018-1364-5

M. Lazar and F. Jarre (2016), 'Calibration by optimization without using derivatives', *Optim. Engng* **17**, 833–860.

S. Le Digabel (2011), 'Algorithm 909: NOMAD: Nonlinear optimization with the MADS algorithm', *ACM Trans. Math. Software* **37**, 44:1–44:15.

S. Le Digabel and S. M. Wild (2015), A taxonomy of constraints in black-box simulation-based optimization. arXiv:1505.07881

L. Le Gratiet and C. Cannamela (2015), 'Cokriging-based sequential design strategies using fast cross-validation techniques for multi-fidelity computer codes', *Technometrics* **57**, 418–427.

P. L'Ecuyer and G. Yin (1998), 'Budget-dependent convergence rate of stochastic approximation', *SIAM J. Optim.* **8**, 217–247.

H. Lee, R. B. Gramacy, C. Linkletter and G. A. Gray (2011), 'Optimization subject to hidden constraints via statistical emulation', *Pacific J. Optim.* **7**, 467–478.

C. Lemarechal and R. Mifflin, eds (1978), *Nonsmooth Optimization: Proceedings of an IIASA Workshop*, Pergamon Press.

K. Levenberg (1944), 'A method for the solution of certain non-linear problems in least squares', *Quart. Appl. Math.* **2**, 164–168.

R. M. Lewis and V. Torczon (1999), 'Pattern search algorithms for bound constrained minimization', *SIAM J. Optim.* **9**, 1082–1099.

R. M. Lewis and V. Torczon (2000), 'Pattern search methods for linearly constrained minimization', *SIAM J. Optim.* **10**, 917–941.

R. M. Lewis and V. Torczon (2002), 'A globally convergent augmented Lagrangian pattern search algorithm for optimization with general constraints and simple bounds', *SIAM J. Optim.* **12**, 1075–1089.

R. M. Lewis and V. Torczon (2010), A direct search approach to nonlinear programming problems using an augmented Lagrangian method with explicit treatment of the linear constraints. Technical report WM-CS-2010-01, Department of Computer Science, College of William and Mary.

R. M. Lewis, V. Torczon and M. W. Trosset (2000), 'Direct search methods: Then and now', *J. Comput. Appl. Math.* **124**, 191–207.

S. Leyffer (2015), 'It's to solve problems: An interview with Roger Fletcher', *Optima* **99**, 1–6.

D.-H. Li and M. Fukushima (2000), 'A derivative-free line search and global convergence of Broyden-like method for nonlinear equations', *Optim. Methods Software* **13**, 181–201.

Q. Li and D.-H. Li (2011), 'A class of derivative-free methods for large-scale nonlinear monotone equations', *IMA J. Numer. Anal.* **31**, 1625–1635.

Q. Liu, J. Zeng and G. Yang (2015), 'MrDIRECT: A multilevel robust DIRECT algorithm for global optimization problems', *J. Global Optim.* **62**, 205–227.

S. Liu, B. Kailkhura, P.-Y. Chen, P. Ting, S. Chang and L. Amini (2018), Zeroth-order stochastic variance reduction for nonconvex optimization. In *Advances in Neural Information Processing Systems 31* (S. Bengio *et al.*, eds), Curran Associates, pp. 3731–3741.

G. Liuzzi and S. Lucidi (2009), 'A derivative-free algorithm for inequality constrained nonlinear programming via smoothing of an ℓ_∞ penalty function', *SIAM J. Optim.* **20**, 1–29.

G. Liuzzi, S. Lucidi and F. Rinaldi (2011), 'Derivative-free methods for bound constrained mixed-integer optimization', *Comput. Optim. Appl.* **53**, 505–526.

G. Liuzzi, S. Lucidi and F. Rinaldi (2015), 'Derivative-free methods for mixed-integer constrained optimization problems', *J. Optim. Theory Appl.* **164**, 933–965.

G. Liuzzi, S. Lucidi and F. Rinaldi (2016), 'A derivative-free approach to constrained multiobjective nonsmooth optimization', *SIAM J. Optim.* **26**, 2744–2774.

G. Liuzzi, S. Lucidi and F. Rinaldi (2018), An algorithmic framework based on primitive directions and nonmonotone line searches for black box problems with integer variables. Report 6471, Optimization Online. http://www.optimization-online.org/DB_HTML/2018/02/6471.html

G. Liuzzi, S. Lucidi and M. Sciandrone (2006), 'A derivative-free algorithm for linearly constrained finite minimax problems', *SIAM J. Optim.* **16**, 1054–1075.

G. Liuzzi, S. Lucidi and M. Sciandrone (2010), 'Sequential penalty derivative-free methods for nonlinear constrained optimization', *SIAM J. Optim.* **20**, 2614–2635.

S. Lucidi and M. Sciandrone (2002*a*), 'A derivative-free algorithm for bound constrained optimization', *Comput. Optim. Appl.* **21**, 119–142.

S. Lucidi and M. Sciandrone (2002*b*), 'On the global convergence of derivative-free methods for unconstrained optimization', *SIAM J. Optim.* **13**, 97–116.

S. Lucidi, M. Sciandrone and P. Tseng (2002), 'Objective-derivative-free methods for constrained optimization', *Math. Program.* **92**, 37–59.

J. Ma and X. Zhang (2009), 'Pattern search methods for finite minimax problems', *J. Appl. Math. Comput.* **32**, 491–506.

K. Madsen (1975), Minimax solution of non-linear equations without calculating derivatives. In *Nondifferentiable Optimization* (M. L. Balinski and P. Wolfe), Vol. 3 of Mathematical Programming Studies, Springer, pp. 110–126.

A. Maggiar, A. Wächter, I. S. Dolinskaya and J. Staum (2018), 'A derivative-free trust-region algorithm for the optimization of functions smoothed via Gaussian convolution using adaptive multiple importance sampling', *SIAM J. Optim.* **28**, 1478–1507.

M. Marazzi and J. Nocedal (2002), 'Wedge trust region methods for derivative free optimization', *Math. Program.* **91**, 289–305.

A. March and K. Willcox (2012), 'Provably convergent multifidelity optimization algorithm not requiring high-fidelity derivatives', *AIAA J.* **50**, 1079–1089.

D. W. Marquardt (1963), 'An algorithm for least-squares estimation of nonlinear parameters', *J. Soc. Ind. Appl. Math.* **11**, 431–441.

J. M. Martínez and F. N. C. Sobral (2012), 'Constrained derivative-free optimization on thin domains', *J. Global Optim.* **56**, 1217–1232.

J. Matyas (1965), 'Random optimization', *Automat. Remote Control* **26**, 246–253.

J. H. May (1974), Linearly constrained nonlinear programming: A solution method that does not require analytic derivatives. PhD thesis, Yale University.

J. H. May (1979), 'Solving nonlinear programs without using analytic derivatives', *Oper. Res.* **27**, 457–484.

R. L. McKinney (1962), 'Positive bases for linear spaces', *Trans. Amer. Math. Soc.* **103**, 131–131.

K. I. M. McKinnon (1998), 'Convergence of the Nelder–Mead simplex method to a nonstationary point', *SIAM J. Optim.* **9**, 148–158.

M. Menickelly (2017), Random models in nonlinear optimization. PhD thesis, Lehigh University.

M. Menickelly and S. M. Wild (2019), 'Derivative-free robust optimization by outer approximations', *Math. Program.* doi:10.1007/s10107-018-1326-9

A. G. Mersha and S. Dempe (2011), 'Direct search algorithm for bilevel programming problems', *Comput. Optim. Appl.* **49**, 1–15.

R. Mifflin (1975), 'A superlinearly convergent algorithm for minimization without evaluating derivatives', *Math. Program.* **9**, 100–117.

J. Mockus (1989), *Bayesian Approach to Global Optimization: Theory and Applications*, Springer.

J. J. Moré (1978), The Levenberg–Marquardt algorithm: Implementation and theory. In *Numerical Analysis: Dundee 1977* (G. A. Watson, ed.), Vol. 630 of Lecture Notes in Mathematics, Springer, pp. 105–116.

J. J. Moré and D. C. Sorensen (1983), 'Computing a trust region step', *SIAM J. Sci. Statist. Comput.* **4**, 553–572.

J. J. Moré and S. M. Wild (2009), 'Benchmarking derivative-free optimization algorithms', *SIAM J. Optim.* **20**, 172–191.

J. J. Moré and S. M. Wild (2011), 'Estimating computational noise', *SIAM J. Sci. Comput.* **33**, 1292–1314.

J. J. Moré and S. M. Wild (2012), 'Estimating derivatives of noisy simulations', *ACM Trans. Math. Software* **38**, 19:1–19:21.

J. J. Moré and S. M. Wild (2014), 'Do you trust derivatives or differences?', *J. Comput. Phys.* **273**, 268–277.

B. Morini, M. Porcelli and P. L. Toint (2018), 'Approximate norm descent methods for constrained nonlinear systems', *Math. Comput.* **87** (311), 1327–1351.

J. Müller (2016), 'MISO: Mixed-integer surrogate optimization framework', *Optim. Engng* **17**, 177–203.

J. Müller and M. Day (2019), 'Surrogate optimization of computationally expensive black-box problems with hidden constraints', *INFORMS J. Comput.*, to appear.

J. Müller and J. D. Woodbury (2017), 'GOSAC: Global optimization with surrogate approximation of constraints', *J. Global Optim.* **69**, 117–136.

J. Müller, C. A. Shoemaker and R. Piché (2013*a*), 'SO-I: A surrogate model algorithm for expensive nonlinear integer programming problems including global optimization applications', *J. Global Optim.* **59**, 865–889.

J. Müller, C. A. Shoemaker and R. Piché (2013*b*), 'SO-MI: A surrogate model algorithm for computationally expensive nonlinear mixed-integer black-box global optimization problems', *Comput. Oper. Res.* **40**, 1383–1400.

R. Munos (2011), Optimistic optimization of a deterministic function without the knowledge of its smoothness. In *Advances in Neural Information Processing Systems 24* (J. Shawe-Taylor *et al.*, eds), Curran Associates, pp. 783–791.

S. G. Nash (2000), 'A multigrid approach to discretized optimization problems', *Optim. Methods Software* **14**, 99–116.

L. Nazareth and P. Tseng (2002), 'Gilding the lily: A variant of the Nelder–Mead algorithm based on golden-section search', *Comput. Optim. Appl.* **22**, 133–144.

J. A. Nelder and R. Mead (1965), 'A simplex method for function minimization', *Comput. J.* **7**, 308–313.

Y. Nesterov (2004), *Introductory Lectures on Convex Optimization: A Basic Course*, Vol. 87 of Applied Optimization, Springer.

Y. Nesterov and V. Spokoiny (2017), 'Random gradient-free minimization of convex functions', *Found. Comput. Math.* **17**, 527–566.

A. Neumaier (2004), Complete search in continuous global optimization and constraint satisfaction. In *Acta Numerica*, Vol. 13, Cambridge University Press, pp. 271–369.

A. Neumaier, H. Fendl, H. Schilly and T. Leitner (2011), 'VXQR: Derivative-free unconstrained optimization based on QR factorizations', *Soft Comput.* **15**, 2287–2298.

E. Newby and M. M. Ali (2015), 'A trust-region-based derivative free algorithm for mixed integer programming', *Comput. Optim. Appl.* **60**, 199–229.

R. Oeuvray (2005), Trust-region methods based on radial basis functions with application to biomedical imaging. PhD thesis, EPFL.

R. Oeuvray and M. Bierlaire (2009), 'BOOSTERS: A derivative-free algorithm based on radial basis functions', *Internat. J. Model. Simul.* **29**, 26–36.

P.-M. Olsson (2014), Methods for network optimization and parallel derivative-free optimization. PhD thesis, Linköping University.

E. O. Omojokun (1989), Trust region algorithms for optimization with nonlinear equality and inequality constraints. PhD thesis, University of Colorado at Boulder.

C. Paquette and K. Scheinberg (2018), A stochastic line search method with convergence rate analysis. arXiv:1807.07994

R. Pasupathy (2010), 'On choosing parameters in retrospective-approximation algorithms for stochastic root finding and simulation optimization', *Oper. Res.* **58**, 889–901.

R. Pasupathy, P. Glynn, S. Ghosh and F. S. Hashemi (2018), 'On sampling rates in simulation-based recursions', *SIAM J. Optim.* **28**, 45–73.

N. R. Patel, R. L. Smith and Z. B. Zabinsky (1989), 'Pure adaptive search in Monte Carlo optimization', *Math. Program.* **43**, 317–328.

G. Peckham (1970), 'A new method for minimising a sum of squares without calculating gradients', *Comput. J.* **13**, 418–420.

V. Picheny, R. B. Gramacy, S. M. Wild and S. Le Digabel (2016), Bayesian optimization under mixed constraints with a slack-variable augmented Lagrangian. In *Advances in Neural Information Processing Systems 29* (D. D. Lee *et al.*, eds), Curran Associates, pp. 1435–1443.

T. D. Plantenga (2009), HOPSPACK 2.0 user manual. Technical report SAND-2009-6265, Sandia National Laboratories.

E. Polak (1971), *Computational Methods in Optimization: A Unified Approach*, Academic Press.

E. Polak and M. Wetter (2006), 'Precision control for generalized pattern search algorithms with adaptive precision function evaluations', *SIAM J. Optim.* **16**, 650–669.

B. T. Polyak (1987), *Introduction to Optimization*, Optimization Software.

M. Porcelli and P. L. Toint (2017), 'BFO: A trainable derivative-free brute force optimizer for nonlinear bound-constrained optimization and equilibrium computations with continuous and discrete variables', *ACM Trans. Math. Software* **44**, 1–25.

T. Pourmohamad (2016), Combining multivariate stochastic process models with filter methods for constrained optimization. PhD thesis, UC Santa Cruz.

M. J. D. Powell (1964), 'An efficient method for finding the minimum of a function of several variables without calculating derivatives', *Comput. J.* **7**, 155–162.

M. J. D. Powell (1965), 'A method for minimizing a sum of squares of non-linear functions without calculating derivatives', *Comput. J.* **7**, 303–307.

M. J. D. Powell (1975), 'A view of unconstrained minimization algorithms that do not require derivatives', *ACM Trans. Math. Software* **1**, 97–107.

M. J. D. Powell (1994), A direct search optimization method that models the objective and constraint functions by linear interpolation. In *Advances in Optimization and Numerical Analysis* (S. Gomez and J. P. Hennart, eds), Vol. 275 of Mathematics and its Applications, Springer, pp. 51–67.

M. J. D. Powell (1997), Trust region calculations revisited. In *Numerical Analysis 1997* (D. F. Griffiths *et al.*, eds), Vol. 380 of Pitman Research Notes in Mathematics, Addison Wesley Longman, pp. 193–211.

M. J. D. Powell (1998*a*), Direct search algorithms for optimization calculations. In *Acta Numerica*, Vol. 7, Cambridge University Press, pp. 287–336.

M. J. D. Powell (1998*b*), The use of band matrices for second derivative approximations in trust region algorithms. In *Advances in Nonlinear Programming* (Y. Yuan, ed.), Vol. 14 of Applied Optimization, Springer, pp. 3–28.

M. J. D. Powell (2001), 'On the Lagrange functions of quadratic models that are defined by interpolation', *Optim. Methods Software* **16**, 289–309.

M. J. D. Powell (2002), 'UOBYQA: Unconstrained optimization by quadratic approximation', *Math. Program.* **92**, 555–582.

M. J. D. Powell (2003), 'On trust region methods for unconstrained minimization without derivatives', *Math. Program.* **97**, 605–623.

M. J. D. Powell (2004*a*), 'Least Frobenius norm updating of quadratic models that satisfy interpolation conditions', *Math. Program.* **100**, 183–215.

M. J. D. Powell (2004*b*), 'On the use of quadratic models in unconstrained minimization without derivatives', *Optim. Methods Software* **19**, 399–411.

M. J. D. Powell (2004*c*), On updating the inverse of a KKT matrix. Technical report DAMTP 2004/NA01, University of Cambridge.

M. J. D. Powell (2006), The NEWUOA software for unconstrained optimization without derivatives. In *Large-Scale Nonlinear Optimization* (G. Di Pillo and M. Roma, eds), Vol. 83 of Nonconvex Optimization and its Applications, Springer, pp. 255–297.

M. J. D. Powell (2007), A view of algorithms for optimization without derivatives. Technical report DAMTP 2007/NA03, University of Cambridge.

M. J. D. Powell (2008), 'Developments of NEWUOA for minimization without derivatives', *IMA J. Numer. Anal.* **28**, 649–664.

M. J. D. Powell (2009), The BOBYQA algorithm for bound constrained optimization without derivatives. Technical report DAMTP 2009/NA06, University of Cambridge.

M. J. D. Powell (2010), 'On the convergence of a wide range of trust region methods for unconstrained optimization', *IMA J. Numer. Anal.* **30**, 289–301.

M. J. D. Powell (2012), 'On the convergence of trust region algorithms for unconstrained minimization without derivatives', *Comput. Optim. Appl.* **53**, 527–555.

M. J. D. Powell (2013), 'Beyond symmetric Broyden for updating quadratic models in minimization without derivatives', *Math. Program.* **138**, 475–500.

M. J. D. Powell (2015), 'On fast trust region methods for quadratic models with linear constraints', *Math. Program. Comput.* **7**, 237–267.

W. H. Press, S. A. Teukolsky, W. T. Vetterling and B. P. Flannery (2007), *Numerical Recipes in Fortran: The Art of Scientific Computing*, third edition, Cambridge University Press.

C. J. Price and P. L. Toint (2006), 'Exploiting problem structure in pattern search methods for unconstrained optimization', *Optim. Methods Software* **21**, 479–491.

C. J. Price, I. D. Coope and D. Byatt (2002), 'A convergent variant of the Nelder–Mead algorithm', *J. Optim. Theory Appl.* **113**, 5–19.

M. L. Ralston and R. I. Jennrich (1978), 'Dud: A derivative-free algorithm for nonlinear least squares', *Technometrics* **20**, 7–14.

K. Rashid, S. Ambani and E. Cetinkaya (2012), 'An adaptive multiquadric radial basis function method for expensive black-box mixed-integer nonlinear constrained optimization', *Engng Optim.* **45**, 185–206.

L. A. Rastrigin (1963), 'The convergence of the random search method in the extremal control of many-parameter system', *Automat. Remote Control* **24**, 1337–1342.

S. J. Reddi, A. Hefny, S. Sra, B. Poczos and A. Smola (2016), Stochastic variance reduction for nonconvex optimization. In *33rd International Conference on Machine Learning* (M. F. Balcan and K. Q. Weinberger, eds). *Proc. Mach. Learn. Res.* **48**, 314–323.

J. Regier, M. I. Jordan and J. McAuliffe (2017), Fast black-box variational inference through stochastic trust-region optimization. In *Advances in Neural Information Processing Systems 30* (I. Guyon *et al.*, eds), Curran Associates, pp. 2402–2411.

R. G. Regis (2013), 'Constrained optimization by radial basis function interpolation for high-dimensional expensive black-box problems with infeasible initial points', *Engng Optim.* **46**, 218–243.

R. G. Regis (2015), 'The calculus of simplex gradients', *Optim. Lett.* **9**, 845–865.

R. G. Regis (2016), 'On the properties of positive spanning sets and positive bases', *Optim. Engng* **17**, 229–262.

R. G. Regis and C. A. Shoemaker (2007), 'A stochastic radial basis function method for the global optimization of expensive functions', *INFORMS J. Comput.* **19**, 457–509.

R. G. Regis and S. M. Wild (2017), 'CONORBIT: Constrained optimization by radial basis function interpolation in trust regions', *Optim. Methods Software* **32**, 552–580.

A. H. G. Rinnooy Kan and G. T. Timmer (1987*a*), 'Stochastic global optimization methods, I: Clustering methods', *Math. Program.* **39**, 27–56.

A. H. G. Rinnooy Kan and G. T. Timmer (1987*b*), 'Stochastic global optimization methods, II: Multi level methods', *Math. Program.* **39**, 57–78.

L. M. Rios and N. V. Sahinidis (2013), 'Derivative-free optimization: A review of algorithms and comparison of software implementations', *J. Global Optim.* **56**, 1247–1293.

H. Robbins (1952), 'Some aspects of the sequential design of experiments', *Bull. Amer. Math. Soc.* **58**, 527–536.

H. Robbins and S. Monro (1951), 'A stochastic approximation method', *Ann. Math. Statist.* **22**, 400–407.

R. Rockafellar and R. J.-B. Wets (2009), *Variational Analysis*, Springer.

H. H. Rosenbrock (1960), 'An automatic method for finding the greatest or least value of a function', *Comput. J.* **3**, 175–184.

D. Ruppert (1991), Stochastic approximation. In *Handbook of Sequential Analysis* (B. K. Ghosh and P. K. Sen, eds), Vol. 118 of Statistics: A Series of Textbooks and Monographs, CRC Press, pp. 503–529.

D. Russo and B. Van Roy (2016), 'An information-theoretic analysis of Thompson sampling', *J. Mach. Learn. Res.* **17**, 2442–2471.

A. S. Rykov (1980), 'Simplex direct search algorithms', *Automat. Remote Control* **41**, 784–793.

J. H. Ryu and S. Kim (2014), 'A derivative-free trust-region method for biobjective optimization', *SIAM J. Optim.* **24**, 334–362.

J. Sacks (1958), 'Asymptotic distribution of stochastic approximation procedures', *Ann. Math. Statist.* **29**, 373–405.

A. Saha and A. Tewari (2011), Improved regret guarantees for online smooth convex optimization with bandit feedback. In *14th International Conference on Artifical Intelligence and Statistics (AISTATS)*. Proc. Mach. Learn. Res. **15**, 636–642.

P. R. Sampaio and P. L. Toint (2015), 'A derivative-free trust-funnel method for equality-constrained nonlinear optimization', *Comput. Optim. Appl.* **61**, 25–49.

P. R. Sampaio and P. L. Toint (2016), 'Numerical experience with a derivative-free trust-funnel method for nonlinear optimization problems with general nonlinear constraints', *Optim. Methods Software* **31**, 511–534.

S. Sankaran, C. Audet and A. L. Marsden (2010), 'A method for stochastic constrained optimization using derivative-free surrogate pattern search and collocation', *J. Comput. Phys.* **229**, 4664–4682.

K. Scheinberg and P. L. Toint (2010), 'Self-correcting geometry in model-based algorithms for derivative-free unconstrained optimization', *SIAM J. Optim.* **20**, 3512–3532.

B. Shahriari, K. Swersky, Z. Wang, R. P. Adams and N. de Freitas (2016), 'Taking the human out of the loop: A review of Bayesian optimization', *Proc. IEEE* **104**, 148–175.

O. Shamir (2013), On the complexity of bandit and derivative-free stochastic convex optimization. In *26th Annual Conference on Learning Theory (COLT 2013)* (S. Shalev-Shwartz and I. Steinwart, eds). *Proc. Mach. Learn. Res.* **30**, 3–24.

O. Shamir (2017), 'An optimal algorithm for bandit and zero-order convex optimization with two-point feedback', *J. Mach. Learn. Res.* **18** (52), 1–11.

S. Shashaani, F. S. Hashemi and R. Pasupathy (2018), 'ASTRO-DF: A class of adaptive sampling trust-region algorithms for derivative-free stochastic optimization', *SIAM J. Optim.* **28**, 3145–3176.

S. Shashaani, S. R. Hunter and R. Pasupathy (2016), ASTRO-DF: Adaptive sampling trust-region optimization algorithms, heuristics, and numerical experience. In *2016 Winter Simulation Conference (WSC)*, IEEE.

C. A. Shoemaker and R. G. Regis (2003), MAPO: using a committee of algorithm-experts for parallel optimization of costly functions. In *Proceedings of the ACM Symposium on Parallel Algorithms and Architectures*, ACM, pp. 242–243.

J. C. Spall (1992), 'Multivariate stochastic approximation using a simultaneous perturbation gradient approximation', *IEEE Trans. Automat. Control* **37**, 332–341.

J. C. Spall (2005), *Introduction to Stochastic Search and Optimization: Estimation, Simulation, and Control*, Wiley.

W. Spendley (1969), Nonlinear least squares fitting using a modified simplex minimization method. In *Optimization* (R. Fletcher, ed.), Academic Press, pp. 259–270.

W. Spendley, G. R. Hext and F. R. Himsworth (1962), 'Sequential application of simplex designs in optimisation and evolutionary operation', *Technometrics* **4**, 441–461.

N. Srebro, K. Sridharan and A. Tewari (2011), On the universality of online mirror descent. In *Advances in Neural Information Processing Systems 24* (J. Shawe-Taylor *et al.*, eds), Curran Associates, pp. 2645–2653.

T. A. Sriver, J. W. Chrissis and M. A. Abramson (2009), 'Pattern search ranking and selection algorithms for mixed variable simulation-based optimization', *European J. Oper. Res.* **198**, 878–890.

S. U. Stich, C. L. Müller and B. Gärtner (2013), 'Optimization of convex functions with random pursuit', *SIAM J. Optim.* **23**, 1284–1309.

M. Taddy, H. K. H. Lee, G. A. Gray and J. D. Griffin (2009), 'Bayesian guided pattern search for robust local optimization', *Technometrics* **51**, 389–401.

V. Torczon (1991), 'On the convergence of the multidirectional search algorithm', *SIAM J. Optim.* **1**, 123–145.

V. Torczon (1997), 'On the convergence of pattern search algorithms', *SIAM J. Optim.* **7**, 1–25.

A. Törn and A. Žilinskas (1989), *Global Optimization*, Springer.

A. Tröltzsch (2016), 'A sequential quadratic programming algorithm for equality-constrained optimization without derivatives', *Optim. Lett.* **10**, 383–399.

P. Tseng (1999), 'Fortified-descent simplicial search method: A general approach', *SIAM J. Optim.* **10**, 269–288.

M. Valko, A. Carpentier and R. Munos (2013), Stochastic simultaneous optimistic optimization. In *30th International Conference on Machine Learning (ICML 13)*. *Proc. Mach. Learn. Res.* **28**, 19–27.

B. Van Dyke and T. J. Asaki (2013), 'Using QR decomposition to obtain a new instance of mesh adaptive direct search with uniformly distributed polling directions', *J. Optim. Theory Appl.* **159**, 805–821.

F. Vanden Berghen (2004), CONDOR: A constrained, non-linear, derivative-free parallel optimizer for continuous, high computing load, noisy objective functions. PhD thesis, Université Libre de Bruxelles.

F. Vanden Berghen and H. Bersini (2005), 'CONDOR: A new parallel, constrained extension of Powell's UOBYQA algorithm: Experimental results and comparison with the DFO algorithm', *J. Comput. Appl. Math.* **181**, 157–175.

A. Verdério, E. W. Karas, L. G. Pedroso and K. Scheinberg (2017), 'On the construction of quadratic models for derivative-free trust-region algorithms', *EURO J. Comput. Optim.* **5**, 501–527.

L. N. Vicente (2013), 'Worst case complexity of direct search', *EURO J. Comput. Optim.* **1**, 143–153.

L. N. Vicente and A. Custódio (2012), 'Analysis of direct searches for discontinuous functions', *Math. Program.* **133**, 299–325.

K. K. Vu, C. D'Ambrosio, Y. Hamadi and L. Liberti (2016), 'Surrogate-based methods for black-box optimization', *Internat. Trans. Oper. Res.* **24**, 393–424.

Y. Wang, S. S. Du, S. Balakrishnan and A. Singh (2018), Stochastic zeroth-order optimization in high dimensions. In *21st International Conference on Artificial Intelligence and Statistics* (A. Storkey and F. Perez-Cruz, eds). *Proc. Mach. Learn. Res.* **84**, 1356–1365.

H. Wendland (2005), *Scattered Data Approximation*, Cambridge Monographs on Applied and Computational Mathematics, Cambridge University Press.

S. M. Wild (2008*a*), Derivative-free optimization algorithms for computationally expensive functions. PhD thesis, Cornell University.

S. M. Wild (2008*b*), MNH: A derivative-free optimization algorithm using minimal norm Hessians. In *Tenth Copper Mountain Conference on Iterative Methods*. http://grandmaster.colorado.edu/~copper/2008/SCWinners/Wild.pdf

S. M. Wild (2017), Solving derivative-free nonlinear least squares problems with POUNDERS. In *Advances and Trends in Optimization with Engineering Applications* (T. Terlaky *et al.*, eds), SIAM, pp. 529–540.

S. M. Wild and C. A. Shoemaker (2011), 'Global convergence of radial basis function trust region derivative-free algorithms', *SIAM J. Optim.* **21**, 761–781.

S. M. Wild and C. A. Shoemaker (2013), 'Global convergence of radial basis function trust-region algorithms for derivative-free optimization', *SIAM Review* **55**, 349–371.

S. M. Wild, R. G. Regis and C. A. Shoemaker (2008), 'ORBIT: Optimization by radial basis function interpolation in trust-regions', *SIAM J. Sci. Comput.* **30**, 3197–3219.

D. H. Winfield (1969), Function and functional optimization by interpolation in data tables. PhD thesis, Harvard University.

D. H. Winfield (1973), 'Function minimization by interpolation in a data table', *J. Inst. Math. Appl.* **12**, 339–347.

D. J. Woods (1985), An interactive approach for solving multi-objective optimization problems. PhD thesis, Rice University.

M. H. Wright (1995), Direct search methods: Once scorned, now respectable. In *Numerical Analysis 1995* (D. F. Griffiths and G. A. Watson, eds), Vol. 344 of Pitman Research Notes in Mathematics, Addison Wesley Longman, pp. 191–208.

S. Xiong, P. Z. G. Qian and C. F. J. Wu (2013), 'Sequential design and analysis of high-accuracy and low-accuracy computer codes', *Technometrics* **55**, 37–46.

J. Xu and L. Zikatanov (2017), Algebraic multigrid methods. In *Acta Numerica*, Vol. 26, Cambridge University Press, pp. 591–721.

W.-C. Yu (1979), 'Positive basis and a class of direct search techniques', *Scientia Sinica, Special Issue of Mathematics* **1**, 53–67.

Y.-X. Yuan (1985), 'Conditions for convergence of trust region algorithms for nonsmooth optimization', *Math. Program.* **31**, 220–228.

Z. B. Zabinsky and R. L. Smith (1992), 'Pure adaptive search in global optimization', *Math. Program.* **53**, 323–338.

D. Zhang and G.-H. Lin (2014), 'Bilevel direct search method for leader–follower problems and application in health insurance', *Comput. Oper. Res.* **41**, 359–373.

H. Zhang and A. R. Conn (2012), 'On the local convergence of a derivative-free algorithm for least-squares minimization', *Comput. Optim. Appl.* **51**, 481–507.

H. Zhang, A. R. Conn and K. Scheinberg (2010), 'A derivative-free algorithm for least-squares minimization', *SIAM J. Optim.* **20**, 3555–3576.

L. Zhang, T. Yang, R. Jin and Z.-H. Zhou (2015), Online bandit learning for a special class of non-convex losses. In *29th AAAI Conference on Artificial Intelligence (AAAI '15)*, AAAI Press, pp. 3158–3164.

Z. Zhang (2014), 'Sobolev seminorm of quadratic functions with applications to derivative-free optimization', *Math. Program.* **146**, 77–96.

A. A. Zhigljavsky (1991), *Theory of Global Random Search*, Springer.

Acta Numerica (2019), pp. 405–539
doi:10.1017/S0962492919000047

Numerical methods for Kohn–Sham density functional theory

Lin Lin[*]
Department of Mathematics,
University of California, Berkeley,
and Computational Research Division,
Lawrence Berkeley National Laboratory, Berkeley, CA 94720, USA
E-mail: linlin@math.berkeley.edu

Jianfeng Lu[†]
Department of Mathematics,
Department of Physics, and Department of Chemistry,
Duke University, Durham, NC 27708, USA
E-mail: jianfeng@math.duke.edu

Lexing Ying[‡]
Department of Mathematics and
Institute for Computational and Mathematical Engineering,
Stanford University, Stanford, CA 94305, USA
E-mail: lexing@stanford.edu

Kohn–Sham density functional theory (DFT) is the most widely used electronic structure theory. Despite significant progress in the past few decades, the numerical solution of Kohn–Sham DFT problems remains challenging, especially for large-scale systems. In this paper we review the basics as well as state-of-the-art numerical methods, and focus on the unique numerical challenges of DFT.

[*] Partially supported by the US National Science Foundation via grant DMS-1652330, and by the US Department of Energy via grants DE-SC0017867 and DE-AC02-05CH11231.

[†] Partially supported by the US National Science Foundation via grants DMS-1454939 and ACI-1450280 and by the US Department of Energy via grant DE-SC0019449.

[‡] Partially supported by the US Department of Energy, Office of Science, Office of Advanced Scientific Computing Research, the 'Scientific Discovery through Advanced Computing (SciDAC)' programme and the US National Science Foundation via grant DMS-1818449.

CONTENTS

1. Introduction

Kohn–Sham density functional theory (Hohenberg and Kohn 1964, Kohn and Sham 1965) (DFT) is today the most widely used electronic structure theory in chemistry, materials science and other related fields, and was recognized by the Nobel Prize in Chemistry awarded to Walter Kohn in 1998. Despite being developed in the mid-1960s, Kohn–Sham DFT was not a popular approach until the beginning of the 1980s, when practical exchange-correlation functionals obtained from quantum Monte Carlo calculations (Ceperley and Alder 1980, Perdew and Zunger 1981) became available. Together with the development of more advanced exchange-correlation functionals (*e.g.* Lee, Yang and Parr 1988, Becke 1993, Perdew, Burke and Ernzerhof 1996*a*, Heyd, Scuseria and Ernzerhof 2003), Kohn–Sham DFT has been demonstrated to be capable of predicting chemical and material properties for a wide range of systems from *first principles*. More specifically, under the Born–Oppenheimer approximation that the nuclei are treated as classical particles, in principle the only input needed for Kohn–Sham DFT calculations is the most basic information concerning the system: atomic species and atomic positions.

The success of Kohn–Sham DFT cannot be separated from the vast improvement in numerical algorithms and computer architecture, particularly during the 1980s and 1990s. Some relatively basic ideas, such as using the conjugate gradient method for solving eigenvalue problems (Payne *et al.* 1992), efficient use of the planewave basis set (*i.e.* pseudospectral methods) (Kresse and Furthmüller 1996), and density matrix based methods (*i.e.* computing matrix functions) (Goedecker 1999), have been cited thousands to tens of thousands of times. Unfortunately, to a large extent applied mathematicians have missed this 'golden era' of developing numerical methods for Kohn–Sham DFT, so most numerical methods used in practical calculations were developed by physicists and chemists. By the middle of the 2000s, many numerical methods widely used today had been implemented in commercial or open source software packages, such as ABINIT (Gonze *et al.* 2016), CASTEP (Clark *et al.* 2005), FHI-aims (Blum *et al.* 2009), Molpro (Werner *et al.* 2012), NWChem (Valiev *et al.* 2010), Qbox (Gygi 2008), QChem (Shao *et al.* 2015), Quantum ESPRESSO (Giannozzi *et al.* 2017), SIESTA (Soler *et al.* 2002) and VASP (Kresse and Furthmüller 1996), to name just a few. In fact, the availability of these robust and mature software packages is *the* driving force behind the wide range of applicability of Kohn–Sham DFT for practical physical and chemical systems.

On the other hand, as Kohn–Sham DFT is being applied to ever larger and more complex quantum systems, new numerical challenges have emerged and thus provided new opportunities for further algorithmic development. For instance, the cost of orthogonalizing Kohn–Sham orbitals is typically not noticeable for small systems. However, the computational cost of this step scales cubically with respect to the system size, and quickly dominates the computational cost for large systems, often known as the 'cubic scaling wall'. Due to the cubic scaling wall, most practical electronic structure calculations are performed for systems with tens to hundreds of atoms, and can occasionally reach a few thousand atoms. Since the orthogonalization step is so inherently wired into all eigensolvers, any algorithm with reduced asymptotic complexity must necessarily not seek for an eigen-decomposition of the Kohn–Sham Hamiltonian. It should be noted that most electronic structure software packages used today are designed *around* the eigensolver. Hence avoiding the eigensolver entirely implies major changes for the algorithmic and software flow. With massively parallel high performance computers (HPC) becoming more widely available, it is also crucial to develop scalable numerical algorithms that can be efficiently scaled to tens of thousands of computational processors to tackle challenging electronic structure problems within a reasonable amount of wall clock time. With the development of low complexity and scalable new algorithms in the past decade, it is now possible to carry out electronic structure calculations for systems with tens of thousands of atoms, even for difficult metallic systems (*i.e.* systems without an energy band gap).

Another trend in recent developments of DFT is to move towards more complex exchange-correlation functionals. This is because simple exchange-correlation functionals, such as those based on local density approximation (LDA) and generalized gradient approximation (GGA), cannot reach 'chemical accuracy' (error of the energy below 1 kcal/mol) needed for quantitative predictability for many important systems, such as transition metal oxides. To this end, more involved functionals on the fourth and fifth rungs of the 'Jacob's ladder' of exchange-correlation functionals (see the discussion in Section 2.5) become useful, and sometimes indispensable, for treating such systems. These functionals not only have a more complex formulation but may also fundamentally change the mathematical and hence algorithmic structure of the Kohn–Sham DFT problem. For instance, Kohn–Sham DFT with rung-1 to rung-3 functionals can be solved as a nonlinear eigenvalue problem with differential operators. The operator becomes an integro-differential operator when rung-4 functionals (also called hybrid functionals) are used, and the very notion of self-consistency becomes difficult to define properly when rung-5 functionals are used. We remark that practical electronic structure calculations with such functionals have only been pursued for the past decade in the physics community, at least when a large basis set such as the planewave basis set is concerned. The computational scaling of hybrid functional calculations is also cubic with respect to the system size, the same as that of the LDA and GGA functionals. However, hybrid functionals involve the Fock exchange operator, and the computational time of hybrid functional calculations can often be more than 20 times larger than that of LDA and GGA calculations. Therefore hybrid functional calculations are typically done for systems with tens to hundreds of atoms. Rung-5 functional calculations are even more expensive. This brings new opportunities for designing better algorithms to overcome the computational bottleneck with more complex functionals. With the development of new algorithms, the cost of hybrid functional calculations can now be reduced to within twice the cost of GGA calculations, thus enabling hybrid functional calculations for systems with thousands of atoms.

As a final example, the surge of 'high-throughput computing', such as that driven by the materials genome initiative (https://www.mgi.gov), requires hundreds of thousands of calculations to be performed automatically to form a vast database. This requires *all steps* of numerical algorithms to be performed in a robust fashion. Many numerical algorithms in electronic structure calculations were not designed to meet this criterion, as they often require many 'knobs', *i.e.* tuning parameters to reach convergence. Sometimes the tuning parameters require heavy human intervention and expert knowledge of the system. Therefore, the design of numerical methods to perform the same task, but in a more *robust* and *automatic* fashion, also raises new challenges.

There have been numerous advances in the past two decades towards addressing such new challenges, including our own attempts to reduce computational complexity, improve parallel scalability and design robust algorithms in a number of directions. This review aims to introduce both basic concepts and recent developments in numerical methods for solving Kohn–Sham DFT to a mathematical audience. Although we will try to cover the whole landscape of DFT algorithms, the discussion is *heavily biased* towards our own work and view of the field, and some omissions are inevitable.

For simplicity, unless otherwise specified, we will consider isolated, charge-neutral, spinless systems throughout this review. Most methods we discuss can be easily generalized to periodic systems with Γ-point sampling of the Brillouin zone (*i.e.* systems with periodic boundary conditions), and often further to periodic systems with general **k**-point sampling strategies, charged systems, as well as spin-polarized systems. We will also focus on the so-called 'single-shot' DFT calculations, *i.e.*, calculations for a single given atomic configuration. The algorithms can then be used naturally when calculations for multiple – or even massively many – atomic configurations are needed, such as in the context of geometry optimization and *ab initio* molecular dynamics simulation (Car and Parrinello 1985, Marx and Hutter 2009). For a more general introduction to density functional theory, we refer readers to books by Parr and Yang (1989), Dreizler and Gross (1990), Eschrig (1996), Kaxiras (2003), Martin (2008) and Lin and Lu (2019).

The basic theory and formalism of Kohn–Sham DFT will be introduced in Section 2. The rest of the paper discusses components of the DFT calculations: numerical discretization in Section 3, evaluation of the Kohn–Sham maps in Sections 5 and 6, and self-consistent iteration in Section 7. In addition, Section 4 is devoted to several algorithmic tools that have proved to be helpful in a number of contexts in electronic structure calculations. Section 8 concludes with an outlook for trends and future developments. The list of notations is given at the end of the paper for readers' convenience.

2. Theory and formulation

2.1. Quantum many-body problem

In principle, all particles in a physical system, including both nuclei and electrons, are quantum particles and should be treated using quantum mechanics. Since the mass of the lightest element in the periodic table (hydrogen) is around 2000 times larger than that of the electron, the commonly used Born–Oppenheimer approximation assumes that the nuclei can be described by classical mechanics. This is often an accurate approximation, and will be assumed throughout this review.

In this review, we are mostly concerned with isolated systems surrounded by a vacuum in $\mathbb{R}^3$. This is also called the 'free space' boundary condition. Other settings such as the Dirichlet boundary condition in a finite-sized box can be discussed similarly. We use atomic units (a.u.), *i.e.* $m_e = e = \hbar = 1/(4\pi\epsilon_0) = k_B = 1$, where m_e is the mass of an electron, e is the unit charge, $\hbar$ is the reduced Planck constant, $1/(4\pi\epsilon_0)$ is the Coulomb constant, and k_B is the Boltzmann constant. Under the Born–Oppenheimer approximation, the many-body Hamiltonian with M nuclei and N electrons in $\mathbb{R}^3$ is

$$H = \sum_{i=1}^{N} -\frac{1}{2}\Delta_{\mathbf{r}_i} + \sum_{i=1}^{N} V_{\text{ext}}(\mathbf{r}_i; \{\mathbf{R}_I\}) + \sum_{i<j}^{N} \frac{1}{|\mathbf{r}_i - \mathbf{r}_j|} + \sum_{I<J}^{M} \frac{Z_I Z_J}{|\mathbf{R}_I - \mathbf{R}_J|}$$

$$\equiv \hat{T} + \hat{V}_{\text{ext}} + \hat{V}_{\text{ee}} + E_{\text{II}}, \tag{2.1}$$

where the first three terms on the second line are the kinetic energy, external potential and electron–electron interactions respectively:

$$\hat{T} = \sum_{i=1}^{N} -\frac{1}{2}\Delta_{\mathbf{r}_i}, \quad \hat{V}_{\text{ext}} = \sum_{i=1}^{N} V_{\text{ext}}(\mathbf{r}_i; \{\mathbf{R}_I\}), \quad \hat{V}_{\text{ee}} = \sum_{i<j}^{N} \frac{1}{|\mathbf{r}_i - \mathbf{r}_j|}.$$

For a given atomic configuration $\{\mathbf{R}_I\}$, the ion–ion interaction simply adds a constant shift to the Hamiltonian with

$$E_{\text{II}} = \sum_{I<J}^{M} \frac{Z_I Z_J}{|\mathbf{R}_I - \mathbf{R}_J|}.$$

We remark that the treatment of periodic systems (more specifically, with a Γ-point only sampling strategy of the Brillouin zone) is very similar to that for isolated systems. Even for isolated charge-neutral systems, periodic boundary conditions are often used, since they give reasonably accurate approximations as long as certain care is taken with the electrostatic energy (Makov and Payne 1995). Periodic boundary conditions are also more natural for certain numerical methods, such as planewave methods, even for molecular systems.

The ground state is often the most important state. This is because the energy gap $E_1 - E_0$ for many electron systems is of the order of electron volts (eV), or 10^4 Kelvin when the unit of measurement is $k_B T$ (k_B is the Boltzmann constant). This is much greater than room temperature (300 Kelvin). According to the Boltzmann distribution, the probability of the quantum state E_i being occupied is $\mathrm{e}^{-\beta E_i}/Z$, where β is the inverse temperature, and Z is a normalization factor called the partition function. Hence the ground state is often the dominating state. Even for certain systems where $E_1 - E_0$ is small or zero, the ground state can still be very important. This paper focuses on the calculation of the ground-state energy. In the discussion below, we will simply refer to E_0 as E for notational simplicity.

Let $\mathbf{x}_i = (\mathbf{r}_i, \sigma_i)$ denote both the spatial and spin degrees of freedom for the ith electron (though electrons are indistinguishable quantum particles). The many-body ground-state wavefunction $\Psi(\mathbf{x}_1, \ldots, \mathbf{x}_N)$ is associated with the smallest eigenvalue E of the linear eigenvalue problem

$$H\Psi = E\Psi. \tag{2.2}$$

Due to the Pauli exclusion principle for the identical electrons, Ψ is a totally anti-symmetric function:

$$\Psi(\mathbf{x}_1,\ldots,\mathbf{x}_i,\ldots,\mathbf{x}_j,\ldots,\mathbf{x}_N) = -\Psi(\mathbf{x}_1,\ldots,\mathbf{x}_j,\ldots,\mathbf{x}_i,\ldots,\mathbf{x}_N). \qquad (2.3)$$

The Courant–Fisher minimax theorem states that eigenvalue problem (2.2) can also be viewed as an optimization problem,

$$E = \inf_{|\Psi\rangle\in\mathcal{A}_N,\langle\Psi|\Psi\rangle=1} \langle\Psi|H|\Psi\rangle, \qquad (2.4)$$

where $\mathcal{A}_N$ is the set of totally anti-symmetric functions. Equation (2.4) is the *variational principle* for the ground state in quantum mechanics. Here

$$\langle\Psi|\Psi\rangle := \int \Psi^*(\mathbf{x}_1,\ldots,\mathbf{x}_N)\Psi(\mathbf{x}_1,\ldots,\mathbf{x}_N)\,\mathrm{d}\mathbf{x}_1\cdots\mathbf{x}_N = 1 \qquad (2.5)$$

defines the normalization condition for the many-body wavefunction, and the integration with respect to a multi-index $\mathbf{x}$ should be interpreted as

$$\int f(\mathbf{x})\,\mathrm{d}\mathbf{x} := \sum_{\sigma\in\{\uparrow,\downarrow\}} \int_{\mathbb{R}^3} f(\mathbf{r},\sigma)\,\mathrm{d}\mathbf{r}. \qquad (2.6)$$

In order to solve the quantum many-body problem, it appears at first sight that one has to know the quantum many-body wavefunction. Surprisingly, this is at least formally not the case if we are only interested in the ground state. First established by Hohenberg and Kohn (1964), density functional theory (DFT) discovered that the many-body electron density, defined by

$$\rho(\mathbf{r}) = N \sum_{\sigma\in\{\uparrow,\downarrow\}} \int |\Psi((\mathbf{r},\sigma),\mathbf{x}_2,\ldots,\mathbf{x}_n)|^2\,\mathrm{d}\mathbf{x}_2\cdots\mathrm{d}\mathbf{x}_N, \qquad (2.7)$$

is all one needs to determine the quantum many-body ground state. More specifically, assuming the ground state is non-degenerate, there is a one-to-one correspondence between the electron density ρ and the external potential V_{ext}, so that knowledge of ρ is sufficient to reconstruct the many-body wavefunction Ψ. Shortly thereafter, Mermin (1965) proposed the DFT formulation in the finite temperature set-up to include thermal effects as well. The most widely used form of DFT, called Kohn–Sham DFT, was proposed by Kohn and Sham (1965). Below we introduce DFT from the perspective of constrained minimization, which was first proposed by Levy (1979) and then rigorously established by Lieb (1983). We also remark that there is as yet no analogous mathematically rigorous theory of DFT for excited states.

2.2. Kohn–Sham density functional theory

According to the variational principle (2.4), the ground-state energy can be written as a constrained minimization problem:

$$E = \inf_{\Psi \in \mathcal{A}_N, \langle \Psi | \Psi \rangle = 1} \langle \Psi | H | \Psi \rangle = \inf_{\rho \in \mathcal{J}_N} \left\{ \inf_{\substack{\Psi \in \mathcal{A}_N \\ \Psi \mapsto \rho}} \langle \Psi | H | \Psi \rangle \right\}. \tag{2.8}$$

Here, on the right-hand side, a constrained minimization given the density ρ is first carried out for the wavefunction Ψ. The resulting value is then optimized as a functional of ρ within the function space

$$\mathcal{J}_N = \left\{ \rho \geq 0 : \int \rho(\mathbf{r}) \, \mathrm{d}\mathbf{r} = N, \quad \nabla \sqrt{\rho} \in L^2(\mathbb{R}^3) \right\}. \tag{2.9}$$

Here the condition $\nabla \sqrt{\rho} \in L^2(\mathbb{R}^3)$ is implied by the finiteness of the kinetic energy, since

$$\frac{1}{2} \int |\nabla \sqrt{\rho}(\mathbf{r})|^2 \, \mathrm{d}\mathbf{r} \leq \frac{N}{2} \sum_{\sigma \in \{\uparrow, \downarrow\}} \int |\nabla_{\mathbf{r}} \Psi((\mathbf{r}, \sigma), \mathbf{x}_2, \ldots, \mathbf{x}_n)|^2 \, \mathrm{d}\mathbf{x}_1 \cdots \mathrm{d}\mathbf{x}_N$$
$$= \langle \Psi | \hat{T} | \Psi \rangle. \tag{2.10}$$

It can also be proved that $\rho \in \mathcal{J}_N$ implies that there exists at least one $\Psi \in \mathcal{A}_N$ with finite kinetic energy such that $\Psi \mapsto \rho$ (Lieb 1983).

Therefore the constrained minimization procedure is well-defined, and we have

$$E = \inf_{\rho \in \mathcal{J}_N} \left\{ \inf_{\substack{\Psi \in \mathcal{A}_N \\ \Psi \mapsto \rho}} \langle \Psi | (\hat{T} + \hat{V}_{\mathrm{ee}}) | \Psi \rangle + \int \rho(\mathbf{r}) V_{\mathrm{ext}}(\mathbf{r}) \, \mathrm{d}\mathbf{r} \right\} + E_{\mathrm{II}} \tag{2.11}$$

$$= \inf_{\rho \in \mathcal{J}_N} \left\{ \mathcal{F}_{\mathrm{LL}}[\rho] + \int \rho(\mathbf{r}) V_{\mathrm{ext}}(\mathbf{r}) \, \mathrm{d}\mathbf{r} \right\} + E_{\mathrm{II}}. \tag{2.12}$$

Here, together with equation (2.7), we have used

$$\langle \Psi | \hat{V}_{\mathrm{ext}} | \Psi \rangle = N \int |\Psi(\mathbf{x}_1, \ldots, \mathbf{x}_N)|^2 V_{\mathrm{ext}}(\mathbf{r}_1) \, \mathrm{d}\mathbf{x}_1 \cdots \mathrm{d}\mathbf{x}_N = \int \rho(\mathbf{r}) V_{\mathrm{ext}}(\mathbf{r}) \, \mathrm{d}\mathbf{r}. \tag{2.13}$$

The functional $\mathcal{F}_{\mathrm{LL}}[\rho]$ is *universal*, since it depends only on the kinetic and electron–electron repulsion but not on the external potential V_{ext}, which is specified by the atomic configuration. Another important consequence of density functional theory, often known as the Hohenberg–Kohn theorem (Hohenberg and Kohn 1964), is that if $\rho^\star$ is the minimizer of (2.12) and if $\Psi^\star$ is the unique minimizer that results in $\mathcal{F}_{\mathrm{LL}}[\rho^\star]$, then the many-body ground-state wavefunction $\Psi^\star$ is determined by the electron density $\rho^\star$. In other words, if the ground state is non-degenerate, then there is a one-to-one mapping between the ground-state electron density and the many-body ground-state wavefunction. All that remains to make DFT useful is thus a good explicit expression that approximates the functional $\mathcal{F}_{\mathrm{LL}}[\rho]$.

Since the very early days of quantum mechanics, physicists have been seeking the approximation to $\mathcal{F}_{\mathrm{LL}}[\rho]$ (before it was connected to density functional theory) pioneered by Thomas (1927) and Fermi (1927). Until the 1960s, efforts were mainly restricted to uniform electron gas, that is, the system is in a periodic box with a constant external potential field, where many calculations can be done analytically. Despite significant progress in the past few decades, modelling $\mathcal{F}_{\mathrm{LL}}[\rho]$ remains a very difficult task. To appreciate the difficulty, just recall the atomic shell structure from the eigenfunctions of the Hamiltonian operator of a hydrogen atom. It is already highly non-trivial to find such a mapping for a single atom. Furthermore, in chemistry and materials science, the absolute value of the energy is usually not the most important quantity. It is the *relative* energy difference that determines whether or not a chemical process should occur. This often requires the ground-state energy to be calculated with 99.9% accuracy or higher.

The breakthrough of density functional theory is generally attributed to Kohn and Sham (1965), who proposed combining DFT with single-particle orbital structure. Given N one-particle orthonormal spin-orbitals $\{\psi_i(\mathbf{x})\}_{i=1}^{N}$, that is,

$$\langle \psi_i | \psi_j \rangle := \int \psi_i^*(\mathbf{x})\psi_j(\mathbf{x})\,\mathrm{d}\mathbf{x} = \delta_{ij}, \quad i,j = 1,\ldots,N, \qquad (2.14)$$

the associated Slater determinant is given by

$$\Psi(\mathbf{x}_1,\ldots,\mathbf{x}_N) = \frac{1}{\sqrt{N!}}\det\begin{pmatrix} \psi_1(\mathbf{x}_1) & \psi_1(\mathbf{x}_2) & \cdots & \psi_1(\mathbf{x}_N) \\ \psi_2(\mathbf{x}_1) & \psi_2(\mathbf{x}_2) & \cdots & \psi_2(\mathbf{x}_N) \\ \vdots & \vdots & \ddots & \vdots \\ \psi_N(\mathbf{x}_1) & \psi_N(\mathbf{x}_2) & \cdots & \psi_N(\mathbf{x}_N) \end{pmatrix}, \qquad (2.15)$$

which is anti-symmetric by the property of determinants. The computation of the electron density (2.7) with respect to a Slater determinant can be simplified as

$$\rho(\mathbf{r}) = \sum_{i=1}^{N} \sum_{\sigma \in \{\uparrow,\downarrow\}} |\psi_i(\mathbf{r},\sigma)|^2. \qquad (2.16)$$

Using constrained minimization over Slater determinants, the Kohn–Sham proposal can be interpreted as

$$\mathcal{F}_{\mathrm{LL}}[\rho] = \inf_{\substack{\Psi \in \mathcal{A}_N^0 \\ \Psi \mapsto \rho}} \langle \Psi | \hat{T} | \Psi \rangle + \frac{1}{2}\int \frac{\rho(\mathbf{r})\rho(\mathbf{r}')}{|\mathbf{r}-\mathbf{r}'|}\,\mathrm{d}\mathbf{r}\,\mathrm{d}\mathbf{r}' + E_{\mathrm{xc}}[\rho], \qquad (2.17)$$

Figure 2.1. (Credit: Burke 2012.) A 'zoo' of exchange-correlation functionals in DFT.

where $\mathcal{A}_N^0$ is the set of Slater determinants with N orbitals. It turns out that for any $\rho \in \mathcal{J}_N$ there exists at least one $\Psi \in \mathcal{A}_N^0$ that gives the density ρ, and the constrained minimization of the kinetic energy term is well-defined (Lieb 1983). In equation (2.17), the first term corresponds to the kinetic energy from N non-interacting single-particle orbitals. The second term is the Hartree energy, which characterizes the electron–electron repulsion energy at the mean-field level. The last term $E_{\mathrm{xc}}[\rho]$, called the exchange-correlation functional, at first glance simply defines whatever we do not know about $\mathcal{F}_{\mathrm{LL}}[\rho]$. The insight from Kohn and Sham (1965) is that the kinetic and Hartree terms often account for more than 95% of the total energy. Therefore the approximation to $E_{\mathrm{xc}}[\rho]$, while still very difficult, is a much easier task than approximating $\mathcal{F}_{\mathrm{LL}}[\rho]$ directly. As a result, the approximation to the ground-state energy in the Kohn–Sham DFT is given by minimizing over the Kohn–Sham energy functional:

$$E^{\mathrm{KS}} = \inf_{\Psi \in \mathcal{A}_N^0} \left\{ \langle \Psi | \hat{T} | \Psi \rangle + \frac{1}{2} \int \frac{\rho(\mathbf{r})\rho(\mathbf{r}')}{|\mathbf{r} - \mathbf{r}'|} \, \mathrm{d}\mathbf{r} \, \mathrm{d}\mathbf{r}' \right.$$
$$\left. + E_{\mathrm{xc}}[\rho] + \int \rho(\mathbf{r}) V_{\mathrm{ext}}(\mathbf{r}) \, \mathrm{d}\mathbf{r} \right\} + E_{\mathrm{II}}, \qquad (2.18)$$

with ρ given by the Slater determinant Ψ. Writing the kinetic energy of the Slater determinant more explicitly, we have

$$E^{\mathrm{KS}} = \inf_{\{\psi_i\}_{i=1}^N, \langle \psi_i | \psi_j \rangle = \delta_{ij}} \mathcal{F}^{\mathrm{KS}}(\{\psi_i\}) + E_{\mathrm{II}} \qquad (2.19)$$

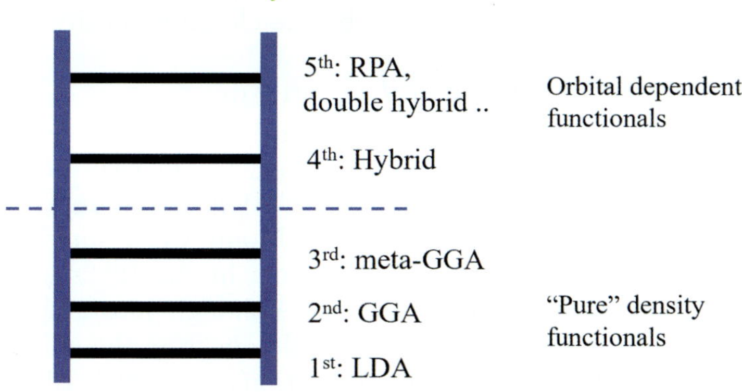

Figure 2.2. The 'Jacob's ladder' of exchange-correlation functionals.

with

$$\mathcal{F}^{\mathrm{KS}}(\{\psi_i\}) = \frac{1}{2} \sum_{i=1}^{N} \int |\nabla_{\mathbf{r}} \psi_i(\mathbf{x})|^2 \, d\mathbf{x} + \int \rho(\mathbf{r}) V_{\mathrm{ext}}(\mathbf{r}) \, d\mathbf{r}$$

$$+ \frac{1}{2} \int \frac{\rho(\mathbf{r})\rho(\mathbf{r}')}{|\mathbf{r} - \mathbf{r}'|} \, d\mathbf{r} \, d\mathbf{r}' + E_{\mathrm{xc}}[\rho]. \tag{2.20}$$

The Kohn–Sham DFT would in principle be exact if we had access to the exact exchange-correlation functional $E_{\mathrm{xc}}[\rho]$. The exchange-correlation functional is also universal, *i.e.* independent of the external potential V_{ext} and hence the atomic configuration. In order to use Kohn–Sham DFT in practice, the exchange-correlation functional E_{xc} must be approximated. Starting from the local density approximation proposed by Kohn and Sham, a 'zoo' of exchange-correlation functionals has been proposed: see an incomplete list in Figure 2.1 by Burke (2012).

According to Perdew (Perdew and Schmidt 2001, Perdew 2013), these exchange-correlation functionals can be organized using a 'Jacob's ladder' for exchange-correlation functionals (Figure 2.2). When no exchange-correlation functional is used, Kohn–Sham DFT is essentially a Hartree approximation (with Pauli's exclusion principle). In such a case, DFT is significantly less accurate than Hartree–Fock theory (Szabo and Ostlund 1989), and the latter is often regarded as the starting point of quantum chemistry. This is referred to as 'Hartree's hell'. Moving up the ladder, the accuracy of the DFT calculation *generally* improves towards the 'heaven of chemical accuracy' of 1 kcal/mol (or 1.6×10^{-3} Hartree per atom) when compared to experimental results. Correspondingly, the functional forms become increasingly more complex, which leads to higher computational costs.

On the first rung of the ladder, we have the local density approximation (LDA), where E_{xc} is modelled locally by the electron density

$$E_{\mathrm{xc}}[\rho] = \int \epsilon_{\mathrm{xc}}(\rho(\mathbf{r}))\,\mathrm{d}\mathbf{r}. \qquad (2.21)$$

Note that the integrand $\epsilon_{\mathrm{xc}}(\rho(\mathbf{r}))$ depends only on the electron density at $\mathbf{r}$. The most widely used LDA exchange-correlation functional is obtained by parametrizing the result from the quantum Monte Carlo simulation (QMC) obtained for the uniform electron gas system in the 1980s (Ceperley and Alder 1980, Perdew and Zunger 1981). Although most, if not all, real chemical and materials systems are very different from the uniform electron gas system, the Kohn–Sham DFT calculations with such LDA exchange-correlation functionals already perform surprisingly well for many systems.

In order to improve the accuracy of the exchange-correlation functional, on the next rung of the ladder we have the generalized gradient approximation (GGA) (Lee, Yang and Parr 1988, Becke 1988, Perdew, Burke and Ernzerhof 1996a), which in addition depends on information on the gradient of electron density,

$$E_{\mathrm{xc}}[\rho] = \int \epsilon_{\mathrm{xc}}(\rho(\mathbf{r}), \sigma(\mathbf{r}))\,\mathrm{d}\mathbf{r}, \qquad (2.22)$$

where the functional depends only on the norm of the gradient

$$\sigma(\mathbf{r}) := |\nabla\rho(\mathbf{r})|^2$$

due to local rotational symmetry. The GGA functionals are currently the most widely used functionals since they achieve a balance between accuracy and computational cost.

The third rung of the ladder is the meta-GGA approximation (Staroverov, Scuseria, Tao and Perdew 2003, Sun, Ruzsinszky and Perdew 2015), where second-order derivative information is also added to the approximation, in particular the kinetic energy density

$$\tau(\mathbf{r}) := \frac{1}{2} \sum_{\sigma\in\{\uparrow,\downarrow\}} \sum_{i=1}^{N} |\nabla_{\mathbf{r}}\psi_i(\mathbf{r}, \sigma)|^2.$$

The meta-GGA energy functional then takes the form

$$E_{\mathrm{xc}}[\{\psi_i\}] = \int \epsilon_{\mathrm{xc}}(\rho(\mathbf{r}), \sigma(\mathbf{r}), \tau(\mathbf{r}))\,\mathrm{d}\mathbf{r}. \qquad (2.23)$$

Some other meta-GGA functionals also involve second-order information on the density, *i.e.* $\nabla^2\rho(\mathbf{r})$, and for simplicity we omit treatment of such terms. Since τ is the local kinetic energy of the orbitals, it may seem that the meta-GGA functional is no longer strictly a density functional. It turns out that the Euler–Lagrange equation associated with the meta-GGA functional can still be written in the same form as that of the LDA and GGA energy functionals, as we will see in the next section. Hence, the LDA, GGA and meta-GGA functionals are known as *semi-local functionals*, and their numerical treatment is very similar, as will be discussed in Section 5.

Let us remark that the exchange-correlation functionals discussed so far do not explicitly depend on spin degrees of freedom (*i.e.* they are only functionals of the electron density rather than the spin-dependent electron density). Spin-dependent exchange-correlation functionals, such as local-spin-density approximation (LSDA) (von Barth and Hedin 1972), are developed for spin-polarized systems. From the perspective of numerical algorithms they are rather similar to the exchange-correlation functionals without spin dependence, so we will not go into the details.

The exchange-correlation functionals beyond the third rung are called non-local functionals, and will be discussed in Section 2.5. Before diving into these more complicated exchange-correlation functionals, we first introduce the Kohn–Sham equations, *i.e.* the Euler–Lagrange equation for Kohn–Sham density functional theory.

2.3. Kohn–Sham equations for semi-local functionals

In order to minimize the Kohn–Sham energy functional (2.20), we need to find the stationary point of the Lagrangian. Let us consider the case of the LDA exchange-correlation functional first. Note that

$$\frac{1}{2}\frac{\delta \mathcal{F}^{\mathrm{KS}}(\{\psi_i\})}{\delta \psi_i^*(\mathbf{x})} = \left(-\frac{1}{2}\Delta_{\mathbf{r}} + V_{\mathrm{ext}} + V_H[\rho] + V_{\mathrm{xc}}[\rho]\right)\psi_i(\mathbf{x}) =: H^{\mathrm{KS}}[\rho]\psi_i(\mathbf{x}),$$

$$(2.24)$$

where the effective Kohn–Sham Hamiltonian $H^{\mathrm{KS}}[\rho]$ is an operator acting on the orbitals which depends on ρ, given by the orbitals as in equation (2.16). The Hartree potential V_H is given by the Coulomb kernel via

$$V_H[\rho](\mathbf{r}) = (v_C \rho)(\mathbf{r}) := \int \frac{\rho(\mathbf{r}')}{|\mathbf{r} - \mathbf{r}'|}\, d\mathbf{r}', \qquad (2.25)$$

and the exchange-correlation potential is defined to be the functional derivative of E_{xc}, that is,

$$V_{\mathrm{xc}}[\rho] = \frac{\delta E_{\mathrm{xc}}[\rho]}{\delta \rho}, \qquad (2.26)$$

which for now can be thought of as a ρ-dependent potential acting on orbitals. V_{xc} of more general forms will be discussed further and made more explicit at the end of this section.

Taking into account the orthonormality condition (2.14), we obtain the Euler–Lagrange equations

$$H^{\mathrm{KS}}[\rho]\psi_i(\mathbf{x}) = \left(-\frac{1}{2}\Delta_{\mathbf{r}} + V_{\mathrm{ext}} + V_H[\rho] + V_{\mathrm{xc}}[\rho]\right)\psi_i(\mathbf{x}) = \sum_{j=1}^{N}\psi_j(\mathbf{x})\lambda_{ij},$$

$$(2.27)$$

where the λ_{ij} are Lagrange multipliers. Let us further simplify the Euler–Lagrange equations to reveal the structure of the problem. First note that due to the orthonormality of $\{\psi_i\}$ and the self-adjointness of $H^{\mathrm{KS}}[\rho]$, we have

$$\lambda_{ij} = \langle\psi_j|H^{\mathrm{KS}}[\rho]|\psi_i\rangle = \langle\psi_i|H^{\mathrm{KS}}[\rho]|\psi_j\rangle^* = \lambda_{ji}^*. \tag{2.28}$$

Therefore $\Lambda = (\lambda_{ij})$ is a Hermitian matrix, and we may assume an eigen-decomposition of Λ as

$$\Lambda = U\,\mathrm{diag}(\varepsilon_1,\ldots,\varepsilon_n)U^*, \tag{2.29}$$

where U is a unitary matrix.

Now consider a rotation of the orbitals, that is,

$$\varphi_i(\mathbf{x}) = \sum_{j=1}^{N}\psi_j(\mathbf{x})U_{ji}. \tag{2.30}$$

Since U is a unitary matrix, the transformation preserves the electron density and hence the effective Kohn–Sham Hamiltonian. The equations for the rotated orbitals $\{\varphi_i\}$ become

$$H^{\mathrm{KS}}[\rho]\varphi_i(\mathbf{x}) = H^{\mathrm{KS}}[\rho]\sum_{j=1}^{N}\psi_j(\mathbf{x})U_{ji} = \sum_{k,j=1}^{N}\psi_k(\mathbf{x})\lambda_{kj}U_{ji}$$

$$= \sum_{k=1}^{N}\psi_k(\mathbf{x})U_{ki}\varepsilon_i = \varphi_i(\mathbf{x})\varepsilon_i. \tag{2.31}$$

Therefore, without loss of generality (up to a rotation of orbitals), it suffices to consider the Euler–Lagrange equations of the form

$$\left(-\frac{1}{2}\Delta_{\mathbf{r}} + V_{\mathrm{ext}} + V_H[\rho] + V_{\mathrm{xc}}[\rho]\right)\psi_i(\mathbf{x}) = \varepsilon_i\psi_i(\mathbf{x}), \quad i = 1,\ldots,N. \tag{2.32}$$

Since the operator H^{KS} depends on the orbitals $\{\psi_i\}$ via the electron density ρ, this is a set of nonlinear eigenvalue problems, known as the *Kohn–Sham equations*. Here the Hamiltonian is nonlinear with respect to the eigenvectors, rather than the eigenvalues.

The Kohn–Sham equations (2.32) must be solved self-consistently with respect to the electron density ρ. For a given electron density ρ, the Hamiltonian $H^{\mathrm{KS}}[\rho] = -\frac{1}{2}\Delta_{\mathbf{r}} + V_{\mathrm{eff}}(\mathbf{r})$ is a self-adjoint linear operator, where the effective potential induced by ρ is

$$V_{\mathrm{eff}}(\mathbf{r}) = V_{\mathrm{ext}}(\mathbf{r}) + V_H[\rho](\mathbf{r}) + V_{\mathrm{xc}}[\rho](\mathbf{r}). \tag{2.33}$$

The Kohn–Sham orbitals $\{\psi_i\}$ are thus eigenfunctions of $H^{\mathrm{KS}}[\rho]$. It should be noted that *a priori* there is no guarantee that $\{\psi_i\}_{i=1}^N$ should correspond to the lowest N eigenvalues (counting multiplicity) of $H^{\mathrm{KS}}[\rho]$ to achieve the global minimum of the Kohn–Sham energy functional (2.18). In practice, this is often assumed in solving the Kohn–Sham equations, known as the *aufbau principle*. It is known that the aufbau principle holds for non-interacting systems, as well as certain Hartree–Fock models. However, it can be violated for more complex models such as Kohn–Sham DFT. Assuming the aufbau principle, the first N eigenfunctions $\{\psi_i\}_{i=1}^N$ are called *occupied orbitals*, while the eigenfunctions of $H^{\mathrm{KS}}[\rho]$ with higher eigenvalues are called *virtual* or *unoccupied orbitals*.

For Kohn–Sham DFT with semi-local exchange-correlation functionals, the effective Kohn–Sham potential $V_{\mathrm{eff}}[\rho]$, and hence the Hamiltonian matrix, depends only on the density ρ through the Kohn–Sham potential. We refer to the mapping from V_{eff} to ρ as the *Kohn–Sham map*, denoted by

$$\rho = \mathcal{F}_{\mathrm{KS}}[V_{\mathrm{eff}}]. \tag{2.34}$$

The electron density ρ can be evaluated from the Kohn–Sham map by solving a linear eigenvalue problem, or using density matrix techniques, to be discussed in Section 5. Hence ρ and V_{eff} should be iteratively determined by each other until convergence. This is called the self-consistent field (SCF) iteration.

The numerical solution for solving the Kohn–Sham equations can thus be divided into three subproblems, as illustrated in Figure 2.3: discretization of the Kohn–Sham Hamiltonian using a finite basis set, evaluation of the Kohn–Sham map, and iteration until reaching self-consistency. The content of this review is thus largely organized around these three topics. Note that during the self-consistent iteration, one also needs to form the Kohn–Sham Hamiltonian by evaluating the Hartree and exchange-correlation potential. With some abuse of terminology, we consider this step as part of the discretization problem in Figure 2.3. We remark that the evaluation of the Hartree potential requires solving a Poisson equation. Depending on the choice of the basis set, the solution can be efficiently obtained using the fast Fourier transform or multigrid methods (see *e.g.* Brandt 1977, Brandt, McCormick and Ruge 1985, Briggs, Henson and McCormick 2000, Fattebert and Bernholc 2000).

Let us now come back to the exchange-correlation potential, given by the functional derivative of E_{xc} with respect to the density. For LDA, the

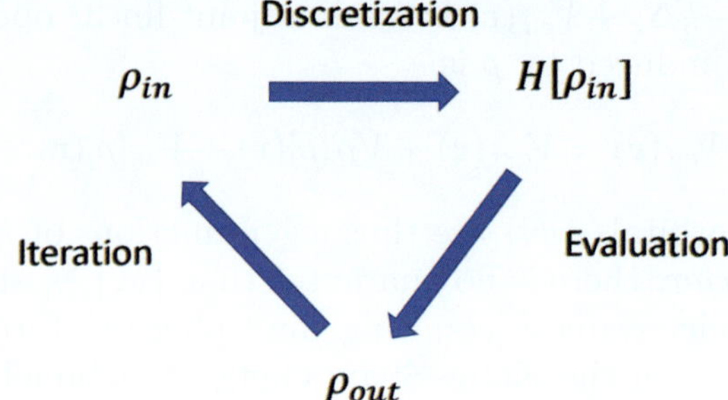

Figure 2.3. Three subproblems for the self-consistent solution of the Kohn–Sham equations. With some abuse of terminology, the step of forming the Kohn–Sham Hamiltonian is considered to be part of the discretization step in this figure.

variation of E_{xc} is given by

$$\delta E_{\mathrm{xc}} = \int \frac{\partial \epsilon_{\mathrm{xc}}}{\partial \rho}(\rho(\mathbf{r}))\delta\rho(\mathbf{r})\,\mathrm{d}\mathbf{r}.$$

Hence

$$V_{\mathrm{xc}}[\rho](\mathbf{r}) = \frac{\delta E_{\mathrm{xc}}}{\delta \rho(\mathbf{r})} = \frac{\partial \epsilon_{\mathrm{xc}}}{\partial \rho}(\rho(\mathbf{r})).$$

For GGA, the variation with respect to the density gives

$$\delta E_{\mathrm{xc}} = \int \frac{\partial \epsilon_{\mathrm{xc}}}{\partial \rho}\delta\rho + \frac{\partial \epsilon_{\mathrm{xc}}}{\partial \sigma}\delta\sigma\,\mathrm{d}\mathbf{r}.$$

Using the chain rule,

$$\delta\sigma = |\nabla(\rho + \delta\rho)|^2 - |\nabla\rho|^2 = 2\nabla\rho \cdot \nabla\delta\rho + O(|\delta\rho|^2),$$

we have

$$\delta E_{\mathrm{xc}} = \int \frac{\partial \epsilon_{\mathrm{xc}}}{\partial \rho}\delta\rho + 2\frac{\partial \epsilon_{\mathrm{xc}}}{\partial \sigma}(\nabla\rho \cdot \nabla\delta\rho)\,\mathrm{d}\mathbf{r} = \int \frac{\partial \epsilon_{\mathrm{xc}}}{\partial \rho}\delta\rho - 2\nabla \cdot \left(\frac{\partial \epsilon_{\mathrm{xc}}}{\partial \sigma}\nabla\rho\right)\delta\rho\,\mathrm{d}\mathbf{r}.$$

Thus

$$V_{\mathrm{xc}}[\rho] = \frac{\partial \epsilon_{\mathrm{xc}}}{\partial \rho} - 2\nabla \cdot \left(\frac{\partial \epsilon_{\mathrm{xc}}}{\partial \sigma}\nabla\rho\right).$$

The derivation of the exchange-correlation potential is slightly different for meta-GGA, since the kinetic energy density $\tau(\mathbf{r})$ explicitly involves the Kohn–Sham orbitals $\{\psi_i\}$. We still start with

$$\delta E_{\mathrm{xc}} = \int \frac{\partial \epsilon_{\mathrm{xc}}}{\partial \rho}\delta\rho + \frac{\partial \epsilon_{\mathrm{xc}}}{\partial \sigma}\delta\sigma + \frac{\partial \epsilon_{\mathrm{xc}}}{\partial \tau}\delta\tau\,\mathrm{d}\mathbf{r}.$$

Since

$$\delta\tau(\mathbf{r}) = \sum_{i=1}^{N} \frac{1}{2}\nabla\psi_i^*(\mathbf{r}) \cdot \nabla\delta\psi_i(\mathbf{r}) + \frac{1}{2}\nabla\delta\psi_i^*(\mathbf{r}) \cdot \nabla\psi_i(\mathbf{r}),$$

we can use integration by parts to obtain

$$\int \frac{\partial\epsilon_{\mathrm{xc}}}{\partial\tau}\delta\tau\,\mathrm{d}\mathbf{r} = -\frac{1}{2}\sum_{i=1}^{N}\int \delta\psi_i^*\nabla\cdot\left(\frac{\partial\epsilon_{\mathrm{xc}}}{\partial\tau}\nabla\right)\psi_i\,\mathrm{d}\mathbf{r}$$
$$-\frac{1}{2}\sum_{i=1}^{N}\int \psi_i^*\nabla\cdot\left(\frac{\partial\epsilon_{\mathrm{xc}}}{\partial\tau}\nabla\right)\delta\psi_i\,\mathrm{d}\mathbf{r}. \tag{2.35}$$

Recall that the Kohn–Sham equation is obtained by variation with respect to $\delta\psi_i^*$, thus the exchange-correlation 'potential' $V_{\mathrm{xc}}[\rho]$ applied to the occupied orbital ψ_i can be viewed as

$$V_{\mathrm{xc}}[\rho]\psi_i = \left[\frac{\partial\epsilon_{\mathrm{xc}}}{\partial\rho} - 2\nabla\cdot\left(\frac{\partial\epsilon_{\mathrm{xc}}}{\partial\sigma}\nabla\rho\right) - \frac{1}{2}\nabla\cdot\left(\frac{\partial\epsilon_{\mathrm{xc}}}{\partial\tau}\nabla\right)\right]\psi_i. \tag{2.36}$$

Therefore, the exchange-correlation functional $V_{\mathrm{xc}}[\rho]$ is still independent of the orbitals, and can be defined only from the electron density as

$$V_{\mathrm{xc}}[\rho] = \frac{\partial\epsilon_{\mathrm{xc}}}{\partial\rho} - 2\nabla\cdot\left(\frac{\partial\epsilon_{\mathrm{xc}}}{\partial\sigma}\nabla\rho\right) - \frac{1}{2}\nabla\cdot\left(\frac{\partial\epsilon_{\mathrm{xc}}}{\partial\tau}\nabla\right). \tag{2.37}$$

Strictly speaking this is no longer a potential, as it involves a differential operator $\nabla\cdot((\partial\epsilon_{\mathrm{xc}}/\partial\tau)\nabla)$ acting on the orbitals. On the other hand, this is still a local operator, and can be treated in a similar way to our previous discussions.

For simplicity, for the rest of the paper we will neglect the spin degrees of freedom and consider the 'spinless' electrons. Hence all single-particle orbitals can be written as $\psi_i(\mathbf{r})$ instead of $\psi_i(\mathbf{x})$. The numerical algorithms can be easily adapted to take into account the spin degrees of freedom.

2.4. Density matrix formulation

Consider a (one-body) Hamiltonian operator $H = -\frac{1}{2}\Delta + V$. Assume that H has a discrete spectrum and denote the eigenpairs of H as $\{(\varepsilon_i, \psi_i)\}$:

$$H\psi_i = \varepsilon_i\psi_i. \tag{2.38}$$

Assume that the system has N electrons that occupy the first N eigenstates according to Pauli's exclusion principle. Here, for definiteness, we assume that $\varepsilon_N < \varepsilon_{N+1}$ to avoid degeneracy.

The Kohn–Sham energy functionals are invariant with respect to unitary rotations of the orbitals. The unitary rotation matrix is often referred to as the *gauge* degrees of freedom. Hence, the physical quantity is the subspace spanned by the occupied orbitals, instead of the individual eigenfunctions. This subspace $\mathrm{span}\{\psi_i\}_{i=1,\ldots,N}$ is known as the Kohn–Sham *occupied subspace* and can be represented by the *density matrix*

$$P = \sum_{i=1}^{N} |\psi_i\rangle\langle\psi_i|. \tag{2.39}$$

Since the $\{\psi_i\}$ are orthonormal, we have

$$P^2 = \sum_{i,j=1}^{N} |\psi_i\rangle\langle\psi_i|\psi_j\rangle\langle\psi_j| = \sum_{i=1}^{N} |\psi_i\rangle\langle\psi_i| = P. \tag{2.40}$$

Hence P is self-adjoint and idempotent, and is the projection operator onto the occupied space. The kernel of P, viewed as an integral operator, is given by

$$P(\mathbf{r}, \mathbf{r}') = \sum_{i=1}^{N} \psi_i(\mathbf{r})\psi_i^*(\mathbf{r}'). \tag{2.41}$$

In particular, we observe that the diagonal part of the kernel is just the electron density

$$P(\mathbf{r}, \mathbf{r}) = \sum_{i=1}^{N} \psi_i(\mathbf{r})\psi_i^*(\mathbf{r}) = \sum_{i=1}^{N} |\psi_i(\mathbf{r})|^2 = \rho(\mathbf{r}). \tag{2.42}$$

Moreover the trace of P is equal to the number of electrons:

$$\mathrm{Tr}\, P = \int P(\mathbf{r}, \mathbf{r})\, \mathrm{d}\mathbf{r} = N. \tag{2.43}$$

The Kohn–Sham equations can be reformulated in terms of the density matrix. Compared with the orbital representation, the density matrix is more intrinsic, as it is invariant with respect to the unitary rotation of the orbitals. In terms of numerical algorithms, use of the density matrix can also be advantageous, especially for large-scale problems, as we will discuss in Section 5.

Note that P can be represented independent of the orbitals. Since H is a self-adjoint operator, for any Borel-measurable function f on the real line, a matrix function $f(H)$ can be defined using its spectral decomposition as

$$f(H) = \sum_{i} f(\varepsilon_i)|\psi_i\rangle\langle\psi_i|. \tag{2.44}$$

Recall that for simplicity we have assumed that the spectrum of H is discrete. Thus, assuming that $\varepsilon_N < \varepsilon_{N+1}$ and the parameter $\mu \in \mathbb{R}$ satisfies

$\varepsilon_N \leq \mu < \varepsilon_{N+1}$, we have

$$\mathbb{1}_{(-\infty,\mu]}(H) = \sum_i \mathbb{1}_{(-\infty,\mu]}(\varepsilon_i)|\psi_i\rangle\langle\psi_i| = \sum_{i:\varepsilon_i\leq\mu} |\psi_i\rangle\langle\psi_i| = \sum_{i=1}^{N} |\psi_i\rangle\langle\psi_i| = P.$$

(2.45)

We conclude with

$$P = \mathbb{1}_{(-\infty,\mu]}(H), \qquad (2.46)$$

where the right-hand side is the spectral projection onto the interval $(-\infty, \mu]$. Here μ is known as the *Fermi level* or the *chemical potential*.

2.5. Non-local functionals

Let us now introduce the non-local exchange-correlation functionals, which are on the fourth and fifth rungs of Perdew's ladder above the semi-local ones introduced in Section 2.2. These more complex density functionals can improve the fidelity of DFT, but they may significantly increase the computational cost at the same time.

On the fourth rung are the so-called hybrid density functionals, such as B3LYP (Becke 1993), PBE0 (Perdew, Ernzerhof and Burke 1996*b*) and HSE (Heyd, Scuseria and Ernzerhof 2003). They have been shown to improve the accuracy of Kohn–Sham DFT calculations, such as the computation of adsorption energies for molecules on surfaces, and molecular frontier level alignment relative to metals and semiconductors. This is achieved by incorporating a fraction of the Hartree–Fock exact exchange or screened exchange operator into the Kohn–Sham Hamiltonian. The hybrid energy functionals depend not only on the density but also on the density matrix, and hence, strictly speaking, they are no longer 'pure' density functionals. A hybrid functional typically takes the following form of hybridization based on a density-based exchange energy and the exact exchange energy (Becke 1993, Heyd *et al.* 2003):

$$E_{\mathrm{xc}}[P] = E_{\mathrm{c}}[\rho] + (1-\alpha)E_{\mathrm{x}}[\rho] + \alpha E_{\mathrm{x}}^{\mathrm{EX}}[P]. \qquad (2.47)$$

Here α is a parameter for the fraction of the non-local exchange contribution, and E_{x} and E_{c} are the exchange and correlation parts (see *e.g.* Martin 2008 for the separation of the exchange-correlation functional into the exchange and correlation components, respectively) from semi-local functionals, such as GGA functionals. $E_{\mathrm{x}}^{\mathrm{EX}}[P]$ is the (screened) Hartree–Fock exchange energy corresponding to the density matrix P:

$$E_{\mathrm{x}}^{\mathrm{EX}}[P] = -\frac{1}{2}\iint |P(\mathbf{r},\mathbf{r}')|^2 K(\mathbf{r},\mathbf{r}')\,\mathrm{d}\mathbf{r}\,\mathrm{d}\mathbf{r}', \qquad (2.48)$$

where $K(\mathbf{r},\mathbf{r}')$ is the kernel for the electron–electron interaction. For example, for Hartree–Fock exchange, $K(\mathbf{r},\mathbf{r}') = v_C(\mathbf{r},\mathbf{r}') = 1/|\mathbf{r}-\mathbf{r}'|$ is the Coulomb kernel. In theories with screened Fock exchange interactions (Heyd *et al.* 2003), K can be a screened Coulomb kernel with $K(\mathbf{r}, \mathbf{r}') = \mathrm{erfc}(\alpha_s |\mathbf{r} - \mathbf{r}'|)/|\mathbf{r} - \mathbf{r}'|$, where α_s is the inverse screening length parameter.

On the fifth rung of the ladder, we have functionals that depend not only on the density and occupied Kohn–Sham orbitals but also on other quantities such as the eigenvalues and virtual orbitals of the Kohn–Sham effective Hamiltonian. They are sometimes referred to as *orbital functionals*. Examples of such functionals include the double hybrid functionals (Grimme 2006, Zhang, Xu and Goddard III 2009, Goerigk and Grimme 2014), van der Waals functionals (Dion *et al.* 2004), random phase approximation (RPA) functionals (Ren, Rinke, Joas and Scheffler 2012*b*, Chen *et al.* 2017), and other functionals based on many-body perturbation theory (see *e.g.* Mori-Sánchez, Wu and Yang 2005, Ren, Rinke, Scuseria and Scheffler 2013, Zhang, Rinke and Scheffler 2016).

For an example of rung-5 functionals, let us consider the RPA functional, for which the exchange-correlation $E_{\mathrm{xc}} = E_{\mathrm{x}} + E_{\mathrm{c}}$ consists of the exact exchange and the random phase approximation (Bohm and Pines 1953, Gell-Mann and Brueckner 1957) of the correlation energy E_{c} given by

$$E_{\mathrm{c}}^{\mathrm{RPA}} = \frac{1}{2\pi} \int_0^\infty \mathrm{Tr}[\ln(1 - \hat{\chi}^0(\mathrm{i}\omega)v_C) + \hat{\chi}^0(\mathrm{i}\omega)v_C]\,\mathrm{d}\omega, \tag{2.49}$$

where v_C is the Coulomb kernel in (2.25), and $\hat{\chi}^0$ is the dynamic polarizability operator

$$\hat{\chi}^0(\mathbf{r}, \mathbf{r}', \mathrm{i}\omega) = \sum_i^{\mathrm{occ}} \sum_a^{\mathrm{vir}} \frac{\psi_i^*(\mathbf{r})\psi_a(\mathbf{r})\psi_a^*(\mathbf{r}')\psi_i(\mathbf{r}')}{\varepsilon_i - \varepsilon_a - \mathrm{i}\omega} + \mathrm{c.c.} \tag{2.50}$$

Here 'occ' and 'vir' denote the set of occupied and virtual orbitals, respectively, and c.c. stands for the complex conjugate of the previous term. Approximation of the correlation is obtained by applying random phase approximation combined with the adiabatic connection (Langreth and Perdew 1975, Gunnarsson and Lundqvist 1976, Ren *et al.* 2012*b*). Note that equation (2.50) contains the virtual orbitals and also the Kohn–Sham orbital energies, which result from the dynamic linear response of the system.

The Kohn–Sham equations of non-local functionals involve non-local operators (hence the name non-local functionals). For rung-4 functionals, the Kohn–Sham equations take the form

$$H[P]\psi_i = \left(-\frac{1}{2}\Delta + V_{\mathrm{ext}} + V_{\mathrm{Hxc}}[\rho] + \alpha V_{\mathrm{x}}^{\mathrm{EX}}[P]\right)\psi_i = \varepsilon_i\psi_i,$$

$$\int \psi_i^*(\mathbf{r})\psi_j(\mathbf{r})\,\mathrm{d}\mathbf{r} = \delta_{ij}, \quad P(\mathbf{r}, \mathbf{r}') = \sum_{i=1}^N \psi_i(\mathbf{r})\psi_i^*(\mathbf{r}'). \tag{2.51}$$

Here V_{Hxc} is the Hartree and exchange-correlation contribution from the electron density ρ only, and $V_x^{EX}[P]$ is derived from the exact exchange functional $E_x^{EX}[P]$, with kernel

$$V_x^{EX}[P](\mathbf{r}, \mathbf{r}') = -P(\mathbf{r}, \mathbf{r}')K(\mathbf{r}, \mathbf{r}'). \qquad (2.52)$$

$V_x^{EX}[P]$ is often called the Fock exchange operator, and is negative semi-definite. Equation (2.51) is called the Hartree–Fock-like equation. In this case the Kohn–Sham map in equation (2.34) becomes the *generalized Kohn–Sham map* from V_{eff} to the density matrix P. The numerical treatment of the exact exchange term, as well as the Hartree–Fock-like equation, will be discussed in detail in Section 6.

The self-consistency for RPA functionals is much more complicated, since the functional involves virtual orbitals and orbital energies. Most current strategies are to first obtain the effective Hamiltonian and corresponding Kohn–Sham orbitals based on a semi-local or hybrid functional, and then calculate the RPA correlation in a post-processing step. Self-consistency calculations have been performed using the optimized effective potential (OEP) framework (Godby, Schlüter and Sham 1986, Godby, Schlüter and Sham 1988, Fukazawa and Akai 2015) or more recently the generalized optimized effective potential approach (Jin *et al.* 2017), but they come with considerable numerical effort. Finding an efficient approach for self-consistent treatment of RPA functionals is an active research area, and in this review we will not discuss self-consistency for rung-5 functionals in detail.

2.6. Finite temperature density functional theory

The Kohn–Sham density functional theory for ground-state quantum systems discussed so far can be extended to systems at finite temperature (Mermin 1965). Instead of a variational principle for the ground-state energy, the functional for free energy is minimized with respect to the density matrix, that is,

$$\mathcal{F}_\beta[\rho] = \inf_{\substack{P \in \mathcal{D} \\ P \mapsto \rho}} \left(\frac{1}{2}\operatorname{Tr}((-\Delta)P) + \beta^{-1}\operatorname{Tr}(P\ln P + (I - P)\ln(I - P)) \right.$$

$$\left. + \frac{1}{2}\int \frac{\rho(\mathbf{r})\rho(\mathbf{r}')}{|\mathbf{r} - \mathbf{r}'|}\,\mathrm{d}\mathbf{r}\,\mathrm{d}\mathbf{r}' + E_{\text{xc},\beta}[\rho] \right), \qquad (2.53)$$

where $\operatorname{Tr}(P\ln P + (I - P)\ln(I - P))$ is the Fermi–Dirac entropy (Parr and Yang 1989) and $\mathcal{D}$ is the set of all one-particle density matrices for an N-electron system:

$$\mathcal{D} = \{P \in \mathcal{B}(L^2(\mathbb{R}^3)) : P = P^*, 0 \preceq P \preceq I, \operatorname{Tr} P = N\}. \qquad (2.54)$$

Here $\mathcal{B}(L^2(\mathbb{R}^3))$ stands for the bounded operators on $L^2(\mathbb{R}^3)$ and the constraint $0 \preceq P \preceq I$ comes from the Pauli exclusion principle. To see this, let the eigen-decomposition of P be

$$P = \sum_i f_i |\psi_i\rangle\langle\psi_i|. \tag{2.55}$$

The constraints $0 \preceq P \preceq I$ and $\operatorname{Tr} P = N$ imply that the occupation number $\{f_i\}$ satisfies

$$0 \leq f_i \leq 1 \quad \text{and} \quad \sum_i f_i = N. \tag{2.56}$$

Thus no state is occupied by more than one electron. Here the summation is performed over all single-particle orbitals, rather than only the occupied ones.

We can write the variational problem more explicitly in terms of the (f_i, ψ_i) as

$$F = \inf_{\substack{\{f_i\},\{\psi_i\} \\ 0 \leq f_i \leq 1,\ \sum_i f_i = N \\ \langle\psi_i|\psi_j\rangle = \delta_{ij}}} F_\beta^{\mathrm{KS}}(\{f_i\}, \{\psi_i\}), \tag{2.57}$$

with

$$\begin{aligned}
F_\beta^{\mathrm{KS}}(\{f_i\}, \{\psi_i\}) \\
= \frac{1}{2}\sum_i \int f_i |\nabla\psi_i|^2 \, \mathrm{d}\mathbf{r} + \beta^{-1}\sum_i (f_i \ln f_i + (1 - f_i)\ln(1 - f_i)) \\
+ \int \rho(\mathbf{r}) V_{\mathrm{ext}}(\mathbf{r}) \, \mathrm{d}\mathbf{r} + \frac{1}{2}\int \frac{\rho(\mathbf{r})\rho(\mathbf{r}')}{|\mathbf{r} - \mathbf{r}'|} \, \mathrm{d}\mathbf{r}\,\mathrm{d}\mathbf{r}' + E_{\mathrm{xc},\beta}[\rho],
\end{aligned} \tag{2.58}$$

and the density is given by

$$\rho(\mathbf{r}) = \sum_i f_i |\psi_i(\mathbf{r})|^2. \tag{2.59}$$

In principle, the exchange-correlation functional for the finite temperature $E_{\mathrm{xc},\beta}[\rho]$ depends on β and also has a ladder of approximation schemes like the zero temperature case. However, most finite temperature DFT calculations in practice still use the temperature-independent exchange-correlation functional at the semi-local level. The zero and finite temperature functional approximations share the same mathematical structure. Hence, for the purpose of our discussions, we will not explicitly distinguish them below.

We now consider the Kohn–Sham equations for the finite temperature functional (2.53). As the functional involves minimization with respect to the density matrix P, it is more convenient to consider the combined min-imization of density and the associated density matrix:

$$\inf_{P \in \mathcal{D}} \left(\frac{1}{2} \mathrm{Tr}((-\Delta)P) + \beta^{-1}\, \mathrm{Tr}(P \ln P + (I - P) \ln(I - P)) \right.$$
$$\left. + \frac{1}{2} \int \frac{\rho(\mathbf{r})\rho(\mathbf{r}')}{|\mathbf{r} - \mathbf{r}'|}\, \mathrm{d}\mathbf{r}\, \mathrm{d}\mathbf{r}' + E_{\mathrm{xc},\beta}[\rho] \right), \tag{2.60}$$

where ρ is given by the diagonal of the density matrix P. Taking the variation with respect to P, we obtain the Euler–Lagrange equation

$$H_\beta^{\mathrm{KS}}[\rho] + \beta^{-1}(\ln P - \ln(I - P)) - \mu = 0, \tag{2.61}$$

where μ is the Lagrange multiplier associated with the constraint $\mathrm{Tr}\, P = N$ and the effective Hamiltonian is given as in the zero temperature case (*cf.* (2.24)), that is,

$$H_\beta^{\mathrm{KS}}[\rho] = -\frac{1}{2}\Delta + V_{\mathrm{ext}} + V_H[\rho] + V_{\mathrm{xc},\beta}[\rho]. \tag{2.62}$$

Solving (2.61) for P, we get

$$P = [I + \exp(\beta(H_\beta^{\mathrm{KS}}[\rho] - \mu))]^{-1}. \tag{2.63}$$

Letting f_β be the *Fermi–Dirac distribution* function

$$f_\beta(\varepsilon) = \frac{1}{1 + \exp(\beta\varepsilon)}, \tag{2.64}$$

we arrive at the self-consistent equation

$$P = f_\beta(H_\beta^{\mathrm{KS}}[\rho] - \mu) = \sum_i f_\beta(\varepsilon_i - \mu)|\psi_i\rangle\langle\psi_i|, \tag{2.65}$$

where ε_i and ψ_i are the eigenvalue and associated eigenfunction of $H_\beta^{\mathrm{KS}}[\rho]$ respectively. We see that the occupation number is given by

$$f_i = f_\beta(\varepsilon_i - \mu) = \frac{1}{1 + \exp(\beta(\varepsilon_i - \mu))}. \tag{2.66}$$

Thus $f_i \in (0, 1)$, and all eigenstates are occupied with a fractional number. Also, notice that as $\beta \to \infty$ (zero temperature limit), f_β converges to the function f_∞:

$$f_\infty(\varepsilon) = \begin{cases} 1 & \varepsilon < 0, \\ 1/2 & \varepsilon = 0, \\ 0 & \varepsilon > 0. \end{cases} \tag{2.67}$$

Let μ_β denote the Lagrange multiplier for the inverse temperature β, where we make the dependence on β explicit. In the limit, we can show that (as before, we assume $\varepsilon_N < \varepsilon_{N+1}$)

$$\lim_{\beta \to \infty} \mu_\beta = \frac{1}{2}(\varepsilon_N + \varepsilon_{N+1}) =: \mu_\infty. \tag{2.68}$$

Therefore, the density matrix of finite temperature is consistent with the definition of the chemical potential at zero temperature. Compared to equation (2.46), we may also write

$$P = f_\infty(H - \mu_\infty).$$

2.7. Pseudopotential approximation

In typical electronic structure calculations, not all electrons play the same role. In a single-particle picture, it is well known that the core electrons (*i.e.* orbitals with relatively low energies) are barely affected by the chemical environment. It is therefore desirable to remove the core electrons from the actual computation, and solve the Kohn–Sham equations for valence electrons (*i.e.* orbitals with relatively high energies) only. Another difficulty in all-electron calculations lies in the treatment of the singular Coulomb interaction between nuclei and electrons, which introduces a cusp in the Kohn–Sham orbitals at each atomic position. The electron–nucleus cusp makes uniform basis functions such as planewaves (to be discussed in Section 3.1) very inefficient for all-electron calculations.

Hence the pseudopotential approximation is introduced to solve both issues simultaneously, which is widely used especially in solid-state physics and materials science. Although this is not a systematically controlled approximation, numerical results indicate that the error introduced by the use of the pseudopotential can be much smaller compared to other sources of error in Kohn–Sham DFT calculations. Simply speaking, the use of pseudopotentials has the following benefits.

(1) The number of Kohn–Sham orbitals depends only on the number of valence electrons. This is particularly important for heavy elements, where each atom involves several tens of electrons, while the number of valence electrons can be much smaller.

(2) The valence electron orbitals are typically smoother than the core electron orbitals, and hence require a smaller number of basis functions such as planewaves to resolve.

(3) After removing the core electrons, the resulting pseudo-valence electron orbitals have fewer nodes near the nuclei. This further enhances the smoothness of the orbitals and reduces computational cost.

Below we introduce the basic idea of the norm-conserving pseudopotential (Hamann, Schlüter and Chiang 1979, Troullier and Martins 1991) – the earliest and still the most widely used form of pseudopotential. With some abuse of notation, in this section we use Cartesian coordinates $(r_1, r_2, r_3)^\top$ and spherical coordinates $(r, \theta, \phi)^\top$ for the same electron position $\mathbf{r}$ interchangeably. When a function $f(\mathbf{r})$ depends only on the radial distance r, we will not distinguish between $f(\mathbf{r})$ and $f(r)$. Similarly, if $f(\mathbf{r})$ depends only on the angular variables θ and ϕ, we will not distinguish between $f(\mathbf{r})$ and $f(\theta, \phi)$.

For a single atom at the origin $\mathbf{R} = \mathbf{0}$, the electron–nucleus interaction becomes $V_{\mathrm{ext}}(r) = -Z/r$, where Z is the charge of the nucleus. Consider the atomic Kohn–Sham equation

$$\left(-\frac{1}{2}\Delta + V_{\mathrm{ext}}(r) + V_{\mathrm{Hxc}}[\rho](r) \right) \psi_i(\mathbf{r}) = \varepsilon_i \psi_i(\mathbf{r}), \quad i = 1, \ldots, N,$$

$$\rho(\mathbf{r}) = \sum_{i=1}^{N} |\psi_i(\mathbf{r})|^2. \tag{2.69}$$

Here V_{Hxc} includes the contribution from the Hartree and exchange-correlation interactions, which also depend only on the radial distance r if the electron density $\rho(\mathbf{r}) = \rho(r)$ satisfies spherical symmetry. However, some single-particle orbital $\psi_i(\mathbf{r})$ might not be spherically symmetric and thus still depend on all components of $\mathbf{r}$.

Let us separate the N Kohn–Sham orbitals into two groups: core electron orbitals $\{\psi_i\}_{i=1}^{N_c}$ and valence electron orbitals $\{\psi_i\}_{i=N_c+1}^{N}$. This also defines the number of valence orbitals $N_v := N - N_c$. The goal of the pseudopotential approximation is to find an operator V_{ps} which is defined to satisfy the following modified equation:

$$\left(-\frac{1}{2}\Delta + V_{\mathrm{ps}} + V_{\mathrm{Hxc}}[\widetilde{\rho}](r) \right) \widetilde{\psi}_i(\mathbf{r}) = \widetilde{\varepsilon}_i \widetilde{\psi}_i(\mathbf{r}), \quad i = 1, \ldots, N_v,$$

$$\widetilde{\rho}(\mathbf{r}) = \sum_{i=1}^{N_v} |\widetilde{\psi}_i(\mathbf{r})|^2. \tag{2.70}$$

The orbitals $\{\widetilde{\psi}_i\}$ are called pseudo-valence orbitals, and the eigenvalues $\{\widetilde{\varepsilon}_i\}$ are called pseudo-valence eigenvalues, respectively. Note that we need only solve N_v instead of N orbitals. Furthermore, we require that the solution to equation (2.70) satisfies the following conditions, which are called the Hamann–Schlüter–Chiang (HSC) conditions (Hamann *et al.* 1979).

(1) The pseudo-valence eigenvalues agree with the real valence eigenvalues:

$$\widetilde{\varepsilon}_i = \varepsilon_{i+N_c}, \quad i = 1, \ldots, N_v. \tag{2.71}$$

(2) The pseudo-valence orbitals agree with the real valence orbitals outside a given radius r_c:

$$\widetilde{\psi}_i(\mathbf{r}) = \psi_{i+N_c}(\mathbf{r}), \quad r \geq r_c, \quad i = 1, \ldots, N_v. \tag{2.72}$$

(3) The pseudo-valence orbitals are normalized, that is,

$$\langle \widetilde{\psi}_i | \widetilde{\psi}_i \rangle = 1. \tag{2.73}$$

In particular, condition (3) is called the norm-conservation condition, which leads to the name of this class of pseudopotential approximation. Although the pseudo-valence orbitals should form an orthonormal set of functions, in the discussion below we will associate the orbitals with spherical harmonic functions that are orthogonal to each other. Hence the orthogonality condition will be automatically satisfied, and only the normalization condition is imposed here.

The choice of pseudopotential satisfying the HSC conditions is by no means unique. In order to construct a V_{ps} satisfying the HSC conditions, for simplicity let us first consider the case where $N_v = 1$. Note that the pseudo-valence orbital $\widetilde{\psi}_1$ now becomes the ground state of equation (2.70), and must therefore be a nodeless function (Lieb and Loss 2001). On the other hand, the real valence orbital ψ_1 should be orthogonal to all core orbitals, and therefore must have nodes within the core region $r \leq r_c$. Hence the real and pseudo-valence orbitals have qualitatively different shapes within the core region, but this does not affect the shape of the valence orbitals outside the core region due to HSC condition (2). When $N_v = 1$, the valence orbital is expected to be spherically symmetric (called an s-orbital):

$$\psi_{N_c+1}(\mathbf{r}) = \varphi(r), \tag{2.74}$$

where we use the notation $\varphi(r)$ to emphasize that the valence orbital depends only on the radial distance r and is real. Similarly the pseudo-valence orbital should also be spherically symmetric and we denote

$$\widetilde{\psi}_1(\mathbf{r}) = \widetilde{\varphi}(r). \tag{2.75}$$

In such a case, V_{ps} can be chosen to be a local potential that depends only on r, denoted by $V_{\mathrm{loc}}(r)$. To see this, we may simply choose a pseudo-valence orbital $\widetilde{\varphi}(r)$ such that HSC conditions (2) and (3) are satisfied. In order to satisfy HSC condition (1), we may invert the Schrödinger equation (2.70) as

$$V_{\mathrm{loc}}(r) = \varepsilon_1 + \frac{\Delta_{\mathbf{r}}\widetilde{\psi}_1(\mathbf{r})}{2\widetilde{\psi}_1(\mathbf{r})} - V_{\mathrm{Hxc}}[\widetilde{\rho}](r) = \varepsilon_1 + \frac{1}{2r^2}\frac{\partial}{\partial r}\left(r^2 \frac{\partial \widetilde{\varphi}}{\partial r}\right) - V_{\mathrm{Hxc}}[\widetilde{\rho}](r).$$

$$\tag{2.76}$$

Note that $V_{\mathrm{loc}}(r)$ depends only on the radial distance due to the assumptions of $\widetilde{\psi}_1$ and $\widetilde{\rho}$. Figure 2.4 shows an example comparing the valence 2s orbital of the fluorine atom (F) in an all-electron calculation, and the

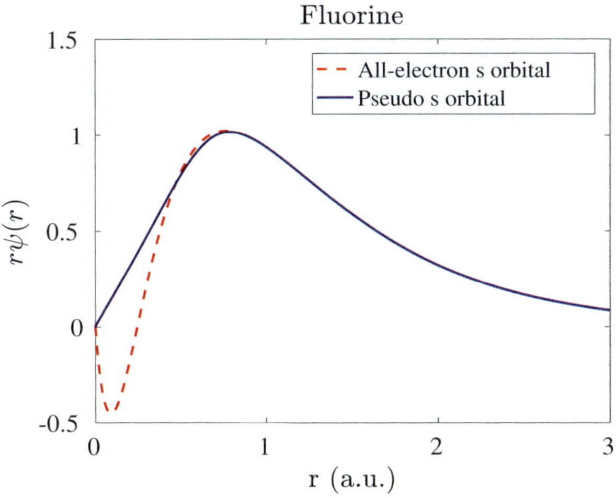

Figure 2.4. Comparison of the 2s orbital of the fluorine atom in an all-electron and a pseudopotential calculation. The cut-off radius r_c is set to 1.0 a.u.

pseudo-valence orbital using a Troullier–Martins pseudopotential (Troullier and Martins 1991). We find that the orbital becomes significantly smoother in the pseudopotential approximation. In particular, the pseudo-valence orbital is nodeless in the core region.

However, such a strategy becomes problematic when $N_v > 1$. To see this, following the notation of solutions to hydrogen-like atoms (Landau and Lifshitz 1991), we first relabel the real and pseudo-valence orbitals as (here $\hat{\mathbf{r}} = \mathbf{r}/r$)

$$\psi_{lm}(\mathbf{r}) = \varphi_l(r)Y_{lm}(\hat{\mathbf{r}}), \quad \widetilde{\psi}_{lm}(\mathbf{r}) = \widetilde{\varphi}_l(r)Y_{lm}(\hat{\mathbf{r}}). \tag{2.77}$$

The integers l, m are called the azimuthal and magnetic quantum numbers, respectively. For each azimuthal quantum number l, the admissible integer m should satisfy $-l \leq m \leq l$. Compared to the complete solution for hydrogen-like atoms, the choice of 'valence electron' fixes the principal quantum number implicitly. The radial parts $\varphi_l, \widetilde{\varphi}_l$ are real. $Y_{lm}(\hat{\mathbf{r}})$ is a spherical harmonic function defined on $\mathbb{S}^2$, which can also be viewed as a function $Y_{lm}(\theta, \varphi)$ that depends only on the angular degrees of freedom in spherical coordinates. The spherical harmonic functions satisfy the ortho-normality condition

$$\int_{\mathbb{S}^2} Y_{lm}^*(\hat{\mathbf{r}})Y_{l'm'}(\hat{\mathbf{r}})\, \mathrm{d}\hat{\mathbf{r}}$$
$$= \int_0^\pi \int_0^{2\pi} Y_{lm}^*(\theta, \varphi)Y_{l'm'}(\theta, \varphi)\sin^2\theta\, \mathrm{d}\varphi\, \mathrm{d}\theta = \delta_{ll'}\delta_{mm'}. \tag{2.78}$$

Now if we choose a radial function $\widetilde{\varphi}_l$ for the pseudo-valence orbital $\widetilde{\psi}_{lm}$, we may invert the Schrödinger equation and obtain

$$
\begin{aligned}
V_{\mathrm{loc},l}(r) &= \varepsilon_l + \frac{\Delta_{\mathbf{r}}\widetilde{\psi}_{lm}(\mathbf{r})}{2\widetilde{\psi}_{lm}(\mathbf{r})} - V_{\mathrm{Hxc}}[\widetilde{\rho}](r) \\
&= \varepsilon_l + \frac{1}{2r^2}\frac{\partial}{\partial r}\left(r^2\frac{\partial\widetilde{\varphi}_l}{\partial r}\right) - \frac{l(l+1)}{2r^2}\widetilde{\varphi}_l(r) - V_{\mathrm{Hxc}}[\widetilde{\rho}](r).
\end{aligned}
\tag{2.79}
$$

Due to the spherical symmetry, ε_l, $V_{\mathrm{loc},l}$ depend only on l but not on m. Again, $V_{\mathrm{loc},l}(r)$ is only a function of the radial distance r. Clearly V_{ps} cannot be equal to $V_{\mathrm{loc},l}(r)$ for all different choices of l.

Below we demonstrate that HSC condition (1) can be satisfied if we allow V_{ps} to be a *non-local potential*, *i.e.* an integral operator. We first choose the local potential to be V_{loc,l_0} for a particular angular momentum l_0. Then let the kernel of V_{ps} take the following form:

$$
V_{\mathrm{ps}}(\mathbf{r},\mathbf{r}') = V_{\mathrm{loc},l_0}(r)\delta_{\mathbf{r},\mathbf{r}'} + \sum_{lm} Y_{lm}(\hat{\mathbf{r}})(V_{\mathrm{loc},l}(r) - V_{\mathrm{loc},l_0}(r))\delta_{r,r'}Y_{lm}^*(\hat{\mathbf{r}}')/r^2.
\tag{2.80}
$$

Applying equation (2.80) to a pseudo-valence orbital $\widetilde{\psi}_{lm}$, we have

$$
\begin{aligned}
(V_{\mathrm{ps}}\widetilde{\psi}_{lm})(\mathbf{r}) &= \int V_{\mathrm{ps}}(\mathbf{r},\mathbf{r}')\widetilde{\psi}_{lm}(\mathbf{r}')\,\mathrm{d}\mathbf{r}' \\
&= V_{\mathrm{loc},l_0}(r)\widetilde{\psi}_{lm}(\mathbf{r}) + Y_{lm}(\hat{\mathbf{r}})(V_{\mathrm{loc},l}(r) - V_{\mathrm{loc},l_0}(r))\widetilde{\varphi}_l(r) \\
&= V_{\mathrm{loc},l}(r)\widetilde{\psi}_{lm}(\mathbf{r}).
\end{aligned}
\tag{2.81}
$$

Here we have used the form (2.77) and the orthonormality condition of spherical harmonics. Combined with equation (2.79), we find that HSC condition (1) is satisfied for any valence orbital $\widetilde{\psi}_{lm}$ under consideration.

Since the kernel of V_{ps} is local with respect to the radial variable, and non-local with respect to angular variables, equation (2.80) is called the semi-local form of the pseudopotential. However, for a general atomic configuration, the semi-local form cannot be treated efficiently other than being discretized as a dense matrix. Therefore the semi-local pseudopotential is rarely used in practice. In order to reduce the computational cost, we may note that it is only necessary for equation (2.81) to hold. Define

$$
\delta V_{\mathrm{loc},l}(r) := V_{\mathrm{loc},l}(r) - V_{\mathrm{loc},l_0}(r).
\tag{2.82}
$$

We note that

$$
\begin{aligned}
\langle \delta V_{\mathrm{loc},l}\widetilde{\psi}_{lm}|\widetilde{\psi}_{l'm'}\rangle &= \int \delta V_{\mathrm{loc},l}(r)\widetilde{\varphi}_l^*(r)\widetilde{\varphi}_{l'}(r)Y_{lm}^*(\hat{\mathbf{r}})Y_{l'm'}(\hat{\mathbf{r}})\,\mathrm{d}\mathbf{r} \\
&= \delta_{ll'}\delta_{mm'}\int_0^\infty r^2\delta V_{\mathrm{loc},l}(r)|\widetilde{\varphi}_l(r)|^2\,\mathrm{d}r.
\end{aligned}
\tag{2.83}
$$

Here we have used the orthonormality condition of the spherical harmonics. Now define the pseudopotential as

$$V_{\mathrm{ps}} = V_{\mathrm{loc},l_0} + \sum_{lm} \frac{1}{\langle \delta V_{\mathrm{loc},l} \widetilde{\psi}_{lm} | \widetilde{\psi}_{lm} \rangle} | \delta V_{\mathrm{loc},l} \widetilde{\psi}_{lm} \rangle \langle \delta V_{\mathrm{loc},l} \widetilde{\psi}_{lm} |. \qquad (2.84)$$

We may readily verify that

$$V_{\mathrm{ps}} \widetilde{\psi}_{lm} = V_{\mathrm{loc},l_0} \widetilde{\psi}_{lm} + \delta V_{\mathrm{loc},l} \widetilde{\psi}_{lm} = V_{\mathrm{loc},l} \widetilde{\psi}_{lm}, \qquad (2.85)$$

which means that V_{ps} also satisfies HSC condition (1). Compared to the semi-local form, the pseudopotential (2.84) consists of a local component, as well as a non-local component that can be stored as a low-rank matrix. The rank is equal to the number of pseudo-valence orbitals under consideration. This is called the Kleinman–Bylander form of non-local pseudopotential (Kleinman and Bylander 1982), and is used in almost all modern electronic structure software packages using norm-conserving pseudopotentials.

Let us define

$$b_{lm}(\mathbf{r}) = \frac{1}{\sqrt{|\langle \delta V_{\mathrm{loc},l} \widetilde{\psi}_{lm} | \widetilde{\psi}_{lm} \rangle|}} \delta V_{\mathrm{loc}}(r) \widetilde{\varphi}_l(r) Y_{lm}(\hat{\mathbf{r}}), \qquad (2.86)$$

which is called a projection vector. From HSC condition (2) and equation (2.79), we may readily find that $b_{lm}(\mathbf{r}) = 0$ if $r \geq r_c$, and hence is localized in real space. Defining $\gamma_l = \mathrm{sign}(\langle \delta V_{\mathrm{loc},l} \widetilde{\psi}_{lm} | \widetilde{\psi}_{lm} \rangle)$ and $V_{\mathrm{loc}}(\mathbf{r}) := V_{\mathrm{loc},l_0}(r)$, we may rewrite the kernel of the Kleinman–Bylander form as

$$V_{\mathrm{ps}}(\mathbf{r}, \mathbf{r}') = V_{\mathrm{loc}}(\mathbf{r}) \delta_{\mathbf{r}, \mathbf{r}'} + \sum_{lm} \gamma_l b_{lm}(\mathbf{r}) b_{lm}^*(\mathbf{r}'). \qquad (2.87)$$

Finally, the pseudopotential (2.87) only comes from one atom centred at the origin. For a general atomic configuration $\{\mathbf{R}_I\}_{I=1}^M$, the kernel of the non-local pseudopotential takes the form

$$V_{\mathrm{ps}}(\mathbf{r}, \mathbf{r}'; \{\mathbf{R}_I\}) = \left\{ \sum_{I=1}^M V_{\mathrm{loc},I}(\mathbf{r} - \mathbf{R}_I) \right\} \delta_{\mathbf{r}, \mathbf{r}'}$$

$$+ \left\{ \sum_{I=1}^M \sum_{\ell=1}^{L_I} \gamma_{\ell,I} b_{\ell,I}(\mathbf{r} - \mathbf{R}_I) b_{\ell,I}^*(\mathbf{r}' - \mathbf{R}_I) \right\}$$

$$= V_{\mathrm{loc}}(\mathbf{r}; \{\mathbf{R}_I\}) \delta_{\mathbf{r}, \mathbf{r}'} + V_{\mathrm{nl}}(\mathbf{r}, \mathbf{r}'; \{\mathbf{R}_I\}). \qquad (2.88)$$

Here we have combined lm into a multi-index ℓ. Furthermore, $V_{\mathrm{loc},I}, b_{\ell,I}$ and the number of projection vectors L_I depend on the atom type and hence have the added index I. In this notation, V_{loc} and V_{nl} denote the collection of local and non-local parts of the pseudopotential, respectively. Equation (2.88) is the general form of norm-conserving pseudopotential and will be assumed throughout this review.

More explicitly, the Kohn–Sham energy under the pseudopotential approximation becomes

$$E^{\mathrm{KS}} = \inf_{\{\psi_i\}_{i=1}^N,\,\langle\psi_i|\psi_j\rangle=\delta_{ij}} \tag{2.89}$$

$$\left\{ \frac{1}{2}\sum_{i=1}^N \int |\nabla\psi_i(\mathbf{r})|^2 \, \mathrm{d}\mathbf{r} + \sum_{i=1}^N \int \psi_i^*(\mathbf{r})V_{\mathrm{nl}}(\mathbf{r},\mathbf{r}';\{\mathbf{R}_I\})\psi_i(\mathbf{r}') \, \mathrm{d}\mathbf{r}\,\mathrm{d}\mathbf{r}' \right.$$

$$\left. + \int \rho(\mathbf{r})V_{\mathrm{loc}}(\mathbf{r};\{\mathbf{R}_I\}) \, \mathrm{d}\mathbf{r} + \frac{1}{2}\int \frac{\rho(\mathbf{r})\rho(\mathbf{r}')}{|\mathbf{r}-\mathbf{r}'|} \, \mathrm{d}\mathbf{r}\,\mathrm{d}\mathbf{r}' + E_{\mathrm{xc}}[\rho] \right\} + E_{\mathrm{II}}.$$

The corresponding Kohn–Sham Hamiltonian becomes

$$H^{\mathrm{KS}}[\rho] = -\frac{1}{2}\Delta + V_{\mathrm{loc}} + V_{\mathrm{nl}} + V_H[\rho] + V_{\mathrm{xc}}[\rho]. \tag{2.90}$$

When the electron density is given, the effective potential (2.33) becomes

$$V_{\mathrm{eff}}(\mathbf{r}) = V_{\mathrm{loc}}(\mathbf{r}) + V_H[\rho](\mathbf{r}) + V_{\mathrm{xc}}[\rho](\mathbf{r}), \tag{2.91}$$

where we have separated out the contribution from the non-local pseudopotential V_{nl}.

Remarks

Although not discussed here, another major reason for using pseudopotential is to take into account relativistic effects, which are non-negligible for heavy elements such as lead and gold. Since relativistic effects mainly affect the behaviour of core electrons, we may solve the relativistic Dirac–Kohn–Sham equation (Thaller 1992, Belpassi, Tarantelli, Sgamellotti and Quiney 2005) for a single atom to obtain the relativistically corrected valence orbitals ψ_{lm} and energies ε_l. The rest of the procedure for generating the pseudopotential is the same as above.

There are a number of widely used norm-conserving pseudopotentials, such as the Troullier–Martins (TM) pseudopotential (Troullier and Martins 1991), the Hartwigsen–Goedecker–Hutter (HGH) pseudopotential (Hartwigsen, Goedecker and Hutter 1998) and the optimized norm-conserving Vanderbilt (ONCV) pseudopotential (Hamann 2013), to name just a few. All these pseudopotentials can be written in the form (2.88), though the *interpretation* of V_{loc}, b_ℓ and even the number of projectors L may be different. Recently the semi-local form of the pseudopotential has been analysed mathematically (Cancès and Mourad 2016). However, the Kleinman–Bylander form of pseudopotential may occasionally introduce 'ghost states', which are unphysical states with artificially low energies. In practice such ghost states can often be removed by choosing the local potential to have a different angular momentum l, and hence they are often not considered to be a problem in most electronic structure calculations. However, the existence of ghost states introduces difficulty into the mathematical analysis.

It may appear that the norm-conservation condition should be naturally imposed since the pseudo-valence orbitals $\tilde{\psi}_{lm}$ should be eigenfunctions of a modified Schrödinger operator. However, it has been found that by relaxing the norm-conservation condition, one can further improve the smoothness of the pseudopotential and hence reduce the computational cost. The most well-known example is the Vanderbilt ultrasoft pseudopotential (Vanderbilt 1990). Moreover, the widely used projected augmented wave (PAW) method (Blöchl 1994) can also be viewed as a pseudopotential, which also violates the norm-conservation condition. Compared to norm-conserving pseudopotentials, one drawback of such approaches is that the Kohn–Sham equations become inherently a generalized eigenvalue problem even when the basis set is orthonormal (see Section 3).

2.8. Physical quantities of interest

Once the Kohn–Sham equations have converged, the total energy can be evaluated readily from equation (2.89). It can also be evaluated from the Kohn–Sham eigenvalues as

$$E^{\mathrm{KS}} = \sum_{i=1}^{N} \varepsilon_i - \frac{1}{2} \int \frac{\rho(\mathbf{r})\rho(\mathbf{r}')}{|\mathbf{r} - \mathbf{r}'|}\, \mathrm{d}\mathbf{r}\, \mathrm{d}\mathbf{r}' - \int \rho(\mathbf{r}) V_{\mathrm{xc}}[\rho](\mathbf{r})\, \mathrm{d}\mathbf{r} + E_{\mathrm{xc}}[\rho] + E_{\mathrm{II}}.$$

$$(2.92)$$

Here the summation of the eigenvalues $\sum_{i=1}^{N} \varepsilon_i$ is called the band energy. The difference between the total energy and the band energy (other than the nuclei interaction energy E_{II}) is called the *double-counting* term, which is due to the nonlinearity of the Hartree energy and the exchange-correlation energy functionals. When finite temperature effects are included, the entropy also needs to be evaluated in order to compute the free energy.

In Kohn–Sham DFT, besides the total energy and the electron density, we are often interested in computing the atomic force, which is necessary for performing geometry relaxation and *ab initio* molecular dynamics simulation. Once the SCF iteration reaches convergence, the force on the Ith atom can be computed as the negative derivative of the total energy with respect to the atomic position $\mathbf{R}_I$:

$$\mathbf{F}_I = -\frac{\partial E^{\mathrm{KS}}(\{\mathbf{R}_I\})}{\partial \mathbf{R}_I}.$$

$$(2.93)$$

The required derivative can be computed directly, for example via finite differences. However, even for first-order accuracy, the number of energy evaluations for a system containing M atoms is $3M + 1$, that is, the Kohn–Sham equations must be solved $3M + 1$ times independently. This approach becomes prohibitively expensive as the system size increases. The cost of the force calculation is greatly reduced via the Hellmann–Feynman theorem (Martin 2008), which states that, at self-consistency, the partial derivative $\partial/\partial\mathbf{R}_I$ only needs to be applied to terms in equation (2.89) which depend *explicitly* on the atomic position $\mathbf{R}_I$. The Hellmann–Feynman (HF) force is then given by

$$\mathbf{F}_I = -\int \frac{\partial V_{\mathrm{loc}}}{\partial\mathbf{R}_I}(\mathbf{r};\{\mathbf{R}_I\})\rho(\mathbf{r})\,\mathrm{d}\mathbf{r} - \sum_{i=1}^{N}\int \psi_i^*(\mathbf{r})\frac{\partial V_{\mathrm{nl}}}{\partial\mathbf{R}_I}(\mathbf{r},\mathbf{r}';\{\mathbf{R}_I\})\psi_i(\mathbf{r}')\,\mathrm{d}\mathbf{r}\,\mathrm{d}\mathbf{r}'$$

$$+ \sum_{J\neq I}\frac{Z_I Z_J}{|\mathbf{R}_I - \mathbf{R}_J|^3}(\mathbf{R}_I - \mathbf{R}_J). \tag{2.94}$$

Note that equation (2.88) gives

$$\frac{\partial V_{\mathrm{loc}}}{\partial\mathbf{R}_I}(\mathbf{r};\{\mathbf{R}_I\}) = \frac{\partial V_{\mathrm{loc},I}}{\partial\mathbf{R}_I}(\mathbf{r} - \mathbf{R}_I) = -\nabla_\mathbf{r}V_{\mathrm{loc},I}(\mathbf{r} - \mathbf{R}_I),$$

and

$$\frac{\partial V_{\mathrm{nl}}}{\partial\mathbf{R}_I}(\mathbf{r},\mathbf{r}';\{\mathbf{R}_I\})$$

$$= \sum_{\ell=1}^{L_I}\gamma_{I,\ell}\left(\frac{\partial b_{I,\ell}}{\partial\mathbf{R}_I}(\mathbf{r} - \mathbf{R}_I)b_{I,\ell}^*(\mathbf{r}' - \mathbf{R}_I) + b_{I,\ell}(\mathbf{r} - \mathbf{R}_I)\frac{\partial b_{I,\ell}^*}{\partial\mathbf{R}_I}(\mathbf{r}' - \mathbf{R}_I)\right)$$

$$= -\sum_{\ell=1}^{L_I}\gamma_{I,\ell}(\nabla_\mathbf{r}b_{I,\ell}(\mathbf{r} - \mathbf{R}_I)b_{I,\ell}^*(\mathbf{r}' - \mathbf{R}_I) + b_{I,\ell}(\mathbf{r} - \mathbf{R}_I)\nabla_{\mathbf{r}'}b_{I,\ell}^*(\mathbf{r}' - \mathbf{R}_I)).$$

Then the Hellmann–Feynman force in equation (2.94) can be written as

$$\mathbf{F}_I = \int \nabla_\mathbf{r}V_{\mathrm{loc},I}(\mathbf{r} - \mathbf{R}_I)\rho(\mathbf{r})\,\mathrm{d}\mathbf{r}$$

$$+ 2\mathrm{Re}\sum_{i=1}^{N}\sum_{\ell=1}^{L_I}\gamma_{I,\ell}\left(\int \psi_i^*(\mathbf{r})\nabla_\mathbf{r}b_{I,\ell}(\mathbf{r} - \mathbf{R}_I)\,\mathrm{d}\mathbf{r}\right)\left(\int b_{I,\ell}^*(\mathbf{r}' - \mathbf{R}_I)\psi_i(\mathbf{r}')\,\mathrm{d}\mathbf{r}'\right)$$

$$+ \sum_{J\neq I}\frac{Z_I Z_J}{|\mathbf{R}_I - \mathbf{R}_J|^3}(\mathbf{R}_I - \mathbf{R}_J). \tag{2.95}$$

From the computational cost point of view, if we let N_g denote the number of grid points to discretize quantities such as $\rho(\mathbf{r})$ in the global domain to perform a quadrature, then the cost of computing the force due to the local potential $\int \nabla_\mathbf{r}V_{\mathrm{loc},I}(\mathbf{r} - \mathbf{R}_I)\rho(\mathbf{r})\,\mathrm{d}\mathbf{r}$ is $O(N_g)$, since $V_{\mathrm{loc},I}(\mathbf{r} - \mathbf{R}_I)$ is a delocalized quantity in the global domain. On the other hand, each non-local projector $b_{I,\ell}(\mathbf{r} - \mathbf{R}_I)$ is localized around $\mathbf{R}_I$, and the cost of evaluating the integral $(\int \psi_i^*(\mathbf{r})\nabla_\mathbf{r}b_{I,\ell}(\mathbf{r} - \mathbf{R}_I)\,\mathrm{d}\mathbf{r})$ or $(\int b_{I,\ell}^*(\mathbf{r}' - \mathbf{R}_I)\psi_i(\mathbf{r}')\,\mathrm{d}\mathbf{r}')$ is a constant N_l independent of the global number of grid points N_g. The computation of the last term,

$$\sum_{J \neq I} \frac{Z_I Z_J}{|\mathbf{R}_I - \mathbf{R}_J|^3}(\mathbf{R}_I - \mathbf{R}_J),$$

involves only scalar operations, and its cost is usually negligibly small in electronic structure calculations. N_g and N_I are proportional to the number of electrons N. Hence, neglecting constant terms independent of N, we have that the computational cost of the Hellmann–Feynman force on each atom is $O(N_g + NL_I N_l) \sim O(N)$, and the cost of computing forces on all atoms is $O(N^2)$.

Remarks

The computational cost of evaluating equation (2.95) can scale as $O(N^3)$ if implemented straightforwardly. This is because, for each atom, the associated non-local pseudopotential needs to be applied to all orbitals via an integral. The cost is thus $MNN_g \sim O(N^3)$. Note that each projection vector $b_{\ell,I}$ has (at least approximately) compact support in real space and can thus be stored as a sparse vector, using the real-space representation (see Section 3.1.2). The number of non-zeros in the sparse vector is independent of the number of quadrature points in the global domain. The cost is then reduced to $O(N^2)$ as discussed above. We remark that use of the real-space representation may result in higher numerical error than the Fourier space representation (see Section 3.1.1), especially if the real-space grid is not dense enough. Hence several electronic structure software packages such as Quantum ESPRESSO still prefer the Fourier space representation, even though the cost scales as $O(N^3)$. In certain contexts, the computational cost can be further reduced to $O(N)$ using density matrix formalism (see Section 3.3).

The atomic force evaluated as in equation (2.95) uses the Hellmann–Feynman theorem, and hence is called the Hellmann–Feynman force. When numerical discretization as in Section 3 is under consideration, the derivative of the Kohn–Sham energy may involve the derivative with respect to a basis function. If the basis set depends on the atomic configuration, such as in the case of Gaussian-type orbitals to be discussed in Section 3.2, then the Hellmann–Feynman force does not agree with the negative derivative of the Kohn–Sham energy. The remaining difference is called the Pulay force (Pulay 1969). On the other hand, if the discretization is independent of atomic configuration, such as in the case of the planewave basis set in Section 3.1, the Pulay force vanishes and the Hellmann–Feynman force becomes exact. We also remark that even if the basis set depends on the atomic configuration, the magnitude of the Pulay force will systematically decrease as the basis set approaches the complete basis set limit.

3. Numerical discretization

In order to solve Kohn–Sham DFT in practice, the Hamiltonian operator must first be discretized, for instance by a finite-sized basis set. The efficiency of a discretization scheme can be measured in terms of the number of degrees of freedom per atom, or the dimension of the discretized Hamiltonian matrix. Besides this standard metric, one special feature of electronic structure calculations is that the choice of the discretization scheme can directly affect the effectiveness of subsequent numerical methods to evaluate the Kohn–Sham map, which is to be discussed in detail in Sections 5 and 6. In the following discussion, we divide the numerical discretization schemes roughly into three categories: large basis sets (Section 3.1), small basis sets (Section 3.2) and adaptive basis sets (Section 3.3).

3.1. Large basis sets

In electronic structure calculations, a large basis set typically requires $100 \sim 10\,000$ basis functions per atom to achieve chemical accuracy, even when pseudopotentials are used. The size of the resulting Hamiltonian matrix is usually of the order of $10^3 \sim 10^6$. Hence it is not numerically efficient, or even feasible at all, to diagonalize the Hamiltonian matrix. In such a case, iterative algorithms should be used to compute the occupied orbitals, which will be discussed in Section 4.2. Most standard basis sets used to solve PDEs numerically fall into this category, for instance the planewave method (Payne *et al.* 1992, Kresse and Furthmüller 1996) (also known as the Fourier basis set, or more precisely, the pseudospectral method), the finite element method (Tsuchida and Tsukada 1995, Suryanarayana *et al.* 2010, Bao, Hu and Liu 2012, Chen *et al.* 2014) and the wavelet method (Genovese *et al.* 2008), to name just a few. Strictly speaking, the finite difference method (Chelikowsky, Troullier and Saad 1994) does not use a basis set. However, the number of degrees of freedom needed by the finite difference method is approximately in the same range, and hence finite difference can also be regarded as a large basis set method. The main advantage of using a large basis set is that physical quantities such as energies and forces can converge systematically with respect to refinement of the basis set, by tuning one or a few parameters.

3.1.1. Fourier basis set

For charge-neutral systems, periodic boundary conditions are often sufficiently accurate when the Coulomb energy is treated with care (Makov and Payne 1995). One of the key advantages of working with periodic boundary conditions is that they allow use of the Fourier basis set, which is arguably the most widely used large basis set for electronic structure calculations. In the Fourier basis set, the basis functions are of the form $\exp(i\mathbf{g}\cdot\mathbf{r})$, where the parameter $\mathbf{g}$ is chosen from a specific set of frequencies. The Fourier basis set has several key advantages. First, the discretization of the Kohn–Sham Hamiltonian takes a rather simple form under this basis set. Second, as it is independent of the atom configurations, the Pulay force vanishes, thus making the force calculation rather straightforward. Third, by leveraging the fast Fourier transform (FFT), computation of the matrix–vector multiplication with the Kohn–Sham Hamiltonian can be carried out efficiently, and this allows for efficient iterative diagonalization while evaluating the Kohn–Sham map. However, the Fourier basis set also comes with a noticeable disadvantage: as a fixed basis set, it lacks the flexibility of local refinement, which can be a serious issue for all-electron calculations. Fortunately, introduction of the pseudopotentials addresses this problem by eliminating the singularity at the nuclei, as well as the highly localized core electrons. This allows one to focus solely on the smooth valence electrons. In the rest of this subsection, we shall work with the Kohn–Sham Hamiltonian under the pseudopotentials approximation as in equation (2.90).

Let us consider for simplicity a rectangular computational domain $\Omega = [0, L_1] \times [0, L_2] \times [0, L_3]$ with periodic boundary conditions. It is natural to associate with Ω a reciprocal lattice in frequency space given by

$$\mathbb{L}^* = \left\{ \mathbf{g} = \left(\frac{2\pi}{L_1} i_1, \frac{2\pi}{L_2} i_2, \frac{2\pi}{L_3} i_3 \right) : (i_1, i_2, i_3) \in \mathbb{Z}^3 \right\},$$

and the Fourier basis functions are the complex exponentials

$$\phi_{\mathbf{g}}(\mathbf{r}) = \frac{1}{\sqrt{|\Omega|}} \exp(i\mathbf{g} \cdot \mathbf{r})$$

indexed by each $\mathbf{g} \in \mathbb{L}^*$. In order to obtain a finite set of basis functions, one typically introduces an energy cut-off E_{cut} and restricts to only the Fourier modes indexed by

$$\mathbb{G}_{\mathrm{cut}} := \left\{ \mathbf{g} \in \mathbb{L}^* : \frac{1}{2} |\mathbf{g}|^2 \leq E_{\mathrm{cut}} \right\} \subset \mathbb{L}^*.$$

We will use N_b to denote the cardinality of $\mathbb{G}_{\mathrm{cut}}$, $i.e.$ the number of basis functions used within the energy cut-off.

In order to present the Kohn–Sham Hamiltonian under this basis, it is convenient to introduce two Cartesian grids, one in frequency space and one in real space. The one in frequency space is

$$\mathbb{G} = \left\{ \mathbf{g} = \left(\frac{2\pi}{L_1} i_1, \frac{2\pi}{L_2} i_2, \frac{2\pi}{L_3} i_3 \right) : \right.$$
$$\left. -\frac{n_1}{2} \leq i_1 < \frac{n_1}{2}, -\frac{n_2}{2} \leq i_2 < \frac{n_2}{2}, -\frac{n_3}{2} \leq i_3 < \frac{n_3}{2} \right\},$$

where n_1, n_2, and n_3 are chosen in such a way that the width of $\mathbb{G}$ in each dimension is typically twice the width of $\mathbb{G}_{\mathrm{cut}}$. The reason for such a choice

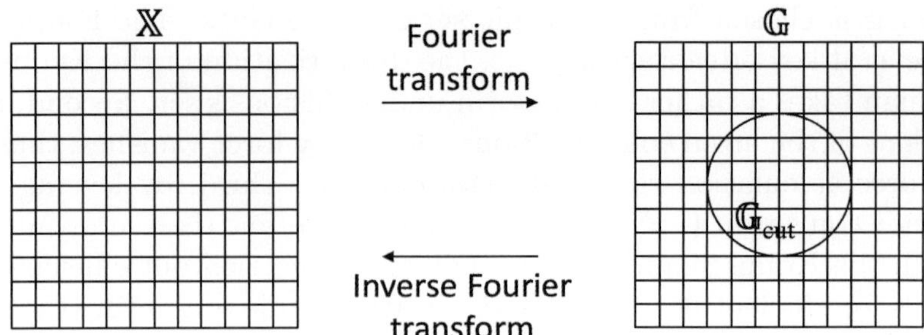

Figure 3.1. Illustration of the grids used for the Fourier basis set: $\mathbb{X}$, $\mathbb{G}$ and $\mathbb{G}_{\mathrm{cut}}$.

will be discussed below. The Cartesian grid in real space is denoted by

$$\mathbb{X} = \left\{ \mathbf{r} = \left(\frac{i_1}{n_1} L_1, \frac{i_2}{n_2} L_2, \frac{i_3}{n_3} L_3 \right) : 0 \le i_1 < n_1, 0 \le i_2 < n_2, 0 \le i_3 < n_3 \right\}.$$

Note that the number of grid points in both grids is equal to $N_g = n_1 n_2 n_3$.

For a function $\psi(\mathbf{r})$ in the span of $\{\phi_{\mathbf{g}}(\mathbf{r}) : \mathbf{g} \in \mathbb{G}_{\mathrm{cut}}\}$ with spanning coefficients $\{\hat{\psi}_{\mathbf{g}} : \mathbf{g} \in \mathbb{G}_{\mathrm{cut}}\}$, its value at any point $\mathbf{r} \in \mathbb{X}$ is given by

$$\psi(\mathbf{r}) = \sum_{\mathbf{g} \in \mathbb{G}_{\mathrm{cut}}} \hat{\psi}_{\mathbf{g}} \phi_{\mathbf{g}}(\mathbf{r}) = \sum_{\mathbf{g} \in \mathbb{G}_{\mathrm{cut}}} \hat{\psi}_{\mathbf{g}} \frac{1}{\sqrt{|\Omega|}} \exp(i \mathbf{g} \cdot \mathbf{r}). \tag{3.1}$$

Suppose now that $\psi(\mathbf{r})$ is a sufficiently smooth function. Then the basis coefficients $\{\hat{\psi}_{\mathbf{g}} : \mathbf{g} \in \mathbb{G}\}$ can be approximated using the samples of $\psi(\mathbf{r})$ at $\mathbb{X}$, computed via the inverse Fourier transform as

$$\hat{\psi}_{\mathbf{g}} = \int_\Omega \psi(\mathbf{r}) \phi_{\mathbf{g}}^*(\mathbf{r}) \, \mathrm{d}\mathbf{r}$$

$$= \int_\Omega \psi(\mathbf{r}) \frac{1}{\sqrt{|\Omega|}} \exp(-i \mathbf{g} \cdot \mathbf{r}) \, \mathrm{d}\mathbf{r} \approx \sum_{\mathbf{r} \in \mathbb{X}} \frac{\sqrt{|\Omega|}}{N_g} \psi(\mathbf{r}) \exp(-i \mathbf{g} \cdot \mathbf{r}). \tag{3.2}$$

Note that the last approximation is exact if $\psi(\mathbf{r})$ belongs to the span of $\{\phi_{\mathbf{g}}(\mathbf{r}) : \mathbf{g} \in \mathbb{G}_{\mathrm{cut}}\}$.

Equations (3.1) and (3.2) show how to transform between the real-space and frequency-space representations for a function $\psi(\mathbf{r})$. Given the spanning coefficients $\{\hat{\psi}_{\mathbf{g}} : \mathbf{g} \in \mathbb{G}_{\mathrm{cut}}\}$, the values $\{\psi(\mathbf{r}) : \mathbf{r} \in \mathbb{X}\}$ are computed by first extending $\hat{\psi}_{\mathbf{g}}$ from $\mathbb{G}_{\mathrm{cut}}$ to $\mathbb{G}$ with zero-padding and then applying a discrete inverse Fourier transform. In the opposite direction, given the real-space values $\{\psi(\mathbf{r}) : \mathbf{r} \in \mathbb{X}\}$, the coefficients $\{\hat{\psi}_{\mathbf{g}} : \mathbf{g} \in \mathbb{G}_{\mathrm{cut}}\}$ are computed by first applying a discrete Fourier transform and then restricting the result from $\mathbb{G}$ to $\mathbb{G}_{\mathrm{cut}}$. In terms of computational efficiency, it is essential that both directions can be accelerated using the fast Fourier transform. Figure 3.1 illustrates the grids involved in these conversions.

The Fourier basis set typically results in a dense matrix for the Kohn–Sham Hamiltonian in (2.89). Direct diagonalization of such dense matrices is often computationally too expensive. As a result, iterative diagonalization is often used instead (see Section 4.2). The key step of an iterative diagonalization algorithm is the application of the Kohn–Sham Hamiltonian to a given function represented in the Fourier basis set.

More precisely, for a function ψ given in terms of its Fourier coefficients $\{\hat{\psi}_\mathbf{g} : \mathbf{g} \in \mathbb{G}_{\mathrm{cut}}\}$, the goal is to evaluate the Fourier coefficients of $H^{\mathrm{KS}}\psi$ in $\mathbb{G}_{\mathrm{cut}}$ effectively. Note from the definition (2.90) that the Kohn–Sham Hamiltonian H^{KS} can be decomposed into three parts:

- the kinetic energy part $-\frac{1}{2}\Delta$,

- the local potential part, that is,

$$V_{\mathrm{loc}} + V_H[\rho] + V_{\mathrm{xc}}[\rho],$$

  where V_{loc} is the local part of the pseudopotential approximation,

- the non-local part of the pseudopotential approximation V_{nl}.

The overall strategy is to treat each part individually, either in real space or in frequency space depending on computational convenience.

Before describing in detail how each of these three parts is treated, we note that the Kohn–Sham Hamiltonian depends on the electron density $\rho(\cdot)$. In the actual numerical computation, one needs access to the values of $\rho(\cdot)$ at each $\mathbf{r} \in \mathbb{X}$. $\rho(\mathbf{r})$ is formed from the current copies of the Kohn–Sham orbitals $\{\psi_i(\mathbf{r})\}$ with $i = 1, \ldots, N$. In the Fourier basis method, $\{\psi_i(\mathbf{r})\}$ are represented in the frequency domain $\hat{\psi}_{\mathbf{g};i}$ with $\mathbf{g} \in \mathbb{G}_{\mathrm{cut}}$. The computation of $\rho(\mathbf{r})$ involves evaluating $\psi_i(\mathbf{r})$ for each i (*i.e.* one fast Fourier transform), squaring them, and summing the results over $i = 1, \ldots, N$. In order to avoid the aliasing error coming from the squaring step, it is typically required that the sidelength of $\mathbb{G}$ is significantly larger than the diameter of the grid $\mathbb{G}_{\mathrm{cut}}$ (for details see *e.g.* Fornberg 1998). In this case, the conventional choice in the electronic structure community is to set the sidelength of $\mathbb{G}$ to be at least twice the diameter of $\mathbb{G}_{\mathrm{cut}}$, as mentioned above.

With $\rho(\mathbf{r})$ ready at each $\mathbf{r} \in \mathbb{X}$, let us now go through each of the three parts when applying the Kohn–Sham Hamiltonian H^{KS} to a function ψ. First, consider the kinetic energy part $-\frac{1}{2}\Delta\psi$. It is well known that a differential operator in real space is equivalent to a pointwise multiplication operator in frequency space. Therefore, applying the kinetic part $-\frac{1}{2}\Delta$ can be carried out by multiplying $\hat{\psi}_\mathbf{g}$ by $\frac{1}{2}|\mathbf{g}|^2$ for each $\mathbf{g} \in \mathbb{G}_{\mathrm{cut}}$.

Second, for the computation of the local potential part, one first applies an inverse Fourier transform on $\{\hat{\psi}_\mathbf{g} : \mathbf{g} \in \mathbb{G}_{\mathrm{cut}}\}$ to obtain the real-space representation $\{\psi(\mathbf{r}) : \mathbf{r} \in \mathbb{X}\}$. As mentioned above, the local potential consists of several parts,

$$(V_{\mathrm{loc}}(\mathbf{r}) + V_H[\rho](\mathbf{r}) + V_{\mathrm{xc}}[\rho](\mathbf{r}))\psi(\mathbf{r}), \tag{3.3}$$

The Hartree potential involves convolution in real space, and therefore it is convenient to treat it in the Fourier domain. More precisely, one first applies the Fourier transform to the electron density $\rho(\mathbf{r})$ with $\mathbf{r} \in \mathbb{X}$ to obtain $\{\widehat{\rho}_{\mathbf{g}} : \mathbf{g} \in \mathbb{G}_{\mathrm{cut}}\}$. Then each $\widehat{\rho}_{\mathbf{g}}$ is multiplied by $1/|\mathbf{g}|^2$, except that at $\mathbf{g} = 0$ it is kept at zero. This is due to the charge neutrality assumption, which allows the formally divergent contribution from the Fourier mode $\mathbf{g} = 0$ to be cancelled by the contribution from the electron–ion interaction. Finally, an inverse Fourier transform brings the result back to real space. When all the parts are ready, we compute the sum and multiply the result pointwise with $\psi(\mathbf{r})$. Note that the above procedure of computing derivative terms in frequency space and the multiplication terms in real space is the *pseudo-spectral method* for the numerical solution of partial differential equations (Fornberg 1998).

Third, the computation of the non-local potential part

$$\int \left(\sum_{I=1}^{M} \sum_{\ell=1}^{L_I} \gamma_{\ell,I} b_{\ell,I}(\mathbf{r} - \mathbf{R}_I) b_{\ell,I}^*(\mathbf{r}' - \mathbf{R}_I) \right) \psi(\mathbf{r}') \, \mathrm{d}\mathbf{r}' \tag{3.4}$$

can be treated either in frequency or in real space. When the basis functions $b_{\ell,I}(\mathbf{r}')$ are localized, the treatment in real space is often preferred due to its simplicity. More precisely, one evaluates the integral $\int b_{\ell,I}^*(\mathbf{r}')\psi(\mathbf{r}') \, \mathrm{d}\mathbf{r}'$ for each pair (ℓ, I) with a discrete sum over $\mathbb{X}$ and then scales $b_{\ell,I}(\mathbf{r})$ with this integral value before summing them up over (ℓ, I).

At this point, one holds the contribution from both local and non-local potentials in the spatial grid $\mathbb{X}$. In order to obtain its coefficients in frequency space on the grid $\mathbb{G}_{\mathrm{cut}}$, one simply applies a discrete fast Fourier transform.

3.1.2. Real-space representation

The above discussion is given in terms of the Fourier basis set. Most of the computation can be carried out almost equivalently in terms of a real-space basis function set. Therefore, for each $\mathbf{r}' \in \mathbb{X}$ we define the coefficients

$$\hat{\varphi}_{\mathbf{g};\mathbf{r}'} = \frac{1}{\sqrt{N_g}} \, \mathrm{e}^{-\mathrm{i}\mathbf{g}\cdot\mathbf{r}'}$$

with $\mathbf{g} \in \mathbb{G}$. Its inverse Fourier transform is a real-space function

$$\varphi_{\mathbf{r}'}(\mathbf{r}) = \sum_{\mathbf{g}\in\mathbb{G}} \hat{\varphi}_{\mathbf{g};\mathbf{r}'} \phi_{\mathbf{g}}(\mathbf{r}) = \frac{1}{\sqrt{N_g|\Omega|}} \sum_{\mathbf{g}\in\mathbb{G}} \exp(\mathrm{i}\mathbf{g} \cdot (\mathbf{r} - \mathbf{r}')). \tag{3.5}$$

This is called the periodic sinc function (Skylaris, Haynes, Mostofi and Payne 2005), or psinc function for short. In particular, the psinc functions can be viewed as the numerical δ-function on the discrete set $\mathbb{X}$:

$$\varphi_{\mathbf{r}'}(\mathbf{r}) = \sqrt{\frac{N_g}{|\Omega|}}\delta_{\mathbf{r},\mathbf{r}'}, \quad \mathbf{r},\mathbf{r}' \in \mathbb{X}. \tag{3.6}$$

Hence a smooth function $\psi(\mathbf{r})$ can be expanded as

$$\psi(\mathbf{r}) \approx \sum_{\mathbf{r}'\in\mathbb{X}} \varphi_{\mathbf{r}'}(\mathbf{r})\psi(\mathbf{r}'). \tag{3.7}$$

The basis set $\{\varphi_{\mathbf{r}'}\}$ is often called the psinc basis set, or the planewave dual basis set. For a given function, the nodal representation (3.7) allows us to identify its function values evaluated at Cartesian grid points with the expansion coefficients under the psinc basis set. This is particularly convenient for describing many numerical algorithms below, such as the selected columns of the density matrix method (Section 4.3) and the interpolative separable density fitting method (Section 4.4). If a discretization scheme allows such a nodal representation, it is often referred to as a *real-space representation*. In such a case, the number of basis functions N_b can be identified with the number of grid points N_g. The finite difference discretization can also be viewed as a real-space representation. However, with some abuse of notation we shall use the words *psinc basis set* and *real-space representation* interchangeably in the following discussion. When the real-space representation is used, we may also slightly abuse the notation by using $\mathbf{r}$ to denote an element from a *discrete set* of points $\{\mathbf{r}_i\}_{i=1}^{N_g}$.

3.2. Small basis set

In electronic structure calculations, a small basis set typically only requires $10 \sim 100$ basis functions per atom to achieve chemical accuracy. This can be the case even for all-electron calculations in the absence of pseudopotentials. The reason why such a small basis set can be achieved without significant deterioration of the accuracy is that in quantum chemistry the electron orbitals usually do not vary arbitrarily in the presence of a chemical environment. Hence useful information can be extracted from the atomic limit, which gives rise to the atomic basis set. The size of the discretized Hamiltonian matrix is often of the order of $10^2 \sim 10^4$, which makes direct diagonalization of the Hamiltonian matrix a viable approach (and often the fastest option for small systems too) to solve Kohn–Sham DFT. For certain discretization schemes and exchange-correlation functionals, the discretized Hamiltonian is even a sparse matrix due to the spatial localization of the basis set. This permits the use of efficient numerical methods to reduce the asymptotic complexity.

In order to understand the atomic basis set, let us first recall the one-body Schrödinger equation for hydrogen-like atoms, given in spherical coordinates

(r, θ, φ) by

$$E\psi(\mathbf{r}) = \tag{3.8}$$

$$-\frac{1}{2}\left(\frac{1}{r^2}\frac{\partial}{\partial r}\left(r^2\frac{\partial}{\partial r}\right) + \frac{1}{r^2\sin\theta}\frac{\partial}{\partial\theta}\left(\sin\theta\frac{\partial}{\partial\theta}\right) + \frac{1}{r^2\sin^2\theta}\frac{\partial^2}{\partial\varphi^2}\right)\psi(\mathbf{r}) - \frac{Z}{r}\psi(\mathbf{r}).$$

The spherical part of the operator is given by

$$-\frac{1}{r^2}\mathbf{L}^2 := -\frac{1}{r^2}\left(\frac{1}{\sin\theta}\frac{\partial}{\partial\theta}\left(\sin\theta\frac{\partial}{\partial\theta}\right) + \frac{1}{\sin^2\theta}\frac{\partial^2}{\partial\varphi^2}\right), \tag{3.9}$$

where $\mathbf{L}^2$ is the spherical Laplacian operator, whose eigenfunctions are given by the spherical harmonics $Y_{lm}(\theta, \varphi)$ parametrized by the azimuthal quantum number l and the magnetic quantum number m. For each non-negative integer l, the admissible m are given by $-l, -l+1, \ldots, l$. The eigenfunction of the one-body Schrödinger operator for hydrogen-like atoms is then given by

$$\psi_i(\mathbf{r}) = \frac{u_i(r)}{r}Y_{lm}(\hat{\mathbf{r}}), \tag{3.10}$$

where the term r in the denominator is used to capture the singularity of the solution near the origin $r = 0$. For notational simplicity we let l, m depend implicitly on the orbital index i.

The idea of the atomic basis set is then to use functions of the type (3.10) centred at each atom in the system to discretize the Kohn–Sham equations. Given the nuclei positions $\{\mathbf{R}_I\}$, the basis set is given by

$$\left\{\varphi_i(\mathbf{r} - \mathbf{R}_I) = \frac{u_i(|\mathbf{r} - \mathbf{R}_I|)}{|\mathbf{r} - \mathbf{R}_I|}Y_{lm}\left(\frac{\mathbf{r} - \mathbf{R}_I}{|\mathbf{r} - \mathbf{R}_I|}\right) : i = 1, \ldots, n_I, \, I = 1, \ldots, M\right\},$$
$$\tag{3.11}$$

where n_I atomic orbitals are used for the Ith nucleus. Such discretization has the clear advantage that the basis set is able to capture the singularity of the solutions to the Kohn–Sham problem near the nuclei. Hence only a small number of degrees of freedom are needed to discretize the Kohn–Sham orbitals. Moreover, the atomic basis set can be used even without the pseudopotential approximation.

Based on different choices of the radial part of the basis functions u_i, some widely used small basis sets are as follows.

(1) *Slater-type orbital (STO)*. The radial part takes the form

$$u(r) = Cr^n e^{-\zeta r}, \tag{3.12}$$

where the non-negative integer n plays the role of principal quantum number, ζ is a constant related to the effective charge of the nucleus, and C is the normalization factor. The physical motivation of STO is clear, as (3.12) gives the radial part of the eigenfunctions of hydrogen-like atoms.

(2) *Gaussian-type orbitals (GTO).* Numerical integration involving STO can be difficult (we will further discuss the quadrature issues below). Gaussian-type orbitals – Gaussian functions or Gaussians multiplied by polynomials – are proposed as basis functions to avoid this difficulty, since many integrals can then be calculated explicitly. Many GTO basis sets have been proposed (Jensen 2013) in the quantum chemistry literature, starting from the idea of fitting STO using a few Gaussians (known as the STO-nG minimal basis), to the most widely used correlation-consistent basis sets (Dunning 1989).

(3) *Numerical atomic-orbitals (NAO).* Instead of using a predetermined analytical form of the basis functions, the idea of NAO is to obtain u_i by numerically solving a Schrödinger-like radial equation (after writing wavefunctions as products of the radial part and spherical harmonics, as in (2.77)). Thus we have

$$\left(-\frac{1}{2}\frac{\mathrm{d}^2}{\mathrm{d}r^2} + \frac{l(l+1)}{r^2} + v_i(r) + v_{\mathrm{cut}}(r) \right) u_i(r) = \epsilon_i u_i(r), \qquad (3.13)$$

where $(l(l+1))/r^2$ comes from the spherical Laplacian and the choice of spherical harmonics with azimuthal quantum number l, $v_i(r)$ is a radial potential chosen to control the main behaviour of u_i, and $v_{\mathrm{cut}}(r)$ is a confining potential to ensure that u_i decays rapidly beyond a certain radius and can be treated as a compactly supported function. We refer readers to Blum *et al.* (2009) for details. The compact support ensures the locality of the resulting discrete Hamiltonian and facilitates numerical computations based on the NAO.

Let us denote the collection of atomic orbital basis functions by

$$\{\phi_p(\mathbf{r}) : p = 1, \ldots, N_b\}.$$

In general, the basis set is not orthonormal. Therefore, by the usual Galerkin projection, the Kohn–Sham equation becomes a (nonlinear) generalized eigenvalue problem,

$$Hc_i = \varepsilon_i S c_i, \qquad (3.14)$$

where H is the discrete Hamiltonian matrix

$$H_{pq} = \langle \phi_p | H | \phi_q \rangle, \qquad (3.15)$$

and S is the overlap matrix (*i.e.* Gram matrix)

$$S_{pq} = \langle \phi_p | \phi_q \rangle. \qquad (3.16)$$

Since the basis functions $\{\phi_p\}$ are by construction either compactly supported or decay rapidly, the resulting H and S matrices are sparse, which enables efficient algorithms for solving the Kohn–Sham eigenvalue problems, as will be discussed in Section 5.

After solving (3.14), the Kohn–Sham orbitals are

$$\psi_i(\mathbf{r}) = \sum_p \phi_p(\mathbf{r}) c_{p,i}, \tag{3.17}$$

and thus the electron density is given by

$$\rho(\mathbf{r}) = \sum_{i=1}^{N} |\psi_i(\mathbf{r})|^2 = \sum_{i=1}^{N} \sum_{p,q=1}^{N_b} c_{p,i}^* c_{q,i} \phi_p^*(\mathbf{r}) \phi_q(\mathbf{r}). \tag{3.18}$$

Thus, to obtain the contribution to the matrix elements H_{ij} from the Hartree potential, the two-electron repulsion integral (we adopt the physicists' notation, as opposed to the chemists' notation) of the form

$$\langle pq|rs \rangle = \iint \frac{\phi_p^*(\mathbf{r}) \phi_q^*(\mathbf{r}') \phi_r(\mathbf{r}) \phi_s(\mathbf{r}')}{|\mathbf{r} - \mathbf{r}'|} \, \mathrm{d}\mathbf{r} \, \mathrm{d}\mathbf{r}' \tag{3.19}$$

is needed. In the case of Gaussian orbitals, the above integral can be obtained analytically by reducing the problem to integrals of Gaussians using the integral representation of the Coulomb kernel

$$\frac{1}{r} = \frac{1}{\pi^{1/2}} \int_{-\infty}^{+\infty} \exp(-r^2 \xi^2) \, \mathrm{d}\xi. \tag{3.20}$$

For other basis functions, numerical quadratures are needed, for example by using the density fitting technique (see Section 4.4). We also refer readers to Reine, Helgaker and Lindh (2012) for a review of multi-electron integrals.

As discussed in Section 2.8, since small basis sets are typically constructed to be centred around each atom, the Hellmann–Feynman force may not be accurate enough for the atomic force, unless a large number of basis functions are used in the calculation. In such a case, the Pulay force involving derivative quantities $\{\partial \phi_p / \partial \mathbf{R}_I\}$ is needed. We refer readers to Soler *et al.* (2002), for example, for details of the implementation in the context of numerical atomic orbitals.

3.3. Adaptive basis set

As discussed in Section 3.1, a large basis set allows one to systematically improve the accuracy of the numerical discretization by tuning one or a handful of parameters (*e.g.* the kinetic energy cut-off E_{cut} in the planewave method and the grid spacing in the finite difference method). The disadvantage of a large basis set is the much higher number of degrees of freedom per atom. On the other hand, as shown in Section 3.2, a small basis set can significantly reduce the number of degrees of freedom per atom, but it can be more difficult to improve its quality in a systematic fashion, and the improvement may rely heavily on the practitioner's experience with the underlying chemical system. An adaptive basis set aims to combine the best of both worlds, namely to achieve systematic improvability while significantly reducing the number of degrees of freedom per atom.

While adaptive basis sets are not yet as widely used as large or small basis sets, there have been a number of developments along this direction in the past few decades. Examples include the non-orthogonal generalized Wannier function (NGWF) approach in the ONETEP package (Skylaris *et al.* 2005), the adaptive minimal basis approach in the BigDFT package (Mohr *et al.* 2014), the filter diagonalization approach (Rayson and Briddon 2009), the localized spectrum slicing approach (Lin 2017) and the adaptive local basis set approach (Lin, Lu, Ying and E 2012*a*), to name just a few. Below we briefly introduce the adaptive local basis set (ALB) (Lin *et al.* 2012*a*) approach, which uses a discontinuous basis set for electronic structure calculations.

The basic idea of ALB is to partition the global domain into a number of subdomains (called elements), and to generate basis functions that are adapted to the Kohn–Sham solutions for each element. This allows the resulting basis functions to capture the structure of atomic orbitals but also effects from the chemical environment. Each basis function is continuous inside its associated supporting element but discontinuous across element boundaries. Then the discontinuous Galerkin (DG) method (see *e.g.* Babuška and Zlámal 1973, Arnold 1982, Cockburn, Karniadakis and Shu 2000) is used to construct a finite-dimensional projected Kohn–Sham Hamiltonian matrix with significantly reduced dimension. The projected Kohn–Sham problem can be solved efficiently in a similar way to the case when a small basis set is used.

Let Ω denote the global computational domain with periodic boundary conditions, and let $\mathcal{K}$ be a collection of quasi-uniform rectangular partitions of Ω into non-overlapping elements:

$$\mathcal{K} = \{\kappa_1, \kappa_2, \ldots, \kappa_M\}. \tag{3.21}$$

For $\kappa \in \mathcal{K}$, we let $\overline{\kappa}$ denote the closure of κ. The periodic boundary condition on Ω implies that the partition is regular across the boundary $\partial\Omega$.

We let $H^1(\kappa)$ denote the standard Sobolev space of $L^2(\kappa)$-functions such that the first partial derivatives are also in $L^2(\kappa)$. We denote the set of piecewise H^1-functions by

$$H^1(\mathcal{K}) = \{v \in L^2(\Omega) : v|_\kappa \in H^1(\kappa), \text{ for all } \kappa \in \mathcal{K}\},$$

which is also referred to as the broken Sobolev space. For $v, w \in H^1(\mathcal{K})$, we define the L^2-inner product as

$$(v, w)_{\mathcal{K}} = \sum_{\kappa \in \mathcal{K}} (v, w)_\kappa := \sum_{\kappa \in \mathcal{K}} \int_\kappa v^*(\mathbf{r}) w(\mathbf{r}) \, d\mathbf{r}, \tag{3.22}$$

which induces a norm $\|v\|_{\mathcal{K}} = (v, v)_{\mathcal{K}}^{1/2}$. For $v, w \in H^1(\mathcal{K})$ and $\kappa, \kappa' \in \mathcal{K}$, define the average and jump operators on a face $\overline{\kappa} \cap \overline{\kappa}'$ by

$$\langle v \rangle = \frac{1}{2}(v|_\kappa + v|_{\kappa'}), \quad \langle \nabla v \rangle = \frac{1}{2}(\nabla v|_\kappa + \nabla v|_{\kappa'}), \tag{3.23}$$

and

$$[\![v]\!] = v|_\kappa \mathbf{n}_\kappa + v|_{\kappa'} \mathbf{n}_{\kappa'}, \quad [\![\nabla v]\!] = \nabla v|_\kappa \cdot \mathbf{n}_\kappa + \nabla v|_{\kappa'} \cdot \mathbf{n}_{\kappa'}, \tag{3.24}$$

where $\mathbf{n}_\kappa$ denotes the exterior unit normal of the element κ.

In order to solve Kohn–Sham DFT in the broken Sobolev space $H^1(\mathcal{K})$, we also need to identify a basis set which spans a subspace of $H^1(\mathcal{K})$. Let N_κ be the number of DOFs on κ, and the total number of DOFs is $N_b = \sum_{\kappa \in \mathcal{K}} N_\kappa$. Let $\mathbb{V}(\kappa) = \text{span} \{\phi_{\kappa,j}\}_{j=1}^{N_\kappa}$, where each $\phi_{\kappa,j}$ is a function defined on Ω compactly supported in κ. Hence $\mathbb{V}(\kappa)$ is a subspace of $H^1(\mathcal{K})$ and is associated with a finite-dimensional approximation for $H^1(\kappa)$. Then $\mathbb{V} = \bigoplus_{\kappa \in \mathcal{K}} \mathbb{V}(\kappa)$ is a finite-dimensional approximation to $H^1(\mathcal{K})$. We also assume all functions $\{\phi_{\kappa,j}\}$ form an orthonormal set in the sense that

$$(\phi_{\kappa,j}, \phi_{\kappa',j'})_\mathcal{K} = \delta_{\kappa,\kappa'}\delta_{j,j'}, \quad \text{for all } \kappa, \kappa' \in \mathcal{K}, 1 \le j \le N_\kappa, 1 \le j' \le N_{\kappa'}. \tag{3.25}$$

For the moment, we assume that the basis set $\mathbb{V}$ has been given. For $w, v \in \mathbb{V}$, we introduce the following bilinear form:

$$a(w, v) = \sum_{\kappa \in \mathcal{K}} \left[\frac{1}{2}(\nabla w, \nabla v)_\kappa + ((V_{\text{eff}} + V_{\text{nl}})w, v)_\kappa \right.$$

$$\left. - (\nabla w, [\![\psi_v]\!])_{\partial\kappa} - ([\![w]\!], \nabla v)_{\partial\kappa} + \alpha([\![w]\!], [\![v]\!])_{\partial\kappa} \right]. \tag{3.26}$$

Here the first term on the right-hand side is the kinetic energy, and the second term is the effective local potential and the non-local potential as in equation (2.91). The third and fourth terms are obtained from integration by parts of the Laplacian operator, which cures the ill-defined operation of applying the Laplacian operator to discontinuous functions on the global domain. The last term is a penalty term to guarantee the stability condition (Arnold, Brezzi, Cockburn and Marini 2002). The penalty parameter needs to be sufficiently large, and the value of α depends on the choice of basis set $\mathbb{V}$.

In order to find the approximate Kohn–Sham orbitals $\{\psi_i\}_{i=1}^N \subset \mathbb{V}$, that is,

$$\psi_i(\mathbf{r}) = \sum_\kappa \sum_{j=1}^{N_\kappa} c_{\kappa,j;i}\phi_{\kappa,j}(\mathbf{r}), \tag{3.27}$$

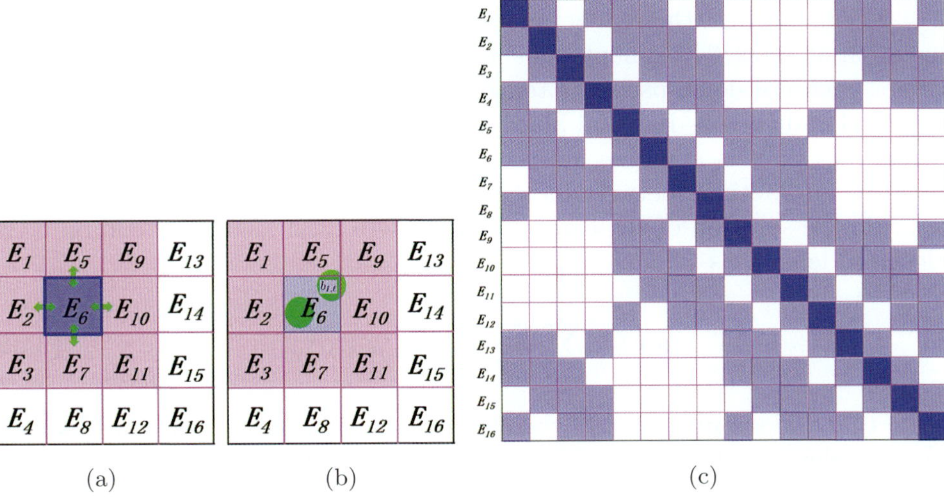

Figure 3.2. (Credit: Hu, Lin and Yang 2015.) A model system in 2D partitioned into 16 (4 × 4) equal-sized elements. The elements are denoted by E_i.

we minimize the following energy functional:

$$\mathcal{E}^{\mathrm{DG}}(\{\psi_i\}) = \sum_{i=1}^{N} a(\psi_i, \psi_i), \tag{3.28}$$

subject to the orthonormal constraint for $\{\psi_i\}$. The associated Euler–Lagrange equation corresponding to this minimization problem gives rise to the following eigenvalue problem:

$$\sum_{\kappa',j'} H^{\mathrm{DG}}_{\kappa,j;\kappa',j'} c_{\kappa',j';i} = \varepsilon_i^{\mathrm{DG}} c_{\kappa,j;i}, \quad i = 1, \ldots, N, \tag{3.29}$$

where the projected Hamiltonian matrix is given by

$$H^{\mathrm{DG}}_{\kappa,j;\kappa',j'} = a(\phi_{\kappa,j}, \phi_{\kappa',j'}). \tag{3.30}$$

The H^{DG} matrix can be naturally partitioned into matrix blocks as sketched in Figure 3.2. The kinetic energy and the local pseudopotential can only contribute to the diagonal matrix blocks. The terms from boundary integrals can contribute to both the diagonal and the off-diagonal blocks of H^{DG} corresponding to adjacent elements. Each boundary term involves only two neighbouring elements by definition, as plotted in Figure 3.2(a). For the non-local pseudopotential, since a projection vector of the non-local pseudopotential is spatially localized, we require the dimension of every element along each direction (usually around $6 \sim 8$ Bohr) to be larger than the size of the non-zero support of each projection vector (usually around $2 \sim 4$ Bohr). Thus, the non-zero support of each projection vector can overlap with at most 2^d elements as shown in Figure 3.2(b) ($d = 1, 2, 3$). As a result, each non-local pseudopotential term may also contribute to both the diagonal and the off-diagonal blocks corresponding to adjacent elements. In summary, H^{DG} is a sparse matrix and the non-zero matrix blocks correspond to interactions between neighbouring elements (Figure 3.2(c)).

From the solution of the eigenvalue problem (3.29), we may compute the electron density as

$$\rho(\mathbf{r}) = \sum_{i=1}^{N} \sum_{\kappa} \left| \sum_{j} c_{\kappa,j;i} \phi_{\kappa,j}(\mathbf{r}) \right|^2 . \tag{3.31}$$

Note that the computation of the electron density can be performed locally in each element due to the locality of the basis functions $\phi_{\kappa,j}$. Then the electron density can be fed back to the next step of the SCF iteration.

There are several methods to generate the basis set $\mathbb{V}$ in the broken Sobolev space. Here we introduce the adaptive local basis set in Lin, Lu, Ying and E (2012a), which is obtained by solving local Kohn–Sham problems. For each element κ, we form an *extended element* $\widetilde{\kappa}$ around κ, and we refer to $\widetilde{\kappa} \backslash \kappa$ as the buffer region for κ. On $\widetilde{\kappa}$ we solve the eigenvalue problem

$$\left(-\frac{1}{2}\Delta + V_{\mathrm{eff}}^{\widetilde{\kappa}} + V_{\mathrm{nl}}^{\widetilde{\kappa}} \right) \widetilde{\phi}_{\kappa,j} = \lambda_{\kappa,j} \widetilde{\phi}_{\kappa,j}. \tag{3.32}$$

with certain boundary conditions on $\partial \widetilde{\kappa}$. Here we define $V_{\mathrm{eff}}^{\widetilde{\kappa}} = V_{\mathrm{eff}}|_{\widetilde{\kappa}}$ to be the restriction of the effective potential at the current SCF step to $\widetilde{\kappa}$, and $V_{\mathrm{nl}}^{\widetilde{\kappa}} = V_{\mathrm{nl}}|_{\widetilde{\kappa}}$ to be the restriction of the non-local potential to $\widetilde{\kappa}$. This eigenvalue problem can be solved using a standard basis set such as finite difference, finite elements or planewaves. Note that the size of the extended element $\widetilde{\kappa}$ is independent of the size of the global domain, and so is the number of basis functions per element. We then restrict $\{\widetilde{\phi}_{\kappa,j}\}$ from $\widetilde{\kappa}$ to κ. The truncated functions are not necessarily orthonormal. Therefore, we apply a singular value decomposition (SVD) to obtain $\{\phi_{\kappa,j}\}$. The SVD procedure can ensure the orthonormal constraint of the basis functions inside each element, as well as eliminating the approximately linearly dependent functions in the basis set. We then extend each $\phi_{k,j}$ to the global domain by setting it to zero outside κ to generate the basis set $\mathbb{V}$. As a result, the overlap matrix corresponding to the adaptive local basis set is an identity matrix.

There are a number of possible ways to set the boundary conditions for the local problem (3.32). In practice, we use the periodic boundary condition for all eigenfunctions $\{\widetilde{\phi}_{\kappa,j}\}$ in $\widetilde{\kappa}$. In some sense, the details of the boundary condition do not affect the accuracy of the adaptive local basis set much as the buffer size increases. The periodic boundary condition permits the use of Fourier basis sets for solving the local problem. Note that while the eigenfunctions are required to satisfy periodic boundary conditions, the potential is obtained from the restriction of the global Kohn–Sham potential to the extended element. The size of each extended element should be chosen to balance between the effectiveness of the basis functions and the computational cost of obtaining them. For a typical choice used in practice, the elements are chosen to be of the same size, and each element contains on average a few atoms. The partition does not need to be updated when the atomic configuration is changed, as in the case of structure optimization and molecular dynamics.

Once the SCF iteration reaches convergence, we may evaluate the atomic forces. Recall the discussion in Section 2.8 that the atomic force generally consists of two components: the Hellmann–Feynman force and the Pulay force. Since each ALB depends on the atomic configuration, the Pulay force does not vanish as in the planewave basis set. Compared to atomic orbitals, the Pulay force for the adaptive basis set is much more difficult to evaluate. Nonetheless, the adaptive basis sets are constructed to be aware of the environment of each atom, and can be much closer to a complete basis set for describing occupied states than numerical atomic orbitals. Numerical results indicate that the Pulay force for the atomic basis set is indeed small, and use of the Hellman–Feynman formula can readily achieve chemical accuracy (Zhang *et al.* 2017). We also remark that using the spatial locality of the basis set $\mathbb{V}$, the computational cost of evaluating the force for all atoms can be reduced from $O(N^2)$ to $O(N)$. Compared to equation (2.95), the main reduction of the computational cost comes from (i) the localized pseudocharge associated with the local pseudopotential $V_{\mathrm{loc},I}$ (the pseudo-charge is defined as $-\Delta V_{\mathrm{loc},I}/4\pi$), and (ii) the density matrix formulation for the contribution from the non-local pseudopotential. We refer readers to Zhang *et al.* (2017) for more details.

The adaptive local basis set and the DG formulation have been implemented in the DGDFT (discontinuous Galerkin density functional theory) software package. DGDFT is a massively parallel electronic structure software package designed for large-scale DFT calculations involving up to tens of thousands of atoms (Hu, Lin and Yang 2015). As an illustrating example, Figure 3.3 shows the ALBs of a 2D phosphorene monolayer with 140 phosphorus atoms (P_{140}). This is a two-dimensional system, and the global domain is partitioned into 64 equal-sized elements along the Y and Z directions, respectively. We show the isosurfaces of the first three ALB functions for the element denoted by E_{10} in Figure 3.3(a–c). Each ALB function shown is strictly localized inside E_{10} and is therefore discontinuous across the boundary of elements. On the other hand, each ALB function is delocalized across a few atoms inside the element since they are obtained from eigenfunctions of local Kohn–Sham Hamiltonian. Although the basis

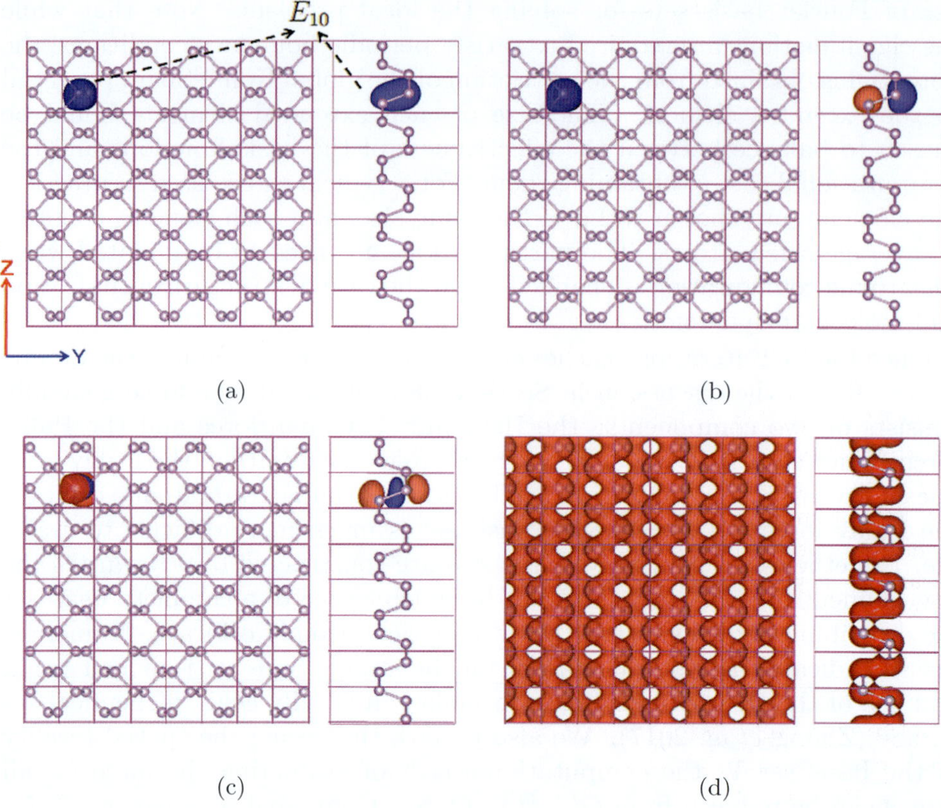

Figure 3.3. (Credit: Hu, Lin and Yang 2015.) The P_{140} system, with the isosurfaces (0.04 Hartree/Bohr^3) of the first three ALB functions belonging to the tenth element E_{10}: (a) ϕ_1, (b) ϕ_2, (c) ϕ_3 and (d) the electron density ρ in the top and side views in the global domain in the example of P_{140}. The red and blue regions indicate positive and negative isosurfaces, respectively. There are 64 elements and 80 ALB functions in each element, corresponding to 37 basis functions per atom.

functions are discontinuous, the electron density is well-defined and is very close to being a continuous function in the global domain (Figure 3.3(d)).

Figure 3.4 shows the convergence of the total energy and atomic forces for the quasi-1D and 3D Si systems with an increasing number of ALB functions per atom. First, we see that chemical accuracy is obtained with less than ten basis functions per atom for the quasi-1D system and a few tens of basis functions per atom for the 3D system. Furthermore, for the quasi-1D Si system, we find that the total energy error can be as small as 2.78×10^{-8} Ha/atom and the maximum error of the atomic forces can be as small as 8.47×10^{-7} Ha/Bohr when 22.5 ALB functions per atom are used.

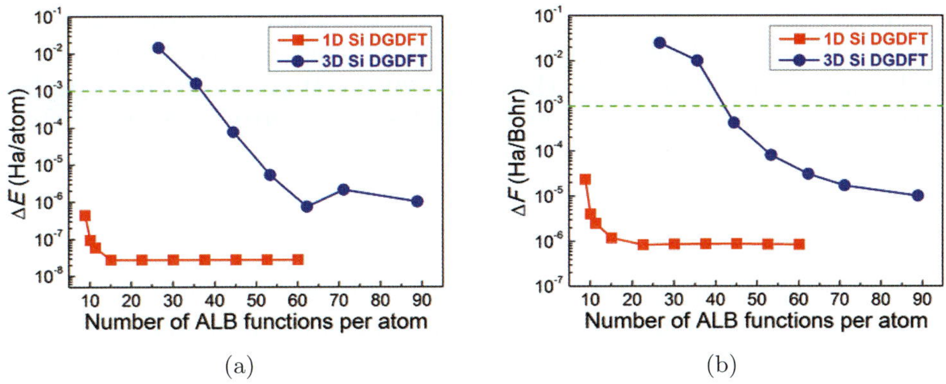

(a)   (b)

Figure 3.4. (Credit: Zhang *et al.* 2017.) Convergence of DGDFT total energy and forces for quasi-1D and 3D Si systems to reference planewave results with increasing number of ALB functions per atom. (a) Total energy error per atom ΔE (Ha/atom). (b) Maximum atomic force error ΔF (Ha/Bohr). The dashed green line corresponds to chemical accuracy. We set $E_{\mathrm{cut}} = 60$ Ha and penalty parameter $\alpha = 40$. When the number of basis functions per atom is sufficiently large, the error is well below the target accuracy (green dashed line).

For the bulk 3D Si system, the total energy error can be as small as 2.10×10^{-6} Ha/atom and the maximum error in atomic forces can be as small as 1.70×10^{-5} Ha/Bohr when 71.11 ALB functions per atom are used.

Remarks

The adaptive basis set is closely related to the 'hybrid basis set', which directly combines the large basis set and the small basis set. The idea of such a combination dates back to Slater in his augmented planewave method (APW) (Slater 1937), which was later improved to become the linearized APW method (Andersen 1975). The idea is to use an atomic basis set near the atom to resolve the singularity due to the electron–nucleus interaction in all-electron calculations, and use smooth functions such as planewaves in the interstitial region. Since planewaves are used without compression, the number of degrees of freedom per atom in the hybrid basis set is still relatively large. Recently there have been further developments along this direction using a partition of unity approach (Cai, Bai, Pask and Sukumar 2013) and the discontinuous Galerkin approach (Lu, Cai, Xin and Guo 2013).

The interior penalty formulation of DG requires setting a penalty parameter. For general non-polynomial basis functions, the value of the penalty parameter is not known *a priori*. We have demonstrated that the adjustable penalty parameter is mainly used to ensure the stability of the numerical scheme, and has relatively little effect on the accuracy of the scheme when it takes a large range of values (Lin *et al.* 2012*a*, Hu *et al.* 2015). Another possible solution can be found in Lin and Stamm (2016), which provides a formula for evaluating α on the fly for general non-polynomial basis sets based on the solution of eigenvalue problems restricted to each element κ. This further leads to rigorous *a posteriori* error estimates for using non-polynomial basis functions in the DG setting to solve linear PDEs and eigenvalue problems (Kaye, Lin and Yang 2015, Lin and Stamm 2016, Lin and Stamm 2017). This allows adaptive refinement of the basis set to improve its efficiency for heterogeneous systems.

The adaptive local basis set has been improved in a number of directions. As in the NGWF approach, the adaptive local basis set can be obtained via numerical optimization (Lin, Lu, Ying and E 2012*b*), which rigorously eliminates the Pulay force when the minimizer is achieved. The basis functions can also be obtained from solutions to linear systems on the global domain, which removes the need to impose artificial boundary conditions on the extended element and simultaneously improves the accuracy of the basis set (Li and Lin 2019). The adaptive local basis set can also be constructed by partitioning the discretized basis indices instead of the continuous domain (Xu, Suryanarayana and Pask 2018). This is called the discrete discontinuous basis projection method. Formally, this is a Galerkin method instead of a discontinuous Galerkin approach, and the advantage of this method is that it removes the need to impose penalty conditions on the basis set.

4. Algorithmic tools

This section introduces a few algorithmic tools that are useful in electronic structure calculations. We introduce direct and iterative eigensolvers in Sections 4.1 and 4.2, respectively. These are the standard methods for solving the eigenvalue problem obtained in Kohn–Sham DFT after discretization. In Section 4.3 we discuss how to compress information on the Kohn–Sham orbitals through localization. Localized representation of the occupied space forms the basis for linear scaling algorithms for insulating systems to be discussed in Section 5.2. Then in Section 4.4 we introduce the interpolative separable density fitting (ISDF) technique to compress the information stored in the pair products of Kohn–Sham orbitals, which will become useful in reducing the computational cost of Kohn–Sham DFT calculations with rung-4 and rung-5 functionals. We remark that eigensolvers, localization and density fitting are generic techniques that are useful in a number of other contexts of scientific computing. Hence we summarize these techniques separately in this section.

4.1. Direct diagonalization methods

When a small basis set is used and the system size is not too large, it is most efficient to solve the standard or generalized eigenvalue problem (3.14) using a generic eigensolver for dense matrices. It can be performed in serial using the LAPACK software package (Anderson *et al.* 1999) or in parallel using the ScaLAPACK software package (Blackford *et al.* 1997). Recently dense solvers specialized for solving large-scale Hermitian eigenvalue problems, such as ELPA (Marek *et al.* 2014), have been demonstrated to be more efficient and to scale massively in parallel to more than 100 000 computational cores. As a rule of thumb, dense eigensolvers are most efficient when the matrix dimension is 10 000 or less, but are not commonly used when the matrix dimension is beyond 100 000, due to the steep increase of the computational cost that scales as $O(N_b^3)$.

4.2. Iterative diagonalization methods

When a large basis set is used, iterative diagonalization methods that rely on the matrix–vector product of $H^{\mathrm{KS}}[\rho]$ with arbitrary vectors become particularly attractive, especially when the matrix–vector product can be performed efficiently. The two common cases are when $H^{\mathrm{KS}}[\rho]$ is sparse and when $H^{\mathrm{KS}}[\rho]$ is constructed using the Fourier basis set so that the fast Fourier transform can be used. Among the iterative diagonalization methods, we focus on Krylov subspace methods here. The discussion of another class of methods called filtering methods will be deferred to Section 5.1.

The key idea of Krylov subspace methods is to form a low-dimensional subspace and approximate the eigenpairs with those of the projected Hamiltonian into this subspace. Of these methods, the Davidson–Liu method (Davidson 1975, Liu 1978) and the LOBPCG method (Knyazev 2001) are widely used due to their simplicity and efficiency.

Consider for simplicity the standard eigenvalue problem

$$HX = X\Lambda,$$

where the columns of the $N_b \times N$ matrix X are the eigenvectors and the diagonal entries of the $N \times N$ diagonal matrix Λ give the eigenvalues.

In its simplest version, the Davidson–Liu method solves this problem by iteratively minimizing $\mathrm{Tr}[X^*HX]$ over a matrix X that satisfies the orthogonality constraint $X^*X = I$ and takes its columns from a subspace spanned by a collection of $2N$ vectors. These $2N$ vectors are set to be the columns of the matrix X from the previous step and another $N_b \times N$ matrix W, which is a preconditioned residual of X. More precisely, in each iteration one updates X via

$$X \leftarrow XC_X + WC_W, \quad W = TR := T(HX - X(X^*HX)),$$

Algorithm 1: Davidson–Liu method for solving the Kohn–Sham DFT eigenvalue problems $H\psi_i = \varepsilon_i \psi_i$

Input: Hamiltonian matrix H and initial wavefunctions $\{\psi_i^0\}_{i=1}^N$.
Output: Eigenvalues $\{\varepsilon_i\}_{i=1}^N$ and wavefunctions $\{\psi_i\}_{i=1}^N$.
1:   Initialize X by $\{\psi_i^0\}_{i=1}^N$ and orthonormalize X.
2:   **while** convergence not reached **do**
3:      Compute the preconditioned residual $W \leftarrow T(HX - X(X^*HX))$, where T is a preconditioner.
4:      Update the trial subspace $S \leftarrow [X, W]$.
5:      Solve the projected eigenvalue problem $S^*HSC = S^*SC\Lambda$ and obtain the coefficients $C = \begin{bmatrix} C_X \\ C_W \end{bmatrix}$.
6:      Compute $X \leftarrow SC$.
7:   **end while**
8:   Update $\{\psi_i\}_{i=1}^N \leftarrow X$.

---

where $R = HX - X(X^*HX)$ is the residual and T is a preconditioner that is ideally close to $(H - \lambda I)^{-1}$ with a certain shift λ. The original version of the Davidson–Liu method allows one to use many copies of the history; here, in order to simplify the discussion, we keep only the most recent copy of X and W. The coefficients C_X and C_W are obtained by computing the lowest N eigenpairs of the projected $2N \times 2N$ generalized eigenvalue problem

$$S^*HSC = S^*SC\Lambda, \quad C = \begin{bmatrix} C_X \\ C_W \end{bmatrix} \in \mathbb{R}^{2N \times N},$$

where $S = [X, W] \in \mathbb{R}^{N_b \times 2N}$ represents the trial subspace and C contains the optimal coefficients. The Davidson–Liu method is summarized in Algorithm 1.

The LOBPCG algorithm improves on the Davidson–Liu algorithm by introducing a conjugate search direction. More precisely, it solves $HX = X\Lambda$ by iteratively minimizing $\text{Tr}[X^*HX]$ over X that satisfies the orthogonality constraint $X^*X = I$ and takes its columns from a subspace spanned by $3N$ vectors. These $3N$ vectors are the columns of the matrix X of the previous step, its preconditioned residual W and an extra matrix P that consists of certain conjugate directions. The coefficients C_X, C_W and C_P are obtained by computing the lowest N eigenpairs of the projected $3N \times 3N$ generalized eigenvalue problem

$$S^*HSC = S^*SC\Lambda, \quad C = \begin{bmatrix} C_X \\ C_W \\ C_P \end{bmatrix} \in \mathbb{R}^{3N \times N},$$

Algorithm 2: LOBPCG method for solving the Kohn–Sham DFT eigenvalue problems $H\psi_i = \varepsilon_i \psi_i$

Input: Hamiltonian matrix H and initial wavefunctions $\{\psi_i^0\}_{i=1}^N$.
Output: Eigenvalues $\{\varepsilon_i\}_{i=1}^N$ and wavefunctions $\{\psi_i\}_{i=1}^N$.
1: Initialize X by $\{\psi_i^0\}_{i=1}^N$ and orthonormalize X.
2: **while** convergence not reached **do**
3:    Compute the preconditioned residual $W \leftarrow T(HX - X(X^*HX))$, where T is a preconditioner.
4:    Update the trial subspace $S \leftarrow [X, W, P]$.
5:    Solve the projected eigenvalue problem $S^*HSC = S^*SC\Lambda$ and obtain the coefficients $C = \begin{bmatrix} C_X \\ C_W \\ C_P \end{bmatrix}$.
6:    Compute the conjugate gradient direction $P \leftarrow WC_W + PC_P$.
7:    Compute $X \leftarrow XC_X + P$.
8: **end while**
9: Update $\{\psi_i\}_{i=1}^N \leftarrow X$.

---

where $S = [X, W, P]$ is the trial subspace. The LOBPCG method is outlined in Algorithm 2.

Note that when N is relatively small ($N \sim 10$–1000), the computational cost of the $2N \times 2N$ or $3N \times 3N$ projected eigenvalue problem in the Rayleigh–Ritz step is negligible. However, when N is relatively large ($N \sim 1000$–$10\,000$), the cost of solving such projected eigenvalue problems can become dominant. Furthermore, even with recent advances in dense eigensolvers such as ELPA (Marek *et al.* 2014), the Rayleigh–Ritz step can still limit the parallel scalability and dominate the wall clock time of the Davidson–Liu and LOBPCG algorithms when a large number of processors are used.

More recently, the projected preconditioned conjugate gradient (PPCG) (Vecharynski, Yang and Pask 2015) algorithm has been proposed to reduce the cost of the Rayleigh–Ritz step. The main idea of PPCG is to replace the $3N \times 3N$ projected eigenvalue problem of LOBPCG by N subproblems, each of size 3×3. Compared to the solution of the subspace problem from a full Rayleigh–Ritz step, PPCG relaxes the optimality of the search direction at each step, but the advantage is that all N subproblems can be solved independently. The computational cost of this step scales linearly with N, and is negligibly small even when N is large. The orbitals still need to be orthogonalized at each step to avoid degeneracy. We remark that the idea of the PPCG method is closely related to another variant of the LOBPCG method called LOBPCG II (Knyazev 2001).

Remarks

Algorithms 1 and 2 update all columns of X simultaneously. This is often referred to as a block eigensolver, which can efficiently utilize BLAS 3 level matrix–matrix multiplication operations, and is suitable for parallelization. The disadvantage of a block eigensolver is that the cost of the Rayleigh–Ritz step can be expensive when N becomes large. Different eigenvectors often converge with different rates, and the subspace problem may thus become degenerate. In such a case, the converged eigenvectors should be deflated from the unconverged ones to avoid instability. Another strategy is to solve the eigenpairs one by one, starting from the smallest eigenvalue. Such a method is often referred to as a single-band method (Payne *et al.* 1992) in electronic structure calculation. Since each eigenvector is directly orthogonalized with all previously converged eigenvectors, the single-band method can be numerically more stable. Besides the eigensolvers mentioned in this section, the linear eigenvalue problem can also be solved via optimization methods under orthogonality constraints (Payne *et al.* 1992, Edelman, Arias and Smith 1998, Yang, Meza and Wang 2006, Wen and Yin 2013). We also remark that optimization techniques can be used to minimize the nonlinear Kohn–Sham energy functional directly with respect to the Kohn–Sham orbitals.

The choice of the preconditioner T can significantly affect the convergence rate of eigensolvers. In the context of the planewave basis set, the most commonly used preconditioner (Teter, Payne and Allan 1989) is diagonal in frequency space, $T = F^{-1}\hat{T}F$, where F is the discrete Fourier transform matrix, with the coefficients $\hat{T}$ given by

$$\hat{T}(\mathbf{g}, \mathbf{g}') = \delta_{\mathbf{g},\mathbf{g}'} \left(\frac{27 + 18E_{\mathbf{g}} + 12E_{\mathbf{g}}^2 + 8E_{\mathbf{g}}^3}{27 + 18E_{\mathbf{g}} + 12E_{\mathbf{g}}^2 + 8E_{\mathbf{g}}^3 + 16E_{\mathbf{g}}^4} \right), \tag{4.1}$$

where $E_{\mathbf{g}} = |\mathbf{g}|^2/2$ is the kinetic energy associated with the Fourier mode $\mathbf{g}$. The purpose of the choice (4.1) is that the Fourier modes of long wavelengths (*i.e.* $|\mathbf{g}|$ is small) remain approximately unchanged, while Fourier modes of short wavelengths (*i.e.* $|\mathbf{g}|$ is large) are dampened by $1/E_{\mathbf{g}}$, which is approximately $2(-\Delta)^{-1}$. The rationale is that the Laplacian operator is an unbounded operator, while the pseudopotential, Hartree and exchange-correlation potentials are usually bounded operators. As the planewave basis set is being refined, the Laplacian operator dominates the spectral radius of the Hamiltonian, especially along the Fourier modes with short wavelengths. Hence, along such modes, the preconditioner T approximates H^{-1} well.

The preconditioner (4.1) can be efficiently implemented in the context of the planewave basis set using only two FFTs per orbital, but it becomes considerably more difficult in other large basis sets such as the finite element method, and the finite difference method. In such a case, one often performs preconditioning by directly taking $T = -\Delta^{-1}$, and applying the preconditioner to the residual vector r amounts to solving a Poisson equation.

4.3. Localization methods

Localized representations of electronic wavefunctions have a wide range of applications in quantum physics, chemistry and materials science. As they require significantly less cost for both storage and computation, localized representations are the foundation of so-called 'linear scaling methods' (Kohn 1996, Goedecker 1999, Bowler and Miyazaki 2012) for solving quantum problems (see more discussion in Section 5.2). They can also be used to analyse the chemical bonding in complex materials, interpolate the band structure of crystals, accelerate ground and excited state electronic structure calculations, and form reduced-order models for strongly correlated many-body systems (Marzari *et al.* 2012).

The localization problem can be stated as follows. The Kohn–Sham orbitals $\{\psi_i\}_{i=1}^N$ are eigenfunctions of a Hamiltonian matrix, and are generally delocalized across the entire system, *i.e.* with significant magnitude in large portions of the computational domain. Nonetheless, if we apply a unitary rotation $U \in \mathbb{C}^{N \times N}$ to the Kohn–Sham orbitals

$$w_i(\mathbf{r}) = \sum_{j=1}^N \psi_j(\mathbf{r}) U_{ji}, \qquad (4.2)$$

the density matrix and hence the electron density are invariant with respect to such rotation. However, for certain systems there exist unitary transformations for which the resulting orthonormal functions $\{w_i\}_{i=1}^N$ are approximately only supported on a small portion of the computational domain. For isolated systems, $\{w_i\}_{i=1}^N$ are often called Boys orbitals (Foster and Boys 1960), while for periodic systems they are often called Wannier functions (Wannier 1937). Below we will simply refer to the localized orbitals as Wannier functions.

For isolated systems surrounded by a vacuum, the occupied orbitals always decay exponentially to zero as $|\mathbf{r}| \to \infty$ (Lieb and Loss 2001). Hence qualitatively, both the Kohn–Sham orbitals ψ_i and the localized orbitals w_i decay exponentially, and the localization is defined only in a quantitative sense.

On the other hand, for periodic crystals, the Kohn–Sham orbitals satisfy the Bloch boundary condition on each unit cell, and hence are delocalized on a macroscopic scale. However, for insulating systems with a finite band gap, the kernel of the density matrix decays exponentially along the off-diagonal direction (Kohn 1959, Blount 1962, Nenciu 1983, E and Lu 2011, Benzi,

Boito and Razouk 2013, Lin and Lu 2016), in the sense that

$$|P(\mathbf{r}, \mathbf{r}')| \lesssim e^{-C|\mathbf{r}-\mathbf{r}'|}, \quad |\mathbf{r} - \mathbf{r}'| \to \infty. \tag{4.3}$$

This is related to the *near-sightedness principle* in the physics literature (Kohn 1996, Prodan and Kohn 2005).

If the system is also topologically trivial, then one may further find a proper choice of the gauge to construct a set of exponentially localized Wannier functions (Kohn 1959, Blount 1962, Kivelson 1982, Nenciu 1983, Brouder *et al.* 2007, E, Li and Lu 2010, Panati and Pisante 2013). Therefore, localized representation can make a qualitative difference. In the past few decades, many numerical algorithms have been designed to compute such localized orbitals (Marzari and Vanderbilt 1997, Koch and Goedecker 2001, Gygi 2009, Damle, Lin and Ying 2015, Mustafa, Coh, Cohen and Louie 2015, Cancès, Levitt, Panati and Stoltz 2017, Damle, Lin and Ying 2017, Damle and Lin 2018).

For isolated systems, the localized representation can be identified by minimizing the following *spread functional* (Foster and Boys 1960, Marzari and Vanderbilt 1997):

$$\inf_{U} \; \Omega[\{w_i\}_{i=1}^{N}] = \sum_{i=1}^{N} \left[\langle w_i | \mathbf{r}^2 | w_i \rangle - (\langle w_i | \mathbf{r} | w_i \rangle)^2 \right]$$

$$\text{subject to } w_i = \sum_{j=1}^{N} \psi_j U_{ji}, \; U^* U = I_N. \tag{4.4}$$

In three-dimensional space, the formula for the functional Ω should be interpreted as

$$\sum_{\alpha=1}^{3} [\langle w_i | \mathbf{r}_\alpha^2 | w_i \rangle - (\langle w_i | \mathbf{r}_\alpha | w_i \rangle)^2] := \langle w_i | (\mathbf{r} - \langle \mathbf{r} \rangle_i) \cdot (\mathbf{r} - \langle \mathbf{r} \rangle_i) | w_i \rangle,$$

where $\langle \mathbf{r} \rangle_i := \langle w_i | \mathbf{r} | w_i \rangle$. Hence the spread functional can be written compactly as

$$\Omega[\{w_i\}_{i=1}^{N}] = \sum_{i=1}^{N} \langle w_i | (\mathbf{r} - \langle \mathbf{r} \rangle_i)^2 | w_i \rangle.$$

The functional Ω characterizes the total spatial spread of the rotated orbitals $\{w_i\}$, in terms of the second moment around each centre $\langle \mathbf{r} \rangle_i$. A smaller spread value indicates a more localized representation of the Kohn–Sham occupied subspace. Numerically, the localization problem (4.4) can be solved as a constrained minimization problem. One main drawback of the localization procedure above is the reliance on a nonlinear optimization subproblem, and in practice such nonlinear optimization algorithms can frequently get stuck at local minima, which can lead to a qualitatively different physical interpretation.

The selected column of the density matrix (SCDM) method (Damle *et al.* 2015) is a recently developed procedure to overcome this problem. For simplicity we focus on the case of isolated, insulating systems below, and the formulation can be generalized to periodic crystals as well as metallic systems (Damle and Lin 2018). We also assume that the real-space representation is used, and we do not distinguish a continuous vector $\psi(\mathbf{r})$ and the corresponding discretized vector $\{u(\mathbf{r}_i)\}_{i=1}^{N_g}$, where the grid point $\mathbf{r}_i$ also serves as a basis index in the real-space representation.

For an isolated system, the density matrix P is a spectral projector of rank N to the occupied space. The kernel of the density matrix remains the same under the unitary transformation

$$P(\mathbf{r}, \mathbf{r}') = \sum_{i=1}^{N} w_i(\mathbf{r}) w_i^*(\mathbf{r}').$$

Thus, if there *exists* a set of localized orbitals $\{w_i\}$, $P(\mathbf{r}, \mathbf{r}')$ also decays rapidly as $|\mathbf{r} - \mathbf{r}'| \to \infty$. Intuitively, if we can select a set of N points $\mathcal{C} := \{\hat{\mathbf{r}}_\mu\}_{\mu=1}^{N}$ so that the corresponding column vectors of the kernel $\{P(\mathbf{r}, \hat{\mathbf{r}}_\mu)\}_{i=1}^{N}$ are the 'most representative' and well-conditioned column vectors of P, these vectors almost form the desired Wannier functions up to the orthonormality condition.

In order to select the set $\mathcal{C}$, using a real-space representation, we let $\Psi = [\psi_1, \ldots, \psi_N] \in \mathbb{C}^{N_g \times N}$ be a matrix with orthonormal columns. The corresponding discretized density matrix kernel, still denoted by P, is given by $P = \Psi\Psi^*$. From a numerical linear algebra point of view, the most representative column vectors can be identified via a QR factorization with column-pivoting (QRCP) (Golub and Van Loan 2013) applied to P, that is,

$$P\Pi = \widetilde{Q}\widetilde{R}. \tag{4.5}$$

Here $\Pi \in \mathbb{R}^{N_g \times N_g}$ is a permutation matrix, $\widetilde{Q} \in \mathbb{C}^{N_g \times N_g}$ is a unitary matrix, and $\widetilde{R} \in \mathbb{C}^{N_g \times N_g}$ is an upper triangular matrix. The QRCP decomposition is not unique, but the absolute value of the diagonal entries $|\widetilde{R}_{ii}|$ should always follow a non-increasing order, *i.e.* $|\widetilde{R}_{11}| \geq |\widetilde{R}_{22}| \geq \cdots$. Note that P is a rank-N matrix, and hence we must have $\widetilde{R}_{n,n} = 0, n > N$. The row indices of the non-zero entries in the first N columns of Π directly give the set $\mathcal{C}$.

One drawback of this method is that the cost of QRCP applied to P scales as $O(N_g^3)$, or at least $O(N_g^2 N)$. This becomes prohibitively expensive when a large basis set representation is used. The SCDM method (Damle *et al.* 2015) proposes that the set $\mathcal{C}$ can be equivalently computed via the

QRCP of the matrix Ψ^* as

$$\Psi^*\Pi = QR \equiv Q[R_1, R_2]. \tag{4.6}$$

Here $Q \in \mathbb{C}^{N \times N}$ is a unitary matrix of reduced dimension, $R_1 \in \mathbb{C}^{N \times N}$ is an upper triangular matrix, and $R_2 \in \mathbb{C}^{N \times (N_g - N)}$ is a rectangular matrix. The cost of the factorization (4.6) only scales as $O(N_g N^2)$, which is similar to the cost of the matrix orthogonalization step in iterative eigensolvers as discussed in Section 4.2. Once the factorization (4.6) is obtained, we readily have

$$P\Pi = \Psi\Psi^*\Pi = (\Psi Q)[R_1, R_2].$$

Hence QRCP applied to Ψ^* *implicitly* provides a QRCP for the density matrix P.

Having chosen $\mathcal{C}$, we must now orthonormalize the localized column vectors $\{P(\mathbf{r}, \hat{\mathbf{r}}_\mu)\}_{\mu=1}^N$ without destroying their locality. Note that

$$P(\mathbf{r}, \hat{\mathbf{r}}_\mu) = \sum_{i=1}^N \psi_i(\mathbf{r})\Xi_{i,\mu},$$

where $\Xi \in \mathbb{C}^{N \times N}$ has matrix elements

$$\Xi_{i,\mu} = \psi_i^*(\hat{\mathbf{r}}_\mu). \tag{4.7}$$

One way to enforce the orthogonality is

$$w_\mu(\mathbf{r}) = \sum_{i=1}^N \psi_i(\mathbf{r})U_{i,\mu}, \quad U = \Xi(\Xi^*\Xi)^{-1/2}. \tag{4.8}$$

It is clear that $U \in \mathbb{C}^{N \times N}$ is a unitary matrix, and U is in fact the minimizer of the orthogonal Procrustes problem:

$$U = \arg\min_{V^*V=I} \|V - \Xi\|_F. \tag{4.9}$$

In the quantum chemistry literature, the solution of the orthogonal Procrustes problem via the matrix square-root transformation in equation (4.8) is called the Löwdin transformation (Löwdin 1950).

Note that

$$(\Xi^*\Xi)_{\mu,\nu} = \sum_{i=1}^N \psi_i(\hat{\mathbf{r}}_\mu)\psi_i^*(\hat{\mathbf{r}}_\nu) = P(\hat{\mathbf{r}}_\mu, \hat{\mathbf{r}}_\nu). \tag{4.10}$$

The decay properties of the matrix P imply that $\{P(\hat{\mathbf{r}}_\mu, \hat{\mathbf{r}}_\nu)\}$ is a localized matrix, but of smaller size $N \times N$. If the eigenvalues $(\Xi^*\Xi)^{-1/2}$ are bounded away from 0, then $(\Xi^*\Xi)^{-1/2}$ will itself be localized (E and Lu 2011, Benzi *et al.* 2013), and consequently $\{w_i\}_{i=1}^N$ will be localized, orthonormal Wannier functions. Pseudocode for the SCDM algorithm is given in Algorithm 3.

Algorithm 3: Selected columns of the density matrix (SCDM) algorithm

---

Input: Kohn–Sham orbitals $\Psi \in \mathbb{C}^{N_g \times N}$
Output: Localized orbitals $W \in \mathbb{C}^{N_g \times N}$
  1: Compute the selected columns set $\mathcal{C} := \{\hat{\mathbf{r}}_\mu\}_{\mu=1}^N$ using QRCP via (4.6).
  2: Evaluate the Ξ matrix via (4.7).
  3: Perform orthogonalization via (4.8).

Below we demonstrate the performance of the SCDM method by computing localized basis functions for two three-dimensional systems. Figure 4.1(a) shows one of the orthogonalized SCDM orbitals, obtained from a silicon crystal with 512 atoms consisting of $4\times4\times4$ unit cells with a diamond structure, and the dimension of each unit cell is $10.26 \times 10.26 \times 10.26$ a.u. Figure 4.1(b) shows one of the orthogonalized SCDM orbitals, obtained from a water system with 64 molecules in a cubic supercell with dimension $22.08 \times 22.08 \times 22.08$ a.u. The orbitals are very localized in real space and resemble the shape of maximally localized Wannier functions (Marzari and Vanderbilt 1997). In fact, in this case the method automatically finds the centres of all localized orbitals, which for the silicon crystal are in the middle of the Si–Si bond, and for water is closer to the oxygen atoms than to the hydrogen atoms.

Remarks
Besides using Löwdin transformation, the gauge matrix U can also be directly identified from the QRCP (4.6) via a simple choice $U = Q$. This in fact corresponds to another orthogonalization method called the Cholesky-QR factorization, and in this context the Cholesky decomposition can be performed explicitly or implicitly (Golub and Van Loan 2013, Damle *et al.* 2017). Numerical results indicate that the two orthogonalization methods have comparable performance. However, the Löwdin transformation can be better when the system has certain spatial symmetries, and can be more easily generalized to periodic crystals.

The numerical methods for finding localized orbitals from the set of occupied orbitals of an insulating systems can be directly generalized to any set of eigenvalues in a given energy interval $\mathcal{I}$ satisfying the band isolation condition:

$$\inf_{\varepsilon_i \in \mathcal{I}, \varepsilon_{i'} \notin \mathcal{I}} |\varepsilon_i - \varepsilon_{i'}| := \varepsilon_g > 0. \qquad (4.11)$$

When this condition is violated (*i.e.* $\varepsilon_g = 0$), the eigenvalues in $\mathcal{I}$ become *entangled*. Entangled eigenvalues appear ubiquitously in metallic systems,

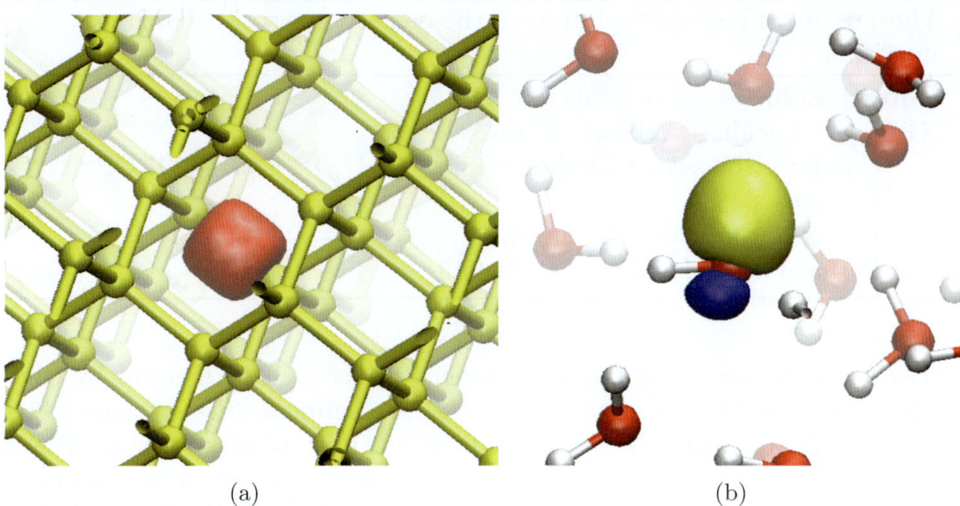

(a)   (b)

Figure 4.1. (Credit: Damle, Lin and Ying 2015.) Isosurface for an orthogonalized SCDM orbital. (a) A silicon crystal with 512 Si atoms (yellow balls). The red isosurface characterizes the localized orbitals located between two Si atoms. (b) A water system with 64 O atoms (red balls) and 128 H atoms (white balls). The yellow and blue isosurfaces characterize the positive and negative portions of localized orbitals.

and also in insulating systems when conduction bands or a selected range of valence bands are considered. The problem now becomes significantly more difficult: to identify a subspace $\mathcal{V}_w$ that admits a localized basis, and to construct such a basis.

The most widely used method to construct localized functions in this scenario is a *disentanglement* procedure (Souza, Marzari and Vanderbilt 2001). However, the result of the disentanglement procedure may depend sensitively on the initial guess. Often, detailed knowledge of the underlying physical system is required to obtain physically meaningful results. The SCDM method has been extended to handle systems with an entangled band structure in a more robust way (Damle and Lin 2018, Damle, Levitt and Lin 2019), using QRCP for a set of 'quasi' density matrices. Figure 4.2 demonstrates that the extended SCDM algorithm can accurately interpolate the band structure of graphene, even when zooming in on the region near the Dirac point.

4.4. Interpolative separable density fitting method

Kohn–Sham DFT calculations with rung-4 functionals and beyond often require computing quantities related to the *pair product of orbitals*, in the form $\{\varphi_i^*(\mathbf{r})\psi_j(\mathbf{r})\}_{i,j=1}^N$. Below we present a technique called interpolative

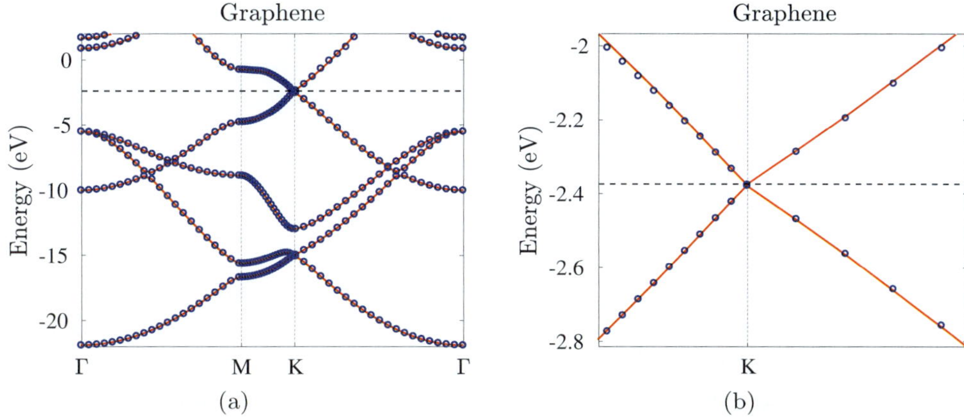

Figure 4.2. (Credit: Damle and Lin 2018.) Wannier interpolation with SCDM for the band structure for graphene (a) below the Fermi energy and (b) near the Dirac point. Direct calculation (red lines) and SCDM-based Wannier interpolation (blue circles). Here the chemical potential $\mu = -2.5$ eV, $\sigma = 4.0$ eV, and we use a $12 \times 12 \times 1$ **k**-grid to construct the Wannier functions.

separable density fitting (ISDF) (Lu and Ying 2015, Lu and Ying 2016), which allows us to reduce the number of pairs from $O(N^2)$ to $O(N)$ in general. This can be used to reduce the cost of rung-4 functional calculations with a large basis set (Hu, Lin and Yang 2017a, Dong, Hu and Lin 2018) (see Section 6.1), as well as RPA correlation energy calculations (Lu and Thicke 2017a) (see Section 6.3). It has also been shown to be useful in other contexts such as large-scale phonon calculations (Lin, Xu and Ying 2017).

To see why it is possible to find a compressed representation of the pair products, let us consider a real-space representation. Note that the number of grid points N_g to represent the orbital pairs scales linearly with respect to N. Hence, as N increases, the number of orbital pairs will eventually exceed N_g. This suggests that the numerical rank of $\{\varphi_i^* \psi_j\}$, viewed as a matrix of size $N_g \times N^2$, must scale asymptotically as $O(N)$. On the analytic level, consider the first N eigenfunctions $\{\psi_i\}$, $i = 1, \ldots, N$ of an effective Hamiltonian on a compact domain or closed manifold. Lu, Sogge and Steinerberger (2018) proved that for any ϵ there exists a subspace B_N of dimension $O_\delta(\epsilon^{-\delta} N^{1+\delta})$ such that

$$\|\psi_i \psi_j - \Pi_{B_N}(\psi_i \psi_j)\|_{L^2} \le \epsilon, \tag{4.12}$$

where δ is an arbitrary positive constant. In other words, we can approximate the N^2 pair products of eigenfunctions with almost $O(N)$ auxiliary basis functions, regardless of the discretization level.

Motivated by the above observation, the ISDF method uses the following compression format:

$$\varphi_i^*(\mathbf{r})\psi_j(\mathbf{r}) \approx \sum_{\mu=1}^{N_\mu} \varphi_i^*(\hat{\mathbf{r}}_\mu)\psi_j(\hat{\mathbf{r}}_\mu)\zeta_\mu(\mathbf{r}). \tag{4.13}$$

Here $\{\hat{\mathbf{r}}_\mu\}_{\mu=1}^{N_\mu}$ is a subset of real-space grid points $\{\mathbf{r}_i\}_{i=1}^{N_g}$ on which the orbitals are evaluated. We will refer to $\{\hat{\mathbf{r}}_\mu\}_{\mu=1}^{N_\mu}$ as the interpolation points, and $\{\zeta_\mu(\mathbf{r})\}_{\mu=1}^{N_\mu}$ sampled on $\{\mathbf{r}_i\}_{i=1}^{N_g}$ the interpolation vectors.

The ISDF decomposition can also be understood from the perspective of interpolation. Let $\{\hat{\mathbf{r}}_\mu\}$ denote a set of grid points in real space, and let $\zeta_\mu(\mathbf{r})$ be the Lagrange interpolation function on these grid points satisfying

$$\zeta_\mu(\hat{\mathbf{r}}_{\mu'}) = \begin{cases} 1 & \mu = \mu', \\ 0 & \text{otherwise.} \end{cases} \tag{4.14}$$

Then the ISDF decomposition would become sufficiently accurate as one systematically refines the set $\{\hat{\mathbf{r}}_\mu\}_{\mu=1}^{N_\mu}$. In the worst case, all grid points need to be selected and $N_\mu = N_g$. However, numerical evidence and analytical results suggest that the decomposition often becomes sufficiently accurate when $N_\mu \ll N_g$, especially when both $\{\varphi_i(\mathbf{r})\}$ and $\{\psi_i(\mathbf{r})\}$ consist of functions that are sufficiently smooth.

We first assume that the interpolation points $\{\hat{\mathbf{r}}_\mu\}$ are given, and discuss how to find the interpolation vectors. Define

$$\{Z_{ij}(\mathbf{r}) := \varphi_i^*(\mathbf{r})\psi_j(\mathbf{r})\}_{1\leq i\leq N, 1\leq j\leq N}, \tag{4.15}$$

and equation (4.13) can be written as

$$Z = \Theta C, \tag{4.16}$$

where each column of Z is defined by equation (4.15) sampled on real-space grids $\{\mathbf{r}_i\}_{i=1}^{N_g}$. $\Theta = [\zeta_1, \zeta_2, \ldots, \zeta_{N_\mu}]$ contains the interpolating vectors, and each column of C with a multi-index (i, j) is given by

$$[\varphi_i^*(\hat{\mathbf{r}}_1)\psi_j(\hat{\mathbf{r}}_1), \ldots, \varphi_i^*(\hat{\mathbf{r}}_\mu)\psi_j(\hat{\mathbf{r}}_\mu), \ldots, \varphi_i^*(\hat{\mathbf{r}}_{N_\mu})\psi_j(\hat{\mathbf{r}}_{N_\mu})]^\top.$$

Equation (4.16) is an over-determined linear systems with respect to the interpolation vectors Θ. One possible way to solve the over-determined system is to impose the Galerkin condition

$$ZC^* = \Theta CC^*. \tag{4.17}$$

It follows that the interpolating vectors can be obtained from

$$\Theta = ZC^*(CC^*)^{-1}. \tag{4.18}$$

Note that the solution given by equation (4.18) is a least-squares approximation to the solution of equation (4.13).

Algorithm 4: Interpolative separable density fitting (ISDF) method

---

Input: Functions $\{\varphi_i(\mathbf{r})\}$ and $\{\psi_j(\mathbf{r})\}$
Output: Interpolation points $\{\hat{\mathbf{r}}_\mu\}$, C, and Θ
  1: Form a matrix Z of size $N_g \times N^2$ with column ij given by
     $\{Z_{ij}(\mathbf{r}) := \varphi_i^*(\mathbf{r})\psi_j(\mathbf{r})\}_{1 \leq i \leq N, 1 \leq j \leq jN}$,
  2: Perform QRCP to Z^* to obtain $Z^*\Pi = QR$. Here Q is an $N^2 \times N_g$
     matrix that has orthonormal columns, R is an upper triangular
     matrix, and Π is a permutation matrix. Although a naive
     implementation takes $O(N^2 N_g^2)$ steps, randomized projection
     using the pair product structure of Z can reduce the complexity to
     $O(N N_g^2)$.
  3: The locations of the non-zero entries of the first N_μ columns of Π
     give $\{\hat{\mathbf{r}}_\mu\}$ and C.
  4: $\Theta = ZC^*(CC^*)^{-1}$.

It may appear that the cost of matrix–matrix multiplications ZC^* and CC^* scales as $O(N^4)$, because the size of Z is $N_g \times N^2$ and the size of C is $N_\mu \times N^2$. However, both multiplications can be carried out with fewer operations due to the separable structure of Z and C, and the computational complexity of computing the interpolation vectors is therefore $O(N^3)$.

In order to optimize the set of interpolation points, a general strategy is to use the QR factorization with column pivoting (QRCP), as in the case of the SCDM method in Section 4.3. Pseudocode for the ISDF algorithm is given in Algorithm 4.

To demonstrate the interpolation points selected from the QRCP procedure, we consider a water molecule, with $N_\mu = 8$ interpolation points distributed in real space (see Figure 4.3). The choice of interpolation points agrees with chemical intuition: the locations of these points are consistent with the distribution of the electron density, and the points are relatively well separated, so that the set of interpolation vectors do not become linearly dependent.

The disadvantage of using equation (4.16) directly is that the storage requirement for the matrix Z is $O(N^3)$ and the computational cost associated with a standard QRCP procedure is $O(N^4)$. One possibility is to lower the cost of QRCP by using a random matrix to subsample columns of the matrix Z to form a smaller matrix $\tilde{Z}$ of size $N_g \times \tilde{N}_\mu$, where $\tilde{N}_\mu$ is only slightly larger than N_μ (Lu and Ying 2015, Lu and Ying 2016). The reduced matrix size allows the computational cost of the QRCP procedure to be reduced to $O(N^3)$. Since the QRCP algorithm has been implemented in standard linear algebra software packages such as LAPACK and ScaLAPACK, the implementation and parallelization of ISDF is then relatively straightforward.

(a)   (b)

Figure 4.3. (Credit: Hu, Lin and Yang 2017a.) (a) The electron density (yellow isosurfaces) and (b) the interpolation points (green squares) $\{\hat{\mathbf{r}}_\mu\}_{\mu=1}^{N_\mu}$ ($N_\mu = 8$) selected from the real-space grid points $\{\mathbf{r}_i\}_{i=1}^{N_g}$ ($N_g = 66^3$) for a water molecule in a 10 Å × 10 Å × 10 Å box. The white and red balls denote hydrogen and oxygen atoms, respectively.

Another possibility for choosing interpolation points is to use a heuristic strategy. Note that an effective choice of the set of interpolation points should satisfy the following two conditions.

(i) The distribution of the interpolation points should roughly follow the distribution of the electron density. In particular, there should be more points when the electron density is high, and fewer or even zero points if the electron density is very low.

(ii) The interpolation points should not be very close to each other. Otherwise matrix rows represented by the interpolation points are nearly linearly dependent, and the matrix formed by the interpolation vectors become highly ill-conditioned.

The QRCP procedure satisfies both (i) and (ii) simultaneously. On the other hand, the conditions above can also be satisfied via a much simpler centroidal Voronoi tessellation (CVT) procedure applied to a weight vector, such as the electron density. More specifically, in the centroidal Voronoi tessellation (CVT) approach, we may partition the grid points in the global domain into N_μ Voronoi cells. The interpolation points can then be simply chosen to be the centroids corresponding to each cell. The CVT procedure can be effectively implemented through a K-Means algorithm (MacQueen 1967). Besides reduction of the computational cost, the use of a K-Means algorithm also produces a smoother potential energy surface, particularly in the context of *ab initio* molecular dynamics. We refer to Dong, Hu and Lin (2018) for more details of this approach.

Remarks

As in the SCDM method in Section 4.3, the ISDF method also relies on the real-space representation and smoothness of the orbitals. When small basis sets such as Gaussian-type orbitals are used, one possible way of using ISDF is to first interpolate the orbitals onto a real-space grid. However, in the context of all-electron calculations, the grid size needed to reach chemical accuracy can be exceedingly large. Hence in the quantum chemistry literature, the compression is often performed using the density fitting method (a.k.a. resolution of identity) (Weigend 2002, Ren *et al.* 2012*a*), applied to the basis functions $\{\phi_p\}_{p=1}^{N_b}$ directly. More specifically, density fitting seeks the following decomposition:

$$\phi_p^*(\mathbf{r})\phi_q(\mathbf{r}) \approx \sum_{\mu=1}^{N_\mu} C_\mu^{pq} \zeta_\mu(\mathbf{r}), \tag{4.19}$$

where the number of auxiliary functions N_μ is expected to be much smaller than the number of pairs N_b^2. The density fitting method can be viewed as an approximate solution to the singular value decomposition (SVD) of the matrix formed by the product of pairs of orbitals.

We remark that there is an important technical difference between ISDF and standard density fitting methods. In ISDF, the interpolation points $\{\hat{\mathbf{r}}_\mu\}$ are chosen first, and the auxiliary functions $\{\zeta_\mu\}$ are decided accordingly via a least-squares procedure. The interpolation coefficients in ISDF are obtained directly with a factorized form

$$C_\mu^{pq} = \phi_p^*(\hat{\mathbf{r}}_\mu)\phi_q(\hat{\mathbf{r}}_\mu). \tag{4.20}$$

In standard density fitting methods, the auxiliary functions are often chosen *a priori*, and part of the reason is to facilitate the computation of the two-electron integrals in subsequent calculations. The interpolation coefficients $\{C_\mu^{pq}\}$ are then determined via a least-squares procedure. Since $\{C_\mu^{pq}\}$ is generally a dense three-way tensor, the computational cost of density fitting scales generally as $O(N^4)$ unless additional locality constraints are enforced (Ren *et al.* 2012*a*).

Another technique to reduce the cost of the density fitting method is the tensor hypercontraction (THC) method (Parrish, Hohenstein, Martínez and Sherrill 2012, Parrish, Hohenstein, Martínez and Sherrill 2013), which is closely related to the ISDF method. Conceptually, THC aims at compressing the two-electron integrals directly, without going through intermediate decomposition of the form (4.19) with separable coefficients as in (4.20). However, since the two-electron integrals is a four-way tensor to start with, the computational cost associated with the THC decomposition can be $O(N^5)$. Once the decomposition is obtained, the THC format can then be used to reduce the cost of post-Hartree–Fock calculations, such as MP2 and coupled cluster theories.

5. Evaluation of the Kohn–Sham map: semi-local functional

When solving Kohn–Sham DFT using the self-consistent field iteration, the most computationally expensive step is the evaluation of the Kohn–Sham map. For a given effective Hamiltonian at the current iteration, one needs to determine the electronic structure, in the form of occupied orbitals or the electron density. In this section we discuss the evaluation of the Kohn–Sham map for semi-local functionals. Given an effective Hamiltonian as a matrix, the straightforward way to obtain the corresponding density is to use conventional direct diagonalization (see Section 4.1) or iterative diagonalization (see Section 4.2) algorithms to obtain the low-lying eigenfunctions and then the electron density. These methods inherently display cubic scaling with respect to the system size, and become computationally expensive or even infeasible for large-scale systems.

To reduce the computational cost, it is important to realize that we require only the electron density in the Kohn–Sham map, but not the eigenfunctions or eigenvalues. More efficient algorithms can thus be developed for the Kohn–Sham map. In Section 5.1, we review filtering methods that are iterative methods *not* trying to converge to each individual eigenvector, but instead to the occupied subspace, which can often be much more efficient than traditional iterative solvers for eigenvalue problems.

Since we do not need each individual eigenvector, we can use an alternative representation for the occupied subspace, such as the density matrix or localized orbitals. By exploiting the sparsity, it is even possible to achieve linear scaling for insulating systems. We will review such methods in Section 5.2. While linear scaling methods are in principle desirable, it is often challenging to control their accuracy unless the system has a relatively large gap. The pole expansion and selected inversion method (PEXSI) has been developed as a reduced scaling method that yields accurate approximation for general systems, which will be discussed in Section 5.3.

5.1. Filtering methods

Instead of using Krylov-type methods to approximate the low-lying eigenmodes, one can also identify the low eigenmodes by evaluating an appropriate polynomial of the Hamiltonian H. The main criterion for choosing such a polynomial $p(z)$ is that it should maximize the magnitudes at the eigenvalues $\lambda_1, \ldots, \lambda_N$ and at the same time minimize the magnitudes at the eigenvalues $\lambda_{N+1}, \ldots, \lambda_{N_b}$. One of the most popular choices is the shifted-and-scaled Chebyshev polynomial $P_k(z)$ that maps the standard interval $[-1, 1]$ to $[\lambda_{N+1}, \lambda_{N_b}]$, where the subscript k denotes the degree of this Chebyshev polynomial (Zhou, Saad, Tiago and Chelikowsky 2006). The absolute values of Chebyshev polynomials are bounded by 1 within the interval $[-1, 1]$ and grow rapidly outside this interval. Hence the shifted-and-scaled

---

Algorithm 5: Chebyshev filtering method for solving the Kohn–Sham DFT eigenvalue problems $H\psi_i = \varepsilon_i\psi_i$

Input: Hamiltonian matrix H and an orthonormal matrix $X \in \mathbb{C}^{N_b \times N_s}$.
Output: Eigenvalues $\{\varepsilon_i\}_{i=1}^N$ and wavefunctions $\{\psi_i\}_{i=1}^N$.
  1: Estimate λ_{N+1} and λ_{N_b} using a few steps of the Lanczos algorithm.
  2: **while** convergence not reached **do**
  3:     Apply the Chebyshev polynomial $P_k(H)$ to X: $Y = P_k(H)X$.
  3:     Orthonormalize columns of Y.
  4:     Compute the projected Hamiltonian matrix $\tilde{H} = Y^*HY$ and solve the eigenproblem $\tilde{H}\tilde{\Psi} = \tilde{\Psi}\tilde{D}$.
  5:     Subspace rotation $X = Y\tilde{\Psi}$.
  6: **end while**
  7: Update $\{\psi_i\}_{i=1}^N$ from the first N columns of X.

Chebyshev polynomial $P_k(z)$ only amplifies the occupied states of the H. In practice, one does not know *a priori* the values of λ_{N+1} and λ_{N_b}. These values need to be estimated from the Kohn–Sham Hamiltonian (*e.g.* applying a few steps of the Lanczos iteration (Zhou, Chelikowsky and Saad 2014)), and also updated on the fly. When applying the Chebyshev filtering algorithm, we also often choose the number of states N_s to be slightly larger than the number of occupied orbitals N, to accelerate the convergence. Pseudocode for the Chebyshev filtering algorithm is shown in Algorithm 5.

The Chebyshev filtering algorithm is presented here as a diagonalization procedure, that is, one iterates until the low-lying eigenvalues converge. However, in many implementations, due to the existence of the outer SCF iteration, one often performs a handful of iterations (applying Chebyshev polynomials and orthonormalization) and accepts whatever improvement it produces. Sometimes even a single iteration per outer SCF iteration may be sufficient. An equivalent view is that it combines the outer SCF loop with the inner diagonalization loop, without performing accurate diagonalization. For many problems, such a combination does not increase the number of outer SCF iterations and hence may significantly reduce the overall running time.

Application of the Chebyshev polynomial to all occupied states scales as $O(N_bN)$. For large problems, the most expensive step is the solution of the Rayleigh–Ritz problem in step 4 (also called the subspace diagonalization step) of Algorithm 5, which scales as $O(N^3)$. Although the cubic scaling cannot be avoided in the filtering-based approaches, the prefactors can be significantly reduced by the following complementary subspace method (Banerjee *et al.* 2018).

In the Chebyshev filtering approach, the density matrix is computed as

$$P = Y \tilde{P} Y^*, \tag{5.1}$$

where $\tilde{P} \in \mathbb{R}^{N_s \times N_s}$ is the projected density matrix, with its eigenvalues given by the occupation numbers $\{f_i\}_{i=1}^{N_s}$, evaluated according to the Fermi–Dirac distribution in equation (2.66), for example (we switch to the finite temperature case here). Since N_s is only slightly larger than N, most of the occupation numbers are approximately equal to 1. We denote states 1 to N_1 as those with occupation numbers equal to 1 within numerical tolerance. The remaining states, from $N_1 + 1$ to N_s, have occupation numbers less than 1. Let N_t be the number of these fractionally occupied states, *i.e.* $N_t = N_s - N_1$. The eigenvectors of the projected density matrix are $\{\tilde{\psi}_i\}_{i=1}^{N_s}$, and we may rewrite the expression for the projected density matrix as

$$\begin{aligned}
\tilde{P} &= \sum_{i=1}^{N_s} f_i \tilde{\psi}_i \tilde{\psi}_i^* \\
&= \sum_{i=1}^{N_s} \tilde{\psi}_i \tilde{\psi}_i^* - \sum_{i=N_1+1}^{N_s} \tilde{\psi}_i \tilde{\psi}_i^* + \sum_{i=N_1+1}^{N_s} f_i \tilde{\psi}_i \tilde{\psi}_i^* \\
&= \tilde{\mathcal{I}} - \sum_{i=N_1+1}^{N_s} (1 - f_i) \tilde{\psi}_i \tilde{\psi}_i^* .
\end{aligned} \tag{5.2}$$

In equation (5.2) above, $\tilde{\mathcal{I}}$ is the identity matrix of dimension $N_s \times N_s$. That the first term of equation (5.2) is the identity matrix follows from the resolution of the identity. Hence, if the N_t top eigenvectors $\tilde{\psi}_i$ and corresponding occupation numbers f_i are known, the projected density matrix $\tilde{P}$ may be computed. Thus, instead of determining the full $N_s \times N_s$ set of vectors, we need to determine only an extremal block of vectors (of dimension $N_s \times N_t$), corresponding to the states $i = N_1 + 1$ to N_s.

Moreover, physical quantities such as the electron density, energy, entropy and atomic force can all be computed by knowing only the top eigenstates. These eigenstates can be efficiently evaluated via several Chebyshev polynomial filtering steps applied to the projected Hamiltonian matrix $\tilde{H}$. Using this complementary subspace strategy with two levels of Chebyshev filtering (CS2CF), for insulating systems, the Rayleigh–Ritz step may be avoided altogether. For metallic systems, the cost of the Rayleigh–Ritz step can also be significantly reduced. Table 5.1 shows that the CS2CF strategy can be applied to insulating and metallic systems with $O(10^4)$ atoms and efficiently parallelized over $O(10^4)$ computational cores. The wall clock time to solution can be more than an order of magnitude faster than parallel dense diagonalization methods such as ELPA for large systems.

Table 5.1. (Credit: Banerjee *et al.* 2018.) Wall clock times for one SCF iteration large systems using the CS2CF strategy in DGDFT.

System	No. of atoms	No. of electrons	Comput. cores	CS2CF time (s)	CS2CF subspace time (s)	ELPA (s)
Electrolyte $3D_{3\times3\times3}$	8 586	29 808	3 456	34	19	647
SiDiamond $3D_{10\times10\times10}$	8 000	32 000	3 456	40	24	648
Graphene $2D_{8\times8}$	11 520	23 040	4 608	35	27	262
CuFCC $3D_{10\times10\times10}$	4 000	44 000	3 000	75	46	199
LiBCC $3D_{12\times12\times12}$	27 648	82 944	12 960	180	165	5844

Remarks

The idea of filtering the Hamiltonian matrix also appears in the spectral slicing approach (Zhang, Smith, Sternberg and Zapol 2007, Polizzi 2009, Schofield, Chelikowsky and Saad 2012, Aktulga *et al.* 2014). The basic idea is to project matrix functions $f_i(H)$ to a set of orbitals, where $f_i(\cdot)$ can be polynomials or rational functions approximately supported only on a small segment $\mathcal{I}_i$. Then one may perform diagonalization methods to compute the eigenvalues restricted to the segment. Finally, the eigenvalues obtained from different segments of the spectrum are merged together. The spectral splicing approach is naturally suited to massive parallelization. Each core or multiple cores together may focus on calculations associated with one segment, and may significantly reduce the wall clock time. Note that spectral slicing methods may involve multiple parameters, such as the partitioning strategy of the spectrum, the degrees of filtering polynomials, and the merging strategy of different segments of the spectrum. Finding an efficient and robust strategy to determine these parameters is still an active research area.

5.2. Linear scaling methods

Direct diagonalization and iterative methods for solving the Kohn–Sham eigenvalue problem scale cubically with system size. The cubic scaling is the bottleneck for electronic structure calculations of large systems. When a set of localized basis functions such as atomic orbitals and real-space representation are used, the resulting Hamiltonian matrix from the Kohn–Sham eigenvalue problem is localized. For insulating systems, the density matrix is localized and there exist localized representations – Wannier functions – for the subspace spanned by the occupied orbitals, as discussed in Section 4.3. One might hope to obtain the electronic structure, in the form of either a density matrix or localized orbitals, with a computation cost that scales linearly with system size. Such algorithms are referred to as 'linear scaling methods'. Many algorithms have been proposed and investigated over the past 30 years.

To achieve linear scaling, one needs to bypass the diagonalization of the Kohn–Sham Hamiltonian matrix. The crucial observation is that in the self-consistent iteration, we do not need full information on the eigenvectors. Thus, it is possible to reformulate the Kohn–Sham eigenvalue problem in a way that allows us to exploit the locality of the density matrix or localized orbitals to reduce computational complexity. In the following, we review a few representative linear scaling methods. While the list does not cover all algorithms developed in the literature, the discussion covers most of the existing ideas and strategies to achieve linear scaling. We refer readers to Goedecker (1999) and Bowler and Miyazaki (2012) for a more extensive review of linear scaling methods.

5.2.1. Divide-and-conquer method

The divide-and-conquer method proposed by Weitao Yang (Yang 1991a, Yang 1991b) is the first linear scaling method proposed in the literature for Kohn–Sham DFT. The idea of divide-and-conquer is to divide the physical system into several subsystems. The effective Hamiltonian for each subsystem is then solved separately, and the electron density of the whole system is obtained by merging the results of the subsystems together.

The divide-and-conquer method for each SCF step is described in Algorithm 6. Here a real-space representation is assumed. The divide-and-conquer method was originally developed for discretization using localized basis functions (Yang 1991a, Yang 1991b), in which case the algorithm is slightly different but shares the same spirit.

The divide-and-conquer method relies on the near-sightedness of the electron matter, as formulated by Kohn (1996). Roughly speaking, the dependence of electron density at $\mathbf{r}$ on the effective potential at $\mathbf{r}'$ is negligible when $|\mathbf{r} - \mathbf{r}'|$ is large. Hence, in the interior of the domain Ω_κ, the electron density corresponding to the Hamiltonian H is well approximated by that of the truncated Hamiltonian H_κ, since H agrees with H_κ inside Ω_κ. To make this work, besides requiring that the original system H is insulating, we also require that the truncated system H_κ does not destroy the energy gap (which could happen in practice, $e.g.$ for heterogeneous systems). See

---

Algorithm 6: Divide-and-conquer method for evaluating the Kohn–Sham map

---

Input: Hamiltonian matrix H.
Output: Electron density ρ corresponding to H.

1: Construct an overlapping partition $\{\Omega_\kappa\}$ of the whole computational domain Ω and the associated partition of unity $\{p_\kappa\}$ satisfying

$$\sum_\kappa p_\kappa(\mathbf{r}) = 1, \quad \text{for all } \mathbf{r} \in \Omega \text{ and } \operatorname{supp} p_\kappa \subset \Omega_\kappa.$$

2: $H_\kappa = H|_{\Omega_\kappa}$: restrict the Hamiltonian on the subdomain Ω_κ with appropriate boundary conditions.

3: Solve the eigenvalue problem in each subsystem

$$H_\kappa \psi_j^\kappa(\mathbf{r}) = \varepsilon_j^\kappa \psi_j^\kappa(\mathbf{r}), \quad \mathbf{r} \in \Omega_\kappa.$$

4: Determine the Fermi energy μ by fixing the total number of electrons:

$$N = \sum_\kappa \sum_j f_\beta(\varepsilon_j^\kappa - \mu) \int_{\Omega_\kappa} p_\kappa(\mathbf{r})|\psi_j^\kappa(\mathbf{r})|^2 \, \mathrm{d}\mathbf{r},$$

where f_β is the Fermi–Dirac function.

5: Construct the electron density as

$$\rho(\mathbf{r}) = \sum_\kappa p_\kappa(\mathbf{r}) \sum_j f_\beta(\varepsilon_j^\kappa - \mu)|\psi_j^\kappa(\mathbf{r})|^2.$$

Chen and Lu (2016) for a mathematical analysis of the divide-and-conquer method, based on the analysis tools developed in E and Lu (2011).

Further developments of the divide-and-conquer method and related domain decomposition type method can be found in Yang and Lee (1995), Wang, Zhao and Meza (2008), Zhao, Meza and Wang (2008), Barrault, Cancès, Hager and Le Bris (2007) and Bencteux *et al.* (2008). A great advantage of the method lies in the intrinsic parallelism of the computation for each subsystem, which has been utilized for large-scale calculations with more than 10^6 atoms and 10^{12} electronic degrees of freedom (Kobayashi and Nakai 2009, Ohba *et al.* 2012, Shimojo, Kalia, Nakano and Vashishta 2008, Shimojo *et al.* 2011).

The divide-and-conquer method is based on separating the total computational domain into smaller ones and solving the eigenvalue problem for each subdomain. There are other linear scaling methods based on reformulations of the Kohn–Sham problems to bypass the eigenvalue problem, as we will discuss below.

5.2.2. Orbital minimization methods

According to the Courant–Fisher variational principle, the first N eigenvectors of the discretized Hamiltonian matrix H can be found by solving

$$E = \min_{X,\, X^*X=I} \operatorname{Tr}(X^*HX), \tag{5.3}$$

where $X \in \mathbb{C}^{N_b \times N}$ is the matrix of discretized orbitals. The orthogonality constraint $X^*X = I$ is a bottleneck for linear scaling, as the cost of both the Gram–Schmidt orthogonalization and the Rayleigh–Ritz step scale cubically with respect to the system size.

The orbital minimization method (OMM) (Mauri, Galli and Car 1993, Ordejón, Drabold, Grumbach and Martin 1993, Mauri and Galli 1994, Ordejón, Drabold, Martin and Grumbach 1995) is based instead on an unconstrained variational formulation,

$$E = \min_{X} \operatorname{Tr}((2I - X^*X)(X^*(H - \varepsilon_{\max})X)), \tag{5.4}$$

where $\varepsilon_{\max}$ is an upper bound of the eigenvalue of H. The shift of the Hamiltonian by $-\varepsilon_{\max}$ makes all its eigenvalues negative, so the objective function is bounded from below, while it does not affect the eigenvectors of H.

While the OMM variational problem looks rather different from (5.3), it can be shown that the global minimizer X of (5.4) spans the eigenspace of the lowest N eigenvalues of H (assuming non-degeneracy) (Mauri *et al.* 1993, Pfrommer, Demmel and Simon 1999). In fact, somewhat surprisingly, while the objective function (5.4) is non-convex, it is proved in Lu and Thicke (2017*b*) that every local minimum of (5.4) is in fact also a global minimum. Thus it suffices for an optimization algorithm to converge locally.

The OMM minimization can be used as an alternative to direct diagonalization, as a way to obtain the Kohn–Sham map. This is implemented in the library libOMM (Corsetti 2014) with the preconditioned conjugate gradient algorithm.

More commonly, OMM is combined with truncation of the iterates of X to achieve linear scaling by only keeping $O(1)$ entries for each column of X during the iteration (Mauri *et al.* 1993, Ordejón *et al.* 1993). The hope is that the algorithm converges to the localized representation of the occupied space for insulating systems. However, with truncation, the optimization procedure might get stuck. Strategies to alleviate the problem have been proposed, for example in Kim, Mauri and Galli (1995), Tsuchida (2007), Gao and E (2009) and Lu and Thicke (2017*b*). The OMM is used in the SIESTA package (Soler *et al.* 2002) to achieve linear scaling.

As in OMM, localization and truncation steps can also be combined with other iterative algorithms that only rely on the subspace (rather than individual eigenvectors) to achieve linear scaling. One such algorithm is the localized subspace iteration algorithm (E *et al.* 2010, Garcia-Cervera, Lu, Xuan and E 2009), based on the Chebyshev filtering method (Algorithm 5). The Rayleigh–Ritz step (*i.e.* step 4 of Algorithm 5) is replaced by a localization and truncation step to maintain sparsity.

5.2.3. Density matrix minimization methods

Besides using orbitals, one can also use the density matrix in the variational approach for linear scaling algorithms. The starting point is the following variational principle in terms of density matrices:

$$E = \min_{P} \operatorname{Tr}(PH)$$

$$\text{subject to } P^2 = P, \ P = P^*, \ \operatorname{Tr} P = N. \tag{5.5}$$

In the above minimization, the idempotency constraint $P^2 = P$ is not easy to deal with, similar to the orthogonality constraint for orbitals. The idea is to circumvent the constraint by using some alternative variational principles. One such variational principle is

$$E = \min_{P} E^{\mathrm{DMM}}[P] := \min_{P} \operatorname{Tr}((3P^2 - 2P^3)(H - \mu))$$

$$\text{subject to } P = P^*, \tag{5.6}$$

where μ is the chemical potential assumed to be known, whose determination can be done by solving the constraint that $\operatorname{Tr} P = N$ (we will further discuss how to determine the chemical potential at the end of Section 5.2.4). Note that in (5.6) we have replaced P in the trace term by the polynomial $3P^2 - 2P^3$ (this is known as the McWeeny purification polynomial, which will be discussed in more detail in Section 5.2.4) and used a constant shift of the Hamiltonian by the chemical potential. To minimize (5.6), one can typically start with an initial guess (*e.g.* $P_{(0)} = \frac{1}{2}I$) and employ the conjugate gradient method to minimize the energy.

The energy (5.6) is in fact not bounded from below. Let ε_i be one of the eigenvalues of H that is larger than μ; we can take $P = \sum_i f_i |\psi_i\rangle\langle\psi_i|$ and send $f_i \to \infty$ to make the energy go to $-\infty$. On the other hand, convergence can be proved with particular choices of initial conditions and minimization algorithms. The variation of (5.6) can be computed as

$$\delta E^{\mathrm{DMM}} = \operatorname{Tr}\left[\frac{\delta E^{\mathrm{DMM}}}{\delta P}\delta P\right], \tag{5.7}$$

where

$$\frac{\delta E^{\mathrm{DMM}}}{\delta P} = 3(PH + HP - 2\mu P) - 2(P^2 H + PHP + HP^2 - 3\mu P^2). \tag{5.8}$$

When P commutes with H, equation (5.8) can be simplified as

$$\frac{\delta E^{\mathrm{DMM}}}{\delta P} = (6P - 6P^2)(H - \mu) = 6P(I - P)(H - \mu). \tag{5.9}$$

Equation (5.9) suggests that DMM may have many stationary points. In fact, any 'pure' density matrix satisfying $P^2 = P$ is a stationary point. Nonetheless, if the initial guess P_0 is not a stationary point, and can be represented as

$$P_0 = \sum_i f_i \psi_i \psi_i^*,$$

where $0 < f_i < 1$ for all i, that is, all the states are fractionally occupied. Then

$$\frac{\delta E^{\mathrm{DMM}}}{\delta P_0} = \sum_i 6(\varepsilon_i - \mu) f_i (1 - f_i) \psi_i \psi_i^*. \tag{5.10}$$

Thus, if $\varepsilon_i < \mu$, the gradient direction will increase f_i, and *vice versa*. The steepest descent algorithm will then converge to the density matrix corresponding to H, if the chemical potential μ is correctly chosen. A common choice of initial condition is $P_0 = \frac{1}{2}I$, which satisfies the condition $0 < f_i < 1$ in general.

We may also understand the OMM functional (5.4) from the perspective of density matrices. Setting $P = XX^*$, we can rewrite (5.4) as

$$E = \min_P \mathrm{Tr}((2P - P^2)(H - E_{\mathrm{max}}))$$

$$\text{subject to } P = P^*,\ 0 \preceq P \preceq I,\ \mathrm{rank}\, P \leq N. \tag{5.11}$$

The semi-definite constraint $0 \preceq P \preceq I$ is not easy to deal with (see Lai, Lu and Osher 2015, Lai and Lu 2016 for some algorithms for zero and finite temperature extensions), and hence in practice the OMM formulation in terms of orbitals is preferred.

Density matrix based algorithms provide a possible route to reduce the computational complexity for the evaluation of the Kohn–Sham map. Note that the gradient of the DMM energy (5.8) only involves matrix–matrix multiplication operations. Hence, if each matrix can be approximated by a sparse matrix, as in the case of insulating systems (see Section 4.3), the computational cost may be significantly reduced. For such systems, the number of non-zero entries of the density matrix only increases linearly with respect to N. Therefore with a proper implementation, all matrix–matrix multiplication operations can be carried out with $O(N)$ cost, which leads to linear scaling. The DMM algorithm is used by the ONETEP package (Skylaris *et al.* 2005) together with optimization of the localized basis functions.

5.2.4. Density matrix purification

Besides density matrix minimization, another class of density matrix based algorithms is density matrix purification, which works for zero temperature systems, as it directly uses the idempotency of the density matrix.

Let us first consider the problem of constructing an idempotent matrix, starting from a given Hermitian matrix with eigenvalues close to 0 and 1 (but not exactly 0 and 1). One strategy is McWeeny purification (McWeeny 1960), which recursively applies the function $f_{\mathrm{McW}}(x) = 3x^2 - 2x^3$ to the matrix, starting from the initial Hermitian matrix P_0:

$$P_{n+1} = f_{\mathrm{McW}}(P_n) = 3P_n^2 - 2P_n^3. \tag{5.12}$$

To see why McWeeny purification works, note that the iterate remains Hermitian throughout the iteration and the eigenfunctions remain the same; it hence suffices to keep track of the eigenvalues. By focusing on a specific eigenvalue λ_0 of P_0, we have

$$\lambda_{n+1} = f_{\mathrm{McW}}(\lambda_n) = 3\lambda_n^2 - 2\lambda_n^3. \tag{5.13}$$

The fixed point of this map is given by the solution to

$$x = f_{\mathrm{McW}}(x) = 3x^2 - 2x^3, \tag{5.14}$$

whose three roots are given by $0, 1/2$ and 1. Calculating the derivative of f_{McW} shows that

$$f'_{\mathrm{McW}}(x) = 6x - 6x^2 = \begin{cases} 0 & x = 0, \\ 3/2 & x = 1/2, \\ 0 & x = 1. \end{cases} \tag{5.15}$$

Therefore, 0 and 1 are the stable fixed points, while $1/2$ is unstable. We also observe that

$$\begin{aligned} x &< f_{\mathrm{McW}}(x) \quad \text{for } x \in (0, 1/2), \\ x &> f_{\mathrm{McW}}(x) \quad \text{for } x \in (1/2, 1). \end{aligned}$$

Hence, the iteration converges to 0 if the initial condition lies in $[0, 1/2)$, while it converges to 1 if the initial condition lies in $(1/2, 1]$. (See Figure 5.1. In fact, the iteration converges to 0 if starting within $[-1/2, 1/2)$ and to 1 if starting within $(1/2, 3/2]$.)

Given these observations, let us now consider how to use purification to get the density matrix

$$P = \mathbb{1}_{(-\infty, 0]}(H - \mu). \tag{5.16}$$

Note that P shares the same eigenfunctions as H, so this fits into the purification framework. We want to make all the eigenvalues of H below the

---

Algorithm 7: McWeeny purification algorithm for density matrix

---

Input: Hamiltonian matrix H and chemical potential μ.
Output: Density matrix P.

1: Estimate lower and upper bounds of eigenvalues $\varepsilon_{\min}$ and $\varepsilon_{\max}$, such that $\mathrm{spec}(H) \subset [\varepsilon_{\min}, \varepsilon_{\max}]$.
2: Set initial density matrix

$$P \leftarrow \frac{\alpha}{2}(\mu - H) + \frac{1}{2}I,$$

with $\alpha = \min\{(\varepsilon_{\max} - \mu)^{-1}, (\mu - \varepsilon_{\min})^{-1}\}$.
3: **while** convergence not reached **do**
4:　　$P \leftarrow f_{\mathrm{McW}}(P)$
5: **end while**

---

chemical potential μ converge to 1 and all the eigenvalues of H above converge to 0. Algorithm 7 starts from a rescaled version of the matrix $\mu - H$ and applies purification.

The initial density matrix

$$P_0 = \frac{\alpha}{2}(\mu - H) + \frac{1}{2}I$$

with the proper choice of α guarantees convergence. Observe that at the mth iteration, the density matrix is approximated by

$$P \approx P_m = f_{\mathrm{McW}} \circ f_{\mathrm{McW}} \circ \cdots \circ f_{\mathrm{McW}}(P_0). \tag{5.17}$$

This is a polynomial of the Hamiltonian H and gives a recursive polynomial approximation to the density matrix. Due to the recursion, the degree of the polynomial becomes quite high: in the mth iteration, the polynomial has highest degree monomial x^{3^m} where $2m$ matrix–matrix multiplication is needed. This is useful since our goal is to obtain a polynomial approximation of a Heaviside function (Fermi–Dirac function at zero temperature), hence a high-order polynomial is required to reduce the effect of the Gibbs phenomenon. In fact, the McWeeny purification method can be understood as the Newton–Schultz algorithm for the matrix sign function (Higham 2008), and thus gives a fast iterative scheme for the density matrix.

One drawback of McWeeny purification and the density matrix minimization discussed above is that they require the chemical potential μ as input. Extensions that do not rely on input chemical potential have been developed, such as the canonical purification (Paler and Manolopoulos 1998), trace correcting purification (Niklasson 2002), trace resetting purification (Niklasson, Tymczak and Challacombe 2003) and generalized canonical purification (Truflandier, Dianzinga and Bowler 2016). These algorithms are used in large-scale parallel implementations (Chow, Liu, Smelyanskiy

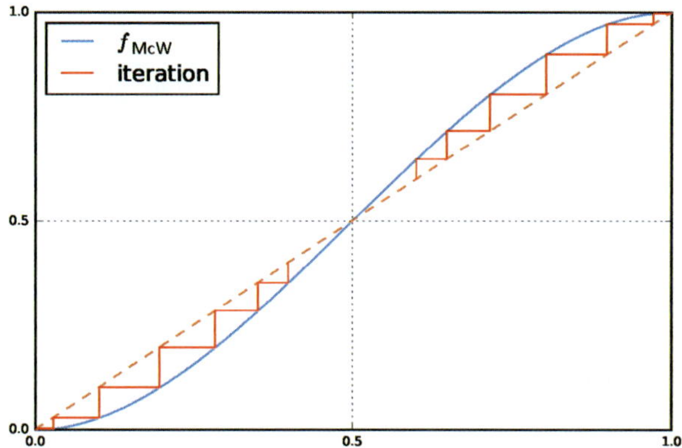

Figure 5.1. Shape of f_{McW} for McWeeny purification.

and Hammond 2015, Dawson and Nakajima 2018). We refer interested readers to Niklasson (2011) for a recent review on these methods.

5.2.5. Fermi operator expansion

Another approach to yielding linear scaling algorithms is the Fermi operator expansion (FOE) method. Consider the density matrix at the finite temperature

$$P = f_\beta(H - \mu). \tag{5.18}$$

The right-hand side is a matrix function with respect to the Hamiltonian matrix H. Instead of diagonalizing H and evaluating the matrix function using the eigen-decomposition, the basic idea of FOE is to expand the Fermi–Dirac function $f_\beta(\cdot)$ into an m-term expansion as

$$f_\beta(\varepsilon) \approx f_{\beta,m}(\varepsilon) = \sum_{n=1}^{m} g_n(\varepsilon). \tag{5.19}$$

The corresponding matrix function approximation is

$$f_\beta(H - \mu) \approx f_{\beta,m}(H - \mu) = \sum_{n=1}^{m} g_n(H - \mu). \tag{5.20}$$

The above formulation is general. We only require each term $g_n(H - \mu)$ to be a simple function, so that the corresponding matrix function can be evaluated directly without diagonalizing the matrix. For instance, g_n can be chosen to be a polynomial function or a rational function.

The approximation error of the matrix form (5.20) is directly related to that of the scalar form (5.19). Given the eigendecomposition

$$H\psi_i = \varepsilon_i \psi_i,$$

we have for any vector v

$$(f_\beta(H-\mu)-f_{\beta,m}(H-\mu))v = \sum_i (f_\beta(\varepsilon_i-\mu)-f_{\beta,m}(\varepsilon_i-\mu))\psi_i\langle\psi_i|v\rangle. \quad (5.21)$$

Thus,

$$\begin{aligned}
\|(f_\beta(H-\mu)&-f_{\beta,m}(H-\mu))v\|_2^2 \\
&= \sum_i |f_\beta(\varepsilon_i-\mu)-f_{\beta,m}(\varepsilon_i-\mu)|^2|\langle\psi_i|v\rangle|^2 \\
&\le \sup_i |f_\beta(\varepsilon_i-\mu)-f_{\beta,m}(\varepsilon_i-\mu)|^2 \sum_i |\langle\psi_i|v\rangle|^2 \\
&= \sup_{t\in\mathrm{spec}(H-\mu)} |f_\beta(t)-f_{\beta,m}(t)|^2\|v\|_2^2.
\end{aligned} \quad (5.22)$$

The equation above can be rewritten as

$$\|f_\beta(H-\mu)-f_{\beta,m}(H-\mu)\|_2 \le \|f_\beta(\cdot)-f_{\beta,m}(\cdot)\|_\infty, \quad (5.23)$$

where the left-hand side is the operator norm for matrices and the right-hand side is the L^∞-norm for scalar functions. Thus the error of the Fermi operator expansion (5.20) will be small as long as the corresponding approximation is small in the sense of expansions for scalar functions (5.19).

One example of FOE is to expand the Fermi–Dirac function into polynomials (Goedecker and Colombo 1994):

$$f_\beta(\varepsilon) \approx \sum_{n=1}^{m} c_n \varepsilon^{n-1}. \quad (5.24)$$

The corresponding matrix function version is

$$f_\beta(H-\mu) \approx \sum_{n=1}^{m} c_n (H-\mu)^{n-1}. \quad (5.25)$$

Note that each term of equation (5.25) is simply a matrix power $(H-\mu)^n$, which can be evaluated using only matrix–matrix multiplication recursively, without diagonalizing the matrix H. In order to implement the FOE (5.25) for a high-order polynomial, it is more efficient and stable to expand $f_\beta(t)$ using Chebyshev polynomials. For insulators, the number of terms needed in equation (5.25) scales as $\log \epsilon^{-1}$ to reach target accuracy ϵ (Trefethen 2008). In particular, the number of terms is independent of the system size and can be treated as a constant. When H is a sparse matrix, this means that the polynomial approximation to $f_\beta(H-\mu)$ is a sparse matrix as well, and the number of non-zeros only scales linearly with respect to the system size. Hence the FOE method (5.25) is a linear scaling algorithm.

Besides expansion using polynomials, another possibility is to approximate the Fermi–Dirac function using rational functions. A rational function can be decomposed into a linear combination of terms of the form $(\varepsilon - z)^{-p}$, where $z \in \mathbb{C}$ and $p \geq 1$. In particular, if all terms use $p = 1$ the resulting expansion is called a simple pole expansion, or just the pole expansion. Compared to polynomial expansion, there are two main advantages in using the rational expansion. First, the number of terms needed for the rational expansion can be much smaller than that needed for polynomial expansion to achieve the same accuracy. This is particularly the case for systems with small gaps. Second, the use of the pole expansion can yield fast algorithms with reduced complexity even for metallic systems. This is called the pole expansion and selected inversion algorithm (PEXSI), which will be discussed in Section 5.3. To our knowledge, PEXSI is so far *the only* algorithm allowing such complexity reduction.

5.3. Pole expansion and selected inversion method

While linear scaling algorithms in principle yield fast algorithms for the evaluation of Kohn–Sham maps, their accuracy often crucially depends on the decay of orbitals or density matrices, and they are usually only suitable for insulating systems with a large gap. Another practical drawback is that they often require user input on support of truncation and other tuning parameters to achieve a balance between efficiency and accuracy. The pole expansion and selected inversion method (PEXSI) (Lin *et al.* 2009*b*, Lin, Chen, Yang and He 2013) is a reduced scaling algorithm with computational scaling at most $O(N^2)$ and smaller for lower-dimensional systems. While it has a worse computational scaling than linear, the PEXSI algorithm can be applied to general systems and gives accurate results.

The PEXSI algorithm is one type of FOE method, and uses the following pole expansion to approximate the Fermi–Dirac distribution:

$$P \approx \sum_{l=1}^{m} \omega_l (H - z_l)^{-1}. \tag{5.26}$$

If the band gap is small or zero, the number of terms needed in order for the polynomial expansion to reach a certain target accuracy ϵ scales as $O(\beta \Delta E)$, where ΔE is the spectral radius of the shifted operator $H - \mu$ (Goedecker 1999). Although the number of terms of a straightforward construction of the pole expansion also scales as $O(\beta \Delta E)$ (Baroni and Giannozzi 1992), it has subsequently been improved to $O((\beta \Delta E)^{1/2})$ (Ozaki 2007), $O((\beta \Delta E)^{1/3})$ (Ceriotti, Kühne and Parrinello 2008), and finally $O(\log(\beta \Delta E))$ (Lin, Lu, Ying and E 2009*a*).

In order to obtain such a pole expansion, one possibility is to use the Cauchy contour integral formulation. Note that the Fermi–Dirac function $f_\beta(\varepsilon) = 1/(1 + e^{\beta \varepsilon})$ is a meromorphic function in $\mathbb{C}$, and the only poles are at $\varepsilon = (2n + 1)\mathrm{i}\pi/\beta, n \in \mathbb{Z}$. Furthermore, $f_\beta(\varepsilon)$ can be expanded using the

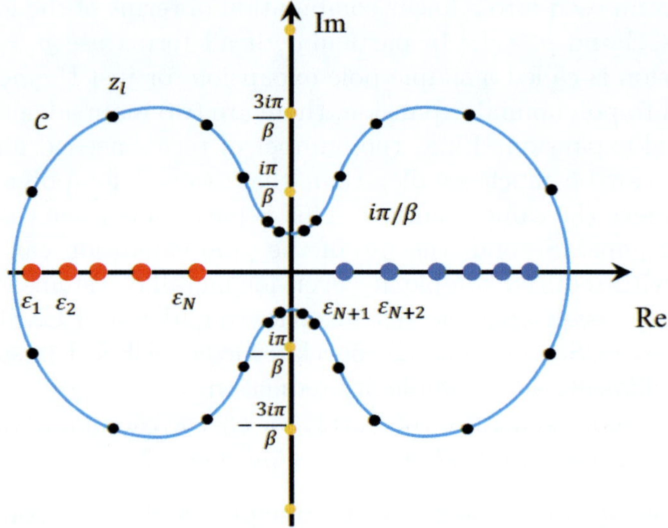

Figure 5.2. Contour integral representation and pole expansion for the density matrix at finite temperature.

following Matsubara expansion (Mahan 2000):

$$f_\beta(\varepsilon) = \frac{1}{2} - \frac{1}{\beta} \sum_{n \in \mathbb{Z}} \frac{1}{\varepsilon - (2n+1)\mathrm{i}\pi/\beta}. \tag{5.27}$$

Note that equation (5.27) only converges conditionally, and the infinite summation must be performed symmetrically with respect to the positive and negative choices of n. The number of terms needed in the direct truncation of the Matsubara series naturally scales as $O(\beta\Delta E)$.

The efficiency of the pole expansion can be improved by using a contour integral formulation:

$$f_\beta(H - \mu) = \frac{1}{2\pi\mathrm{i}} \oint_{\mathscr{C}} f_\beta(z)((z + \mu)I - H)^{-1}\,\mathrm{d}z. \tag{5.28}$$

Here the contour $\mathscr{C}$ should be chosen so that it includes all the (real) eigenvalues of $H - \mu$, without including any poles of $f_\beta(z)$, i.e. $(2n+1)\mathrm{i}\pi/\beta, n \in \mathbb{Z}$. This leads to the 'dumbbell-shaped' contour used in Lin, Lu, Ying and E (2009a) (Figure 5.2). The contour is symmetric with respect to the chemical potential μ. The discretization points are chosen to be denser around μ to resolve the sharp transition of the Fermi–Dirac function at μ. At finite temperature, the contour integral formulation remains well-defined for gapless systems, $i.e.$ $\varepsilon_N = \varepsilon_{N+1}$.

Each term in the pole expansion corresponds to a matrix inverse, or Green's function $(z_l - H)^{-1}$, which can be evaluated directly without diagonalizing the matrix H. Equation (5.26) converts the problem of computing P to the problem of computing m Green's functions. In order to find the Kohn–Sham map, we do not need the entire density matrix P, but only the electron density which corresponds to the diagonal entries of P (again for simplicity we assume that the basis set is orthogonal and the overlap matrix S is an identity matrix). This amounts to the question of finding the diagonal entries of a Green's function. Note that even if H is a sparse matrix, the matrix inverse $G_l := (z_l - H)^{-1}$ can be a fully dense matrix. One naive method is to first evaluate each Green's function and extract its diagonal elements. However, when H is a sparse matrix, the computation of diagonal entries, and more generally the entries of G_l corresponding to the sparsity pattern of H, can be evaluated much more efficiently by means of the selected inversion method (Erisman and Tinney 1975, Lin *et al.* 2009*b*, Lin *et al.* 2011, Jacquelin, Lin and Yang 2016).

Although we assume H to be a general Hermitian matrix throughout the paper, for the simplicity of discussion in this section, we assume H to be a real symmetric matrix. This makes $A = z_l - H \in \mathbb{C}^{N_b \times N_b}$ a complex symmetric, non-singular matrix. For such a matrix, the standard approach to computing A^{-1} is to first decompose A as

$$A = LDL^\top, \tag{5.29}$$

where L is a unit lower triangular matrix and D is a diagonal or a block-diagonal matrix. Equation (5.29) is often known as the $LDL^\top$ factorization of A. Given such a factorization, one can obtain $A^{-1} = (x_1, x_2, \ldots, x_{N_b})$ by solving a number of triangular systems,

$$Ly = e_j, \quad Dw = y, \quad L^\top x_j = w, \tag{5.30}$$

for $j = 1, 2, \ldots, N_b$, where e_j is the jth column of the identity matrix I. The computational cost of such algorithms is generally $O(N_b^3)$. However, when A is sparse, we can exploit the sparsity structure of L and e_j to reduce the complexity of computing selected components of A^{-1}.

The selected inversion algorithm can be heuristically understood as follows (Lin *et al.* 2011). Let A be partitioned into the 2×2 block form

$$A = \begin{pmatrix} \alpha & b^\top \\ b & \widetilde{A} \end{pmatrix}. \tag{5.31}$$

The first step of an $LDL^\top$ factorization produces a decomposition of A that can be expressed by

$$A = \begin{pmatrix} 1 & \\ \ell & I \end{pmatrix} \begin{pmatrix} \alpha & \\ & \widetilde{A} - b\alpha^{-1}b^\top \end{pmatrix} \begin{pmatrix} 1 & \ell^\top \\ & I \end{pmatrix}, \tag{5.32}$$

where α is often referred to as a pivot, $\ell = b\alpha^{-1}$ and $S = \widetilde{A} - b\alpha^{-1}b^\top$ is known as the *Schur complement*. The same type of decomposition can be applied recursively to the Schur complement S until its dimension becomes 1. The product of lower triangular matrices produced from the recursive procedure, which all have the form

$$\begin{pmatrix} I & & \\ & 1 & \\ & \ell^{(i)} & I \end{pmatrix},$$

where $\ell^{(1)} = \ell = b\alpha^{-1}$, yields the final L factor. At this last step the matrix in the middle becomes diagonal, which is the D matrix.

From equation (5.32), A^{-1} can be expressed by

$$A^{-1} = \begin{pmatrix} \alpha^{-1} + \ell^\top S^{-1}\ell & -\ell^\top S^{-1} \\ -S^{-1}\ell & S^{-1} \end{pmatrix}. \tag{5.33}$$

This expression suggests that once α and ℓ are known, the task of computing A^{-1} can be reduced to that of computing S^{-1}. Because a sequence of Schur complements is produced recursively in the $LDL^\top$ factorization of A, the computation of A^{-1} can be organized in a recursive fashion too. Clearly, the reciprocal of the last entry of D is the (N_b, N_b)th entry of A^{-1}. Starting from this entry, which is also the 1×1 Schur complement produced in the $(N_b - 1)$th step of the $LDL^\top$ factorization procedure, we can construct the inverse of the 2×2 Schur complement produced at the $(N_b - 2)$th step of the factorization procedure, using the recipe given by (5.33). This 2×2 matrix is the trailing 2×2 block of A^{-1}. As we proceed from the lower right corner of L and D towards their upper left corner, more and more elements of A^{-1} are recovered. Finally, we may recover all the diagonal entries of A^{-1} *exactly*. In fact, given the factorization $A = LDL^\top$, the selected inversion algorithm can be used to efficiently compute all entries

$$\{A_{i,j}^{-1} : (L + L^\top)_{i,j} \neq 0\}.$$

The validity of the selected inversion algorithm can be verified by the following statement. For any $1 \leq k < N_b$, define

$$\mathcal{C} = \{i : L_{i,k} \neq 0, \ i > k\}. \tag{5.34}$$

Then all entries $\{A_{i,k}^{-1} : i \in \mathcal{C}\}$, $\{A_{k,j}^{-1} : j \in \mathcal{C}\}$, and $A_{k,k}^{-1}$ can be computed using only the L, D factors and

$$\{A_{i,j}^{-1} : (L + L^\top)_{i,j} \neq 0, \ i, j > k\}.$$

Algorithm 8: Selected inversion algorithm based on $LDL^\top$ factorization

---

Input: $LDL^\top$ factorization of a symmetric matrix $A \in \mathbb{C}^{N_b \times N_b}$.
Output: Selected elements of A^{-1}, i.e. $\{A_{i,j}^{-1} : (L + L^\top)_{i,j} \neq 0\}$.

1: Calculate $A_{N_b,N_b}^{-1} \leftarrow (D_{N_b,N_b})^{-1}$.
2: **for** $k = N_b - 1, \ldots, 1$ **do**
3:      Find the collection of indices $\mathcal{C} = \{i \mid i > k, L_{i,k} \neq 0\}$.
4:      Calculate $A_{\mathcal{C},k}^{-1} \leftarrow -A_{\mathcal{C},\mathcal{C}}^{-1} L_{\mathcal{C},k}$.
5:      Calculate $A_{k,\mathcal{C}}^{-1} \leftarrow (A_{\mathcal{C},k}^{-1})^\top$.
6:      Calculate $A_{k,k}^{-1} \leftarrow (D_{k,k})^{-1} - A_{k,\mathcal{C}}^{-1} L_{\mathcal{C},k}$.
7: **end for**

To see why this is the case, consider $\{A_{i,k}^{-1} : i \in \mathcal{C}\}$. As in equation (5.33) we can derive

$$A_{i,k}^{-1} = -\sum_{j=k+1}^{N_b} A_{i,j}^{-1} L_{j,k}, \quad i \in \mathcal{C}. \tag{5.35}$$

If $L_{j,k} \neq 0$, then $A_{i,j}^{-1}$ is needed in the sum. Since we are only interested in computing $A_{i,k}^{-1}$ for $i \in \mathcal{C}$, the i and j indices are constrained to satisfy the conditions $L_{j,k} \neq 0$ and $L_{i,k} \neq 0$. Due to the non-zero fill-in pattern in the $LDL^\top$ factorization, we have $(L + L^\top)_{i,j} \neq 0$ and the statement is proved for $\{A_{i,k}^{-1} : i \in \mathcal{C}\}$. The argument is the same for $\{A_{k,j}^{-1} : j \in \mathcal{C}\}$ due to symmetry. Finally for the diagonal entry, we have

$$A_{k,k}^{-1} = D_{k,k}^{-1} - \sum_{i=k+1}^{N_b} L_{i,k} A_{i,k}^{-1}, \tag{5.36}$$

which can be readily computed given $\{A_{i,k}^{-1} : i \in \mathcal{C}\}$ is available. This proves the statement above.

In order to understand the asymptotic complexity of the selected inversion algorithm, without loss of generality we assume the sparsity pattern of H is similar to that obtained by the second-order central difference discretization of a Laplace operator. The computational cost associated with the $LDL^\top$ factorization, as well as the selected inversion algorithm, scales as $O(N_b)$, $O(N_b^{1.5})$ and $O(N_b^2)$ for one-, two- and three-dimensional systems, respectively (Lin *et al.* 2009*a*). We remark that such a complexity count is robust to changes of the discretization scheme as long as local basis sets are used. Hence for quasi-1D systems (such as nanotubes) and quasi-2D systems (such as monolayer systems and surfaces), the computational cost also scales as $O(N_b)$ and $O(N_b^{1.5})$, respectively. Pseudocode for the selected inversion algorithm is given in Algorithm 8.

In practice, a column-based sparse factorization and selected inversion algorithm as illustrated in Algorithm 8 may not be efficient due to the lack of level 3 BLAS operations. For a sparse matrix A, the columns of A and the L factor can be partitioned into supernodes. A supernode is a set of contiguous columns $\mathcal{J} = \{j, j+1, \ldots, j+s\}$ of the L factor that have the same or similar non-zero sparsity structure below the $(j+s)$th row (Ashcraft and Grimes 1989). This allows use of matrix–matrix multiplications, which can significantly improve the efficiency.

The pole expansion and selected inversion (PEXSI) method (Lin *et al.* 2009*a*, Lin *et al.* 2011, Jacquelin *et al.* 2016) therefore combines the pole expansion and the selected inversion, and evaluates the Kohn–Sham map without solving any eigenvalues or eigenfunctions. The selected inversion method is an exact method if exact arithmetic is used, that is, the only error in the selected inversion method is due to round-off errors. Hence the accuracy of the PEXSI method is determined by the pole expansion, which can be systematically improved by increasing the number of poles m. The PEXSI method is ideally suited to massively parallel computers. The treatment of the poles can be parallelized in a straightforward fashion, with communication only needed at the end to construct the density matrix. The selected inversion method itself can also be massively parallelizable to thousands of processors (Jacquelin *et al.* 2016), and the total number of processors that can be efficiently used by PEXSI can be over 100 000.

The PEXSI software package (available at http://www.pexsi.org, distributed under the BSD license) has now been integrated into electronic structure packages such as BigDFT, CP2K, DGDFT, FHI-aims, QuantumWise ATK and SIESTA, and is part of the 'Electronic Structure Infrastructure' (ELSI) package (Yu *et al.* 2018).

In addition to computing the charge density at a reduced computational complexity in each SCF iteration, we can also use PEXSI to compute energy, free energy and the atomic forces efficiently without diagonalizing the Kohn–Sham Hamiltonian, using *the same set of poles* as those used for computing the charge density (Lin *et al.* 2013). Another numerical issue associated with the PEXSI technique, as well as the Fermi operator expansion techniques in general, is to determine the chemical potential, so that the condition

$$N = N_\beta(\mu) := \mathrm{Tr}[P] \tag{5.37}$$

is satisfied. Note that $N_\beta(\cdot)$ is a non-decreasing function of μ. Hence the chemical potential can be determined via a bisection strategy, or Newton's method. When Newton's method is used, the chemical potential converges rapidly near its correct value. However, the standard Newton's method may not be robust enough when the initial guess is far away from the correct chemical potential. It may give, for example, too large a correction when

$N'_\beta(\mu)$ is close to zero, such as when μ is near the edge or in the middle of a band gap.

One way to overcome the above difficulty is to efficiently approximate the function $N_\beta(\varepsilon)$ to narrow down the region in which the correct μ must lie. This function can be seen effectively as a (temperature smeared) cumulative density of states, counting the number of eigenvalues in the interval $(-\infty, \varepsilon)$. We can evaluate such a zero temperature limit, denoted by $N_\infty(\varepsilon)$, again without computing any eigenvalues of H. Instead, we perform a matrix decomposition of the shifted matrix $H - \varepsilon = LDL^\top$, where L is unit lower triangular and D is diagonal. It follows from Sylvester's law of inertia (Sylvester 1852) that the inertia (the number of negative, zero and positive eigenvalues) of a real symmetric matrix does not change under a congruent transform, that D has the same inertia as that of $H - \varepsilon$. Hence, we can obtain $N_\infty(\varepsilon)$ by simply counting the number of negative entries in D. Note that the matrix decomposition $H - \varepsilon = LDL^\top$ can be computed efficiently by using a sparse $LDL^\top$ or LU factorization in real arithmetic. It requires fewer floating point operations than the complex arithmetic direct sparse factorization used in PEXSI.

We apply the parallel PEXSI method to the DG_Graphene_2048 and DG_Graphene_8192 systems, which are disordered graphene systems with 2048 and 8192 atoms, respectively, and compare its performance with a standard approach that requires a partial diagonalization of (H, S). We use a ScaLAPACK subroutine pdsyevr (Vömel 2010), which is based on the multiple relatively robust representations (MRRR) algorithm, to perform the diagonalization.

Figure 5.3 shows that for graphene problems with 2048 and 8192 processors, the PEXSI technique is nearly two orders of magnitude faster than the ScaLAPACK routine pdsyevr, and can be scalable to a much larger number of processors. The advantage of PEXSI becomes even clearer for a disordered graphene system with 32 768 atoms (see Figure 5.4). For this case, the diagonalization routine is no longer feasible, while the time to solution for the PEXSI technique can be as small as 241 s.

Remarks

Unlike most linear scaling algorithms discussed in this section, the PEXSI method does not rely on the decay properties of the density matrix. Hence the efficiency of PEXSI is roughly the same for metallic, semiconducting and insulating systems. For insulating systems with a relatively large energy gap, PEXSI may become slightly more efficient due to the potentially smaller number of poles needed to approximate the density matrix. The computational cost of PEXSI scales as $O(N_b^\alpha)$, where $\alpha = 1, 3/2, 2$ for quasi-1D, 2D and 3D systems, respectively, where N_b is the number of basis functions.

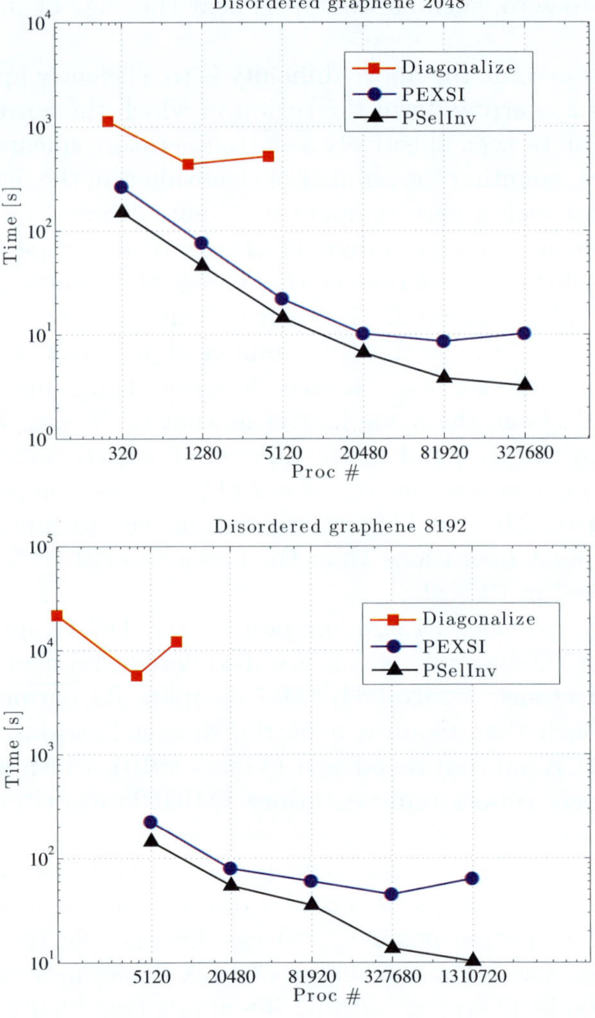

Figure 5.3. (Credit: Jacquelin, Lin and Yang 2016.) Wall clock time versus number of cores for a graphene system.

Compared to dense eigensolvers which scale as $O(N_b^3)$, the PEXSI method can be highly advantageous for large systems, as demonstrated in numerical results above. On the other hand, recall that iterative diagonalization methods discussed in Section 4.2 scale as $O(N_b N^2)$; we find that the cross-over point between PEXSI and iterative diagonalization methods is roughly at $N \sim N_b^{(\alpha-1)/2}$. This implies that PEXSI is most efficient when small basis sets such as Gaussian-type orbitals are used, and becomes less efficient for large basis sets such as finite elements. This statement is in fact true for most linear scaling methods discussed in Section 5.2.

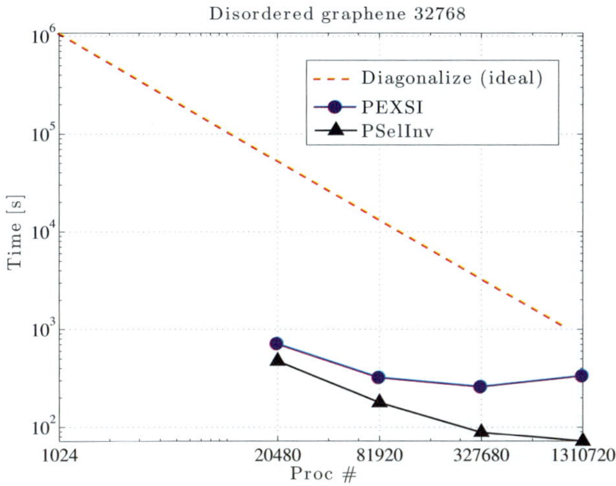

Figure 5.4. (Credit: Jacquelin, Lin and Yang 2016.) Wall clock time versus number of cores for a graphene system with 32 768 atoms.

In the discussion above, we have assumed that H is a real symmetric matrix, and hence $A = z_l - H$ is complex symmetric. This assumption is valid for molecular systems, as well as Γ point sampling of periodic systems. In a more general situation, such as in the context of **k**-point sampling for periodic systems, or when the system is magnetic, H is a Hermitian matrix, and $A = z_l - H$ is merely a *structurally symmetric* matrix. In such a case, the $LDL^\top$ factorization should be replaced by the LU factorization, and the non-symmetric version of the selected inversion algorithm should be used instead (Jacquelin, Lin and Yang 2018). Besides PEXSI, general-purpose selected inversion algorithms are also available in other software packages such as PARDISO (Schenk and Gartner 2006) and MUMPS (Amestoy, Duff, L'Excellent and Koster 2001). The selected inversion-type algorithms are also used in other contexts such as quantum transport calculations (Li, Ahmed, Klimeck and Darve 2008, Petersen *et al.* 2009).

Recently it has been shown that the pre-constant of the logarithmic scaling factor can be further reduced to be numerically near-optimal (Moussa 2016). This makes the pole expansion highly efficient for metallic systems, and the number of poles needed in practical calculations is typically only $10 \sim 40$.

During the self-consistency field iteration, the zero temperature limit $N_\infty(\varepsilon)$ evaluated from the inertia-counting procedure may be used to construct upper and lower bounds on the chemical potential. Then, coupled with a strategy to evaluate $N_\beta(\varepsilon)$ accurately using PEXSI over multiple energy points ε, we may accurately determine the chemical potential within the window specified by the inertia-counting procedure. In particular, it is possible to perform PEXSI calculations over multiple energy points *only once per SCF iteration*, without sacrificing accuracy at convergence (Jia and Lin 2017).

6. Evaluation of the Kohn–Sham map: non-local functional

Kohn–Sham DFT calculations with non-local functionals, such as rung-4 functionals (hybrid functionals) and rung-5 functionals, can be considerably more costly than calculations with exchange-correlation functionals from the first three rungs of the ladder. More specifically, Kohn–Sham equations with local and semi-local functionals can be viewed as eigenvalue problems corresponding to differential operators. When a rung-4 functional is used, the Kohn–Sham Hamiltonian operator becomes an integro-differential operator due to the Fock exchange term. For rung-5 functionals, the self-consistency is computationally rather challenging, and most calculations are done as a post-processing step to obtain the correlation energy as a perturbation to the self-consistent solution for a semi-local functional.

Throughout the discussion in this section, we assume that a large basis set such as planewaves or real-space representation is used. We first discuss numerical methods for evaluating the Fock exchange operator in Section 6.1. We introduce an acceleration technique called the adaptive compression method in Section 6.2. We discuss a method for evaluating the RPA correlation energy as an example of the rung-5 functionals in Section 6.3.

6.1. Fock exchange operator

The Fock exchange operator $V_{\mathrm{x}}^{\mathrm{EX}}[P]$ is an integral operator, introduced in equation (2.52), recalled here for readers' convenience:

$$V_{\mathrm{x}}^{\mathrm{EX}}[P](\mathbf{r}, \mathbf{r}') = -P(\mathbf{r}, \mathbf{r}')K(\mathbf{r}, \mathbf{r}'),$$

where

$$P(\mathbf{r}, \mathbf{r}') = \sum_{i=1}^{N} \psi_i(\mathbf{r})\psi_i^*(\mathbf{r}').$$

Here we assume the system is an insulating system for simplicity. The kernel of $V_{\mathrm{x}}^{\mathrm{EX}}[P]$ is not of low rank due to the Hadamard product (*i.e.* element-wise product) between the kernels of P and K. When a small basis set $\Phi = [\phi_1, \ldots, \phi_{N_b}]$ is used, each occupied orbital is expanded as

$$\psi_i(\mathbf{r}) = \sum_p \phi_p(\mathbf{r})c_{p,i}, \quad i = 1, \ldots, N. \tag{6.1}$$

The matrix elements of the Fock exchange operator become

$$
\begin{aligned}
(V_{\mathrm{x}}^{\mathrm{EX}}[P])_{pq} &= -\int \phi_p^*(\mathbf{r}) P(\mathbf{r}, \mathbf{r}') K(\mathbf{r}, \mathbf{r}') \phi_q(\mathbf{r}')\, \mathrm{d}\mathbf{r}\, \mathrm{d}\mathbf{r}' \\
&= -\sum_{i=1}^{N} \sum_{r,s} c_{r,i} c_{s,i}^* \int \phi_p^*(\mathbf{r}) \phi_r(\mathbf{r}) \phi_s^*(\mathbf{r}') \phi_q(\mathbf{r}') K(\mathbf{r}, \mathbf{r}')\, \mathrm{d}\mathbf{r}\, \mathrm{d}\mathbf{r}' \\
&:= -\sum_{i=1}^{N} \sum_{r,s} c_{r,i} c_{s,i}^* \langle ps | rq \rangle.
\end{aligned}
\tag{6.2}
$$

Equation (6.2) defines the two-electron integral tensor $\langle ps | rq \rangle$, which is a fourth-order tensor. Hence the computational cost of constructing the Fock exchange matrix $(V_{\mathrm{x}}^{\mathrm{EX}}[P])_{pq}$ typically scales as $O(N^4)$.

It turns out that when a large basis set is used, the asymptotic complexity associated with the Fock exchange operator can be reduced to $O(N^3)$ *without approximation*. Note that in such a case, it is often prohibitively expensive to explicitly construct or to store the Fock exchange operator directly. It is only viable to apply it to an occupied orbital ψ_j as

$$
(V_{\mathrm{x}}^{\mathrm{EX}}[P]\psi_j)(\mathbf{r}) = -\sum_{i=1}^{N} \varphi_i(\mathbf{r}) \int K(\mathbf{r}, \mathbf{r}') \varphi_i^*(\mathbf{r}') \psi_j(\mathbf{r}')\, \mathrm{d}\mathbf{r}',
\tag{6.3}
$$

for

$$
P = \sum_{i=1}^{N} |\varphi_i\rangle \langle \varphi_i|.
$$

Here we deliberately distinguish the orbitals in the density matrix ($\{\varphi_i\}$) and the orbitals $V_{\mathrm{x}}^{\mathrm{EX}}$ acts on ($\{\psi_j\}$) to emphasize that they may correspond to different density matrices before self-consistency is achieved. The integral $\int K(\mathbf{r}, \mathbf{r}') \varphi_i^*(\mathbf{r}') \psi(\mathbf{r}')\, \mathrm{d}\mathbf{r}'$ can be implemented by solving N Poisson-like equations. Let N_g denote the grid size, and the computational cost of solving each Poisson-like equation scales as $O(N_g \log N_g)$ when fast Fourier transform can be used. When applied to all pairs of occupied orbitals, the computational cost scales as $O(N_g \log N_g N^2)$ to $O(N^3)$. Note that the preconstant of this cubic scaling component can be very large. In practical Hartree–Fock calculations, the application of the Fock exchange operator can often take more than 95% of the overall computational time.

The Hartree–Fock-like equations require the density matrix P to be computed self-consistently. A common strategy is to solve the linearized Hartree–Fock equation by fixing the density matrix P so that $H[P]$ becomes a fixed operator. Then one solves a nonlinear fixed-point problem to obtain the self-consistent P. The most time-consuming step is to solve the linearized

Hartree–Fock equation. We will describe the strategy for achieving self-consistency in Section 7.4.

In order to reduce the computational cost of evaluating the Fock exchange operator, one strategy is to compute in parallel with a large number of cores. Note that the N^2 Poisson-like equations associated with the pair products $\varphi_i^*(\mathbf{r})\psi_j(\mathbf{r})$ are independent of each other. Hence one may employ in principle $O(N^2)$ processors to distribute the work (Duchemin and Gygi 2010, Valiev *et al.* 2010). The disadvantage of this approach is that the rest of the components of a Kohn–Sham solver often cannot scale to such a large number of processors, and hence the excessive number of computational cores may not be used efficiently.

In Section 4.3, we have introduced localization techniques to find localized representations of the Kohn–Sham subspace via the density matrix and Wannier functions. These techniques can be used to reduce the computational cost of evaluating the Kohn–Sham map as in Section 5. They can also be used to reduce the cost of applying exchange operators to $O(N)$. More specifically, let $\{\varphi_i\}$, $\{\psi_j\}$ be exponentially localized Wannier functions after a proper rotation operation, and the pair product $\varphi_i^*(\mathbf{r})\psi_j(\mathbf{r})$ would vanish if the support sets of φ_i and ψ_j do not overlap with each other. This reduces the number of pairs from $O(N^2)$ to $O(N)$. Furthermore, each Poisson-like equation only needs to be solved on a computational domain that is independent of the system size. Taking both factors into account, we find that the cost of equation (6.3) is reduced to $O(N)$. We refer readers to Wu, Selloni and Car (2009) and Dawson and Gygi (2015), for example, for more details of linear scaling hybrid functional calculations. As in the discussion in Section 5.2, linear scaling algorithms typically have a large prefactor, and hence they become advantageous to the cubic scaling methods only for systems of relatively large sizes.

On the other hand, for metallic and semi-conducting systems, the decay rate of the density matrix along the off-diagonal direction can be very slow. This may also introduce a significantly larger number of non-zero entries in the Hamiltonian matrix. Hence Green's function-based methods, such as the PEXSI method, become less efficient in the context of hybrid functional calculations.

However, the interpolative separable density fitting (ISDF) (Lu and Ying 2015) technique introduced in Section 4.4 can be applicable to insulating, semiconducting and metallic systems. Following the ISDF decomposition equation (4.13), the Poisson-like equation only needs to be solved with $\{\zeta_\mu(\mathbf{r})\}$ on the right-hand side. This reduces the number of Poisson-like equations from $O(N^2)$ to $N_\mu \sim O(N)$. Of course, the solutions of these equations need to be reassembled through linear algebra operations to obtain $\{V_{\mathrm{x}}^{\mathrm{EX}}[P]\psi_j\}_{j=1}^N$. The cost of these linear algebra steps is still $O(N^3)$ but the preconstant can be effectively reduced (Hu *et al.* 2017*a*, Dong *et al.* 2018). Hence ISDF is effective when it is relatively costly to solve Poisson-like equations, such as in the context of planewave methods, and even more so for finite difference and finite element methods.

6.2. Adaptive compression method

When a large basis set is used, the standard method for solving Hartree–Fock-like equations self-consistently is the two-level nested SCF method in Section 7.4. The motivation for this method is that the Fock exchange energy is only a small fraction (usually 5% or less) of the total energy, although such a contribution is sufficient to result in qualitatively different results for the study of many chemical systems. The main idea is then to separate the self-consistent field (SCF) iteration into two sets of SCF iterations. In the *inner SCF iteration*, the exchange operation $V_{\mathrm{x}}^{\mathrm{EX}}$ defined by a set of orbitals $\{\varphi_i\}_{i=1}^N$ is fixed, and the Hamiltonian operator depends only on the density $\rho(\mathbf{r})$. This allows the use of efficient charge mixing schemes to reduce the number of SCF iterations. In other words, the SCF iteration then proceeds as in Kohn–Sham DFT calculations with a fixed exchange operator. In the *outer SCF iteration*, the orbitals defining the exchange operator are updated via a fixed-point iteration. Furthermore, each inner SCF iteration requires the solution of a linear eigenvalue problem, which is to be performed with iterative diagonalization methods discussed in Section 4.2 and requires repeated application of the Hamiltonian operator to occupied orbitals. The application of the Fock exchange operator to occupied orbitals according to equation (6.3) appears in this innermost loop, which makes electronic structure calculations with rung-4 functionals much more costly than those with rung-1 to rung-3 functionals.

The adaptively compressed exchange operator (ACE) method (Lin 2016) accelerates hybrid functional calculations by reducing the frequency of applications of the Fock exchange operator without compromising the accuracy of the self-consistent solution. This is achieved by constructing a low-rank surrogate operator, denoted by $\widetilde{V}_{\mathrm{x}}^{\mathrm{EX}}$, to approximate the Fock exchange operator $V_{\mathrm{x}}^{\mathrm{EX}}$. Note that $V_{\mathrm{x}}^{\mathrm{EX}}$ is generally a dense, full-rank operator. Hence the low-rank surrogate cannot be expected to be accurate when applied to an arbitrary vector. Instead we only require $\widetilde{V}_{\mathrm{x}}^{\mathrm{EX}}$ to be accurate when applied to all occupied orbitals.

More specifically, for a given set of orbitals $\{\varphi_i\}_{i=1}^N$ defining the density matrix implicitly and thus the exchange operator, we first compute the application of the exchange operator to $\{\varphi_i\}_{i=1}^N$ using

$$W_i(\mathbf{r}) = (V_{\mathrm{x}}^{\mathrm{EX}}[\{\varphi\}]\varphi_i)(\mathbf{r}), \quad i = 1, \ldots, N. \tag{6.4}$$

The adaptively compressed exchange operator, denoted by $\widetilde{V}_{\mathrm{x}}^{\mathrm{EX}}$, should satisfy the conditions

$$(\widetilde{V}_{\mathrm{x}}^{\mathrm{EX}}\varphi_i)(\mathbf{r}) = W_i(\mathbf{r}) \quad \text{and} \quad \widetilde{V}_{\mathrm{x}}^{\mathrm{EX}}(\mathbf{r}, \mathbf{r}') = (\widetilde{V}_{\mathrm{x}}^{\mathrm{EX}}(\mathbf{r}', \mathbf{r}))^*. \tag{6.5}$$

The choice of this surrogate operator is not unique. One possible choice satisfying the conditions (6.5) is given by

$$\widetilde{V}_{\mathrm{x}}^{\mathrm{EX}}(\mathbf{r}, \mathbf{r}') = \sum_{i,j=1}^{N} W_i(\mathbf{r}) B_{ij} W_j^*(\mathbf{r}'), \tag{6.6}$$

where $B = M^{-1}$ is a negative definite matrix, and

$$M_{kl} = \int \varphi_k^*(\mathbf{r}) W_l(\mathbf{r}) \, \mathrm{d}\mathbf{r}.$$

Perform Cholesky factorization for $-M$, *i.e.* $M = -LL^*$, where L is a lower triangular matrix. Then we get $B = -L^{-*}L^{-1}$. Define the projection vector in the ACE formulation as

$$\xi_k(\mathbf{r}) = \sum_{i=1}^{N} W_i(\mathbf{r})(L^{-*})_{ik}. \tag{6.7}$$

Then the adaptively compressed exchange operator is given by

$$\widetilde{V}_{\mathrm{x}}^{\mathrm{EX}}(\mathbf{r}, \mathbf{r}') = -\sum_{k=1}^{N} \xi_k(\mathbf{r})\xi_k(\mathbf{r}'). \tag{6.8}$$

The ACE can be readily used to reduce the computational cost of the exchange energy, without the need to solve any extra Poisson equations, using

$$E_{\mathrm{x}}^{\mathrm{EX}} = \frac{1}{2} \sum_{i=1}^{N} \iint \psi_i^*(\mathbf{r})\widetilde{V}_{\mathrm{x}}^{\mathrm{EX}}(\mathbf{r}, \mathbf{r}')\psi_i(\mathbf{r}') \, \mathrm{d}\mathbf{r} \, \mathrm{d}\mathbf{r}'$$

$$= -\frac{1}{2} \sum_{i,k=1}^{N} \left| \int \psi_i^*(\mathbf{r})\xi_k(\mathbf{r}) \, \mathrm{d}\mathbf{r} \right|^2. \tag{6.9}$$

Note that once ACE is constructed, the cost of applying $\widetilde{V}_{\mathrm{x}}^{\mathrm{EX}}$ to any orbital ψ is similar to the application of a non-local pseudopotential operator, thanks to its low-rank structure. ACE only needs to be constructed once per outer iteration, and can be repeatedly used for all the subsequent inner SCF iterations for the electron density, and each iterative step for solving the linear eigenvalue problem. The ACE formulation has been integrated in electronic structure software packages such as Quantum ESPRESSO.

Table 6.1 demonstrates the accuracy of the ACE formulation for Kohn–Sham DFT calculations with the HSE06 (Heyd *et al.* 2003) hybrid functional, for silicon systems ranging from 64 to 1000 atoms. The ACE formu-

Table 6.1. Comparison between the conventional hybrid DFT calculations and ACE-enabled hybrid DFT calculations in terms of the HF energy E_x^{EX} (Hartree) and the energy gap E_{gap} (Hartree) for the Si_{64}, Si_{216}, Si_{512} and Si_{1000} systems. The corresponding relative errors of the HF energy are shown in parentheses.

| Methods | ACE HSE06 | | Conventional HSE06 | |
Systems	E_x^{EX}	E_{gap}	E_x^{EX}	E_{gap}
Si_{64}	-13.541616 (10^{-6})	1.488335	-13.541629	1.488352
Si_{216}	-45.471192 (10^{-7})	1.449790	-45.471190	1.449790
Si_{512}	-107.698011 (10^{-7})	1.324901	-107.698016	1.324902
Si_{1000}	-210.300628 (10^{-6})	1.289162	-210.300524	1.289128

Table 6.2. Comparison between the conventional hybrid DFT calculation and an ACE-enabled hybrid DFT calculation in terms of the number of inner SCF iterations and wall clock time spent in each outer SCF iteration for Si_{1000} on 2000 cores.

| Methods | ACE HSE06 | | Conventional HSE06 | |
#Outer SCF	#Inner SCF	Time (s)	#Inner SCF	Time (s)
1	6	356	6	2518
2	5	320	5	2044
3	5	308	4	1665

lation can evaluate the HF energy and energy gap accurately (total energy difference is under 10^{-4} Hartree) even for large systems, and the remaining difference is in fact due to the tolerance of the SCF iteration. Table 6.2 demonstrates that ACE can perform hybrid functional calculations at a fraction of the cost of conventional methods (Hu *et al.* 2017*c*).

The efficiency of the ACE formulation also rests on the assumption that the magnitude of the Fock exchange operator is relatively small compared to other components of the Hamiltonian. To be more specific, we demonstrate below that in the context of linearized Hartree–Fock-like equations, the ACE formulation leads to desirable convergence properties compared to standard iterative solvers.

To make the discussion more general, we consider the following linear eigenvalue problem (linearized Hartree–Fock-like equations take this form):

$$(A + B)v_i = \lambda_i v_i, \quad i = 1, \ldots, N. \tag{6.10}$$

Here $A, B \in \mathbb{C}^{N_b \times N_b}$ are Hermitian matrices. The eigenvalues $\{\lambda_i\}$ are real and ordered non-decreasingly. Due to the Pauli exclusion principle we need to compute the eigenpairs (λ_i, v_i) corresponding to the lowest N eigenvalues, which are assumed to be separated from the rest of the eigenvalues by a positive spectral gap $\lambda_g := \lambda_{N+1} - \lambda_N$. The matrix A in (6.10) is obtained by discretizing the Hamiltonian operator excluding the exchange operator, while B as a discretization of the exchange operator is negative definite.

In order to reduce the number of matrix–vector multiplication operations Bv, the simplest idea is to fix $w_i := Bv_i$ at some stage, and to replace Bv_i with w_i for a number of iterations. This leads to the following subproblem:

$$Av_i + w_i = \lambda_i v_i, \quad i = 1, \ldots, N. \tag{6.11}$$

Note that equation (6.11) is not an eigenvalue problem: if v_i is a solution to (6.11), then v_i multiplied by a constant c is typically not a solution. Equation (6.11) could be solved using optimization-based methods, but such a problem is typically more difficult than a Hermitian eigenvalue problem. In practice, software packages for solving Hartree–Fock-like equations are typically built around eigensolvers, which is another important factor that makes subproblem (6.11) undesirable.

The adaptive compression method re-uses the information in $\{w_i\}$ in a different way, retaining the structure of the eigenvalue problem (6.10). Define $V = [v_1, \ldots, v_N]$, $W = [w_1, \ldots, w_N]$, so $V, W \in \mathbb{C}^{N_b \times N}$, and construct

$$\widetilde{B}[V] = W(W^*V)^{-1}W^*. \tag{6.12}$$

Since $B \prec 0$, $W^*V \equiv V^*BV$ has only negative eigenvalues and is invertible. $\widetilde{B}[V]$ is Hermitian of rank N, and agrees with B when applied to V as

$$\widetilde{B}[V]V = W(W^*V)^{-1}W^*V = W = BV. \tag{6.13}$$

We shall refer to the operation from B to $\widetilde{B}[V]$ as an *adaptive compression*. It turns out that the adaptive compression $\widetilde{B}[V]$ is the unique rank-N Hermitian matrix that agrees with B on span V. Furthermore, $B \preceq \widetilde{B}[V] \preceq 0$ (Lin and Lindsey 2019).

Note that the subspace span V is precisely the solution for (6.10) and is not known *a priori*. Therefore $\widetilde{B}$ needs to be constructed in an adaptive manner. Starting from some initial guess $V^{(0)}$, we will obtain a sequence $V^{(k)}$ and corresponding compressed operators $\widetilde{B}[V^{(k)}]$. More specifically, ACE uses a fixed-point iteration given by

$$(A + \widetilde{B}[V^{(k)}])v_i^{(k+1)} = \lambda_i^{(k+1)} v_i^{(k+1)}, \quad i = 1, \ldots, N. \tag{6.14}$$

In each iteration, after $\widetilde{B}[V^{(k)}]$ is constructed, (6.14) can be solved via *any* iterative eigensolver to obtain $V^{(k+1)}$. The iterative eigensolver only requires application of A and the low-rank matrix $\widetilde{B}$ to vectors, and does not require any additional application of B until $V^{(k+1)}$ is obtained. If span $V^{(k)}$

Algorithm 9: Adaptive compression method for solving linear eigenvalue problem

---

1: Initialize $V^{(0)}$ by solving $Av_i^{(0)} = \lambda_i^{(0)} v_i^{(0)}, \quad i = 1, \ldots, N$.
2: **while** convergence not reached **do**
3:     Compute $W^{(k)} = BV^{(k)}$.
4:     Evaluate $[(W^{(k)})^* V^{(k)}]^{-1}$ to construct $\widetilde{B}[V^{(k)}]$ implicitly.
5:     Solve (6.14) to obtain $V^{(k+1)}$.
6:     Set $k \leftarrow k + 1$.
7: **end while**

---

converges to span V, then the consistency condition $\widetilde{B}[V]V = BV$ is satisfied, and the adaptive compression method is numerically exact. The adaptive compression method for solving the linear eigenvalue problem (6.10) is given in Algorithm 9, where we initialize $V^{(0)}$ by solving the eigenvalue problem in the absence of B.

At first glance, the advantage of converting a linear eigenvalue problem (6.10) into a nonlinear eigenvalue problem (6.14) is unclear. The advantage of the adaptive compression method comes from the decoupling of the matrix–vector multiplication operations Av and Bv, and asymptotically the number of Bv operations is independent of the spectral radius $\|A\|_2$. More specifically, starting from an initial density matrix $P^{(0)}$, the asymptotic convergence rate measured by the convergence of the density matrix at the kth iteration is given by

$$\|P - P^{(k)}\|_2 \lesssim \gamma^k \|P - P^{(0)}\|_2, \quad \text{where} \quad \gamma \leq \frac{\|B\|_2}{\|B\|_2 + \lambda_g}. \tag{6.15}$$

Furthermore, one may prove that the adaptive compression method converges globally starting from *almost every* initial guess. We refer readers to Lin and Lindsey (2019) for more details.

Remarks

An alternative way to adaptively reduce the rank of the exchange operator is via a projector-based compression of the exchange operator (Duchemin and Gygi 2010, Boffi, Jain and Natan 2016). Compared to the discussion in Section 2.7, we can also find that the Kleinman–Bylander form of the pseudopotential follows the same spirit as that in the ACE formulation to construct a low-rank approximation to the semi-local pseudopotential. However, in the case of the pseudopotential, the orbitals are fixed and are not computed adaptively. The adaptive compression method is also related to the Nyström sketching method in numerical optimization, and can be used to accelerate the convergence rate of optimization-based electronic structure solvers as well (Hu *et al.* 2018).

6.3. RPA correlation energy

After hybrid functionals that involve the exact exchange operator, the next family of approximations to the exchange-correlation functional involves virtual orbitals, such as the random phase approximation to the correlation functional discussed in Section 2.5; see (2.49). The approximate functional form on the fifth rung is still under active development and thus the efficient numerical algorithm for such functionals is also an active research field with many recent ideas and on-going work. It is beyond the scope of this work to review all these developments, and we will restrict our focus to recent work on cubic scaling algorithms for the computation of RPA correlation energy (Lu and Thicke 2017a) based on the interpolative separable density fitting, as discussed in Section 4.4.

Several cubic scaling methods to calculate the RPA correlation energy have been developed. Recall that the dynamic polarizability operator is defined as

$$\hat{\chi}^0(\mathbf{r}, \mathbf{r}', \mathrm{i}\omega) = \sum_{i}^{\mathrm{occ}} \sum_{a}^{\mathrm{vir}} \frac{\psi_i^*(\mathbf{r})\psi_a(\mathbf{r})\psi_a^*(\mathbf{r}')\psi_i(\mathbf{r}')}{\varepsilon_i - \varepsilon_a - \mathrm{i}\omega} + \mathrm{c.c.} \qquad (6.16)$$

The general idea is to split up the i and a dependence in the computation of $\hat{\chi}^0$ by introducing a new integral. Two different integrals have been utilized for this purpose. The first is

$$-\int_0^\infty \mathrm{e}^{\varepsilon_i t}\, \mathrm{e}^{-\varepsilon_a t}\, \mathrm{e}^{-\mathrm{i}\omega t}\, \mathrm{d}t = \frac{1}{\varepsilon_i - \varepsilon_a - \mathrm{i}\omega}, \qquad (6.17)$$

where $\varepsilon_a > \varepsilon_i$ since the former corresponds to a virtual orbital and the latter to an occupied orbital. Using this integral, one separates the dependence of i and a in (6.16) into a product of exponentials inside the integral. This leads to the Laplace transform cubic scaling methods. This idea was first applied to RPA calculations by Kaltak, Klimeš and Kresse (2014a, 2014b), using a projector augmented wave (PAW) basis. It was later extended to atomic orbitals (Schurkus and Ochsenfeld 2016, Luenser, Schurkus and Ochsenfeld 2017, Wilhelm, Seewald, Del Ben and Hutter 2016). Another integral decomposition that can be used to break up the i and a dependence is

$$\frac{1}{2\pi\mathrm{i}} \oint_{\mathscr{C}} \frac{1}{(\lambda - \varepsilon_i + \mathrm{i}\omega)(\lambda - \varepsilon_a)}\, \mathrm{d}\lambda = \frac{1}{\varepsilon_i - \varepsilon_a - \mathrm{i}\omega}, \qquad (6.18)$$

where $\mathscr{C}$ is a positively oriented closed contour that encloses $\varepsilon_i - \mathrm{i}\omega$, but not ε_a: an example is shown in Figure 6.1. This idea was first used in the context of cubic scaling RPA in Moussa (2014), and combined with the ISDF approach in Lu and Thicke (2017a) to further reduce the computational cost by taking advantage of the 'low-rank' nature of $\hat{\chi}^0$. If the usual density fitting approximation is used, the $\hat{\chi}^0$ cannot be constructed in $O(N^3)$ since

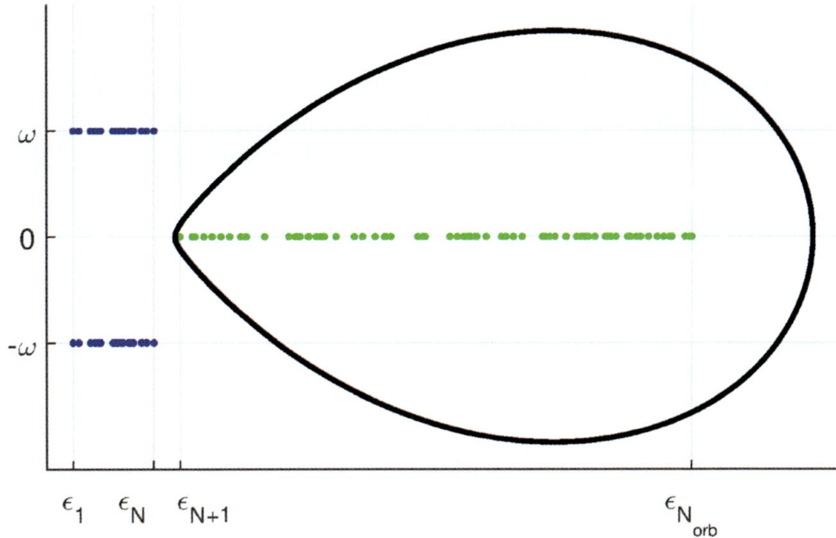

Figure 6.1. (Credit: Lu and Thicke 2017a.) An example of contour $\mathscr{C}$. The blue points represent $\{\varepsilon_i \pm \mathrm{i}\omega\}_{i=1}^{N}$ (for a particular choice of ω) and the green points represent $\{\varepsilon_a\}_{a=N+1}^{N_{\mathrm{orb}}}$.

i and a are coupled in the coefficients of the density fitting method. The solution to this is provided by the interpolative separable density fitting (ISDF) method (Lu and Ying 2015) which keeps the dependence on i and a separate. We remark that density fitting has been widely used to improve the computational efficiency in electronic structure theory, including rung-5 functionals (Feyereisen, Fitzgerald and Komornicki 1993, Weigend, Häser, Patzelt and Ahlrichs 1998, Ren $et\ al.$ 2012a); the ISDF has the additional benefit of keeping indices separated to make it more flexible.

After we choose the contour $\mathscr{C}$ and apply Cauchy's integral formula to split up the dependence of i and a in the denominator of (6.16), we obtain a reformulation for $\hat{\chi}^0$ which can be computed in cubic time complexity,

$$\langle \mathbf{r}|\hat{\chi}^0(\mathrm{i}\omega)|\mathbf{r}'\rangle = \sum_{i}^{\mathrm{occ}} \sum_{a}^{\mathrm{vir}} \frac{\psi_i^*(\mathbf{r})\psi_a(\mathbf{r})\psi_a^*(\mathbf{r}')\psi_i(\mathbf{r}')}{\varepsilon_j - \varepsilon_k - \mathrm{i}\omega} + \mathrm{c.c.} \tag{6.19}$$

$$= \frac{1}{2\pi\mathrm{i}} \oint_{\mathscr{C}} \left(\sum_{i}^{\mathrm{occ}} \frac{\psi_i^*(\mathbf{r})\psi_i(\mathbf{r}')}{\lambda - \varepsilon_i + \mathrm{i}\omega} \right) \left(\sum_{a}^{\mathrm{vir}} \frac{\psi_a(\mathbf{r})\psi_a^*(\mathbf{r}')}{\lambda - \varepsilon_a} \right) \mathrm{d}\lambda + \mathrm{c.c.}$$

Similar to that discussed in the case of PEXSI (see Section 5.3), the contour integral can be discretized and the number of quadrature points is logarithmic in $(\varepsilon_{N_{\mathrm{orb}}} - \varepsilon_N)/(\varepsilon_{N+1} - \varepsilon_N)$; we refer readers to Lu and Thicke (2017a) for details.

Note that the formula (6.19) already provides a cubic scaling method for calculating $\chi^0(i\omega)$. In particular, ignoring logarithmic factors, $\chi^0(i\omega)$ can be calculated with cost $O(N_{\mathrm{orb}} N_g^2)$, assuming we keep N_{orb} total orbitals in the summation (known as the virtual orbital cut-off). However, the number of grid points N_g could be much larger than the number of orbitals N_{orb}. This motivates the use of ISDF to reduce the prefactor in the computational cost. We approximate the $\hat\chi^0$ operator by an operator $\tilde\chi^0$ by using the ISDF approximation:

$$
\langle \mathbf{r} | \hat\chi^0(i\omega) | \mathbf{r}' \rangle
$$
$$
= \frac{1}{2\pi i} \oint_{\mathscr{C}} \left(\sum_i^{\mathrm{occ}} \frac{\psi_i^*(\mathbf{r})\psi_i(\mathbf{r}')}{\lambda - \varepsilon_i + i\omega} \right) \left(\sum_a^{\mathrm{vir}} \frac{\psi_a(\mathbf{r})\psi_a^*(\mathbf{r}')}{\lambda - \varepsilon_a} \right) d\lambda + \text{c.c.}
$$
$$
\approx \sum_{\mu\nu} \frac{1}{2\pi i} \oint_{\mathscr{C}} \left(\sum_i^{\mathrm{occ}} \frac{\psi_i^*(\hat{\mathbf{r}}_\mu)\psi_i(\hat{\mathbf{r}}_\nu)}{\lambda - \varepsilon_i + i\omega} \right) \left(\sum_a^{\mathrm{vir}} \frac{\psi_a(\hat{\mathbf{r}}_\mu)\psi_a^*(\hat{\mathbf{r}}_\nu)}{\lambda - \varepsilon_a} \right) d\lambda\, \zeta_\mu(\mathbf{r})\zeta_\nu(\mathbf{r}') + \text{c.c.}
$$
$$
=: \sum_{\mu\nu} \chi^0_{\mu\nu}(i\omega)\zeta_\mu(\mathbf{r})\zeta_\nu(\mathbf{r}') =: \langle \mathbf{r} | \tilde\chi^0(i\omega) | \mathbf{r}' \rangle, \tag{6.20}
$$

where the last line defines $\chi^0_{\mu\nu}$ and $\tilde\chi^0$. We note that in the above the separability of the ISDF coefficients into the i and a components is crucial. Without this separability (*e.g.* if a conventional density fitting were used) we would not be able to calculate $\tilde\chi^0$ in cubic time since the sums over i and a would not decouple.

We now return our attention to the RPA correlation energy under random phase approximation (2.49). Given $\tilde\chi^0$ sufficiently close to $\hat\chi^0$, we can approximate $\mathrm{Tr}[\ln(I - \hat\chi^0 v_C)]$ using (with a suitable choice of constant c for convergence of the series)

$$
\ln(I - \tilde\chi^0 v_C) = \ln[cI - ((c-1)I + \tilde\chi^0 v_C)]
$$
$$
= \ln(c)I + \ln\left[I - \frac{1}{c}((c-1)I + \tilde\chi^0 v_C) \right]
$$
$$
= \ln(c)I - \sum_{\ell=1}^{\infty} \frac{[(c-1)I + \tilde\chi^0 v_C]^\ell}{\ell c^\ell}
$$
$$
= \ln(c)I - \sum_{\ell=1}^{\infty} \frac{1}{\ell c^\ell} \sum_{p=0}^{\ell} \binom{\ell}{p}(c-1)^{\ell-p}(\tilde\chi^0 v_C)^p, \tag{6.21}
$$

and represent each the polynomial in terms of $\tilde\chi^0 v_C$ on the right-hand side in the auxiliary basis. After some manipulation, this results in our final desired approximation,

$$
\mathrm{Tr}[\ln(1 - \hat\chi^0(i\omega)v_C) + \hat\chi^0(i\omega)v_C] \approx \mathrm{Tr}[\ln(1 - \chi^0(i\omega)v) + \chi^0(i\omega)v], \tag{6.22}
$$

where χ^0 and v are the matrix elements in the auxiliary basis.

Algorithm 10: Cubic scaling calculation of the RPA correlation energy

Input: Kohn–Sham orbitals $\{\psi_k\}$ and corresponding energies $\{\varepsilon_k\}$.

Output: $E_{\mathrm{c}}^{\mathrm{RPA}}$

1: Use $\{\psi_k\}_{k=1}^{N_{\mathrm{orb}}}$ as the input to ISDF to obtain $\{\hat{\mathbf{r}}_\mu\}_{k=1}^{N_\mu}$ and $\{\zeta_\mu\}_{k=1}^{N_\mu}$.

2: Compute the matrix $v_{\mu,\nu} = \langle \zeta_\mu | v_C | \zeta_\nu \rangle$.

3: For each quadrature point ω_m for the contour integral on $\mathscr{C}$:

    (a) Compute $\chi_{\mu,\nu}^0(i\omega_m)$ as defined in (6.20).

    (b) Compute $\dfrac{1}{2\pi} \mathrm{Tr}[\ln(1 - \chi^0(i\omega_m)v) + \chi^0(i\omega_m)v]$.

4: Calculate

$$E_{\mathrm{c}}^{\mathrm{RPA}} = \frac{1}{2\pi} \int_0^\infty \mathrm{Tr}[\ln(1 - \chi^0(i\omega)v) + \chi^0(i\omega)v]\, \mathrm{d}\omega$$

via numerical quadrature.

We present the cubic scaling algorithm for the calculation of the RPA correlation energy as in Algorithm 10. Details of the algorithm can be found in Lu and Thicke (2017a).

Our numerical results use the following test problem. Our two-dimensional spatial grid is $10 \times 10N$ equally spaced points. First, we solve for the Kohn–Sham orbitals of the periodic system with an external potential consisting of N Gaussian potential wells, the centres of which are randomly perturbed from the centres of their respective 10×10 box of grid points. Then the eigenvectors of H are used as the orbitals in the calculation of the RPA correlation energy.

We investigate the cubic scaling behaviour of the algorithm in Figure 6.2. The quartic scaling method using traditional density fitting is also plotted for comparison. In this example, we choose the virtual orbital cut-off $N_{\mathrm{orb}} = 0.2N_g$ and the system size is scaled up to $N = 160$. We observe that the cubic scaling algorithm greatly outperforms the quartic scaling algorithm for large system sizes.

7. Self-consistent field iteration

In this section we discuss numerical methods for performing self-consistent field (SCF) iterations. In Sections 7.1 and 7.2 we introduce basic SCF iteration techniques such as the fixed-point iteration and the simple mixing method, as well as acceleration techniques based on Newton and quasi-Newton methods. We introduce these techniques in the context of semi-local functionals. In Section 7.3 we discuss preconditioning techniques for SCF

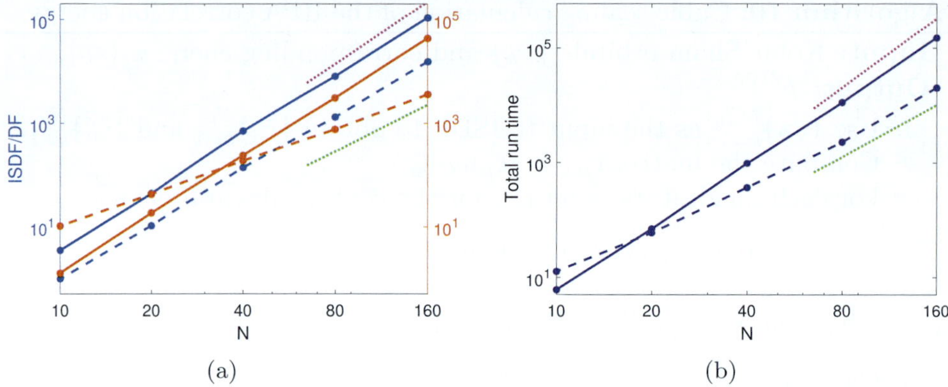

Figure 6.2. (Credit: Lu and Thicke 2017*a*.) Timing results for the quartic scaling method are plotted with solid lines, and results for the cubic scaling method are plotted with dashed lines. For reference, the purple and green dotted lines represent the slopes of N^4 and N^3 respectively. (a) Comparison of the time required to calculate χ^0 and the time to perform the respective density fitting schemes for each method. (b) The total running time to calculate E_c^{RPA} for each method.

iterations specific to semi-local functionals using a large basis set. Similarly Section 7.4 introduces SCF techniques specifically for non-local functionals. Throughout the discussion, we will assume a large basis set is used.

7.1. Fixed-point iteration and simple mixing

For a fixed atomic configuration, the self-consistent iteration starts from certain initial electron density denoted by ρ_0. We let ρ_k, V_k denote the electron density and the effective potential V_{eff} at the kth SCF iteration, respectively. When pseudopotentials are used, since the non-local pseudopotential is independent of ρ, V_{eff} is still a local potential as in equation (2.91). Then the flow of the SCF iteration becomes

$$\cdots \to \rho_k \to V_k = V_{\mathrm{eff}}[\rho_k] \to \rho_{k+1} \to V_{k+1} = V_{\mathrm{eff}}[\rho_{k+1}] \to \cdots. \qquad (7.1)$$

Depending on the starting point, the relation (7.1) can be viewed as a mapping from ρ_k to ρ_{k+1}, or from V_k to V_{k+1}. The former viewpoint is called *density mixing*, while the latter is called *potential mixing*. There is no qualitative difference between the two types of mixing schemes. However, the density mixing has the extra constraint that the density must be non-negative everywhere, and must be normalized to have the correct number of electrons. In practice, this constraint can be easily satisfied by setting all the negative entries of ρ (usually these entries have very small magnitude) to 0 or a small positive number, followed by a normalization step. On the other hand, potential mixing is formally free of such constraints, and hence we will consider potential mixing below. We remark that both density mixing and potential mixing strategies are widely used in electronic structure software packages, and the algorithm below for potential mixing can be used for density mixing as well.

When self-consistency is reached, the converged effective potential is denoted by $V_\star$ and satisfies the nonlinear equation

$$V_\star = V_{\mathrm{eff}}[\mathcal{F}_{\mathrm{KS}}[V_\star]]. \tag{7.2}$$

The simplest version of the SCF iteration is the fixed-point iteration, where the potential at the $(k+1)$th step is directly given by the output potential at the kth step:

$$V_{k+1} = V_{\mathrm{eff}}[\mathcal{F}_{\mathrm{KS}}[V_k]]. \tag{7.3}$$

Since the exchange-correlation functional is neither convex nor concave with respect to the electron density, rigorous study of the global convergence properties of SCF schemes in Kohn–Sham DFT calculations is difficult. Hence we consider the linear response regime, where we assume the initial effective potential V_0 is already close to $V_\star$. Let

$$e_k := V_k - V_\star$$

be the error of the potential at the kth iteration. In order to study the propagation of the error in the fixed-point iteration (7.3), we apply the chain rule

$$e_{k+1} = f_{\mathrm{Hxc}}\chi_0 e_k + O(\|e_k\|^2). \tag{7.4}$$

Here $f_{\mathrm{Hxc}} = \delta V_{\mathrm{eff}}/\delta\rho$ is the kernel for the Hartree and exchange-correlation contributions, and $\chi_0 = \delta\mathcal{F}_{\mathrm{KS}}/\delta V$ is called the independent particle polarizability operator. In the linear response regime, we assume the $O(\|e_k\|^2)$ term is small and it is omitted in the following discussion. Then after k steps

$$e_k \approx (f_{\mathrm{Hxc}}\chi_0)^k e_0. \tag{7.5}$$

Hence in the linear response regime, the convergence of the fixed-point iteration requires that the spectral radius of the operator, denoted by $r_\sigma(f_{\mathrm{Hxc}}\chi_0)$, is smaller than 1. Unfortunately, such a spectral radius is generally much larger than 1, and the error in the fixed-point iteration will therefore diverge, even if the initial potential is already very close to the self-consistent potential.

In order to achieve the self-consistent solution, the simplest practically usable scheme is the simple mixing method, which introduces a slight modification of the fixed-point iteration, shown in Algorithm 11. The iteration can also be written equivalently as

$$V_{k+1} = \alpha V_{\mathrm{eff}}[\mathcal{F}_{\mathrm{KS}}[V_k]] + (1-\alpha)V_k. \tag{7.6}$$

Algorithm 11: Simple mixing method

Input: Initial guess V_0 of the potential and relaxation parameter α.

1: **for** $k = 0, \ldots$, until convergence **do**
2:     Apply the Kohn–Sham map to compute the density $\rho_k = \mathcal{F}_{\mathrm{KS}}[V_k]$.
3:     Form the residual error $r_k = V_k - V_{\mathrm{eff}}[\rho_k]$.
4:     $V_{k+1} = V_k - \alpha r_k$.
5: **end for**

In the simple mixing method, the error propagation follows

$$e_{k+1} = (I - \alpha\epsilon_d)e_k + O(\|e_k\|^2). \tag{7.7}$$

Here $\epsilon_d = I - f_{\mathrm{Hxc}}\chi_0$ is called the dielectric operator. Equation (7.7) can be iterated, yielding

$$e_k \approx (I - \alpha\epsilon_d)^k e_0. \tag{7.8}$$

When the exchange-correlation functional contribution is dropped from f_{Hxc}, the resulting approximation is called the random phase approximation (RPA).[1] Then ϵ_d can be transformed using a similarity transformation to a positive definite matrix. In particular, ϵ_d is diagonalizable with real positive eigenvalues. In order to achieve convergence in the linear response regime, we need each eigenvalue of ϵ_d, denoted by λ, to satisfy

$$|1 - \alpha\lambda| < 1,$$

or equivalently

$$0 < \alpha < \frac{2}{\lambda}.$$

Let $\lambda_{\min}$ and $\lambda_{\max}$ denote the smallest and largest eigenvalues of ϵ_d, respectively, and the spectral radius $r_\sigma(\epsilon_d)$ is then given by $\lambda_{\max}$. If

$$0 < \alpha < \frac{2}{r_\sigma(\epsilon_d)} \tag{7.9}$$

is satisfied, the simple mixing would converge.

It remains to determine the *optimal* choice of α satisfying the constraint (7.9). This requires the solution of the following minimax problem:

$$\min_\alpha \max_\lambda |1 - \alpha\lambda|. \tag{7.10}$$

---

[1] Here the meaning of 'RPA' is related to the RPA functional in Section 6.3, in the sense that the exchange-correlation kernel is dropped in both cases, but the usage has different origins.

The optimal choice of α satisfies

$$1 - \alpha\lambda_{\min} = \alpha\lambda_{\max} - 1, \tag{7.11}$$

or

$$\alpha = \frac{2}{\lambda_{\min} + \lambda_{\max}}. \tag{7.12}$$

Substituting this choice of α into (7.7), we find that the optimal convergence rate of simple mixing is

$$\max_{\lambda}|1 - \alpha\lambda| = \frac{\lambda_{\max} - \lambda_{\min}}{\lambda_{\max} + \lambda_{\min}} = \frac{\kappa(\epsilon_d) - 1}{\kappa(\epsilon_d) + 1}. \tag{7.13}$$

Here $\kappa(\varepsilon_d) = \lambda_{\max}/\lambda_{\min}$ is the condition number of the dielectric operator.

In practical calculations, the condition number might be very large and the simple mixing method converges with a very slow rate. Thus the simple mixing method is rarely used directly in practical electronic structure calculations.

7.2. Newton and quasi-Newton type methods

The convergence of the simple mixing method often requires a rather small mixing constant α. Hence the SCF procedure may take many iterations to converge. One possible acceleration can be achieved by using Newton's method, which can be written as

$$V_{k+1} = V_k - \mathcal{J}_k^{-1} r_k. \tag{7.14}$$

Here $\mathcal{J}_k$ is the Jacobian matrix for the residual map

$$\delta V \mapsto \delta V - V_{\mathrm{eff}}\big[\mathcal{F}_{\mathrm{KS}}[V_k + \delta V]\big].$$

Note that at converged potential $V_\star$, the Jacobian matrix $\mathcal{J}_\star = \epsilon_d$. Hence the simple mixing method can also be interpreted as approximating the inverse of the Jacobian matrix $\mathcal{J}_k^{-1}$ simply by αI in the evaluation of the Jacobian matrix for the composition map $V_{\mathrm{eff}} \circ \mathcal{F}_{\mathrm{KS}}$. For a system with N electrons, the evaluation of the Jacobian matrix requires in principle $O(N)$ evaluations of the Kohn–Sham map, which is prohibitively expensive.

The Jacobian-free Newton–Krylov method replaces the need for the explicit evaluation of the Jacobian matrix by solving a linear equation

$$\mathcal{J}_k \delta V_k = -r_k \tag{7.15}$$

to obtain the Newton update δV_k. This can be done using iterative methods for solving linear equations, such as the generalized minimal residual method (GMRES) (Saad and Schultz 1986). In order to compute the matrix–vector

multiplication related to the Jacobian matrix, one can use the finite difference approximation

$$\mathcal{J}_k \delta V_k \approx \delta V_k - \left(V_{\text{eff}}\left[\mathcal{F}_{\text{KS}}[V_k + \delta V_k]\right] - V_{\text{eff}}\left[\mathcal{F}_{\text{KS}}[V_k]\right] \right). \tag{7.16}$$

The finite difference calculation requires at least one additional function
evaluation of $\mathcal{F}_{\text{KS}}(V_k + \delta V)$ per iteration step. Therefore, even though
Newton's method may exhibit local quadratic convergence, each Newton
iteration may take many inner iterations in order to solve the linear equation (7.15).

A widely used alternative to Newton's method is the class of quasi-Newton
methods, which replace $\mathcal{J}_k^{-1}$ with an approximate matrix C_k that is easy
to compute and apply. Then the updating strategy becomes

$$V_{k+1} = V_k - C_k r_k. \tag{7.17}$$

Using Broyden's techniques (Nocedal and Wright 1999), one can sys-
tematically approximate $\mathcal{J}_k$ or $\mathcal{J}_k^{-1}$. In Broyden's second method, C_k is
obtained by performing a sequence of low-rank modifications to some initial
approximation C_0 of the Jacobian inverse using a recursive formula (Fang
and Saad 2009, Marks and Luke 2008). At each step, C_k is obtained by
solving the following constrained optimization problem:

$$\min_C \frac{1}{2}\|C - C_{k-1}\|_F^2$$
$$\text{subject to } S_k = CY_k, \tag{7.18}$$

where C_{k-1} is the approximation to the Jacobian constructed in the $(k-1)$th
Broyden iteration. The matrices S_k and Y_k above are defined as

$$S_k = (s_k, s_{k-1}, \ldots, s_{k-\ell}), \quad Y_k = (y_k, y_{k-1}, \ldots, y_{k-\ell}), \tag{7.19}$$

where s_j and y_j are defined by $s_j = V_j - V_{j-1}$ and $y_j = r_j - r_{j-1}$ respectively.
The number ℓ is the length of history used in Broyden's method, which is
typically set to $5 \sim 20$ in practical calculations.

Equation (7.18) is a constrained optimization problem and can be solved
using the method of Lagrange multipliers, given by

$$C_k = C_{k-1} + (S_k - C_{k-1}Y_k)Y_k^\dagger. \tag{7.20}$$

Here $Y_k^\dagger$ denotes the Moore–Penrose pseudo-inverse of Y_k, that is, $Y_k^\dagger = (Y_k^* Y_k)^{-1} Y_k^*$. We remark that in practice $Y_k^\dagger$ is not constructed explicitly
since we need only apply $Y_k^\dagger$ to a residual vector r_k. This operation can
be carried out by solving a linear least-squares problem with appropriate
regularization (*e.g.* through a truncated singular value decomposition).

A variant of Broyden's method is Anderson's method (Anderson 1965),
which is widely used in electronic structure software packages. Anderson's

---

Algorithm 12: Anderson's method

Input: Initial guess V_0 of the potential, relaxation parameter α, history length ℓ

1: **for** $k = 0, \ldots,$ until convergence **do**
2:     Apply the Kohn–Sham map to compute the density $\rho_k = \mathcal{F}_{\mathrm{KS}}[V_k]$.
3:     Form the residual error $r_k = V_k - V_{\mathrm{eff}}[\rho_k]$.
4:     Define $S_k = (s_k, s_{k-1}, \ldots, s_{k-\ell})$ with $s_j = V_j - V_{j-1}$ and $Y_k = (y_k, y_{k-1}, \ldots, y_{k-\ell})$ with $y_j = r_j - r_{j-1}$. For $k < \ell$, keep only the vectors starting from index 0 for both S_k and Y_k matrices.
5:     Form $d_k = Y_k^\dagger r_k$ by solving a linear least-squares problem with appropriate regularization.
6:     $V_{k+1} = V_k - \alpha(r_k - Y_k d_k) - S_k d_k$.
7: **end for**

---

method fixes C_{k-1} to the initial approximation C_0 when solving (7.18) during each iteration. It follows from equation (7.17) that Anderson's method updates the potential as

$$V_{k+1} = V_k - C_0(I - Y_k Y_k^\dagger)r_k - S_k Y_k^\dagger r_k. \tag{7.21}$$

In particular, if C_0 is set to αI, we obtain Anderson's method,

$$V_{k+1} = V_k - \alpha(I - Y_k Y_k^\dagger)r_k - S_k Y_k^\dagger r_k,$$

commonly used in Kohn–Sham DFT solvers. Pseudocode for Anderson's method is shown in Algorithm 12.

An alternative way to derive Broyden's method is through a technique called direct inversion of iterative subspace (DIIS). The technique was originally developed by Pulay to accelerate Hartree–Fock calculations (Pulay 1980). Hence it is often referred to as Pulay mixing. The motivation of Pulay's method is to minimize the residual $V - V_{\mathrm{eff}}[\mathcal{F}_{\mathrm{KS}}(V)]$ within the subspace spanned by $\{V_{k-\ell-1}, \ldots, V_k\}$. In Pulay's original work (Pulay 1980), the optimal approximation to V is expressed as

$$V_{\mathrm{opt}} = \sum_{j=k-\ell-1}^{k} \alpha_j V_j,$$

where V_j $(j = k - \ell - 1, \ldots, k)$ are previous approximations to V, and the coefficients α_j are chosen to satisfy the constraint $\sum_{j=k-\ell-1}^{k} \alpha_j = 1$.

When the V_j are all sufficiently close to the fixed-point solution,

$$V_{\mathrm{eff}}\left[\mathcal{F}_{\mathrm{KS}}\left(\sum_j \alpha_j V_j\right)\right] \approx \sum_j \alpha_j V_{\mathrm{eff}}[\mathcal{F}_{\mathrm{KS}}(V_j)]$$

holds approximately. Hence we may obtain α_j (and consequently V_{opt}) by solving the following quadratic programming problem:

$$\min_{\{\alpha_j\}} \left\| \sum_{j=k-\ell-1}^{k} \alpha_j r_j \right\|_2^2$$

$$\text{subject to} \quad \sum_{j=k-\ell-1}^{k} \alpha_j = 1, \tag{7.22}$$

where $r_j = V_j - V_{\text{eff}}[\mathcal{F}_{\text{KS}}(V_j)]$.

Note that (7.22) can be reformulated as an unconstrained minimization problem if V_{opt} is required to take the form

$$V_{\text{opt}} = V_k + \sum_{j=k-\ell}^{k} \beta_j (V_j - V_{j-1}),$$

where β_j can be any unconstrained real number. Again, if we assume $V_{\text{eff}}[\mathcal{F}_{\text{KS}}(V)]$ is approximately linear around V_j and let $b = (\beta_{k-\ell}, \dots, \beta_k)^\top$, minimizing $\|V_{\text{opt}} - V_{\text{eff}}[\mathcal{F}_{\text{KS}}(V_{\text{opt}})]\|$ with respect to $\{\beta_j\}$ yields $b = -Y_k^\dagger r_k$, where Y_k is the same as that defined in (7.19). Then Pulay's method for updating V is thus $V_{k+1} = V_{\text{opt}}$. More generally, we may introduce a C_0 matrix as

$$V_{k+1} = V_{\text{opt}} - C_0(V_{\text{opt}} - V_{\text{eff}}[\mathcal{F}_{\text{KS}}(V_{\text{opt}})]). \tag{7.23}$$

Substituting $V_{\text{opt}} = V_k - S_k Y_k^\dagger r_k$ into (7.23), together with the linearity assumption of $V_{\text{eff}}[\mathcal{F}_{\text{KS}}(V)]$, yields exactly Anderson's updating formula (7.21).

The matrix C_0 plays the role of a preconditioner, and hence a better C_0 can be chosen to accelerate the convergence of Anderson's method in practical electronic structure calculations, as will be discussed in Section 7.3.

It is worthwhile remarking that a variant of the Pulay method above can be derived by taking other definitions of the residual vector r. This leads to the commutator-DIIS (C-DIIS) method (Pulay 1982), which defines the residual as the commutator of the Hamiltonian operator and the density matrix. This is the most widely used method in quantum chemistry software packages for achieving self-consistency. On the other hand, the C-DIIS method requires explicit storage of the density matrix and the Hamiltonian matrix for a few iterations, and hence is only feasible for calculations with small basis sets.

In the C-DIIS method, the residual, now written as R, is the commutator between $H[P]$ and P, that is,

$$R[P] = H[P]P - PH[P]. \tag{7.24}$$

Note that the Hamiltonian matrix $H[P]$ and the effective potential $V_{\text{eff}}[P]$

contain the same information. Let H_k denote the approximate Hamiltonian produced at step k. We define a new Hamiltonian $\tilde{H}_{k+1}$ at step $(k+1)$ as a linear combination of previous approximations to the Hamiltonian, that is,

$$\tilde{H}_{k+1} = \sum_{j=k-\ell-1}^{k} \alpha_j H_j, \tag{7.25}$$

where α_j satisfies the constraint $\sum_{j=k-\ell-1}^{k} \alpha_j = 1$. Each Hamiltonian matrix H_j defines a density matrix P_j via the generalized Kohn–Sham map. Before self-consistency is reached, the residual $R_j = R[P_j]$ defined by equation (7.24) is non-zero. However, when all the Hamiltonian matrices $\{H_j\}$ are close to the self-consistent Hamiltonian operator $H_\star$, it is reasonable to expect the residual associated with $\tilde{H}_{k+1}$ to be well approximated by $\tilde{R}_{k+1} \equiv \sum_{j=k-\ell-1}^{k} \alpha_j R_j$. The C-DIIS method determines $\{\alpha_j\}$ by minimizing $\tilde{R}_{k+1}$, that is, we solve the following constrained minimization problem in each C-DIIS iteration:

$$\min_{\{\alpha_j\}} \left\| \sum_{j=k-\ell-1}^{k} \alpha_j R_j \right\|_F^2$$

$$\text{subject to} \quad \sum_{j=k-\ell-1}^{k} \alpha_j = 1. \tag{7.26}$$

Here the Frobenius norm is defined as $\|A\|_F^2 = \mathrm{Tr}[A^*A]$.

As in the comparison between Anderson's method and Pulay's method above, the constraint in (7.26) can be eliminated by rewriting equation (7.25) as

$$\tilde{H}_{k+1} = H_k + \sum_{j=k-\ell}^{k} \beta_j (H_{j-1} - H_j). \tag{7.27}$$

Define $Y_j = R_{j-1} - R_j$ for $k - \ell + 1 \le j \le k$. Then the constraint minimization problem (7.26) becomes an unconstrained minimization problem:

$$\min_{\{\beta_j\}} \left\| R_k + \sum_{j=k-\ell+1}^{k} \beta_j Y_j \right\|_F^2. \tag{7.28}$$

As a result, equation (7.28) has an analytic solution,

$$\beta = -M^{-1}b, \tag{7.29}$$

where the $\ell \times \ell$ matrix M and the vector b are defined as

$$M_{ij} = \mathrm{Tr}[Y_i^* Y_j], \quad b_j = \mathrm{Tr}[Y_j^* R_k], \tag{7.30}$$

respectively.

7.3. Preconditioning techniques

From the analysis of the simple mixing method we observe that the convergence rate is determined by the condition number of $\epsilon_d = \mathcal{J}_\star$. Hence we should examine the dependence of $\kappa(\mathcal{J}_\star)$ with respect to the system size. Here we mainly examine periodic systems, of which the system size is often characterized by the number of unit cells in the computational domain. To simplify our discussion, we assume the unit cell to be a simple cubic cell with a lattice constant L. For non-periodic systems such as molecules, we can construct a fictitious (cubic) supercell that encloses the molecule and periodically extend the supercell so that properties of the system can be analysed via Fourier analysis. In both cases, we assume the number of atoms in each supercell is proportional to L^3.

When $\mathcal{J}_\star$ is similar to a positive definite matrix, the smallest eigenvalue $\lambda_{\min}$ is often independent of the system size, but the largest eigenvalue $\lambda_{\max}$ may depend sensitively on the system size. For instance, for the jellium system (or uniform electron gas), $\mathcal{J}_\star$ becomes a diagonal matrix in the Fourier space, and the eigenvalues can be evaluated explicitly as (Ziman 1979)

$$\lambda_{\mathbf{q}} = 1 + \left(\frac{4\pi}{q^2} + \kappa^2 \right) \gamma F_L(q), \quad \mathbf{q} \in \mathbb{R}^3, \tag{7.31}$$

where $q = |\mathbf{q}|$, γ, κ are constants, and $F_L(q)$ is known as the *Lindhard* response function, which satisfies

$$\lim_{q \to 0} F_L(q) = 1, \quad \lim_{q \to \infty} F_L(q) = 0. \tag{7.32}$$

The above formulation is formally valid when the system size is infinite. Note that the smallest possible value q is $2\pi/L$, and then $\lambda_{\max}$ is bounded by $1 + \gamma(L^2/\pi + \kappa^2)$. As a result, the convergence of SCF iteration schemes becomes slower as the system size increases. If the mixing parameters are not adjusted properly, the long wavelength modes (corresponding to small q) tend to be amplified as the SCF iteration proceeds. This phenomenon is called 'charge sloshing'.

In order to accelerate the convergence, one can replace the scalar α in the simple mixing method with a matrix C_0, and the convergence rate is then determined by the condition number of the matrix $C_0 \mathcal{J}_\star$. As discussed in Section 7.2, the choice of the matrix C_0 in Anderson's method plays the role of a preconditioner. The ideal preconditioner is $C_0 = \mathcal{J}_\star^{-1}$ as $\kappa(C_0 \mathcal{J}_\star) = 1$. This corresponds to Newton's method. To overcome charge sloshing, a widely used preconditioner is the Kerker preconditioner (Kerker 1981), which is much easier to apply. It assigns a smaller weight to the long wavelength Fourier modes in order to attenuate charge sloshing. More specifically, the Kerker preconditioner is a diagonal matrix in the Fourier space, and its eigenvalues are given by

$$\lambda_{\mathbf{q}}^{\text{Kerker}} = 1 + \frac{4\pi\widetilde{\gamma}}{q^2}, \tag{7.33}$$

where $\widetilde{\gamma} > 0$ is an adjustable parameter. For the uniform electron gas, the matrix $C_0 \mathcal{J}_\star$ is a diagonal matrix in the Fourier space, and its eigenvalues are

$$\widetilde{\lambda}_{\mathbf{q}} = \frac{q^2 + \gamma F_L(q)(4\pi + \kappa^2 q^2)}{q^2 + 4\pi\widetilde{\gamma}}. \tag{7.34}$$

The eigenvalue $\widetilde{\lambda}_{\mathbf{q}}$ becomes approximately 1 for large q and $F_L(q)\gamma/\widetilde{\gamma}$ for small q. Hence, after preconditioning, the eigenvalues can be bounded by constants independent of the system size. In practice it is found that the Kerker preconditioner is an effective preconditioner for simple metals such as sodium (Na) or aluminium (Al).

However, the Kerker preconditioner is not an appropriate preconditioner for insulating systems. Although in general the Jacobian associated with the insulating system cannot be diagonalized by the Fourier basis, it can be shown that $e^{i\mathbf{q}\cdot\mathbf{r}}$ is an approximate eigenfunction of the independent particle polarizability operator χ_0 with the corresponding eigenvalue $-\xi q^2$ (Pick, Cohen and Martin 1970, Ghosez, Gonze and Godby 1997) for small q, where $\xi > 0$ is a constant factor. If we neglect the contribution from the exchange-correlation kernel, $e^{i\mathbf{q}\cdot\mathbf{r}}$ is also an approximate eigenfunction of $\mathcal{J}_\star$ with the corresponding eigenvalue $1 + (4\pi/q^2)q^2\xi = 1 + 4\pi\xi$ for small q. If C_0 is chosen to be the Kerker preconditioner, then the corresponding eigenvalue of $C_0\mathcal{J}_\star$ for small q is approximately

$$\widetilde{\lambda}_{\mathbf{q}} = \frac{q^2}{q^2 + 4\pi\widetilde{\gamma}}(1 + 4\pi\xi).$$

As the system size L increases, the smallest eigenvalue $\widetilde{\lambda}_{\mathbf{q}} \sim L^{-2}$, and thus the condition number $\kappa(C_0\mathcal{J}_\star)$ increases as the system size increases. For simple insulating systems, one good preconditioner is simply a constant, as the convergence of the simple mixing method is already independent of the system size.

As in the above discussion, simple insulating and metallic systems call for different types of preconditioners to accelerate the convergence of a fixed-point iteration for solving the Kohn–Sham problem. A natural question one may ask is how we should construct a preconditioner for a complex material that may contain both insulating and metallic components or metal surfaces. This leads to the elliptic preconditioner (Lin and Yang 2013) as one promising strategy.

Note that under RPA (neglecting the exchange-correlation kernel), we may approximate $\mathcal{J}_\star^{-1}$ as

$$\widetilde{\mathcal{J}}_\star^{-1} = (v_C^{-1} - \chi_0)^{-1} v_C^{-1}.$$

Since the Coulomb kernel $v_C^{-1} = -\Delta/(4\pi)$, applying $\widetilde{\mathcal{J}}_\star^{-1}$ to a residual vector r_k simply amounts to solving the equation

$$(-\Delta - 4\pi\chi_0)\tilde{r}_k = -\Delta r_k. \tag{7.35}$$

To construct a preconditioner, we will replace χ_0 with a simpler operator. In many cases, we can choose the approximation to be a local (diagonal) operator defined by a function $b(\mathbf{r})$, although other types of more sophisticated operators are possible. To compensate for the simplification of χ_0, we replace the Laplacian operator on the left of (7.35) by $-\nabla \cdot (a(\mathbf{r})\nabla)$ for some appropriately chosen function $a(\mathbf{r})$. This additional change yields the following elliptic PDE:

$$(-\nabla \cdot (a(\mathbf{r})\nabla) + 4\pi b(\mathbf{r}))\tilde{r}_k = -\Delta r_k. \tag{7.36}$$

Because our construction of the preconditioner involves solving an elliptic equation, equation (7.36) is called an elliptic preconditioner.

The elliptic preconditioner is naturally compatible with previous preconditioners for simple metals or insulators. For example, for uniform electron gas, setting $a(\mathbf{r}) = 1$ and $b(\mathbf{r}) = -\tilde{\gamma}$ for some constant $\tilde{\gamma} > 0$ yields

$$(-\Delta + 4\pi\tilde{\gamma})\tilde{r}_k = -\Delta r_k. \tag{7.37}$$

The solution of the above equation is exactly the same as that produced by the Kerker preconditioner. For simple insulating system, setting $a(\mathbf{r}) = \alpha^{-1}$ and $b(\mathbf{r}) = 0$ yields

$$-\alpha^{-1}\Delta\tilde{r}_k = -\Delta r_k.$$

The solution to the above equation is simply

$$\tilde{r}_k = \alpha r_k. \tag{7.38}$$

Such a solution corresponds to simple mixing with a mixing parameter α.

For a complex system that consists of both insulating and metallic components, it is desirable to choose approximations to $a(\mathbf{r})$ and $b(\mathbf{r})$ that are spatially dependent. The asymptotic behaviour of χ with respect to the sizes of both insulating and metallic systems suggests that $a(\mathbf{r})$ and $b(\mathbf{r})$ should be chosen to satisfy $a(\mathbf{r}) \geq 1$ and $b(\mathbf{r}) \geq 0$.

The implementation of the elliptic preconditioner only requires solving an elliptic equation, for instance through an iterative linear solver, such as the conjugate gradient method. In particular, fast algorithms such as multigrid (Brandt 1977), the fast multipole method (FMM) (Greengard and Rokhlin 1987), hierarchical matrix solvers (Hackbusch 1999) and hierarchical semi-separable (HSS) matrix solvers (Chandrasekaran, Gu and Pals 2006) can be used to solve (7.36) in $O(N)$ arithmetic operations.

Algorithm 13: Two-level nested SCF method for solving Hartree–Fock-like equations

---

1: **while** Exchange energy is not converged **do**
2:     **while** Electron density ρ is not converged **do**
3:         Solve the linear eigenvalue problem $H\psi_i = \varepsilon_i \psi_i$ with any iterative eigensolver.
4:         Update $\rho^{\text{out}}(\mathbf{r}) \leftarrow \sum_{i=1}^{N} |\psi_i(\mathbf{r})|^2$.
5:         Update ρ using ρ^{out} and charge densities computed and saved from previous iterations using a charge mixing scheme.
6:     **end while**
7:     Compute the exchange energy.
8:     Update $\{\varphi_i\}_{i=1}^{N} \leftarrow \{\psi_i\}_{i=1}^{N}$.
9: **end while**

7.4. Rung-4 functionals

As discussed in Section 7.2, the C-DIIS method is the most widely used method for Kohn–Sham DFT calculations with non-local functionals using a small basis set, such as Gaussian-type orbitals and atomic orbitals. However, it requires the explicit storage of the density matrix and the Hamiltonian matrix, and thus cannot be used in the setting of a large basis set such as planewaves. On the other hand, one cannot simply take the output Kohn–Sham orbitals from one SCF iteration and use them as the input Kohn–Sham orbitals for the next SCF iteration. As analysed in Section 7.1, such a fixed-point iteration is vulnerable to large eigenvalues of the Jacobian matrix and can suffer from the 'charge sloshing' problem. In practice, the most commonly used method for converging Kohn–Sham DFT calculations with a large basis set is a two-level nested SCF procedure. This is implemented, for instance, in the Quantum ESPRESSO software package. The motivation for using a two-level SCF procedure is to apply advanced charge mixing schemes to the electron density in the inner iteration to mitigate charge sloshing, and to use a fixed-point iteration to update the Kohn–Sham orbitals and consequently the exchange potential in the outer iteration. The update of the exchange potential is more costly, even though its contribution to total energy is typically much smaller.

The two-level nested SCF method is summarized in Algorithm 13. In each outer iteration, the exchange operator $V_X[P]$ is updated. This is implicitly done by updating a set of orbitals $\{\varphi_i\}_{i=1}^{N}$ defining the density matrix as $P = \sum_{i=1}^{N} \varphi_i \varphi_i^*$. We remark that this set of orbitals may be different from the Kohn–Sham orbitals in the inner SCF iteration. The update is done through a fixed-point iteration, that is, $\{\varphi_i\}_{i=1}^{N}$ are given by the output Kohn–Sham orbitals in the previous outer iteration. In the inner SCF iteration, with the exchange operator fixed, the Hamiltonian H depends only on the electron density ρ. Charge mixing schemes for ρ can be performed in a similar fashion to what is done in a standard Kohn–Sham DFT calculation without the exchange operator in the inner SCF iteration. Finally, within each inner iteration, with both P and ρ fixed, the Hartree–Fock-like equation becomes a linear eigenvalue problem and can be solved by an iterative eigensolver discussed in Section 4.2. The outer SCF iteration continues until convergence is reached, which can be monitored using the change in the exchange energy, for example.

Recently, a new method, called the projected commutator DIIS (PC-DIIS) method, has been introduced so that the C-DIIS method can be used for Kohn–Sham DFT calculations with non-local functionals performed with a large basis set (Hu, Lin and Yang 2017b). Since it is not possible to explicitly store or mix the density matrices in such calculations, it is tempting to perform the DIIS procedure on the $N_b \times N$ orbital matrix $\Psi = [\psi_1, \ldots, \psi_N]$. However, one key difference between the density matrix and the orbital matrix is that the former is gauge-invariant. That is, if we replace Ψ with ΨU, where U is an $N \times N$ unitary gauge matrix, the density matrix $P = \Psi \Psi^* = \Psi U U^* \Psi^*$ does not change. Therefore, it is completely safe to combine two density matrices constructed from Ψ that differ by a gauge transformation, as the total energy is gauge-invariant.

However, since the orbital matrix Ψ is not gauge-invariant, combining successive approximations to Ψ that differ by a gauge transformation may hinder the stability of the SCF iteration. To overcome this difficulty, we introduce an auxiliary orbital matrix Φ that spans the same subspace spanned by Ψ. This orbital matrix is obtained by applying the orthogonal projection operator associated with Ψ to a reference orbital matrix Φ_{ref} to be specified later. That is, Φ is chosen to be

$$\Phi = P\Phi_{\mathrm{ref}} = \Psi(\Psi^*\Phi_{\mathrm{ref}}). \tag{7.39}$$

We require Φ_{ref} to be fixed throughout the entire SCF procedure. Note that Φ is invariant to any gauge transformation applied to Ψ. Therefore, the auxiliary orbital matrices obtained in successive SCF iterations can be safely combined to produce a better approximation to the desired invariant subspace. The columns of Φ are generally not orthogonal to each other. However, as long as the columns of Φ are not linearly dependent, both Ψ and Φ span the range of the density matrix P, which can also be written as

$$P = \Phi(\Phi^*\Phi)^{-1}\Phi^*. \tag{7.40}$$

The PC-DIIS method constructs a new approximation to Φ in the kth SCF iteration by taking a linear combination of the auxiliary orbital matrices

$\Phi_{k-\ell}, \ldots, \Phi_k$ obtained in the most recent $\ell + 1$ iterations, that is,

$$\tilde{\Phi}_{k+1} = \sum_{j=k-\ell}^{k} \alpha_j \Phi_j. \tag{7.41}$$

The coefficients $\{\alpha_j\}$ in (7.41) are determined by minimizing the residual associated with $\tilde{\Phi}_{k+1}$, which, under the assumption that Φ_j are sufficiently close to the solution of the Kohn–Sham equations, is well approximated by $R \equiv \sum_{j=k-\ell}^{k} \alpha_j R_{\Phi_j}$, where the residual associated with an auxiliary orbital matrix Φ is defined by

$$R_\Phi = H[P]P\Phi_{\text{ref}} - PH[P]\Phi_{\text{ref}} = (H[P]\Psi)(\Psi^*\Phi_{\text{ref}}) - \Psi((H[P]\Psi)^*\Phi_{\text{ref}}). \tag{7.42}$$

Note that evaluation of the residual in equation (7.42) only requires multiplying $H[P]$ by Ψ and the multiplications of matrices of sizes $N_b \times N$ and $N \times N$ only. These operations are already used in iterative methods for computing the desired eigenvectors Ψ of H. The PC-DIIS algorithm does not require $P, H[P]$ or R to be constructed or stored explicitly.

An interesting observation is that if $\Phi_{\text{ref}} = \Psi$ then $\Psi^* H[P]\Psi$ is a diagonal matrix denoted by Λ. Consequently, the projected commutator takes the form

$$R_\Phi = H[P]\Psi - \Psi\Lambda. \tag{7.43}$$

This expression coincides with the standard definition of the residual associated with an approximate eigenpair (Λ, Ψ). Hence the PC-DIIS method can also be viewed as an extension of an iterative eigensolver for nonlinear problems.

As in the reformulation of the constrained minimization problem into an unconstrained minimization problem in the C-DIIS method, the constrained minimization problem

$$\min_{\{\alpha_j\}} \left\| \sum_{j=k-\ell}^{k} \alpha_j R_{\Phi_j} \right\|_F^2$$

$$\text{subject to} \quad \sum_{j=k-\ell}^{k} \alpha_j = 1, \tag{7.44}$$

to be solved in the PC-DIIS method can also be reformulated as an unconstrained minimization problem. Using the same change of variable as that presented in Section 7.2, we can write

$$\tilde{\Phi}_{k+1} = \Phi_k + \sum_{j=k-\ell+1}^{k} \beta_j(\Phi_{j-1} - \Phi_j). \tag{7.45}$$

If we let $Y_{\Phi_j} = R_{\Phi_{j-1}} - R_{\Phi_j}$, the coefficients β_j in (7.45) can be retrieved

---

Algorithm 14: The PC-DIIS method for solving Hartree–Fock-like equations

---

Input: Reference orbitals Φ_{ref}.
Output: Approximate solution $\Psi = \{\psi_i\}$, $i = 1, 2 \ldots, N$.
1: Construct the initial Hamiltonian H and evaluate the exchange energy using Φ_{ref}.
2: **while** Exchange energy is not converged **do**
3:     Solve the linear eigenvalue problem $H[P]\psi_i = \varepsilon_i \psi_i$ using an iterative eigensolver.
4:     Evaluate Φ, R_Φ according to (7.39), (7.42).
5:     Perform the DIIS procedure according to (7.45) to obtain the new $\tilde{\Phi}$ which implicitly defines a density matrix P via (7.40).
6:     Update the Hamiltonian $H[P]$.
7:     Compute the exchange energy.
8: **end while**

---

from the vector $\beta = -(M^\Phi)^{-1} b^\Phi$, where

$$M_{ij}^\Phi = \mathrm{Tr}[Y_{\Phi_i}^* Y_{\Phi_j}], \quad b_j^\Phi = \mathrm{Tr}[Y_{\Phi_j}^* R_{\Phi_k}]. \tag{7.46}$$

Once $\tilde{\Phi}_{k+1}$ is obtained, a density matrix associated with this orbital matrix is implicitly defined through equation (7.40). This implicitly defined density matrix allows us to construct a new Hamiltonian from which a new set of Kohn–Sham orbitals Ψ_{k+1} and auxiliary orbitals Φ_{k+1} can be computed.

We now discuss how to choose the gauge-fixing matrix Φ_{ref}. Note that in hybrid functional calculations, the contribution from the exchange operator is relatively small. Hence the density matrix associated with Kohn–Sham orbitals obtained from a DFT calculation that uses a local or semi-local exchange-correlation functional is already a good initial guess for the density matrix required in a hybrid functional calculation. Therefore, we may use these orbitals as Φ_{ref}. Compared to the two-level nested loop structure, the PC-DIIS method only requires one level of SCF iteration. The PC-DIIS method is summarized in Algorithm 14.

The discussion above is applicable when Ψ only contains the occupied orbitals. When Ψ also involves the virtual orbitals, we use the fact that the density matrix defining the Fock exchange operator only involves the occupied orbitals, and we need only apply the PC-DIIS method to the occupied orbitals. We also remark that the PC-DIIS method is not yet applicable for finite temperature calculations with fractionally occupied orbitals.

8. Conclusion and future directions

In this paper, we have reviewed some basic aspects and recent developments of numerical strategies for solving Kohn–Sham DFT. Most of the numerical methods focus on large systems, either due to the presence of a large number of electrons (thousands to tens of thousands), or due to the use of a large basis set (such as the planewave basis set). In this sense, the numerical algorithms in this paper are more suitable for applications in quantum physics and materials science, which often favour a large basis set and relatively large system sizes. On the other hand, the quantum chemistry literature often favours a small basis set (such as Gaussian-type orbitals) and relatively small system sizes, partly due to their focus on post-Hartree–Fock and post-DFT methods, which can be much more accurate but also more costly than DFT calculations. Even so, we remark that there has been growing interest in large system sizes and large basis sets in quantum chemistry (Sun, Berkelbach, McClain and Chan 2017, Stoudenmire and White 2017, Mardirossian, McClain and Chan 2018), and the methods or ideas reviewed in this paper may then become applicable as well.

Although many advanced numerical methods have been implemented in mature DFT software packages, there are still many outstanding challenges for solving Kohn–Sham DFT. Below we provide our own perspectives organized according to the contents of this review.

(1) *Numerical discretization.* The pseudopotential approximation greatly facilitated the efficient discretization of the Kohn–Sham Hamiltonian. Nonetheless, it is not a systematic approximation, and is regarded by many as one of the 'artistic' components for solving Kohn–Sham DFT. Efficient discretizations that allow efficient all-electron calculations to be performed accurately, such as the numerical atomic basis sets, wavelets, and recent development of the Gausslet basis set (White 2017), are still very much of interest. A key challenge would be to effectively control the number of degrees of freedom, so that calculations can still be performed efficiently for large systems.

(2) *Evaluation of the Kohn–Sham map with semi-local functionals.* The cross-over point of linear scaling methods and reduced scaling methods over conventional cubic scaling methods is still large, especially for three-dimensional bulk systems. Note that some key factors affecting the cross-over point, such as the decay rate of the density matrix, are determined by physical systems under consideration rather than numerical algorithms. Hence one cannot realistically expect the cross-over point to be reduced to below several hundreds of atoms in general. Nonetheless, many physical processes, such as the battery degradation process, require large simulation of systems with thousands to tens

of thousands of atoms. Therefore further improvement of numerical algorithms and their parallel scalability, particularly for metallic systems, may enable a wide range of applications beyond reach today.

(3) *Evaluation of the Kohn–Sham map with non-local functionals.* The applicability range of Kohn–Sham DFT is ultimately determined by the choice of exchange-correlation functionals, and an increasing number of numerical calculations are now performed using rung-4 and rung-5 functionals. In some sense, the precise form of these functionals is still under development and somewhat debated in the literature. For instance, the self-consistently screened hybrid functional (Brawand, Vörös, Govoni and Galli 2016) was only developed in recent years, and there is as yet no consensus on the formulation of self-consistent rung-5 functionals. In particular, rung-5 functionals are closely related to Green's function methods such as those in the many-body perturbation theory (MBPT) (Hedin 1965, Onida, Reining and Rubio 2002), which are often treated as post-DFT methods. For strongly correlated systems, novel exchange-correlation functionals such as those based on the strictly correlated electron (SCE) limit (Seidl, Gori-Giorgi and Savin 2007, Malet and Gori-Giorgi 2012) offer new perspectives on the design of exchange-correlation functionals, though they have yet to be shown to be effective for practically relevant chemical systems.

(4) *Self-consistent field iterations.* For large-scale heterogeneous systems with a small or zero energy gap, the SCF iteration often converges slowly, or fails to converge at all. This is particularly problematic for geometry optimization and molecular dynamics simulation, where a large number of iterations need to be performed at each step. It remains challenging to design efficient and robust approaches for self-consistency for large-scale systems. From this perspective, optimization based algorithms with guaranteed reduction of energy at each step could be more effective and robust.

(5) *Numerical analysis.* Due to the nonlinearity of the exchange-correlation functional, numerical analysis of Kohn–Sham DFT can be very challenging. Much progress has been made for simplified models such as the Thomas–Fermi-type models and reduced Hartree–Fock models (see *e.g.* Cancès, Deleurence and Lewin 2008, Cancès and Lewin 2010, Cancès and Mourad 2014). Even so, there are still many basic questions from numerical analysis that remain to be answered, such as the convergence analysis of self-consistent field iterations beyond the linear response regime.

Finally, we would like to emphasize again the fact that although 'Kohn–Sham DFT' has become synonymous with 'electronic structure theory' in many contexts, it is only *one* of the many branches of electronic structure theories. We have not touched upon most of the wavefunction methods in the quantum chemistry literature, nor the methods for excited states or time-dependent problems. The fact that Kohn–Sham DFT has become so widely used today is because it strikes the right balance between efficiency, accuracy and general applicability *so far*. This balance has been constantly reshaped together with the development of many other theories, and will continue to be further reshaped in the future. Let us mention two possibilities.

Recently, there has been rapid surge of interest in applying machine learning tools to electronic structure theories. In the past few years there has been considerable progress in building inter-atomic potentials, for which the quality can be comparable to a first-principles simulation (see *e.g.* Bartók, Payne and Csányi 2010, Zhang *et al.* 2018). Although machine learning based methods will still require electronic structure calculations to generate the training data, it is possible that in future we will not need to perform a monolithic, long-time first-principles molecular dynamics simulation for a large system. Instead it may be efficient enough to perform long molecular dynamics calculations for many small- to medium-sized systems in parallel, or for large systems but at many discontinuous snapshots. This may enable DFT calculations on a massively parallel scale, and many large systems beyond reach today might then become feasible. It may also be possible to use machine learning tools to build better density functionals or to more efficiently solve DFT. The interaction between physical modelling, data and numerical algorithms promises to further improve the predictive power of electronic structure theories for materials and chemical systems.

Another possibility is the fusion of DFT methods with other physical theories. In the quantum physics and chemistry literature, such multiscale-like theories are called 'quantum embedding theories'. Kohn–Sham DFT can be combined with more coarse-grained theories such as molecular mechanics to simulate large-scale systems such as biological systems. These 'quantum mechanics / molecular mechanics' (QM/MM) methods began in the 1970s (Warshel and Levitt 1976) and were recognized by the Nobel Prize in Chemistry in 2013. Kohn–Sham DFT can also be combined with more accurate post-DFT methods, such as coupled cluster methods, density matrix renormalization group methods, quantum Monte Carlo methods and exact diagonalization methods (Sun and Chan 2016, Georges, Kotliar, Krauth and Rozenberg 1996, Kotliar *et al.* 2006, Knizia and Chan 2012, Manby, Stella, Goodpaster and Miller III 2012) to solve strongly correlated systems. As of today, these methods certainly still lack the ease of use and general applicability compared to Kohn–Sham DFT. There are challenges as well as opportunities for applied mathematicians to make contributions.

Notation

General conventions	
i	imaginary unit
z^*	complex conjugate of the complex number z
N	number of electrons
M	number of nuclei
β	inverse temperature
$\oint_{\mathscr{C}} \mathrm{d}\lambda$	contour integral
$\langle\psi\rvert,\ \lvert\psi\rangle,\ \langle\psi\rvert\varphi\rangle$	bra vector, ket vector and bra–ket in Dirac notation

Coordinates	
$\mathbf{r},\ r_\alpha$	single electron spatial coordinate and its Cartesian components, $\alpha = x, y, z$ or $1, 2, 3$
$\mathbf{p},\ p_\alpha$	single electron momentum coordinate and its Cartesian components
$\mathbf{x}_i = (\mathbf{r}_i, \sigma_i)$	space-spin coordinates of the ith electron
Z_I	charge of the Ith nuclei
$\mathbf{R}_I$	spatial coordinate of the Ith nuclei

DFT-related	
Ψ or $\lvert\Psi\rangle$	N-electron wavefunction
P	single-particle density matrix
$\rho(\mathbf{r})$	electron density
$\psi_i(\mathbf{r})$ or $\varphi_i(\mathbf{r})$	ith single electron spatial orbital
ε_i	eigenvalue of the ith orbital
f_i	occupation number for the ith orbital
i, j	occupied eigenvalue index
a, b	unoccupied eigenvalue index
μ	chemical potential
$V(\mathbf{r})$	single-particle potential
$V_{\mathrm{ext}}(\mathbf{r})$	external potential
$V_{\mathrm{H}}, V_{\mathrm{x}}, V_{\mathrm{c}}, V_{\mathrm{xc}}, V_{\mathrm{Hxc}}$	Hartree, exchange, correlation, exchange-correlation and Hartree-exchange-correlation potentials
$\mathcal{F}_{\mathrm{KS}}[V]$	Kohn–Sham map from potential to density

Notation for matrix representation	
$A^{\top}$	transpose of A
A^* or $A^{\dagger}$	Hermitian transpose / adjoint of A
N_g	number of grid points / degrees of freedom
N_b	number of basis functions

Continued on next page

Continued from previous page

$\Psi = [\psi_1, \ldots, \psi_N]$	a matrix collecting N single-particle orbitals
$\Phi = [\phi_1, \ldots, \phi_{N_b}]$	a matrix collecting N_b basis functions, usually of size $N_g \times N_b$
H, S	discretized Hamiltonian and overlap matrices
G	discretized Green's function
I	identity matrix

Other quantities

$\mathbb{1}_{(-\infty,0)}$	indicator function
f_β	finite temperature Fermi–Dirac function
f_∞	zero temperature Fermi–Dirac function, the same as an indicator function
η	a small positive quantity approaching 0

REFERENCES[2]

H. M. Aktulga, L. Lin, C. Haine, E. G. Ng and C. Yang (2014), 'Parallel eigenvalue calculation based on multiple shift–invert Lanczos and contour integral based spectral projection method', *Parallel Comput.* **40**, 195–212.

P. Amestoy, I. Duff, J.-Y. L'Excellent and J. Koster (2001), 'A fully asynchronous multifrontal solver using distributed dynamic scheduling', *SIAM J. Matrix Anal. Appl.* **23**, 15–41.

O. K. Andersen (1975), 'Linear methods in band theory', *Phys. Rev.* B **12**, 3060–3083.

D. G. Anderson (1965), 'Iterative procedures for nonlinear integral equations', *J. Assoc. Comput. Mach.* **12**, 547–560.

E. Anderson, Z. Bai, C. Bischof, S. Blackford, J. Demmel, J. Dongarra, J. Du Croz, A. Greenbaum, S. Hammarling, A. McKenney and D. Sorensen (1999), *LAPACK Users' Guide*, third edition, SIAM.

D. N. Arnold (1982), 'An interior penalty finite element method with discontinuous elements', *SIAM J. Numer. Anal.* **19**, 742–760.

D. N. Arnold, F. Brezzi, B. Cockburn and L. D. Marini (2002), 'Unified analysis of discontinuous Galerkin methods for elliptic problems', *SIAM J. Numer. Anal.* **39**, 1749–1779.

C. Ashcraft and R. Grimes (1989), 'The influence of relaxed supernode partitions on the multifrontal method', *ACM Trans. Math. Software* **15**, 291–309.

I. Babuška and M. Zlámal (1973), 'Nonconforming elements in the finite element method with penalty', *SIAM J. Numer. Anal.* **10**, 863–875.

A. S. Banerjee, L. Lin, P. Suryanarayana, C. Yang and J. E. Pask (2018), 'Two-level Chebyshev filter based complementary subspace method for pushing the

---

[2] The URLs cited in this work were correct at the time of going to press, but the publisher and the authors make no undertaking that the citations remain live or are accurate or appropriate.

envelope of large-scale electronic structure calculations', *J. Chem. Theory Comput.* **14**, 2930–2946.

G. Bao, G. Hu and D. Liu (2012), 'An *h*-adaptive finite element solver for the calculations of the electronic structures', *J. Comput. Phys.* **231**, 4967–4979.

S. Baroni and P. Giannozzi (1992), 'Towards very large-scale electronic-structure calculations', *Europhys. Lett.* **17**, 547–552.

M. Barrault, E. Cancès, W. Hager and C. Le Bris (2007), 'Multilevel domain decomposition for electronic structure calculations', *J. Comput. Phys.* **222**, 86–109.

A. P. Bartók, M. C. Payne and G. Csányi (2010), 'Gaussian approximation potentials: The accuracy of quantum mechanics, without the electrons', *Phys. Rev. Lett.* **104**, 1–4.

A. D. Becke (1988), 'Density-functional exchange-energy approximation with correct asymptotic behavior', *Phys. Rev. A* **38**, 3098–3100.

A. D. Becke (1993), 'Density functional thermochemistry, III: The role of exact exchange', *J. Chem. Phys.* **98**, 5648–5652.

L. Belpassi, F. Tarantelli, A. Sgamellotti and H. M. Quiney (2005), 'Computational strategies for a four-component Dirac–Kohn–Sham program: Implementation and first applications', *J. Chem. Phys.* **122**, 184109.

G. Bencteux, M. Barrault, E. Cancès, W. W. Hager and C. Le Bris (2008), Domain decomposition and electronic structure computations: A promising approach. In *Numerical Analysis and Scientific Computing for PDEs and Their Challenging Applications* (R. Glowinski and P. Neittaanmäki, eds), Vol. 16 of Computational Methods in Applied Sciences, Springer, pp. 147–164.

M. Benzi, P. Boito and N. Razouk (2013), 'Decay properties of spectral projectors with applications to electronic structure', *SIAM Rev.* **55**, 3–64.

L. S. Blackford, J. Choi, A. Cleary, E. D'Azevedo, J. Demmel, I. Dhillon, S. Hammarling, G. Henry, A. Petitet, K. Stanley, D. Walker and R. C. Whaley (1997), *ScaLAPACK Users' Guide*, SIAM.

P. E. Blöchl (1994), 'Projector augmented-wave method', *Phys. Rev. B* **50**, 17953–17979.

E. I. Blount (1962), 'Formalisms of band theory', *Solid State Phys.* **13**, 305–373.

V. Blum, R. Gehrke, F. Hanke, P. Havu, V. Havu, X. Ren, K. Reuter and M. Scheffler (2009), '*Ab initio* molecular simulations with numeric atom-centered orbitals', *Comput. Phys. Commun.* **180**, 2175–2196.

N. M. Boffi, M. Jain and A. Natan (2016), 'Efficient computation of the Hartree–Fock exchange in real-space with projection operators', *J. Chem. Theory Comput.* **12**, 3614–3622.

D. Bohm and D. Pines (1953), 'A collective description of electron interactions, III: Coulomb interactions in a degenerate electron gas', *Phys. Rev.* **92**, 609–625.

D. R. Bowler and T. Miyazaki (2012), '$O(N)$ methods in electronic structure calculations', *Rep. Prog. Phys.* **75**, 036503.

A. Brandt (1977), 'Multi-level adaptive solutions to boundary-value problems', *Math. Comp.* **31**, 333–390.

A. Brandt, S. McCormick and J. Ruge (1985), Algebraic multigrid (AMG) for sparse matrix equations. In *Sparsity and its Applications*, Cambridge University Press, pp. 257–284.

N. P. Brawand, M. Vörös, M. Govoni and G. Galli (2016), 'Generalization of dielectric-dependent hybrid functionals to finite systems', *Phys. Rev. X* **6**, 041002.

W. Briggs, V. E. Henson and S. F. McCormick (2000), *A Multigrid Tutorial*, second edition, SIAM.

C. Brouder, G. Panati, M. Calandra, C. Mourougane and N. Marzari (2007), 'Exponential localization of Wannier functions in insulators', *Phys. Rev. Lett.* **98**, 046402.

K. Burke (2012), 'Perspective on density functional theory', *J. Chem. Phys.* **136**, 150901.

Y. Cai, Z. Bai, J. E. Pask and N. Sukumar (2013), 'Hybrid preconditioning for iterative diagonalization of ill-conditioned generalized eigenvalue problems in electronic structure calculations', *J. Comput. Phys.* **255**, 16–30.

E. Cancès and M. Lewin (2010), 'The dielectric permittivity of crystals in the reduced Hartree–Fock approximation', *Arch. Rational Mech. Anal.* **197**, 139–177.

E. Cancès and N. Mourad (2014), 'A mathematical perspective on density functional perturbation theory', *Nonlinearity* **27**, 1999.

E. Cancès and N. Mourad (2016), 'Existence of a type of optimal norm-conserving pseudopotentials for Kohn–Sham models', *Commun. Math. Sci.* **14**, 1315–1352.

E. Cancès, A. Deleurence and M. Lewin (2008), 'A new approach to the modeling of local defects in crystals: The reduced Hartree–Fock case', *Commun. Math. Phys.* **281**, 129–177.

E. Cancès, A. Levitt, G. Panati and G. Stoltz (2017), 'Robust determination of maximally localized Wannier functions', *Phys. Rev. B* **95**, 075114.

R. Car and M. Parrinello (1985), 'Unified approach for molecular dynamics and density-functional theory', *Phys. Rev. Lett.* **55**, 2471–2474.

D. M. Ceperley and B. J. Alder (1980), 'Ground state of the electron gas by a stochastic method', *Phys. Rev. Lett.* **45**, 566–569.

M. Ceriotti, T. Kühne and M. Parrinello (2008), 'An efficient and accurate decomposition of the Fermi operator', *J. Chem. Phys.* **129**, 024707.

S. Chandrasekaran, M. Gu and T. Pals (2006), 'A fast ULV decomposition solver for hierarchically semiseparable representations', *SIAM J. Matrix Anal. Appl.* **28**, 603–622.

J. Chelikowsky, N. Troullier and Y. Saad (1994), 'Finite-difference-pseudopotential method: Electronic structure calculations without a basis', *Phys. Rev. Lett.* **72**, 1240–1243.

G. P. Chen, V. K. Voora, M. M. Agee, S. G. Balasubramani and F. Furche (2017), 'Random-phase approximation method', *Ann. Rev. Phys. Chem.* **68**, 421–445.

H. Chen, X. Dai, X. Gong, L. He and A. Zhou (2014), 'Adaptive finite element approximations for Kohn–Sham models', *Multiscale Model. Simul.* **12**, 1828–1869.

J. Chen and J. Lu (2016), 'Analysis of the divide-and-conquer method for electronic structure calculations', *Math. Comp.* **85**, 2919–2938.

E. Chow, X. Liu, M. Smelyanskiy and J. R. Hammond (2015), 'Parallel scalability of Hartree–Fock calculations', *J. Chem. Phys.* **142**, 104103.

S. J. Clark, M. D. Segall, C. J. Pickard, P. J. Hasnip, M. J. Probert, K. Refson and M. C. Payne (2005), 'First principles methods using CASTEP', *Z. Kristallographie* **220**, 567–570.

B. Cockburn, G. Karniadakis and C.-W. Shu (2000), *Discontinuous Galerkin methods: Theory, Computation and Applications*, Vol. 11 of Lecture Notes in Computational Science and Engineering, Springer.

F. Corsetti (2014), 'The orbital minimization method for electronic structure calculations with finite-range atomic basis sets', *Comput. Phys. Commun.* **185**, 873–883.

A. Damle and L. Lin (2018), 'Disentanglement via entanglement: A unified method for Wannier localization', *Math. Model. Simul.* **16**, 1392–1410.

A. Damle, A. Levitt and L. Lin (2019), 'Variational formulation for Wannier functions with entangled band structure', *SIAM Multiscale Model. Simul.* **17**, 167–191.

A. Damle, L. Lin and L. Ying (2015), 'Compressed representation of Kohn–Sham orbitals via selected columns of the density matrix', *J. Chem. Theory Comput.* **11**, 1463–1469.

A. Damle, L. Lin and L. Ying (2017), 'SCDM-k: Localized orbitals for solids via selected columns of the density matrix', *J. Comput. Phys.* **334**, 1–15.

E. R. Davidson (1975), 'The iterative calculation of a few of the lowest eigenvalues and corresponding eigenvectors of large real symmetric matrices', *J. Comput. Phys.* **17**, 87–94.

W. Dawson and F. Gygi (2015), 'Performance and accuracy of recursive subspace bisection for hybrid DFT calculations in inhomogeneous systems', *J. Chem. Theory Comput.* **11**, 4655–4663.

W. Dawson and T. Nakajima (2018), 'Massively parallel sparse matrix function calculations with NTPoly', *Comput. Phys. Commun.* **225**, 154–165.

M. Dion, H. Rydberg, E. Schröder, D. C. Langreth and B. I. Lundqvist (2004), 'Van der Waals density functional for general geometries', *Phys. Rev. Lett.* **92**, 246401.

K. Dong, W. Hu and L. Lin (2018), 'Interpolative separable density fitting through centroidal Voronoi tessellation with applications to hybrid functional electronic structure calculations', *J. Chem. Theory Comput.* **14**, 1311–1320.

R. M. Dreizler and E. K. U. Gross (1990), *Density Functional Theory*, Springer.

I. Duchemin and F. Gygi (2010), 'A scalable and accurate algorithm for the computation of Hartree–Fock exchange', *Comput. Phys. Commun.* **181**, 855–860.

T. H. Dunning (1989), 'Gaussian basis sets for use in correlated molecular calculations, I: The atoms boron through neon and hydrogen', *J. Chem. Phys.* **90**, 1007–1023.

W. E and J. Lu (2011), 'The electronic structure of smoothly deformed crystals: Wannier functions and the Cauchy–Born rule', *Arch. Ration. Mech. Anal.* **199**, 407–433.

W. E, T. Li and J. Lu (2010), 'Localized bases of eigensubspaces and operator compression', *Proc. Nat. Acad. Sci.* **107**, 1273–1278.

A. Edelman, T. A. Arias and S. T. Smith (1998), 'The geometry of algorithms with orthogonality constraints', *SIAM J. Matrix Anal. Appl.* **20**, 303–353.

A. Erisman and W. Tinney (1975), 'On computing certain elements of the inverse of a sparse matrix', *Comm. Assoc. Comput. Mach.* **18**, 177–179.

H. Eschrig (1996), *The Fundamentals of Density Functional Theory*, Springer.

H.-R. Fang and Y. Saad (2009), 'Two classes of multisecant methods for nonlinear acceleration', *Numer. Linear Algebra Appl.* **16**, 197–221.

J. L. Fattebert and J. Bernholc (2000), 'Towards grid-based $O(N)$ density-functional theory methods: Optimized nonorthogonal orbitals and multigrid acceleration', *Phys. Rev. B* **62**, 1713–1722.

E. Fermi (1927), 'Un metodo statistico per la determinazione di alcune prioprietà dell'atomo', *Rend. Accad. Naz. Lincei.* **6**, 602–607.

M. Feyereisen, G. Fitzgerald and A. Komornicki (1993), 'Use of approximate integrals in *ab initio* theory: An application in MP2 energy calculations', *Chem. Phys. Lett.* **208**, 359–363.

B. Fornberg (1998), *A Practical Guide to Pseudospectral Methods*, Cambridge Monographs on Applied and Computational Mathematics, Cambridge University Press.

J. M. Foster and S. F. Boys (1960), 'Canonical configurational interaction procedure', *Rev. Mod. Phys.* **32**, 300–302.

T. Fukazawa and H. Akai (2015), 'Optimized effective potential method and application to static RPA correlation', *J. Phys. Condens. Matter* **27**, 115502.

W. Gao and W. E (2009), 'Orbital minimization with localization', *Discrete Contin. Dyn. Syst.* **23**, 249–264.

C. J. Garcia-Cervera, J. Lu, Y. Xuan and W. E (2009), 'Linear-scaling subspace-iteration algorithm with optimally localized nonorthogonal wave functions for Kohn–Sham density functional theory', *Phys. Rev. B* **79**, 115110.

M. Gell-Mann and K. A. Brueckner (1957), 'Correlation energy of an electron gas at high density', *Phys. Rev.* **106**, 364–368.

L. Genovese, A. Neelov, S. Goedecker, T. Deutsch, S. A. Ghasemi, A. Willand, D. Caliste, O. Zilberberg, M. Rayson, A. Bergman and R. Schneider (2008), 'Daubechies wavelets as a basis set for density functional pseudopotential calculations', *J. Chem. Phys.* **129**, 014109.

A. Georges, G. Kotliar, W. Krauth and M. J. Rozenberg (1996), 'Dynamical mean-field theory of strongly correlated fermion systems and the limit of infinite dimensions', *Rev. Mod. Phys.* **68**, 13–125.

P. Ghosez, X. Gonze and R. W. Godby (1997), 'Long-wavelength behavior of the exchange-correlation kernel in the Kohn–Sham theory of periodic systems', *Phys. Rev. B* **56**, 12811–12817.

P. Giannozzi, O. Andreussi, T. Brumme, O. Bunau, M. B. Nardelli, M. Calandra, R. Car, C. Cavazzoni, D. Ceresoli, M. Cococcioni, N. Colonna, I. Carnimeo, A. D. Corso, S. de Gironcoli, P. Delugas, R. A. DiStasio Jr, A. Ferretti, A. Floris, G. Fratesi, G. Fugallo, R. Gebauer, U. Gerstmann, F. Giustino, T. Gorni, J. Jia, M. Kawamura, H.-Y. Ko, A. Kokalj, E. Küçükbenli, M. Lazzeri, M. Marsili, N. Marzari, F. Mauri, N. L. Nguyen, H.-V. Nguyen, A. O. de-la Roza, L. Paulatto, S. Poncé, D. Rocca, R. Sabatini, B. Santra, M. Schlipf, A. P. Seitsonen, A. Smogunov, I. Timrov, T. Thonhauser, P. Umari, N. Vast, X. Wu and S. Baroni (2017), 'Advanced capabilities for materials modelling with QUANTUM ESPRESSO', *J. Phys. Condens. Matter* **29**, 465901.

R. W. Godby, M. Schlüter and L. J. Sham (1986), 'Accurate exchange-correlation potential for Silicon and its discontinuity on addition of an electron', *Phys. Rev. Lett.* **56**, 2415–2418.

R. W. Godby, M. Schlüter and L. J. Sham (1988), 'Self-energy operators and exchange-correlation potentials in semiconductors', *Phys. Rev. B* **37**, 10159–10175.

S. Goedecker (1999), 'Linear scaling electronic structure methods', *Rev. Mod. Phys.* **71**, 1085–1123.

S. Goedecker and L. Colombo (1994), 'Efficient linear scaling algorithm for tight-binding molecular dynamics', *Phys. Rev. Lett.* **73**, 122–125.

L. Goerigk and S. Grimme (2014), 'Double-hybrid density functionals', *WIREs Comput. Mol. Sci.* **4**, 576–600.

G. H. Golub and C. F. Van Loan (2013), *Matrix Computations*, fourth edition, Johns Hopkins University Press.

X. Gonze, F. Jollet, F. Abreu Araujo, D. Adams, B. Amadon, T. Applencourt, C. Audouze, J.-M. Beuken, J. Bieder, A. Bokhanchuk, E. Bousquet, F. Bruneval, D. Caliste, M. Côté, F. Dahm, F. Da Pieve, M. Delaveau, M. Di Gennaro, B. Dorado, C. Espejo, G. Geneste, L. Genovese, A. Gerossier, M. Giantomassi, Y. Gillet, D. Hamann, L. He, G. Jomard, J. Laflamme Janssen, S. Le Roux, A. Levitt, A. Lherbier, F. Liu, I. Lukačević, A. Martin, C. Martins, M. Oliveira, S. Poncé, Y. Pouillon, T. Rangel, G.-M. Rignanese, A. Romero, B. Rousseau, O. Rubel, A. Shukri, M. Stankovski, M. Torrent, M. Van Setten, B. Van Troeye, M. Verstraete, D. Waroquiers, J. Wiktor, B. Xu, A. Zhou and J. Zwanziger (2016), 'Recent developments in the ABINIT software package', *Comput. Phys. Commun.* **205**, 106–131.

L. Greengard and V. Rokhlin (1987), 'A fast algorithm for particle simulations', *J. Comput. Phys.* **73**, 325–348.

S. Grimme (2006), 'Semiempirical hybrid density functional with perturbative second-order correlation', *J. Chem. Phys.* **124**, 034108.

O. Gunnarsson and B. I. Lundqvist (1976), 'Exchange and correlation in atoms, molecules, and solids by the spin-density-functional formalism', *Phys. Rev. B* **13**, 4274–4298.

F. Gygi (2008), 'Architecture of Qbox: A scalable first-principles molecular dynamics code', *IBM J. Res. Dev.* **52**, 137–144.

F. Gygi (2009), 'Compact representations of Kohn–Sham invariant subspaces', *Phys. Rev. Lett.* **102**, 166406.

W. Hackbusch (1999), 'A sparse matrix arithmetic based on $\mathcal{H}$-matrices, I: Introduction to $\mathcal{H}$-matrices', *Computing* **62**, 89–108.

D. R. Hamann (2013), 'Optimized norm-conserving Vanderbilt pseudopotentials', *Phys. Rev. B* **88**, 085117.

D. R. Hamann, M. Schlüter and C. Chiang (1979), 'Norm-conserving pseudopotentials', *Phys. Rev. Lett.* **43**, 1494–1497.

C. Hartwigsen, S. Goedecker and J. Hutter (1998), 'Relativistic separable dual-space Gaussian pseudopotentials from H to Rn', *Phys. Rev. B* **58**, 3641–3662.

L. Hedin (1965), 'New method for calculating the one-particle Green's function with application to the electron-gas problem', *Phys. Rev. A* **139**, 796–823.

J. Heyd, G. E. Scuseria and M. Ernzerhof (2003), 'Hybrid functionals based on a screened Coulomb potential', *J. Chem. Phys.* **118**, 8207–8215.

N. Higham (2008), *Functions of Matrices: Theory and Computation*, SIAM.

P. Hohenberg and W. Kohn (1964), 'Inhomogeneous electron gas', *Phys. Rev. B* **136**, 864–871.

J. Hu, B. Jiang, L. Lin, Z. Wen and Y. Yuan (2018), Structured quasi-Newton methods for optimization with orthogonality constraints. arXiv:1809.00452

W. Hu, L. Lin and C. Yang (2015), 'DGDFT: A massively parallel method for large scale density functional theory calculations', *J. Chem. Phys.* **143**, 124110.

W. Hu, L. Lin and C. Yang (2017*a*), 'Interpolative separable density fitting decomposition for accelerating hybrid density functional calculations with applications to defects in silicon', *J. Chem. Theory Comput.* **13**, 5420–5431.

W. Hu, L. Lin and C. Yang (2017*b*), 'Projected commutator DIIS method for accelerating hybrid functional electronic structure calculations', *J. Chem. Theory Comput.* **13**, 5458–5467.

W. Hu, L. Lin, A. Banerjee, E. Vecharynski and C. Yang (2017*c*), 'Adaptively compressed exchange operator for large scale hybrid density functional calculations with applications to the adsorption of water on silicene', *J. Chem. Theory Comput.* **13**, 1188–1198.

M. Jacquelin, L. Lin and C. Yang (2016), 'PSelInv: A distributed memory parallel algorithm for selected inversion: The symmetric case', *ACM Trans. Math. Software* **43**, 21.

M. Jacquelin, L. Lin and C. Yang (2018), 'PSelInv: A distributed memory parallel algorithm for selected inversion: The non-symmetric case', *Parallel Comput.* **74**, 84–98.

F. Jensen (2013), 'Atomic orbital basis sets', *WIREs Comput. Mol. Sci.* **3**, 273–295.

W. Jia and L. Lin (2017), 'Robust determination of the chemical potential in the pole expansion and selected inversion method for solving Kohn–Sham density functional theory', *J. Chem. Phys.* **147**, 144107.

Y. Jin, D. Zhang, Z. Chen, N. Q. Su and W. Yang (2017), 'Generalized optimized effective potential for orbital functionals and self-consistent calculation of random phase approximation', *J. Phys. Chem. Lett.* **8**, 4746–4751.

M. Kaltak, J. Klimeš and G. Kresse (2014*a*), 'Cubic scaling algorithm for the random phase approximation: Self-interstitials and vacancies in Si', *Phys. Rev. B* **90**, 054115.

M. Kaltak, J. Klimeš and G. Kresse (2014*b*), 'Low scaling algorithms for the random phase approximation: Imaginary time and Laplace transformations', *J. Chem. Theory Comput.* **10**, 2498–2507.

E. Kaxiras (2003), *Atomic and Electronic Structure of Solids*, Cambridge University Press.

J. Kaye, L. Lin and C. Yang (2015), '*A posteriori* error estimator for adaptive local basis functions to solve Kohn–Sham density functional theory', *Commun. Math. Sci.* **13**, 1741–1773.

G. P. Kerker (1981), 'Efficient iteration scheme for self-consistent pseudopotential calculations', *Phys. Rev. B* **23**, 3082–3084.

J. Kim, F. Mauri and G. Galli (1995), 'Total-energy global optimization using nonorthogonal localized orbitals', *Phys. Rev. B* **52**, 1640–1648.

S. Kivelson (1982), 'Wannier functions in one-dimensional disordered systems: Application to fractionally charged solitons', *Phys. Rev. B* **26**, 4269–4277.

L. Kleinman and D. M. Bylander (1982), 'Efficacious form for model pseudopotentials', *Phys. Rev. Lett.* **48**, 1425–1428.

G. Knizia and G. Chan (2012), 'Density matrix embedding: A simple alternative to dynamical mean-field theory', *Phys. Rev. Lett.* **109**, 186404.

A. V. Knyazev (2001), 'Toward the optimal preconditioned eigensolver: Locally optimal block preconditioned conjugate gradient method', *SIAM J. Sci. Comput.* **23**, 517–541.

M. Kobayashi and H. Nakai (2009), 'Divide-and-conquer-based linear-scaling approach for traditional and renormalized coupled cluster methods with single, double, and noniterative triple excitations', *J. Chem. Phys.* **131**, 114108.

E. Koch and S. Goedecker (2001), 'Locality properties and Wannier functions for interacting systems', *Solid State Commun.* **119**, 105–109.

W. Kohn (1959), 'Analytic properties of Bloch waves and Wannier functions', *Phys. Rev.* **115**, 809–821.

W. Kohn (1996), 'Density functional and density matrix method scaling linearly with the number of atoms', *Phys. Rev. Lett.* **76**, 3168–3171.

W. Kohn and L. Sham (1965), 'Self-consistent equations including exchange and correlation effects', *Phys. Rev. A* **140**, 1133–1138.

G. Kotliar, S. Y. Savrasov, K. Haule, V. S. Oudovenko, O. Parcollet and C. A. Marianetti (2006), 'Electronic structure calculations with dynamical mean-field theory', *Rev. Mod. Phys.* **78**, 865–951.

G. Kresse and J. Furthmüller (1996), 'Efficient iterative schemes for *ab initio* total-energy calculations using a plane-wave basis set', *Phys. Rev. B* **54**, 11169–11186.

R. Lai and J. Lu (2016), 'Localized density matrix minimization and linear scaling algorithms', *J. Comput. Phys.* **315**, 194–210.

R. Lai, J. Lu and S. Osher (2015), 'Density matrix minimization with ℓ_1 regularization', *Commun. Math. Sci.* **13**, 2097–2117.

L. Landau and E. Lifshitz (1991), *Quantum Mechanics: Non-Relativistic Theory*, Butterworth-Heinemann.

D. C. Langreth and J. P. Perdew (1975), 'The exchange-correlation energy of a metallic surface', *Solid State Commun.* **17**, 1425–1429.

C. Lee, W. Yang and R. G. Parr (1988), 'Development of the Colle–Salvetti correlation-energy formula into a functional of the electron density', *Phys. Rev. B* **37**, 785–789.

M. Levy (1979), 'Universal variational functionals of electron densities, first-order density matrices, and natural spin-orbitals and solution of the *v*-representability problem', *Proc. Nat. Acad. Sci.* **76**, 6062–6065.

S. Li, S. Ahmed, G. Klimeck and E. Darve (2008), 'Computing entries of the inverse of a sparse matrix using the FIND algorithm', *J. Comput. Phys.* **227**, 9408–9427.

Y. Li and L. Lin (2019), 'Globally constructed adaptive local basis set for spectral projectors of second order differential operators', *Multiscale Model. Simul.* **17**, 92–116.

E. H. Lieb (1983), 'Density functionals for Coulomb systems', *Int. J. Quantum Chem.* **24**, 243–277.

E. H. Lieb and M. Loss (2001), *Analysis*, Vol. 14 of Graduate Studies in Mathematics, AMS.

L. Lin (2016), 'Adaptively compressed exchange operator', *J. Chem. Theory Comput.* **12**, 2242–2249.

L. Lin (2017), 'Localized spectrum slicing', *Math. Comp.* **86**, 2345–2371.

L. Lin and M. Lindsey (2019), 'Convergence of adaptive compression methods for Hartree–Fock-like equations', *Commun. Pure Appl. Math.* **72**, 451–499.

L. Lin and J. Lu (2016), 'Decay estimates of discretized Green's functions for Schrödinger type operators', *Sci. China Math.* **59**, 1561–1578.

L. Lin and J. Lu (2019), *A Mathematical Introduction to Electronic Structure Theory*, SIAM, to appear.

L. Lin and B. Stamm (2016), '*A posteriori* error estimates for discontinuous Galerkin methods using non-polynomial basis functions, I: Second order linear PDE', *Math. Model. Numer. Anal.* **50**, 1193–1222.

L. Lin and B. Stamm (2017), '*A posteriori* error estimates for discontinuous Galerkin methods using non-polynomial basis functions, II: Eigenvalue problems', *Math. Model. Numer. Anal.* **51**, 1733–1753.

L. Lin and C. Yang (2013), 'Elliptic preconditioner for accelerating self consistent field iteration in Kohn–Sham density functional theory', *SIAM J. Sci. Comp.* **35**, S277–S298.

L. Lin, M. Chen, C. Yang and L. He (2013), 'Accelerating atomic orbital-based electronic structure calculation via pole expansion and selected inversion', *J. Phys. Condens. Matter* **25**, 295501.

L. Lin, J. Lu, L. Ying and W. E (2009*a*), 'Pole-based approximation of the Fermi–Dirac function', *Chin. Ann. Math.* B **30**, 729–742.

L. Lin, J. Lu, L. Ying and W. E (2012*a*), 'Adaptive local basis set for Kohn–Sham density functional theory in a discontinuous Galerkin framework, I: Total energy calculation', *J. Comput. Phys.* **231**, 2140–2154.

L. Lin, J. Lu, L. Ying and W. E (2012*b*), 'Optimized local basis function for Kohn–Sham density functional theory', *J. Comput. Phys.* **231**, 4515–4529.

L. Lin, J. Lu, L. Ying, R. Car and W. E (2009*b*), 'Fast algorithm for extracting the diagonal of the inverse matrix with application to the electronic structure analysis of metallic systems', *Commun. Math. Sci.* **7**, 755–777.

L. Lin, Z. Xu and L. Ying (2017), 'Adaptively compressed polarizability operator for accelerating large scale *ab initio* phonon calculations', *Multiscale Model. Simul.* **15**, 29–55.

L. Lin, C. Yang, J. Meza, J. Lu, L. Ying and W. E (2011), 'SelInv: An algorithm for selected inversion of a sparse symmetric matrix', *ACM. Trans. Math. Software* **37**, 40.

B. Liu (1978), The simultaneous expansion method for the iterative solution of several of the lowest eigenvalues and corresponding eigenvectors of large real-symmetric matrices. Report LBL-8158, Lawrence Berkeley Laboratory, University of California, Berkeley.

P.-O. Löwdin (1950), 'On the non-orthogonality problem connected with the use of atomic wave functions in the theory of molecules and crystals', *J. Chem. Phys.* **18**, 365–375.

J. Lu and K. Thicke (2017*a*), 'Cubic scaling algorithm for RPA correlation using interpolative separable density fitting', *J. Comput. Phys.* **351**, 187–202.

J. Lu and K. Thicke (2017*b*), 'Orbital minimization method with ℓ^1 regularization', *J. Comput. Phys.* **336**, 87–103.

J. Lu and L. Ying (2015), 'Compression of the electron repulsion integral tensor in tensor hypercontraction format with cubic scaling cost', *J. Comput. Phys.* **302**, 329–335.

J. Lu and L. Ying (2016), 'Fast algorithm for periodic density fitting for Bloch waves', *Ann. Math. Sci. Appl.* **1**, 321–339.

J. Lu, C. D. Sogge and S. Steinerberger (2018), Approximating pointwise products of Laplacian eigenfunctions. arXiv:1811.10447

T. Lu, W. Cai, J. Xin and Y. Guo (2013), 'Linear scaling discontinuous Galerkin density matrix minimization method with local orbital enriched finite element basis: 1-D lattice model system', *Commun. Comput. Phys.* **14**, 276–300.

A. Luenser, H. F. Schurkus and C. Ochsenfeld (2017), 'Vanishing-overhead linear-scaling Random Phase Approximation by Cholesky decomposition and an attenuated Coulomb-metric', *J. Chem. Theory Comput.* **13**, 1647–1655.

J. MacQueen (1967), Some methods for classification and analysis of multivariate observations. In *Proc. Fifth Berkeley Symposium on Mathematical Statistics and Probability*, Vol. 1, pp. 281–297, University of California Press.

G. Mahan (2000), *Many-Particle Physics*, Plenum.

G. Makov and M. C. Payne (1995), 'Periodic boundary conditions in *ab initio* calculations', *Phys. Rev. B* **51**, 4014–4022.

F. Malet and P. Gori-Giorgi (2012), 'Strong correlation in Kohn–Sham density functional theory', *Phys. Rev. Lett.* **109**, 246402.

F. R. Manby, M. Stella, J. D. Goodpaster and T. F. Miller III (2012), 'A simple, exact density-functional-theory embedding scheme', *J. Chem. Theory Comput.* **8**, 2564–2568.

N. Mardirossian, J. D. McClain and G. Chan (2018), 'Lowering of the complexity of quantum chemistry methods by choice of representation', *J. Chem. Phys.* **148**, 044106.

A. Marek, V. Blum, R. Johanni, V. Havu, B. Lang, T. Auckenthaler, A. Heinecke, H.-J. Bungartz and H. Lederer (2014), 'The ELPA library: Scalable parallel eigenvalue solutions for electronic structure theory and computational science', *J. Phys. Condens. Matter* **26**, 213201.

L. D. Marks and D. R. Luke (2008), 'Robust mixing for *ab initio* quantum mechanical calculations', *Phys. Rev. B* **78**, 075114–075125.

R. Martin (2008), *Electronic Structure: Basic Theory and Practical Methods*, Cambridge University Press.

D. Marx and J. Hutter (2009), *Ab Initio Molecular Dynamics: Basic Theory and Advanced Methods*, Cambridge University Press.

N. Marzari and D. Vanderbilt (1997), 'Maximally localized generalized Wannier functions for composite energy bands', *Phys. Rev. B* **56**, 12847–12865.

N. Marzari, A. A. Mostofi, J. R. Yates, I. Souza and D. Vanderbilt (2012), 'Maximally localized Wannier functions: Theory and applications', *Rev. Mod. Phys.* **84**, 1419–1475.

F. Mauri and G. Galli (1994), 'Electronic-structure calculations and molecular-dynamics simulations with linear system-size scaling', *Phys. Rev. B* **50**, 4316–4326.

F. Mauri, G. Galli and R. Car (1993), 'Orbital formulation for electronic-structure calculations with linear system-size scaling', *Phys. Rev. B* **47**, 9973–9976.

R. McWeeny (1960), 'Some recent advances in density matrix theory', *Rev. Mod. Phys.* **32**, 335–369.

N. Mermin (1965), 'Thermal properties of the inhomogeneous electron gas', *Phys. Rev. A* **137**, 1441–1443.

S. Mohr, L. E. Ratcliff, P. Boulanger, L. Genovese, D. Caliste, T. Deutsch and S. Goedecker (2014), 'Daubechies wavelets for linear scaling density functional theory', *J. Chem. Phys.* **140**, 204110.

P. Mori-Sánchez, Q. Wu and W. Yang (2005), 'Orbital-dependent correlation energy in density-functional theory based on a second-order perturbation approach: Success and failure', *J. Chem. Phys.* **123**, 062204.

J. E. Moussa (2014), 'Cubic-scaling algorithm and self-consistent field for the random-phase approximation with second-order screened exchange', *J. Chem. Phys.* **140**, 014107.

J. E. Moussa (2016), 'Minimax rational approximation of the Fermi–Dirac distribution', *J. Chem. Phys.* **145**, 164108.

J. I. Mustafa, S. Coh, M. L. Cohen and S. G. Louie (2015), 'Automated construction of maximally localized Wannier functions: Optimized projection functions method', *Phys. Rev. B* **92**, 165134.

G. Nenciu (1983), 'Existence of the exponentially localised Wannier functions', *Comm. Math. Phys.* **91**, 81–85.

A. M. N. Niklasson (2002), 'Expansion algorithm for the density matrix', *Phys. Rev. B.* **66**, 155115.

A. M. N. Niklasson (2011), Linear-scaling techniques in computational chemistry and physics. In *Challenges and Advances in Computational Chemistry and Physics* (R. Zalesny et al., eds), Springer, pp. 439–473.

A. M. N. Niklasson, C. J. Tymczak and M. Challacombe (2003), 'Trace resetting density matrix purification in $\mathcal{O}(N)$ self-consistent-field theory', *J. Chem. Phys.* **118**, 8611–8620.

J. Nocedal and S. J. Wright (1999), *Numerical Optimization*, Springer.

N. Ohba, S. Ogata, T. Kouno, T. Tamura and R. Kobayashi (2012), 'Linear scaling algorithm of real-space density functional theory of electrons with correlated overlapping domains', *Comput. Phys. Commun.* **183**, 1664–1673.

G. Onida, L. Reining and A. Rubio (2002), 'Electronic excitations: Density-functional versus many-body Green's-function approaches', *Rev. Mod. Phys.* **74**, 601–659.

P. Ordejón, D. A. Drabold, M. P. Grumbach and R. M. Martin (1993), 'Unconstrained minimization approach for electronic computations that scales linearly with system size', *Phys. Rev. B* **48**, 14646–14649.

P. Ordejón, D. A. Drabold, R. M. Martin and M. P. Grumbach (1995), 'Linear system-size scaling methods for electronic-structure calculations', *Phys. Rev. B* **51**, 1456–1476.

T. Ozaki (2007), 'Continued fraction representation of the Fermi–Dirac function for large-scale electronic structure calculations', *Phys. Rev. B* **75**, 035123.

A. H. R. Paler and D. E. Manolopoulos (1998), 'Canonical purification of the density matrix in electronic-structure theory', *Phys. Rev. B* **58**, 12704–12711.

G. Panati and A. Pisante (2013), 'Bloch bundles, Marzari–Vanderbilt functional and maximally localized Wannier functions', *Commun. Math. Phys.* **322**, 835–875.

R. Parr and W. Yang (1989), *Density Functional Theory of Atoms and Molecules*, Oxford University Press.

R. M. Parrish, E. G. Hohenstein, T. J. Martínez and C. D. Sherrill (2012), 'Tensor hypercontraction, II: Least-squares renormalization', *J. Chem. Phys.* **137**, 224106.

R. M. Parrish, E. G. Hohenstein, T. J. Martínez and C. D. Sherrill (2013), 'Discrete variable representation in electronic structure theory: Quadrature grids for least-squares tensor hypercontraction', *J. Chem. Phys.* **138**, 194107.

M. C. Payne, M. P. Teter, D. C. Allen, T. A. Arias and J. D. Joannopoulos (1992), 'Iterative minimization techniques for *ab initio* total energy calculation: Molecular dynamics and conjugate gradients', *Rev. Mod. Phys.* **64**, 1045–1097.

J. P. Perdew (2013), 'Climbing the ladder of density functional approximations', *MRS Bull.* **38**, 743–750.

J. P. Perdew and K. Schmidt (2001), Jacob's ladder of density functional approximations for the exchange-correlation energy. In *AIP Conference Proceedings*, Vol. 577, pp. 1–20.

J. P. Perdew and A. Zunger (1981), 'Self-interaction correction to density-functional approximations for many-electron systems', *Phys. Rev. B* **23**, 5048–5079.

J. P. Perdew, K. Burke and M. Ernzerhof (1996a), 'Generalized gradient approximation made simple', *Phys. Rev. Lett.* **77**, 3865–3868.

J. P. Perdew, M. Ernzerhof and K. Burke (1996b), 'Rationale for mixing exact exchange with density functional approximations', *J. Chem. Phys.* **105**, 9982–9985.

D. E. Petersen, S. Li, K. Stokbro, H. H. B. Sørensen, P. C. Hansen, S. Skelboe and E. Darve (2009), 'A hybrid method for the parallel computation of Green's functions', *J. Comput. Phys.* **228**, 5020–5039.

B. Pfrommer, J. Demmel and H. Simon (1999), 'Unconstrained energy functionals for electronic structure calculations', *J. Comput. Phys.* **150**, 287–298.

R. Pick, M. Cohen and R. Martin (1970), 'Microscopic theory of force constants in the adiabatic approximation', *Phys. Rev. B* **1**, 910–920.

E. Polizzi (2009), 'Density-matrix-based algorithm for solving eigenvalue problems', *Phys. Rev. B* **79**, 115112–115117.

E. Prodan and W. Kohn (2005), 'Nearsightedness of electronic matter', *Proc. Nat. Acad. Sci.* **102**, 11635–11638.

P. Pulay (1969), '*Ab initio* calculation of force constants and equilibrium geometries in polyatomic molecules, I: Theory', *Mol. Phys.* **17**, 197–204.

P. Pulay (1980), 'Convergence acceleration of iterative sequences: The case of SCF iteration', *Chem. Phys. Lett.* **73**, 393–398.

P. Pulay (1982), 'Improved SCF convergence acceleration', *J. Comput. Chem.* **3**, 54–69.

M. J. Rayson and P. R. Briddon (2009), 'Highly efficient method for Kohn–Sham density functional calculations of 500–10 000 atom systems', *Phys. Rev. B* **80**, 205104.

S. Reine, T. Helgaker and R. Lindh (2012), 'Multi-electron integrals', *WIREs Comput. Mol. Sci.* **2**, 290–303.

X. Ren, P. Rinke, V. Blum, J. Wieferink, A. Tkatchenko, A. Sanfilippo, K. Reuter and M. Scheffler (2012a), 'Resolution-of-identity approach to Hartree–Fock, hybrid density functionals, RPA, MP2 and GW with numeric atom-centered orbital basis functions', *New J. Phys.* **14**, 053020.

X. Ren, P. Rinke, C. Joas and M. Scheffler (2012b), 'Random-phase approximation and its applications in computational chemistry and materials science', *J. Mater. Sci.* **47**, 7447–7471.

X. Ren, P. Rinke, G. E. Scuseria and M. Scheffler (2013), 'Renormalized second-order perturbation theory for the electron correlation energy: Concept, implementation, and benchmarks', *Phys. Rev. B* **88**, 035120.

Y. Saad and M. H. Schultz (1986), 'GMRES: A generalized minimal residual algorithm for solving nonsymmetric linear systems', *SIAM J. Sci. Statist. Comput.* **7**, 856–869.

O. Schenk and K. Gartner (2006), 'On fast factorization pivoting methods for symmetric indefinite systems', *Elec. Trans. Numer. Anal.* **23**, 158–179.

G. Schofield, J. R. Chelikowsky and Y. Saad (2012), 'A spectrum slicing method for the Kohn–Sham problem', *Comput. Phys. Commun.* **183**, 497–505.

H. F. Schurkus and C. Ochsenfeld (2016), 'Communication: An effective linear-scaling atomic-orbital reformulation of the random-phase approximation using a contracted double-Laplace transformation', *J. Chem. Phys.* **144**, 031101.

M. Seidl, P. Gori-Giorgi and A. Savin (2007), 'Strictly correlated electrons in density-functional theory: A general formulation with applications to spherical densities', *Phys. Rev. A* **75**, 042511.

Y. Shao, Z. Gan, E. Epifanovsky, A. T. Gilbert, M. Wormit, J. Kussmann, A. W. Lange, A. Behn, J. Deng, X. Feng, D. Ghosh, M. Goldey, P. R. Horn, L. D. Jacobson, I. Kaliman, R. Z. Khaliullin, T. Kuś, A. Landau, J. Liu, E. I. Proynov, Y. M. Rhee, R. M. Richard, M. A. Rohrdanz, R. P. Steele, E. J. Sundstrom, H. L. Woodcock III, P. M. Zimmerman, D. Zuev, B. Albrecht, E. Alguire, B. Austin, G. J. O. Beran, Y. A. Bernard, E. Berquist, K. Brandhorst, K. B. Bravaya, S. T. Brown, D. Casanova, C.-M. Chang, Y. Chen, S. H. Chien, K. D. Closser, D. L. Crittenden, M. Diedenhofen, R. A. DiStasio Jr, H. Do, A. D. Dutoi, R. G. Edgar, S. Fatehi, L. Fusti-Molnar, A. Ghysels, A. Golubeva-Zadorozhnaya, J. Gomes, M. W. Hanson-Heine, P. H. Harbach, A. W. Hauser, E. G. Hohenstein, Z. C. Holden, T.-C. Jagau, H. Ji, B. Kaduk, K. Khistyaev, J. Kim, J. Kim, R. A. King, P. Klunzinger, D. Kosenkov, T. Kowalczyk, C. M. Krauter, K. U. Lao, A. D. Laurent, K. V. Lawler, S. V. Levchenko, C. Y. Lin, F. Liu, E. Livshits, R. C. Lochan, A. Luenser,

P. Manohar, S. F. Manzer, S.-P. Mao, N. Mardirossian, A. V. Marenich, S. A. Maurer, N. J. Mayhall, E. Neuscamman, C. M. Oana, R. Olivares-Amaya, D. P. O'Neill, J. A. Parkhill, T. M. Perrine, R. Peverati, A. Prociuk, D. R. Rehn, E. Rosta, N. J. Russ, S. M. Sharada, S. Sharma, D. W. Small, A. Sodt, T. Stein, D. Stück, Y.-C. Su, A. J. Thom, T. Tsuchimochi, V. Vanovschi, L. Vogt, O. Vydrov, T. Wang, M. A. Watson, J. Wenzel, A. White, C. F. Williams, J. Yang, S. Yeganeh, S. R. Yost, Z.-Q. You, I. Y. Zhang, X. Zhang, Y. Zhao, B. R. Brooks, G. K. Chan, D. M. Chipman, C. J. Cramer, W. A. Goddard III, M. S. Gordon, W. J. Hehre, A. Klamt, H. F. Schaefer III, M. W. Schmidt, C. D. Sherrill, D. G. Truhlar, A. Warshel, X. Xu, A. Aspuru-Guzik, R. Baer, A. T. Bell, N. A. Besley, J.-D. Chai, A. Dreuw, B. D. Dunietz, T. R. Furlani, S. R. Gwaltney, C.-P. Hsu, Y. Jung, J. Kong, D. S. Lambrecht, W. Liang, C. Ochsenfeld, V. A. Rassolov, L. V. Slipchenko, J. E. Subotnik, T. V. Voorhis, J. M. Herbert, A. I. Krylov, P. M. Gill and M. Head-Gordon (2015), 'Advances in molecular quantum chemistry contained in the Q-Chem 4 program package', *Mol. Phys.* **113**, 184–215.

F. Shimojo, R. K. Kalia, A. Nakano and P. Vashishta (2008), 'Divide-and-conquer density functional theory on hierarchical real-space grids: Parallel implementation and applications', *Phys. Rev. B* **77**, 085103.

F. Shimojo, S. Ohmura, A. Nakano, R. Kalia and P. Vashishta (2011), 'Large-scale atomistic simulations of nanostructured materials based on divide-and-conquer density functional theory', *Eur. Phys. J. Spec. Top.* **196**, 53–63.

C. Skylaris, P. Haynes, A. Mostofi and M. Payne (2005), 'Introducing ONETEP: Linear-scaling density functional simulations on parallel computers', *J. Chem. Phys.* **122**, 084119.

J. C. Slater (1937), 'Wave functions in a periodic potential', *Phys. Rev.* **51**, 846–851.

J. M. Soler, E. Artacho, J. D. Gale, A. García, J. Junquera, P. Ordejón and D. Sánchez-Portal (2002), 'The SIESTA method for *ab initio* order-N materials simulation', *J. Phys. Condens. Matter* **14**, 2745–2779.

I. Souza, N. Marzari and D. Vanderbilt (2001), 'Maximally localized Wannier functions for entangled energy bands', *Phys. Rev. B* **65**, 035109.

V. N. Staroverov, G. E. Scuseria, J. Tao and J. P. Perdew (2003), 'Comparative assessment of a new nonempirical density functional: Molecules and hydrogen-bonded complexes', *J. Chem. Phys.* **119**, 12129–12137.

E. M. Stoudenmire and S. R. White (2017), 'Sliced basis density matrix renormalization group for electronic structure', *Phys. Rev. Lett.* **119**, 046401.

J. Sun, A. Ruzsinszky and J. P. Perdew (2015), 'Strongly constrained and appropriately normed semilocal density functional', *Phys. Rev. Lett.* **115**, 036402.

Q. Sun and G. K.-L. Chan (2016), 'Quantum embedding theories', *Acc. Chem. Res.* **49**, 2705–2712.

Q. Sun, T. C. Berkelbach, J. D. McClain and G. Chan (2017), 'Gaussian and plane-wave mixed density fitting for periodic systems', *J. Chem. Phys.* **147**, 164119.

P. Suryanarayana, V. Gavani, T. Blesgen, K. Bhattacharya and M. Ortiz (2010), 'Non-periodic finite-element formulation of Kohn–Sham density functional theory', *J. Mech. Phys. Solids* **58**, 258–280.

J. J. Sylvester (1852), 'A demonstration of the theorem that every homogeneous quadratic polynomial is reducible by real orthogonal substitutions to the form of a sum of positive and negative squares', *Philos. Mag.* **4**, 138–142.

A. Szabo and N. Ostlund (1989), *Modern Quantum Chemistry: Introduction to Advanced Electronic Structure Theory*, McGraw-Hill.

M. Teter, M. Payne and D. Allan (1989), 'Solution of Schrödinger's equation for large systems', *Phys. Rev. B* **40**, 12255–12263.

B. Thaller (1992), *The Dirac Equation*, Springer.

L. H. Thomas (1927), 'The calculation of atomic fields', *Proc. Camb. Phil. Soc.* **23**, 542–548.

L. N. Trefethen (2008), 'Is Gauss quadrature better than Clenshaw–Curtis?', *SIAM Rev.* **50**, 67–87.

N. Troullier and J. L. Martins (1991), 'Efficient pseudopotentials for plane-wave calculations', *Phys. Rev. B* **43**, 1993–2006.

L. A. Truflandier, R. M. Dianzinga and D. R. Bowler (2016), 'Communication: Generalized canonical for density matrix minimization', *J. Chem. Phys.* **144**, 091102.

E. Tsuchida (2007), 'Augmented orbital minimization method for linear scaling electronic structure calculations', *J. Phys. Soc. Japan* **76**, 034708.

E. Tsuchida and M. Tsukada (1995), 'Electronic-structure calculations based on the finite-element method', *Phys. Rev. B* **52**, 5573–5578.

M. Valiev, E. J. Bylaska, N. Govind, K. Kowalski, T. P. Straatsma, H. J. J. Van Dam, D. Wang, J. Nieplocha, E. Apra, T. L. Windus and W. De Jong (2010), 'NWChem: A comprehensive and scalable open-source solution for large scale molecular simulations', *Comput. Phys. Commun.* **181**, 1477–1489.

D. Vanderbilt (1990), 'Soft self-consistent pseudopotentials in a generalized eigenvalue formalism', *Phys. Rev. B* **41**, 7892–7895.

E. Vecharynski, C. Yang and J. E. Pask (2015), 'A projected preconditioned conjugate gradient algorithm for computing many extreme eigenpairs of a Hermitian matrix', *J. Comput. Phys.* **290**, 73–89.

C. Vömel (2010), 'ScaLAPACK's MRRR algorithm', *ACM Trans. Math. Software* **37**, 1.

U. von Barth and L. Hedin (1972), 'A local exchange-correlation potential for the spin polarized case', *J. Phys. C Solid State Phys.* **5**, 1629–1642.

L.-W. Wang, Z. Zhao and J. Meza (2008), 'Linear-scaling three-dimensional fragment method for large-scale electronic structure calculations', *Phys. Rev. B* **77**, 165113.

G. H. Wannier (1937), 'The structure of electronic excitation levels in insulating crystals', *Phys. Rev.* **52**, 191–197.

A. Warshel and M. Levitt (1976), 'Theoretical studies of enzymic reactions: Dielectric, electrostatic and steric stabilization of the carbonium ion in the reaction of lysozyme', *J. Mol. Biol.* **103**, 227–249.

F. Weigend (2002), 'A fully direct RI-HF algorithm: Implementation, optimised auxiliary basis sets, demonstration of accuracy and efficiency', *Phys. Chem. Chem. Phys.* **4**, 4285–4291.

F. Weigend, M. Häser, H. Patzelt and R. Ahlrichs (1998), 'RI-MP2: Optimized auxiliary basis sets and demonstration of efficiency', *Chem. Phys. Lett.* **294**, 143–152.

Z. Wen and W. Yin (2013), 'A feasible method for optimization with orthogonality constraints', *Math. Program.* **142**, 397–434.

H. Werner, P. J. Knowles, G. Knizia, F. R. Manby and M. Schütz (2012), 'Molpro: A general-purpose quantum chemistry program package', *WIREs Comput. Mol. Sci.* **2**, 242–253.

S. R. White (2017), 'Hybrid grid/basis set discretizations of the Schrödinger equation', *J. Chem. Phys.* **147**, 244102.

J. Wilhelm, P. Seewald, M. Del Ben and J. Hutter (2016), 'Large-scale cubic-scaling random phase approximation correlation energy calculations using a Gaussian basis', *J. Chem. Theory Comput.* **12**, 5851–5859.

X. Wu, A. Selloni and R. Car (2009), 'Order-N implementation of exact exchange in extended insulating systems', *Phys. Rev. B* **79**, 085102.

Q. Xu, P. Suryanarayana and J. E. Pask (2018), 'Discrete discontinuous basis projection method for large-scale electronic structure calculations', *J. Chem. Phys.* **149**, 094104.

C. Yang, J. Meza and L. Wang (2006), 'A constrained optimization algorithm for total energy minimization in electronic structure calculations', *J. Comput. Phys.* **217**, 709–721.

W. Yang (1991*a*), 'Direct calculation of electron density in density-functional theory', *Phys. Rev. Lett.* **66**, 1438–1441.

W. Yang (1991*b*), 'Direct calculation of electron density in density-functional theory: Implementation for benzene and a tetrapeptide', *Phys. Rev. A* **44**, 7823–7826.

W. Yang and T.-S. Lee (1995), 'A density-matrix divide-and-conquer approach for electronic structure calculations of large molecules', *J. Chem. Phys.* **103**, 5674–5678.

V. W.-z. Yu, F. Corsetti, A. García, W. P. Huhn, M. Jacquelin, W. Jia, B. Lange, L. Lin, J. Lu, W. Mi, A. Seifitokaldani, A. Vazquez-Mayagoitia, C. Yang, H. Yang and V. Blum (2018), 'ELSI: A unified software interface for Kohn–Sham electronic structure solvers', *Comput. Phys. Commun.* **222**, 267–285.

G. Zhang, L. Lin, W. Hu, C. Yang and J. E. Pask (2017), 'Adaptive local basis set for Kohn–Sham density functional theory in a discontinuous Galerkin framework, II: Force, vibration, and molecular dynamics calculations', *J. Comput. Phys.* **335**, 426–443.

H. Zhang, B. Smith, M. Sternberg and P. Zapol (2007), 'SIPs: Shift-and-invert parallel spectral transformations', *ACM Trans. Math. Software* **33**, 9–19.

I. Y. Zhang, P. Rinke and M. Scheffler (2016), 'Wave-function inspired density functional applied to the H_2 / H_2^+ challenge', *New J. Phys.* **18**, 073026.

L. Zhang, J. Han, H. Wang, R. Car and W. E (2018), 'Deep potential molecular dynamics: A scalable model with the accuracy of quantum mechanics', *Phys. Rev. Lett.* **120**, 143001.

Y. Zhang, X. Xu and W. A. Goddard III (2009), 'Doubly hybrid density functional for accurate descriptions of nonbond interactions, thermochemistry, and thermochemical kinetics', *Proc. Nat. Acad. Sci. USA* **106**, 4963–4968.

Z. Zhao, J. Meza and L.-W. Wang (2008), 'A divide-and-conquer linear scaling three-dimensional fragment method for large scale electronic structure calculations', *J. Phys. Condens. Matter* **20**, 294203.

Y. Zhou, J. R. Chelikowsky and Y. Saad (2014), 'Chebyshev-filtered subspace iteration method free of sparse diagonalization for solving the Kohn–Sham equation', *J. Comput. Phys.* **274**, 770–782.

Y. Zhou, Y. Saad, M. L. Tiago and J. R. Chelikowsky (2006), 'Self-consistent-field calculations using Chebyshev-filtered subspace iteration', *J. Comput. Phys.* **219**, 172–184.

J. M. Ziman (1979), *Principles of the Theory of Solids*, Cambridge University Press.

Acta Numerica (2019), pp. 541–633
doi:10.1017/S0962492919000035
© Cambridge University Press, 2019
Printed in the United Kingdom

Approximation algorithms in combinatorial scientific computing

Alex Pothen[*]

*Department of Computer Science, Purdue University,
West Lafayette, IN 47907, USA*
E-mail: apothen@purdue.edu

S. M. Ferdous[†]

*Department of Computer Science, Purdue University,
West Lafayette, IN 47907, USA*
E-mail: sferdou@purdue.edu

Fredrik Manne

*Department of Informatics, University of Bergen,
N-5020 Bergen, Norway*
E-mail: fredrikm@ii.uib.no

We survey recent work on approximation algorithms for computing degree-constrained subgraphs in graphs and their applications in combinatorial scientific computing. The problems we consider include maximization versions of cardinality matching, edge-weighted matching, vertex-weighted matching and edge-weighted b-matching, and minimization versions of weighted edge cover and b-edge cover. Exact algorithms for these problems are impractical for massive graphs with several millions of edges. For each problem we discuss theoretical foundations, the design of several linear or near-linear time approximation algorithms, their implementations on serial and parallel computers, and applications. Our focus is on practical algorithms that yield good performance on modern computer architectures with multiple threads and interconnected processors. We also include information about the software available for these problems.

[*] The work of the first two authors was supported in part by US NSF grant CCF-1637534; the US Department of Energy through grant DE-FG02-13ER26135; and the Exascale Computing Project (17-SC-20-SC), a collaborative effort of the DOE Office of Science and the NNSA.

CONTENTS

1. Introduction

We discuss recent progress in the design of approximation algorithms for two problems on graphs, with their applications to combinatorial scientific computing (CSC). The problems involve the computation of degree-constrained subgraphs of a graph that might represent its significant subgraphs. Computing these subgraphs reduce the computational costs and memory required of algorithms that obtain information from the graph, such as semi-supervised classification in machine learning, or the solution of sparse systems of linear equations. These subgraphs also help remove noise from the data so that machine learning algorithms perform better in classification tasks.

The first problem we consider is the classical one of computing a matching in a graph. A matching is a subset of vertex-disjoint edges; hence there is *at most* one edge of the subset incident on each vertex in the graph. Here we could seek to maximize the cardinality of a matching, or when weights are assigned to edges, maximize the sum of the weights of edges in a matching. We will also discuss a less studied variant where the weights are on the vertices instead of the edges. A generalization of matching is the b-matching problem, where we are given natural numbers $b(v)$ for each vertex v in the graph, and are required to choose at most $b(v)$ matched edges incident on v. When weights are assigned to the edges, we seek to maximize the sum of weights of the matched edges.

The second problem we consider is edge cover, where we are required to choose *at least* one edge incident on each vertex to belong to the edge cover. Here we seek to minimize the cardinality of the edges in the cover, or the sum of weights of the edges in the cover. The generalization of an edge cover leads to the b-edge cover problem, where given natural numbers $b(v)$ for each vertex v, we are required to choose at least $b(v)$ edges incident on v

to belong to the edge cover. Again, we seek to minimize the sum of weights of the edges in the cover. Our work on this problem was motivated by an application to a data privacy problem called adaptive anonymity.

Both of these problems and their variants have polynomial time algorithms to solve them; however, the asymptotic run time is larger than the product of the number of edges times the square root of the number of vertices, and this is too high to be practical for graphs with millions or billions of vertices and edges. Furthermore, exact algorithms for these problems have little concurrency. Hence we turn to the design of approximation algorithms that have near-linear time complexity in the size of the graph. We also design approximation algorithms that possess high concurrency, so that they can be implemented efficiently on parallel computers.

An *exact* algorithm for an optimization problem computes the optimum value of its objective function. An *approximation algorithm* for an optimization problem computes a value that is within some factor α (a constant or a function of the problem size) of the optimal value for all problem instances. For a maximization problem (as in matching), the ratio of the value computed by the approximation algorithm to the maximum value is at least $\alpha < 1$ for all instances; for a minimization problem (as in edge cover), the ratio of the value computed by the approximation algorithm to the optimal value is at most $\alpha > 1$, again for all instances. We say that this is an α-approximation algorithm for the problem, and that the approximation ratio of the algorithm is α. Note that this worst-case approximation ratio is obtained analytically by an *a priori* argument, and the approximation ratio for a specific instance might be much better than α. For many optimization problems, known exact algorithms might not have polynomial time complexity. Also, for many problems, for any $\epsilon > 0$ if an algorithm with approximation ratio $n^{(1-\epsilon)}$ exists then $P = NP$, which suggests that a polynomial time approximation algorithm might not exist. An algorithm for an optimization problem for which we cannot obtain an approximation ratio is called a *heuristic algorithm*. This is the situation for many problems in CSC, and we can evaluate an algorithm by empirically comparing the value of the objective function it computes with other algorithms on a collection of test problems.

There are several advantages that approximation algorithms have over exact algorithms, which we enumerate as follows.

- Approximation algorithms have lower run time complexity relative to exact algorithms, often linear or nearly linear in the size of the problem. Exact algorithms with polynomial run times can be too slow for massive graphs, but the faster approximation algorithms may be practical. Furthermore, in practice these approximation algorithms compute solutions that are nearly optimal.

- Approximation algorithms are conceptually simpler than exact algorithms, and their proofs of correctness could also be simpler.

- Approximation algorithms are easier to implement when compared to the more sophisticated exact algorithms, which is practically an important reason for their widespread use.

- Approximation algorithms can be designed to have more concurrency than exact algorithms. The simplicity of implementation of approximation algorithms is even more important for algorithms to be implemented on parallel computers. This is a major motivating factor for our work on matchings and edge covers.

- Often a matching or edge cover algorithm is used at each step of an approximation algorithm or a heuristic to solve another problem. In this case, exact matchings are not required, and its use increases the run time of the algorithm, limiting the size of the problems that could be solved. This is the case for network alignment (Khan, Gleich, Pothen and Halappanavar 2012), for adaptive anonymity (Khan *et al.* 2018*a*), k-nearest neighbour graph construction (Ferdous, Pothen and Khan 2018), ontology alignment (Kolyvakis, Kalousis, Smith and Kiritsis 2018), *etc.*

Approximation algorithms have been studied in the discrete mathematics, theoretical computer science and operations research communities. Books discussing approximation algorithms include Hochbaum (1997), Vazirani (2003), Williamson and Shmoys (2011) and Du, Ko and Hu (2012). An earlier survey on approximation algorithms for the matching problem was provided by Hougardy (2009). Our discussion of the matching problem is fairly disjoint from this survey; we have included more recent algorithms such as the SUITOR and b-SUITOR algorithms, and problems such as b-matching and the vertex-weighted matching problem, as well as parallel algorithms. We are not aware of an earlier survey of the edge cover and b-edge cover problems. Goemans and Williamson (1997) survey the primal–dual method for designing approximation algorithms and apply it to several network design problems, including vertex cover and edge cover.

We conclude this section with a brief discussion of the research area and community of combinatorial scientific computing (CSC). Research in CSC focuses on the design, theoretical analysis, computational evaluation and deployment of combinatorial algorithms to solve problems in computational science and engineering. The CSC community has its roots in the research areas of sparse matrix computations, algorithmic differentiation and parallel computing. This community was formally organized in the early 2000s, and biennial workshops on CSC have been held under the auspices of the Society for Industrial and Applied Mathematics

(SIAM) since 2004. Information about these meetings and proceedings published by SIAM is available at www.csc-research.org. Snapshots of recent research in CSC are included in the collection of articles edited by Naumann and Schenk (2012); the first chapter by Hendrickson and Pothen provides more detail on the historical development and emerging research areas of CSC. More recently the CSC community has joined with other computational discrete mathematicians to organize a SIAM Activity Group on Applied and Computational Discrete Algorithms. Additional information is available at www.siam.org/membership/Activity-Groups/detail/applied-and-computational-discrete-algorithms.

2. Maximum cardinality matching

2.1. Elementary definitions and concepts

We begin with notation and concepts needed for discussing matching and edge cover problems. We consider a simple, loopless, undirected graph $G = (V, E)$, where V is the set of vertices, E is the set of edges, and $|V| \equiv n$ and $|E| \equiv m$. The neighbours of a vertex u will be denoted by $\mathrm{adj}(u)$ or $N(u)$; the vertex u itself is not a member of the adjacency set. The cardinality of the adjacency set is the degree of the vertex, denoted $\deg(u)$. The maximum degree of a vertex in the graph will be denoted Δ.

An edge $e = (u, v)$ has the vertices u and v as its endpoints. We say that e is an edge incident on u (and v), and that u and v are adjacent vertices. Two edges e and f are adjacent or are neighbours if they share a common endpoint. The set of edges adjacent to $e = (u, v)$ consists of other edges in G with u or v as an endpoint.

A *path* in a graph is sequence of edges $\{(v_1, v_2), (v_2, v_3), \dots (v_k, v_{k+1})\}$, where the vertices are distinct and consecutive edges share an endpoint. The *length* of the path is the number of edges k. A *cycle* is the concatenation of a path and an edge joining its first and last vertices, (v_1, v_{k+1}).

A matching in the graph G is a subset of vertex-disjoint edges M. Thus at most one edge in M is incident on each vertex in V. We say that an edge in M is a matched edge, and the endpoints of a matched edge are matched vertices. If an edge belongs to $E \setminus M$ then it is unmatched, and similarly for unmatched vertices.

Let M be a matching. Then a path or cycle P is *alternating* if it consists of edges drawn alternately from M and $E \setminus M$. An alternating path P is an M-*augmenting path* if it begins and ends with an unmatched edge. An augmenting path has one more unmatched edge than matched edges. By exchanging matched edges with unmatched edges along the augmenting path, *i.e.* by computing $M \oplus P$, we obtain a matching M' with one more matched edge than M. (Here $A \oplus B = (A \setminus B) \cup (B \setminus A)$.) In some situations, we consider two matchings M_1 and M_2, and consider the symmetric difference $M_1 \oplus M_2$. The symmetric difference consists of isolated vertices (an edge belonging to both matchings), and alternating paths and cycles; here the edges belong alternately to M_1 and M_2, and thus vertices on such paths have degree 1 or 2 in the subgraph induced by the two matchings. Such paths could be used to augment the cardinality of the matching M_2 if M_1 has more edges on the path. Alternating cycles have the same number of edges from M_1 and M_2, and cannot augment the cardinality of either matching. Augmenting paths and cycles with respect to weights on edges will be discussed in Section 3.2.

A maximum cardinality matching in a graph could be obtained by an algorithm that begins with the empty matching and at each step finds an augmenting path from an unmatched vertex. Hopcroft and Karp (1973) obtained a $O(\sqrt{n}m)$-time algorithm for finding maximum cardinality matchings in bipartite graphs, and this was extended to an algorithm for finding maximum matchings in non-bipartite graphs by Micali and Vazirani (1980).

Matching algorithms are discussed in many books on graphs and graph algorithms as well as specialized books on matchings, for example Burkard, Dell'Amico and Martello (2009), Lovász and Plummer (2009) and Schrijver (2003).

2.2. Augmenting path-based approximation

We begin by proving a lemma obtained by Hopcroft and Karp (1973).

Lemma 2.1. If all augmenting paths with respect to a matching M in a graph G have length greater than or equal to $2k - 1$, then the matching is a $(k - 1)/k$-approximation of a maximum cardinality matching M^*.

Proof. We consider the symmetric difference of a maximum cardinality matching M^* and the given matching M. It consists of vertices of degree zero, one or two, since at most one edge in M^* and one edge in M can be incident on any given vertex. Isolated vertices and alternating cycles have the same number of edges from M and M^*, and the ratio $|M|/|M^*|$ is one. There are $|M^*| - |M|$ vertex-disjoint M-augmenting paths in the symmetric difference that account for the differences in cardinalities of the two matchings. Each augmenting path P has at least $k - 1$ edges from M and k edges from M^*, and hence for the path the ratio $|M \cap P|/|M^* \cap P| \geq (k - 1)/k$. Since the inequality is true for every augmenting path, the approximation ratio is $|M|/|M^*| \geq (k - 1)/k$. $\square$

An important special case of the above lemma is when the augmenting path has length at least three, that is, we find a maximal matching in G. Then we have a 1/2-approximation to the maximum cardinality matching. This observation is used in many practical matching codes as an initialization step before commencing searches for longer augmenting paths. By increasing the lengths of the augmenting paths, one obtains an approximation ratio as close to one as possible; however, with higher augmenting path lengths, the algorithm more closely resembles the exact algorithm due to Micali and Vazirani (1980), which requires $O(\sqrt{n})$ phases, where each phase constructs a maximal, vertex-disjoint set of shortest augmenting paths. Bast, Mehlhorn, Schäfer and Tamaki (2006) have shown that on an Erdős–Rényi random graph on n vertices with the probability of an edge equal to $33/n$, $G(n, 33/n)$, the Micali–Vazirani algorithm requires at most $O(\log n)$ phases. A similar result was obtained earlier by Motwani (1994). Hougardy (2009) showed that an approximation ratio of $4/5$ is achievable in $O(m)$ time for the maximum cardinality matching problem by finding augmenting paths of length less than or equal to seven.

2.3. Randomized approximation algorithms

We now consider two randomized algorithms that compute approximations better than $1/2$ by first scaling the adjacency matrix of the bipartite graph to a doubly stochastic matrix, when the graph has the property that every edge belongs to some maximum cardinality matching. Such graphs are said to have *total support*. (Another way of stating this condition is that the Dulmage–Mendelsohn decomposition of the bipartite graph consists of subgraphs that have (1) a perfect matching, or (2) a row-perfect matching, or (3) a column-perfect matching, and no edge joins two distinct components to each other. The Dulmage–Mendelsohn decomposition is discussed briefly in Section 3.4.1 and in more detail in Pothen and Fan (1990).) One of the advantages of these algorithms is that they are highly concurrent, and hence can be easily implemented on parallel architectures. These algorithms were designed and implemented by Dufossé, Kaya and Uçar (2015), and we follow their discussion.

We consider a bipartite graph $G = (V_1, V_2, E)$ with the same number of vertices in the two sets V_1 and V_2, so that its adjacency matrix is a square matrix with elements belonging to $\{0, 1\}$. Every edge joins some vertex $i \in V_1$ with some vertex in $j \in V_2$, and the element (i, j) in the adjacency matrix is then 1; it would be 0 if there is no such edge. The scaling algorithm was designed by Sinkhorn and Knopp (1967), and it alternates in scaling the matrix with a diagonal matrix of the reciprocal of the column sums and another diagonal matrix with the reciprocal of the row sums. When this process is iterated on a bipartite graph that has total support, the scaled adjacency matrix converges to a doubly stochastic matrix, *i.e.* a matrix whose column sums and row sums are equal to one. The values of the matrix elements are used in a randomized algorithm to determine the probability of matching an edge.

Algorithm 1 Sinkhorn–Knopp (A, ϵ)

Input: An $n \times n$ adjacency matrix A with total support, and an error threshold ϵ.

Output: Row scaling array d_r and a column scaling array d_c.

  1: Initialize $d_r(i) = 1$, $d_c(i) = 1$, for $i = 1, \ldots, n$.
  2: Initialize $\mathrm{csum}(j) = \sum_i A_{i,j}$, for $j = 1, \ldots, n$.
  3: **while** $\max_j |1 - \mathrm{csum}(j)| > \epsilon$ **do**
  4:      $d_c(j) = \mathrm{csum}(j)$, for $j = 1, \ldots, n$
  5:      **for** $i = 1$ **to** n **do**
  6:          $\mathrm{rsum}(i) = \sum_j A_{i,j} \times d_c(j)$
  7:          $d_r(i) = 1/\mathrm{rsum}(i)$
  8:      **end for**
  9:      **for** $j = 1$ **to** n **do**
10:          $\mathrm{csum}(j) = \sum_i A_{i,j} \times d_r(i)$
11:          $d_c(j) = 1/\mathrm{csum}(j)$
12:      **end for**
13: **end while**
14: **return** d_r, d_c

Algorithm 1 describes the procedure to convert an adjacency matrix with total support to a doubly stochastic matrix. Here ϵ is an upper bound on the permissible distance from a column sum of one. Other scaling algorithms could also be used for this purpose, but the Sinkhorn–Knopp algorithm is more concurrent. Also Dufossé, Kaya and Uçar (2015) report that five to ten iterations of this scaling algorithm suffice to compute approximate matchings.

Algorithm 2 describes how a random matching is computed from the doubly stochastic matrix S derived from the bipartite graph. The variable $\mathrm{Diag}(d_r)$ denotes a diagonal matrix obtained with the elements of the vector d_r on the diagonal. The algorithm uses the matrix element s_{ij} to determine the probability with which a column j is matched to a row i. In this algorithm a row could attempt to match to a column that is already matched to another row; in this case, one of the rows, say the last, succeeds, and the other row gets unmatched.

Dufossé *et al.* (2015) proved that this random matching will match $n(1 - 1/e)$ vertices with high probability, where e is the base of natural logarithm, the Euler number.

Theorem 2.2. Let A be an $n \times n$ adjacency matrix corresponding to a bipartite graph $G = (V_1, V_2, E)$ with total support. Then the random matching obtained by Algorithm 2 has expected cardinality $n(1 - 1/e)$ as $n \to \infty$.

Algorithm 2 $(1 - 1/e)$-APPROXIMATE MAXIMUM CARDINALITY MATCHING (A, ϵ)

Input: An $n \times n$ matrix A with total support.
Output: An array cmatch(.) of the rows matched to columns.
1: $(d_r, d_c) = $ Sinkhorn–Knopp(A, ϵ)
2: $S = \mathrm{Diag}(d_r)\, A\, \mathrm{Diag}(d_c)$
3: **for** $i = 1$ **to** n **do**
4:      Pick a random column $j \in \mathrm{adj}(i)$ with probability s_{ij};
5:      cmatch$(j) = i$
6: **end for**
7: **return** cmatch

Proof. The probability that a column j is not matched to any of the rows in $\mathrm{adj}(j)$ is $\prod_{i \in \mathrm{adj}(j)}(1 - s_{ij})$. Let d_j denote the degree of column j, *i.e.* $|\mathrm{adj}(j)|$. From the inequality that the geometric mean is less than or equal to the arithmetic mean, we have

$$\left(\prod_{i \in \mathrm{adj}(j)} (1 - s_{ij}) \right)^{1/d_j} \leq \frac{d_j - \sum_{i \in \mathrm{adj}(j)} s_{ij}}{d_j}.$$

Since S is doubly stochastic, the sum on the right-hand side of the inequality is one; hence after taking the d_jth power, the right-hand side simplifies to

$$(1 - 1/d_j)^{d_j} \leq (1/e).$$

Thus the probability that column j is matched is at least $1 - 1/e$, and the expected size of the matching is at least $n(1 - 1/e)$. $\qquad\square$

The algorithm we have described is a $((1 - 1/e) \approx 0.632)$-approximation algorithm for maximum cardinality matching on bipartite graphs with total support. We mention that an online algorithm for maximum cardinality matching has the same approximation ratio. This latter algorithm was originally designed by Karp, Vazirani and Vazirani (1990), and a simpler proof of its correctness and approximation ratio was provided by Birnbaum and Mathieu (2008).

A better approximation ratio can be obtained from a variant 'two-sided' matching algorithm. In order to describe this algorithm, we need to discuss an algorithm due to Karp and Sipser (1981) for computing a random matching in a graph. The Karp–Sipser algorithm matches a vertex of degree one to its only neighbour, since it must be matched in any maximum cardinality matching. Then it deletes the endpoints of the matched edge, and the edges incident on them from the graph. This may create new vertices of degree one. The algorithm repeatedly matches vertices of degree one, until none is left. At this stage, if all vertices have been matched, then the algorithm terminates. If not, it chooses a random vertex in the graph and matches it to one of its neighbours, and then deletes the endpoints and the edges incident on them. The algorithm iterates until all vertices are matched.

In the two-sided matching algorithm, vertices in both the vertex sets V_1 and V_2 choose at random a single neighbour, with the probability given by the matrix elements in the doubly stochastic matrix S. The subgraph induced by the chosen edges has at most $2n$ edges. It could be fewer than $2n$, if two vertices belonging to different vertex sets choose each other. Each connected component of this induced graph can have at most the same number of edges as the number of vertices, and hence it is a tree or a graph with one cycle. In this graph, we run the Karp–Sipser algorithm to compute a matching. Since each connected component is either a tree or a unicyclic graph, the Karp–Sipser algorithm computes a maximum cardinality matching in the graph. When the initial graph has total support, Dufossé, Kaya and Uçar (2015) have proved that the two-sided algorithm leads to a 0.866-approximation for maximum cardinality matching as $n \to \infty$ with high probability.

3. Edge-weighted matching

In this section we present approximation algorithms for the weighted matching problem on an undirected graph $G = (V, E, w)$ with $w : E \to \mathbb{R}_{\geq 0}$. The objective is to find a matching M such that the sum of the weights of the edges in M is as large as possible. We denote such a maximum weight matching by M^*. The fastest exact algorithm for the weighted matching problem has run time $O(mn + n^2 \log n)$ (Gabow 2018), and the run time of computing an optimal solution can get prohibitively large even for moderate sized graphs, it is of interest to investigate fast approximation algorithms. As we will explain, some of these algorithms also lend themselves well to parallel execution.

Definitions

Let M be a matching. An alternating path or cycle P is an *augmentation* if $M \oplus P$ is also a matching. The weight of a set of edges S is $w(S) = \sum_{e \in S} w(e)$. The *gain* of an alternating path or cycle P is $g(P) = w(P \setminus M) - w(P \cap M)$.

For $k \geq 1$, a *k-augmentation* is an augmentation containing *at most k* edges not in M. Figure 3.1(a) shows all possible 1-augmentations. In the figure a solid line denotes an edge in the current matching and a dotted line an edge not in the matching. By including the paths in Figure 3.1(b) we get all possible 2-augmentations that are paths, while Figure 3.1(c) shows the only cycle possible in a 2-augmentation. Together these graphs show all possible 2-augmentations. In terms of cardinality a k-augmentation that

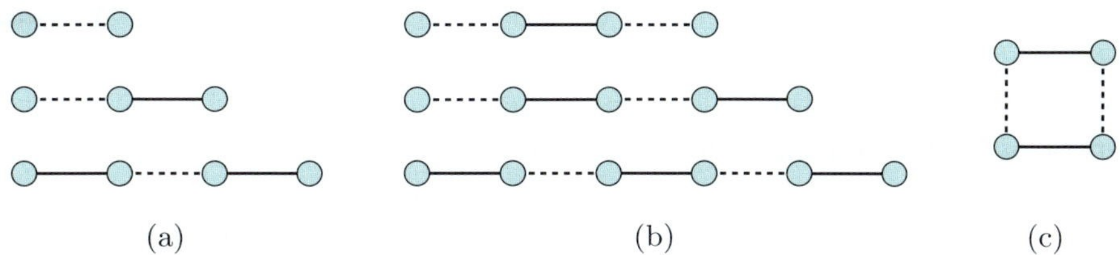

(a)   (b)   (c)

Figure 3.1. 2-augmenting paths and cycles.

has $l \leq k$ unmatched edges contains either $l-1$, l or $l+1$ matched edges. The gain of an augmentation is positive if the sum of the weights of the unmatched edges is greater than the sum of the weights of the matched edges. (Note that an alternating path with one more matched edge than unmatched edges could have positive gain, but would not be an augmenting path for the cardinality of the matching.)

3.1. Approximation lemma

It appears to be well known that if a matching does not admit positive gain $(k-1)$-augmentations then it must have a weight of at least a factor $(k-1)/k$ times the weight of a maximum matching $w(M^*)$. Still, to the best of our knowledge a proof of this does not appear to have been written down previously. We therefore include a formal proof here. This is a generalization of the proof given for Theorem 1 in Drake and Hougardy (2003*a*) which shows the result for $k = 2$.

Lemma 3.1. Let M be a matching on G and k an integer greater than 1 such that M does not admit any positive gain $(k-1)$-augmentations. Then

$$\frac{k-1}{k} w(M^*) \leq w(M),$$

where M^* is a maximum weight matching.

Proof. Let $S = M \cap M^*$, that is, S consists of the edges of G that are common to both matchings M and M^*. Further, let $R = M \setminus S$ and $R^* = M^* \setminus S$. Then $M = S \cup R$ and $M^* = S \cup R^*$. We will show that

$$w(R) \geq \frac{k-1}{k} \cdot w(R^*).$$

The result will then follow since

$$w(M) = w(S) + w(R) \geq \frac{k-1}{k} \cdot (w(S) + w(R^*)) = \frac{k-1}{k} \cdot w(M^*).$$

Since R and R^* are matchings, the subgraph induced by the edges $T = R \cup R^*$ consists of vertices of degree one or two. Thus the edges in T can be split into a set of even length cycles C and a set of paths P. We show the result separately for each cycle in C and each path in P.

Consider a cycle $C_i \in C$. If C_i contains at most $2k - 2$ edges then by the assumption of the lemma it follows that

$$w(C_i \cap M) \geq w(C_i \cap M^*) \geq \frac{k-1}{k} \cdot w(C_i \cap M^*).$$

Assume therefore that $|C_i| = 2r$ where $2r \geq 2k$. We denote by m_i edges of C_i that belong to the matching M, and by m_i^* edges that belong to M^*. Let $m_1, m_1^*, m_2, m_2^*, \ldots, m_{k-1}, m_{k-1}^*, m_k$ be a clockwise path of $2k - 1$ consecutive distinct edges from C_i, where each $m_j \in M$ and each $m_j^* \in M^*$. Then by the assumption of the lemma, it follows that $\sum_{j=1}^{k} w(m_j) \geq \sum_{j=1}^{k-1} w(m_j^*)$. There are r distinct starting edges belonging to M for such a path, each one giving rise to a different inequality. In these inequalities, every $e \in C_i \cap M$ will appear once in every position m_j, where $1 \leq j \leq k$. Thus each $e \in C_i \cap M$ appears in exactly k of these inequalities. Similarly, each $f \in C_i \cap M^*$ appears in exactly $k - 1$ inequalities. It follows that if we sum the left and right sides of the inequalities we get

$$k \cdot w(C_i \cap M) \geq (k - 1) \cdot w(C_i \cap M^*),$$

which leads to the desired inequality for each $C_i \in C$.

Next, consider a path $P_i \in P$. Note first that $|P_i| \geq 2$ and thus contains at least one edge from each of M and M^*. If this were not the case then either M would contain an augmenting path of length one, or M^* would not be maximal. Now append two paths, each containing $2k - 2$ dummy edges of weight 0, to the first and last vertex of P_i respectively. As in the case for C_i, repeatedly overlay a path $m_1, m_1^*, m_2, m_2^*, \ldots, m_{k-1}, m_{k-1}^*, m_k$ of $2k - 1$ edges onto P_i such that each $e \in P_i \cap M$ appears exactly once for each m_j, where $1 \leq j \leq k$. Note that dummy vertices may be assigned to the start and the end of such paths. For each path the inequality

$$\sum_{j=1}^{k} w(m_j) \geq \sum_{j=1}^{k-1} w(m_j^*)$$

still holds true as any assigned dummy edges have weight 0. Since every $e \in P_i \cap M$ appears in exactly k inequalities, while each $f \in P_j \cap M^*$ appears in exactly $k - 1$ inequalities, it follows that by taking the sum of the inequalities we get

$$w(P_i \cap M) \geq \frac{k-1}{k} \cdot w(P_i \cap M^*). \qquad \square$$

Note that the statement of Lemma 3.1 is slightly more restrictive than that of Lemma 2.1 in that the former lemma does not permit the existence of any weight-increasing path of length $2k - 1$ edges with respect to the

---

Algorithm 3 LOCAL GREEDY$(G(V, E, w))$

1: $M = \emptyset$
2: Set every $u \in V$ as unmarked
3: **for** $u \in V$ **do**
4:     **if** u is unmarked **then**
5:         $(u, v) = \arg\max_{x \in N(u)}\{w(u, x) \text{ s.t. } x \text{ is unmarked}\}$
6:         Mark u and v
7:         $M = M \cup (u, v)$
8:     **end if**
9: **end for**
10: **return** M

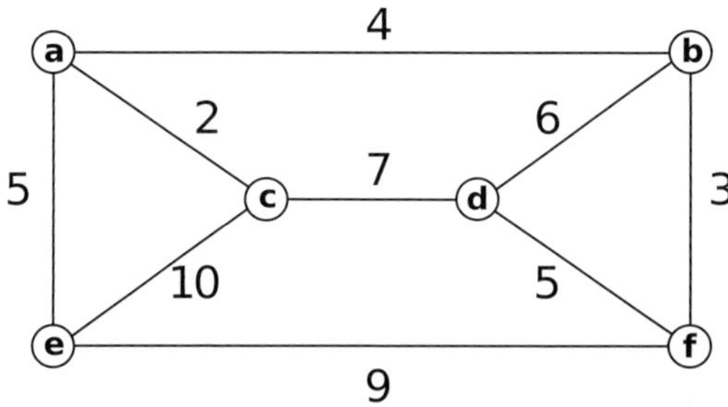

Figure 3.2. Example graph for approximate weighted matching.

matching M, where k edges belong to M and $k - 1$ edges to M^*. Such an augmentation is not permissible for maximum cardinality matching as it would reduce the cardinality of the matching M.

3.2. Edge-weighted approximation algorithms

3.2.1. GREEDY and PATH GROWING algorithms

In the following we give an overview of various efforts at designing efficient linear or close to linear time algorithms for the edge-weighted matching problem.

Perhaps the simplest matching heuristic is to visit each vertex u once and unless u is already incident on a matched edge, add an edge (u, v) of maximum weight to the matching where v is not already incident on a matched edge. The heuristic is shown in Algorithm 3.

As an example consider the graph in Figure 3.2 consisting of six vertices and nine edges with integer weights. If the graph is traversed in lexicographical order then LOCAL GREEDY will compute the matching $M = \{(a, e), (b, d)\}$ of weight 11. Note that the optimal matching is given by $M^* = \{(a, b), (c, d), (e, f)\}$ of weight 20.

Algorithm 4 Greedy($G(V, E, w)$)

1: $M = \emptyset$
2: Set every $u \in V$ as unmarked
3: **for** $(u, v) \in E$ by non-increasing weight **do**
4:     **if** u and v are both unmarked **then**
5:         Mark u and v
6:         $M = M \cup (u, v)$
7:     **end if**
8: **end for**
9: **return** M

While this strategy has linear run time in the size of G, it does not offer any bound on the quality of the obtained solution. To see this consider an even length (number of edges) path $v_1, v_2, \ldots, v_k$ where $w(v_i, v_{i+1}) = \epsilon$ if i is odd, $w(v_i, v_{i+1}) = \delta$ if i is even, and $\epsilon < \delta$. If the Local Greedy algorithm processes the vertices in the given order, it will compute a matching with weight $k/2 \cdot \epsilon$, while the optimal solution has weight $k/2 \cdot \delta$.

One way to obtain a quality guarantee on the computed matching is to visit the edges in a predefined order. This gives rise to the classical greedy matching algorithm as shown in Algorithm 4. Here a matching is computed by considering each edge e by non-increasing weights. Then if e is not adjacent to an edge already in the matching it is added to the matching; otherwise it is discarded.

In the graph in Figure 3.2 Greedy will compute $M = \{(c, e), (b, d)\}$ of weight 16. The following result shows that Greedy always gives a solution of weight at least half of the optimal one.

Lemma 3.2. The Greedy algorithm computes a matching M that is a $1/2$-approximation to a maximum weight matching M^*.

Proof. We show that M does not admit a 1-augmentation, and the result will then follow from Lemma 3.1. For every $e \in E \setminus M$ there must be either one or two edges in M adjacent to e, otherwise e would have been added to M. Let $e' \in M$ be an edge adjacent to e of maximum weight. If $w(e) > w(e')$ then e would have been considered before any of its adjacent edges in M and then added to M. It follows that $w(e) \leq w(e')$ and that e cannot be part of a 1-augmentation with respect to M. $\square$

It is clear that the Greedy algorithm runs in time $O(m \log n)$ due to the sorting of the edges. However, it is possible to compute a $1/2$-approximation in linear time. One such is the Path Growing algorithm (Drake and Hougardy 2003b), shown in Algorithm 5. In line 9 of the algorithm, we concatenate the edge (u, v) to the path P.

Algorithm 5 Path Growing($G(V, E, w)$)

1: $M = \emptyset$
2: Mark each $u \in V$ as unvisited
3: **for** $u \in V$ **do**
4:     **if** u is unvisited **then**
5:         $P = \{\}$
6:         Mark u as visited
7:         **while** u has unvisited neighbours **do**
8:             $v = \arg\max_{x \in N(u)}\{w(u, x), x \text{ is unvisited}\}$
9:             $P = P \circ (u, v)$
10:             $u = v$
11:             Mark u as visited
12:         **end while**
13:     **end if**
14:     Add the heavier set of the odd or even numbered edges of P to M
15: **end for**
16: **return** M

In its original form the algorithm starts at an arbitrary vertex u and then grows a path P from u by traversing a heaviest edge incident on u that leads to an unvisited vertex v. The process is then continued from u and terminates when a vertex is reached such that there is no edge leading to an unvisited vertex. The result is a path $P = e_0, e_1, \ldots, e_k$. This is then repeated starting from every unvisited vertex. For each such path one selects the heavier set of the odd or even numbered edges to add to M.

In the graph in Figure 3.2 starting from vertex a Path Growing would follow the path a, e, c, d, b, f. The edges of this path would then be divided into sets $\{(a, e), (c, d), (b, f)\}$ (odd) of weight 15 and $\{(c, e), (b, d)\}$ (even) of weight 16. The final solution consists of the even set.

The run time of Path Growing is clearly $O(n + m)$ as the search for a path is only started once from each vertex and each edge is only considered twice (once from each of its endpoints). Note that the resulting matching might contain 1-augmentations. This is the case for a path of length four, e_0, e_1, e_2, e_3 with edge weights 3, 1, 2 and 3. Then the resulting matching will consist of the edges e_0 and e_2; however, e_2 and e_3 form a 1-augmentation. Thus it is not possible to use Lemma 3.1 to show the performance guarantee. Still, it is not hard to prove that the resulting matching is a 1/2-approximation.

Lemma 3.3. The Path Growing algorithm computes a matching M that is a 1/2-approximation to a maximum weight matching M^*.

Proof. Let $K = P_0, P_1, \ldots, P_k$ be the non-empty paths generated by Path Growing. We show that $\sum_i w(P_i) \geq w(M^*)$. The result then follows since

$$w(M) \geq \sum_i \frac{1}{2} \cdot w(P_i) \geq \frac{1}{2} \cdot w(M^*).$$

Consider an edge $(u, v) \in M^*$. Then if (u, v) is not part of some P_i at least one of u and v must be part of some path P_j, otherwise Path Growing would have started a new path from either u or v. Assume therefore that u is part of P_i and that if v is part of P_j then $i < j$. This ensures that u is visited before v irrespective of if v is part of some P_j or not. Thus when Path Growing visits u it is possible to choose the edge (u, v). Since this is not done it must choose an edge (u, x) to add to P_i such that $w(u, x) \geq w(u, v)$. Note that u can only be incident on at most one edge in M^*. It therefore follows that for every edge $e \in M^*$ either e is in some P_i or there is a unique edge $e' \in P_i$ such that $w(e) \leq w(e')$. $\qquad\square$

The Path Growing algorithm can be improved further without increasing its asymptotic run time, by noting that one can compute an optimal matching for each P_i instead of just choosing between the even and odd numbered edges. This is done using dynamic programming as follows. Let $P = e_0, e_1, \ldots, e_k$ be the edges on a selected path and let $\mathrm{Opt}(i)$ denote the weight of an optimal matching on $e_0, e_1, \ldots, e_{i-1}$. Then the weight of an optimal matching on P can be computed in linear time using the recursion

$$\mathrm{Opt}(i) = \max\{\mathrm{Opt}(i-1), \mathrm{Opt}(i-2) + w(e_{i-1})\},$$

with $\mathrm{Opt}(0) = 0$ and $\mathrm{Opt}(1) = w(e_0)$. The actual matching can be calculated by storing a Boolean value for each i to indicate if e_{i-1} is used in the computation of $\mathrm{Opt}(i)$. We denote this algorithm as Path Growing-DP.

The first linear time $1/2$-approximation algorithm for weighted matching was given by Preis (1999), and it is based on the notion of a *dominating* edge. An edge e is dominating if it is heavier than all of its incident edges. This Locally Dominant Edge algorithm repeatedly locates and adds dominant edges to the matching. When an edge is added to the matching, all adjoining edges are removed from the graph before the next dominant edge is found. The outline of this strategy is shown in Algorithm 6.

To be able to break ties we rank edges of equal weight first by the ID of the higher-numbered endpoint and then by the lower-numbered one. In practice we will only be comparing edges of equal weight incident on the same vertex and thus we always break ties by comparing the ID of the opposing vertex.

To achieve the linear time run time Algorithm 6 performs a carefully designed depth-first search (DFS) of G, locating and removing dominating edges as it progresses. Since the actual algorithm is quite involved, we

---

Algorithm 6 LOCALLY DOMINANT EDGE($G(V, E, w)$)

1: $M = \emptyset$
2: Mark each $u \in V$ as unvisited
3: **while** E is not empty **do**
4:     Locate a dominating edge (u, v)
5:     $M = M \cup (u, v)$
6:     Remove all edges incident on u and v from E
7: **end while**
8: **return** M

---

do not list the details. Moreover, a DFS search is inherently linear, thus prohibiting parallel execution. Hence we present variants of Algorithm 6 that have worse theoretical run times, but which are faster in practice and that also lend themselves better to parallel execution.

Lemma 3.4. If tie-breaking is done consistently, Algorithm 6 computes the same matching as the GREEDY algorithm.

Proof. Let M and M' be the matchings computed by Algorithms 4 and 6, respectively. Let $S = e_0, e_1, \ldots, e_{m-1}$ be the edges in E by non-increasing weight. The proof is by induction on the edges in S with the induction hypothesis being that M and M' consists of exactly the same edges from $e_0, e_1, \ldots, e_k$.

For $k = 0$ we only consider the heaviest edge in G. Obviously this will be included in the matchings computed by both algorithms. Assume therefore that the induction hypothesis holds for all $l < k$, where $k \geq 1$, and consider edge e_k. If e_k is not used by GREEDY then it must be adjacent to some edge e_l where $l < k$ and $e_l \in M$. By assumption $e_l \in M'$, and since it is not possible for two adjacent edges to be picked by Algorithm 6, it follows that $e_k \notin M'$.

If $e_k \in M$ then it cannot be adjacent to any edge $e_l \in M$ where $l < k$. The only way that Algorithm 6 can avoid including e_k in M' is if it is removed because some adjacent edge e_l is included in M'. For e_l to be dominant before e_k is removed requires that $w(e_l) > w(e_k)$. Since the edges are considered by non-increasing weight this implies that $l < k$, and thus by the induction hypothesis $e_l \notin M'$. It follows that $e_k \in M'$ if and only if $e_k \in M$. $\qquad\square$

The simplest algorithm based on discovering dominating edges is the LOCAL MAX algorithm of Birn *et al.* (2013). The algorithm operates in stages. At each stage a maximal set of dominating edges are discovered and added to the matching. These edges, along with their vertices, are then removed from the graph before the process is repeated. The algorithm terminates when the graph is empty. It is not hard to see that the run time of each stage is linear in the size of the remaining graph. It is possible that only a constant number of vertices and edges are removed at each stage, thus leading to a worst-case run time of $O(nm)$. However, it can be shown that if the edge weights are drawn from a uniform random distribution, then half of the remaining edges will be removed at each iteration (Birn *et al.* 2013). Thus in this case the expected number of rounds is $O(\log n)$ and the expected run time is $O(m)$.

3.2.2. *The* Suitor *algorithm*

The Suitor algorithm, designed by Manne and Halappanavar (2014), is another variant of Algorithm 6 that also computes the same matching as Greedy. The algorithm is based on considering the vertices as players that are making bids to match with one of their neighbours. The value of a bid from a vertex u to a vertex v is $w(u, v)$. We denote the highest bid a vertex u has received as its suitor value, denoted by $s(u)$.

The algorithm works by each vertex u proposing to its neighbour v such that $w(u, v)$ is maximized under the constraint that v has not already received a higher offer than $w(u, v)$. (We say that a vertex v that satisfies this property is an eligible neighbour of u.) If at a later stage in the algorithm v receives a better offer, then the bid from u to v is annulled, and u has to make a new bid to its next heaviest eligible neighbour. The algorithm terminates when every vertex u is either a suitor of one of its neighbours or when u has no more vertices to propose to. At this point the edges where $s(u) = v$ and $s(v) = u$ constitute the matching.

There is no restriction on the order in which the vertices in V are processed nor on the order in which rejected suitors are processed. This leaves considerable freedom in how the algorithm is implemented. Two of the most natural ways to handle this is to use either a queue or a stack to organize the vertices that still need to be processed. If a queue is used then all vertices are initially put in the queue and the next considered vertex is always taken as the head of the queue. Any vertex that gets dislodged as a suitor because its chosen neighbour receives a better offer will be added at the end of the queue. The algorithm terminates when the queue is empty.

When using a stack all vertices are initially put on the stack and the next vertex to process is taken from the top of the stack. Any vertex that gets dislodged as a suitor is put on the top of the stack. Note that this is equivalent to processing each vertex exactly once and then immediately starting to process any vertex that gets dislodged. This variant of the Suitor algorithm is shown in Algorithm 7. In the algorithm, if a vertex x does not have a suitor, *i.e.* $s(x) = \text{NULL}$, then $w(u, s(x)) = 0$.

As to correctness consider any initial dominating edge (u, v), for instance the heaviest edge in E. Then the first time u is processed, the value of

Algorithm 7 Suitor$(G(V, E, w))$

```
1: for each u ∈ V do
2:     s(u) = NULL
3: end for
4: for each u ∈ V do
5:     repeat
6:         v = arg max_{x∈N(u)}{w(u, x) : w(u, x) > w(x, s(x))}
7:         t = u
8:         u = s(v)
9:         if v ≠ NULL then
10:            s(v) = t
11:        end if
12:    until u == NULL
13: end for
14: return M
```

$s(v)$ will be set to u. This follows since u has no better alternative and it cannot be prevented from doing so by any other neighbour $x \in N(v)$ as $w(u, v) > w(v, x)$. Using the same argument it follows that $s(u)$ will be set to v when v is initially processed. The suitor values of u and v will remain unchanged throughout the execution for the rest of the algorithm, again because (u, v) is dominant. It follows that for the final result of the vertices in $V - \{u, v\}$, the net effect is the same as if u and v were removed initially along with their adjacent edges. The only difference is that some vertices might become temporary suitors of u or v, but when they are dislodged, as they will be, they will behave as if u and v had been removed initially. Thus Algorithm 7 follows the same pattern as Algorithm 6 in computing the same matching as Greedy.

For the time complexity note first that a vertex can only propose once to each of its neighbours. Thus it follows that there can at most be $2m$ proposals. The time complexity therefore depends on how the next vertex to propose to is determined. If a linear search is performed each time, the time complexity is $O(n + m\Delta)$. If the weights of the edges incident on each vertex are sorted initially, then the rest of the algorithm runs in $O(n + m)$ time. The run time will then be dominated by the sorting, which takes time $O(m \log \Delta)$. Other strategies include using a partial Quicksort partitioning to find the heaviest neighbours, and then only when these do not contain any more potential candidates to match with are more candidates found.

In order to say something about the expected run time of Suitor we note that there is a strong relationship between computing a greedy matching and the stable marriage problem (SM).

In the SM problem we are given two equal-sized sets $L = \{l_1, l_2, \ldots, l_n\}$ and $R = \{r_1, r_2, \ldots, r_n\}$, typically referred to as 'men' and 'women' respectively. Every man and woman has a total ranking of all the members of the opposite sex. These give the 'desirability' for each participant to match with a member of the other set. The object is to find a complete matching M (*i.e.* a pairing) between the entries in L and R such that no two $l_i \in L$ and $r_j \in R$ would both obtain a higher-ranked partner if they were to abandon their current partner in M and rematch with each other. Any such solution is *stable*.

The main differences between computing a matching in a bipartite graph and an SM instance is that for SM one only considers how attractive two partners are relative to each other and also that attractiveness might not be symmetric. Thus one can have cycles such as l_1, r_1, l_2, r_2, l_1 where each one prefers to match with the next one in the list. This does not happen in the matching problem. Moreover, the overall objective is to find any stable solution, whereas in a matching instance possible solutions can be ranked.

Gale and Shapley (1962) formulated the stable marriage problem and also proposed the first algorithm for solving it. The algorithm operates in rounds as follows. In the first round each man in L proposes to his most preferred woman in R. Each woman will then reject all proposals she has received in this round except the one that is the highest in her ranking. In subsequent rounds each man that was rejected in the previous round will again propose to the woman that he has ranked highest, but now disregarding any woman that he has already proposed to in previous rounds. Gale and Shapley showed that this process will terminate with each man in L being matched to a woman in R and that this solution is stable. The algorithm also converges to a stable solution even if each participant has only ranked a subset of the opposing participants. For a recent discussion of stable matching and generalizations such as stable room-mates, stable fixtures, *etc.*, see the book by Manlove (2013).

Manne, Naim, Lerring and Halappanavar (2016) show that a greedy matching on a graph G can be computed by reformulating the problem as an appropriately constructed instance of SM and then solving this using the Gale–Shapley algorithm. In fact, the Gale–Shapley algorithm has a strong resemblance to the Suitor algorithm when using a queue. Similarly, there is a variation of the Gale–Shapley algorithm due to McVitie and Wilson (1971) corresponding to the Suitor algorithm when using a stack.

Wilson (1972) showed that for any profile of women's preferences, if the men's preferences are random, then the expected sum of men's rankings of their mates as assigned by the Gale–Shapley algorithm is bounded above by nH_n, where H_n is the nth harmonic number. Knoblauch (2007) showed that this is also an approximate lower bound in the sense that the ratio of the expected sum of men's rankings of their assigned mates and $(n+1)H_n - n$

has limit 1 as $n \to \infty$. Thus if the men's preferences are random then this sum is $\Theta(n \ln n)$ for large n. However, it is not hard to design instances where this sum is $\Theta(n^2)$. One such case is when the men have identical preferences. These results carry over to the SUITOR algorithm in that for a graph with random weights the expected number of proposals made is also bounded by $O(n \ln n)$, and if the graph is sufficiently dense by $\Theta(n \ln n)$.

The depth of a parallel algorithm is the length of the critical path in the algorithm, and the work in the algorithm is the total number of operations performed by the algorithm. Blelloch, Fineman and Shun (2012) have shown that a maximal matching in a graph can be computed with $O(m)$ work and $O(\log m \log \Delta)$ depth. Their proof technique was adapted by Khan, Pothen and Ferdous (2018b) to show that the parallel SUITOR algorithm has $O(m)$ work and $O(\log m \log \Delta)$ depth when the edge weights are chosen uniformly at random.

Among the algorithms presented thus far, experimental evaluations have shown that the highest-weight matching is obtained by PATH GROWING-DP. This is due to the use of dynamic programming to select an optimal solution on the discovered paths. The GLOBAL PATHS ALGORITHM by Maue and Sanders (2007) further expands this idea, by trying to get a heavier set of edges on which to perform the dynamic programming. The algorithm starts by sorting the edges and then considers each edge by order of non-increasing weight, similar to the GREEDY algorithm. An edge is kept for further consideration in a set S if it either connects at most two paths already contained in S, or if it completes a cycle of even length contained in S. The algorithm then uses dynamic programming on the paths and cycles in S to obtain the final matching. Even though the GLOBAL PATHS ALGORITHM in most instances gives higher weight matchings compared to PATH GROWING-DP, it does not improve the approximation ratio beyond half.

The idea of computing optimal matchings on restricted subgraphs was also used by Manne and Halappanavar (2014) in the M1M2 algorithm. This algorithm is based on the observation that the union of the edges from two matchings always consists of paths and even length cycles. The M1M2 algorithm first computes a greedy matching M_1 on a graph $G(V, E, w)$. It then computes a second greedy matching M_2 on $G(V, E \setminus M_1, w)$ before performing dynamic programming on $M_1 \cup M_2$ in the same way as was done in GLOBAL PATHS ALGORITHM. This gave results of a similar quality to GLOBAL PATHS ALGORITHM. Further use of this idea was shown in Idelberger and Manne (2014) where the dynamic programming was expanded to maximum weight spanning trees, while also performing mergers of multiple matchings following a tree-like structure.

While the algorithms presented thus far tend to perform quite well in practice, often producing optimal or close to optimal solutions, they all have in common that they cannot guarantee a performance ratio higher than 1/2. To do so requires that one considers augmentations containing more than one unmatched edge.

Drake and Hougardy (2005) presented the first such algorithm which computes a $(2/3 - \epsilon)$-approximation in time $O(m\epsilon^{-1})$. The algorithm is based on increasing the weight of a maximal matching by repeatedly performing 2-augmentations. This algorithm was subsequently simplified by Pettie and Sanders (2004) who also improved the run time to $m \log \epsilon^{-1}$. Their algorithm is based on finding the best 2-augmentation *centred* in a node u. For a matching M and 2-augmenting path P, a centre node $u \in P$ has the property that each edge $(x, y) \in P \setminus M$ is either incident on u or on the matching partner v of u. For a particular vertex u one can then find the best 2-augmenting path centred in u in time $O(\deg(u) + \deg(v))$.

Pettie and Sanders presented both a simple randomized algorithm RAMA and a slightly more involved deterministic one. In RAMA one picks a random vertex u and then finds the best 2-augmentation centred at v and augments if this gives a positive gain. By repeating this $(1/3) \log \epsilon^{-1}$ times the algorithm has an expected run time of $O(m \log \epsilon^{-1})$ and an expected performance ratio of $2/3 - \epsilon$ (Pettie and Sanders 2004). The algorithm can either start with an empty matching or from an existing one. A variant of RAMA, named ROMA was later determined to be more effective in practice (Maue and Sanders 2007). In ROMA the algorithm iterates multiple times through all vertices in a random order and performs the same augmentation step as in RAMA. This algorithm computed a higher weight matching when initialized with a matching given by Global Paths Algorithm, albeit at the cost of higher run times. Subsequent developments have given $(3/4 - \epsilon)$-approximation algorithms running in time $O(m \log n \log \epsilon^{-1})$ (Duan and Pettie 2010, Hanke and Hougardy 2010). However, these have not been tested in practice.

Finally, Duan and Pettie (2014) presented a $(1 - \epsilon)$-approximation algorithm that runs in time $O(m\epsilon^{-1} \log \epsilon^{-1})$. This algorithm employs the scaling approach to computing weighted matchings, with the subproblem at each scale solved by a primal–dual linear programming formulation of matching. The feasibility and complementary slackness conditions are enforced approximately, with the violation of these conditions dynamically dependent on the scale of the computation.

3.3. Experiments on edge-weighted matching algorithms

The data set consists of ten graphs taken from the SuiteSparse collection (Davis and Hu 2011). Structural properties of these graphs are shown in Table 3.1.

Table 3.1. Structural properties of graphs used for weighted matching, sorted in increasing order of edges.

Problems	Vertices	Edges
af_shell10	1 508 065	25 582 130
Serena	1 391 349	31 570 176
audikw_1	943 695	38 354 076
channel-500x100x100-b050	4 802 000	42 681 372
delaunay_n24	16 777 216	50 331 601
europe_osm	50 912 018	54 054 660
Flan_1565	1 564 794	57 920 625
Cube_Coup_dt6	2 164 760	62 520 692
kron_g500-logn21	2 097 152	91 040 932
nlpkkt200	16 240 000	215 992 816

3.3.1. Comparison of exact and approximation algorithms

We begin by comparing the performance of exact algorithms for maximum edge-weighted matching with several approximation algorithms. We report on the exact algorithm implemented in LEDA (Mehlhorn and Näher 1999). We report on three variant algorithms: LEDA1 uses no initialization, LEDA2 employs a greedy 1/2-approximation matching, and LEDA3 uses a fractional matching initialization. The latter initialization computes a fractional $\{0, 1/2, 1\}$ solution to a linear programming formulation of the edge-weighted matching problem; the solution is computed by a combinatorial algorithm, not an LP solver. The fractional solution is then rounded to obtain an initial matching. The $(2/3 - \epsilon)$-approximation algorithm ROMA is initialized with the Global Paths Algorithm and the Suitor algorithms. We report results for the RAMA algorithm without any initialization since the initialized versions computed lower weights than the corresponding variants of the ROMA algorithm; however, it is faster than the latter. The Duan and Pettie $(1 - \epsilon)$-approximation algorithm for weighted matching was implemented by Al-Herz and Pothen (2019), and we report results for it as well.

For the experiments we used an Intel Xeon E5-2660 processor-based system (part of the Purdue University Community Cluster), called *Rice*.[1] The machine consists of two processors, each with ten cores running at 2.6 GHz

---

[1] https://www.rcac.purdue.edu/compute/rice/

(20 cores in total) with 25 MB unified L3 cache and 64 GB of memory. The operating system is Red Hat Enterprise Linux release 6.9. All code was developed using C++ and compiled using the g++ compiler (version: 4.4.7) using the –O3 flag.

Run times and relative performances are reported in Table 3.2. Note that LEDA1 does not terminate on the last problem in the set, whereas LEDA2 and LEDA3 do. For the other nine problems, there is not much difference between the uninitialized version and the greedy initialization, whereas the fractional matching initialization is more effective, about four times faster than the uninitialized version. The $(2/3 - \epsilon)$-algorithms and the $(1 - \epsilon)$-approximation algorithms are 20–30 times faster than LEDA1, while the Suitor algorithm is about 1700 times faster, in geometric mean. For the last problem, we use LEDA2 as the baseline algorithm, the Suitor algorithm is 2000 times faster than LEDA2, while for the other approximation algorithms the factor is in the range 35–50.

Now we compare the weights computed by the algorithms. All the exact algorithmic variants compute the same weight, and this is reported in Table 3.3. For the approximation algorithms we report the gap to optimality expressed as a percentage. This value is computed as the ratio of the difference in weight between the maximum and approximate weights and the maximum weight. The Suitor algorithm computes more than 94% of the maximum weight in geometric mean, with the lowest value being 78%. ROMA obtains a lower gap than RAMA, and the ROMA algorithm initialized with the Suitor algorithm computes the highest weight among the $(2/3 - \epsilon)$-approximation algorithms, but the $(1 - \epsilon)$-algorithm with $\epsilon = 1/4$ lowers the gap further to about 1.6%. Note that all these algorithms obtain much better weights than their worst-case approximation ratios.

Finally we compare the cardinalities of the maximum weight matching and approximate matchings in Table 3.4. Note that the cardinality of the exact algorithm is not necessarily equal to that of the maximum cardinality matching. The gap in cardinality is computed analogously to the gap in weights. The cardinality of the approximation algorithms is lower than that of the exact algorithm. The gap in cardinalities is about 3% for Suitor, while it is less than 0.5% for the other approximation algorithms.

The relative performance of these algorithms has a strong dependence on the weights. When we use these algorithms to solve vertex-weighted matchings (with vertex weights chosen uniformly at random, and edge weights computed by adding the vertex weights on the endpoints of an edge), there are problems for which the exact algorithms do not terminate in hundreds of hours. For these problems the $(2/3 - \epsilon)$-algorithms tend to obtain better weights than the $(1 - \epsilon)$-approximation algorithm. We discuss these in Section 5 on vertex-weighted matchings.

Table 3.2. Comparing the run times of exact and approximation algorithms for maximum edge-weighted matching. Edge weights are from a uniform random distribution in the range [1 1000]. Run times of LEDA without any initialization are reported in LEDA1, and for all other algorithms the ratios of the run time of LEDA1 to the other algorithm are presented. For the last problem in the set, LEDA1 did not complete in four hours, and LEDA with greedy initialization (LEDA2) is the baseline. LEDA3 is the LEDA matching algorithm with fractional matching initialization.

	Time (s)			Relative performance					
	Exact	Exact	Exact	$1 - \epsilon$		RAMA	ROMA	Suitor, ROMA	Suitor
	LEDA1	LEDA2	LEDA3	Scaling			$2/3 - \epsilon$		
				$\epsilon = 1/4$	$\epsilon = 1/3$		$\epsilon = 0.01$		
af_shell10	310	1.10	1.70	18.00	24.00	24	23.00	23	1200
Serena	610	1.10	2.10	49.00	63.00	37	36.00	35	1400
audikw_1	990	1.00	2.50	120.00	150.00	62	62.00	58	2300
channel-500x100x100-b050	610	1.30	2.80	13.00	14.00	17	14.00	15	920
delaunay_n24	720	1.60	4.40	2.60	2.90	12	7.90	11	610
europe_osm	3200	3.30	12.00	4.80	6.00	31	9.20	26	1600
Flan_1565	940	1.10	4.80	49.00	80.00	35	33.00	32	1500
Cube_Coup_dt6	1400	1.20	4.70	71.00	94.00	43	38.00	38	1800
kron_g500-logn21	3500	2.00	15.00	41.00	51.00	64	36.00	53	3800
Geometric mean		1.40	4.30	24	30	32	24	29	1500
	Time (s)			Relative performance					
nlpkkt200		7100	2.30	44	50	39	40	36	1700

Table 3.3. Comparison of weights obtained by the exact and approximation algorithms for maximum edge-weighted matching. For each algorithm, we report the difference between one and the ratio of the weight computed by the approximation algorithm and the maximum weight obtained by the exact algorithm, expressed as a percentage.

	Weight	Gap to optimal weight (%)					
	Exact	$1 - \epsilon$		RAMA	ROMA	Suitor, ROMA	Suitor
					$2/3 - \epsilon$		
	LEDA	Scaling					
		$\epsilon = 1/4$	$\epsilon = 1/3$		$\epsilon = 0.01$		
af_shell10	7.20E+08	1.30	1.70	3.90	3.60	2.00	5.10
Serena	6.70E+08	1.20	1.70	4.20	3.90	2.10	4.90
audikw_1	4.60E+08	1.40	1.80	3.70	3.50	2.00	4.70
channel-500x100x100-b050	2.20E+09	1.80	2.40	4.80	4.30	2.70	6.90
delaunay_n24	6.40E+09	2.60	3.60	3.00	2.30	2.10	7.90
europe_osm	1.50E+10	1.30	1.70	1.50	0.50	0.53	3.70
Flan_1565	7.70E+08	1.10	1.40	3.50	3.20	1.60	3.60
Cube_Coup_dt6	1.10E+09	1.10	1.40	3.80	3.50	1.80	3.90
kron_g500-logn21	3.50E+08	4.70	6.10	2.80	1.80	1.70	22.00
nlpkkt200	7.60E+09	1.40	1.90	4.10	3.60	2.10	5.70
Geometric mean		1.60	2.10	3.40	2.70	1.70	5.80

Table 3.4. Comparison of cardinalities obtained by the maximum edge-weighted matching algorithm and the approximation algorithms. Note that this is not the cardinality obtained by a maximum cardinality matching. For each algorithm, we report the difference between one and the ratio of the cardinality of the matching computed by the approximation algorithm and the cardinality of the maximum edge-weighted matching, expressed as a percentage.

	Cardinality	Gap to cardinality of maximum weight matching (%)					
	Exact	$1 - \epsilon$		RAMA	ROMA	SUITOR, ROMA	SUITOR
	LEDA	Scaling		$2/3 - \epsilon$			
		$\epsilon = 1/4$	$\epsilon = 1/3$	$\epsilon = 0.01$			
af_shell10	754 032	0.16	0.24	0.18	0.16	0.19	1.80
Serena	695 674	0.08	0.14	0.09	0.08	0.09	1.80
audikw_1	471 847	0.01	0.03	0.04	0.04	0.04	1.60
channel-500x100x100-b050	2 400 980	0.76	0.99	0.48	0.42	0.41	3.60
delaunay_n24	8 217 157	4.10	4.60	2.60	2.00	2.20	8.20
europe_osm	22 996 323	3.50	3.80	1.90	0.83	1.00	5.90
Flan_1565	782 397	0.01	0.02	0.03	0.03	0.03	1.00
Cube_Coup_dt6	1 082 380	0.02	0.03	0.04	0.04	0.04	1.20
kron_g500-logn21	442 107	8.30	9.60	4.20	3.20	3.20	24.00
nlpkkt200	8 000 000	0.33	0.47	0.14	0.13	0.16	2.80
Geometric mean		0.25	0.36	0.26	0.21	0.23	3.10

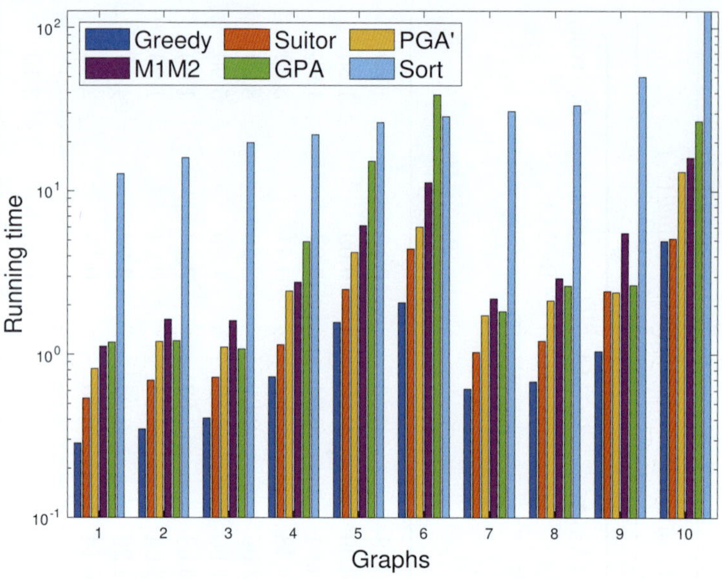

Figure 3.3. Run times for five 1/2-approximation algorithms.

3.3.2. Comprehensive comparison of approximation algorithms

Now we compare the various approximation algorithms using original edge weights from the graphs rather than the random integer weights used in the previous experiments. We first compare the 1/2-approximation algorithms. Figure 3.3 shows the run time of the Global Paths Algorithm (GPA), and the Greedy, Suitor, Path Growing-DP and M1M2 algorithms for the ten graphs. For Greedy and Path Growing-DP the time to sort the edges is shown separately. This is done using the standard *qsort* routine in C. Thus for an application where the edges are not already sorted, then this time must be added to get the total run time.

For these experiments we used a Dell computer running GNU/Linux equipped with four 10 core 2.00 GHz Intel Xeon E7-4850 processors with a total of 128 GB memory. Algorithms were implemented in C using OpenMP for parallelization and compiled with `gcc` using the –O3 flag.

As can be seen from the figure, for all but one graph the time to sort dominates the time taken by all other algorithms, in many cases by as much as an order of magnitude. When one excludes the time to sort, Greedy is the fastest algorithm followed by Suitor and Path Growing-DP. These again outperform M1M2 and the Global Paths Algorithm. If one includes the time to sort the edges then Suitor outperforms Greedy while M1M2 outperforms the Global Paths Algorithm.

To measure the quality of the solutions given by these algorithms we compute the average performance ratio for each algorithm compared to the best algorithm for each graph. This shows that Greedy, Path Growing-DP, M1M2 and Path Growing algorithms have an average performance ratio of 2.44%, 1.67%, 0.21% and 0.04%, respectively, higher than the best algorithm for each graph. Thus it follows that the more time-consuming algorithms also gives better results.

Next, we compare the use of either RAMA or ROMA for post-processing a solution. We first note that in terms of run time RAMA is about 10% slower than ROMA. Each application of ROMA is on average 4.9 times slower than the corresponding edge sorting. Thus there is a fairly large cost for using this post-processing. As there is also a slight advantage of using ROMA over RAMA in terms of quality of solution, we only present results from using ROMA.

How much improvement ROMA gives depends on the starting configuration. The average improvement for using Suitor followed by ROMA was 4.9% while both M1M2 and Global Paths Algorithm were improved by 2.6%. In terms of absolute quality of the final solution Suitor followed by ROMA always gave the best solution. On average M1M2 was 0.19% worse while Global Paths Algorithm was 2.11%. Thus, in terms of both run time and quality, the preferred strategy is to use Suitor followed by ROMA.

In Figure 3.4 we show the speedup curves for Suitor, M1M2 and Local Max algorithms when run on the graph nlpkkt200. For both Suitor and M1M2 the speedups increase almost linearly as long as there are available threads. Although the CPUs can perform hyper-threading, this does not give any additional speedup when using 50 threads. The speedup for Local Max drops off earlier than Suitor. The sequential run time for Local Max is about 70% higher than that of Suitor. Combined with the lower speedup it follows that the parallel run time of Local Max is even worse.

Approximation algorithms for edge-weighted matchings have been implemented on GPUs by several authors. We list some of these papers: Cohen and Castonguay (2012), Fagginger Auer and Bisseling (2012), Halappanavar et al. (2012) and Naim et al. (2015).

3.4. Applications of matchings

3.4.1. Maximum cardinality matching

Maximum cardinality matchings in bipartite graphs have been employed in sparse matrix algorithms for several purposes. A maximum matching can be used to permute a sparse matrix to place the largest number of non-zeros on the diagonal. This number is the structural rank of the matrix, which is an upper bound on the numerical rank.

In the following we say that a matrix has a row-perfect matching to mean that its bipartite graph consisting of row vertices and column vertices has a

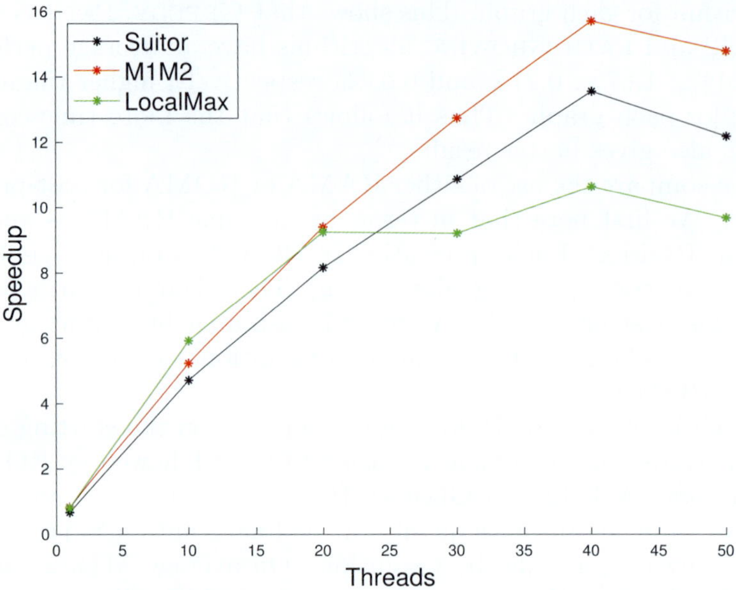

Figure 3.4. Speedup for SUITOR, M1M2 and LOCAL MAX algorithms on the nlpkkt200 graph.

matching in which all rows are matched. The concept of a column perfect matching is similar, and a perfect matching is both row-perfect and column-perfect.

A maximum matching can be used to compute the block (lower) triangular form (BTF) of a sparse matrix, which has the block matrix structure

$$A = \begin{pmatrix} A_{11} & A_{12} & A_{13} \\ 0 & A_{22} & A_{23} \\ 0 & 0 & A_{33} \end{pmatrix},$$

where A_{11} has fewer rows than columns and has a row-perfect matching, A_{22} is a square matrix with a perfect matching, and A_{33} has fewer columns than rows and has a column-perfect matching. The submatrices A_{12}, A_{13} and A_{23} could be zero. The submatrix A_{11} has the *strong Hall property* with respect to its rows (Coleman, Edenbrandt and Gilbert 1986, Pothen and Fan 1990). This property states that every set of k rows has non-zeros in at least $(k+1)$ columns. The submatrix A_{33} has the strong Hall property with respect to its columns, and A_{22} has the strong Hall property with respect to both rows and columns (in this last case $k < n$, where n is the dimension of A_{22}).

The BTF of a sparse matrix is derived from the Dulmage–Mendelsohn decomposition of the bipartite graph of the matrix, which could be computed from a maximum matching in the graph. The subgraph corresponding to A_{11} is obtained by following all alternating paths from unmatched columns, and including all rows and columns reached in the subgraph. The subgraph corresponding to A_{33} is obtained by following all alternating paths from unmatched rows, and including all rows and columns reached. The submatrix corresponding to A_{22} consists of all rows and columns not reached thus far, and they must be matched to each other since all unmatched columns and rows have been accounted for. (There is a finer decomposition for A_{22} which we do not describe here.) The decomposition is unique for a bipartite graph and is independent of the specific maximum matching used to induce it. Further details may be found in Pothen and Fan (1990).

The BTF can be used to reduce the work in solving sparse linear systems of equations by a factorization-based algorithm, since only the diagonal blocks in the BTF need to be factored. The BTF has been used to solve Kirchoff's equations in circuit design within the Xyce circuit simulation code (Keiter *et al.* 2011), and it is also used to solve nonlinear systems of equations in systems modelling software such as Modelica (Fritzson 2014). The strong Hall matrices in the BTF could be used to correctly predict the non-zero structures of orthogonal-triangular (QR) factors of sparse matrices (Coleman *et al.* 1986, Pothen 1993). Matchings have also been used to compute sparse bases for the null space of sparse, under-determined matrices, as also sparse bases for the column space of such matrices (Coleman and Pothen 1987, Pinar, Chow and Pothen 2006).

In applications such as the BTF, one needs to compute a maximum cardinality matching in bipartite graphs, and an approximation will not suffice. Duff, Kaya and Uçar (2011) have addressed the several choices one needs to obtain an efficient practical algorithm. The Karp–Sipser algorithm is used as an initialization, and then there are choices to be made on if the augmenting path searches should use breadth-first search (BFS) or depth-first search (DFS) or some combination of them. It is also clear from their work and other authors, that the theoretically less efficient $O(nm)$ time algorithms are faster than the Hopcroft–Karp algorithm with $O(n^{1/2}m)$ time complexity. Azad, Buluç and Pothen (2017) have designed efficient shared-memory parallel algorithms for this problem. A distributed-memory parallel algorithm scaling to hundred-thousand cores has been implemented using sparse matrix vector products using the Graph-BLAS operations by Azad and Buluç (2016).

3.4.2. Maximum edge-weighted matchings

An important application we consider is sparse Gaussian elimination (LU factorization). Here, permuting the sparse matrix to have large elements on the diagonal before the numerical factorization makes it less likely that numerical pivoting (row permutations) will be required during the computation to prevent large element growth in the factors. Olschowka and Neumaier (1996) described a pivoting strategy using a primal–dual algorithm for the assignment problem, which was then adapted and implemented for sparse bipartite graphs by Duff and Koster (2001). A perfect matching that has the maximum product of matched elements is computed in this context, and the resulting code, MC64, is widely used. Hogg and Scott (2013) discuss several pivoting strategies, including matching-based orderings, to solve symmetric, indefinite systems of equations.

As parallel computers are able to solve systems of linear equations with millions of rows and columns, one challenge that has remained here is that the primal–dual algorithm used to compute perfect matchings does not have much concurrency. Hogg and Scott (2015) have employed the auction algorithm on shared-memory machines to compute the matching in parallel, but scaling to the large numbers of cores on distributed-memory machines remained an open problem. This raised the possibility that approximation algorithms for matching could be employed in this context. In recent work Azad *et al.* (2018) have described an approach that scales to 17 000 cores, and we consider this next.

Azad *et al.* (2018) first compute a perfect matching in the bipartite graph and then seek to increase its weight. (A perfect matching must exist if the matrix is non-singular.) The perfect matching is computed by a distributed-memory parallel algorithm that recasts the problem in terms of matrix–vector products in a suitable semi-ring using the Graph-BLAS operations (Azad and Buluç 2016). The increase in weight is accomplished by means of the $(2/3 - \epsilon)$-approximation algorithm of Pettie and Sanders (2004), which can be simplified here because the graph is bipartite and the matching is perfect. Hence they search for weight-increasing cycles with four edges (see Figure 3.1), finding for each vertex a cycle that leads to the largest increase in matching weight. From the set of cycles obtained, a maximal set of vertex-disjoint cycles is chosen using a greedy algorithm, and the matching weight is increased without affecting the cardinality of the perfect matching. The approximation guarantee of this algorithm is $(2/3 - \epsilon)$ and it is computed in $O(m \log \epsilon^{-1})$ time.

In a distributed-memory parallel implementation, it is not easy to find a maximal set of vertex-disjoint cycles without extensive communication, and hence these authors instead use only local comparisons to obtain a heavy set of vertex-disjoint cycles. Thus the parallel algorithm does not have the approximation guarantee, but it computes a heavy-weight matching that performs well in practice; it runs in $O((m/p)(1 + \beta) + \alpha p)$ time, where p is the number of processors, α is the latency and β is the inverse bandwidth. The authors report results for the algorithm on distributed-memory machines with 17 000 cores. Their code has been incorporated into the distributed-memory SuperLU solver for sparse, unsymmetric, linear systems of equations (Li and Demmel 2003). Additional details on the applications of matchings in sparse matrix algorithms is provided in Duff and Uçar (2012).

Exact algorithms for maximum cardinality matchings and maximum edge-weighted matchings in bipartite and non-bipartite graphs are available in the LEDA library (Mehlhorn and Näher 1999). We compare the performance of approximation algorithms for vertex- and edge-weighted matchings with the LEDA codes in Section 5.

3.5. Matching software

In Table 3.5 we tabulate software libraries available for several matching problems. Software for b-matching, edge cover and b-edge cover will be discussed in their respective sections. An earlier discussion of matching algorithms and software is provided by Burkard *et al.* (2009).

The MC64 code in the Harwell Subroutine Library (HSL) computes a maximum product edge-weighted perfect matching in a bipartite graph in order to place large elements on the diagonal of a sparse, square matrix (Duff and Koster 2001). The primal–dual Hungarian algorithm is employed with a logarithmic transformation of the weights to compute a maximum product matching. Code for computing a maximum sum edge-weighted perfect matching is also included. This code is widely used in the preprocessing step in solving sparse systems of linear equations.

LEDA (Mehlhorn and Näher 1999) has codes for computing maximum edge-weighted matchings in both bipartite graphs and non-bipartite graphs. In our experience, this code is the fastest of the exact matching codes available for this problem. The algorithm uses a fractional matching initialization for non-bipartite graphs, which speeds it up over a greedy initialization. It is commercial software. The LEMON library (Dezső, Jüttner and Kovács 2011) contains an exact maximum edge-weighted matching code. BLOSSOM V is another code for minimum weight perfect matching in non-bipartite graphs by Kolmogorov (2009).

The colleagues and students of the first author have created a software library called Matchbox, which contains a number of exact and approximation algorithms for several variants of the matching problem. Included are exact algorithms for maximum cardinality matching in bipartite and non-bipartite graphs, and a 1/2-approximation algorithm for bipartite graphs. For edge-weighted matching, it has exact algorithms for maximum edge weight, perfect matching with maximum edge weight, and maximum weight semi-perfect matching, all for bipartite graphs. It also has a 1/2-approximate greedy algorithm for both bipartite and non-bipartite graphs. For vertex-weighted matchings, it has exact algorithms for both bipartite and non-

Table 3.5. Matching software libraries.

Library/Code	URL	Algorithms included
HSL/MC64	http://www.hsl.rl.ac.uk/catalogue/mc64.html	Max. product (and sum) edge-weighted perfect matching in bipartite graphs
Blossom V	http://pub.ist.ac.at/~vnk/software.html	Min. edge-weighted perfect matching
LEDA	http://www.algorithmic-solutions.com/leda	Max. edge-weighted matching in bipartite and general graphs
LEMON	http://lemon.cs.elte.hu	Max. edge-weighted matching in general graphs
Matchbox	https://github.com/CSCsw/matchbox	*Cardinality* Max. matching in general graphs; 1/2-approximate matching in bipartite graphs *Edge weight* Max. edge weight, perfect matching with max. edge weight, and max. edge-weighted semi-perfect matching, all in bipartite graphs; 1/2-approximate edge-weighted matching (greedy) for general graphs *Vertex weight* Max. matching for general graphs; 2/3-, $(1-\epsilon)$- and variants of 1/2-approximation algorithms (Greedy, Locally Dominant Edge, Path Growing-DP, Suitor) for max. matching in general graphs

Table 3.5 continued.

Library/Code	URL	Algorithms included
SUITOR	http://www.ii.uib.no/~fredrikm/matching/	Several 1/2-approximation edge-weighted matching algorithms including SUITOR
Matchmaker	http://people.sabanciuniv.edu/kaya/software.html	Ten variant bipartite max. cardinality matching algorithms
Matchmaker2	http://tda.gatech.edu/software/matchmaker2/	GPU code for bipartite max. cardinality matching
MS-BFS-GRAFT	https://bitbucket.org/azadcse/ms-bfs-graft	Multi-threaded code for 1/2-approximate bipartite max. cardinality matching
CombBLAS 2.0	https://bitbucket.org/berkeleylab/combinatorial-blas-2.0, directory CombBLAS/Applications/BipartiteMatchings/	Distributed-memory algorithms for bipartite matching (max. cardinality, 1/2-approximate cardinality, heavy-weight perfect matching) via matrix vector multiplication in a semi-ring

bipartite graphs. The collection of approximation algorithms includes 2/3-, $(1 - \epsilon)$- and several variants of 1/2-approximation algorithms (Greedy, Locally Dominant Edge, Path Growing-DP, Suitor), for both bipartite and non-bipartite graphs. (The bipartite graph algorithms for vertex-weighted matching make use of the Mendelsohn–Dulmage theorem to make them more efficient.)

Codes for several 1/2-approximation algorithms for edge-weighted matching, including the Suitor algorithm, will be available from the third author's website. Included are Suitor, Local Max, Locally Dominant Edge, Path Growing, Path Growing-DP, Global Paths Algorithm and M1M2 algorithms. Multi-threaded parallel implementations of some of these algorithms are also included.

Implementations of ten variants of exact algorithms for maximum cardinality matching in bipartite graphs are provided by Kamer Kaya in a code called Matchmaker. The algorithms are based on breadth-first search, depth-first search and push–relabel techniques. MATLAB interfaces to the codes are also included. A GPU code for this problem, Matchmaker2, is also available (Deveci, Kaya, Uçar and Çatalyürek 2013).

A parallel multi-threaded code for computing 1/2-approximate maximum cardinality matchings in bipartite graphs has been provided by Azad. Distributed-memory parallel codes for exact maximum cardinality matching, 1/2-approximate cardinality matching and heavy-weight perfect matching, all in bipartite graphs, are also available. These codes employ Combinatorial BLAS (matrix operations in a suitably defined semi-ring) (Buluç and Gilbert 2011) to compute the matchings.

4. The b-matching problem

4.1. Background on b-matching

We turn to a generalization of the matching problem. Given a graph $G = (V, E)$ and a function $b(.)$ that maps each vertex to a natural number, a b-matching is a subset of edges M such that at most $b(v)$ edges in M are incident on each vertex v. We assume that $b(v) \leq \deg(v)$ for all v. If $b(v)$ is identically equal to one, then we have the matching problem. If the edges have weights w, then the weight of a b-matching is the sum of the weights of the matched edges, and we seek to compute a b-matching of maximum weight. We denote $b(V) = \sum_{v \in V} b(v)$ and $\beta = \max_{v \in V} b(v)$.

An exact algorithm for a maximum weight b-matching was first designed by Edmonds (1965), and Pulleyblank (1973) later gave a pseudo-polynomial time algorithm with complexity $O(mnb(V))$. Anstee (1987) proposed a three-stage algorithm where the b-matching problem is solved by transforming it to a Hitchcock transportation problem, rounding the solution to integer values, and finally invoking Pulleyblank's algorithm. Derigs and Metz (1986) and Miller and Pekny (1995) improved the Anstee algorithm further. Padberg and Rao (1982) developed another algorithm using the branch and cut approach, and Grötschel and Holland (1985) solved the problem using the cutting plane technique. A survey of exact algorithms for b-matchings was provided by Müller-Hannemann and Schwartz (2000). More recently, Huang and Jebara (2011) proposed an exact b-matching algorithm based on belief propagation. The algorithm assumes that the solution is unique, and otherwise it does not guarantee convergence.

We now describe work on approximate b-matching. Mestre (2006) showed that a b-matching is a relaxation of a matroid called a 2-extendible system, and hence that the greedy algorithm gives a 1/2-approximation for a maximum weight b-matching. We describe his proof in the next subsection. Mestre also generalized the path-growing algorithm of Drake and Hougardy (2003b) to obtain an $O(\beta m)$ time 1/2-approximation algorithm. These algorithms are slower in practice than a serial b-SUITOR algorithm that generalizes the SUITOR algorithm (Khan *et al.* 2016b). Since the PATH GROWING algorithm is inherently sequential, it is not a good candidate for parallelization. Additionally, Mestre (2006) generalized a randomized algorithm for matching to obtain a $(2/3 - \epsilon)$-approximation algorithm with expected run time $O(\beta m \log(1/\epsilon))$. De Francisci Morales, Gionis and Sozio (2011) have adapted the GREEDY algorithm and an integer linear program (ILP) based algorithm to the MapReduce environment to compute b-matchings in bipartite graphs. b-matching algorithms have also been developed using linear programming (Koufogiannakis and Young 2011, Manshadi *et al.* 2013), but these methods are orders of magnitude slower than the b-SUITOR algorithm. Georgiadis and Papatriantafilou (2013) have developed a distributed algorithm based on adding locally dominating edges to the b-matching, which leads to a LOCALLY DOMINANT EDGE algorithm. Our recent work on b-MATCHING algorithms have focused on developing the LOCALLY DOMINANT EDGE and b-SUITOR algorithms, and implementing them efficiently on serial computers and shared-memory and distributed-memory multiprocessors (Khan *et al.* 2016a, 2016b). We will discuss the b-SUITOR algorithm later in this section.

4.2. Half-approximation algorithms for b-matching

4.2.1. The GREEDY algorithm

Algorithm 8 describes the GREEDY algorithm for computing a b-matching in a graph G.

The GREEDY algorithm is a 1/2-approximation algorithm for the maximum weight b-matching problem. We prove this by an argument that employs matroid theory due to Mestre (2006).

Algorithm 8 GREEDY($G(V, E, w, b)$)

1: Sort edges in non-increasing order of weights
2: $M = \emptyset$
3: **for** $e = (u, v) \in E$ in order **do**
4:     **if** $M + e$ is a b-matching **then**
5:         $M = M + e$
6:     **end if**
7: **end for**
8: **return** M

Let E be a set of elements, and $\mathcal{I}$ a collection of subsets of E. The tuple $(E, \mathcal{I})$ is a *matroid* if

(i) for all $A \subseteq B$, if $B \in \mathcal{I}$, then $A \in \mathcal{I}$, and

(ii) for all $A, B \in \mathcal{I}$ with $|A| < |B|$, there exists an element $x \in B \setminus A$ such that $A + x \in \mathcal{I}$.

The sets in $\mathcal{I}$ are called the *independent* sets of the matroid, and by the first property, the empty set is independent. Maximal independent sets are called the bases of the matroid, and by the second property, all bases have the same cardinality. The reader unfamiliar with matroids may find two examples helpful. If E is the set of columns of a matrix, and independence corresponds to linear independence in a vector space, then we have a matric matroid. If E is the set of edges of a connected undirected graph without loops, and a subset of edges is defined to be independent if the edges do not induce a cycle, then we have a graphic matroid. A basis here corresponds to a spanning tree of the graph.

Assign every element in E a non-negative weight, define the weight of a set as the sum of the weights of its elements, and consider the problem of finding an independent set of largest weight. The GREEDY algorithm begins with the empty set, and at each step adds an element of largest weight if its addition will preserve independence, and rejects it otherwise. The GREEDY algorithm finds an independent set (a basis) of maximum weight if and only if the tuple $(E, \mathcal{I})$ is a matroid.

It is well known that a matching does not correspond to a matroid, but a matching in a bipartite graph is the intersection of two matroids. We now consider a more general system called a k-extendible system related to a matroid. A k-extendible system is a tuple $(E, \mathcal{I})$ that satisfies property (i) of a matroid, but property (ii) is replaced by

(ii$'$) for all $A \subseteq B \in \mathcal{I}$ and $x \in E$, if $A + x \in \mathcal{I}$ but $B + x \notin \mathcal{I}$, then there exists $Y \subseteq B \setminus A$, such that $B \setminus Y + x \in \mathcal{I}$ and $|Y| \leq k$.

In words, this means that if we can augment a smaller independent set A with a new element x and preserve its independence, then it should be possible to remove a subset of size at most k from any superset of A that is independent, add x to the new set and still preserve independence. Maximal independent sets do not necessarily have unique maximum cardinality for k-extendible systems.

As a concrete example, let us consider an undirected graph $G = (V, E, w)$ with vertex set V, edge set E and non-negative weights on its edges W. Further, we are given a set of natural numbers $b(v) \leq \deg(v)$ for each $v \in V$. Let the collection of independent sets $\mathcal{I}$ be defined as subsets of edges that consist of a b-matching in G, i.e. subsets of edges M such that $\deg_M(v) \leq b(v)$ for all $v \in V$. Let A be a b-matching in G and let $B \supset A$ be another b-matching. Let $x = (u, v)$ be an edge that could be added to A but not to B to obtain a larger b-matching. The reason that this edge cannot be added to B is that the values of one or both of $b(u)$ and $b(v)$ would be exceeded. By removing one edge incident on u and another edge incident on v from B, we would be able to add the edge (u, v) to B and preserve a b-matching. Thus we have shown that a b-matching is a 2-extendible system.

We now provide some concepts and prove a preliminary lemma in order to prove that the Greedy algorithm computes a $1/2$-approximation algorithm. We prove a more general result for k-extendible systems. An *extension* of an independent set $A \in \mathcal{I}$ is a superset B of A with $B \in \mathcal{I}$. We denote an extension of maximum weight of a set A by $\mathrm{OPT}(A)$.

Let the Greedy algorithm choose elements $x_1, x_2, \ldots, x_l$ on a maximization problem on a k-independence system $(E, \mathcal{I})$. We denote the set of chosen elements at the end of the ith iteration of the Greedy algorithm by S_i, with $S_0 = \emptyset$ and $S_i = S_{i-1} + x_i$. Note that $\mathrm{OPT}(\emptyset)$ is the optimal solution to the maximization problem, and $\mathrm{OPT}(S_l) = S_l$, since the set S_l is maximal. Furthermore, we have

$$w(\mathrm{OPT}(S_0)) \geq w(\mathrm{OPT}(S_1)) \geq \cdots \geq w(\mathrm{OPT}(S_l)),$$

since with increasing index i, S_i has fewer extensions available than S_{i-1}.

Lemma 4.1. If $(E, \mathcal{I})$ is a k-extendible system, then the element x_i chosen by the Greedy algorithm at the ith step satisfies

$$w(\mathrm{OPT}(S_{i-1})) \leq w(\mathrm{OPT}(S_i)) + (k - 1)\, w(x_i).$$

Proof. By the definition of k-extendibility, $S_{i-1} \in \mathcal{I}$ and $\mathrm{OPT}(S_{i-1}) \in \mathcal{I}$. Now since the Greedy algorithm chooses the element x_i at the ith step, it does not belong to S_{i-1}, but it could belong to $\mathrm{OPT}(S_{i-1})$. If it does, then $\mathrm{OPT}(S_{i-1}) = \mathrm{OPT}(S_i)$, and the lemma holds trivially. Hence consider the situation when the element $x_i \notin \mathrm{OPT}(S_{i-1})$. By k-extendibility, there exists $Y \in \mathrm{OPT}(S_{i-1}) \setminus S_{i-1}$ such that $\mathrm{OPT}(S_{i-1}) \setminus Y + x_i \in \mathcal{I}$, where

$|Y| \leq k$. We have the relationships

$$w(\mathrm{OPT}(S_{i-1}) = w(\mathrm{OPT}(S_{i-1}) \setminus Y + x_i) + w(Y) - w(x_i),$$
$$\leq w(\mathrm{OPT}(S_i)) + w(Y) - w(x_i).$$

The first line of the equation above follows since Y belongs to $\mathrm{OPT}(S_{i-1})$ but x_i does not. The second follows because $\mathrm{OPT}(S_{i-1}) \setminus Y + x_i$ is an extension of the set $S_i = S_{i-1} + x_i$ by the choice of Y, and $\mathrm{OPT}(S_i)$ is an extension of maximum weight of the set S_i.

Now we show that any element $y \in Y$ is not heavier than the element x_i, i.e. $w(y) \leq w(x_i)$. Assume for the sake of contradiction that $w(y) > w(x_i)$. If y is heavier than x_i, then since $y \notin S_{i-1}$ by the choice of Y, the element y must have been considered by the Greedy algorithm before x_i was included in S_i, but was not included in the greedy solution. Thus there exists $j \leq i$ such that $S_j + y \notin \mathcal{I}$, but $S_j + y \subseteq \mathrm{OPT}(S_{i-1}) \in \mathcal{I}$. But this contradicts property (i) of a k-independence system. Thus $w(y) \leq w(x_i)$, and since all weights are positive, $w(Y) \leq kw(x_i)$. $\qquad\square$

Theorem 4.2 (Mestre). If $(E, \mathcal{I})$ is a k-extendible system, then the Greedy algorithm is a $1/k$-approximation algorithm for computing an independent set of maximum weight.

Proof. Let $\{x_i : i = 1, 2, \ldots, l\}$ be the elements chosen by the Greedy algorithm, and the corresponding sets of chosen elements be denoted by $\{S_i : i = 1, 2, \ldots, l\}$. We apply Lemma 4.1 l times, beginning with the empty set S_0, to get

$$w(\mathrm{OPT}(S_0)) \leq w(\mathrm{OPT}(S_1)) + (k-1)\, w(x_1)$$

$$\leq w(\mathrm{OPT}(S_l)) + (k-1) \sum_{i=1}^{l} w(x_i)$$

$$= w(S_l) + (k-1)\, w(S_l) = k\, w(S_l).$$

The first term in last line of the equation follows from the fact that the set S_l is maximal, and the second term is obtained by the summing the weight of the elements in S_l. Since $\mathrm{OPT}(S_0)$ is an independent set of maximum weight, we obtain the $1/k$-approximation property of the Greedy algorithm. $\qquad\square$

4.2.2. *The b-Suitor algorithm*

The b-Suitor algorithm, designed by Khan *et al.* (2016*b*), computes a $1/2$-approximate b-matching; indeed it computes a matching identical to the one obtained by the Greedy algorithm, provided ties in weights are broken consistently in both algorithms. This algorithm is based on proposals, much like the algorithms for the stable marriage problem and its variants (here the stable fixtures problem), and is a generalization of the SUITOR algorithm. Vertices can propose to their heaviest neighbours, and these proposals may be reciprocated or annulled by other vertices. Two vertices are matched when both propose to each other. Unlike the GREEDY algorithm, the b-SUITOR algorithm does not need to process edges in non-increasing order of weights; instead, it can process vertices in any order, although each vertex has to make proposals to its available neighbours in non-increasing order of weights.

The pseudo-code for the b-SUITOR algorithm is described in Algorithm 9. It maintains a priority queue S to track the proposals made in the algorithm. The queue $S(u)$ consists of the suitors of a vertex u, $i.e.$ those neighbours of u that currently have an active proposal to u. The operation $S(u).insert(v)$ adds a vertex v to the priority queue $S(u)$, and $S(u).remove(v)$ removes it. The value $r(u)$ is the number of proposals that a vertex u currently has made to its neighbours. We keep track of the lowest weight of a proposal received by a vertex u (made by a suitor of u) in $S(u)$.last. If u has received fewer than $b(u)$ proposals, this value is NULL.

In each iteration of the outer **while** loop, the algorithm processes vertices that have their current $b(.)$ values unsatisfied, and makes the requisite number of proposals to satisfy the $b(.)$ values; these vertices are stored in a set Q. During the iteration it collects vertices whose $b(.)$ values may be decremented in a set Q' so that they could be processed in the next iteration. For each vertex $u \in Q$, while its $b(u)$ value is not satisfied and adj(u) has not been exhaustively searched, the algorithm finds eligible neighbours to propose to. A neighbour v is an *eligible neighbour* of u if it holds fewer than $b(v)$ proposals, or u can beat the lowest offer that v currently has, which is $S(v)$.last. If u cannot beat this offer, then it considers its next heaviest neighbour, and this prevents proposals being extended which at the current stage of the algorithm have no chance of success, saving work. But if v has fewer than $b(v)$ proposals, or if u can beat the lowest offer that v has, then u proposes to v, and becomes a suitor of v. In the latter case, u annuls the lowest-weight proposal of v, say from a vertex y, and y has to make another proposal in the next iteration. The value of $S(v)$.last is also updated.

Algorithm 9 shows that when a proposal made by a vertex y is annulled, it is placed into a queue to be processed in the next iteration. This corresponds to the Gale–Shapley algorithm for stable marriage which processes proposals in rounds. Instead, one could consider the vertex y making a proposal immediately, and this would correspond to the McVitie–Wilson algorithm. It is the possibility of proposals being annulled that permits vertices to be processed in any order, thus increasing the concurrency in the algorithm. We have shown that if the edge weights are chosen uniformly at random, then the total expected work in the algorithm is linear in the number of edges in the graph, so that the algorithm does not create a lot of unnecessary work

---

Algorithm 9 b-SUITOR ALGORITHM

Input: Graph $G = (V, E, w, b)$.
Output: A $1/2-$approximate edge-weighted b-matching M.
Data Structures: Q is the set of vertices that propose in each iteration of the outer **while** loop, and Q' is the set of vertices that need to make proposals in the next iteration. $S(u)$ is the set of suitors of u, $r(u)$ is the number of outstanding proposals that u currently has made.

```
 1: procedure b-SUITOR(G, b)
 2:     Q = V; Q' = ∅;
 3:     Initialize array S to be empty and r to zero;
 4:     while Q ≠ ∅ do
 5:         for vertices u ∈ Q in any order do
 6:             while r(u) < b(u) and adj(u) ≠ exhausted do
 7:                                            ▷ Make a proposal from u
 8:                 Let v ∈ N(u) be the heaviest eligible neighbour of u;
 9:                 if v ≠ NULL then
10:                                            ▷ Make u a suitor of v
11:                     Insert u into S(v);
12:                     r(u) = r(u) + 1;
13:                     if u annuls the proposal of a vertex y then
14:                         Remove y from S(v);
15:                         Add y to Q'; r(y) = r(y) - 1;
16:                     end if
17:                 else adj(u) = exhausted;
18:                 end if
19:             end while
20:         end for
21:         Q = Q'; Q' = ∅;
22:     end while
23: end procedure
```

(Khan *et al.* 2018b). This paper also shows that the depth (the length of the critical path) is $O(\log m \log \Delta)$.

4.3. *Implementation and applications of b-matchings*

The b-SUITOR algorithm was implemented on serial, shared-memory and distributed-memory computers, and it is currently the fastest practical algorithm that we know of (Khan *et al.* 2016a, Khan *et al.* 2016b). It has been compared with the GREEDY algorithm and the LOCALLY DOMINANT EDGE algorithm (LDE), and it outperforms both these algorithms with regard to run times. All of these algorithms compute the same b-matching provided ties in weights are broken in a consistent manner. The b-Suitor algorithm has the desirable property that the parallel algorithms and the serial algorithm compute the same b-matching. This algorithm was scaled to more than $12\,000$ cores on a distributed-memory parallel computer. An implementation on GPUs was reported by Naim and Manne (2018).

b-matchings have been applied to a number of problems such as finite element mesh refinement (Müller-Hannemann and Schwartz 2000), median location problems (Tamir and Mitchell 1998), spectral data clustering (Jebara and Shchogolev 2006), *etc.* An important application of b-matching is in graph-based semi-supervised machine learning, where a b-matching has been used to replace the well-known k-nearest neighbour graph construction (Jebara, Wang and Chang 2009). Both of these constructions are discussed in this context by Subramanya and Talukdar (2014). We will discuss this matter in more detail when we consider applications of b-edge cover, but we state here that an approximate b-matching construction reduces the time complexity of this approach from $O(b(V)m \log n)$ to $O(m \log \beta)$, without any discernible loss in quality in the classification. Recently, Choromanski, Jebara and Tang (2013) used b-matching to solve a data privacy problem called adaptive anonymity. Again, we will discuss this problem as an application of b-edge cover.

An implementation of b-Suitor is available at https://github.com/ExaGraph. Parallel versions of this code will be included in this repository.

5. Vertex-weighted matching

We consider a variant of the matching problem that has not been well studied. We are given a graph $G = (V, E)$, and a weight function $w : V \mapsto \mathbb{R}_{\geq 0}$ that assigns non-negative real-valued weights to vertices. The weight of a matching in G is now the sum of the weights on the matched vertices. The problem is to compute a matching of maximum (vertex-) weight in the graph G (MVM).

By summing the weights on the endpoints of an edge and assigning it to the edge, we can transform the vertex-weighted matching problem into an edge-weighted matching problem. So at first blush it appears that we can solve MVM problems with edge-weighted matching algorithms. But at least for exact algorithms, this can increase the run times of the edge-weighted matching algorithms by three or four orders of magnitude (Dobrian, Halappanavar, Pothen and Al-Herz 2018). Additionally the MVM problem has a rich structure that leads to simpler algorithms than those employed for the MEM problem. We can say that the MVM problem is closer to the maximum cardinality matching problem than the maximum edge-weighted matching problem. Dobrian *et al.* (2018) have designed both exact and approximation algorithms for this problem.

An MVM can be characterized in two different ways. The first characterization is in terms of augmenting paths and weight-increasing paths. An augmenting path is defined as earlier, a path that has alternating unmatched edges and matched edges, with one more unmatched edge than matched edges. By augmenting a matching using this path, we add the weights of the unmatched endpoints of the path to the matching, and hence the cardinality of the matching increases while the weight of the augmented matching cannot decrease since vertex weights are non-negative. A *reversing path* is an alternating path with an even number of edges that begins with a matched edge and ends with an unmatched edge. A reversing path is *weight-increasing* if the matched endpoint of the path has lower weight than the unmatched endpoint. In this case, by switching the matched and unmatched edges on the path we increase the weight of the matching. Dobrian *et al.* (2018) proved the following result.

Theorem 5.1. A matching M is an MVM if and only if there is neither an augmenting path nor a weight-increasing path with respect to M.

The second characterization is in terms of weight vectors. For any matching M, consider a weight vector which lists the weights of the matched vertices in non-increasing order. Now we can compare two matchings by comparing their weight vectors lexicographically. The following result may also be found in Dobrian *et al.* (2018).

Theorem 5.2. A matching M is an MVM if and only if its weight vector is lexicographically maximum among all weight vectors of matchings.

These results lead to two different exact algorithms for the MVM problem.

One of these algorithms processes vertices in non-increasing order of their weights, and from each unmatched vertex u searches for a heaviest unmatched vertex v it can reach by augmenting paths. If the augmenting path search is successful, then the matching is augmented. If it is not successful, we discard this unmatched vertex u (we will not find an augmenting path from u in the future steps of the algorithm). In both cases, we process the next heaviest unmatched vertex, terminating when we have processed or matched all the vertices. In this algorithm by the choice of vertices matched at each step, there will be no weight-increasing path, and it suffices to search for augmenting paths.

A second algorithm would first compute a maximum cardinality matching (of arbitrary weight) in the graph. Then we search for weight-increasing paths of even length from each unmatched vertex. If we succeed, then we switch matched to unmatched edges and *vice versa* on this path, and increase the matching weight. If we do not succeed, then we process the next unmatched vertex. Note that in this case, by construction, there cannot be an augmenting path. This algorithm is attractive practically for several reasons. The first is that this algorithm does not need to process vertices in non-increasing order of weights, and hence there is more concurrency in this algorithm, making an implementation on a parallel computer feasible. The second is that it is attractive if we compute the Gallai–Edmonds decomposition (Lovász and Plummer 2009), since this decomposition identifies a subgraph that has a perfect matching in any maximum cardinality matching. We can remove this subgraph from further consideration, since all vertices in the subgraph are matched and every maximum cardinality matching has the same weight. Hence we need to run the MVM algorithm only on the residual graph.

The first of these algorithms was designed by Dobrian *et al.* (2018) and has time complexity $O(nm)$. Spencer and Mayr (1984) have designed an exact MVM algorithm with lower $O(m\sqrt{n}\log n)$ time complexity, which employs recursion to compute the matching. As far as we know, this algorithm has not been implemented, and it is not clear if it is practical.

An important reason for designing the exact algorithm discussed in the previous paragraph is that by restricting the length of the augmenting path to at most three, we may obtain a 2/3-approximation algorithm. This result was first obtained for bipartite graphs by Dobrian *et al.* (2018). Here one can split the MVM problem into two 'one-side-weighted' subproblems. This means given a bipartite graph $G = (V_1, V_2, E)$ in the first subproblem we ignore the weights on V_2, and in the second subproblem we ignore the weights on V_1. The advantage is that now we can find any augmenting path from an unmatched vertex in V_1 (V_2) for the first (second) subproblem. The vertices still need to be processed in non-increasing order of weights as in the exact algorithm, and we limit the search for augmenting paths to length at most three. Once the two matchings M_1 and M_2 are obtained, the Mendelsohn and Dulmage (1958) theorem can be invoked to find a matching M in which all the V_1 vertices (the weighted vertices) in M_1 and the V_2 vertices (again the weighted vertices) in M_2 are matched. The new matching M has the same weight as the sum of the matchings M_1 and M_2, and it is a 2/3-approximation to the maximum vertex-weighted matching. The time complexity of this algorithm is $O(m + n\log n)$, which is linear, except for the sorting step. While this algorithm is easy to state, its proof of correctness is somewhat involved. The proof works by finding, for each vertex that is not matched in the approximation algorithm but matched by the exact algorithm (failed vertices), two matched vertices in the approximate matching that are at least as heavy as the failed vertex.

Al-Herz and Pothen (2019) have extended this result to non-bipartite graphs. Here, since there is no bipartition of the vertices, we cannot invoke the Mendelsohn–Dulmage theorem, and new concepts about augmenting paths are needed. The 2/3-approximation algorithm is described in Algorithm 10; it processes vertices in non-increasing order of weights, and

---

Algorithm 10 Two-Third Approximate Matching $(G = (V, E, w))$

Input: A graph G with weights w on the vertices.
Output: A 2/3-approximate vertex-weighted matching M.

1:   $M \leftarrow \emptyset$; $Q \leftarrow V$
2: **while** $Q \neq \emptyset$ **do**
3:      Let u be a heaviest vertex in Q
4:      Let v denote a heaviest unmatched vertex reachable from u by an augmenting path P of length at most three
5:      **if** P is found **then**
6:          $M \leftarrow M \oplus P$; delete v from Q
7:      **end if**
8:      Delete u from Q
9: **end while**
10: **return** M

---

matches each vertex if possible to a heaviest unmatched vertex reachable by augmenting paths of length at most three. The time complexity of this algorithm is $O(m \log \Delta + n \log n)$. The proof of the approximation ratio involves charging the weight of each failed vertex to two distinct matched vertices in the approximate matching such that they have equal or higher weight relative to the failure. These vertices are found by examining the augmenting paths in the approximation algorithm, where we distinguish between a vertex from which an augmenting path search begins (an origin), and a vertex where it ends (a terminus).

Al-Herz and Pothen (2019) have implemented the 2/3-approximation algorithm for MVM and compare it with an exact maximum edge-weighted matching algorithm in LEDA (Mehlhorn and Näher 1999), the $(2/3 - \epsilon)$-approximation algorithms of Maue and Sanders (2007), the $(1 - \epsilon)$-algorithm of Duan and Pettie (2014) and a greedy 1/2-approximation algorithm for MVM. The LEDA algorithm benefits from the fractional matching initialization; it is about twelve times faster relative to the LEDA algorithm without any initialization, on problems where both terminated. There are a number of problems where LEDA without initialization did not terminate in four hours. On problems where LEDA with fractional initialization terminated in at most four hours but where the algorithm with no initialization did not, we use the former algorithm as the baseline. The exact MVM algorithm (it does not use any initialization) was slower than the LEDA algorithm by a factor less than two. The $(1 - \epsilon)$-approximation algorithm with $\epsilon = 1/3$ was faster than LEDA by a factor of seven for these problems; the $(2/3 - \epsilon)$-approximation algorithm with a Global Paths Algorithm initialization was about nine times faster; the 2/3-approximate MVM was 140 times faster than LEDA; the 1/2-approximation algorithm for MVM was 400 times faster. There was also a problem (nlpkkt200) where none of the exact algorithms terminated in four hours, but the 2/3-approximate MVM computed a matching in under a minute. In terms of weights, the 2/3-approximate MVM and the $(2/3-\epsilon)$-approximate MEM computed more than 99.99% of the maximum weight; the $(1-\epsilon)$-approximation algorithm was about 1% off the optimal, and the 1/2-approximate algorithm was 3% off the optimal.

Vertex-weighted matchings have been applied to the design of network switches (Tabatabaee, Georgiadis and Tassiulas 2001), and scheduling of astronaut training sessions (Bell 1994). More recently, online algorithms for vertex-weighted matchings in bipartite graphs have been surveyed by Mehta (2012) due to applications in internet advertising. Our interest in this problem was spurred by its application to computing sparse null space and column space bases of sparse, under-determined matrices (Coleman and Pothen 1987, Pinar *et al.* 2006).

6. The edge cover problem

An edge cover in a graph $G = (V, E)$ is a set of edges C such that there is at least one edge belonging to C incident on each vertex in V. The set of edges E is a trivial edge cover of the graph G. We consider the problem of finding an edge cover with minimum cardinality. The following relationship between a maximum matching and a minimum edge cover is due to Gallai (1959) and Norman and Rabin (1959), and is described in Theorems 19.1 and 19.2 of Schrijver (2003).

Theorem 6.1. Every maximum cardinality matching of a graph $G = (V, E)$ is contained in a minimum cardinality edge cover, and every minimum cardinality edge cover contains a maximum cardinality matching. Moreover if we have a maximum cardinality matching in G, we can find a minimum cardinality edge cover in $O(m)$ time, and *vice versa.*

The $O(m)$ time in the last part of Theorem 6.1 comes from a simple algorithm due to Norman and Rabin (1959). Given a maximum cardinality matching, we add the matched edges to the edge cover; additionally we add one of the edges incident on each unmatched vertex to the edge cover. This results in a minimum cardinality edge cover. Further, a minimum cardinality edge cover can be computed in $O(\sqrt{n}m)$ time, the time needed for a maximum cardinality matching.

Now we consider a weighted graph $G = (V, E, w)$, where $w : E \mapsto \mathbb{R}_{\geq 0}$ is a weight function that maps each edge to a non-negative weight. Recall that the weight of a set S is the sum of all the weights of the edges in S. We seek to minimize the weight of an edge cover of G. In the graph in

Figure 3.2, there are two minimum weight edge covers of weight 15: the set of edges $\{(a, e), (b, f), (c, d)\}$ and $\{(a, c), (a, e), (b, f), (d, f)\}$.

The simplest algorithm for computing a minimum weight edge cover selects an edge of minimum weight incident on each vertex and adds it to the cover. This Nearest Neighbour algorithm (and many other algorithms that we consider) does not compute a minimal edge cover; that is, there could be redundant edges that decrease the weight of an edge cover when they are removed, while retaining the property of being an edge cover of the graph. We will show later in the context of the more general b-edge cover problem that this algorithm computes a 2-approximation to an edge cover of minimum weight, even without the removal of the redundant edges.

Next we describe an approximation-preserving reduction of the minimum weight edge cover problem to the maximum weight matching problem. The reduction is mentioned in Chapter 27 of Schrijver (2003), and Huang and Pettie (2017) proved that it is approximation-preserving. We proceed to describe this reduction next.

For each vertex $v \in V$ we define $\mu(v)$ to be the minimum weight of any edge incident on v. Now we compute a transformed weight w' for each edge $e = (u, v)$ as

$$w'(u, v) = \mu(u) + \mu(v) - w(u, v).$$

(Note that we write $w(u, v)$ instead of $w((u, v))$.) Consider how the weight of an edge is transformed under this mapping. There are three cases.

Case 1. $\mu(u) = \mu(v) = w(u, v)$. We call such an edge locally subdominant, since this edge is of minimum weight among all of its neighbouring edges. The reader can verify that the transformation does not change the weight of such an edge; thus $w'(u, v) = w(u, v)$.

Case 2. $\mu(u) = w((u, v)) > \mu(v)$. The reader can verify that $w'(u, v) = \mu(v) < w(u, v)$.

Case 3. $w(u, v) > \mu(v) > \mu(u)$. It is easily verified that $w'(u, v) < \mu(u) < w(u, v)$.

The other cases may be obtained from the symmetry of u and v. In all cases, we see that the transformed weight of an edge is no larger than the minimum weight among its neighbouring edges.

Assume that we are given a matching M with the transformed weights on the edges. Define $V(M)$ to be the set of matched vertices in M, and let $e(v)$ denote an edge of minimum weight incident on v. We can compute an edge cover C as follows:

$$C = M \cup \{e(v) : v \in V \setminus V(M)\}.$$

Hence the edge cover consists of the matched edges and a set of minimum weight edges incident on the unmatched vertices.

If M^* is a maximum weight matching with respect to the weights w', then the resulting edge cover C^* is an edge cover of minimum weight with respect to the weights w. Furthermore, Huang and Pettie showed that this reduction is approximation-preserving.

Theorem 6.2. Let M be $(1-\epsilon)$-approximate matching obtained with the transformed weights w' from a graph $G = (V, E, w)$. Then the edges in M together with a set of minimum weight edges incident on the unmatched vertices constitute a $(1 + \epsilon)$-approximate edge cover C of the graph G with respect to the original weights w.

Proof. Let C^* denote an optimal edge cover with respect to the weights w and let M^* denote an optimal matching with respect to the weights w'. Since we have shown that $w'(u, v) \leq w(u, v)$ for all edges (u, v), we have that

$$w'(M^*) = \sum_{(u,v)\in M^*} w'(u, v) \leq \sum_{(u,v)\in M^*} w(u, v) = w(M^*) \leq w(C^*).$$

We make use of this result in the following.

From the construction of C and the definition of M, we have

$$\begin{aligned}
w(C) &= w(M) + \mu(V \setminus V(M)) \\
&= \mu(V(M)) - w'(M) + \mu(V \setminus V(M)) \\
&= \mu(V) - w'(M).
\end{aligned}$$

Similarly we obtain

$$w(C^*) = \mu(V) - w'(M^*). \tag{6.1}$$

Making use of the approximation ratio of the matching algorithm that computed M, we obtain

$$\begin{aligned}
w(C) &= \mu(V) - w'(M) \\
&\leq \mu(V) - (1 - \epsilon)w'(M^*) \\
&= \mu(V) - w'(M^*) + \epsilon\, w'(M^*) \\
&= w(C^*) + \epsilon\, w'(M^*) \\
&\leq (1 + \epsilon)\, w(C^*).
\end{aligned}$$

In the third line we used (6.1), and in the fourth line we made use of the inequality from the first paragraph. This completes the proof. $\square$

We can conclude that the time complexity of computing a minimum weight edge cover is the same as that of computing a maximum weight matching, $O(mn + n^2 \log n)$ (Gabow 2018). Furthermore, we can compute an approximate minimum weight edge cover using one of the approximate maximum weight matching algorithms.

We now turn to a reduction of an edge cover to a minimum weight perfect matching problem, described in Schrijver (2003). We make a second copy of the graph G which we call G'. We join each vertex v in G with its corresponding vertex v' in G' by an edge with weight set to twice the minimum weight edge incident on v. We call these latter edges the linking edges. By construction the doubled graph has a perfect matching. Furthermore, we can choose the matched edges in G' to correspond to copies of the matched edges in G. We replace each linking edge (v, v') in the matching by a minimum weight edge incident on the vertex v, and add it to the edge cover. To these edges we add the matched edges from G, to obtain a minimum weight edge cover of G. This reduction of an edge cover to a perfect matching was modified to compute a prize-collecting variant of the minimum weight edge cover (in which vertices might not be covered by paying a cost) by Azad et $al.$ (2010) to compute a dissimilarity measure for high-dimensional proteomic data.

A third reduction to matching, also mentioned in Schrijver (2003), computes $b'(v) = \deg(v) - 1$, and then computes a b'-matching of maximum weight. A minimum weight edge cover is the set of edges complementary to the matching. We will discuss this algorithm in more detail when we consider the b-edge cover problem. We will also describe some computational results on the edge cover problem after we discuss the latter problem.

7. The b-edge cover problem

Given a graph $G = (V, E)$ and a function $b(v)$ that maps each vertex $V \in V$ to a natural number, a b-edge cover is a subset of edges C such that *at least* $b(v)$ edges in C are incident on v. We assume that $b(v) \leq \deg(v)$. If $b(v)$ is identically equal to one for all vertices v, then we have the edge cover problem. If the edges are weighted by a non-negative function $w(.)$, then the weight of a b-edge cover is the sum of weights of the edges in the cover. The problem we consider is to compute a b-edge cover of minimum weight. Unfortunately the weight transformation approach we used to compute a minimum weight edge cover from a maximum weight matching does not work for b-edge cover. An exact algorithm for the problem has polynomial time complexity, since it can be computed as the complement of a maximum weight b'-matching, where $b'(v) = \deg(v) - b(v)$ for all $v \in V$. We seek to design approximation algorithms that have near-linear time complexity and have more concurrency.

7.1. b-EDGE COVER *algorithms*

The simplest algorithm that one could think of for the b-edge cover problem is for each vertex v to independently choose $b(v)$ edges to add to the cover. This is the b-NEAREST NEIGHBOUR algorithm (bNN) described in Algo-

---

Algorithm 11 b-NEAREST NEIGHBOUR$(G = (V, E, w, b))$

1: $C = \emptyset$
2: **for** each $v \in V$ **do**
3:     $E_v = b(v)$ lightest edges incident on v
4:     $C = C \cup E_v$
5: **end for**
6: **return** C

---

rithm 11, and we show in Section 7.5 that this leads to a 2-approximation algorithm for the minimum weight b-edge cover problem.

Practically the weight of the edge cover could be reduced by removing redundant edges. An edge (u, v) is *redundant* if there are more than $b(u)$ edges from the cover incident on u and there are more than $b(v)$ edges from the cover incident on v. We say that such a vertex u or v is over-saturated; if there are exactly $b(u)$ edges incident on a vertex u it is saturated; and if fewer than $b(u)$ edges are incident on u (this can happen only during the computation of an edge cover since it violates the requirements of an edge cover), then it is unsaturated.

The next algorithm one could think of is a greedy algorithm, except that unlike the matching problem, here the weights have to be dynamically updated. This algorithm is inspired by a greedy algorithm for the set multicover problem, in which we are given a collection of subsets of a set, and we are required to choose subsets from the collection so that each element v in the set is covered $b(v)$ times. The GREEDY algorithm for the set cover problem is described by Chvatal (1979). The b-edge cover problem is a special case of the set multicover problem where the elements in the set correspond to vertices, and subsets to edges, which consist of pairs of vertices.

We define the *effective weight* of an edge as the weight of the edge divided by the number of its unsaturated endpoints. The GREEDY algorithm for minimum weight edge cover works as follows. Initially, no vertices are covered, and the effective weights of all the edges are half of the edge weights. At each iteration the algorithm chooses an edge of minimum effective weight and adds it to the cover. It then decrements the $b(.)$ values of the endpoints of this edge by one, and updates the effective weights of its neighbouring edges. For the effective weight update, there are three possibilities for each edge: (i) none of its endpoints is saturated, and there is no change in its effective weight, (ii) one of the endpoints is saturated, and its effective weight doubles, or (iii) both endpoints are saturated, its effective weight becomes infinite, and the edge is marked as deleted. The algorithm iterates until all vertices are saturated. The algorithm is described in Algorithm 12.

We can see from an analysis of a PRIMAL DUAL algorithm for this problem (provided in Section 7.4) that this algorithm produces an edge cover whose

---

Algorithm 12 GREEDY-b-EDGE COVER $(G = (V, E, w, b))$

1: $C = \emptyset$
2: Compute effective weight of all edges $e \in E$
3: **while** there exists an unsaturated vertex **do**
4:     Add an edge of minimum effective weight $e = (u, v)$ to C
5:     Delete e
6:     Decrement $b(u)$ and $b(v)$ by one
7:     **for** $x \in \{u, v\}$ **do**
8:         **if** $b(x) = 0$ **then** update effective weights of edges incident on x
9:             Delete any edge (x, y) with $b(y) = 0$
10:         **end if**
11:     **end for**
12: **end while**
13: **return** C

---

weight is at most $3/2$ the minimum weight. The worst-case time complexity of the GREEDY algorithm is $O(\beta m \log n)$.

The next algorithm we consider is the LAZY GREEDY algorithm. The effective weight of an edge can only increase during the GREEDY algorithm, and we exploit this observation to design a faster variant. The idea is to delay updating effective weights of edges, which is the most expensive step in the algorithm, until they are needed. If the edges are maintained in non-increasing order of weights in a heap, then we update the effective weight of only the top edge; if after the update, its effective weight is no larger than the effective weight of the next edge in the heap, then we could add the top edge to the cover as well. A similar property of greedy algorithms has been exploited in submodular optimization, where this algorithm is known as the LAZY GREEDY algorithm (Minoux 1978). Its time complexity is $O(m \log n)$, which is better than the GREEDY algorithm, and in practice it performs even better. We refer the reader to Ferdous, Pothen and Khan (2018) for more details on this algorithm.

Yet another variant of the GREEDY algorithm is the LOCALLY SUB-DOMINANT EDGE (LSE) algorithm. An edge is *locally subdominant* if it has minimum effective weight among all of its neighbouring edges. The LSE algorithm identifies locally subdominant edges, adds them to the cover, and updates the effective weight of its neighbouring edges. This process iterates until no unsaturated vertices remain. This algorithm was proposed by Khan *et al.* (2016a), and they proved that it is also a $3/2$-approximation algorithm.

We briefly consider an algorithm that obtains a b-edge cover from the complement of a b'-matching. For each vertex v, define $b'(v) = \deg(v) - b(v)$, and then compute a b'-matching of maximum weight. The complement of the matching is a b-edge cover of minimum weight. What is interesting is that if we use a 1/2-approximation algorithm that matches locally dominant edges such as the GREEDY b'-MATCHING algorithm or the b'-SUITOR algorithm, then we can show that the resulting b-edge cover is a 2-approximation to a minimum weight b-edge cover. This algorithm, called the MATCHING COMPLEMENT EDGE COVER (MCE) algorithm, was designed by Khan *et al.* (2018b). The MCE algorithm works with static edge weights since it computes a matching, not an edge cover, and hence it has much more concurrency than the other algorithms described thus far except for the b-NEAREST NEIGHBOUR algorithm. An important feature of the MCE algorithm is that it computes a maximal b'-matching, and hence the complement is a minimal b-edge cover, which implies that there are no redundant edges to remove.

There are interesting relationships among the algorithms that have been described in this section. The GREEDY, LAZY GREEDY and LSE algorithms compute the same b-edge cover of a graph if (1) ties in weights are broken consistently, and (2) redundant edges are removed greedily. By this we mean that we consider the subgraph induced by the over-saturated vertices, compute a maximum weight matching in this graph, and remove these edges as redundant. These results are obtained in Khan and Pothen (2016) and Ferdous, Pothen and Khan (2018). The PRIMAL DUAL algorithm we describe in the next section will also compute the same b-edge cover under the two conditions mentioned above. Since we prove that this last algorithm has the approximation ratio 3/2, we can conclude that all the algorithms mentioned in this paragraph have the same approximation ratio.

Finally we discuss the $(1 + \epsilon)$-approximation algorithm for b-edge cover designed by Huang and Pettie (2017). One way to solve a minimum weight b-edge cover is to reduce it to a capacitated b-matching problem and then solve the latter problem (Schrijver 2003). Unfortunately this reduction is not approximation-preserving. However, the authors show that if a b-matching is constructed maintaining a certain approximate complementary slackness condition, then the complement graph, a b-edge cover, would also satisfy an approximate complimentary slackness condition, which results in the desired approximation preservation. They describe a linear time algorithm to find such a b-matching. This linear time scaling algorithm is a generalization of the approximation algorithm for maximum weight matching designed by Duan and Pettie. As a by-product they also obtain an $O(m\sqrt{b(V)})$-time algorithm for the cardinality version of these problems.

7.2. *A linear programming formulation*

Now we turn to a linear programming formulation of the b-edge cover problem to describe primal–dual algorithms for the b-edge cover problem. Using the primal–dual framework, we will describe a 3/2-approximation algorithm, analyse the bNN algorithm and show that it is a 2-approximation algorithm, and describe a Δ-approximation algorithm.

We begin by describing the primal and dual linear programming (LP) formulations of the minimum weight b-edge cover problem. Recall that we consider a graph $G = (V, E, w, b)$, where $w(.)$ denotes the non-negative weights on the edges, and we need to choose at least $b(v)$ edges incident on each vertex v. Recall that a vertex v is *uncovered* or *unsaturated* if a current b-edge cover has fewer than $b(v)$ edges in it.

Define a vector $\mathbf{x} \in \{0, 1\}^m$ with the intent that $x(e) = 1$ if the edge is in the cover, and 0 otherwise. Denote the set of edges incident on a vertex v by $\delta(v)$ (*i.e.* the set of edges one of whose endpoints is v). The integer linear program (ILP) formulation of the minimum weight edge cover problem is as follows:

$$\min \sum_{e \in E} w(e)x(e), \text{ subject to } \sum_{e \in \delta(v)} x(e) \geq b(v), \text{ for all } v \in V,$$

$$x(e) \in \{0, 1\}, \text{ for all } e \in E. \tag{7.1}$$

The linear programming relaxation of the ILP is obtained by relaxing the binary constraint $x(e) \in \{0, 1\}$ to $0 \leq x(e) \leq 1$. The relaxation is as follows:

$$\min \sum_{e \in E} w(e)x(e), \text{ subject to } \sum_{e \in \delta(v)} x(e) \geq b(v), \text{ for all } v \in V,$$

$$x_e \geq 0, \; -x(e) \geq -1, \text{ for all } e \in E. \tag{7.2}$$

The final constraint, equivalent to $x(e) \leq 1$, indicates that we may use each edge e at most once in the cover. Any $\mathbf{x}$ satisfying the constraints of this linear program is a feasible solution.

To construct the dual program of the relaxed linear program we define a dual variable for each of the constraints. We define two sets of variables, namely $\mathbf{y} \in \mathbb{R}^n_{\geq 0}$ and $\mathbf{z} \in \mathbb{R}^m_{\geq 0}$. The dual is as follows:

$$\max \sum_{v \in V} b(v)y(v) - \sum_{e \in E} z(e),$$

$$\text{subject to } y(i) + y(j) - z(e) \leq w(e), \text{ for all } e = (i, j) \in E,$$

$$y(v), z(e) \geq 0, \text{ for all } v \in V \text{ and } e \in E. \tag{7.3}$$

Again, any $\mathbf{y}$ and $\mathbf{z}$ satisfying the constraints of the dual program are dual feasible.

For 1-edge cover, the upper bound constraints on the primal variables for the LP relaxation shown in (7.2), *i.e.* $x(e) \leq 1$ for all $e \in E$, are not necessary. If in the solution there is any $x(e) > 1$, we can change it to $x(e) = 1$, producing a feasible solution with a lower objective value. For the b-edge cover formulation, without the upper bound constraints we cannot guarantee that the relaxation produces a solution in $[0, 1]$. Consider the dual problem, (7.3). The dual variable corresponding to each upper bound constraint on $x(e)$ in the primal is $z(e)$. These variables serve the same purpose as its counterpart constraints in the primal, by helping the program to choose distinct edges in the cover by providing enough slack to make some of the dual constraints feasible. We will discuss the role of $z(e)$ variables in our analysis of the algorithm.

Let EC^* be the objective value of an *optimum* b-edge cover solution, EC_{LP} the objective value of an *optimum* solution of the relaxed LP shown in (7.2), and EC_{dual} the objective value of a *feasible* solution of the dual problem shown in (7.3). Then from linear programming and weak duality theory we have

$$EC_{dual} \leq EC_{LP} \leq EC^*. \tag{7.4}$$

7.3. Dual fitting algorithms

Now let us assume $\mathbf{x}$ is a feasible *integral* solution to the primal linear program, and let $\mathbf{y^a}$, $\mathbf{z^a}$ be the *approximate dual* solutions to the corresponding dual program. We say these are *approximate dual* variables as they may not necessarily satisfy the dual constraints.

Suppose we have a hypothetical algorithm that satisfies the following two properties.

Property 7.1 (paid in full). The algorithm finds these primal and approximate dual variables that maintain the equality of primal and approximate dual objective values, that is,

$$\sum_{e \in E} w(e)x(e) = \sum_{v \in V} b(v)y^a(v) - \sum_{e \in E} z^a(e). \tag{7.5}$$

Property 7.2 (shrinking factor). Let $\alpha > 0$ be a constant such that $\mathbf{y} = \mathbf{y^a}/\alpha$ and $\mathbf{z} = \mathbf{z^a}/\alpha$ become dual feasible variables.

We can prove that this hypothetical algorithm guarantees an α-approximation. Replacing $\mathbf{y^a}$ and $\mathbf{z^a}$ in (7.5), we have

$$\sum_{e \in E} w(e)x(e) = \alpha \cdot \left(\sum_{v \in V} b(v)y(e) - \sum_{e \in E} z(e) \right). \tag{7.6}$$

Since $\mathbf{y}$ and $\mathbf{z}$ are dual feasible, from (7.4) and (7.6) we have

$$\sum_{e \in E} w(e)x(e) = \alpha \cdot \left(\sum_{v \in V} b(v)y(v) - \sum_{e \in E} z(e) \right) \leq \alpha \cdot EC_{LP} \leq \alpha \cdot EC^*. \tag{7.7}$$

This proves the required α-approximation guarantee. We now show how to instantiate this hypothetical algorithm to obtain 3/2- and 2-approximation algorithms for b-edge cover.

7.4. A 3/2-approximation algorithm

Rajagopalan and Vazirani (1993) have employed dual fitting to design an algorithm for set multicover. The Primal Dual algorithm that we present for 3/2-approximation of b-edge cover is motivated by this algorithm. It also generalizes a primal–dual edge cover algorithm discussed in Ferdous, Pothen and Khan (2018). Our aim is to formulate *approximate dual* variables such that the two properties mentioned earlier are satisfied. We first define a few concepts and variables required to understand the algorithm and its analysis.

- An *unsaturated* vertex v is *covered* by one of its incident edges e if during the execution of the algorithm e is selected to cover that vertex. This edge e is called a *covering edge* of the vertex v. Note that after covering a vertex v by e it may still be *unsaturated*. Note also that a b-edge cover might include edges incident on v that are not covering edges of v, since an edge (u, v) may have been chosen as a covering edge of u but not v. We denote by S_v the set of covering edges of v.

- In general during the run of the algorithm an edge e is *available* if it can cover at least one of its endpoints. We define a set Q_e to denote the endpoints that e covers, and hence $0 \le |Q_e| \le 2$. The set C includes the edges in the cover.

- The *effective weight* of an edge, eff_weight(e), is defined as the ratio of the weight of the edge and the number of its *unsaturated* endpoints. The effective weight of an edge can be thought of as the *price* the algorithm needs to pay to cover its unsaturated endpoints. Hence we define price(v, e) as the effective weight of e where v is an unsaturated endpoint of e. When an edge e is included in the cover, we fix the price(v, e) value(s) of the endpoint(s) it covers.

- Let $r(v)$ be a variable defined on each vertex v. We call it the *dynamic requirement of saturation* since this variable will let us know whether the vertex v is already saturated or not. The $r(v)$ values are initialized to the $b(v)$ values.

- We maintain a list $L(v)$ that consists of $r(v)$ edges of lowest effective weight incident on a vertex v. Only these edges will be considered for inclusion in the b-edge cover. If there are two or more vertices with the $r(v)$th lowest effective weight, this list removes the need to process the edges in non-decreasing order of effective weights. Recall that both the effective weights and the values of $r(v)$ are dynamically changing in the algorithm.

The output of the algorithm is a set of edges C. We can derive an integral primal solution from C by setting $x_e = 1$ for all $e \in C$ and $x_e = 0$ for all

---

Algorithm 13 Primal Dual($G = (V, E, w, b)$)

1: $C = \emptyset$
2: **while** there exists an *unsaturated* vertex **do**
3:     Call Price Assignment($G(V, E, w, b)$,max_price)
4:     Call Augment Cover ($G(V, E, w, b)$, max_price, C, r)
5: **end while**
6: **return** C

Algorithm 14 Price Assignment($G = (V, E, w, b)$, max_price)

1: **for** each $v \in V$ **do**
2:     **if** v is *unsaturated* **then**
3:         max_price(v) = {eff_weight(e) : $r(v)$th lowest effective weight of
                            an edge incident on v}
4:         $L(v) = r(v)$ edges of lowest effective weight incident on v
5:     **end if**
6: **end for**

$e \in E \setminus C$. We now introduce max_price(v), a non-negative variable defined on each vertex v, which is equivalent to the approximate dual variable $y^a(v)$. During the execution of the algorithm we set

max_price(v)

$= r(v)$th lowest effective weight among the edges incident on v.

We create another non-negative variable excess(e) equivalent to $z^a(e)$ for each edge e. Unlike the **max_price**, the **excess** variable is not necessary during the run of our algorithm, but it is needed for the proof analysis. Hence we defer its definition till then.

The pseudocode of the algorithm is shown in Algorithm 13. The algorithm iterates until all the vertices become saturated. In each iteration there are two phases: the Price Assignment phase and the Augment Cover phase. The Price Assignment phase computes the max_price(v) values of each unsaturated vertex. These values are used by the Augment Cover phase to add as many edges to the cover as possible.

We provide the pseudocode for the Price Assignment phase in Algorithm 14.

The second phase of the algorithm, the Augment Cover phase, adds vertices to the edge cover using the **max_price** information set by the first phase. The pseudo-code for the Augment Cover phase is presented in Algorithm 15. This phase scans the edges to find eligible ones to add in the cover. An edge e is selected as follows. If the edge $e = (i, j)$ covers both of its endpoints and if its effective weight is less than or equal to the max_price(.) values of both of its endpoints, then it would be included in the cover.

Algorithm 15 AUGMENT COVER$(G = (V, E, w, b), \text{max_price}, \text{price}, C, r)$

1: **for** each $e = (u, v) \in E$ **do**
2:     **if** u and v are both covered **then**
3:         Mark (u, v) as deleted
4:         Continue
5:     **end if**
6:     **if** u and v are both uncovered, $e \in L(u) \cap L(v)$, and
            $\text{eff\_weight}(e) \leq \{\text{max\_price}(u), \text{max\_price}(v)\}$ **then**
7:         Set $\text{price}(u, e)$ and $\text{price}(v, e)$ to $\text{eff\_weight}(e)$
8:         $C = C \cup (u, v)$
9:         Decrease $r(u)$ and $r(v)$ by 1
10:         Mark (u, v) as deleted
11:     **else if** only u is uncovered, $e \in L(u)$, and
            $\text{eff\_weight}(e) \leq \text{max\_price}(u)$ **then**
12:         Set $\text{price}(u, e)$ to $\text{eff\_weight}(e)$
13:         $C = C \cup (u, v)$
14:         Decrease $r(u)$ and $r(v)$ by 1
15:         Mark (u, v) as deleted
16:     **else if** only v is uncovered, $e \in L(v)$, and
            $\text{eff\_weight}(e) \leq \text{max\_price}(v)$ **then**
17:         Set $\text{price}(v, e)$ to $\text{eff\_weight}(e)$
18:         $C = C \cup (u, v)$
19:         Decrease $r(u)$ and $r(v)$ by 1
20:         Mark (u, v) as deleted
21:     **end if**
22: **end for**

Upon finding such an edge e the algorithm fixes the values of $\text{price}(i, e)$ and $\text{price}(j, e)$ to the value of $\text{eff\_weight}(e)$. Note that the equation $\text{price}(i, e) + \text{price}(j, e) = w(e)$ is then satisfied by this edge. On the other hand if the edge e covers only one endpoint u, to be included in the cover its effective weight must be less than or equal to the $\text{max\_price}(u)$ value. In this case we fix $\text{price}(u, e)$ to be $w(e)$ so that the equation $\text{price}(u, e) = w(e)$ holds. Whenever we add an edge to the cover, we mark it as deleted and update the $r(v)$ values of its endpoints. (We update both $r(u)$ and $r(v)$ when an edge (u, v) is deleted for identifying redundant edges in a post-processing.)

Once the algorithm terminates, we have settled the $\text{price}(v, e)$ values for each vertex and covering edge pair. We have also updated $\text{max\_price}(v)$ values for each vertex. For each vertex v there are two kinds of edges incident on v. One kind is the set of *covering* edges, which were the edges used to cover vertex v. The second kind would be other edges incident on v that were necessary to cover the other endpoint w of an edge (v, w). Let

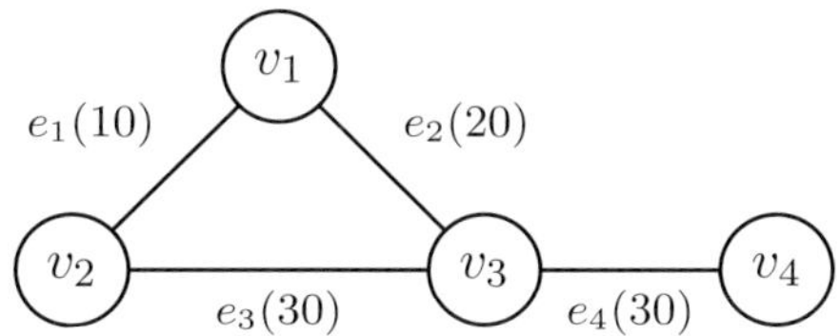

Figure 7.1. A small graph whose b-edge cover is to be computed.

S_v denote the set of covering edges incident on each vertex v, and note that $|S_v| = b(v)$. Observe also that when the algorithm terminates, the value of $\max\_price(v) = \max\{price(v, e) : e \in S_v\}$.

We have not yet defined how we get the other set of approximate dual variables, *i.e.* **excess**. These variables are not necessary for the execution of the algorithm, but they are needed for the proof analysis. We motivate this variable by means of a small example.

Example 7.3. Consider the graph with four vertices v_1, v_2, v_3 and v_4 in Figure 7.1, with the edge labels and weights also shown. The $b(v)$ value is 2 for each vertex v except v_4, for which it is 1. The optimal edge cover is the graph itself. We run the PRIMAL DUAL algorithm on this problem. First, in the PRICE ASSIGNMENT phase, we assign the price values of each vertex and edge pair. Here $price(v_1, e_1) = price(v_2, e_1) = 5$, $price(v_1, e_2) = price(v_3, e_2) = 10$, $price(v_2, e_3) = price(v_3, e_3) = 15$ and $price(v_3, e_4) = price(v_4, e_4) = 15$. The max_price$(v)$ values are as follows: $\max\_price(v_1) = 15$, $\max\_price(v_2) = 10$, $\max\_price(v_3) = 15$ and $\max\_price(v_4) = 15$. In the next phase, suppose we scan through edges in the order e_1, e_2, e_3 and e_4. We select e_1 since the effective weight of this edge is less than the max_price$(.)$ values for both endpoints. We decrease the $r(v_1)$ and $r(v_2)$ values by 1 and mark e_1 as deleted. Similarly we select e_2 and e_3. Note that v_1, v_2 and v_3 are saturated. We cannot add e_4 in this phase because the effective weight of e_4, which is now 30, is greater than max_price(v_4) which is 15. Next we start the second iteration. In the PRICE ASSIGNMENT phase, we set max_price(v_4) as 30 and in the AUGMENT COVER phase we select e_4, since the effective weight of e_4 equals max_price(v_4).

At the termination of the algorithm, the max_price values for our example are as follows: $\max\_price(v_1) = 10$, $\max\_price(v_2) = 15$, $\max\_price(v_3) = 15$ and $\max\_price(v_4) = 30$. Let us consider the dual constraints defined in (7.3). For e_1 the left side of the constraint using the approximate duals is $\max\_price(v_1) + \max\_price(v_2) - excess(e_1)$. But $\max\_price(v_1) + \max\_price(v_2) = 25$, which is much greater than the weight of e_1. So we have a large excess on the summation that we need to balance. This is the purpose of the approximate dual, **excess**. If an edge e was not included in a cover we set $excess(e) = 0$. If an edge $e = (i, j)$ covered both of its endpoints when e was added to the cover, we set

$$\text{excess}(e) = (\text{max\_price}(i) - \text{price}(i, e)) + (\text{max\_price}(j) - \text{price}(j, e)).$$

Otherwise if e covered only one endpoint i when it was added to the cover, then

$$\text{excess}(e) = \text{max\_price}(i) - \text{price}(i, e).$$

We can restate this as

$$\text{excess}(e) = \sum_{q \in Q_e} (\text{max\_price}(q) - \text{price}(q, e)).$$

In the example, the **excess** values of the edges are as follows: $\text{excess}(e_1) = (10-5)+(15-5) = 15$, $\text{excess}(e_2) = (10-10)+(15-10) = 5$, $\text{excess}(e_3) = (15-15)+(15-15) = 0$ and $\text{excess}(e_4) = 30-30 = 0$. Observe that all of the dual constraints now become feasible except for e_4, where the left side of the constraint is $\text{max\_price}(v_3) + \text{max\_price}(v_4) - \text{excess}(e_4) = 15+30-0 = 45$, which is greater than the weight of the edge. We will show that we can scale the approximate duals in such a way that the scaled dual variables always satisfy the constraints.

Lemma 7.4. The approximation ratio of the PRIMAL DUAL algorithm is $3/2$.

Proof. First note that by construction $\text{max\_price}(v)$ is non-negative; since it is the maximum of the price values of incident edges of a vertex, $\text{excess}(e)$ is also non-negative. We need to show that by setting $\alpha = 3/2$, the paid in full and shrinking factor properties defined in Section 7.3 are maintained. Using the approximate duals, **max_price** and **excess**, we first show that the objective value of dual LP defined in (7.3) equals the weight of the cover that we get from the algorithm. Let us first consider the right side of the (approximate) dual objective value, *i.e.* $\sum_{e \in E} \text{excess}(e)$. Note that

$$\sum_{e \in E} \text{excess}(e) = \sum_{e \in C} \text{excess}(e),$$

since $\text{excess}(e) = 0$ if e is not in the cover. Then we have

$$\sum_{e \in C} \text{excess}(e) = \sum_{e \in C} \sum_{q \in Q_e} (\text{max\_price}(q) - \text{price}(q, e))$$

$$= \sum_{e \in C} \sum_{q \in Q_e} \text{max\_price}(q) - \sum_{e \in C} \sum_{q \in Q_e} \text{price}(q, e)$$

$$= \sum_{v \in V} b(v) \cdot \text{max\_price}(v) - \sum_{e \in C} w(e). \tag{7.8}$$

The second term in the last line of (7.8) follows because $\sum_{q \in Q_e} \text{price}(q, e)$ is equal to $w(e)$, an invariant we maintain during the execution of AUGMENT

Cover phase. For the first term note that a particular vertex $v \in V$ will appear exactly $b(v)$ times in the sum.

Replacing $\sum_{e \in E} \text{excess}(e)$ in the objective value of the dual LP from (7.3),

$$\sum_{v \in V} b(v) \cdot \text{max\_price}(v) - \sum_{e \in E} \text{excess}(e)$$

$$= \sum_{v \in V} b(v) \cdot \text{max\_price}(v) - \sum_{v \in V} b(v) \cdot \text{max\_price}(v) + \sum_{e \in C} w(e)$$

$$= \sum_{e \in C} w(e). \tag{7.9}$$

But we cannot substitute **max_price** for **y** and **excess** for **z** because these are not dual feasible. Define $\alpha \equiv 3/2$, and set $\mathbf{y} = \mathbf{max\_price}/\alpha$ and $\mathbf{z} = \mathbf{excess}/\alpha$. We show that the scaled variables **y** and **z** now become feasible. There are two scenarios to consider.

For the first scenario assume that an edge e belongs to the cover. We have two cases.

Case 1. $e = (i, j)$ covers both of its endpoints. Then replacing $y(i)$, $y(j)$ and $z(e)$, the left side of the first constraint in (7.3),

$$\frac{1}{\alpha} \cdot (\text{max\_price}(i) + \text{max\_price}(j) - (\text{max\_price}(i) - \text{price}(i, e))$$

$$- (\text{max\_price}(j) - \text{price}(j, e)))$$

$$= \frac{1}{\alpha}(\text{price}(i, e) + \text{price}(j, e)) \leq \frac{1}{\alpha} w(e) \leq w(e). \tag{7.10}$$

The last line follows from the fact that during the algorithm we maintain $\text{price}(i, e) + \text{price}(j, e) = w(e)$.

Case 2. The edge e covers only one endpoint, say i. Using the definitions of $y(i)$, $y(j)$ and $z(e)$ with $e = (i, j)$, the left side of the constraint in (7.3) becomes

$$\frac{1}{\alpha} \cdot (\text{max\_price}(i) + \text{max\_price}(j) - (\text{max\_price}(i) - \text{price}(i, e)))$$

$$= \frac{1}{\alpha}(\text{max\_price}(j) + \text{price}(i, e)). \tag{7.11}$$

From the algorithm, $\text{price}(i, e) = w(e)$. When vertex j was saturated, vertex i was still unsaturated. We have not picked e as a *covering* edge for j. Since e was in the list of eligible edges $L(j)$, the price values of the covering edges incident on j must be less than or equal to $w(e)/2$. Since max_price of a vertex is the maximum of the price values of the covering edges incident on that vertex, $\text{max\_price}(j) \leq w(e)/2$. So we have

$$\frac{1}{\alpha} \cdot (\text{max\_price}(j) + \text{price}(i, e)) \leq \frac{1}{\alpha} \cdot \frac{3}{2} w(e) \leq w(e).$$

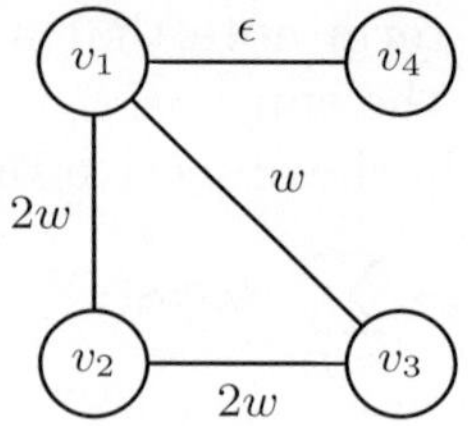

Figure 7.2. A tight example for the Primal Dual algorithm.

We now consider the second scenario when e is not part of the b-edge cover and $\mathrm{excess}(e) = 0$. So the left side of the constraint becomes

$$\frac{1}{\alpha} \cdot (\mathrm{max\_price}(i) + \mathrm{max\_price}(j)).$$

Without loss of generality assume i has become saturated first and then j. This immediately establishes that $\mathrm{max\_price}(i) \leq w(e)/2$ and $\mathrm{max\_price}(j) \leq w(e)$. We have

$$\frac{1}{\alpha} \cdot (\mathrm{max\_price}(i) + \mathrm{max\_price}(j)) \leq w(e).$$

We combine the analysis as follows:

$$\sum_{e \in C} w(e)$$
$$= \sum_{v \in V} b(v) \cdot \mathrm{max\_price}(v) - \sum_{e \in E} \mathrm{excess}(e) \quad \text{(from (7.9))}$$
$$= \alpha \cdot \left(\sum_{v \in V} b(v)y(v) - \sum_{e \in E} z(e) \right) \quad \text{(replacing \textbf{max\_price} and \textbf{excess})}$$
$$= \alpha \cdot EC_{dual} \leq \alpha \cdot EC_{LP} \leq \alpha \cdot EC^* \quad \text{(from (7.6))}.$$

This gives the 3/2-approximation ratio. $\square$

The approximation ratio is tight, and a tight example is the graph shown in Figure 7.2. Suppose $b = 1$ for each vertex. In the first iteration the Primal Dual algorithm will add (v_1, v_4) to the cover, and it cannot add any other edge. In the second iteration since all the remaining edges have the same effective weight, it may add any one of the edges (v_1, v_3), (v_1, v_2) or (v_2, v_3). Suppose it chooses the edge (v_1, v_3). Then to cover the vertex v_2, it has to choose either (v_1, v_2) or (v_2, v_3), resulting in a cover with weight $3w + \epsilon$, whereas the optimal weight is $2w + \epsilon$. So as $\epsilon \to 0$, we get the approximation ratio 3/2.

Next we derive the time complexity of the algorithm. We can do a couple of optimizations on the general algorithm presented in Algorithm 13, which are as follows.

- In the AUGMENT COVER phase, when an edge is selected we mark all the neighbouring vertices of its covered endpoints as potential vertices. During the PRICE ASSIGNMENT phase we need only update the max_price values of the potential vertices.

- In the PRICE ASSIGNMENT phase, when the max_price value changes, we mark all of its incident edges as potential covering edges. In AUGMENT COVER phase we can scan only these edges.

Lemma 7.5. The time complexity of Algorithm 13 is $O(\beta \Delta m)$.

Proof. Initially all the vertices and edges are marked as potential (see the optimization of the algorithm just mentioned). During the execution of the algorithm a vertex v can be marked at most $\deg(v)$ times as potential. Each time it is marked it will have to find an edge incident on it with the $r(v)$th minimum effective weight. One can find such an entry in $O(\beta \deg(v))$ time. Summing over all vertices we obtain $O(\beta \Delta m)$.

Similarly, during the PRICE ASSIGNMENT phase, an edge $e = (i, j)$ can be marked at most $\deg(i) + \deg(j) = O(\Delta)$ times. Summing over m edges we obtain $O(\Delta m)$. Hence the time complexity is $O(\beta \Delta m)$. $\qquad\square$

7.5. A 2-approximation algorithm

We have presented in Algorithm 11 the b-NEAREST NEIGHBOUR algorithm for finding a b-edge cover. It is similar to the popular and well-known k-nearest neighbour graph construction algorithm, used in many domains including machine learning and data mining to represent data by a sparse graph or to sparsify a graph. The difference between these problems is how the values of k and b are defined. In the former case k is constant for all vertices while the latter case is more general, with the option to set user-defined values of $b(v)$ for each vertex in the graph. This algorithm, like many other algorithms for the b-edge cover problem, could have redundant edges in the cover, *i.e.* edges that could be removed while the residual edges form a b-edge cover, thus resulting in an edge cover of lower weight. However, even without removing such edges, we can show that this algorithm gives us an approximate solution to the b-edge cover problem, where the weight of the cover is at most twice the optimal weight. In this section we prove this result using the dual fitting framework developed in Section 7.3.

Lemma 7.6. The approximation ratio of the b-NEAREST NEIGHBOUR algorithm is 2.

Proof. Let $\mathbf{x}$ be the primal integral solution and C a b-edge cover computed by the algorithm; hence $x(e) = 1$ if $e \in C$, and otherwise $x(e) = 0$. Let S_v denote the set of the $b(v)$ lightest edges incident on v. We define a price value for each covering vertex and edge pair, and consider an edge $e = (i, j) \in C$.

If the weight of the edge e is among the lightest $b(i)$ edges incident on the vertex i *and* the lightest $b(j)$ edges incident on j, then we set $\mathrm{price}(i, e) = \mathrm{price}(j, e) = w(e)/2$. In this case we say the edge e covers both of its endpoints. Otherwise if e is only among the lightest $b(i)$ ($b(j)$) edges incident on i (j), we assign $\mathrm{price}(i, e) = w(e)$ ($\mathrm{price}(j, e) = w(e)$). In this case the edge e covers only one of its endpoints. Next we set the approximate dual variables. We define for each vertex v, $\mathrm{max\_price}(v) = \max_{e \in S_v} \mathrm{price}(v, e)$. For each edge $e \in C$, let Q_e denote its covered endpoints. Note that Q_e may contain one or two vertices. We define for each $e \in C$, $\mathrm{excess}(e) = \sum_{q \in Q_e}(\mathrm{max\_price}(q) - \mathrm{price}(q, e))$. If an edge e is not included in the cover then we set $\mathrm{excess}(e) = 0$. Note that since price values of a vertex and edge pair are always non-negative, $\mathrm{max\_price}$ is non-negative. Again, as $\mathrm{max\_price}(v) \geq \mathrm{price}(v, e)$ for all $e \in S_v$, the excess variable is also non-negative.

We now show that with $\alpha = 2$, the two properties mentioned in Section 7.3 are satisfied by the approximate dual variables **max_price** and **excess**.

The first property is the equality of primal objective and approximate dual objective functions, and this follows directly from the corresponding proof in the 3/2-approximation algorithm described in Section 7.4.

For the second property, we show that setting $y(v) = \mathrm{max\_price}(v)/\alpha$ and $z(e) = \mathrm{excess}(e)/\alpha$ for $\alpha = 2$ make these dual feasible. We consider two scenarios for an edge $e \in E$.

In the first scenario e belongs to the cover. Then replacing $y(v)$ and $z(e)$ on the left side of the first constraint of (7.3), we obtain

$$y(i) + y(j) - z(e) = \frac{1}{\alpha}(\mathrm{max\_price}(i) + \mathrm{max\_price}(j)$$

$$- \sum_{q \in Q_e}(\mathrm{max\_price}(q) - \mathrm{price}(q, e))). \qquad (7.12)$$

We have two cases to consider.

Case 1. The edge e covers both of the endpoints, and hence $Q_e = \{i, j\}$. Recall that in this case e is among the lightest $b(i)$ edges incident on i, and the lightest $b(j)$ edges incident on j. The price values are then assigned as $\mathrm{price}(i, e) = \mathrm{price}(j, e) = w(e)/2$. Simplifying, we obtain

$$y(i) + y(j) - z(e) = \frac{1}{\alpha}((\mathrm{max\_price}(i) + \mathrm{max\_price}(j))$$

$$- (\mathrm{max\_price}(i) - \mathrm{price}(i, e)) - (\mathrm{max\_price}(j) - \mathrm{price}(j, e)))$$

$$= \frac{1}{\alpha}(\mathrm{price}(i, e) + \mathrm{price}(j, e))$$

$$= \frac{1}{\alpha}(w(e)/2 + w(e)/2) \leq w(e). \qquad (7.13)$$

Case 2. e covers only one endpoint, say i. In this case we assign $\mathrm{price}(i, e) = w(e)$:

$$
\begin{aligned}
y(i) + y(j) - z(e) &= \frac{1}{\alpha}(\mathrm{max\_price}(i) + \mathrm{max\_price}(j) \\
&\quad - (\mathrm{max\_price}(i) - \mathrm{price}(i, e))) \\
&= \frac{1}{\alpha}(\mathrm{max\_price}(j) + \mathrm{price}(i, e)) \\
&\le \frac{1}{\alpha}(w(e) + w(e)) \le \frac{1}{2}(2w(e)) \le w_e. \qquad (7.14)
\end{aligned}
$$

The last line follows because $\mathrm{max\_price}(j) \le w(e)$, since j is saturated and e is not a covering edge for j.

We now consider the second scenario when e is not part of the cover. In this case $\mathrm{excess}(e) = 0$, and the left side of the constraint becomes

$$
\frac{1}{\alpha} \cdot (\mathrm{max\_price}(i) + \mathrm{max\_price}(j)).
$$

Without loss of generality assume i was saturated first and then j. This establishes that $\mathrm{max\_price}(i) \le w(e)$ and $\mathrm{max\_price}(j) \le w(e)$ since i and j were covered by an edge with lower weight than that of e. We have

$$
\frac{1}{\alpha} \cdot (\mathrm{max\_price}(i) + \mathrm{max\_price}(j)) \le \frac{1}{2} \cdot 2w(e) \le w(e).
$$

We combine these analyses as follows:

$$
\begin{aligned}
\sum_{e \in C} w(e) \\
= \sum_{v \in V} b_v \cdot \mathrm{max\_price}(v) - \sum_{e \in E} \mathrm{excess}(e) \quad &\text{(from (7.9))} \\
= \alpha \cdot \left(\sum_{v \in V} b(v)y(v) - \sum_{e \in E} z(e) \right) \quad &\text{(replacing } \mathbf{max\_price} \text{ and } \mathbf{excess}) \\
= \alpha \cdot EC_{dual} \le \alpha \cdot EC_{LP} \le \alpha \cdot EC^* \quad &\text{(from (7.6)).} \qquad \square
\end{aligned}
$$

As in the 3/2-approximation algorithm we can also show the tightness of the approximation ratio of the b-NEAREST NEIGHBOUR algorithm. We generate a graph with n vertices, where n is odd, as follows. The vertex v_0 is connected to all other vertices $v_1 \ldots v_{(n-1)}$. Each vertex v_i for odd $i > 0$ is connected with $v_{(i+1)}$. All edges have the same weight x. We let $b = 1$ for every vertex. An example with $n = 9$ is shown in Figure 7.3.

The optimal edge cover for this example would have weight $5w$, consisting of the four edges not incident on v_0 and one edge incident on v_0. But the b-NEAREST NEIGHBOUR algorithm could produce an edge cover with weight $8w$ by choosing all edges incident on v_0. For a graph with n vertices, the optimal weight would be $\frac{1}{2}(n-1) * w + w$ but the b-NEAREST NEIGHBOUR

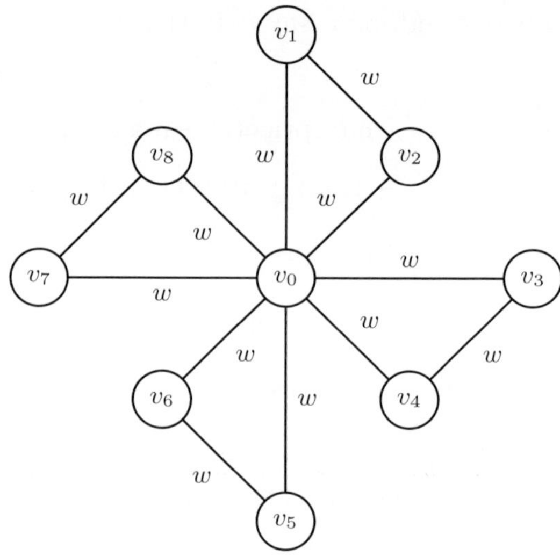

Figure 7.3. A tight example for the b-NEAREST NEIGHBOUR algorithm.

algorithm could produce an edge cover with weight $(n - 1) * w$. Thus the approximation ratio is $(2(n - 1))/(n + 1)$, which as $n \to \infty$ is 2.

Lemma 7.7. The time complexity of the b-NEAREST NEIGHBOUR algorithm is $O(m \log \Delta)$.

Proof. We can sort the adjacency list of a vertex v in $O(\deg(v) \log(\deg(v)))$ time. Summing over all the vertices we obtain $O(m \log \Delta)$. $\qquad\square$

The time complexity can be improved to $O(m)$ if we use the worst-case linear time selection algorithm as described in Cormen, Leiserson, Rivest and Stein (2009) to find the b_ith smallest element in the adjacency list of a vertex.

7.6. Δ-approximation algorithm

Here we present another algorithm based on linear programming duality for b-edge cover, with an approximation ratio of Δ, which is larger than the ratios $3/2$ and 2 for the algorithms we have considered thus far. We do this for several reasons. First, the worst-case approximation ratio does not always determine how well an algorithm does practically. Second, the analysis of this algorithm enables us to present a different technique for designing a primal–dual algorithm. This algorithm is derived from an algorithm for set multicover designed by Hall and Hochbaum (1986), which leads to a better approximation ratio for vertex cover.

Recall that due to weak duality and LP relaxation, the objective value of any feasible solution to the dual problem in (7.3) is a lower bound for the optimum b-edge cover in (7.1). But in general a dual feasible solution does not guarantee an approximation ratio. However, there exists a particular dual feasible solution, a *maximal dual feasible* solution, whose objective value provides a bound on the optimum value. A dual feasible solution (denoted $\bar{\mathbf{y}}$ and $\bar{\mathbf{z}}$) is *maximal* if it satisfies the following three properties.

I There does not exist a feasible solution$(\mathbf{y}, \mathbf{z})$ with $y \geq \bar{y}$, $z \geq \bar{z}$ and

$$\sum_{v \in V} b(v)y(v) - \sum_{e \in E} z(e) > \sum_{v \in V} b(v)\bar{y}(v) - \sum_{e \in E} \bar{z}(e).$$

II $\bar{z}(e) = 0$ whenever $\bar{y}(i) + \bar{y}(j) < w(e)$.

III

$$\sum_{v \in V} \bar{y}(v) \leq \sum_{v \in V} b(v)\bar{y}(v) - \sum_{e \in E} \bar{z}(e).$$

The proposed algorithm is as follows.

(1) Find a *maximal* dual feasible solution $(\bar{\mathbf{y}}, \bar{\mathbf{z}})$.

(2) Output the cover $C = \{e = (i, j) | \bar{y}(i) + \bar{y}(j) - \bar{z}(e) = w(e)\}$.

We first show that such an algorithm would provide a Δ-approximation.

Lemma 7.8. The algorithm is a Δ-approximation algorithm for b-edge cover.

Proof. We first establish that C is a feasible cover. For the sake of contradiction, assume that C is not a feasible cover; hence there exists a vertex v that is not covered by at least $b(v)$ edges. Let

$$\epsilon = \min_{e \in \delta(v)} \{\epsilon_e = w(e) - (\bar{y}(i) + \bar{y}(j) - \bar{z}(e)) \text{ and } \epsilon_e > 0\}.$$

We show that the value of ϵ is well-defined. According to our assumption at most $b(v) - 1$ edges incident on v are included in C. Since $\bar{y}$ and $\bar{z}$ are dual feasible, there must be at least one edge e where $\bar{y}(i) + \bar{y}(j) - \bar{z}(e) < w(e)$, equivalently $w(e) - (\bar{y}(i) + \bar{y}(j) - \bar{z}(e)) > 0$. We set $\bar{y}(v)' = \bar{y}(v) + \epsilon$, and $\bar{z}(e)' = \bar{z}(e) + \epsilon$, for edges $e \in \delta(v)$ and $e \in C$. The variables$(\bar{y}', \bar{z}')$ are dual feasible. But this contradicts the maximality (property I) of $\bar{y}, \bar{z}$, since $\bar{y}'$ increases the first term of the dual objective value by at least $b(v)\epsilon$, and $\bar{z}'$ increases the second term by at most $(b(v) - 1)\epsilon$, with a net increase of at least ϵ. This establishes that C is a feasible cover.

Now since $(\bar{y}, \bar{z})$ is a dual feasible solution, the weak duality theorem applies:

$$\sum_{v \in V} b(v)\bar{y}(v) - \sum_{e \in E} \bar{z}(e) \leq EC_{LP} \leq EC^*. \tag{7.15}$$

Algorithm 16 DUAL FEASIBLE$(G = (V, E, w, b))$

1: Initialize $y_v = 0$ and $z_e = 0$, for all $v \in V$ and $e \in E$
2: Assign $w'(e) = w(e)$, for all $e \in E$
3: **while** there exists an unsaturated vertex $v \in V$ **do**
4:     $f = \arg\min\{w'(e) : e \in \delta(v) - C\}$
5:     $y(v) = y(v) + w'(f)$; $C = C \cup \{f\}$
6:     **for** $e \in \delta(v)$ **do**
7:        $w'(e) = w'(e) - w'(f)$
8:        **if** $w'(e) < 0$ **then**
9:           $z(e) = z(e) - w'(e)$
10:          $w'(e) = 0$
11:        **end if**
12:     **end for**
13:     decrease $r(v)$ by 1
14:     **if** $r(v) = 0$ **then**
15:        Mark v as saturated
16:     **end if**
17: **end while**

Using property III, we obtain

$$\sum_{v \in V} \bar{y}(v) \leq EC^*. \tag{7.16}$$

From the construction of the cover we have

$$\sum_{e \in C} w(e) + \sum_{e \in C} \bar{z}(e) = \sum_{e=(i,j) \in C} \bar{y}(i) + \bar{y}(j)$$

$$= \sum_{v \in V} \sum_{e \in (\delta(v) \cap C)} \bar{y}(v)$$

$$\leq \sum_{v \in V} |\delta(v)| \bar{y}(v)$$

$$\leq \Delta \sum_{v \in V} \bar{y}(v) \leq \Delta \cdot EC^*. \tag{7.17}$$

In the last line we used (7.16). Hence we have established the Δ-approximation ratio for the proposed algorithm. $\qquad\qquad\square$

Next our goal is to design an algorithm that produces a maximal feasible dual solution. One such algorithm is shown in Algorithm 16.

The algorithm first initializes the dual variables $(\mathbf{y}, \mathbf{z})$ to zero. For each edge it maintains a variable, the *residual* weight $\mathbf{w}'$. This is initialized by the weight of the edge. In each iteration, it picks an unsaturated vertex, v, and then it finds an adjacent edge f incident to v with minimum residual weight. Upon finding the edge f, it adds f to the cover, adds $w'(f)$ to $y(v)$, and subtracts $w'(f)$ from the residual weights of all edges incident on v. Note that the iteration in line 6 of the algorithm goes over all edges incident on v, including f and other edges that may have been added to the cover in earlier iterations. If the weight of any residual edge (say e) becomes negative, it subtracts $w'(e)$ from $z(e)$ (thus the value of $z(e)$ increases), and sets $w'(e)$ to zero. It then decreases the requirement $r(v)$ by 1, and marks v as saturated if $r(v) = 0$.

The output of the algorithm is the set C. We now show that the dual vectors derived are maximal and satisfy the three properties.

Claim 7.9. The variables $\mathbf{y}$ and $\mathbf{z}$ are non-negative.

Proof. The initial edge weights are non-negative, and the algorithm maintains the residual weight w' to be non-negative. The variable y is updated by adding w' to it, and $z(e)$ is updated by subtracting $w'(e)$ when its value is negative, and hence these variables are non-negative as well. $\square$

Claim 7.10. All edges e in the cover C satisfy $y(i) + y(j) - z(e) = w(e)$.

Proof. Let i be a vertex at some iteration of the algorithm and $f = (i, j)$ be an available edge with minimum residual weight $w'(f)$. We have $w'(f) \geq 0$ from the previous claim, and then we add $w'(f)$ to $y(i)$. In the **for** loop over edges, we now subtract $w'(f)$ from all edges incident on i, including the edge f. Hence $w'(f)$ is set to zero. Every time the weight of an edge is decreased by some amount in the algorithm, it is transferred to the $y(.)$-variable of one of its endpoints. Hence at this point in the algorithm, $y(i) + y(j) - z(e) = w(f)$, since $z(f)$ is zero as long as $w'(f)$ is non-negative. In future iterations involving other available edges incident on i or j, the invariant $y(i) + y(j) - z(e) = w(e)$ is maintained by increasing the value of $z(e)$. $\square$

Claim 7.11. The inequality $y(i) + y(j) - z(e) \leq w(e)$ holds for all $e \in E \backslash C$.

Proof. Since $w'(e) \geq 0$ for the edges not in the cover, the two endpoints of such edges have absorbed a weight of at most $w(e)$. Note that in this case $z(e) = 0$. $\square$

So $(\mathbf{y}, \mathbf{z})$ is a dual feasible solution, but we need to show that it is also maximal.

Lemma 7.12. The dual vector $(\mathbf{y}, \mathbf{z})$ of the Dual Feasible algorithm is maximal.

Proof. (I) Each vertex v is covered by $b(v)$ edges for which the dual constraints $y(v) + y(u) - z((v, u)) = w(v, u)$ are tight (according to Claim 7.9). Suppose we increase a dual variable $y(v)$ by a non-negative amount ϵ. Now at least $b(v)$ constraints (those corresponding to the covering edges of v)

are violated. (We say at least, since if there is an edge incident on v that is not a covering edge with residual weight less than ϵ, its constraint is also violated.) To compensate for the constraint violations, we need to add ϵ to at least $b(v)$ elements of $\mathbf{z}$. So the increase in objective function is exactly $b(v)\epsilon$ while the decrease is at least $b(v)\epsilon$. Hence the objective function value for (y, z) is not greater than that of $(\bar{y}, \bar{z})$.

(II) According to the construction, the residual weight $w'(e) \geq 0$, for all $e \in E \setminus C$. That means for such edges e we have $y(i) + y(j) \leq w(e)$. Since $w'(e) \geq 0$, line 8 of the algorithm will never be satisfied for e, resulting in $z(e) = 0$.

(III) Since $z(e) = 0$, for all $e \in E \setminus C$, it suffices to show that

$$\sum_{e \in C} z(e) \leq \sum_{v \in V} (b(v) - 1)\, y(v).$$

We will prove this using induction on the number of iterations in the algorithm. Let $y^{(t)}$ and $z^{(t)}$ denote the variable y and z after t iterations, and let the number of iterations in the algorithm be denoted by T. We show that the inequality above is true in every iteration:

$$\sum_{e \in C} z(e)^{(t)} \leq \sum_{v \in V} (b(v) - 1)\, y(v)^{(t)}, \quad t = 1, \ldots, T.$$

For $t = 1$, the left side is zero since the $w'(e)$ values for all the edges are non-negative after the first iteration, and the right side is ≥ 0, since we must have identified an edge incident on a vertex v with the minimum weight and added its weight to $y(v)$. Note that if $b(v)$ equals 1 the right side is zero, and otherwise it is greater than zero. We inductively assume that the inequality holds for $t = 1, \ldots, k - 1$.

Denote the vertex selected at the kth iteration by v, and let f be the minimum weight edge incident on v with weight $w'(f)$. Then the right-hand side of the displayed equation increases by $(b(v) - 1)w'(f)$. Now this could lead to increase in some $z(e)^k$, where e is incident on v. When f is included in the cover, there are at most $(b(v) - 1)$ covering edges already incident on the vertex v. In the current iteration, only the $z(.)$ values of these edges can increase, and hence the net increase on the left side is at most $(b(v) - 1)w'(f)$. Hence the inequality is preserved at the end of this iteration. $\qquad\square$

We can show a tight example of the Δ-approximation algorithm by considering the graph in Figure 7.3 with different weights as follows. The weights of the edges $(v_i, v_{(i+1)})$, where $i = 1 \cdots 7$, are changed to $2\epsilon/\Delta$. Other weights remain the same. The maximum degree in this graph is $n - 1$. Assuming $b = 1$ for every vertex, the optimal cover weight is $\Delta/2 * (2\epsilon/\Delta) + w = \epsilon + w$, whereas if the DUAL FEASIBLE algorithm picks

Table 7.1. Structural properties of our graphs listed in increasing order of edges.

Problems	Vertices	Edges	Mean degree
Fault_639	616 923	5 715 102	19
bone010	986 703	7 861 302	16
Serena	1 382 121	13 716 976	20
mouse_gene	43 126	14 461 095	671
dielFilterV3real	1 102 824	21 583 469	39
Flan_1565	1 564 794	22 636 872	29
kron_g500-logn21	1 544 087	91 040 932	118
hollywood-2011	1 985 306	114 492 816	115
G500_21	1 598 722	118 594 475	148
SSA21	2 089 808	123 097 397	118
eu-2015	10 972 981	257 659 403	47

the first vertex to be v_0, the weight of the edge cover could be Δx. Taking the ratio we have $(\Delta w)/(\epsilon + w)$. As $\epsilon \to 0$, the ratio approaches Δ.

Lemma 7.13. The time complexity of the DUAL FEASIBLE algorithm is $O(\beta\, m)$.

Proof. A vertex v can be selected at most $b(v)$ times. When it is selected it has to find the edge with minimum residual weight, which can be found in $O(\deg(v))$ time. Summing over all vertices we get $\sum_{v \in V} b(v) \cdot \deg(v) = O(\beta\, m)$. $\square$

7.7. Computational results

7.7.1. Experimental set-up

All the experiments were conducted on a Purdue community cluster computer called Rice, described in Section 3.3.

Our test set consists of both real-world and synthetic graphs shown in Table 7.1. We generated two classes of RMAT graphs: (a) G500, representing graphs with skewed degree distributions from the Graph 500 benchmark (Murphy, Wheeler, Barrett and Ang 2010), and (b) SSCA, from the HPCS Scalable Synthetic Compact Applications graph analysis (SSCA#2) benchmark. We used the following parameter settings: (a) $a = 0.57$, $b = c = 0.19$ and $d = 0.05$ for G500, and (b) $a = 0.6$ and $b = c = d = 0.4/3$ for SSCA. Additionally we consider seven problems taken from the SuiteSparse Matrix Collection (Davis and Hu 2011) covering application areas such as medical science, structural engineering and sensor data. We also have a

Table 7.2. Comparison of weights of edge covers computed by approximation algorithms with respect to the exact algorithm.

Problems	OPT weight	Distance from optimality (%)		
		PD	NN	Match
Fault_639	2 475 175	3.21	5.75	1.07
bone010	4 500 059	3.20	5.73	1.04
Serena	5 477 688	3.21	5.62	1.01
mouse_gene	188 517	1.94	3.86	1.24
dielFilterV3real	3 007 336	2.80	4.77	0.82
Flan_1565	4 362 155	3.17	5.62	1.15
kron_g500-logn21	26 301 787	0.12	0.18	0.01
hollywood-2011	9 310 300	1.87	3.69	0.45
G500_21	25 624 663	0.11	0.16	0.01
SSA21	8 243 419	2.85	4.05	0.66
eu-2015	218 514 387	0.49	0.89	0.03
Geometric mean		1.32	2.25	0.28

large web-crawl graph (eu-2015) (Boldi, Marino, Santini and Vigna 2014) and a movie-interaction network (hollywood-2011) (Boldi and Vigna 2004).

All reported results are the average of five runs on graph with random edge weights. For edge cover the uniform random weights are in the range [1 100]. Since LEDA works only with integer weights, the real-valued weights are then rounded to their nearest integers. For the b-edge cover experiment, the uniform random weights are chosen from the range [1 1000].

7.7.2. Edge cover results

We compare four algorithms: the exact algorithm that computes the minimum weight of an edge cover using the weight transformation described in Section 6, the Nearest Neighbour algorithm (NN), the 3/2-approximation Primal Dual algorithm (PD), and a 1/2-approximate maximum weight matching algorithm that with the approximation-preserving weight transformation obtains a 3/2-approximation minimum weight edge cover (Match). We use LEDA's maximum weight matching code to compute the maximum matching with transformed weights, and the Suitor algorithm to compute the approximate matching. We removed redundant edges from the edge covers computed by these algorithms.

Table 7.3. Relative performance of run times of approximation algorithms with respect to the exact algorithm for edge cover.

Problems	Time (s)	Relative performance		
	Exact algorithm	PD	NN	MATCH
Fault_639	8.78	36.61	73.32	44.23
bone010	13.37	37.70	77.23	43.97
Serena	21.48	36.58	74.27	43.28
mouse_gene	14.39	35.74	69.13	40.79
dielFilterV3real	27.30	34.78	69.05	40.74
Flan_1565	27.20	32.91	63.50	39.85
kron_g500-logn21	147.89	37.92	78.83	43.96
hollywood-2011	122.78	33.88	65.43	39.84
G500_21	182.00	36.95	77.53	41.99
SSA21	233.71	40.62	87.41	46.00
eu-2015	408.40	41.22	85.88	48.84
Geometric mean		36.73	74.33	42.97

In Table 7.2 we report the minimum weight computed by the exact algorithm, and the distance to optimality of the other algorithms, computed as $(\text{approx} - \text{opt})/\text{opt} * 100$, where opt and approx are weights from the optimal and approximation algorithms, respectively. Note that all three algorithms perform much better than their worst-case ratios (3/2 for PD and MATCH; 2 for the NN algorithm). The MATCH algorithm is the best performer, followed by PD and then NN.

In Table 7.3 we compare the run times of the algorithms. We report the time taken by the exact algorithm, and report the relative performance of the approximation algorithms as the ratio of the run time of the exact to that of the approximation algorithm. The larger the relative performance, the faster the algorithm. Note that the NN algorithm is the fastest, followed by the MATCH algorithm and then the PD algorithm. But the run times of the approximation algorithms are within a factor of two of each other.

7.7.3. b-edge cover results

We do not have an exact algorithm to compare the approximation algorithms against since the weight transformation reduction does not work for b-edge cover. We compare the PRIMAL DUAL algorithm (PD), the LAZY

Table 7.4. Weight of b-edge covers computed by the PRIMAL DUAL algorithm. 'Increase' is the percentage of increase in weight of edge covers computed by the b-NEAREST NEIGHBOUR algorithm with respect to those from the PRIMAL DUAL algorithm.

Problem	PRIMAL DUAL		Increase (%)	
	Original	Remove redund.	Original	Remove redund.
bone010	3.93E+09	3.93E+09	0.00	0.00
Fault_639	2.86E+09	2.86E+09	0.00	0.00
Serena	6.85E+09	6.84E+09	0.07	0.00
Flan_1565	1.06E+10	1.04E+10	2.14	0.12
G500_21	6.78E+09	6.62E+09	2.20	0.20
kron_g500-logn21	6.00E+09	5.86E+09	3.02	0.36
eu-2015	3.28E+10	3.22E+10	3.63	0.28
mouse_gene	1.11E+08	1.06E+08	5.29	0.45
hollywood-2011	9.80E+09	9.53E+09	7.21	1.65
dielFilterV3real	7.47E+09	7.30E+09	10.50	3.03
SSA21	9.26E+09	8.59E+09	12.88	0.90
Geometric mean			4.77	0.51

GREEDY algorithm (LG) and the b-NEAREST NEIGHBOUR algorithm (bNN). Both the PD and LG algorithms are 3/2-approximate and compute the same b-edge cover, while bNN is 2-approximate. In Table 7.4 we report the weights of the b-edge covers. We include the weight computed by the original algorithm, and then the weight obtained by removing redundant edges. The last two columns show the percentage difference in weights between the bNN and PD algorithms. The results show that the PD algorithm computes smaller weights for the edge cover. The difference in weights between the two original algorithms can be large, up to 13% for these problems, with a geometric mean of about 5%. However, after removing redundant edges, this difference narrows to at most 3%. This implies that more redundant edges are removed from the bNN algorithm.

The run times of the algorithms are plotted in Figure 7.4 as the ratio of the run time of the LG algorithm to the run time of the second algorithm. Values higher than one for an algorithm mean that the algorithm is faster than LG. Note that in geometric mean, the PD algorithm is twice as fast as LG, and the bNN algorithm is about six times as fast as LG.

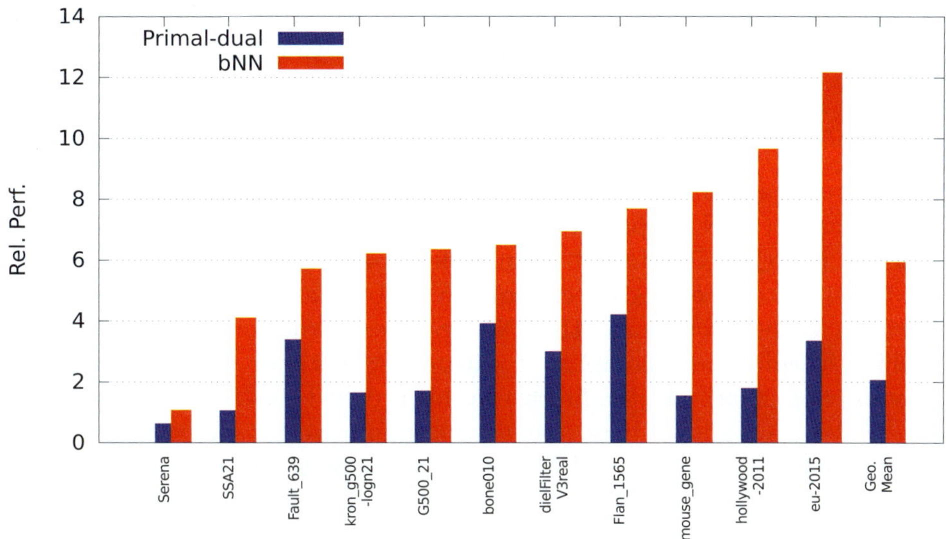

Figure 7.4. Relative performance of run times of the PRIMAL DUAL and b-NEAREST NEIGHBOUR algorithms for b-edge cover with respect to the LAZY GREEDY algorithm.

The Δ-approximation algorithm was implemented by Khan and Pothen (2016) and compared with the LSE algorithm, which has an approximation ratio of $3/2$. The performance of the latter algorithm was highly sensitive to the order in which the vertices are processed, for both weights and run times. Generally the LSE algorithm computed lower weights and it was also faster.

Several of the approximation algorithms for the b-edge cover problem have been implemented on parallel computers. The LSE algorithm and the MATCHING COMPLEMENT EDGE cover (MCE) algorithm have been implemented on shared-memory parallel machines: an IBM Power-8 with 764 cores and an Intel Xeon with 36 cores (Khan *et al.* 2018*b*). The PRIMAL DUAL $3/2$-approximation algorithm has been implemented by us on multiple cores of an Intel Xeon (Ferdous and Pothen, unpublished). The MCE algorithm has also been implemented on 8192 cores of a distributed-memory parallel computer with good speedups (Khan *et al.* 2018*a*), and it has been used to solve the adaptive anonymity problem, which we discuss in the next subsection.

Software for exact (based on the weight transformation to a matching problem) and approximation algorithms for edge cover and b-edge cover (GREEDY, LAZY GREEDY, b-NEAREST NEIGHBOUR and PRIMAL DUAL algorithms) will be included at https://github.com/CSCsw/EdgeCover. Code for the b-SUITOR algorithm from which the MCE b-edge cover could be computed is available at https://github.com/Exa-Graph.

7.8. Applications of b-edge cover

A widely used application of b-edge cover is in computing b-NEAREST NEIGHBOUR graphs (bNN graphs) to construct sparse graphs out of noisy data, although practitioners do not seem to know about the relationship of this construction to the b-edge cover problem. Subramanya and Talukdar (2014) provide a recent discussion of the bNN graph construction in semi-supervised machine learning, and compare it with other methods such as b-matching (Jebara et $al.$ 2009). (In the literature this is called the k-nearest neighbour graph.) We have shown here that the minimum weight b-edge cover formulation leads to a more general formulation where we are not constrained to use an identical value of $b(v)$ for all vertices. This is practically important since choosing $b(v)$ adaptively, $e.g.$ to be proportional to the degree of v, could lead to a graph that more faithfully represents the data. Furthermore, linking the b-nearest neighbour problem to the b-edge cover problem leads to several 3/2-approximation algorithms, while the bNN graph construction leads to a 2-approximation algorithm. We have also shown that the bNN graph construction leads to the inclusion of many redundant edges, which could be removed to reduce the weight of the edge cover, although the worst-case approximation ratio is not reduced by this technique. Use of the bNN graph construction is seen experimentally to lead to highly skewed degree distributions, since several neighbours of a vertex v could include it among its nearest neighbours, whereas the other b-edge cover algorithms provide better control of the degree distribution. This can influence the quality of the classification results obtained in semi-supervised machine learning.

Jebara et $al.$ (2009) have proposed the use of b-matching to solve this problem, using an exact belief propagation-based algorithm for computing perfect b-matchings from a complete graph that represents the data. The use of an exact algorithm makes this approach quite expensive relative to using a 1/2-approximation algorithm to compute the b-matching. Furthermore, working with a complete graph of the data makes the algorithm expensive for large data, whereas with a sparse representation of the data, the b-edge cover formulation ensures that each vertex v has at least $b(v)$ neighbours in the constructed graph (the b-matching only ensures at most $b(v)$ neighbours).

We discuss a problem in data privacy, adaptive anonymity, that stimulated our work on the b-edge cover problem.

Table 7.5 shows a small example to illustrate the adaptive anonymity problem. The top matrix contains binary data measured on six features for six individuals. Each individual is willing to permit their data to be

Table 7.5. A small example illustrating adaptive anonymity. From top to bottom: original input, dissimilarity matrix (Hamming distances) and anonymized output.

Instances	f1	f2	f3	f4	f5	f6
U_1	1	0	1	0	1	0
U_2	1	1	1	1	1	0
U_3	0	1	0	1	0	1
U_4	0	0	0	0	0	1
U_5	1	1	0	0	0	0
U_6	1	1	0	0	0	1

S	U_1	U_2	U_3	U_4	U_5	U_6
U_1	–	**2**	6	4	3	4
U_2		–	4	6	3	4
U_3			–	**2**	4	2
U_4				–	3	2
U_5					–	**1**
U_6						–

Instances	f1	f2	f3	f4	f5	f6
U_1	1	*	1	*	1	0
U_2	1	*	1	*	1	0
U_3	0	*	0	*	0	1
U_4	0	*	0	*	0	1
U_5	1	1	0	0	0	*
U_6	1	1	0	0	0	*

published provided each one is confused with at least one other person in the database. The idea is that a '*' could correspond to a 0 or a 1. The problem is to mask the least number of data items with '*' so that the privacy requirement of every individual is met. The matrix in the middle shows the Hamming distances between pairs of individuals, which is a measure of dissimilarity that indicates in how many entries the data of the two individuals differ. Now we compute a minimum weight 1-edge cover of the complete graph corresponding to the dissimilarity matrix, which includes the edges $(1, 2)$, $(3, 4)$ and $(5, 6)$. Finally, we pair individuals using this edge cover, and introduce '$*$'s in the columns where these pairs of individuals differ. Note that the weight of the edge cover is five, and there are ten '$*$'s in the anonymized output matrix that could be published for machine learning purposes.

More formally, in adaptive anonymity, each individual v expresses a required level of privacy, to be confused with $b(v) - 1$ other individuals. We are given a data set $X \in \mathbb{Z}^{n \times f}$, where n is the number of individuals and f is the number of features. Each row $x_v \in \mathbb{Z}^f$ of X is a contribution of the individual v to the data set and consists of f discrete features. A feature might be race, age, height, weight, income bracket, *etc.*, but not unique identifiers such as social security number. A vector $\mathbf{b}$ of length n, where an element $b(v)$ is a privacy requirement of the vth individual, is also given. The value $b(v)$ specifies that the data of the vth user must be indistinguishable from that of $b(v) - 1$ other users. The output of the algorithm is an anonymized data set $Y \in (\mathbb{Z} \cup \{*\})^{n \times f}$, where the '$*$' symbol indicates that a particular feature has been masked.

The adaptive anonymity problem is NP-hard, but Choromanski *et al.* (2013) proposed an algorithm that finds a good-quality approximate solution. The approximate solution comes from the observation that if we group similar instances together (with respect to their corresponding features) then we need to hide fewer features. The algorithm is a variational optimization method which iterates until some convergence criterion is met or a maximum number of iterations is reached. First, the algorithm creates a complete graph of n vertices corresponding to instances and a weight multiplier matrix initialized to all ones. Within an iteration, the algorithm assigns the weight of an edge between two vertices based on some dissimilarity measure between the two instances, multiplied by the weight multiplier. Next, the algorithm performs a *grouping* step based on the current weight assignment, and then the weight multipliers are adjusted based on the grouping. Thus at each iteration one needs to solve the grouping step.

Choromanski *et al.* (2013) used a perfect b-matching of minimum weight to group similar instances together. This limited the size of the instances they could solve to a few thousand individuals, since the exact b-matching algorithm has $O(b(V) m \log n)$ time complexity. Khan *et al.* (2018a) have shown that one could instead use a minimum weight b-edge cover formulation, and then use an approximation algorithm to compute the b-edge cover. This results in a 2β-approximation algorithm for minimizing the number of stars in the anonymized data, where β is the maximum value of the privacy requirements over individuals. The factor 2 comes from the choice of the algorithm used to compute the b-edge cover, and it could be reduced to $3/2$ with the Greedy or the Primal Dual algorithm. These authors computed the b-edge cover by taking the complement of a $1/2$-approximate b-matching, using the b-SUITOR algorithm. This algorithm was implemented efficiently on shared-memory and distributed-memory parallel computers, to provide the first reported parallel solutions for the adaptive anonymity problem. The work (total number of operations) in the distributed-memory parallel algorithm is $O(b(V)m)$, and its depth is $O(\log \Delta \log m)$. This algorithm also reduced the memory required to solve the problem from quadratic to linear in the number of individuals, since for each row in the dissimilarity matrix we need to store only the top $k(v)$ values (where $k(v)$ is some multiple of $b(v)$) to compute the maximum weight b-matching. If these do not suffice to obtain the b-matching, then the elements that had been processed could be discarded, and an additional set of $k(v)$ elements could be computed. The parallel algorithm solved an anonymity problem involving Medicare/Medicaid physician billing data with more than $700\,000$ individuals and 500 features on a distributed-memory computer with 8192 cores in four minutes. This represents an increase in problem size by three orders of magnitude over the earlier approaches.

8. Other approximation algorithms in CSC

We discuss two other classical problems in CSC where approximation algorithms have been designed or inapproximability results obtained.

8.1. Graph colouring

The graph colouring problem assigns the fewest colours to the vertices of a graph such that adjacent vertices receive different colours. Formally, we find a function $c : V \mapsto \mathbb{N}$ such that for every edge (u, v) we have $c(u) \neq c(v)$, and c uses the minimum number of colours. The *chromatic number* of a graph G is the minimum number of colours needed to colour it. Computing the chromatic number of a graph is an NP-hard problem. Furthermore, an inapproximability result due to Feige and Kilian (1998) states that for any $\epsilon > 0$ no polynomial time algorithm can approximate the chromatic number to within a factor of $n^{1-\epsilon}$ unless $P = NP$.

In CSC, several variants of graph colouring problems are of interest in computing sparse Jacobians and Hessians efficiently. Despite the pessimistic approximability result for colouring, in many contexts in CSC (such as finite element methods for solving partial differential equations) the graphs that need to be coloured are bounded in degree. Any graph can be coloured in at most $\Delta + 1$ colours, and hence the colouring problem is easy for the class of bounded degree graphs. An even tighter upper bound is the colouring number, which is obtained from an ordering computed by iteratively deleting vertices of minimum degree in the graph, updating degrees, and continuing until the graph is empty. (This is the same algorithm that computes the maximum core of a graph.) The maximum value of the minimum degree observed during this process is the colouring number. By colouring the graph in the reverse order in which the vertices are deleted from the graph, one can colour the graph in this many colours. A more detailed discussion is available in Gebremedhin, Manne and Pothen (2005).

For Erdős–Rényi graphs with constant average degree, the chromatic number can be precisely computed, and the greedy colouring algorithm can be shown to colour the graph in at most twice this number of colours. Let $G(n, d/n)$ denote an Erdős–Rényi graph on n vertices with the probability of an edge equal to d/n. We say that a property of a random graph on n vertices A_n holds asymptotically almost surely (a.a.s.) if the probability that A_n is true satisfies $\mathbb{P}(A_n) \to 1$ as $n \to \infty$. Achlioptas and Naor (2005) proved the following result (see also Kang and McDiarmid 2015).

Theorem 8.1. Given $d > 0$, let k_d be the least integer k for which $d < 2(k-1)\ln(k-1)$. Then $\chi(G(n, d/n))$ is $k_d - 1$ or k_d a.a.s. If $d > (2k_d - 3)\ln(k_d - 1)$, then $\chi(G(n, d/n)) = k_d$ a.a.s.

A colouring algorithm that finds maximal independent sets and colours them greedily can be shown to use at most twice the number of colours as the chromatic number (McDiarmid 1984). This paper discusses why this is simultaneously both a good and a bad result.

Now we turn to what is known about approximation algorithms for some of these colouring problems.

A distance-2 colouring of a graph $G = (V, E)$ is a colouring such that a vertex receives a colour distinct from any of its neighbours at distance 2 or less. This variant arises in computing Hessians if one does not take into account the symmetry of the graph. McCormick (1983) described a simple algorithm that computes an $O(\sqrt{n})$-approximation to the distance-2 chromatic number. Two other colouring problems that occur in Hessian computation where we need to compute only one among the two elements H_{ij} or H_{ji} for $i \neq j$, due to the symmetry of the Hessian, are star colouring and acyclic colouring. In star colouring, adjacent vertices receive distinct colours, and also every path on four vertices (P_4) should receive at least three colours. Thus two-coloured subgraphs should be stars. In acyclic colouring, adjacent vertices should receive distinct colours, and every cycle should receive at least three colours. Thus two-coloured subgraphs should be forests. *ZPP* is the class of all problems solvable in zero error probabilistic time. Gebremedhin, Tarafdar, Manne and Pothen (2007) showed that the star chromatic number and the acyclic chromatic number cannot be approximated to within a factor of $n^{(1/3-\epsilon)}$ unless $NP \subseteq ZPP$.

Bicolouring problems arise when we evaluate a Jacobian matrix using both its rows and columns. In this context, we do not need to evaluate every row and every column, since each non-zero in the Jacobian could be computed from its row or its column. Hence we have a colouring problem in which a vertex u could be assigned the colour 0 to indicate that u will not be used to compute any non-zeros in that row or column. The star bicolouring of a bipartite graph $G = (V_1, V_2, E)$ is a function $c : \{V_1 \cup V_2\} \mapsto \mathbb{N} \cup \{0\}$ such that

(1) $c(v_i) \neq c(v_j)$ if $(i, j) \in E$;

(2) the set of non-zero colours (set of 'true' colours) of V_1 is disjoint from the set of non-zero colours of V_2;

(3) two vertices v_i and v_k adjacent to a vertex v_j with $c(v_j) = 0$ receive distinct colours; and

(4) every path on four vertices receives at least three colours.

The acyclic bicolouring problem satisfies all the conditions of the star bicolouring problem, except for the last condition, which is replaced with: every cycle receives at least three colours.

Juedes and Jones (2012) have designed an approximation algorithm for the star bicolouring problem. The algorithm computes distance-2 independent sets of vertices (from either V_1 or V_2) containing a vertex of maximum degree in the current graph using a greedy algorithm, and then assigns them all one colour. The choice of independent sets from V_1 or V_2 is done based on the maximum degree and ratios of $|V_i|/\Delta$. They show that this algorithm has $O(n^{2/3})$ approximation ratio and that its time complexity is $O(m^2/n^{1/3})$. Since every star bicolouring is also an acyclic bicolouring, the approximation result holds for the latter problem as well. The authors have also implemented their algorithms, and showed on graphs with several hundred vertices and thousands of edges that the algorithm is competitive in run time with heuristic colouring algorithms that have been designed for star bicolouring.

8.2. Minimizing fill in sparse Cholesky factorization

The minimum fill problem is one of the most basic problems in the area of sparse matrix computations. Given a sparse, symmetric positive definite matrix A, we wish to permute its rows and columns symmetrically and compute the Cholesky factors of the permuted matrix

$$PAP^T = LL^T,$$

where P is a permutation matrix and L is the lower triangular Cholesky factor, such that the number of non-zeros in L are minimized. A fill element is a matrix element $L_{ij} \neq 0$ such that $A_{ij} = 0$.

In a graph model, we consider the undirected adjacency graph of A, and eliminate its vertices one by one. To eliminate a vertex, we add edges to make all of its neighbours a clique, and then delete the vertex and all edges incident on it. The problem is to find an ordering for eliminating the vertices that minimizes the edges added in this process, *i.e.* the fill edges. This problem was proved to be NP-complete by Yannakakis (1981). It is well known that the graph of the Cholesky factor, the filled graph, is a chordal graph.

Nested dissection, proposed by George (1973) is a technique for solving this problem using the divide and conquer paradigm. It works by finding a vertex separator that divides the graph into roughly two equal-sized (in terms of vertices) subgraphs. The two subgraphs are ordered first followed by the separator. One recurses on the subgraphs to find separators in the subgraphs, and orders them with this process. For planar graphs there exist separators of size $O(n^{1/2})$, and the fill can be bounded by $O(n \log n)$; for graphs (finite element meshes) that can be embedded in three dimensions with good aspect ratios, the separator size is $O(n^{2/3})$, and the fill is $O(n^2)$. Another heuristic which is widely used is the minimum degree algorithm and its many variants.

Unfortunately we cannot prove how close to optimum the solutions from these heuristic fill reduction algorithms are, although they obtained less fill than an approximation algorithm designed by Agrawal, Klein and Ravi (1993) for many matrices. These authors obtained an algorithm with approximation ratio $O(n^{1/2} \log^{3.5} n)$ for the total number of edges in the filled graph. (Their objective function was the sum of the number of edges in the original graph and the fill edges.) Natanzon, Shamir and Sharan (2000) obtained an approximation algorithm for minimizing the fill with approximation ratio eight times the minimum fill size. A polynomial time approximation scheme (PTAS) is a polynomial time algorithm that takes a parameter $\epsilon > 0$ and produces an approximate solution for a minimization problem with approximation ratio $(1 + \epsilon)$. Recently, Cao and Sandeep (2017) proved that if the problems of minimizing the fill or the total number of edges in the filled graph has a PTAS, then $P = NP$.

9. Conclusions

We have surveyed the design and implementation of approximation algorithms for several matching and edge cover problems and their applications in combinatorial scientific computing. Our interest is in algorithms that could be implemented to obtain high performance on modern processor architectures, including serial and parallel computers, both shared-memory and distributed-memory machines. We have viewed approximation as a paradigm for designing parallel algorithms (Khan *et al.* 2018*b*).

The paradigm of designing approximation algorithms for parallelism has been considered in the theoretical computer science community for vertex and set cover problems by Khuller, Vishkin and Young (1994), and for facility location, max cut, set cover and low stretch spanning trees, by Blelloch, Peng and Tangwongsan (2011) and Tangwongsan (2011). The idea underlying many of these parallel algorithms is that a greedy algorithm chooses a most cost-effective element in each iteration, and by allowing a slack, a factor of $(1+\epsilon)$, more elements can be selected at the cost of a slightly worse approximation ratio. These algorithms have poly-logarithmic depth, and although some of them have linear work requirements, there are few parallel implementations that we know of. Blelloch *et al.* (2012) have computed a maximal cardinality matching (which would be a 1/2-approximate algorithm) in parallel.

For matching problems we have discussed new approximation algorithms for maximum cardinality matching. For edge-weighted matching, we have described 1/2-approximation algorithms that employ different paradigms: greedy, path-growing, and proposals (related to the stable matching problem). We have considered $(2/3 - \epsilon)$-approximation algorithms for this problem as well. We have discussed 1/2-approximation algorithms for the b-matching problem, including a matroid-theoretic proof that the greedy algorithm leads to 1/2-approximation. We have also designed 1/2- and 2/3-approximation algorithms for vertex-weighted matching, using techniques that differ from those used for edge-weighted matching. We summarize the matching problems, exact and approximation algorithms to solve them, and their time complexity in Table 9.1.

For the minimum weight edge cover problem we have discussed an approximation-preserving reduction to the maximum weight matching problem, leading to both exact and several approximation algorithms. The b-edge cover problem has a rich collection of approximation algorithms, from the greedy algorithm and variants, and a primal–dual algorithm; these algorithms result in 3/2-approximation. The well-known b-nearest neighbour graph construction leads to a 2-approximation for this problem, as do greedy-like algorithms that do not compute dynamic effective weights, and an algorithm that computes the complement of a suitable 1/2-approximation matching algorithm. For the b-edge cover problem, we discussed a primal–dual linear programming framework that helps establish the approximation ratios of several algorithms. We summarize the edge cover problems, exact and approximation algorithms to solve them, and their time complexity in Table 9.2.

We have considered applications of matching to the solution of sparse systems of linear equations and other matrix computations. For the b-edge cover problem, we have discussed the construction and sparsification of graphs from large, noisy data sets; we have also described the solution of a data privacy problem called adaptive anonymity.

We believe that approximation algorithms represent a fruitful area for progress in designing efficient algorithms for solving problems in CSC, data

Table 9.1. Some of the exact and approximation algorithms for matching problems and their time complexity.

Problem	Algorithm	Reference	Complexity
Maximum matching		Micali and Vazirani (1980)	$O(\sqrt{n}\, m)$
Maximum weight matching		Gabow (2018)	$O(n(m + n \log n))$
1/2-approximation MWM	Path Growing Suitor	Drake and Hougardy (2003b) Manne and Halappanavar (2014)	$O(m)$ $O(m \log \Delta)$
(2/3-ϵ)-approximation MWM		Pettie and Sanders (2004)	$O(m \log \epsilon^{-1})$
(1 $-$ ϵ)-approximation MWM		Duan and Pettie (2014)	$O(m\epsilon^{-1} \log \epsilon^{-1})$
Maximum b-matching		Gabow (1983)	$O(\sqrt{b(V)}\, m)$
Maximum weight b-matching		Gabow (1983)	$O(b(V)\min\{m \log n, n^2\})$
1/2-approx. maximum weight b-matching	b-Suitor	Khan *et al.* (2016b)	$O(m \log \beta)$
Maximum vertex weight matching		Spencer and Mayr (1984)	$O(\sqrt{n}\, m \log n)$
1/2-approximation MVM	Greedy		$O(m + n \log n)$
2/3-approximation MVM		Al-Herz and Pothen (2019)	$O(m \log \Delta + n \log n)$

Table 9.2. Some of the exact and approximation algorithms for edge cover problems and their time complexity.

Problem	Algorithm	Reference	Complexity
Minimum EC		Norman and Rabin (1959), Micali and Vazirani (1980)	$O(\sqrt{n}\,m)$
Minimum weight EC	Weight trans. + MWM	Schrijver (2003), Gabow (2018)	$O(n(m + n \log n))$
3/2-approx. minimum weight EC	Weight trans. + 1/2-MWM	Schrijver (2003), Drake and Hougardy (2003b)	$O(m)$
2-approx. minimum weight EC	Nearest Neighbour	Ferdous $et\ al.$ (2018)	$O(m)$
Minimum b-EC		Huang and Pettie (2017)	$O(\sqrt{b(V)}\,m)$
Minimum weight b-EC	Complement of b'-matching	Schrijver (2003), Gabow (1983)	$O(b'(V)\min\{m \log n, n^2\})$
3/2-approx. min. weight b-EC	Primal Dual Lazy Greedy	Ferdous $et\ al.$ (2018)	$O(\beta\,\Delta m)$ $O(m \log n)$
2-approx. minimum weight b-EC	b-Nearest Neighbour MCE	Khan $et\ al.$ (2018b)	$O(m)$ $O(m \log \beta')$
$(1 + \epsilon)$-approx. min. weight b-EC		Huang and Pettie (2017)	$O(m\epsilon^{-1} \log \epsilon^{-1})$

science, machine learning and other emerging application domains. We expect that the increasing sizes of these problems and the availability of parallel computing resources will demand the development of efficient and concurrent approximation algorithms. We trust that this survey will stimulate further work along these lines.

Acknowledgements

We are grateful to Bora Uçar, ENS Lyon, for his extensive comments on our manuscript. It is a pleasure to thank our colleagues who have collaborated with us on matching and edge cover problems: Arif Khan and Mahantesh Halappanavar, both of Pacific Northwest National Laboratory; Florin Dobrian, Conviva Corporation; Ariful Azad, Indiana University; Ahmed Al-Herz, Purdue University; Johannes Langguth, Simula; Aydin Buluç, Lawrence Berkeley National Laboratory; Mostofa Ali Patwary, NVIDIA; Pradeep Dubey, Intel Corporation; Rob Bisseling, Utrecht University; and Seth Pettie, University of Michigan. We also thank Peter Sanders of the Karlsruhe Institute of Technology and Jens Maue of Zürich for sharing with us their code for the $(2/3 - \epsilon)$-approximation matching algorithms.

REFERENCES[2]

D. Achlioptas and A. Naor (2005), 'The two possible values of the chromatic number of a random graph', *Ann. of Math.* **162**, 1335–1351.

A. Agrawal, P. N. Klein and R. Ravi (1993), Cutting down on fill using nested dissection: Provably good elimination orderings. In *Graph Theory and Sparse Matrix Computations* (A. George, J. R. Gilbert and J. W. H. Liu, eds), Springer, pp. 31–55.

A. Al-Herz and A. Pothen (2019), 'A 2/3-approximation algorithm for vertex-weighted matching', *Discrete Appl. Math.*, under review. arXiv:1902.05877

R. P. Anstee (1987), 'A polynomial algorithm for b-matchings: An alternative approach', *Inform. Process. Lett.* **24**, 153–157.

A. Azad and A. Buluç (2016), Distributed memory algorithms for maximum cardinality matching on bipartite graphs. In *2016 IEEE International Parallel and Distributed Processing Symposium (IPDPS)*, IEEE, pp. 32–42.

A. Azad, A. Buluç and A. Pothen (2017), 'Computing maximum cardinality matchings in parallel on bipartite graphs via tree grafting', *IEEE Trans. Parallel Distrib. Syst.* **28**, 44–59.

A. Azad, A. Buluç, X. S. Li, X. Wang and J. Langguth (2018), 'A distributed memory approximation algorithm for maximum weight perfect bipartite matching', *SIAM J. Sci. Comput.*, under review. arXiv:1801.09809v1

---

[2] The URLs cited in this work were correct at the time of going to press, but the publisher and the authors make no undertaking that the citations remain live or are accurate or appropriate.

A. Azad, J. Langguth, Y. Fang, A. Qi and A. Pothen (2010), Identifying rare cell populations in comparative flow cytometry. In *Algorithms in Bioinformatics: International Workshop on Algorithms in Bioinformatics (WABI)*, Vol. 6293 of Lecture Notes in Bioinformatics, Springer, pp. 162–175.

H. Bast, K. Mehlhorn, G. Schäfer and H. Tamaki (2006), 'Matching algorithms are fast in sparse random graphs', *Theory Comput. Syst.* **39**, 3–14.

C. E. Bell (1994), 'Weighted matching with vertex weights: An application to scheduling training sessions in NASA space shuttle cockpit simulators', *Europ. J. Oper. Res.* **73**, 443–449.

M. Birn, V. Osipov, P. Sanders, C. Schulz and N. Sitchinava (2013), Efficient parallel and external matching. In *Euro-Par 2013 Parallel Processing*, Vol. 8097 of Lecture Notes in Computer Science, Springer, pp. 659–670.

B. Birnbaum and C. Mathieu (2008), 'On-line bipartite matching made simple', *ACM SIGACT News* **39**, 80–87.

G. E. Blelloch, J. T. Fineman and J. Shun (2012), Greedy sequential maximal independent set and matching are parallel on average. In *Proceedings of the 24th Annual ACM Symposium on Parallelism in Algorithms and Architectures (SPAA '12)*, ACM, pp. 308–317.

G. E. Blelloch, R. Peng and K. Tangwongsan (2011), Linear work parallel greedy approximate set cover and variants. In *Proceedings of the 23rd Annual ACM Symposium on Parallelism in Algorithms and Architectures (SPAA '11)*, ACM, pp. 23–32.

P. Boldi and S. Vigna (2004), The WebGraph framework I: Compression techniques. In *Proceedings of the 13th International Conference on World Wide Web (WWW 2004)*, ACM, pp. 595–601.

P. Boldi, A. Marino, M. Santini and S. Vigna (2014), BUbiNG: Massive crawling for the masses. In *Proceedings of the Companion Publication of the 23rd International Conference on World Wide Web*, ACM, pp. 227–228.

A. Buluç and J. R. Gilbert (2011), 'The Combinatorial BLAS: Design, implementation and applications', *Internat. J. High Perf. Comput. Appl.* **25**, 496–509.

R. Burkard, M. Dell'Amico and S. Martello (2009), *Assignment Problems*, SIAM.

Y. Cao and R. B. Sandeep (2017), Minimum fill-in: Inapproximability and almost tight lower bounds. In *Proceedings of the 28th Annual Symposium on Discrete Algorithms (SODA)*, SIAM, pp. 875–880.

K. M. Choromanski, T. Jebara and K. Tang (2013), Adaptive anonymity via *b*-matching. In *Advances in Neural Information Processing Systems (NIPS 2013)* (C. J. C. Burges et al., eds), pp. 3192–3200.

V. Chvatal (1979), 'A greedy heuristic for the set-covering problem', *Math. Oper. Res.* **4**, 233–235.

J. Cohen and P. Castonguay (2012), Efficient graph matching and coloring on GPUs. Presentation available at:
http://on-demand.gputechconf.com/gtc/2012/presentations/S0332-Efficient-Graph-Matching-and-Coloring-on-GPUs.pdf

T. F. Coleman and A. Pothen (1987), 'The null space problem II: Algorithms', *SIAM J. Algebraic Discrete Methods* **8**, 544–563.

T. F. Coleman, A. Edenbrandt and J. R. Gilbert (1986), 'Predicting fill for sparse orthogonal factorization', *J. Assoc. Comput. Mach.* **33**, 517–532.

T. H. Cormen, C. E. Leiserson, R. L. Rivest and C. Stein (2009), *Introduction to Algorithms*, MIT Press.

T. Davis and Y. Hu (2011), 'The University of Florida Sparse Matrix Collection', *ACM Trans. Math. Softw.* **38**, 1:1–1:25.

G. De Francisci Morales, A. Gionis and M. Sozio (2011), 'Social content matching in MapReduce', *Proc. VLDB Endowment* **4**, 460–469.

U. Derigs and A. Metz (1986), 'On the use of optimal fractional matchings for solving the (integer) matching problem', *Computing* **36**, 263–270.

M. Deveci, K. Kaya, B. Uçar and Ü. Çatalyürek (2013), GPU accelerated maximum cardinality matching algorithms for bipartite graphs. In *Proceedings of 19th International Euro-Par Conference on Parallel Processing*, pp. 850—861.

B. Dezső, A. Jüttner and P. Kovács (2011), 'LEMON: An open source C++ graph template library', *Electron. Notes Theoret. Comput. Sci.* **264**, 23–45.

F. Dobrian, M. Halappanavar, A. Pothen and A. Al-Herz (2019), 'A 2/3-approximation algorithm for vertex-weighted matching in bipartite graphs', *SIAM J. Sci. Comput.* **41**, A566–A591.

D. Drake and S. Hougardy (2003a), Linear time local improvements for weighted matchings in graphs. In *Experimental and Efficient Algorithms* (K. Jansen, M. Margraf, M. Mastrolilli and J. Rolim, eds), Vol. 2647 of Lecture Notes in Computer Science, Springer, pp. 107–119.

D. E. Drake and S. Hougardy (2003b), 'A simple approximation algorithm for the weighted matching problem', *Inform. Process. Lett.* **85**, 211–213.

D. E. Drake and S. Hougardy (2005), 'A linear time approximation algorithm for weighted matchings in graphs', *ACM Trans. Algorithms* **1**, 107–122.

D. Du, K. Ko and X. Hu (2012), *Design and Analysis of Approximation Algorithms*, Springer.

R. Duan and S. Pettie (2010), Approximating maximum weight matching in near-linear time. In *2010 IEEE 51st Annual Symposium on Foundations of Computer Science (FOCS '10)*, IEEE, pp. 673–682.

R. Duan and S. Pettie (2014), 'Linear-time approximation for maximum weight matching', *J. Assoc. Comput. Mach.* **61**, 1–23.

I. S. Duff and J. Koster (2001), 'On algorithms for permuting large entries to the diagonal of a sparse matrix', *SIAM J. Matrix Anal. Appl.* **22**, 973–996.

I. S. Duff and B. Uçar (2012), Combinatorial problems in solving linear systems. In *Combinatorial Scientific Computing* (U. Naumann and O. Schenk, eds), CRC, pp. 21–68.

I. S. Duff, K. Kaya and B. Uçar (2011), 'Design, implementation, and analysis of maximum transversal algorithms', *ACM Trans. Math. Softw.* **38**, 13:1–13:31.

F. Dufossé, K. Kaya and B. Uçar (2015), 'Two approximation algorithms for bipartite matching on multicore architectures', *J. Parallel Distrib. Comput.* **85**, 62–78.

J. Edmonds (1965), 'Maximum matching and a polyhedron with 0, 1-vertices', *J. Res. Nat. Bureau Standards* **69B**, 125–130.

B. O. Fagginger Auer and R. H. Bisseling (2012), A GPU algorithm for greedy graph matching. In *Facing the Multicore-Challenge II* (R. Keller, D. Kramer and J. Weiss, eds), Springer, pp. 108–119.

U. Feige and J. Kilian (1998), 'Zero knowledge and the chromatic number', *J. Comput. Systems Sci.* **57**, 187–199.

S. Ferdous, A. Pothen and A. Khan (2018), New approximation algorithms for minimum weighted edge cover. In *2018 Proceedings of the Seventh SIAM Workshop on Combinatorial Scientific Computing*, SIAM, pp. 97–108.

P. Fritzson (2014), *Principles of Object-Oriented Modeling and Simulation with Modelica 3.3: A Cyber-Physical Approach*, Wiley/IEEE.

H. N. Gabow (1983), An efficient reduction technique for degree-constrained subgraph and bidirected network flow problems. In *Proceedings of the 15th Annual ACM Symposium on the Theory of Computing (STOC '83)*, ACM, pp. 448–456.

H. N. Gabow (2018), 'Data structures for weighted matching and extensions to b-matching and f-factors', *ACM Trans. Algorithms* **14**, 39:1–39:80.

D. Gale and L. S. Shapley (1962), 'College admissions and the stability of marriage', *Amer. Math. Monthly* **69**, 9–15.

T. Gallai (1959), 'Über extreme Punkt- und Kantenmengen', *Annales Universitatis Scientiarum Budapestinensis de Rolando Eötvös Nominatae, Sectio Mathematica* **2**, 133–138.

A. H. Gebremedhin, F. Manne and A. Pothen (2005), 'What color is your Jacobian? Graph coloring for computing derivatives', *SIAM Review* **47**, 629–705.

A. H. Gebremedhin, A. Tarafdar, F. Manne and A. Pothen (2007), 'New acyclic and star coloring algorithms with application to computing Hessians', *SIAM J. Sci. Comput.* **29**, 1042–1072.

A. George (1973), 'Nested dissection of a finite element mesh', *SIAM J. Numer. Anal.* **10**, 345–363.

G. Georgiadis and M. Papatriantafilou (2013), 'Overlays with preferences: Distributed, adaptive approximation algorithms for matching with preference lists', *Algorithms* **6**, 824–856.

M. X. Goemans and D. P. Williamson (1997), The primal–dual method for approximation algorithms and its application to network design problems. In *Approximation Algorithms for NP-hard Problems* (D. S. Hochbaum, ed.), PWS Publishing Co., pp. 144–191.

M. Grötschel and O. Holland (1985), 'Solving matching problems with linear programming', *Math. Program.* **33**, 243–259.

M. Halappanavar, J. Feo, O. Villa, F. Dobrian and A. Pothen (2012), 'Approximate weighted matching on emerging manycore and multithreaded architectures', *Internat. J. High Perf. Comput. Appl.* **26**, 413–430.

N. G. Hall and D. S. Hochbaum (1986), 'A fast approximation algorithm for the multicovering problem', *Discrete Appl. Math.* **15**, 35–40.

S. Hanke and S. Hougardy (2010), New approximation algorithms for the weighted matching problem. Research report 101010, Research Institute for Discrete Mathematics, University of Bonn.

D. S. Hochbaum, ed. (1997), *Approximation Algorithms for NP-hard Problems*, PWS Publishing Co.

J. Hogg and J. Scott (2015), 'On the use of suboptimal matchings for scaling and ordering sparse symmetric matrices', *Numer. Linear Algebra Appl.* **22**, 648–663.

J. Hogg and J. Scott (2013), 'Pivoting strategies for tough sparse indefinite systems', *ACM Trans. Math. Softw.* **40**, 4.

J. Hopcroft and R. Karp (1973), 'An $n^{5/2}$ algorithm for maximum matchings in bipartite graphs', *SIAM J. Comput.* **2**, 225–231.

S. Hougardy (2009), Linear time approximation algorithms for degree constrained subgraph problems. In *Research Trends in Combinatorial Optimization* (W. J. Cook, L. Lovász and J. Vygen, eds), Springer, pp. 185–200.

B. C. Huang and T. Jebara (2011), Fast b-matching via sufficient selection belief propagation. In *Proc. 14th International Conference on Artificial Intelligence and Statistics (AISTATS)*, pp. 361–369.

D. Huang and S. Pettie (2017), Approximate generalized matching: f-factors and f-edge covers. arXiv:1706.05761

A. Idelberger and F. Manne (2014), New iterative algorithms for weighted matching. In *Norsk Informatikkonferanse 2014*. www.nik.no/publikasjoner/

T. Jebara and V. Shchogolev (2006), b-matching for spectral clustering. In *Proceedings of the 17th European Conference on Machine Learning (ECML 2006)*, Vol. 4212 of Lecture Notes in Computer Science, Springer, pp. 679–686.

T. Jebara, J. Wang and S.-F. Chang (2009), Graph construction and b-matching for semi-supervised learning. In *Proceedings of the 26th Annual International Conference on Machine Learning (ICML '09)*, ACM, pp. 441–448.

D. Juedes and J. Jones (2012), 'Coloring Jacobians revisited: A new algorithm for acyclic and star bicoloring', *Optim. Methods Softw.* **27**, 295–309.

R. J. Kang and C. McDiarmid (2015), Colouring random graphs. In *Topics in Chromatic Graph Theory* (R. J. Wilson and L. W. Beineke, eds), Cambridge University Press, pp. 199–219.

R. M. Karp and M. Sipser (1981), Maximum matching in sparse random graphs. In *Proceedings of the 22nd Annual Symposium on Foundations of Computer Science (SFCS 1981)*, pp. 364–375.

R. M. Karp, U. Vazirani and V. Vazirani (1990), An optimal algorithm for on-line bipartite matching. In *Proceedings of the 22nd Annual ACM Symposium on Theory of Computing (STOC '90)*, ACM, pp. 352–358.

E. R. Keiter, H. K. Thornquist, R. J. Hoekstra, T. V. Russo, R. L. Schiek and E. L. Rankin (2011), Parallel transistor-level circuit simulation. In *Advanced Simulation and Verification of Electronic and Biological Systems* (P. Li, L. M. Silveira and P. Feldmann, eds), Springer, pp. 1–21.

A. Khan and A. Pothen (2016), A new 3/2-approximation algorithm for the b-edge cover problem. In *2016 Proceedings of the Seventh SIAM Workshop on Combinatorial Scientific Computing*, SIAM, pp. 52–61.

A. Khan, K. Choromanski, A. Pothen, S. Ferdous, M. Halappanavar and A. Tumeo (2018*a*), Adaptive anonymization of data using b-edge cover. In *Proceedings of the International Conference for High Performance Computing, Networking, Storage, and Analysis (SC '18)*, IEEE, pp. 59:1–59:11.

A. M. Khan, D. F. Gleich, A. Pothen and M. Halappanavar (2012), A multithreaded algorithm for network alignment via approximate matching. In *Proceedings of the International Conference on High Performance Computing, Networking, Storage, and Analysis (SC '12)*, IEEE, pp. 64:1–64:11.

A. Khan, A. Pothen and S. M. Ferdous (2018*b*), Parallel algorithms through approximation: *b*-edge cover. In *2018 IEEE International Parallel and Distributed Processing Symposium (IPDPS)*, IEEE, pp. 22–33.

A. Khan, A. Pothen, M. M. Patwary, M. Halappanavar, N. Satish, N. Sundaram and P. Dubey (2016*a*), Designing scalable *b*-matching algorithms on distributed memory multiprocessors by approximation. In *Proceedings of the International Conference for High Performance Computing, Networking, Storage, and Analysis (SC '16)*, IEEE, pp. 773–783.

A. Khan, A. Pothen, M. M. Patwary, N. Satish, N. Sundaram, F. Manne, M. Halappanavar and P. Dubey (2016*b*), 'Efficient approximation algorithms for weighted *b*-matching', *SIAM J. Sci. Comput.* **38**, S593–S619.

S. Khuller, U. Vishkin and N. Young (1994), 'A primal–dual parallel approximation technique applied to weighted set and vertex covers', *J. Algorithms* **17**, 280–289.

V. Knoblauch (2007), Marriage Matching: A conjecture of Donald Knuth. Working papers 2007-15, University of Connecticut, Department of Economics.

V. Kolmogorov (2009), 'BLOSSOM V: A new implementation of a minimum cost perfect matching algorithm', *Math. Prog. Comput.* **1**, 43–67.

P. Kolyvakis, A. Kalousis, B. Smith and D. Kiritsis (2018), 'Biomedical ontology alignment: An approach based on representation learning', *J. Biomed. Semantics* **9**, 21.

C. Koufogiannakis and N. E. Young (2011), 'Distributed algorithms for covering, packing and maximum weighted matching', *Distrib. Comput.* **24**, 45–63.

X. S. Li and J. W. Demmel (2003), 'SuperLU_DIST: A scalable distributed-memory sparse direct solver for unsymmetric linear systems', *ACM Trans. Math. Softw.* **29**, 110–140.

L. Lovász and M. D. Plummer (2009), *Matching Theory*, AMS.

D. F. Manlove (2013), *Algorithmics of Matching Under Preferences*, World Scientific.

F. Manne and M. Halappanavar (2014), New effective multithreaded matching algorithms. In *2014 IEEE 28th International Parallel and Distributed Processing Symposium (IPDPS)*, IEEE, pp. 519–528.

F. Manne, M. Naim, H. Lerring and M. Halappanavar (2016), On stable marriages and greedy matchings. In *2016 Proceedings of the Seventh SIAM Workshop on Combinatorial Scientific Computing*, SIAM, pp. 92–101.

F. M. Manshadi, B. Awerbuch, R. Gemulla, R. Khandekar, J. Mestre and M. Sozio (2013), 'A distributed algorithm for large-scale generalized matching', *Proc. VLDB Endowment* **6**, 613–624.

J. Maue and P. Sanders (2007), Engineering algorithms for approximate weighted matching. In *Experimental Algorithms: 6th International Workshop on Experimental and Efficient Algorithms (WEA 2007)*, Vol. 4525 of Lecture Notes in Computer Science, Springer, pp. 242–255.

S. T. McCormick (1983), 'Optimal approximation of sparse Hessians and its equivalence to a graph coloring problem', *Math. Program.* **26**, 153–171.

C. McDiarmid (1984), 'Colouring random graphs', *Ann. Oper. Res.* **1**, 183–200.

D. G. McVitie and L. B. Wilson (1971), 'The stable marriage problem', *Commun. Assoc. Comput. Mach.* **14**, 486–490.

K. Mehlhorn and S. Näher (1999), LEDA: A platform for combinatorial and geometric computing. www.algorithmic-solutions.com/leda/index.htm

A. Mehta (2012), 'Online matching and ad allocation', *Found. Trends. Theor. Comput. Sci.* **8**, 265–368.

N. S. Mendelsohn and A. L. Dulmage (1958), 'Some generalizations of the problem of distinct representatives', *Canad. J. Math.* **10**, 230–241.

J. Mestre (2006), Greedy in approximation algorithms. In *Algorithms: 14th Annual European Symposium on Algorithms (ESA 2006)*, Vol. 4168 of Lecture Notes in Computer Science, Springer, pp. 528–539.

S. Micali and V. V. Vazirani (1980), An $O(\sqrt{|V|} \cdot |E|)$ algorithm for finding maximum matching in general graphs. In *Proceedings of the 21st Annual Symposium on Foundations of Computer Science (SFCS 1980)*, IEEE, pp. 17–27.

D. L. Miller and J. F. Pekny (1995), 'A staged primal–dual algorithm for perfect b-matching with edge capacities', *ORSA J. Comput.* **7**, 298–320.

M. Minoux (1978), Accelerated greedy algorithms for maximizing submodular set functions. In *Optimization Techniques: Proceedings of the 8th IFIP Conference on Optimization Techniques (IFIP 1977)* (J. Stoer, ed.), Springer, pp. 234–243.

R. Motwani (1994), 'Average-case analysis of algorithms for matchings and related problems', *J. Assoc. Comput. Mach.* **41**, 1329–1356.

M. Müller-Hannemann and A. Schwartz (2000), 'Implementing weighted b-matching algorithms: Insights from a computational study', *J. Exp. Algorithmics* **5**, 8.

R. C. Murphy, K. B. Wheeler, B. W. Barrett and J. A. Ang (2010), Introducing the Graph 500. In *Proceedings of the Cray User's Group Meeting (CUG), 2010*.

M. Naim and F. Manne (2018), Scalable b-matching on GPUs. In *Proceedings of the International Parallel and Distributed Processing Symposium Workshops (IPDPS)*, pp. 637–646.

M. Naim, F. Manne, M. Halappanavar, A. Tumeo and J. Langguth (2015), Optimizing approximate weighted matching on Nvidia Kepler K40. In *IEEE 22nd International Conference on High Performance Computing (HiPC 2015)*, pp. 105–114.

A. Natanzon, R. Shamir and R. Sharan (2000), 'A polynomial approximation for the minimum fill-in problem', *SIAM J. Comput.* **30**, 1067–1079.

U. Naumann and O. Schenk, eds (2012), *Combinatorial Scientific Computing*, CRC Press.

R. Z. Norman and M. O. Rabin (1959), 'An algorithm for a minimum cover of a graph', *Proc. Amer. Math. Soc.* **10**, 315–319.

M. Olschowka and A. Neumaier (1996), 'A new pivoting strategy for Gaussian elimination', *Linear Algebra Appl.* **240** (suppl. C), 131–151.

M. W. Padberg and M. R. Rao (1982), 'Odd minimum cut-sets and b-matchings', *Math. Oper. Res.* **7**, 67–80.

S. Pettie and P. Sanders (2004), 'A simpler linear time $2/3 - \epsilon$ approximation for maximum weight matching', *Inform. Process. Lett.* **91**, 271–276.

A. Pinar, E. Chow and A. Pothen (2006), 'Combinatorial algorithms for computing column space bases that have sparse inverses', *Electron. Trans. Numer. Anal.* **22**, 122–145.

A. Pothen (1993), 'Predicting the structure of sparse orthogonal factors', *Linear Algebra Appl.* **194**, 183–203.

A. Pothen and C.-J. Fan (1990), 'Computing the block triangular form of a sparse matrix', *ACM Trans. Math. Softw.* **16**, 303–324.

R. Preis (1999), Linear time 1/2-approximation algorithm for maximum weighted matching in general graphs. In *1999 Proceedings of 16th Annual Symposium on Theoretical Aspects of Computer Science (STACS 99)*, Vol. 1563 of Lecture Notes in Computer Science, Springer, pp. 259–269.

W. R. Pulleyblank (1973), Faces of matching polyhedra. PhD thesis, Faculty of Mathematics, University of Waterloo.

S. Rajagopalan and V. V. Vazirani (1993), Primal–dual RNC approximation algorithms for (multi)-set (multi)-cover and covering integer programs. In *Proceedings of 1993 IEEE 34th Annual Foundations of Computer Science (SFCS '93)*, IEEE, pp. 322–331.

A. Schrijver (2003), *Combinatorial Optimization: Polyhedra and Efficiency*, Vol. A: *Paths, Flows, Matchings*, Springer.

R. Sinkhorn and P. Knopp (1967), 'Concerning nonnegative matrices and doubly stochastic matrices', *Pacific J. Math.* **21**, 343–348.

T. H. Spencer and E. W. Mayr (1984), Node weighted matching. In *Proceedings of the 11th Colloquium on Automata, Languages, and Programming (ICALP)*, Vol. 172 of Lecture Notes in Computer Science, Springer, pp. 454–464.

A. Subramanya and P. P. Talukdar (2014), *Graph-Based Semi-Supervised Learning*, Vol. 29 of Synthesis Lectures on Artificial Intelligence and Machine Learning, Morgan & Claypool.

V. Tabatabaee, L. Georgiadis and L. Tassiulas (2001), 'QoS provisioning and tracking fluid policies in input queueing switches', *IEEE/ACM Trans. Netw.* **9**, 605–617.

A. Tamir and J. S. B. Mitchell (1998), 'A maximum b-matching problem arising from median location models with applications to the roommates problem', *Math. Program.* **80**, 171–194.

K. Tangwongsan (2011), Efficient parallel approximation algorithms. PhD thesis, Carnegie Mellon University, Pittsburgh, PA.

V. V. Vazirani (2003), *Approximation Algorithms*, Springer.

D. P. Williamson and D. B. Shmoys (2011), *The Design of Approximation Algorithms*, Cambridge University Press.

L. B. Wilson (1972), 'An analysis of the marriage matching assignment algorithm', *BIT* **12**, 569–575.

M. Yannakakis (1981), 'Computing the minimum fill-in is NP-complete', *SIAM J. Algebraic Discrete Methods* **2**, 77–79.

Acta Numerica (2019), pp. 635–711

doi:10.1017/S0962492919000011

Data assimilation: The Schrödinger perspective

Sebastian Reich

Institute of Mathematics,

University of Potsdam,

D-14476 Potsdam, Germany

and

Department of Mathematics and Statistics,

University of Reading,

Reading RG6 6AX, UK

E-mail: sebastian.reich@uni-potsdam.de

Data assimilation addresses the general problem of how to combine model-based predictions with partial and noisy observations of the process in an optimal manner. This survey focuses on sequential data assimilation techniques using probabilistic particle-based algorithms. In addition to surveying recent developments for discrete- and continuous-time data assimilation, both in terms of mathematical foundations and algorithmic implementations, we also provide a unifying framework from the perspective of coupling of measures, and Schrödinger's boundary value problem for stochastic processes in particular.

CONTENTS

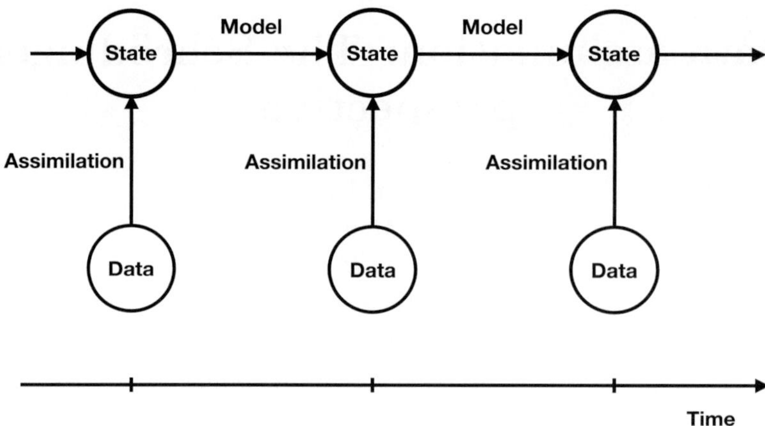

Figure 1.1. Schematic illustration of sequential data assimilation, where model states are propagated forward in time under a given model dynamics and adjusted whenever data become available at discrete instances in time. In this paper, we look at a single transition from a given model state conditioned on all the previous and current data to the next instance in time, and its adjustment under the assimilation of the new data then becoming available.

1. Introduction

This survey focuses on sequential data assimilation techniques for state and parameter estimation in the context of discrete- and continuous-time stochastic diffusion processes. See Figure 1.1. The field itself is well established (Evensen 2006, Särkkä 2013, Law, Stuart and Zygalakis 2015, Reich and Cotter 2015, Asch, Bocquet and Nodet 2017), but is also undergoing continuous development due to new challenges arising from emerging application areas such as medicine, traffic control, biology, cognitive sciences and geosciences.

Data assimilation is typically formulated within a Bayesian framework in order to combine partial and noisy observations with model predictions and their uncertainties with the goal of adjusting model states and model parameters in an optimal manner. In the case of linear systems and Gaussian distributions, this task leads to the celebrated Kalman filter (Särkkä 2013) which even today forms the basis of a number of popular data assimilation schemes and which has given rise to the widely used ensemble Kalman filter (Evensen 2006). Contrary to standard sequential Monte Carlo methods (Doucet, de Freitas and Gordon 2001, Bain and Crisan 2008), the ensemble Kalman filter does not provide a consistent approximation to the sequential filtering problem, while being applicable to very high-dimensional problems. This and other advances have widened the scope of sequential data assimilation and have led to an avalanche of new methods in recent years.

In this review we will focus on probabilistic methods (in contrast to data assimilation techniques based on optimization, such as 3DVar and 4DVar (Evensen 2006, Law *et al.* 2015)) in the form of sequential particle methods. The essential challenge of sequential particle methods is to convert a sample of M particles from a filtering distribution at time t_k into M samples from the filtering distribution at time t_{k+1} without having access to the full filtering distributions. It will also often be the case in practical applications that the sample size will be small to moderate in comparison to the number of variables we need to estimate.

Sequential particle methods can be viewed as a special instance of interacting particle systems (del Moral 2004). We will view such interacting particle systems in this review from the perspective of approximating a certain boundary value problem in the space of probability measures, where the boundary conditions are provided by the underlying stochastic process, the data and Bayes' theorem. This point of view leads naturally to optimal transportation (Villani 2003, Reich and Cotter 2015) and, more importantly for this review, to Schrödinger's problem (Föllmer and Gantert 1997, Leonard 2014, Chen, Georgiou and Pavon 2014), as formulated first by Erwin Schrödinger in the form of a boundary value problem for Brownian motion (Schrödinger 1931).

This paper has been written with the intention of presenting a unifying framework for sequential data assimilation using coupling of measure arguments provided through optimal transportation and Schrödinger's problem. We will also summarize novel algorithmic developments that were inspired by this perspective. Both discrete- and continuous-time processes and data sets will be covered. While the primary focus is on state estimation, the presented material can be extended to combined state and parameter estimation. See Remark 2.2 below.

Remark 1.1. We will primary refer to the methods considered in the survey as particle or ensemble methods instead of the widely used notion of sequential Monte Carlo methods. We will also use the notions of particles, samples and ensemble members synonymously. Since the ensemble size, M, is generally assumed to be small to moderate relative to the number of variables of interest, we will focus on robust but generally biased particle methods.

1.1. Overall organization of the paper

This survey consists of four main parts. We start Section 2 by recalling key mathematical concepts of sequential data assimilation when the data become available at discrete instances in time. Here the underlying dynamic models can be either continuous (*i.e.* generated by a stochastic differential equation) or discrete-in-time. Our initial review of the problem will lead to the identification of three different scenarios of performing sequential data assimilation, which we denote by (A), (B) and (C). While the first two scenarios are linked to the classical importance resampling and optimal proposal densities for particle filtering (Doucet *et al.* 2001), scenario (C) builds upon an intrinsic connection to a certain boundary value problem in the space of joint probability measures first considered by Erwin Schrödinger (1931).

After this initial review, the remaining parts of Section 2 provide more mathematical details on prediction in Section 2.1, filtering and smoothing in Section 2.2, and the Schrödinger approach to sequential data assimilation in Section 2.3. The modification of a given Markov transition kernel via a twisting function will arise as a crucial mathematical construction and will be introduced in Sections 1.2 and 2.1. The next major part of the paper, Section 3, is devoted to numerical implementations of prediction, filtering and smoothing, and the Schrödinger approach as relevant to scenarios (A)–(C) introduced earlier in Section 2. More specifically, this part will cover the ensemble Kalman filter and its extensions to the more general class of linear ensemble transform filters as well as the numerical implementation of the Schrödinger approach to sequential data assimilation using the Sinkhorn algorithm (Sinkhorn 1967, Peyre and Cuturi 2018). Discrete-time stochastic systems with additive Gaussian model errors and stochastic differential equations with constant diffusion coefficient serve as illustrating examples throughout both Sections 2 and 3.

Sections 2 and 3 are followed by two sections on the assimilation of data that arrive continuously in time. In Section 4 we will distinguish between data that are smooth as a function of time and data which have been perturbed by Brownian motion. In both cases, we will demonstrate that the data assimilation problem can be reformulated in terms of so-called mean-field equations, which produce the correct conditional marginal distributions in the state variables. In particular, in Section 4.2 we discuss the feedback particle filter of Yang, Mehta and Meyn (2013) in some detail. The final section of this review, Section 5, covers numerical approximations to these mean-field equations in the form of interacting particle systems. More specifically, the continuous-time ensemble Kalman–Bucy and numerical implementations of the feedback particle filter will be covered in detail. It will be shown in particular that the numerical implementation of the feedback particle filter can be achieved naturally via the approximation of an associated Schrödinger problem using the Sinkhorn algorithm.

In the appendices we provide additional background material on mesh-free approximations of the Fokker–Planck and backward Kolmogorov equations (Appendix A.1), on the regularized Störmer–Verlet time-stepping methods for the hybrid Monte Carlo method, applicable to Bayesian inference problems over path spaces (Appendix A.2), on the ensemble Kalman filter (Appendix A.3), and on the numerical approximation of forward–backward stochastic differential equations (SDEs) (Appendix A.4).

1.2. Summary of essential notations

We typically denote the probability density function (PDF) of a random variable Z by π. Realizations of Z will be denoted by $z = Z(\omega)$.

Realizations of a random variable can also be continuous functions/paths, in which case the associated probability measure on path space is denoted by $\mathbb{Q}$. We will primarily consider continuous functions over the unit time interval and denote the associated random variable by $Z_{[0,1]}$ and its realizations $Z_{[0,1]}(\omega)$ by $z_{[0,t]}$. The restriction of $Z_{[0,1]}$ to a particular instance $t \in [0,1]$ is denoted by Z_t with marginal distribution π_t and realizations $z_t = Z_t(\omega)$.

For a random variable Z having only finitely many outcomes z^i, $i = 1, \ldots, M$, with probabilities p_i, that is,

$$\mathbb{P}[Z(\omega) = z^i] = p_i,$$

we will work with either the probability vector $p = (p_1, \ldots, p_M)^{\mathrm{T}}$ or the empirical measure

$$\pi(z) = \sum_{i=1}^{M} p_i\, \delta(z - z^i),$$

where $\delta(\cdot)$ denotes the standard Dirac delta function.

We use the shorthand

$$\pi[f] = \int f(z)\, \pi(z)\, \mathrm{d}z$$

for the expectation of a function f under a PDF π. Similarly, integration with respect to a probability measure $\mathbb{Q}$, not necessarily absolutely continuous with respect to Lebesgue, will be denoted by

$$\mathbb{Q}[f] = \int f(z)\, \mathbb{Q}(\mathrm{d}z).$$

The notation $\mathbb{E}[f]$ is used if we do not wish to specify the measure explicitly.

The PDF of a Gaussian random variable Z with mean $\bar{z}$ and covariance matrix B will be abbreviated by $\mathrm{n}(z; \bar{z}, B)$. We also use $Z \sim \mathrm{N}(\bar{z}, B)$.

Let $u \in \mathbb{R}^N$, then $D(u) \in \mathbb{R}^{N \times N}$ denotes the diagonal matrix with entries $(D(u))_{ii} = u_i$, $i = 1, \ldots, N$. We also denote the $N \times 1$ vector of ones by $\mathbb{1}_N = (1, \ldots, 1)^{\mathrm{T}} \in \mathbb{R}^N$.

A matrix $P \in \mathbb{R}^{L \times M}$ is called bi-stochastic if all its entries are non-negative, which we will abbreviate by $P \geq 0$, and

$$\sum_{l=1}^{L} q_{li} = p_0, \quad \sum_{i=1}^{M} q_{li} = p_1,$$

where both $p_1 \in \mathbb{R}^L$ and $p_0 \in \mathbb{R}^M$ are probability vectors. A matrix $Q \in \mathbb{R}^{M \times M}$ defines a discrete Markov chain if all its entries are non-negative and

$$\sum_{l=1}^{L} q_{li} = 1.$$

The Kullback–Leibler divergence between two bi-stochastic matrices $P \in \mathbb{R}^{L \times M}$ and $Q \in \mathbb{R}^{L \times M}$ is defined by

$$\mathrm{KL}\,(P\|Q) := \sum_{l,j} p_{lj} \log \frac{p_{lj}}{q_{lj}}.$$

Here we have assumed for simplicity that $q_{lj} > 0$ for all entries of Q. This definition extends to the Kullback–Leibler divergence between two discrete Markov chains.

The transition probability going from state z_0 at time $t = 0$ to state z_1 at time $t = 1$ is denoted by $q_+(z_1|z_0)$. Hence, given an initial PDF $\pi_0(z_0)$ at $t = 0$, the resulting (prediction or forecast) PDF at time $t = 1$ is provided by

$$\pi_1(z_1) := \int q_+(z_1|z_0)\,\pi_0(z_0)\,\mathrm{d}z_0. \tag{1.1}$$

Given a twisting function $\psi(z) > 0$, the twisted transition kernel $q_+^\psi(z_1|z_0)$ is defined by

$$q_+^\psi(z_1|z_0) := \psi(z_1)\,q_+(z_1|z_0)\,\widehat{\psi}(z_0)^{-1} \tag{1.2}$$

provided

$$\widehat{\psi}(z_0) := \int q_+(z_1|z_0)\,\psi(z_1)\,\mathrm{d}z_1 \tag{1.3}$$

is non-zero for all z_0. See Definition 2.8 for more details.

If transitions are characterized by a discrete Markov chain $Q_+ \in \mathbb{R}^M$, then a twisted Markov chain is provided by

$$Q_+^u = D(u)\,Q_+\,D(v)^{-1}$$

for given twisting vector $u \in \mathbb{R}^M$ with positive entries u_i, *i.e.* $u > 0$, and the vector $v \in \mathbb{R}^M$ determined by

$$v = (D(u)\,Q_+)^{\mathrm{T}}\,\mathbb{1}_M.$$

The conditional probability of observing y given z is denoted by $\pi(y|z)$ and the likelihood of z given an observed y is abbreviated by $l(z) = \pi(y|z)$. We will also use the abbreviations

$$\widehat{\pi}_1(z_1) = \pi_1(z_1|y_1)$$

and

$$\widehat{\pi}_0(z_0) = \pi_0(z_0|y_1)$$

to denote the conditional PDFs of a process at time $t = 1$ given data at time $t = 1$ (filtering) and the conditional PDF at time $t = 0$ given data at time $t = 1$ (smoothing), respectively. Finally, we also introduce the evidence

$$\beta := \pi_1[l] = \int p(y_1|z_1)\pi_1(z_1)\,\mathrm{d}z_1$$

of observing y_1 under the given model as represented by the forecast PDF (1.1). A more precise definition of these expressions will be given in the following section.

2. Mathematical foundation of discrete-time DA

Let us assume that we are given partial and noisy observations y_k, $k = 1, \ldots, K$, of a stochastic process in regular time intervals of length $T = 1$. Given a likelihood function $\pi(y|z)$, a Markov transition kernel $q_+(z'|z)$ and an initial distribution π_0, the associated prior and posterior PDFs are given by

$$\pi(z_{0:K}) := \pi_0(z_0) \prod_{k=1}^{K} q_+(z_k|z_{k-1}) \tag{2.1}$$

and

$$\pi(z_{0:K}|y_{1:K}) := \frac{\pi_0(z_0) \prod_{k=1}^{K} \pi(y_k|z_k)\, q_+(z_k|z_{k-1})}{\pi(y_{1:K})}, \tag{2.2}$$

respectively (Jazwinski 1970, Särkkä 2013). While it is of broad interest to approximate the posterior or smoothing PDF (2.2), we will focus on the recursive approximation of the filtering PDFs $\pi(z_k|y_{1:k})$ using sequential particle filters in this paper. More specifically, we wish to address the following computational task.

Problem 2.1. We have M equally weighted Monte Carlo samples z_{k-1}^i, $i = 1, \ldots, M$, from the filtering PDF $\pi(z_{k-1}|y_{1:k-1})$ at time $t = k - 1$ available and we wish to produce M equally weighted samples from the filtering PDF $\pi(z_k|y_{1:k})$ at time $t = k$ having access to the transition kernel $q_+(z_k|z_{k-1})$ and the likelihood $\pi(y_k|z_k)$ only. Since the computational task is exactly the same for all indices $k \geq 1$, we simply set $k = 1$ throughout this paper.

We introduce some notations before we discuss several possibilities of addressing Problem 2.1. Since we do not have direct access to the filtering distribution at time $k = 0$, the PDF at t_0 becomes

$$\pi_0(z_0) := \frac{1}{M} \sum_{i=1}^{M} \delta(z_0 - z_0^i), \qquad (2.3)$$

where $\delta(z)$ denotes the Dirac delta function and z_0^i, $i = 1, \ldots, M$, are M given Monte Carlo samples representing the actual filtering distribution. Recall that we abbreviate the resulting filtering PDF $\pi(z_1|y_1)$ at $t = 1$ by $\widehat{\pi}_1(z_1)$ and the likelihood $\pi(y_1|z_1)$ by $l(z_1)$. Because of (1.1), the forecast PDF is given by

$$\pi_1(z_1) = \frac{1}{M} \sum_{i=1}^{M} q_+(z_1|z_0^i) \qquad (2.4)$$

and the filtering PDF at time $t = 1$ by

$$\widehat{\pi}_1(z_1) := \frac{l(z_1)\,\pi_1(z_1)}{\pi_1[l]} = \frac{1}{\pi_1[l]} \frac{1}{M} \sum_{i=1}^{M} l(z_1)\,q_+(z_1|z_0^i) \qquad (2.5)$$

according to Bayes' theorem.

Remark 2.2. The normalization constant $\pi(y_{1:K})$ in (2.2), also called the evidence, can be determined recursively using

$$\pi(y_{1:k}) = \pi(y_{1:k-1}) \int \pi(y_k, z_{k-1})\,\pi(z_{k-1}|y_{1:k-1})\,\mathrm{d}z_{k-1}$$

$$= \pi(y_{1:k-1}) \int \int \pi(y_k|z_k)\,q_+(z_k|z_{k-1})\,\pi(z_{k-1}|y_{1:k-1})\,\mathrm{d}z_{k-1}\,\mathrm{d}z_k$$

$$= \pi(y_{1:k-1}) \int \pi(y_k|z_k)\,\pi(z_k|y_{1:k-1})\,\mathrm{d}z_k \qquad (2.6)$$

(Särkkä 2013, Reich and Cotter 2015). Since, as for the state estimation problem, the computational task is the same for each index $k \geq 1$, we simply set $k = 1$ and formally use $\pi(y_{1:0}) \equiv 1$. We are then left with

$$\beta := \pi_1[l] = \frac{1}{M} \sum_{i=1}^{M} \int l(z_1)\,q_+(z_1|z_0^i)\,\mathrm{d}z_1 \qquad (2.7)$$

within the setting of Problem 2.1, and β becomes a shorthand for $\pi(y_1)$. If the model depends on parameters, λ, or different models are to be compared, then it is important to evaluate the evidence (2.7) for each parameter value λ or model, respectively. More specifically, if $q_+(z_1|z_0; \lambda)$, then $\beta = \beta(\lambda)$ in (2.7) and larger values of $\beta(\lambda)$ indicate a better fit of the transition kernel to the data for that parameter value. One can then perform Bayesian parameter inference based upon appropriate approximations to the likelihood

$\pi(y_1|\lambda) = \beta(\lambda)$ and a given prior PDF $\pi(\lambda)$. The extension to the complete data set $y_{1:K}$, $K > 1$, is straightforward using (2.6) and an appropriate data assimilation algorithm, *i.e.* algorithms that can tackle Problem 2.1 sequentially.

Alternatively, one can treat a combined state–parameter estimation problem as a particular case of Problem 2.1 by introducing the extended state variable (z, λ) and augmented transition probabilities $Z_1 \sim q_+(\cdot|z_0, \lambda_0)$ and $\mathbb{P}[\Lambda_1 = \lambda_0] = 1$. The state augmentation technique allows one to extend all approaches discussed in this paper for Problem 2.1 to combined state–parameter estimation.

See Kantas *et al.* (2015) for a detailed survey of the topic of combined state and parameter estimation.

The filtering distribution $\widehat{\pi}_1$ at time $t = 1$ implies a smoothing distribution at time $t = 0$, which is given by

$$\widehat{\pi}_0(z_0) := \frac{1}{\beta} \int l(z_1)\, q_+(z_1|z_0)\, \pi_0(z_0)\, \mathrm{d}z_1 = \frac{1}{M} \sum_{i=1}^{M} \gamma^i \delta(z_0 - z_0^i) \qquad (2.8)$$

with weights

$$\gamma^i := \frac{1}{\beta} \int l(z_1)\, q_+(z_1|z_0^i)\, \mathrm{d}z_1. \qquad (2.9)$$

It is important to note that the filtering PDF $\widehat{\pi}_1$ can be obtained from $\widehat{\pi}_0$ using the transition kernels

$$\widehat{q}_+(z_1|z_0^i) := \frac{l(z_1)\, q_+(z_1|z_0^i)}{\beta\,\gamma^i}, \qquad (2.10)$$

that is,

$$\widehat{\pi}_1(z_1) = \frac{1}{M} \sum_{i=1}^{M} \widehat{q}_+(z_1|z_0^i)\, \gamma^i.$$

See Figure 2.1 for a schematic illustration of these distributions and their mutual relationships.

Remark 2.3. The modified transition kernel (2.10) can be seen as a particular instance of a twisted transition kernel (1.2) with $\psi(z) = l(z)/\beta$ and $\widehat{\psi}(z_0^i) = \gamma^i$. Such twisting kernels will play a prominent role in this survey, not only in the context of optimal proposals (Doucet *et al.* 2001, Arulampalam, Maskell, Gordon and Clapp 2002) but also in the context of the Schrödinger approach to data assimilation, *i.e.* scenario (C) below.

The following scenarios of how to tackle Problem 2.1, that is, how to produce the desired samples $\widehat{z}_1^i$, $i = 1, \ldots, M$, from the filtering PDF (2.5), will be considered in this paper.

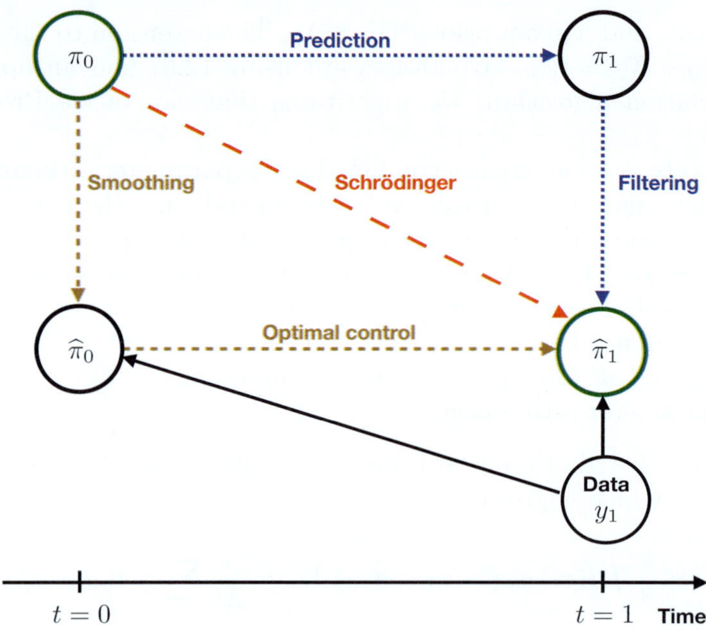

Figure 2.1. Schematic illustration of a single data assimilation cycle. The distribution π_0 characterizes the distribution of states conditioned on all observations up to and including t_0, which we set here to $t = 0$ for simplicity. The predictive distribution at time $t_1 = 1$, as generated by the model dynamics, is denoted by π_1. Upon assimilation of the data y_1 and application of Bayes' formula, one obtains the filtering distribution $\widehat{\pi}_1$. The conditional distribution of states at time t_0 conditioned on all the available data including y_1 is denoted by $\widehat{\pi}_0$. Control theory provides the adjusted model dynamics for transforming $\widehat{\pi}_0$ into $\widehat{\pi}_1$. Finally, the Schrödinger problem links π_0 and $\widehat{\pi}_1$ in the form of a penalized boundary value problem in the space of joint probability measures. Data assimilation scenario (A) corresponds to the dotted lines, scenario (B) to the short-dashed lines, and scenario (C) to the long-dashed line.

Definition 2.4. We define the following three scenarios of how to tackle Problem 2.1.

(A) We first produce samples, z_1^i, from the forecast PDF π_1 and then transform those samples into samples, $\widehat{z}_1^i$, from $\widehat{\pi}_1$. This can be viewed as introducing a Markov transition kernel $q_1(\widehat{z}_1|z_1)$ with the property that

$$\widehat{\pi}_1(\widehat{z}_1) = \int q_1(\widehat{z}_1|z_1)\,\pi_1(z_1)\,\mathrm{d}z_1. \tag{2.11}$$

Techniques from optimal transportation can be used to find appropriate transition kernels (Villani 2003, Villani 2009, Reich and Cotter 2015).

(B) We first produce M samples from the smoothing PDF (2.8) via resampling with replacement and then sample from $\widehat{\pi}_1$ using the smoothing transition kernels (2.10). The resampling can be represented in terms of a Markov transition matrix $Q_0 \in \mathbb{R}^{M \times M}$ such that

$$\gamma = Q_0\, p.$$

Here we have introduced the associated probability vectors

$$\gamma = \left(\frac{\gamma^1}{M}, \dots, \frac{\gamma^M}{M} \right)^{\mathrm{T}} \in \mathbb{R}^M, \quad p = \left(\frac{1}{M}, \dots, \frac{1}{M} \right)^{\mathrm{T}} \in \mathbb{R}^M. \tag{2.12}$$

Techniques from optimal transport will be explored to find such Markov transition matrices in Section 3.

(C) We directly seek Markov transition kernels $q_+^*(z_1|z_0^i)$, $i = 1, \dots, M$, with the property that

$$\widehat{\pi}_1(z_1) = \frac{1}{M} \sum_{i=1}^{M} q_+^*(z_1|z_0^i) \tag{2.13}$$

and then draw a single sample, $\widehat{z}_1^i$, from each kernel $q_+^*(z_1|z_0^i)$. Such kernels can be found by solving a Schrödinger problem (Leonard 2014, Chen *et al.* 2014) as demonstrated in Section 2.3.

Scenario (A) forms the basis of the classical bootstrap particle filter (Doucet *et al.* 2001, Liu 2001, Bain and Crisan 2008, Arulampalam *et al.* 2002) and also provides the starting point for many currently used ensemble-based data assimilation algorithms (Evensen 2006, Reich and Cotter 2015, Law *et al.* 2015). Scenario (B) is also well known in the context of particle filters under the notion of optimal proposal densities (Doucet *et al.* 2001, Arulampalam *et al.* 2002, Fearnhead and Künsch 2018). Recently there has been renewed interest in scenario (B) from the perspective of optimal control and twisting approaches (Guarniero, Johansen and Lee 2017, Heng, Bishop, Deligiannidis and Doucet 2018, Kappen and Ruiz 2016, Ruiz and Kappen 2017). Finally, scenario (C) has not yet been explored in the context of particle filters and data assimilation, primarily because the required kernels q_+^* are typically not available in closed form or cannot be easily sampled from. However, as we will argue in this paper, progress on the numerical solution of Schrödinger's problem (Cuturi 2013, Peyre and Cuturi 2018) turns scenario (C) into a viable option in addition to providing a unifying mathematical framework for data assimilation.

We emphasize that not all existing particle methods fit into these three scenarios. For example, the methods put forward by van Leeuwen (2015) are based on proposal densities which attempt to overcome limitations of scenario (B) and which lead to less variable particle weights, thus attempting to obtain particle filter implementations closer to what we denote here as scenario (C). More broadly speaking, the exploration of alternative proposal densities in the context of data assimilation has started only recently. See, for example, Vanden-Eijnden and Weare (2012), Morzfeld, Tu, Atkins and Chorin (2012), van Leeuwen (2015), Pons Llopis, Kantas, Beskos and Jasra (2018) and van Leeuwen *et al.* (2018).

The accuracy of an ensemble-based data assimilation method can be characterized in terms of its effective sample size M_{eff} (Liu 2001). The relevant effective sample size for scenario (B) is, for example, given by

$$M_{\mathrm{eff}} = \frac{M^2}{\sum_{i=1}^{M}(\gamma^i)^2} = \frac{1}{\|\gamma\|^2}.$$

We find that $M \geq M_{\mathrm{eff}} \geq 1$ and the accuracy of a data assimilation step decreases with decreasing M_{eff}, that is, the convergence rate $1/\sqrt{M}$ of a standard Monte Carlo method is replaced by $1/\sqrt{M_{\mathrm{eff}}}$ (Agapiou, Papaspiliopoulos, Sanz-Alonso and Stuart 2017). Scenario (C) offers a route around this problem by bridging π_0 with $\widehat{\pi}_1$ directly, that is, solving the Schrödinger problem delivers the best possible proposal densities leading to equally weighted particles without the need for resampling.[1]

Example 2.5. We illustrate the three scenarios with a simple example. The prior samples are given by $M = 11$ equally spaced particles $z_0^i \in \mathbb{R}$ from the interval $[-1, 1]$. The forecast PDF π_1 is provided by

$$\pi_1(z) = \frac{1}{M} \sum_{i=1}^{M} \frac{1}{(2\pi)^{1/2}\sigma} \exp\left(-\frac{1}{2\sigma^2}(z - z_0^i)^2\right)$$

with variance $\sigma^2 = 0.1$. The likelihood function is given by

$$\pi(y_1|z) = \frac{1}{(2\pi R)^{1/2}} \exp\left(-\frac{1}{2R}(y_1 - z)^2\right)$$

with $R = 0.1$ and $y_1 = -0.5$. The implied filtering and smoothing distributions can be found in Figure 2.2. Since $\widehat{\pi}_1$ is in the form of a weighted Gaussian mixture distribution, the Markov chain leading from $\widehat{\pi}_0$ to $\widehat{\pi}_1$ can be stated explicitly, that is, (2.10) is provided by

$$\widehat{q}_+(z_1|z_0^i) = \frac{1}{(2\pi)^{1/2}\widehat{\sigma}} \exp\left(-\frac{1}{2\widehat{\sigma}^2}(\bar{z}_1^i - z_1)^2\right) \tag{2.14}$$

with

$$\widehat{\sigma}^2 = \sigma^2 - \frac{\sigma^4}{\sigma^2 + R}, \quad \bar{z}_1^i = z_0^i - \frac{\sigma^2}{\sigma^2 + R}(z_0^i - y_1).$$

---

[1] The kernel (2.10) is called the optimal proposal in the particle filter community. However, the kernel (2.10) is suboptimal in the broader framework considered in this paper.

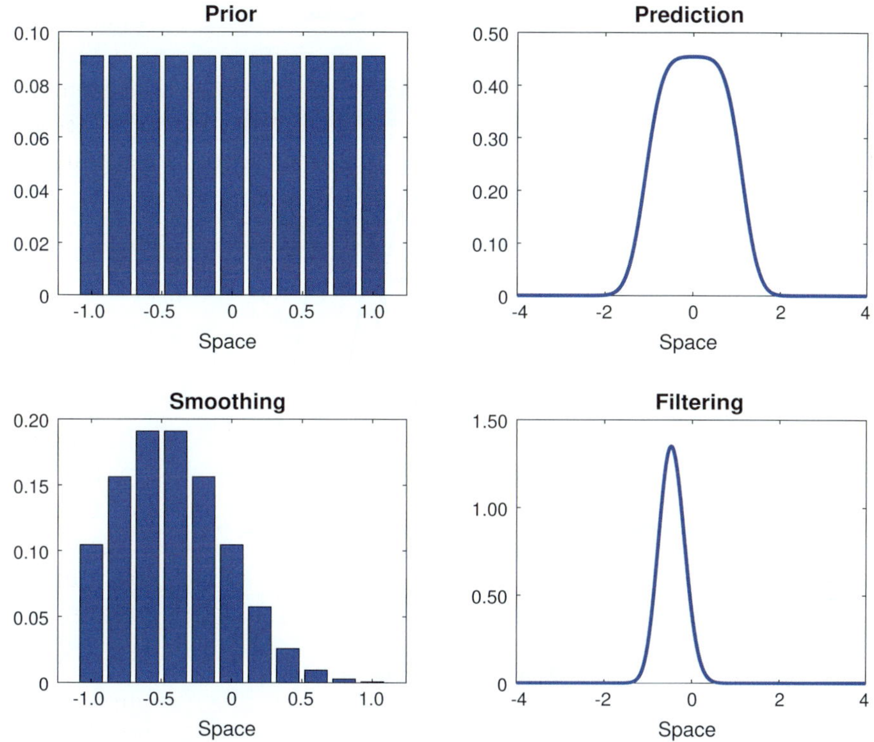

Figure 2.2. The initial PDF π_0, the forecast PDF π_1, the filtering PDF $\widehat{\pi}_1$, and the smoothing PDF $\widehat{\pi}_0$ for a simple Gaussian transition kernel.

The resulting transition kernels are displayed in Figure 2.3 together with the corresponding transition kernels for the Schrödinger approach, which connects π_0 directly with $\widehat{\pi}_1$.

Remark 2.6. It is often assumed in optimal control or rare event simulations arising from statistical mechanics that π_0 in (2.1) is a point measure, that is, the starting point of the simulation is known exactly. See, for example, Hartmann, Richter, Schütte and Zhang (2017). This corresponds to (2.3) with $M = 1$. It turns out that the associated smoothing problem becomes equivalent to Schrödinger's problem under this particular setting since the distribution at $t = 0$ is fixed.

The remainder of this section is structured as follows. We first recapitulate the pure prediction problem for discrete-time Markov processes and continuous-time diffusion processes, after which we discuss the filtering and smoothing problem for a single data assimilation step as relevant for scenarios (A) and (B). The final subsection is devoted to the Schrödinger

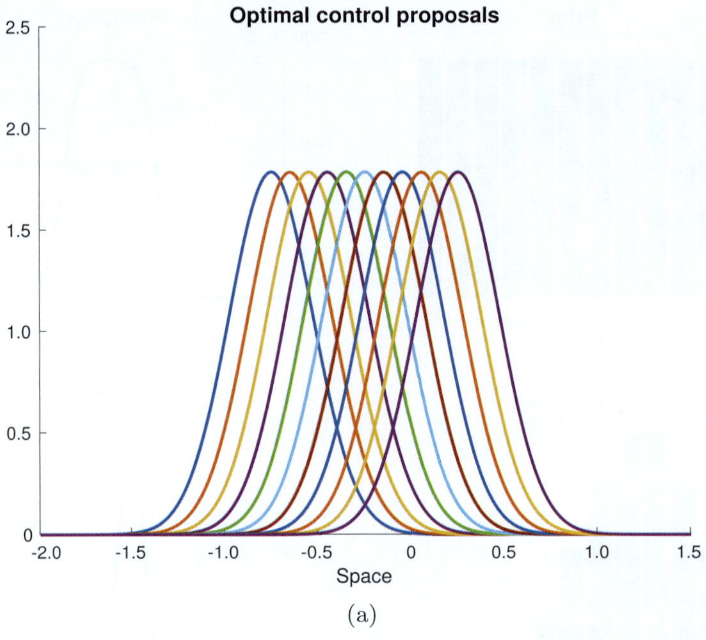

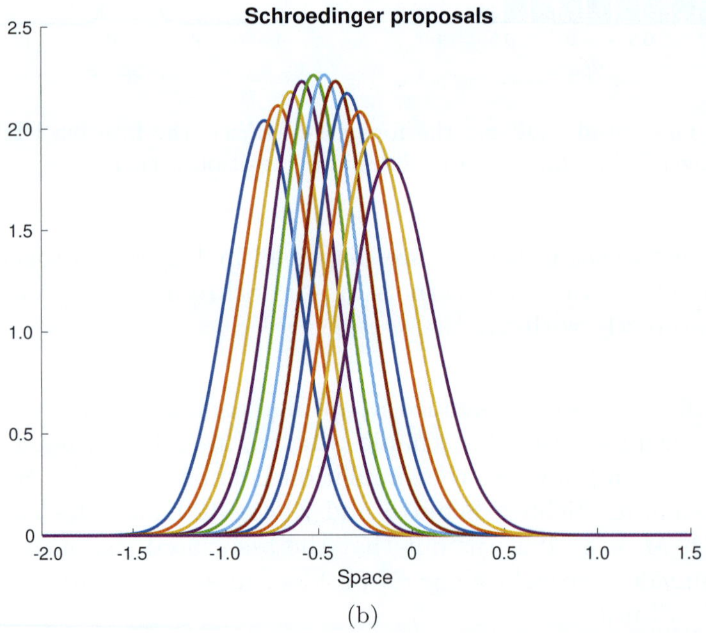

Figure 2.3. (a) The transition kernels (2.14) for the $M = 11$ different particles z_0^i. These correspond to the optimal control path in Figure 2.1. (b) The corresponding transition kernels, which lead directly from π_0 to $\widehat{\pi}_1$. These correspond to the Schrödinger path in Figure 2.1. Details of how to compute these Schrödinger transition kernels, $q_+^*(z_1|z_0^i)$, can be found in Section 3.4.1.

problem (Leonard 2014, Chen *et al.* 2014) of bridging the filtering distribution, π_0, at $t = 0$ directly with the filtering distribution, $\widehat{\pi}_1$, at $t = 1$, thus leading to scenario (C).

2.1. Prediction

We assume under the chosen computational setting that we have access to M samples $z_0^i \in \mathbb{R}^{N_z}$, $i = 1, \ldots, M$, from the filtering distribution at $t = 0$. We also assume that we know (explicitly or implicitly) the forward transition probabilities $q_+(z_1|z_0^i)$ of the underlying Markovian stochastic process. This leads to the forecast PDF π_1 as given by (2.4).

Before we consider two specific examples, we introduce two concepts related to the forward transition kernel which we will need later in order to address scenarios (B) & (C) from Definition 2.4.

We first introduce the backward transition kernel $q_-(z_0|z_1)$, which is defined via the equation

$$q_-(z_0|z_1)\,\pi_1(z_1) = q_+(z_1|z_0)\,\pi_0(z_0).$$

Note that $q_-(z_0|z_1)$ as well as π_0 are not absolutely continuous with respect to the underlying Lebesgue measure, that is,

$$q_-(z_0|z_1) = \frac{1}{M} \sum_{i=1}^{M} \frac{q_+(z_1|z_0^i)}{\pi_1(z_1)}\,\delta(z_0 - z_0^i). \tag{2.15}$$

The backward transition kernel $q_-(z_1|z_0)$ reverses the prediction process in the sense that

$$\pi_0(z_0) = \int q_-(z_0|z_1)\,\pi_1(z_1)\,\mathrm{d}z_1.$$

Remark 2.7.   Let us assume that the detailed balance

$$q_+(z_1|z_0)\,\pi(z_0) = q_+(z_0|z_1)\,\pi(z_1)$$

holds for some PDF π and forward transition kernel $q_+(z_1|z_0)$. Then $\pi_1 = \pi$ for $\pi_0 = \pi$ (invariance of π) and $q_-(z_0|z_1) = q_+(z_1|z_0)$.

We next introduce a class of forward transition kernels using the concept of twisting (Guarniero *et al.* 2017, Heng *et al.* 2018), which is an application of Doob's H-transform technique (Doob 1984).

Definition 2.8.   Given a non-negative twisting function $\psi(z_1)$ such that the modified transition kernel (1.2) is well-defined, one can define the twisted forecast PDF

$$\pi_1^{\psi}(z_1) := \frac{1}{M} \sum_{i=1}^{M} q_+^{\psi}(z_1|z_0^i) = \frac{1}{M} \sum_{i=1}^{M} \frac{\psi(z_1)}{\widehat{\psi}(z_0^i)}\,q_+(z_1|z_0^i). \tag{2.16}$$

The PDFs π_1 and π_1^ψ are related by

$$\frac{\pi_1(z_1)}{\pi_1^\psi(z_1)} = \frac{\sum_{i=1}^M q_+(z_1|z_0^i)}{\sum_{i=1}^M \frac{\psi(z_1)}{\widehat{\psi}(z_0^i)} q_+(z_1|z_0^i)}. \tag{2.17}$$

Equation (2.17) gives rise to importance weights

$$w^i \propto \frac{\pi_1(z_1^i)}{\pi_1^\psi(z_1^i)} \tag{2.18}$$

for samples $z_1^i = Z_1^i(\omega)$ drawn from the twisted forecast PDF, that is,

$$Z_1^i \sim q_+^\psi(\cdot|z_0^i)$$

and

$$\pi_1(z) \approx \frac{1}{M} \sum_{i=1}^M w^i \delta(z - z_1^i)$$

in a weak sense. Here we have assumed that the normalization constant in (2.18) is chosen such that

$$\sum_{i=1}^M w^i = M. \tag{2.19}$$

Such twisted transition kernels will become important when looking at the filtering and smoothing as well as the Schrödinger problem later in this section.

Let us now discuss a couple of specific models which give rise to transition kernels $q_+(z_1|z_0)$. These models will be used throughout this paper to illustrate mathematical and algorithmic concepts.

2.1.1. Gaussian model error

Let us consider the discrete-time stochastic process

$$Z_1 = \Psi(Z_0) + \gamma^{1/2}\Xi_0 \tag{2.20}$$

for given map $\Psi : \mathbb{R}^{N_z} \to \mathbb{R}^{N_z}$, scaling factor $\gamma > 0$, and Gaussian distributed random variable Ξ_0 with mean zero and covariance matrix $B \in \mathbb{R}^{N_z \times N_z}$. The associated forward transition kernel is given by

$$q_+(z_1|z_0) = \mathrm{n}(z_1; \Psi(z_0), \gamma B). \tag{2.21}$$

Recall that we have introduced the shorthand $\mathrm{n}(z; \bar{z}, P)$ for the PDF of a Gaussian random variable with mean $\bar{z}$ and covariance matrix P.

Let us consider a twisting potential ψ of the form

$$\psi(z_1) \propto \exp\left(-\frac{1}{2}(Hz_1 - d)^{\mathrm{T}} R^{-1}(Hz_1 - d)\right)$$

for given $H \in \mathbb{R}^{N_z \times N_d}$, $d \in \mathbb{R}^{N_d}$, and covariance matrix $R \in \mathbb{R}^{N_d \times N_d}$. We define

$$K := BH^{\mathrm{T}}(HBH^{\mathrm{T}} + \gamma^{-1}R)^{-1} \tag{2.22}$$

and

$$\bar{B} := B - KHB, \quad \bar{z}_1^i := \Psi(z_0^i) - K(H\Psi(z_0^i) - d). \tag{2.23}$$

The twisted transition kernels are given by

$$q_+^{\psi}(z_1|z_0^i) = \mathrm{n}(z_1; \bar{z}_1^i, \gamma\bar{B})$$

and

$$\widehat{\psi}(z_0^i) \propto \exp\left(-\frac{1}{2}(H\Psi(z_0^i) - d)^{\mathrm{T}}(R + \gamma HBH^{\mathrm{T}})^{-1}(H\Psi(z_0^i) - d)\right)$$

for $i = 1, \ldots, M$.

2.1.2. SDE models

Consider the (forward) SDE (Pavliotis 2014)

$$\mathrm{d}Z_t^+ = f_t(Z_t^+)\,\mathrm{d}t + \gamma^{1/2}\,\mathrm{d}W_t^+ \tag{2.24}$$

with initial condition $Z_0^+ = z_0$ and $\gamma > 0$. Here W_t^+ stands for standard Brownian motion in the sense that the distribution of $W_{t+\Delta t}^+$, $\Delta t > 0$, conditioned on $w_t^+ = W_t^+(\omega)$ is Gaussian with mean w_t^+ and covariance matrix $\Delta t\, I$ (Pavliotis 2014) and the process Z_t^+ is adapted to W_t^+.

The resulting time-t transition kernels $q_t^+(z|z_0)$ from time zero to time t, $t \in (0, 1]$, satisfy the Fokker–Planck equation (Pavliotis 2014)

$$\partial_t q_t^+(\cdot|z_0) = -\nabla_z \cdot (q_t^+(\cdot|z_0)f_t) + \frac{\gamma}{2}\Delta_z q_t^+(\cdot|z_0)$$

with initial condition $q_0^+(z|z_0) = \delta(z - z_0)$, and the time-one forward transition kernel $q_+(z_1|z_0)$ is given by

$$q_+(z_1|z_0) = q_1^+(z_1|z_0).$$

We introduce the operator $\mathcal{L}_t$ by

$$\mathcal{L}_t g := \nabla_z g \cdot f_t + \frac{\gamma}{2}\Delta_z g$$

and its adjoint $\mathcal{L}_t^{\dagger}$ by

$$\mathcal{L}_t^{\dagger}\pi := -\nabla_z \cdot (\pi\, f_t) + \frac{\gamma}{2}\Delta_z \pi \tag{2.25}$$

(Pavliotis 2014). We call $\mathcal{L}_t^{\dagger}$ the Fokker–Planck operator and $\mathcal{L}_t$ the generator of the Markov process associated to the SDE (2.24).

Solutions (realizations) $z_{[0,1]} = Z_{[0,1]}^+(\omega)$ of the SDE (2.24) with initial conditions drawn from π_0 are continuous functions of time, that is, $z_{[0,1]} \in \mathcal{C} := C([0,1], \mathbb{R}^{N_z})$, and define a probability measure $\mathbb{Q}$ on $\mathcal{C}$, that is,

$$Z_{[0,1]}^+ \sim \mathbb{Q}.$$

We note that the marginal distributions π_t of $\mathbb{Q}$, given by

$$\pi_t(z_t) = \int q_t^+(z_t|z_0)\, \pi_0(z_0)\, \mathrm{d}z_0,$$

also satisfy the Fokker–Planck equation, that is,

$$\partial_t \pi_t = \mathcal{L}_t^\dagger \pi_t = -\nabla_z \cdot (\pi_t\, f_t) + \frac{\gamma}{2}\Delta_z \pi_t \tag{2.26}$$

for given PDF π_0 at time $t = 0$.

Furthermore, we can rewrite the Fokker–Planck equation (2.26) in the form

$$\partial \pi_t = -\nabla_z \cdot (\pi_t(f_t - \gamma\nabla_z \log \pi_t)) - \frac{\gamma}{2}\Delta_z \pi_t, \tag{2.27}$$

which allows us to read off from (2.27) the backward SDE

$$\begin{aligned}
\mathrm{d}Z_t^- &= f_t(Z_t^-)\,\mathrm{d}t - \gamma\nabla_z \log \pi_t\,\mathrm{d}t + \gamma^{1/2}\,\mathrm{d}W_t^-, \\
&= b_t(Z_t^-)\,\mathrm{d}t + \gamma^{1/2}\,\mathrm{d}W_t^-
\end{aligned} \tag{2.28}$$

with final condition $Z_1^- \sim \pi_1$, W_t^- backward Brownian motion, and density-dependent drift term

$$b_t(z) := f_t(z) - \gamma\nabla_z \log \pi_t$$

(Nelson 1984, Chen *et al.* 2014). Here backward Brownian motion is to be understood in the sense that the distribution of $W_{t-\Delta\tau}^-$, $\Delta\tau > 0$, conditioned on $w_t^- = W_t^-(\omega)$ is Gaussian with mean w_t^- and covariance matrix $\Delta\tau\, I$ and all other properties of Brownian motion appropriately adjusted. The process Z_t^- is adapted to W_t^-.

Lemma 2.9. The backward SDE (2.28) induces a corresponding backward transition kernel from time one to time $t = 1 - \tau$ with $\tau \in [0,1]$, denoted by $q_\tau^-(z|z_1)$, which satisfies the time-reversed Fokker–Planck equation

$$\partial_\tau q_\tau^-(\cdot|z_1) = \nabla_z \cdot (q_\tau^-(\cdot|z_1)\, b_{1-\tau}) + \frac{\gamma}{2}\Delta_z q_\tau^-(\cdot|z_1)$$

with boundary condition $q_0^-(z|z_1) = \delta(z - z_1)$ at $\tau = 0$ (or, equivalently, at $t = 1$). The induced backward transition kernel $q_-(z_0|z_1)$ is then given by

$$q_-(z_0|z_1) = q_1^-(z_0|z_1)$$

and satisfies (2.15).

Proof. The lemma follows from the fact that the backward SDE (2.28) implies the Fokker–Planck equation (2.27) and that we have reversed time by introducing $\tau = 1 - t$. $\qquad\square$

Remark 2.10. The notion of a backward SDE also arises in a different context where the driving Brownian motion is still adapted to the past, *i.e.* W_t^+ in our notation, and a final condition is prescribed as for (2.28). See (2.54) below as well as Appendix A.4 and Carmona (2016) for more details.

We note that the mean-field equation,

$$\frac{\mathrm{d}}{\mathrm{d}t} z_t = f_t(z_t) - \frac{\gamma}{2} \nabla_z \log \pi_t(z_t) = \frac{1}{2}(f_t(z_t) + b_t(z_t)), \qquad (2.29)$$

resulting from (2.26), leads to the same marginal distributions π_t as the forward and backward SDEs, respectively. It should be kept in mind, however, that the path measure generated by (2.29) is different from the path measure $\mathbb{Q}$ generated by (2.24).

Please also note that the backward SDE and the mean-field equation (2.29) become singular as $t \to 0$ for the given initial PDF (2.3). A meaningful solution can be defined via regularization of the Dirac delta function, that is,

$$\pi_0(z) \approx \frac{1}{M} \sum_{i=1}^{M} \mathrm{n}(z; z_0^i, \epsilon I),$$

and taking the limit $\epsilon \to 0$.

We will find later that it can be advantageous to modify the given SDE (2.24) by a time-dependent drift term $u_t(z)$, that is,

$$\mathrm{d}Z_t^+ = f_t(Z_t^+)\,\mathrm{d}t + u_t(Z_t^+)\,\mathrm{d}t + \gamma^{1/2}\,\mathrm{d}W_t^+. \qquad (2.30)$$

In particular, such a modification leads to the time-continuous analogue of the twisted transition kernel (1.2) introduced in Section 2.1.

Lemma 2.11. Let $\psi_t(z)$ denote the solutions of the backward Kolmogorov equation

$$\partial_t \psi_t = -\mathcal{L}_t \psi_t = -\nabla_z \psi_t \cdot f_t - \frac{\gamma}{2} \Delta_z \psi_t \qquad (2.31)$$

for given final $\psi_1(z) > 0$ and $t \in [0, 1]$. The controlled SDE (2.30) with

$$u_t(z) := \gamma \nabla_z \log \psi_t(z) \qquad (2.32)$$

leads to a time-one forward transition kernel $q_+^\psi(z_1|z_0)$ which satisfies

$$q_+^\psi(z_1|z_0) = \psi_1(z_1)\, q_+(z_1|z_0)\, \psi_0(z_0)^{-1},$$

where $q_+(z_1|z_0)$ denotes the time-one forward transition kernel of the uncontrolled forward SDE (2.24).

Proof. A proof of this lemma has, for example, been given by Dai Pra (1991, Theorem 2.1). $\qquad\square$

More generally, the modified forward SDE (2.30) with $Z_0^+ \sim \pi_0$ generates a path measure which we denote by $\mathbb{Q}^u$ for given functions $u_t(z)$, $t \in [0, 1]$. Realizations of this path measure are denoted by $z_{[0,1]}^u$. According to Girsanov's theorem (Pavliotis 2014), the two path measures $\mathbb{Q}$ and $\mathbb{Q}^u$ are absolutely continuous with respect to each other, with Radon–Nikodym derivative

$$\frac{\mathrm{d}\mathbb{Q}^u}{\mathrm{d}\mathbb{Q}}\bigg|_{z_{[0,1]}^u} = \exp\left(\frac{1}{2\gamma}\int_0^1 \left(\|u_t\|^2 \,\mathrm{d}t + 2\gamma^{1/2}u_t \cdot \mathrm{d}W_t^+\right)\right), \tag{2.33}$$

provided that the Kullback–Leibler divergence $\mathrm{KL}(\mathbb{Q}^u\|\mathbb{Q})$ between $\mathbb{Q}^u$ and $\mathbb{Q}$, given by

$$\mathrm{KL}(\mathbb{Q}^u\|\mathbb{Q}) := \int \left[\frac{1}{2\gamma}\int_0^1 \|u_t\|^2 \,\mathrm{d}t\right]\mathbb{Q}^u(\mathrm{d}z_{[0,1]}^u), \tag{2.34}$$

is finite. Recall that the Kullback–Leibler divergence between two path measures $\mathbb{P} \ll \mathbb{Q}$ on $\mathcal{C}$ is defined by

$$\mathrm{KL}(\mathbb{P}\|\mathbb{Q}) = \int \log \frac{\mathrm{d}\mathbb{P}}{\mathrm{d}\mathbb{Q}} \,\mathbb{P}(\mathrm{d}z_{[0,1]}).$$

If the modified SDE (2.30) is used to make predictions, then its solutions $z_{[0,1]}^u$ need to be weighted according to the inverse Radon–Nikodym derivative

$$\frac{\mathrm{d}\mathbb{Q}}{\mathrm{d}\mathbb{Q}^u}\bigg|_{z_{[0,1]}^u} = \exp\left(-\frac{1}{2\gamma}\int_0^1 \left(\|u_t\|^2 \,\mathrm{d}t + 2\gamma^{1/2}u_t \cdot \mathrm{d}W_t^+\right)\right) \tag{2.35}$$

in order to reproduce the desired forecast PDF π_1 of the original SDE (2.24).

Remark 2.12. A heuristic derivation of equation (2.33) can be found in Section 3.1.2 below, where we discuss the numerical approximation of SDEs by the Euler–Maruyama method. Equation (2.34) follows immediately from (2.33) by noting that the expectation of Brownian motion under the path measure $\mathbb{Q}^u$ is zero.

2.2. *Filtering and smoothing*

We now incorporate the likelihood

$$l(z_1) = \pi(y_1|z_1)$$

of the data y_1 at time $t_1 = 1$. Bayes' theorem tells us that, given the forecast PDF π_1 at time t_1, the posterior PDF $\widehat{\pi}_1$ is given by (2.5). The distribution $\widehat{\pi}_1$ solves the filtering problem at time t_1 given the data y_1. We also recall the definition of the evidence (2.7). The quantity $\mathcal{F} = -\log \beta$ is called the free energy in statistical physics (Hartmann $et\ al.$ 2017).

An appropriate transition kernel $q_1(\widehat{z}_1|z_1)$, satisfying (2.11), is required in order to complete the transition from π_0 to $\widehat{\pi}_1$ following scenario (A) from Definition 2.4. A suitable framework for finding such transition kernels is via the theory of optimal transportation (Villani 2003). More specifically, let Π_c denote the set of all joint probability measures $\pi(z_1, \widehat{z}_1)$ with marginals

$$\int \pi(z_1, \widehat{z}_1)\, \mathrm{d}\widehat{z}_1 = \pi_1(z_1), \quad \int \pi(z_1, \widehat{z}_1)\, \mathrm{d}z_1 = \widehat{\pi}_1(\widehat{z}_1).$$

We seek the joint measure $\pi^*(z_1, \widehat{z}_1) \in \Pi_c$ which minimizes the expected Euclidean distance between the two associated random variables Z_1 and $\widehat{Z}_1$, that is,

$$\pi^* = \arg \inf_{\pi \in \Pi_c} \int \int \|z_1 - \widehat{z}_1\|^2\, \pi(z_1, \widehat{z}_1)\, \mathrm{d}z_1\, \mathrm{d}\widehat{z}_1. \tag{2.36}$$

The minimizing joint measure is of the form

$$\pi^*(z_1, \widehat{z}_1) = \delta(\widehat{z}_1 - \nabla_z \Phi(z_1))\, \pi_1(z_1) \tag{2.37}$$

with suitable convex potential Φ under appropriate conditions on the PDFs π_1 and $\widehat{\pi}_1$ (Villani 2003). These conditions are satisfied for dynamical systems with Gaussian model errors and typical SDE models. Once the potential Φ (or an approximation) is available, samples z_1^i, $i = 1, \ldots, M$, from the forecast PDF π_1 can be converted into samples $\widehat{z}_1^i$, $i = 1, \ldots, M$, from the filtering distribution $\widehat{\pi}_1$ via the deterministic transformation

$$\widehat{z}_1^i = \nabla_z \Phi(z_1^i). \tag{2.38}$$

We will discuss in Section 3 how to approximate the transformation (2.38) numerically. We will find that many popular data assimilation schemes, such as the ensemble Kalman filter, can be viewed as approximations to (2.38) (Reich and Cotter 2015).

We recall at this point that classical particle filters start from the importance weights

$$w^i \propto \frac{\widehat{\pi}_1(z_1^i)}{\pi_1(z_1^i)} = \frac{l(z_1^i)}{\beta},$$

and obtain the desired samples $\widehat{z}^i$ by an appropriate resampling with replacement scheme (Doucet $et\ al.$ 2001, Arulampalam $et\ al.$ 2002, Douc and Cappe 2005) instead of applying a deterministic transformation of the form (2.38).

Remark 2.13. If one replaces the forward transition kernel $q_+(z_1|z_0)$ with a twisted kernel (1.2), then, using (2.17), the filtering distribution (2.5) satisfies

$$\frac{\widehat{\pi}_1(z_1)}{\pi_1^{\psi}(z_1)} = \frac{l(z_1)\sum_{j=1}^{M} q_+(z_1|z_0^j)}{\beta \sum_{j=1}^{M} \frac{\psi(z_1)}{\widehat{\psi}(z_0^j)} q_+(z_1|z_0^j)}. \tag{2.39}$$

Hence drawing samples z_1^i, $i = 1,\ldots,M$, from π_1^{ψ} instead of π_1 leads to modified importance weights

$$w^i \propto \frac{l(z_1^i)\sum_{j=1}^{M} q_+(z_1^i|z_0^j)}{\beta \sum_{j=1}^{M} \frac{\psi(z_1^i)}{\widehat{\psi}(z_0^j)} q_+(z_1^i|z_0^j)}. \tag{2.40}$$

We will demonstrate in Section 2.3 that finding a twisting potential ψ such that $\widehat{\pi}_1 = \pi_1^{\psi}$, leading to importance weights $w^i = 1$ in (2.40), is equivalent to solving the Schrödinger problem (2.58)–(2.61).

The associated smoothing distribution at time $t = 0$ can be defined as follows. First introduce

$$\psi(z_1) := \frac{\widehat{\pi}_1(z_1)}{\pi_1(z_1)} = \frac{l(z_1)}{\beta}. \tag{2.41}$$

Next we set

$$\widehat{\psi}(z_0) := \int q_+(z_1|z_0)\,\psi(z_1)\,\mathrm{d}z_1 = \beta^{-1}\int q_+(z_1|z_0)\,l(z_1)\,\mathrm{d}z_1, \tag{2.42}$$

and introduce $\widehat{\pi}_0 := \pi_0\,\widehat{\psi}$, that is,

$$\widehat{\pi}_0(z_0) = \frac{1}{M}\sum_{i=1}^{M} \widehat{\psi}(z_0^i)\,\delta(z_0 - z_0^i)$$

$$= \frac{1}{M}\sum_{i=1}^{M} \gamma^i\,\delta(z_0 - z_0^i) \tag{2.43}$$

since $\widehat{\psi}(z_0^i) = \gamma^i$ with γ^i defined by (2.9).

Lemma 2.14. The smoothing PDFs $\widehat{\pi}_0$ and $\widehat{\pi}_1$ satisfy

$$\widehat{\pi}_0(z_0) = \int q_-(z_0|z_1)\,\widehat{\pi}_1(z_1)\,\mathrm{d}z_1 \tag{2.44}$$

with the backward transition kernel defined by (2.15). Furthermore,

$$\widehat{\pi}_1(z_1) = \int \widehat{q}_+(z_1|z_0)\,\widehat{\pi}_0(z_0)\,\mathrm{d}z_0$$

with twisted forward transition kernels

$$\widehat{q}_+(z_1|z_0^i) := \psi(z_1)\,q_+(z_1|z_0^i)\,\widehat{\psi}(z_0^i)^{-1} = \frac{l(z_1)}{\beta\,\gamma^i}q_+(z_1|z_0^i) \tag{2.45}$$

and γ^i, $i = 1, \ldots, M$, defined by (2.9).

Proof. We note that

$$q_-(z_0|z_1)\,\widehat{\pi}_1(z_1) = \frac{\pi_0(z_0)}{\pi_1(z_1)}q_+(z_1|z_0)\widehat{\pi}_1(z_1) = \frac{l(z_1)}{\beta}q_+(z_1|z_0)\,\pi_0(z_0),$$

which implies the first equation. The second equation follows from $\widehat{\pi}_0 = \widehat{\psi}\,\pi_0$ and

$$\int \widehat{q}_+(z_1|z_0)\,\widehat{\pi}_0(z_0)\,\mathrm{d}z_0 = \frac{1}{M}\sum_{i=1}^{M}\frac{l(z_1)}{\beta}\,q_+(z_1|z_0^i).$$

In other words, we have defined a twisted forward transition kernel of the form (1.2). $\qquad\square$

Seen from a more abstract perspective, we have provided an alternative formulation of the joint smoothing distribution

$$\widehat{\pi}(z_0, z_1) := \frac{l(z_1)\,q_+(z_1|z_0)\,\pi_0(z_0)}{\beta} \tag{2.46}$$

in the form of

$$\begin{aligned}
\widehat{\pi}(z_0, z_1) &= \frac{l(z_1)}{\beta}\frac{\psi(z_1)}{\psi(z_1)}q_+(z_1|z_0)\frac{\widehat{\psi}(z_0)}{\widehat{\psi}(z_0)}\pi_0(z_0) \\
&= \widehat{q}_+(z_1|z_0)\,\widehat{\pi}_0(z_0) \tag{2.47}
\end{aligned}$$

because of (2.41). Note that the marginal distributions of $\widehat{\pi}$ are provided by $\widehat{\pi}_0$ and $\widehat{\pi}_1$, respectively.

One can exploit these formulations computationally as follows. If one has generated M equally weighted particles $\widehat{z}_0^j$ from the smoothing distribution (2.43) at time $t = 0$ via resampling with replacement, then one can obtain equally weighted samples $\widehat{z}_1^j$ from the filtering distribution $\widehat{\pi}_1$ using the modified transition kernels (2.45). This is the idea behind the optimal proposal particle filter (Doucet *et al.* 2001, Arulampalam *et al.* 2002, Fearnhead and Künsch 2018) and provides an implementation of scenario (B) as introduced in Definition 2.4.

Remark 2.15. We remark that backward simulation methods use (2.44) in order to address the smoothing problem (2.2) in a sequential forward–backward manner. Since we are not interested in the general smoothing problem in this paper, we refer the reader to the survey by Lindsten and Schön (2013) for more details.

Lemma 2.16. Let $\psi(z) > 0$ be a twisting potential such that

$$l^{\psi}(z_1) := \frac{l(z_1)}{\beta\,\psi(z_1)}\pi_0[\widehat{\psi}]$$

is well-defined with $\widehat{\psi}$ given by (1.3). Then the smoothing PDF (2.46) can be represented as

$$\widehat{\pi}(z_0, z_1) = l^{\psi}(z_1)\, q_+^{\psi}(z_1|z_0)\, \pi_0^{\psi}(z_0), \qquad (2.48)$$

where the modified forward transition kernel $q_+^{\psi}(z_1|z_0)$ is defined by (1.2) and the modified initial PDF by

$$\pi_0^{\psi}(z_0) := \frac{\widehat{\psi}(z_0)\, \pi_0(z_0)}{\pi_0[\widehat{\psi}]}.$$

Proof. This follows from the definition of the smoothing PDF $\widehat{\pi}(z_0, z_1)$ and the twisted transition kernel $q_+^{\psi}(z_1|z_0)$. $\qquad\square$

Remark 2.17. As mentioned before, the choice (2.41) implies $l^{\psi} = \text{const.}$, and leads to the well-known optimal proposal density for particle filters. The more general formulation (2.48) has recently been explored and expanded by Guarniero *et al.* (2017) and Heng *et al.* (2018) in order to derive efficient proposal densities for the general smoothing problem (2.2). Within the simplified formulation (2.48), such approaches reduce to a change of measure from π_0 to π_0^{ψ} at t_0 followed by a forward transition according to q_+^{ψ} and subsequent reweighting by a modified likelihood l^{ψ} at t_1 and hence lead to particle filters that combine scenarios (A) and (B) as introduced in Definition 2.4.

2.2.1. Gaussian model errors (cont.)

We return to the discrete-time process (2.20) and assume a Gaussian measurement error leading to a Gaussian likelihood

$$l(z_1) \propto \exp\left(-\frac{1}{2}(Hz_1 - y_1)^T R^{-1}(Hz_1 - y_1)\right).$$

We set $\psi_1 = l/\beta$ in order to derive the optimal forward kernel for the associated smoothing/filtering problem. Following the discussion from Section 2.1.1, this leads to the modified transition kernels

$$\widehat{q}_+(z_1|z_0^i) := \mathrm{n}(z_1; \bar{z}_1^i, \gamma\bar{B})$$

with $\bar{B}$ and K defined by (2.23) and (2.22), respectively, and

$$\bar{z}_1^i := \Psi(z_0^i) - K(H\Psi(z_0^i) - y_1).$$

The smoothing distribution $\widehat{\pi}_0$ is given by

$$\widehat{\pi}_0(z_0) = \frac{1}{M}\sum_{i=1}^{M}\gamma^i\, \delta(z - z_0^i)$$

with coefficients

$$\gamma^i \propto \exp\left(-\frac{1}{2}(H\Psi(z_0^i) - y_1)^{\mathrm{T}}(R + \gamma H B H^{\mathrm{T}})^{-1}(H\Psi(z_0^i) - y_1)\right).$$

It is easily checked that, indeed,

$$\widehat{\pi}_1(z_1) = \int q_+^\psi(z_1|z_0)\,\widehat{\pi}_0(z_0)\,\mathrm{d}z_0.$$

The results from this subsection have been used in simplified form in Example 2.5 in order to compute (2.14). We also note that a non-optimal, i.e. $\psi(z_1) \neq l(z_1)/\beta$, but Gaussian choice for ψ leads to a Gaussian l^ψ and the transition kernels $q_+^\psi(z_1|z_0^i)$ in (2.48) remain Gaussian as well. This is in contrast to the Schrödinger problem, which we discuss in the following Section 2.3 and which leads to forward transition kernels of the form (2.66) below.

2.2.2. SDE models (cont.)

The likelihood $l(z_1)$ introduces a change of measure over path space $z_{[0,1]} \in \mathcal{C}$ from the forecast measure $\mathbb{Q}$ with marginals π_t to the smoothing measure $\widehat{\mathbb{P}}$ via the Radon–Nikodym derivative

$$\frac{\mathrm{d}\widehat{\mathbb{P}}}{\mathrm{d}\mathbb{Q}}_{\big|z_{[0,1]}} = \frac{l(z_1)}{\beta}. \tag{2.49}$$

We let $\widehat{\pi}_t$ denote the marginal distributions of the smoothing measure $\widehat{\mathbb{P}}$ at time t.

Lemma 2.18. Let ψ_t denote the solution of the backward Kolmogorov equation (2.31) with final condition $\psi_1(z) = l(z)/\beta$ at $t = 1$. Then the controlled forward SDE

$$\mathrm{d}Z_t^+ = (f_t(Z_t^+) + \gamma \nabla_z \log \psi_t(Z_t^+))\,\mathrm{d}t + \gamma^{1/2}\,\mathrm{d}W_t^+, \tag{2.50}$$

with $Z_0^+ \sim \widehat{\pi}_0$ at time $t = 0$, implies $Z_1^+ \sim \widehat{\pi}_1$ at final time $t = 1$.

Proof. The lemma is a consequence of Lemma 2.11 and definition (2.45) of the smoothing kernel $\widehat{q}_+(z_1|z_0)$ with $\psi(z_1) = \psi_1(z_1)$ and $\widehat{\psi}(z_0) = \psi_0(z_0)$. $\square$

The SDE (2.50) is obviously a special case of (2.30) with control law u_t given by (2.32). Note, however, that the initial distributions for (2.30) and (2.50) are different. We will reconcile this fact in the following subsection by considering the associated Schrödinger problem (Föllmer and Gantert 1997, Leonard 2014, Chen *et al.* 2014).

Lemma 2.19. The solution ψ_t of the backward Kolmogorov equation (2.31) with final condition $\psi_1(z) = l(z)/\beta$ at $t = 1$ satisfies

$$\psi_t(z) = \frac{\widehat{\pi}_t(z)}{\pi_t(z)} \tag{2.51}$$

and the PDFs $\widehat{\pi}_t$ coincide with the marginal PDFs of the backward SDE (2.28) with final condition $Z_1^- \sim \widehat{\pi}_1$.

Proof. We first note that (2.51) holds at final time $t = 1$. Furthermore, equation (2.51) implies

$$\partial_t \widehat{\pi}_t = \pi_t \, \partial_t \psi_t - \psi_t \, \partial_t \pi_t.$$

Since π_t satisfies the Fokker–Planck equation (2.26) and ψ_t the backward Kolmogorov equation (2.31), it follows that

$$\partial_t \widehat{\pi}_t = -\nabla_z \cdot (\widehat{\pi}_t(f - \gamma \nabla_z \log \pi_t)) - \frac{\gamma}{2} \Delta_z \widehat{\pi}_t, \tag{2.52}$$

which corresponds to the Fokker–Planck equation (2.27) of the backward SDE (2.28) with final condition $Z_1^- \sim \widehat{\pi}_1$ and marginal PDFs denoted by $\widehat{\pi}_t$ instead of π_t. $\qquad\square$

Note that (2.52) is equivalent to

$$\partial \widehat{\pi}_t = -\nabla_z \cdot (\widehat{\pi}_t(f - \gamma \nabla_z \log \pi_t + \gamma \nabla_z \log \widehat{\pi}_t)) + \frac{\gamma}{2} \Delta_z \widehat{\pi}_t,$$

which in turn is equivalent to the Fokker–Planck equation of the forward smoothing SDE (2.50) since $\psi_t = \widehat{\pi}_t/\pi_t$.

We conclude from the previous lemma that one can either solve the backward Kolmogorov equation (2.31) with $\psi_1(z) = l(z)/\beta$ or the backward SDE (2.28) with $Z_1^- \sim \widehat{\pi}_1 = l\,\pi_1/\beta$ in order to derive the desired control law $u_t(z) = \gamma \nabla_z \log \psi_t$ in (2.50).

Remark 2.20. The notion of a backward SDE used throughout this paper is different from the notion of a backward SDE in the sense of Carmona (2016), for example. More specifically, Itô's formula

$$\mathrm{d}\psi_t = \partial_t \psi_t \, \mathrm{d}t + \nabla_z \psi_t \cdot \mathrm{d}Z_t^+ + \frac{\gamma}{2} \Delta_z \psi_t \, \mathrm{d}t$$

and the fact that ψ_t satisfies the backward Kolmogorov equation (2.31) imply that

$$\mathrm{d}\psi_t = \gamma^{1/2} \, \nabla_z \psi_t \cdot \mathrm{d}W_t^+ \tag{2.53}$$

along solutions of the forward SDE (2.24). In other words, the quantities ψ_t are materially advected along solutions Z_t^+ of the forward SDEs in expectation or, in the language of stochastic analysis, ψ_t is a martingale. Hence, by the martingale representation theorem, we can reformulate the problem of determining ψ_t as follows. Find the solution (Y_t, V_t) of the backward SDE

$$\mathrm{d}Y_t = V_t \cdot \mathrm{d}W_t^+ \tag{2.54}$$

subject to the final condition $Y_1 = l(Z_1^+)/\beta$ at $t = 1$. Here (2.54) has to be understood as a backward SDE in the sense of Carmona (2016), for example, where the solution (Y_t, V_t) is adapted to the forward SDE (2.24), *i.e.* to the past $s \leq t$, whereas the solution Z_t^- of the backward SDE (2.28) is adapted to the future $s \geq t$. The solution of (2.54) is given by $Y_t = \psi_t(Z_t^+)$ and $V_t = \gamma^{1/2}\nabla_z\psi_t(Z_t^+)$ in agreement with (2.53) (compare Carmona 2016, page 42). See Appendix A.4 for the numerical treatment of (2.54).

A variational characterization of the smoothing path measure $\widehat{\mathbb{P}}$ is given by the Donsker–Varadhan principle

$$\widehat{\mathbb{P}} = \arg \inf_{\mathbb{P}\ll\mathbb{Q}} \{-\mathbb{P}[\log(l)] + \mathrm{KL}(\mathbb{P}||\mathbb{Q})\}, \tag{2.55}$$

that is, the distribution $\widehat{\mathbb{P}}$ is chosen such that the expected loss, $-\mathbb{P}[\log(l)]$, is minimized subject to the penalty introduced by the Kullback–Leibler divergence with respect to the original path measure $\mathbb{Q}$. Note that

$$\inf_{\mathbb{P}\ll\mathbb{Q}} \{\mathbb{P}[-\log(l)] + \mathrm{KL}(\mathbb{P}||\mathbb{Q})\} = -\log\beta \tag{2.56}$$

with $\beta = \mathbb{Q}[l]$. The connection between smoothing for SDEs and the Donsker–Varadhan principle has, for example, been discussed by Mitter and Newton (2003). See also Hartmann *et al.* (2017) for an in-depth discussion of variational formulations and their numerical implementation in the context of rare event simulations for which it is generally assumed that $\pi_0(z) = \delta(z - z_0)$ in (2.3), that is, the ensemble size is $M = 1$ when viewed within the context of this paper.

Remark 2.21. One can choose ψ_t differently from the choice made in (2.51) by changing the final condition for the backward Kolmogorov equation (2.31) to any suitable ψ_1. As already discussed for twisted discrete-time smoothing, such modifications give rise to alternative representations of the smoothing distribution $\widehat{\mathbb{P}}$ in terms of modified forward SDEs, likelihood functions and initial distributions. See Kappen and Ruiz (2016) and Ruiz and Kappen (2017) for an application of these ideas to importance sampling in the context of partially observed diffusion processes. More specifically, let u_t denote the associated control law (2.32) for the forward SDE (2.30) with given initial distribution $Z_0^+ \sim q_0$. Then

$$\frac{\mathrm{d}\widehat{\mathbb{P}}}{\mathrm{d}\mathbb{Q}}\Big|_{z_{[0,1]}^u} = \frac{\mathrm{d}\widehat{\mathbb{P}}}{\mathrm{d}\mathbb{Q}^u}\Big|_{z_{[0,1]}^u} \frac{\mathrm{d}\mathbb{Q}^u}{\mathrm{d}\mathbb{Q}}\Big|_{z_{[0,1]}^u},$$

which, using (2.33) and (2.49), implies the modified Radon–Nikodym derivative

$$\frac{\mathrm{d}\widehat{\mathbb{P}}}{\mathrm{d}\mathbb{Q}^u}\bigg|_{z^u_{[0,1]}} = \frac{l(z^u_1)}{\beta}\frac{\pi_0(z^u_0)}{q_0(z^u_0)}\exp\left(-\frac{1}{2\gamma}\int_0^1\left(\|u_t\|^2\,\mathrm{d}t + 2\gamma^{1/2}u_t\cdot\mathrm{d}W^+_t\right)\right). \tag{2.57}$$

Recall from Lemma 2.18 that the control law (2.32) with ψ_t defined by (2.51) together with $q_0 = \widehat{\pi}_0$ leads to $\widehat{\mathbb{P}} = \mathbb{Q}^u$.

2.3. Schrödinger problem

In this subsection we show that scenario (C) from Definition 2.4 leads to a certain boundary value problem first considered by Schrödinger (1931). More specifically, we state the so-called Schrödinger problem and show how it is linked to data assimilation scenario (C).

In order to introduce the Schrödinger problem, we return to the twisting potential approach as utilized in Section 2.2, with two important modifications. These modifications are, first, that the twisting potential ψ is determined implicitly and, second, that the modified transition kernel q^ψ_+ is applied to π_0 instead of the tilted initial density π^ψ_0 as in (2.48). More specifically, we have the following.

Definition 2.22. We seek the pair of functions $\widehat{\psi}(z_0)$ and $\psi(z_1)$ which solve the boundary value problem

$$\pi_0(z_0) = \pi^\psi_0(z_0)\,\widehat{\psi}(z_0), \tag{2.58}$$

$$\widehat{\pi}_1(z_1) = \pi^\psi_1(z_1)\,\psi(z_1), \tag{2.59}$$

$$\pi^\psi_1(z_1) = \int q_+(z_1|z_0)\,\pi^\psi_0(z_0)\,\mathrm{d}z_0, \tag{2.60}$$

$$\widehat{\psi}(z_0) = \int q_+(z_1|z_0)\,\psi(z_1)\,\mathrm{d}z_1, \tag{2.61}$$

for given marginal (filtering) distributions π_0 and $\widehat{\pi}_1$ at $t = 0$ and $t = 1$, respectively. The required modified PDFs π^ψ_0 and π^ψ_1 are defined by (2.58) and (2.59), respectively. The solution $(\widehat{\psi},\psi)$ of the Schrödinger system (2.58)–(2.61) leads to the modified transition kernel

$$q^*_+(z_1|z_0) := \psi(z_1)\,q_+(z_1|z_0)\,\widehat{\psi}(z_0)^{-1}, \tag{2.62}$$

which satisfies

$$\widehat{\pi}_1(z_1) = \int q^*_+(z_1|z_0)\,\pi_0(z_0)\,\mathrm{d}z_0$$

by construction.

The modified transition kernel $q^*_+(z_1|z_0)$ couples the two marginal distributions π_0 and $\widehat{\pi}_1$ with the twisting potential ψ implicitly defined. In other words, q_+^* provides the transition kernel for going from the initial distribution (2.3) at time t_0 to the filtering distribution at time t_1 without the need for any reweighting, *i.e.* the desired transition kernel for scenario (C). See Leonard (2014) and Chen *et al.* (2014) for more mathematical details on the Schrödinger problem.

Remark 2.23. Let us compare the Schrödinger system to the twisting potential approach (2.48) for the smoothing problem from Section 2.2 in some more detail. First, note that the twisting potential approach to smoothing replaces (2.58) with

$$\pi_0^\psi(z_0) = \pi_0(z_0)\,\widehat{\psi}(z_0)$$

and (2.59) with

$$\pi_1^\psi(z_1) = \pi_1(z_1)\,\psi(z_1),$$

where ψ is a given twisting potential normalized such that $\pi_1[\psi] = 1$. The associated $\widehat{\psi}$ is determined by (2.61) as in the twisting approach. In both cases, the modified transition kernel is given by (2.62). Finally, (2.60) is replaced by the prediction step (2.4).

In order to solve the Schrödinger system for our given initial distribution (2.3) and the associated filter distribution $\widehat{\pi}_1$, we make the *ansatz*

$$\pi_0^\psi(z_0) = \frac{1}{M}\sum_{i=1}^{M}\alpha^i\,\delta(z_0 - z_0^i),\qquad \sum_{i=1}^{M}\alpha^i = M.$$

This *ansatz* together with (2.58)–(2.61) immediately implies

$$\widehat{\psi}(z_0^i) = \frac{1}{\alpha^i},\qquad \pi_1^\psi(z_1) = \frac{1}{M}\sum_{i=1}^{M}\alpha^i\,q_+(z_1|z_0^i),$$

as well as

$$\psi(z_1) = \frac{\widehat{\pi}_1(z_1)}{\pi_1^\psi(z_1)} = \frac{l(z_1)}{\beta}\,\frac{\frac{1}{M}\sum_{j=1}^{M}q_+(z_1|z_0^j)}{\frac{1}{M}\sum_{j=1}^{M}\alpha^j\,q_+(z_1|z_0^j)}. \tag{2.63}$$

Hence we arrive at the equations

$$\widehat{\psi}(z_0^i) = \frac{1}{\alpha^i} \tag{2.64}$$

$$= \int \psi(z_1)\,q_+(z_1|z_0^i)\,\mathrm{d}z_1$$

$$= \int \frac{l(z_1)}{\beta}\,\frac{\sum_{j=1}^{M}q_+(z_1|z_0^j)}{\sum_{j=1}^{M}\alpha^j\,q_+(z_1|z_0^j)}\,q_+(z_1|z_0^i)\,\mathrm{d}z_1 \tag{2.65}$$

for $i = 1,\ldots,M$. These M equations have to be solved for the M unknown coefficients $\{\alpha^i\}$. In other words, the Schrödinger problem becomes finite-dimensional in the context of this paper. More specifically, we have the following result.

Lemma 2.24. The forward Schrödinger transition kernel (2.62) is given by

$$q_+^*(z_1|z_0^i) = \frac{\widehat{\pi}_1(z_1)}{\pi_1^\psi(z_1)} q_+(z_1|z_0^i)\,\alpha^i$$

$$= \frac{\alpha^i \sum_{j=1}^M q_+(z_1|z_0^j)}{\sum_{j=1}^M \alpha^j\, q_+(z_1|z_0^j)} \frac{l(z_1)}{\beta}\, q_+(z_1|z_0^i), \qquad (2.66)$$

for each particle z_0^i with the coefficients α^j, $j = 1, \ldots, M$, defined by (2.64)–(2.65).

Proof. Because of (2.64)–(2.65), the forward transition kernels (2.66) satisfy

$$\int q_+^*(z_1|z_0^i)\,\mathrm{d}z_1 = 1 \qquad (2.67)$$

for all $i = 1, \ldots, M$ and

$$\int q_+^*(z_1|z_0)\,\pi_0(z_0)\,\mathrm{d}z = \frac{1}{M} \sum_{i=1}^M q_+^*(z_1|z_0^i)$$

$$= \frac{1}{M} \sum_{i=1}^M \alpha^i q_+(z_1|z_0^i)\, \frac{\widehat{\pi}_1(z_1)}{\frac{1}{M} \sum_{j=1}^M \alpha^j\, q_+(z_1|z_0^j)}$$

$$= \widehat{\pi}_1(z_1), \qquad (2.68)$$

as desired. $\qquad\qquad\qquad\square$

Numerical implementations will be discussed in Section 3. Note that knowledge of the normalizing constant β is not required *a priori* for solving (2.64)–(2.65) since it appears as a common scaling factor.

We note that the coefficients $\{\alpha^j\}$ together with the associated potential ψ from the Schrödinger system provide the optimally twisted prediction kernel (1.2) with respect to the filtering distribution $\widehat{\pi}_1$, that is, we set $\widehat{\psi}(z_0^i) = 1/\alpha^i$ in (2.16) and define the potential ψ by (2.63).

Lemma 2.25. The Schrödinger transition kernel (2.62) satisfies the following constrained variational principle. Consider the joint PDFs given by $\pi(z_0, z_1) := q_+(z_1|z_0)\pi_0(z_0)$ and $\pi^*(z_0, z_1) := q_+^*(z_1|z_0)\pi_0(z_0)$. Then

$$\pi^* = \arg \inf_{\widetilde{\pi}\in\Pi_S} \mathrm{KL}(\widetilde{\pi}||\pi). \qquad (2.69)$$

Here a joint PDF $\widetilde{\pi}(z_0, z_1)$ is an element of Π_S if

$$\int \widetilde{\pi}(z_0, z_1)\,\mathrm{d}z_1 = \pi_0(z_0), \qquad \int \widetilde{\pi}(z_0, z_1)\,\mathrm{d}z_0 = \widehat{\pi}_1(z_1).$$

Proof. See Föllmer and Gantert (1997) for a proof and Remark 2.31 for a heuristic derivation in the case of discrete measures. $\qquad\square$

The constrained variational formulation (2.69) of Schrödinger's problem should be compared to the unconstrained Donsker–Varadhan variational principle

$$\widehat{\pi} = \arg\inf\{-\widetilde{\pi}[\log(l)] + \mathrm{KL}(\widetilde{\pi}||\pi)\} \tag{2.70}$$

for the associated smoothing problem. See Remark 2.27 below.

Remark 2.26. The Schrödinger problem is closely linked to optimal transportation (Cuturi 2013, Leonard 2014, Chen *et al.* 2014). For example, consider the Gaussian transition kernel (2.21) with $\Psi(z) = z$ and $B = I$. Then the solution (2.66) to the associated Schrödinger problem of coupling π_0 and $\widehat{\pi}_1$ reduces to the solution π^* of the associated optimal transport problem

$$\pi^* = \arg\inf_{\widetilde{\pi}\in\Pi_S} \int\int \|z_0 - z_1\|^2\,\widetilde{\pi}(z_0, z_1)\,\mathrm{d}z_0\,\mathrm{d}z_1$$

in the limit $\gamma \to 0$.

2.3.1. SDE models (cont.)

At the SDE level, Schrödinger's problem amounts to continuously bridging the given initial PDF π_0 with the PDF $\widehat{\pi}_1$ at final time using an appropriate modification of the stochastic process $Z_{[0,1]}^+ \sim \mathbb{Q}$ defined by the forward SDE (2.24) with initial distribution π_0 at $t = 0$. The desired modified stochastic process $\mathbb{P}^*$ is defined as the minimizer of

$$\mathcal{L}(\widetilde{\mathbb{P}}) := \mathrm{KL}(\widetilde{\mathbb{P}}||\mathbb{Q})$$

subject to the constraint that the marginal distributions $\widetilde{\pi}_t$ of $\widetilde{\mathbb{P}}$ at time $t = 0$ and $t = 1$ satisfy π_0 and $\widehat{\pi}_1$, respectively (Föllmer and Gantert 1997, Leonard 2014, Chen *et al.* 2014).

Remark 2.27. We note that the Donsker–Varadhan variational principle (2.55), characterizing the smoothing path measure $\widehat{\mathbb{P}}$, can be replaced by

$$\mathbb{P}^* = \arg\inf_{\widetilde{\mathbb{P}}\in\Pi} \{-\widetilde{\pi}_1[\log(l)] + \mathrm{KL}(\widetilde{\mathbb{P}}||\mathbb{Q})\}$$

with

$$\Pi = \{\widetilde{\mathbb{P}} \ll \mathbb{Q} : \widetilde{\pi}_1 = \widehat{\pi}_1,\ \widetilde{\pi}_0 = \pi_0\}$$

in the context of Schrödinger's problem. The associated

$$-\log \beta^* := \inf_{\widetilde{\mathbb{P}} \in \Pi} \left\{ -\widetilde{\pi}_1[\log(l)] + \mathrm{KL}(\widetilde{\mathbb{P}}\|\mathbb{Q}) \right\} = -\widehat{\pi}[\log(l)] + \mathrm{KL}(\mathbb{P}^*\|\mathbb{Q})$$

can be viewed as a generalization of (2.56) and gives rise to a generalized evidence β^*, which could be used for model comparison and parameter estimation.

The Schrödinger process $\mathbb{P}^*$ corresponds to a Markovian process across the whole time domain $[0, 1]$ (Leonard 2014, Chen *et al.* 2014). More specifically, consider the controlled forward SDE (2.30) with initial conditions

$$Z_0^+ \sim \pi_0$$

and a given control law u_t for $t \in [0, 1]$. Let $\mathbb{P}^u$ denote the path measure associated to this process. Then, as discussed in detail by Dai Pra (1991), one can find time-dependent potentials ψ_t with associated control laws (2.32) such that the marginal of the associated path measure $\mathbb{P}^u$ at times $t = 1$ satisfies

$$\pi_1^u = \widehat{\pi}_1$$

and, more generally,

$$\mathbb{P}^* = \mathbb{P}^u.$$

We summarize this result in the following lemma.

Lemma 2.28. The Schrödinger path measure $\mathbb{P}^*$ can be generated by a controlled SDE (2.30) with control law (2.32), where the desired potential ψ_t can be obtained as follows. Let $(\widehat{\psi}, \psi)$ denote the solution of the associated Schrödinger system (2.58)–(2.61), where $q_+(z_1|z_0)$ denotes the time-one forward transition kernel of (2.24). Then ψ_t in (2.32) is the solution of the backward Kolmogorov equation (2.31) with prescribed $\psi_1 = \psi$ at final time $t = 1$.

Remark 2.29. As already pointed out in the context of smoothing, the desired potential ψ_t can also be obtained by solving an appropriate backward SDE. More specifically, given the solution $(\widehat{\psi}, \psi)$ and the implied PDF $\widetilde{\pi}_0^+ := \pi_0^{\psi} = \pi_0/\widehat{\psi}$ of the Schrödinger system (2.58)–(2.61), let $\widetilde{\pi}_t^+$, $t \geq 0$, denote the marginals of the forward SDE (2.24) with $Z_0^+ \sim \widetilde{\pi}_0^+$. Furthermore, consider the backward SDE (2.28) with drift term

$$b_t(z) = f_t(z) - \gamma \nabla_z \log \widetilde{\pi}_t^+(z), \tag{2.71}$$

and final time condition $Z_1^- \sim \widehat{\pi}_1$. Then the choice of $\widetilde{\pi}_0^+$ ensures that $Z_0^- \sim \pi_0$. Furthermore the desired control in (2.30) is provided by

$$u_t = \gamma \nabla_z \log \frac{\widetilde{\pi}_t^-}{\widetilde{\pi}_t^+},$$

where $\widetilde{\pi}_t^-$ denotes the marginal distributions of the backward SDE (2.28) with drift term (2.71) and $\widetilde{\pi}_1^- = \widehat{\pi}_1$. We will return to this reformulation of the Schrödinger problem in Section 3 when considering it as the limit of a sequence of smoothing problems.

Remark 2.30. The solution to the Schrödinger problem for linear SDEs and Gaussian marginal distributions has been discussed in detail by Chen, Georgiou and Pavon (2016b).

2.3.2. Discrete measures

We finally discuss the Schrödinger problem in the context of finite-state Markov chains in more detail. These results will be needed in the following sections on the numerical implementation of the Schrödinger approach to sequential data assimilation.

Let us therefore consider an example which will be closely related to the discussion in Section 3. We are given a bi-stochastic matrix $Q \in \mathbb{R}^{L \times M}$ with all entries satisfying $q_{lj} > 0$ and two discrete probability measures represented by vectors $p_1 \in \mathbb{R}^L$ and $p_0 \in \mathbb{R}^M$, respectively. Again we assume for simplicity that all entries in p_1 and p_0 are strictly positive. We introduce the set of all bi-stochastic $L \times M$ matrices with those discrete probability measures as marginals, that is,

$$\Pi_{\mathrm{s}} := \{ P \in \mathbb{R}^{L \times M} : P \geq 0,\ P^{\mathrm{T}} \mathbb{1}_L = p_0,\ P \mathbb{1}_M = p_1 \}. \tag{2.72}$$

Solving Schrödinger's system (2.58)–(2.61) corresponds to finding two non-negative vectors $u \in \mathbb{R}^L$ and $v \in \mathbb{R}^M$ such that

$$P^* := D(u)\, Q\, D(v)^{-1} \in \Pi_{\mathrm{s}}.$$

In turns out that P^* is uniquely determined and minimizes the Kullback–Leibler divergence between all $P \in \Pi_{\mathrm{s}}$ and the reference matrix Q, that is,

$$P^* = \arg \min_{P \in \Pi_{\mathrm{s}}} \mathrm{KL}\,(P \| Q). \tag{2.73}$$

See Peyre and Cuturi (2018) and the following remark for more details.

Remark 2.31. If one makes the *ansatz*

$$p_{lj} = \frac{u_l q_{lj}}{v_j},$$

then the minimization problem (2.73) becomes equivalent to

$$P^* = \arg \min_{(u,v)>0} \sum_{l,j} p_{lj} (\log u_l - \log v_j)$$

subject to the additional constraints

$$P \mathbb{1}_M = D(u) Q D(v)^{-1} \mathbb{1}_M = p_1, \quad P^{\mathrm{T}} \mathbb{1}_L = D(v)^{-1} Q^{\mathrm{T}} D(u) \mathbb{1}_L = p_0.$$

Note that these constraint determine $u > 0$ and $v > 0$ up to a common scaling factor. Hence (2.73) can be reduced to finding $(u, v) > 0$ such that

$$u^{\mathrm{T}} \mathbb{1}_L = 1, \quad P \mathbb{1}_M = p_1, \quad P^{\mathrm{T}} \mathbb{1}_L = p_0.$$

Hence we have shown that solving the Schrödinger system is equivalent to solving the minimization problem (2.73) for discrete measures. Thus

$$\min_{P \in \Pi_{\mathrm{s}}} \mathrm{KL}\,(P\|Q) = p_1^{\mathrm{T}} \log u - p_0^{\mathrm{T}} \log v.$$

Lemma 2.32. The Sinkhorn iteration (Sinkhorn 1967)

$$u^{k+1} := D(P^{2k} \mathbb{1}_M)^{-1} p_1, \tag{2.74}$$

$$P^{2k+1} := D(u^{k+1}) P^{2k}, \tag{2.75}$$

$$v^{k+1} := D(p_0)^{-1} (P^{2k+1})^{\mathrm{T}} \mathbb{1}_L, \tag{2.76}$$

$$P^{2k+2} := P^{2k+1} D(v^{k+1})^{-1}, \tag{2.77}$$

$k = 0, 1, \ldots$, with initial $P^0 = Q \in \mathbb{R}^{L \times M}$ provides an algorithm for computing P^*, that is,

$$\lim_{k \to \infty} P^k = P^*. \tag{2.78}$$

Proof. See, for example, Peyre and Cuturi (2018) for a proof of (2.78), which is based on the contraction property of the iteration (2.74)–(2.77) with respect to the Hilbert metric on the projective cone of positive vectors. $\square$

It follows from (2.78) that

$$\lim_{k \to \infty} u^k = \mathbb{1}_L, \quad \lim_{k \to \infty} v^k = \mathbb{1}_M.$$

The essential idea of the Sinkhorn iteration is to enforce

$$P^{2k+1} \mathbb{1}_M = p_1, \quad (P^{2k})^{\mathrm{T}} \mathbb{1}_L = p_0$$

at each iteration step and that the matrix P^* satisfies both constraints simultaneously in the limit $k \to \infty$. See Cuturi (2013) for a computationally efficient and robust implementation of the Sinkhorn iteration.

Remark 2.33. One can introduce a similar iteration for the Schrödinger system (2.58)–(2.61). For example, pick $\widehat{\psi}(z_0) = 1$ initially. Then (2.58) implies $\pi_0^{\psi} = \pi_0$ and (2.60) $\pi_1^{\psi} = \pi_1$. Hence $\psi = l/\beta$ in the first iteration. The second iteration starts with $\widehat{\psi}$ determined by (2.61) with $\psi = l/\beta$. We again cycle through (2.58), (2.60) and (2.59) in order to find the next approximation to ψ. The third iteration takes now this ψ and computes the associated $\widehat{\psi}$ from (2.61) *etc.* A numerical implementation of this procedure requires the approximation of two integrals which essentially leads back to a Sinkhorn type algorithm in the weights of an appropriate quadrature rule.

3. Numerical methods

Having summarized the relevant mathematical foundation for prediction, filtering (data assimilation scenario (A)) and smoothing (scenario (B)) and the Schrödinger problem (scenario (C)), we now discuss numerical approximations suitable for ensemble-based data assimilation. It is clearly impossible to cover all available methods, and we will focus on a selection of approaches which are built around the idea of optimal transport, ensemble transform methods and Schrödinger systems. We will also focus on methods that can be applied or extended to problems with high-dimensional state spaces even though we will not explicitly cover this topic in this survey. See Reich and Cotter (2015), van Leeuwen (2015) and Asch *et al.* (2017) instead.

3.1. Prediction

Generating samples from the forecast distributions $q_+(\cdot\,|z_0^i)$ is in most cases straightforward. The computational expenses can, however, vary dramatically, and this impacts on the choice of algorithms for sequential data assimilation. We demonstrate in this subsection how samples from the prediction PDF π_1 can be used to construct an associated finite-state Markov chain that transforms π_0 into an empirical approximation of π_1.

Definition 3.1. Let us assume that we have $L \geq M$ independent samples z_1^l from the M forecast distributions $q_+(\cdot\,|z_0^j)$, $j = 1, \ldots, M$. We introduce the $L \times M$ matrix Q with entries

$$q_{lj} := q_+(z_1^l|z_0^j). \tag{3.1}$$

We now consider the associated bi-stochastic matrix $P^* \in \mathbb{R}^{L \times M}$, as defined by (2.73), with the two probability vectors in (2.72) given by $p_1 = \mathbb{1}_L/L \in \mathbb{R}^L$ and $p_0 = \mathbb{1}_M/M \in \mathbb{R}^M$, respectively. The finite-state Markov chain

$$Q_+ := MP^* \tag{3.2}$$

provides a sample-based approximation to the forward transition kernel $q_+(z_1|z_0)$.

More precisely, the ith column of Q_+ provides an empirical approximation to $q_+(\cdot\,|z_0^i)$ and

$$Q_+ p_0 = p_1 = \frac{1}{L}\mathbb{1}_L,$$

which is in agreement with the fact that the z_1^l are equally weighted samples from the forecast PDF π_1.

Remark 3.2. Because of the simple relation between a bi-stochastic matrix $P \in \Pi_s$ with p_0 in (2.72) given by $p_0 = \mathbb{1}_M/M$ and its associated finite-state Markov chain $Q_+ = MP$, one can reformulate the minimization problem (2.73) in those cases directly in terms of Markov chains $Q_+ \in \Pi_{\mathrm{M}}$ with the definition of Π_{s} adjusted to

$$\Pi_{\mathrm{M}} := \left\{ Q \in \mathbb{R}^{L \times M} : Q \geq 0, \ Q^{\mathrm{T}} \mathbb{1}_L = \mathbb{1}_M, \ \frac{1}{M} Q \mathbb{1}_M = p_1 \right\}. \tag{3.3}$$

Remark 3.3. The associated backward transition kernel $Q_- \in \mathbb{R}^{M \times L}$ satisfies

$$Q_- D(p_1) = (Q_+ D(p_0))^{\mathrm{T}}$$

and is hence given by

$$Q_- = (Q_+ D(p_0))^{\mathrm{T}} D(p_1)^{-1} = \frac{L}{M} Q_+^{\mathrm{T}}.$$

Thus

$$Q_- p_1 = D(p_0) Q_+^{\mathrm{T}} \mathbb{1}_L = D(p_0) \mathbb{1}_M = p_0,$$

as desired.

Definition 3.4. We can extend the concept of twisting to discrete Markov chains such as (3.2). A twisting potential ψ gives rise to a vector $u \in \mathbb{R}^L$ with normalized entries

$$u_l = \frac{\psi(z_1^l)}{\sum_{k=1}^{L} \psi(z_1^k)},$$

$l = 1, \ldots, L$. The twisted finite-state Markov kernel is now defined by

$$Q_+^{\psi} := D(u) Q_+ D(v)^{-1}, \quad v := (D(u) Q_+)^{\mathrm{T}} \mathbb{1}_L \in \mathbb{R}^M, \tag{3.4}$$

and thus $\mathbb{1}_L^{\mathrm{T}} Q_+^{\psi} = \mathbb{1}_M^{\mathrm{T}}$, as required for a Markov kernel. The twisted forecast probability is given by

$$p_1^{\psi} := Q_+^{\psi} p_0$$

with $p_0 = \mathbb{1}_M / M$. Furthermore, if we set $p_0 = v$ then $p_1^{\psi} = u$.

3.1.1. Gaussian model errors (cont.)

The proposal density is given by (2.21) and it is easy to produce $K > 1$ samples from each of the M proposals $q_+(\cdot \,|\, z_0^j)$. Hence we can make the total sample size $L = K M$ as large as desired. In order to produce M samples, $\widetilde{z}_1^j$, from a twisted finite-state Markov chain (3.4), we draw a single realization from each of the M associated discrete random variables $\widetilde{Z}_1^j$, $j = 1, \ldots, M$, with probabilities

$$\mathbb{P}[\widetilde{Z}_1^j(\omega) = z_1^l] = (Q_+^{\psi})_{lj}.$$

We will provide more details when discussing the Schrödinger problem in the context of Gaussian model errors in Section 3.4.1.

3.1.2. SDE models (cont.)

The Euler–Maruyama method (Kloeden and Platen 1992)

$$Z_{n+1}^{+} = Z_n^{+} + f_{t_n}(Z_n^{+})\,\Delta t + (\gamma\Delta t)^{1/2}\,\Xi_n, \quad \Xi_n \sim \mathrm{N}(0, I), \tag{3.5}$$

$n = 0, \ldots, N-1$, will be used for the numerical approximation of (2.24) with step-size $\Delta t := 1/N$, $t_n = n\Delta t$. In other words, we replace $Z_{t_n}^{+}$ with its numerical approximation Z_n^{+}. A numerical approximation (realization) of the whole solution path $z_{[0,1]}$ will be denoted by $z_{0:N} = Z_{0:N}^{+}(\omega)$ and can be computed recursively due to the Markov property of the Euler–Maruyama scheme. The marginal PDFs of Z_n^{+} are denoted by π_n.

For any finite number of time-steps N, we can define a joint PDF $\pi_{0:N}$ on $\mathcal{U}_N = \mathbb{R}^{N_z \times (N+1)}$ via

$$\pi_{0:N}(z_{0:N}) \propto \exp\left(-\frac{1}{2\Delta t}\sum_{n=0}^{N-1}\|\eta_n\|^2\right)\pi_0(z_0) \tag{3.6}$$

with

$$\eta_n := \gamma^{-1/2}(z_{n+1} - z_n - f_{t_n}(z_n)\Delta t) \tag{3.7}$$

and $\eta_n = \Delta t^{1/2}\Xi_n(\omega)$. Note that the joint PDF $\pi_{0:N}(z_{0:N})$ can also be expressed in terms of z_0 and $\eta_{0:N-1}$.

The numerical approximation of SDEs provides an example for which the increase in computational cost for producing $L > M$ samples from the PDF $\pi_{0:N}$ versus $L = M$ is non-trivial, in general.

We now extend Definition 3.1 to the case of temporally discretized SDEs in the form of (3.5).

Definition 3.5.   Let us assume that we have $L = M$ independent numerical solutions $z_{0:N}^{i}$ of (3.5). We introduce an $M \times M$ matrix Q_n for each $n = 1, \ldots, N$ with entries

$$q_{lj} = q_{+}(z_n^{l}|z_{n-1}^{j}) := \mathrm{n}(z_n^{l}; z_{n-1}^{j} + \Delta t f(z_{n-1}^{j}), \gamma\Delta t\, I).$$

With each Q_n we associate a finite-state Markov chain Q_n^{+} as defined by (3.2) for general transition densities q_{+} in Definition 3.1. An approximation of the Markov transition from time $t_0 = 0$ to $t_1 = 1$ is now provided by

$$Q_{+} := \prod_{n=1}^{N} Q_n^{+}. \tag{3.8}$$

Remark 3.6.   The approximation (3.2) can be related to the diffusion map approximation of the infinitesimal generator of Brownian dynamics

$$\mathrm{d}Z_t^{+} = -\nabla_z U(Z_t^{+})\,\mathrm{d}t + \sqrt{2}\,\mathrm{d}W_t^{+} \tag{3.9}$$

with potential $U(z) = -\log\pi^{*}(z)$ in the following sense. First note that

π^* is invariant under the associated Fokker–Planck equation (2.26) with (time-independent) operator $\mathcal{L}^\dagger$ written in the form

$$\mathcal{L}^\dagger \pi = \nabla_z \cdot \left(\pi^* \nabla_z \frac{\pi}{\pi^*} \right).$$

Let z^i, $i = 1, \ldots, M$, denote M samples from the invariant PDF π^* and define the symmetric matrix $Q \in \mathbb{R}^{M \times M}$ with entries

$$q_{lj} = \mathrm{n}(z^l; z^j, 2\Delta t\, I).$$

Then the associated (symmetric) matrix (3.2), as introduced in Definition 3.1, provides a discrete approximation to the evolution of a probability vector $p_0 \propto \pi_0/\pi^*$ over a time-interval Δt and, hence, to the semigroup operator $\mathrm{e}^{\Delta t \mathcal{L}}$ with the infinitesimal generator $\mathcal{L}$ given by

$$\mathcal{L}g = \frac{1}{\pi^*} \nabla_z \cdot (\pi^* \nabla_z g). \tag{3.10}$$

We formally obtain

$$\mathcal{L} \approx \frac{Q_+ - I}{\Delta t} \tag{3.11}$$

for Δt sufficiently small. The symmetry of Q_+ reflects the fact that $\mathcal{L}$ is self-adjoint with respect to the weighted inner product

$$\langle f, g \rangle_{\pi^*} = \int f(z)\, g(z)\, \pi^*(z)\, \mathrm{d}z.$$

See Harlim (2018) for a discussion of alternative diffusion map approximations to the infinitesimal generator $\mathcal{L}$ and Appendix A.1 for an application to the feedback particle filter formulation of continuous-time data assimilation.

We also consider the discretization

$$Z_{n+1}^+ = Z_n^+ + (f_{t_n}(Z_n^+) + u_{t_n}(Z_n^+))\Delta t + (\gamma \Delta t)^{1/2}\, \Xi_n, \tag{3.12}$$

$n = 0, \ldots, N-1$, of a controlled SDE (2.30) with associated PDF $\pi_{0:N}^u$ defined by

$$\pi_{0:N}^u(z_{0:N}^u) \propto \exp\left(-\frac{1}{2\Delta t} \sum_{n=0}^{N-1} \|\eta_n^u\|^2 \right) \pi_0(z_0), \tag{3.13}$$

where

$$\eta_n^u := \gamma^{-1/2}\{z_{n+1}^u - z_n^u - (f_{t_n}(z_n^u) + u_{t_n}(z_n^u))\Delta t\}$$

$$= \eta_n - \frac{\Delta t}{\gamma^{1/2}} u_{t_n}(z_n^u).$$

Here $z_{0:N}^u$ denotes a realization of the discretization (3.12) with control laws

u_{t_n}. We find that

$$\frac{1}{2\Delta t}\|\eta_n^u\|^2 = \frac{1}{2\Delta t}\|\eta_n\|^2 - \frac{1}{\gamma^{1/2}}u_{t_n}(z_n^u)^{\mathrm{T}}\eta_n + \frac{\Delta t}{2\gamma}\|u_{t_n}(z_n^u)\|^2$$
$$= \frac{1}{2\Delta t}\|\eta_n\|^2 - \frac{1}{\gamma^{1/2}}u_{t_n}(z_n^u)^{\mathrm{T}}\eta_n^u - \frac{\Delta t}{2\gamma}\|u_{t_n}(z_n^u)\|^2,$$

and hence

$$\frac{\pi_{0:N}^u(z_{0:N}^u)}{\pi_{0:N}(z_{0:N}^u)} = \exp\left(\frac{1}{2\gamma}\sum_{n=0}^{N-1}\left(\|u_{t_n}(z_n^u)\|^2\Delta t + 2\gamma^{1/2}u_{t_n}(z_n^u)^{\mathrm{T}}\eta_n^u\right)\right), \quad (3.14)$$

which provides a discrete version of (2.33) since $\eta_n^u = \Delta t^{1/2}\,\Xi_n(\omega)$ are increments of Brownian motion over time intervals of length Δt.

Remark 3.7. Instead of discretizing the forward SDE (2.24) in order to produce samples from the forecast PDF π_1, one can also start from the mean-field formulation (2.29) and its time discretization, for example,

$$z_{n+1}^i = z_n^i + (f_{t_n}(z_n^i) + u_{t_n}(z_n^i))\Delta t \qquad (3.15)$$

for $i = 1, \ldots, M$ and

$$u_{t_n}(z) = -\frac{\gamma}{2}\nabla_z \log \widetilde{\pi}_n(z).$$

Here $\widetilde{\pi}_n$ stands for an approximation to the marginal PDF π_{t_n} based on the available samples z_n^i, $i = 1, \ldots, M$. A simple approximation is obtained by the Gaussian PDF

$$\widetilde{\pi}_n(z) = \mathrm{n}(z; \bar{z}_n, P_n^{zz})$$

with empirical mean

$$\bar{z}_n = \frac{1}{M}\sum_{i=1}^{M} z_n^i$$

and empirical covariance matrix

$$P_n^{zz} = \frac{1}{M-1}\sum_{i=1}^{M} z_n^i(z_n^i - \bar{z}_n)^{\mathrm{T}}.$$

The system (3.15) becomes

$$z_{n+1}^i = z_n^i + (f_{t_n}(z_n^i) + \gamma(P_n^{zz})^{-1}(z_n^i - \bar{z}_n))\Delta t,$$

$i = 1, \ldots, M$, and provides an example of an interacting particle approximation. Similar mean-field formulations can be found for the backward SDE (2.28).

3.2. Filtering

Let us assume that we are given M samples, z_1^i, from the forecast PDF using forward transition kernels $q_+(\cdot \,|\, z_0^i)$, $i = 1, \ldots, M$. The likelihood function $l(z)$ leads to importance weights

$$w^i \propto l(z_1^i). \tag{3.16}$$

We also normalize these importance weights such that (2.19) holds.

Remark 3.8. The model evidence β can be estimated from the samples, z_1^i, and the likelihood $l(z)$ as follows:

$$\widetilde{\beta} := \frac{1}{M} \sum_{i=1}^{M} l(z_1^i).$$

If the likelihood is of the form

$$l(z) \propto \exp\left(-\frac{1}{2}(y_1 - h(z))^{\mathrm{T}} R^{-1}(y_1 - h(z)) \right)$$

and the prior distribution in $y = h(z)$ can be approximated as being Gaussian with covariance

$$P^{hh} := \frac{1}{M-1} \sum_{i=1}^{M} h(z_1^i)(h(z_1^i) - \bar{h})^{\mathrm{T}}, \quad \bar{h} := \frac{1}{M} \sum_{i=1}^{M} h(z_1^i),$$

then the evidence can be approximated by

$$\widetilde{\beta} \approx \frac{1}{(2\pi)^{N_y/2}|P^{yy}|^{1/2}} \exp\left(-\frac{1}{2}(y_1 - \bar{h})^{\mathrm{T}}(P^{yy})^{-1}(y_1 - \bar{h}) \right)$$

with

$$P^{yy} := R + P^{hh}.$$

Such an approximation has been used, for example, in Carrassi, Bocquet, Hannart and Ghil (2017). See also Reich and Cotter (2015) for more details on how to compute and use model evidence in the context of sequential data assimilation.

Sequential data assimilation requires that we produce M equally weighted samples $\widehat{z}_1^j \sim \widehat{\pi}_1$ from the M weighted samples $z_1^i \sim \pi_1$ with weights w^i. This is a standard problem in Monte Carlo integration and there are many ways to tackle this problem, among which are multinomial, residual, systematic and stratified resampling (Douc and Cappe 2005). Here we focus on those resampling methods which are based on a discrete Markov chain $P \in \mathbb{R}^{M \times M}$ with the property that

$$w = \frac{1}{M} P \mathbb{1}_M, \quad w = \left(\frac{w^1}{M}, \ldots, \frac{w^M}{M} \right)^{\mathrm{T}}. \tag{3.17}$$

The Markov property of P implies that $P^{\mathrm{T}}\mathbb{1}_M = \mathbb{1}_M$. We now consider the set of all Markov chains Π_{M}, as defined by (3.3), with $p_1 = w$. Any Markov chain $P \in \Pi_{\mathrm{M}}$ can now be used for resampling, but we seek the Markov chain $P^* \in \Pi_{\mathrm{M}}$ which minimizes the expected distance between the samples, that is,

$$P^* = \arg \min_{P \in \Pi_{\mathrm{M}}} \sum_{i,j=1}^{M} p_{ij}\|z_1^i - z_1^j\|^2. \tag{3.18}$$

Note that (3.18) is a special case of the optimal transport problem (2.36) with the involved probability measures being discrete measures. Resampling can now be performed according to

$$\mathbb{P}[\widehat{Z}_1^j(\omega) = z_1^i] = p_{ij}^* \tag{3.19}$$

for $j = 1, \ldots, M$.

Since it is known that (3.18) converges to (2.36) as $M \to \infty$ (McCann 1995) and since (2.36) leads to a transformation (2.38), the resampling step (3.19) has been replaced by

$$\widehat{z}_1^j = \sum_{i=1}^{M} z_1^i\, p_{ij}^* \tag{3.20}$$

in the so-called ensemble transform particle filter (ETPF) (Reich 2013, Reich and Cotter 2015). In other words, the ETPF replaces resampling with probabilities p_{ij}^* by its mean (3.20) for each $j = 1, \ldots, M$. The ETPF leads to a biased approximation to the resampling step which is consistent in the limit $M \to \infty$.

The general formulation (3.20) with the coefficients p_{ij}^* chosen appropriately[2] leads to a large class of so-called ensemble transform particle filters (Reich and Cotter 2015). Ensemble transform particle filters generally result in biased and inconsistent but robust estimates, which have found applications to high-dimensional state space models (Evensen 2006, Vetra-Carvalho et al. 2018) for which traditional particle filters fail due to the 'curse of dimensionality' (Bengtsson, Bickel and Li 2008). More specifically, the class of ensemble transform particle filters includes the popular ensemble Kalman filters (Evensen 2006, Reich and Cotter 2015, Vetra-Carvalho et al. 2018, Carrassi, Bocquet, Bertino and Evensen 2018) and so-called second-order accurate particle filters with coefficients p_{ij}^* in (3.20) chosen such that the weighted ensemble mean

$$\bar{z}_1 := \frac{1}{M} \sum_{i=1}^{M} w^i z_1^i$$

---

[2] The coefficients p_{ij}^* of an ensemble transform particle filter do not need to be nonnegative and only satisfy $\sum_{i=1}^{M} p_{ij}^* = 1$ (Acevedo, de Wiljes and Reich 2017).

and the weighted ensemble covariance matrix

$$\widetilde{P}^{zz} := \frac{1}{M} \sum_{i=1}^{M} w^i (z_1^i - \bar{z}_1)(z_1^i - \bar{z}_1)^{\mathrm{T}}$$

are exactly reproduced by the transformed and equally weighted particles $\widehat{z}_1^j$, $j = 1, \ldots, M$, defined by (3.20), that is,

$$\frac{1}{M} \sum_{j=1}^{M} \widehat{z}_1^j = \bar{z}_1, \quad \frac{1}{M-1} \sum_{j=1}^{M} (\widehat{z}_1^i - \bar{z}_1)(\widehat{z}_1^i - \bar{z}_1)^{\mathrm{T}} = \widetilde{P}^{zz}.$$

See the survey paper by Vetra-Carvalho *et al.* (2018) and the paper by Acevedo *et al.* (2017) for more details. A summary of the ensemble Kalman filter can be found in Appendix A.3.

In addition, hybrid methods (Frei and Künsch 2013, Chustagulprom, Reich and Reinhardt 2016), which bridge between classical particle filters and the ensemble Kalman filter, have recently been successfully applied to atmospheric fluid dynamics (Robert, Leuenberger and Künsch 2018).

Remark 3.9. Another approach for transforming samples, z_1^i, from the forecast PDF π_1 into samples, $\widehat{z}_1^i$, from the filtering PDF $\widehat{\pi}_1$ is provided through the mean-field interpretation

$$\frac{\mathrm{d}}{\mathrm{d}s} \breve{Z}_s = -\nabla_z \log \frac{\breve{\pi}_s(\breve{Z}_s)}{\widehat{\pi}_1(\breve{Z}_s)} \tag{3.21}$$

of the Fokker–Planck equation (2.26) for a random variable $\breve{Z}_s$ with law $\breve{\pi}_s$, drift term $f_s(z) = \nabla_z \log \widehat{\pi}_1$ and $\gamma = 2$, that is,

$$\partial_s \breve{\pi}_s = \nabla_z \cdot \left(\breve{\pi}_s \nabla_z \log \frac{\breve{\pi}_s}{\widehat{\pi}_1} \right)$$

in artificial time $s \geq 0$. It holds under fairly general assumptions that

$$\lim_{s \to \infty} \breve{\pi}_s = \widehat{\pi}_1$$

(Pavliotis 2014) and one can set $\breve{\pi}_0 = \pi_1$. The more common approach would be to solve Brownian dynamics

$$\mathrm{d}\breve{Z}_s = \nabla_z \log \widehat{\pi}_1(Z_s)\,\mathrm{d}t + \sqrt{2}\,\mathrm{d}W_s^+$$

for each sample z_1^i from π_1, *i.e.* $\breve{Z}_0(\omega) = z_1^i$, $i = 1, \ldots, M$, at initial time and

$$\widehat{z}_1^i = \lim_{s \to \infty} \breve{Z}_s(\omega).$$

In other words, formulation (3.21) replaces stochastic Brownian dynamics with a deterministic interacting particle system. See Appendix A.1 and Remark 4.3 for further details.

3.3. Smoothing

Recall that the joint smoothing distribution $\widehat{\pi}(z_0, z_1)$ can be represented in the form (2.47) with modified transition kernel (2.10) and smoothing distribution (2.8) at time t_0 with weights γ^i determined by (2.9).

Let us assume that it is possible to sample from $\widehat{q}_+(z_1|z_0^i)$ and that the weights γ^i are available. Then we can utilize (2.47) in sequential data assimilation as follows. We first resample the z_0^i at time t_0 using a discrete Markov chain $P \in \mathbb{R}^{M \times M}$ satisfying

$$\widehat{p}_0 = \frac{1}{M} P \, \mathbb{1}_M, \quad \widehat{p}_0 := \gamma, \tag{3.22}$$

with γ defined in (2.12). Again optimal transportation can be used to identify a suitable P. More explicitly, we now consider the set of all Markov chains Π_M, as defined by (3.3), with $p_1 = \gamma$. Then the Markov chain P^* arising from the associated optimal transport problem (3.18) can be used for resampling, that is,

$$\mathbb{P}[\widetilde{Z}_0^j(\omega) = z_0^i] = p_{ij}^*.$$

Once equally weighted samples $\widehat{z}_0^i$, $i = 1, \ldots, M$, from $\widehat{\pi}_0$ have been determined, the desired samples $\widehat{z}_1^i$ from $\widehat{\pi}_1$ are simply given by

$$\widehat{z}_1^i := \widehat{Z}_1^i(\omega), \quad \widehat{Z}_1^i \sim \widehat{q}_+(\cdot \,|\widehat{z}_0^i),$$

for $i = 1, \ldots, M$.

The required transition kernels (2.10) are explicitly available for state space models with Gaussian model errors and Gaussian likelihood functions. In many other cases, these kernels are not explicitly available or are difficult to draw from. In such cases, one can resort to sample-based transition kernels.

For example, consider the twisted discrete Markov kernel (3.4) with twisting potential $\psi(z) = l(z)$. The associated vector v from (3.4) then gives rise to a probability vector $\widehat{p}_0 = c\,v \in \mathbb{R}^M$ with $c > 0$ an appropriate scaling factor, and

$$\widehat{p}_1 := Q_+^{\psi} \widehat{p}_0 \tag{3.23}$$

approximates the filtering distribution at time t_1. The Markov transition matrix $Q_+^{\psi} \in \mathbb{R}^{L \times M}$ together with $\widehat{p}_0$ provides an approximation to the smoothing kernel $\widehat{q}_+(z_1|z_0)$ and $\widehat{\pi}_0$, respectively.

The approximations $Q_+^{\psi} \in \mathbb{R}^{L \times M}$ and $\widehat{p}_0 \in \mathbb{R}^M$ can be used to first generate equally weighted samples $\widehat{z}_0^i \in \{z_0^1, \ldots, z_0^M\}$ with distribution $\widehat{p}_0$ via, for example, resampling with replacement. If $\widehat{z}_0^i = z_0^k$ for an index $k = k(i) \in \{1, \ldots, M\}$, then

$$\mathbb{P}[\widehat{Z}_1^i(\omega) = z_1^l] = (Q_+^{\psi})_{lk}$$

for each $i = 1, \ldots, M$. The $\widehat{z}_1^i$ are equally weighted samples from the discrete filtering distribution $\widehat{p}_1$, which is an approximation to the continuous filtering PDF $\widehat{\pi}_1$.

Remark 3.10. One has to take computational complexity and robustness into account when deciding whether to utilize methods from Section 3.2 or this subsection to advance M samples z_0^i from the prior distribution π_0 into M samples $\widehat{z}_1^i$ from the posterior distribution $\widehat{\pi}_1$. While the methods from Section 3.2 are easier to implement, the methods of this subsection benefit from the fact that

$$M > \frac{1}{\|\gamma\|^2} \geq \frac{1}{\|w\|^2} \geq 1,$$

in general, where the importance weights $\gamma \in \mathbb{R}^M$ and $w \in \mathbb{R}^M$ are defined in (2.12) and (3.17), respectively. In other words, the methods from this subsection lead to larger effective sample sizes (Liu 2001, Agapiou *et al.* 2017).

Remark 3.11. We mention that finding efficient methods for solving the more general smoothing problem (2.2) is an active area of research. See, for example, the recent contributions by Guarniero *et al.* (2017) and Heng *et al.* (2018) for discrete-time Markov processes, and Kappen and Ruiz (2016) as well as Ruiz and Kappen (2017) for smoothing in the context of SDEs. Ensemble transform methods of the form (3.20) can also be extended to the general smoothing problem. See, for example, Evensen (2006) and Carrassi *et al.* (2018) for extensions of the ensemble Kalman filter, and Kirchgessner, Tödter, Ahrens and Nerger (2017) for an extension of the nonlinear ensemble transform filter to the smoothing problem.

3.3.1. SDE models (cont.)

After discretization in time, smoothing leads to a change from the forecast PDF (3.6) to

$$\widehat{\pi}_{0:N}(z_{0:N}) := \frac{l(z_N)\pi_{0:N}(z_{0:N})}{\pi_{0:N}[l]}$$

$$\propto \exp\left(-\frac{1}{2\Delta t}\sum_{n=0}^{N-1} \|\xi_n\|^2\right)\pi_0(z_0)\, l(z_N)$$

with ξ_n given by (3.7), or, alternatively,

$$\frac{\widehat{\pi}_{0:N}}{\pi_{0:N}}(z_{0:N}) = \frac{l(z_N)}{\pi_{0:N}[l]}.$$

Remark 3.12. Efficient MCMC methods for sampling high-dimensional smoothing distributions can be found in Beskos *et al.* (2017) and Beskos, Pinski, Sanz-Serna and Stuart (2011). Improved sampling can also be achieved by using regularized Störmer–Verlet time-stepping methods (Reich and Hundertmark 2011) in a hybrid Monte Carlo method (Liu 2001). See Appendix A.2 for more details.

3.4. Schrödinger problem

Recall that the Schrödinger system (2.58)–(2.61) reduces in our context to solving equations (2.64)–(2.65) for the unknown coefficients α^i, $i = 1, \ldots, M$. In order to make this problem tractable we need to replace the required expectation values with respect to $q_+(z_1|z_0^j)$ by Monte Carlo approximations. More specifically, let us assume that we have $L \geq M$ samples z_1^l from the forecast PDF π_1. The associated $L \times M$ matrix Q with entries (3.1) provides a discrete approximation to the underlying Markov process defined by $q_+(z_1|z_0)$ and initial PDF (2.3).

The importance weights in the associated approximation to the filtering distribution

$$\widehat{\pi}_1(z) = \frac{1}{L} \sum_{l=1}^{L} w^l \, \delta(z - z_1^l)$$

are given by (3.16) with the weights normalized such that

$$\sum_{l=1}^{L} w^l = L. \tag{3.24}$$

Finding the coefficients $\{\alpha^i\}$ in (2.64)–(2.65) can now be reformulated as finding two vectors $u \in \mathbb{R}^L$ and $v \in \mathbb{R}^M$ such that

$$P^* := D(u)QD(v)^{-1} \tag{3.25}$$

satisfies $P^* \in \Pi_M$ with $p_1 = w$ in (3.3), that is, more explicitly

$$\Pi_M = \left\{ P \in \mathbb{R}^{L \times M} : p_{lj} \geq 0, \ \sum_{l=1}^{L} p_{lj} = 1, \ \frac{1}{M} \sum_{j=1}^{M} p_{lj} = \frac{w^l}{L} \right\}. \tag{3.26}$$

We note that (3.26) are discrete approximations to (2.67) and (2.68), respectively. The scaling factor $\widehat{\psi}$ in (2.58) is approximated by the vector v up to a normalization constant, while the vector u provides an approximation to ψ in (2.59). Finally, the desired approximations to the Schrödinger transition kernels $q_+^*(z_1|z_0^i)$, $i = 1, \ldots, M$, are provided by the columns of P^*, that is,

$$\mathbb{P}[\widehat{Z}_1^i(\omega) = z_1^l] = p_{li}^*$$

characterizes the desired equally weighted samples $\widehat{z}_1^i$, $i = 1, \ldots, M$, from the filtering distribution $\widehat{\pi}_1$. See the following subsection for more details.

The required vectors u and v can be computed using the iterative Sinkhorn algorithm (2.74)–(2.76) (Cuturi 2013, Peyre and Cuturi 2018).

Remark 3.13. Note that one can replace the forward transition kernel $q_+(z_1|z_0)$ in (3.1) with any suitable twisted prediction kernel (1.2). This results in a modified matrix Q in (3.25) and weights w^l in 3.26. The resulting matrix (3.25) still provides an approximation to the Schrödinger problem.

Remark 3.14. The approximation (3.25) can be extended to an approximation of the Schrödinger forward transition kernels (2.66) in the following sense. We use $\alpha^i = 1/v_i$ in (2.66) and note that the resulting approximation satisfies (2.68) while (2.67) now longer holds exactly. However, since the entries u_l of the vector u appearing in (3.25) satisfy

$$u_l = \frac{w^l}{L} \frac{1}{\sum_{j=1}^{M} q_+(z_1^l|z_0^j)/v_j},$$

it follows that

$$\int q_+^*(z_1|z_0^i)\, dz_1 \approx \frac{1}{L} \sum_{l=1}^{L} \frac{l(z_1^l)}{\beta} \frac{q_+(z_1^l|z_0^i)/v_i}{\sum_{j=1}^{M} q_+(z_1^l|z_0^j)/v_j}$$

$$\approx \frac{1}{L} \sum_{l=1}^{L} w^l \frac{q_+(z_1^l|z_0^i)/v_i}{\sum_{j=1}^{M} q_+(z_1^l|z_0^j)/v_j} = \sum_{l=1}^{L} p_{li}^* = 1.$$

Furthermore, one can use such continuous approximations in combination with Monte Carlo sampling methods which do not require normalized target PDFs.

3.4.1. Gaussian model errors (cont.)

One can easily generate L, $L \geq M$, i.i.d. samples z_1^l from the forecast PDF (2.21), that is,

$$Z_1^l \sim \frac{1}{M} \sum_{j=1}^{M} \mathrm{n}(\,\cdot\,; \Psi(z_0^j), \gamma B),$$

and with the filtering distribution $\hat{\pi}_1$ characterized through the importance weights (3.16).

We define the distance matrix $D \in \mathbb{R}^{L \times M}$ with entries

$$d_{lj} := \frac{1}{2}\|z_1^l - \Psi(z_0^j)\|_B^2, \quad \|z\|_B^2 := z^{\mathrm{T}} B^{-1} z,$$

and the matrix $Q \in \mathbb{R}^{L \times M}$ with entries

$$q_{lj} := \mathrm{e}^{-d_{lj}/\gamma}.$$

The Markov chain $P^* \in \mathbb{R}^{L \times M}$ is now given by

$$P^* = \arg \min_{P \in \Pi_{\mathrm{M}}} \mathrm{KL}(P\|Q)$$

with the set Π_M defined by (3.26).

Once P^* has been computed, the desired Schrödinger transitions from π_0 to $\widehat{\pi}_1$ can be represented as follows. The Schrödinger transition kernels $q_+^*(z_1|z_0^i)$ are approximated for each z_0^i by

$$\tilde{q}_+^*(z_1|z_0^i) := \sum_{l=1}^{L} p_{li}^* \, \delta(z_1 - z_1^l), \quad i = 1, \ldots, M. \tag{3.27}$$

The empirical measure in (3.27) converges weakly to the desired $q_+^*(z_1|z_0^i)$ as $L \to \infty$ and

$$\widehat{\pi}_1(z_1) \approx \frac{1}{M} \sum_{i=1}^{M} \delta(z - \widehat{z}_1^i),$$

with

$$\widehat{z}_1^i = \widehat{Z}_1^i(\omega), \quad \widehat{Z}_1^i \sim \tilde{q}_+^*(\cdot \,|z_0^i), \tag{3.28}$$

provides the desired approximation of $\widehat{\pi}_1$ by M equally weighted particles $\widehat{z}_1^i$, $i = 1, \ldots, M$.

We remark that (3.27) has been used to produce the Schrödinger transition kernels for Example 2.5 and Figure 2.3(b) in particular. More specifically, we have $M = 11$ and used $L = 11\,000$. Since the particles $z_1^l \in \mathbb{R}$, $l = 1, \ldots, L$, are distributed according to the forecast PDF π_1, a function representation of $\tilde{q}_+^*(z_1|z_0^i)$ over all of $\mathbb{R}$ is provided by interpolating p_{li}^* onto $\mathbb{R}$ and multiplication of this interpolated function by $\pi_1(z)$.

For $\gamma \ll 1$, the measure in (3.27) can also be approximated by a Gaussian measure with mean

$$\bar{z}_1^i := \sum_{l=1}^{L} z_1^l p_{li}^*$$

and covariance matrix γB, that is, we replace (3.28) with

$$\widehat{Z}_1^i \sim \mathrm{N}(\bar{z}_1^i, \gamma B)$$

for $i = 1, \ldots, M$.

3.4.2. SDE (cont.)

One can also apply (3.25) in order to approximate the Schrödinger problem associated with SDE models. We typically use $L = M$ in this case and utilize (3.8) in place of Q in (3.25). The set Π_M is still given by (3.26).

Example 3.15. We consider scalar-valued motion of a Brownian particle in a bimodal potential, that is,

$$\mathrm{d}Z_t^+ = Z_t^+ \, \mathrm{d}t - (Z_t^+)^3 \, \mathrm{d}t + \gamma^{1/2} \, \mathrm{d}W_t^+ \tag{3.29}$$

with $\gamma = 0.5$ and initial distribution $Z_0 \sim \mathrm{N}(-1, 0.3)$. At time $t = 2$ we

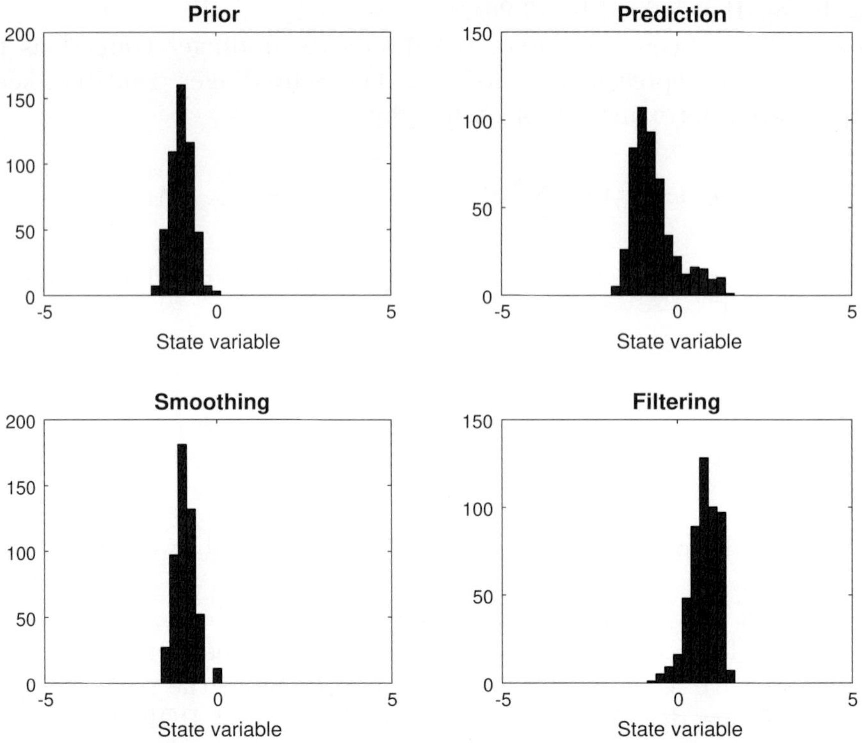

Figure 3.1. Histograms produced from $M = 200$ Monte Carlo samples of the initial PDF π_0, the forecast PDF π_2 at time $t = 2$, the filtering distribution $\widehat{\pi}_2$ at time $t = 2$, and the smoothing PDF $\widehat{\pi}_0$ at time $t = 0$ for a Brownian particle moving in a double well potential.

measure the location $y = 1$ with measurement error variance $R = 0.2$. We simulate the dynamics using $M = 200$ particles and a time-step of $\Delta t = 0.01$ in the Euler–Maruyama discretization (3.5). One can find histograms produced from the Monte Carlo samples in Figure 3.1. The samples from the filtering and smoothing distributions are obtained by resampling with replacement from the weighted distributions with weights given by (3.16). Next we compute (3.8) from the $M = 200$ Monte Carlo samples of (3.29). Eleven out of the 200 transition kernels from π_0 to π_2 (prediction problem) and π_0 to $\widehat{\pi}_2$ (Schrödinger problem) are displayed in Figure 3.2.

The Sinkhorn approach requires relatively large sample sizes M in order to lead to useful approximations. Alternatively we may assume that there is an approximative control term $u_t^{(0)}$ with associated forward SDE

$$\mathrm{d}Z_t^+ = f_t(Z_t)\,\mathrm{d}t + u_t^{(0)}(Z_t^+)\,\mathrm{d}t + \gamma^{1/2}\,\mathrm{d}W_t^+, \quad t \in [0, 1], \qquad (3.30)$$

and $Z_0^+ \sim \pi_0$. We denote the associated path measure by $\mathbb{Q}^{(0)}$. Girsanov's

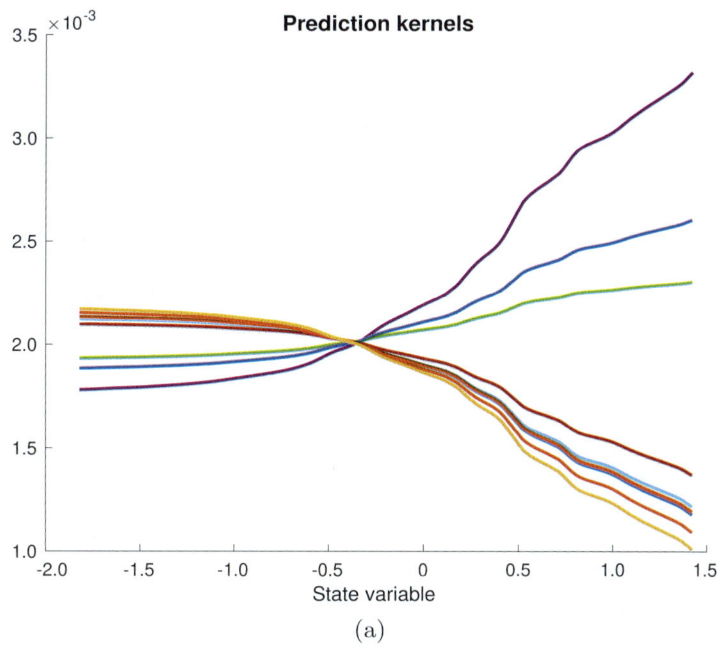

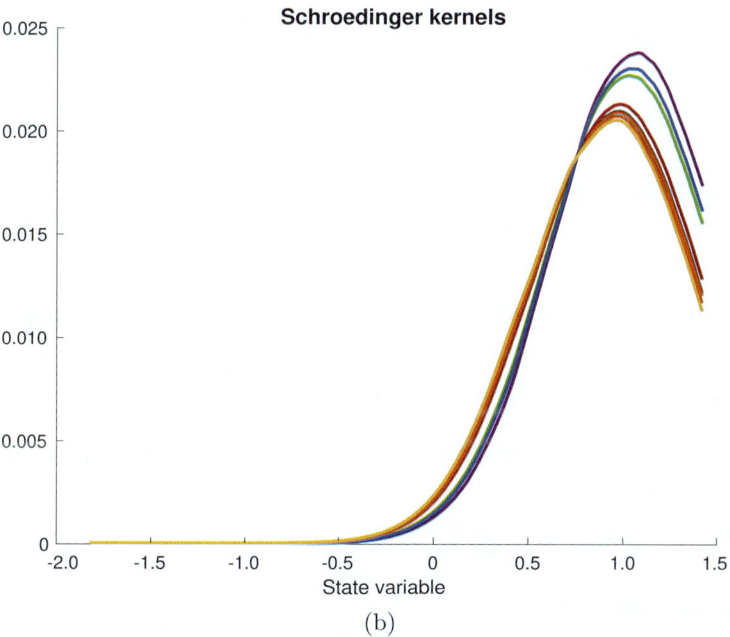

Figure 3.2. (a) Approximations of typical transition kernels from time π_0 to π_2 under the Brownian dynamics model (3.29). (b) Approximations of typical Schrödinger transition kernels from π_0 to $\widehat{\pi}_2$. All approximations were computed using the Sinkhorn algorithm and by linear interpolation between the $M = 200$ data points.

theorem implies that the Radon–Nikodym derivative of $\mathbb{Q}$ with respect to $\mathbb{Q}^{(0)}$ is given by (compare (2.33))

$$\frac{\mathrm{d}\mathbb{Q}}{\mathrm{d}\mathbb{Q}^{(0)}}\Big|_{z_{[0,1]}^{(0)}} = \exp(-V^{(0)}),$$

where $V^{(0)}$ is defined via the stochastic integral

$$V^{(0)} := \frac{1}{2\gamma} \int_0^1 (\|u_t^{(0)}\|^2 \, \mathrm{d}t + 2\gamma^{1/2} u_t^{(0)} \cdot \mathrm{d}W_t^+)$$

along solution paths $z_{[0,1]}^{(0)}$ of (3.30). Because of

$$\frac{\mathrm{d}\widehat{\mathbb{P}}}{\mathrm{d}\mathbb{Q}^{(0)}}\Big|_{z_{[0,1]}^{(0)}} = \frac{\mathrm{d}\widehat{\mathbb{P}}}{\mathrm{d}\mathbb{Q}}\Big|_{z_{[0,1]}^{(0)}} \frac{\mathrm{d}\mathbb{Q}}{\mathrm{d}\mathbb{Q}^{(0)}}\Big|_{z_{[0,1]}^{(0)}} \propto l(z_1^{(0)}) \exp(-V^{(0)}),$$

we can now use (3.30) to importance-sample from the filtering PDF $\widehat{\pi}_1$. The control $u_t^{(0)}$ should be chosen such that the variance in the modified likelihood function

$$l^{(0)}(z_{[0,1]}^{(0)}) := l(z_1^{(0)}) \exp(-V^{(0)}) \tag{3.31}$$

is reduced compared to the uncontrolled case $u_t^{(0)} \equiv 0$. In particular, the filter distribution $\widehat{\pi}_1$ at time $t = 1$ satisfies

$$\widehat{\pi}_1(z_1^{(0)}) \propto l^{(0)}(z_{[0,1]}^{(0)}) \, \pi_1^{(0)}(z_1^{(0)}),$$

where $\pi_t^{(0)}$, $t \in (0,1]$, denote the marginal PDFs generated by (3.30).

We now describe an iterative algorithm for the associated Schrödinger problem in the spirit of the Sinkhorn iteration from Section 2.3.2.

Lemma 3.16. The desired optimal control law can be computed iteratively,

$$u_t^{(k+1)} = u_t^{(k)} + \gamma \nabla_z \log \psi_t^{(k)}, \quad k = 0, 1, \ldots, \tag{3.32}$$

for given $u_t^{(k)}$ and the potential $\psi_t^{(k)}$ obtained as the solutions to the backward Kolmogorov equation

$$\partial_t \psi_t^{(k)} = -\mathcal{L}_t^{(k)} \psi_t^{(k)}, \quad \mathcal{L}_t^{(k)} g := \nabla_z g \cdot (f_t + u_t^{(k)}) + \frac{\gamma}{2}\Delta_z g, \tag{3.33}$$

with final time condition

$$\psi_1^{(k)}(z) := \frac{\widehat{\pi}_1(z)}{\pi_1^{(k)}(z)}. \tag{3.34}$$

Here $\pi_1^{(k)}$ denotes the time-one marginal of the path measure $\mathbb{Q}^{(k)}$ induced by (2.30) with control term $u_t = u_t^{(k)}$ and initial PDF π_0. The recursion

(3.32) is stopped whenever the final time condition (3.34) is sufficiently close to a constant function.

Proof. The extension of the Sinkhorn algorithm to continuous PDFs and its convergence has been discussed by Chen, Georgiou and Pavon (2016a).
$\qquad\square$

Remark 3.17. Note that $\psi_t^{(k)}$ needs to be determined up to a constant of proportionality only since the associated control law is determined from $\psi_t^{(k)}$ by (3.32). One can also replace (3.33) with any other method for solving the smoothing problem associated to the SDE (2.30) with $Z_0^+ \sim \pi_0$, control law $u_t = u_t^{(k)}$, and likelihood function $l(z) = \psi_1^{(k)}(z)$. See Appendix A.4 for a forward–backward SDE formulation in particular.

We need to restrict the class of possible control laws $u_t^{(k)}$ in order to obtain a computationally feasible implementations in practice. For example, a simple class of control laws is provided by linear controls of the form

$$u_t^{(k)}(z) = -B_t^{(k)}(z - m_t^{(k)})$$

with appropriately chosen symmetric positive definite matrices $B_t^{(k)}$ and vectors $m_t^{(k)}$. Such approximations can, for example, be obtained from the smoother extensions of ensemble transform methods mentioned earlier. See also the recent work by Kappen and Ruiz (2016) and Ruiz and Kappen (2017) on numerical methods for the SDE smoothing problem.

4. DA for continuous-time data

In this section we focus on the continuous-time filtering problem over the time interval $[0, 1]$, that is, on the assimilation of data that arrive continuously in time. If one is only interested in transforming samples from the prior distribution at $t = 0$ into samples of the filtering distribution at time $t = 1$, then all methods from the previous sections can be applied once the associated filtering distribution $\widehat{\pi}_1$ is available. However, it is more natural to consider the associated filtering distributions $\widehat{\pi}_t$ for all $t \in (0, 1]$ and to derive appropriate transformations in the form of mean-field equations in continuous time. We distinguish between smooth and non-smooth data y_t, $t \in [0, 1]$.

4.1. Smooth data

We start from a forward SDE model (2.24) with associated path measure $\mathbb{Q}$ over the space of continuous functions $\mathcal{C}$. However, contrary to the previous sections, the likelihood l is defined along a whole solution path $z_{[0,1]}$ as follows:

$$\frac{\mathrm{d}\widehat{\mathbb{P}}}{\mathrm{d}\mathbb{Q}}\bigg|_{z_{[0,1]}} \propto l(z_{[0,1]}), \quad l(z_{[0,1]}) := \exp\left(-\int_0^1 V_t(z_t)\,\mathrm{d}t\right)$$

with the assumption that $\mathbb{Q}[l] < \infty$ and $V_t(z) \geq 0$. A specific example of a suitable V_t is provided by

$$V_t(z) = \frac{1}{2}\|h(z) - y_t\|^2, \tag{4.1}$$

where the data function $y_t \in \mathbb{R}$, $t \in [0, 1]$, is a smooth function of time and $h(z)$ is a forward operator connecting the model states to the observations/data. The associated estimation problem has, for example, been addressed by Mortensen (1968) and Fleming (1997) from an optimal control perspective and has led to what is called the minimum energy estimator. Recall that the filtering PDF $\widehat{\pi}_1$ is the marginal PDF of $\widehat{\mathbb{P}}$ at time $t = 1$.

The associated time-continuous smoothing/filtering problems are based on the time-dependent path measures $\widehat{\mathbb{P}}_t$ defined by

$$\frac{\mathrm{d}\widehat{\mathbb{P}}_t}{\mathrm{d}\mathbb{Q}}\bigg|_{z_{[0,1]}} \propto l(z_{[0,t]}), \quad l(z_{[0,t]}) := \exp\left(-\int_0^t V_s(z_s)\,\mathrm{d}s\right)$$

for $t \in (0, 1]$. We let $\widehat{\pi}_t$ denote the marginal PDF of $\widehat{\mathbb{P}}_t$ at time t. Note that $\widehat{\pi}_t$ is the filtering PDF, *i.e.* the marginal PDF at time t conditioned on all the data available until time t. Also note that $\widehat{\pi}_t$ is different from the marginal (smoothing) PDF of $\widehat{\mathbb{P}}$ at time t.

We now state a modified Fokker–Planck equation which describes the time evolution of the filtering PDFs $\widehat{\pi}_t$.

Lemma 4.1. The marginal distributions $\widehat{\pi}_t$ of $\widehat{\mathbb{P}}_t$ satisfy the modified Fokker–Planck equation

$$\partial_t \widehat{\pi}_t = \mathcal{L}_t^\dagger \widehat{\pi}_t - \widehat{\pi}_t(V_t - \widehat{\pi}_t[V_t]) \tag{4.2}$$

with $\mathcal{L}_t^\dagger$ defined by (2.25).

Proof. This can be seen by setting $\gamma = 0$ and $f_t \equiv 0$ in (2.24) for simplicity and by considering the incremental change of measure induced by the likelihood, that is,

$$\frac{\widehat{\pi}_{t+\delta t}}{\widehat{\pi}_t} \propto \mathrm{e}^{-V_t \delta t} \approx 1 - V_t\,\delta t,$$

and taking the limit $\delta t \to 0$ under the constraint that $\widehat{\pi}_t[1] = 1$ is preserved. $\square$

We now derive a mean-field interpretation of (4.2) and rewrite (4.2) in the form

$$\partial_t \widehat{\pi}_t = \mathcal{L}_t^\dagger \widehat{\pi}_t + \nabla_z \cdot (\widehat{\pi}_t \nabla_z \phi_t), \tag{4.3}$$

where the potential $\phi_t : \mathbb{R}^{N_z} \to \mathbb{R}$ satisfies the elliptic PDE

$$\nabla_z \cdot (\widehat{\pi}_t \nabla_z \phi_t) = -\widehat{\pi}_t (V_t - \widehat{\pi}_t[V_t]). \tag{4.4}$$

Remark 4.2. Necessary conditions for the elliptic PDE (4.4) to be solvable and to lead to bounded gradients $\nabla_z \phi$ for given π_t have been discussed by Laugesen, Mehta, Meyn and Raginsky (2015). It is an open problem to demonstrate that continuous-time data assimilation problems actually satisfy such conditions.

With (4.3) in place, we formally obtain the mean-field equation

$$\mathrm{d}Z_t^+ = \{f_t(Z_t^+) - \nabla_z \phi_t(Z_t^+)\}\,\mathrm{d}t + \gamma^{1/2}\,\mathrm{d}W_t^+, \tag{4.5}$$

and the marginal distributions π_t^u of this controlled SDE agree with the marginals $\widehat{\pi}_t$ of the path measures $\widehat{\mathbb{P}}_t$ at times $t \in (0, 1]$.

The control u_t is not uniquely determined. For example, one can replace (4.4) with

$$\nabla_z \cdot (\pi_t M_t \nabla_z \phi_t) = -\pi_t (V_t - \pi_t[V_t]), \tag{4.6}$$

where M_t is a symmetric positive definite matrix. More specifically, let us assume that π_t is Gaussian with mean $\bar{z}_t$ and covariance matrix P_t^{zz} and that $h(z)$ is linear, *i.e.* $h(z) = Hz$. Then (4.6) can be solved analytically for $M_t = P_t^{zz}$ with

$$\nabla_z \phi_t(z) = \frac{1}{2} H^{\mathrm{T}} (Hz + H\bar{z}_t - 2y_t).$$

The resulting mean-field equation becomes

$$\mathrm{d}Z_t^+ = \left\{ f_t(Z_t^+) - \frac{1}{2} P_t^{zz} H^{\mathrm{T}} (HZ_t^+ + H\bar{z}_t - 2y_t) \right\} \mathrm{d}t + \gamma^{1/2}\,\mathrm{d}W_t^+, \tag{4.7}$$

which gives rise to the ensemble Kalman–Bucy filter upon Monte Carlo discretization (Bergemann and Reich 2012). See Section 5 below and Appendix A.3 for further details.

Remark 4.3. The approach described in this subsection can also be applied to standard Bayesian inference without model dynamics. More specifically, let us assume that we have samples z_0^i, $i = 1, \ldots, M$, from a prior distribution π_0 which we would like to transform into samples from a posterior distribution

$$\pi^*(z) := \frac{l(z)\,\pi_0(z)}{\pi_0[l]}$$

with likelihood $l(z) = \pi(y|z)$. One can introduce a homotopy connecting π_0 with π^*, for example, via

$$\breve{\pi}_s(z) := \frac{l(z)^s\,\pi_0(z)}{\pi_0[l^s]} \tag{4.8}$$

with $s \in [0, 1]$. We find that

$$\frac{\partial \breve{\pi}_s}{\partial s} = \breve{\pi}_s (\log l - \breve{\pi}_s[\log l]). \tag{4.9}$$

We now seek a differential equation

$$\frac{\mathrm{d}}{\mathrm{d}s} \breve{Z}_s = u_s(\breve{Z}_s) \tag{4.10}$$

with $\breve{Z}_0 \sim \pi_0$ such that its marginal distributions $\breve{\pi}_s$ satisfy (4.9) and, in particular, $\breve{Z}_1 \sim \pi^*$. This condition together with Liouville's equation for the time evolution of marginal densities under a differential equation (4.10) leads to

$$-\nabla_z \cdot (\breve{\pi}_s \, u_s) = \breve{\pi}_s (\log l - \breve{\pi}_s[\log l]). \tag{4.11}$$

In order to define u_s in (4.10) uniquely, we make the *ansatz*

$$u_s(z) = -\nabla_z \phi_s(z) \tag{4.12}$$

which leads to the elliptic PDE

$$\nabla_z \cdot (\breve{\pi}_s \, \nabla_z \phi_s) = \breve{\pi}_s (\log l - \breve{\pi}_s[\log l]) \tag{4.13}$$

in the potential ϕ_s. The desired samples from π^* are now obtained as the time-one solutions of (4.10) with 'control law' (4.12) satisfying (4.13) and initial conditions z_0^i, $i = 1, \ldots, M$. There are many modifications of this basic procedure (Daum and Huang 2011, Reich 2011, El Moselhy and Marzouk 2012), some of them leading to explicit expressions for (4.10) such as Gaussian PDFs (Bergemann and Reich 2010) and Gaussian mixture PDFs (Reich 2012). We finally mention that the limit $s \to \infty$ in (4.8) leads, formally, to the PDF $\breve{\pi}_\infty = \delta(z - z_{\mathrm{ML}})$, where z_{ML} denotes the minimizer of $V(z) = -\log \pi(y|z)$, *i.e.* the maximum likelihood estimator, which we assume here to be unique, for example, V is convex. In other words these homotopy methods can be used to solve optimization problems via derivative-free mean-field equations and their interacting particle approximations. See, for example, Zhang, Taghvaei and Mehta (2019) and Schillings and Stuart (2017) as well as Appendices A.1 and A.3 for more details.

4.2. Random data

We now replace (4.1) with an observation model of the form

$$\mathrm{d}Y_t = h(Z_t^+) \, \mathrm{d}t + \mathrm{d}V_t^+,$$

where we set $Y_t \in \mathbb{R}$ for simplicity and V_t^+ denotes standard Brownian motion. The forward operator $h : \mathbb{R}^{N_z} \to \mathbb{R}$ is also assumed to be known. The marginal PDFs $\widehat{\pi}_t$ for Z_t conditioned on all observations y_s with $s \in$

$[0, t]$ satisfy the Kushner–Stratonovitch equation (Jazwinski 1970)

$$\mathrm{d}\widehat{\pi}_t = \mathcal{L}_t^\dagger \widehat{\pi}_t \, \mathrm{d}t + (h - \widehat{\pi}_t[h])(\mathrm{d}Y_t - \widehat{\pi}_t[h] \, \mathrm{d}t) \tag{4.14}$$

with $\mathcal{L}^\dagger$ defined by (2.25). The following observation is important for the subsequent discussion.

Remark 4.4. Consider state-dependent diffusion

$$\mathrm{d}Z_t^+ = \gamma_t(Z_t^+) \circ \mathrm{d}U_t^+, \tag{4.15}$$

in its Stratonovitch interpretation (Pavliotis 2014), where U_t^+ is scalar-valued Brownian motion and $\gamma_t(z) \in \mathbb{R}^{N_z \times 1}$. Here the Stratonovitch interpretation is to be applied to the implicit time-dependence of $\gamma_t(z)$ through Z_t^+ only, that is, the explicit time-dependence of γ_t remains to be Itô-interpreted. The associated Fokker–Planck equation for the marginal PDFs π_t takes the form

$$\partial_t \pi_t = \frac{1}{2} \nabla_z \cdot (\gamma_t \nabla_z \cdot (\pi_t \gamma_t)), \tag{4.16}$$

and expectation values $\bar{g} = \pi_t[g]$ evolve in time according to

$$\pi_t[g] = \pi_0[g] + \int_0^t \pi_s[\mathcal{A}_t g] \, \mathrm{d}s \tag{4.17}$$

with operator $\mathcal{A}_t$ defined by

$$\mathcal{A}_t g = \frac{1}{2} \gamma_t^{\mathrm{T}} \nabla_z (\gamma_t^{\mathrm{T}} \nabla_z g).$$

Now consider the mean-field equation

$$\frac{\mathrm{d}}{\mathrm{d}t} \widetilde{Z}_t = -\frac{1}{2} \gamma_t(\widetilde{Z}_t) \, J_t, \quad J_t := \widetilde{\pi}_t^{-1} \nabla_z \cdot (\widetilde{\pi}_t \gamma_t), \tag{4.18}$$

with $\widetilde{\pi}_t$ the law of $\widetilde{Z}_t$. The associated Liouville equation is

$$\partial_t \widetilde{\pi}_t = \frac{1}{2} \nabla_z \cdot (\widetilde{\pi}_t \gamma_t J_t) = \frac{1}{2} \nabla_z \cdot (\gamma_t \nabla_z \cdot (\gamma_t \widetilde{\pi}_t)).$$

In other words, the marginal PDFs and the associated expectation values evolve identically under (4.15) and (4.18), respectively.

We now state a formulation of the continuous-time filtering problem in terms of appropriate mean-field equations. These equations follow the framework of the feedback particle filter (FPF) as first introduced by Yang *et al.* (2013) and theoretically justified by Laugesen *et al.* (2015). See Crisan and Xiong (2010) and Xiong (2011) for an alternative formulation.

Lemma 4.5. The mean-field SDE

$$\mathrm{d}Z_t^+ = f_t(Z_t^+) \, \mathrm{d}t + \gamma^{1/2} \, \mathrm{d}W_t^+ - K_t(Z_t^+) \circ \mathrm{d}I_t \tag{4.19}$$

with

$$\mathrm{d}I_t := h(Z_t^+)\,\mathrm{d}t - \mathrm{d}Y_t + \mathrm{d}U_t^+,$$

U_t^+ standard Brownian motion, and $K_t := \nabla_z \phi_t$, where the potential ϕ_t satisfies the elliptic PDE

$$\nabla_z \cdot (\pi_t \nabla_z \phi_t) = -\pi_t(h - \pi_t[h]), \tag{4.20}$$

leads to the same evolution of its conditional marginal distributions π_t as (4.14).

Proof. We set $\gamma = 0$ and $f_t \equiv 0$ in (4.19) for simplicity. Then, following (4.16) with $\gamma_t = K_t$, the Fokker–Planck equation for the marginal distributions π_t of (4.19) conditioned on $\{Y_s\}_{s \in [0,t]}$ is given by

$$\begin{aligned}
\mathrm{d}\pi_t &= \nabla_z \cdot (\pi_t K_t(h(z)\,\mathrm{d}t - \mathrm{d}Y_t)) + \nabla_z \cdot (K_t \nabla_z \cdot (\pi_t K_t))\,\mathrm{d}t \tag{4.21}\\
&= (\pi_t[h]\,\mathrm{d}t - \mathrm{d}Y_t)\,\nabla_z \cdot (\pi_t K_t) + \nabla_z \cdot (\pi_t K_t(h(z) - \pi_t[h]))\,\mathrm{d}t \\
&\quad + \nabla_z \cdot (K_t \nabla_z \cdot (\pi_t K_t))\,\mathrm{d}t \tag{4.22}\\
&= \pi_t(h - \pi_t[h])(\mathrm{d}Y_t - \pi_t[h]\,\mathrm{d}t), \tag{4.23}
\end{aligned}$$

as desired, where we have used (4.20) twice to get from (4.22) to (4.23). Also note that both Y_t and U_t^+ contributed to the diffusion-induced final term in (4.21) and hence the factor $1/2$ in (4.16) is replaced by one. $\qquad\square$

Remark 4.6. Using the reformulation (4.18) of (4.15) in Stratonovitch form with $\gamma_t = K_t$ together with (4.20), one can replace $K_t \circ \mathrm{d}U_t^+$ with $\frac{1}{2}K_t(\pi_t[h] - h)\,\mathrm{d}t$, which leads to the alternative

$$\mathrm{d}I_t = \frac{1}{2}(h + \pi_t[h])\,\mathrm{d}t - \mathrm{d}Y_t$$

for the innovation I_t, as originally proposed by Yang *et al.* (2013) in their FPF formulation. We also note that the feedback particle formulation (4.19) can be extended to systems for which the measurement and model errors are correlated. See Nüsken, Reich and Rozdeba (2019) for more details.

The ensemble Kalman–Bucy filter (Bergemann and Reich 2012) with the Kalman gain factor K_t being independent of the state variable z and of the form

$$K_t = P_t^{zh} \tag{4.24}$$

can be viewed as a special case of an FPF. Here P_t^{zh} denotes the covariance matrix between Z_t and $h(Z_t)$ at time t.

5. Numerical methods

In this section we discuss some numerical implementations of the mean-field approach to continuous-time data assimilation. An introduction to standard particle filter implementations can, for example, be found in Bain and Crisan (2008). We start with the continuous-time formulation of the ensemble Kalman filter and state a numerical implementation of the FPF using a Schrödinger formulation in the second part of this section. See also Appendix A.1 for some more details on a particle-based solution of the elliptic PDEs (4.4), (4.13) and (4.20), respectively.

5.1. Ensemble Kalman–Bucy filter

Let us start with the ensemble Kalman–Bucy filter (EnKBF), which arises naturally from the mean-field equations (4.7) and (4.19), respectively, with Kalman gain (4.24) (Bergemann and Reich 2012). We state the EnKBF here in the form

$$\mathrm{d}Z_t^i = f_t(Z_t^i)\,\mathrm{d}t + \gamma^{1/2}\,\mathrm{d}W_t^+ - K_t^M\,\mathrm{d}I_t^i \tag{5.1}$$

for $i = 1, \dots, M$ and

$$K_t^M := \frac{1}{M-1}\sum_{i=1}^M Z_t^i(h(Z_t^i) - \bar{h}_t^M)^{\mathrm{T}}, \quad \bar{h}_t^M := \frac{1}{M}\sum_{i=1}^M h(Z_t^i).$$

The innovations $\mathrm{d}I_t^i$ take different forms depending on whether the data are smooth in time, that is,

$$\mathrm{d}I_t^i = \frac{1}{2}(h(Z_t^i) + \bar{h}_t^M - 2y_t)\,\mathrm{d}t,$$

or contains stochastic contributions, that is,

$$\mathrm{d}I_t^i = \frac{1}{2}(h(Z_t^i) + \bar{h}_t^M)\,\mathrm{d}t - \mathrm{d}y_t, \tag{5.2}$$

or, alternatively,

$$\mathrm{d}I_t^i = h(Z_t^i)\,\mathrm{d}t + \mathrm{d}U_t^i - \mathrm{d}y_t,$$

where U_t^i denotes standard Brownian motion. The SDEs (5.1) can be discretized in time by any suitable time-stepping method such as the Euler–Maruyama scheme (Kloeden and Platen 1992). However, one has to be careful with the choice of the step-size Δt due to potentially stiff contributions from $K_t^M\,\mathrm{d}I_t^i$. See, for example, Amezcua, Kalnay, Ide and Reich (2014) and Blömker, Schillings and Wacker (2018).

Remark 5.1. It is of broad interest to study the stability and accuracy of interacting particle filter algorithms such as the discrete-time EnKF and the continuous-time EnKBF for fixed particle numbers M. On the negative side, it has been shown by Kelly, Majda and Tong (2015) that such algorithms can undergo finite-time instabilities, while it has also been demonstrated (González-Tokman and Hunt 2013, Kelly, Law and Stuart 2014, Tong, Majda and Kelly 2016, de Wiljes, Reich and Stannat 2018) that such algorithms can be stable and accurate under appropriate conditions on the dynamics and measurement process. Asymptotic properties of the EnKF and EnKBF in the limit of $M \to \infty$ have also been studied, for example, by Le Gland, Monbet and Tran (2011), Kwiatowski and Mandel (2015) and de Wiljes *et al.* (2018).

5.2. Feedback particle filter

A Monte Carlo implementation of the FPF (4.19) faces two main obstacles. First, one needs to approximate the potential ϕ_t in (4.20) with the density π_t, which is only available in terms of an empirical measure

$$\pi_t(z) = \frac{1}{M} \sum_{i=1}^{M} \delta(z - z_t^i).$$

Several possible approximations have been discussed by Taghvaei and Mehta (2016) and Taghvaei, de Wiljes, Mehta and Reich (2017). Here we would like to mention in particular an approximation based on diffusion maps, which we summarize in Appendix A.1. Second, one needs to apply suitable time-stepping methods for the SDE (4.19) in Stratonovitch form. Here we suggest using the Euler–Heun method (Burrage, Burrage and Tian 2004),

$$\tilde{z}_{n+1}^i = z_n^i + \Delta t f_{t_n}(z_n^i) + (\gamma \Delta t)^{1/2} \xi_n^i - K_n(z_n^i) \Delta I_n^i,$$

$$z_{n+1}^i = z_n^i + \Delta t f_{t_n}(z_n^i) + (\gamma \Delta t)^{1/2} \xi_n^i - \frac{1}{2}(K_n(z_n^i) + K_n(\tilde{z}_{n+1}^i)) \Delta I_n^i,$$

$i = 1, \ldots, M$, with, for example,

$$\Delta I_n^i = \frac{1}{2}(h(z_n^i) + \bar{h}_n^M) \Delta t - \Delta y_n.$$

While the above implementation of the FPF requires one to solve the elliptic PDE (4.20) twice per time-step, we now suggest a time-stepping approach in terms of an associated Schrödinger problem. Let us assume that we have M equally weighted particles z_n^i representing the conditional filtering distribution at time t_n. We first propagate these particles forward under the drift term alone, that is,

$$\hat{z}_{n+1}^i := z_n^i + \Delta t f_{t_n}(z_n^i), \quad i = 1, \ldots, M.$$

In the next step, we draw $L = KM$ with $K \geq 1$ samples $\tilde{z}_{n+1}^l$ from the forecast PDF

$$\tilde{\pi}(z) := \frac{1}{M} \sum_{i=1}^{M} \mathrm{n}(z; \hat{z}_{n+1}^i, \gamma \Delta t I)$$

and assign importance weights

$$w_{n+1}^l \propto \exp\left(-\frac{\Delta t}{2}(h(\tilde{z}_{n+1}^l))^2 + \Delta y_n h(\tilde{z}_{n+1}^l)\right)$$

with normalization (3.24). Recall that we assumed that $y_t \in \mathbb{R}$ for simplicity and $\Delta y_n := y_{t_{n+1}} - y_{t_n}$. We then solve the Schrödinger problem

$$P^* = \arg \min_{P \in \Pi_{\mathrm{M}}} \mathrm{KL}(P\|Q) \tag{5.3}$$

with the entries of $Q \in \mathbb{R}^{L \times M}$ given by

$$q_{li} = \exp\left(-\frac{1}{2\gamma\Delta t}\|\tilde{z}_{n+1}^l - \hat{z}_{n+1}^i\|^2\right)$$

and the set Π_{M} defined by (3.26). The desired particles $z_{n+1}^i = Z_{n+1}^i(\omega)$ are finally given as realizations of

$$Z_{n+1}^i = \sum_{l=1}^{L} \tilde{z}_{n+1}^l p_{li}^* + (\gamma\Delta t)^{1/2}\Xi_n^i, \quad \Xi_n^i \sim \mathrm{N}(0, I), \tag{5.4}$$

for $i = 1, \ldots, M$.

The update (5.4), with P^* defined by (5.3), can be viewed as data-driven drift correction combined with a standard approximation to the Brownian diffusion part of the underlying SDE model. It remains to be investigated in what sense (5.4) can be viewed as an approximation to the FPF formulation (4.19) as $M \to \infty$ and $\Delta t \to 0$.

Remark 5.2. One can also use the matrix P^* from (5.3) to implement a resampling scheme

$$\mathbb{P}[Z_{n+1}^i(\omega) = \tilde{z}_{n+1}^l] = p_{li}^* \tag{5.5}$$

for $i = 1, \ldots, M$. Note that, contrary to classical resampling schemes based on weighted particles $(\tilde{z}_{n+1}^l, w_{n+1}^l)$, $l = 1, \ldots, L$, the sampling probabilities p_{li}^* take into account the underlying geometry of the forecasts $\hat{z}_{n+1}^i$ in state space.

Example 5.3. We consider the SDE formulation

$$\mathrm{d}Z_t = f(Z_t)\,\mathrm{d}t + \gamma^{1/2}\,\mathrm{d}W_t$$

of a stochastically perturbed Lorenz-63 model (Lorenz 1963, Reich and Cotter 2015, Law *et al.* 2015) with diffusion constant $\gamma = 0.1$. The system is fully observed according to

$$\mathrm{d}Y_t = f(Z_t)\,\mathrm{d}t + R^{1/2}\,\mathrm{d}V_t$$

with measurement error variance $R = 0.1$, and the system is simulated over a time interval $t \in [0, 40\,000]$ with step-size $\Delta t = 0.01$. We implemented a standard particle filter with resampling performed after each time-step and compared the resulting RMS errors with those arising from using (5.4)

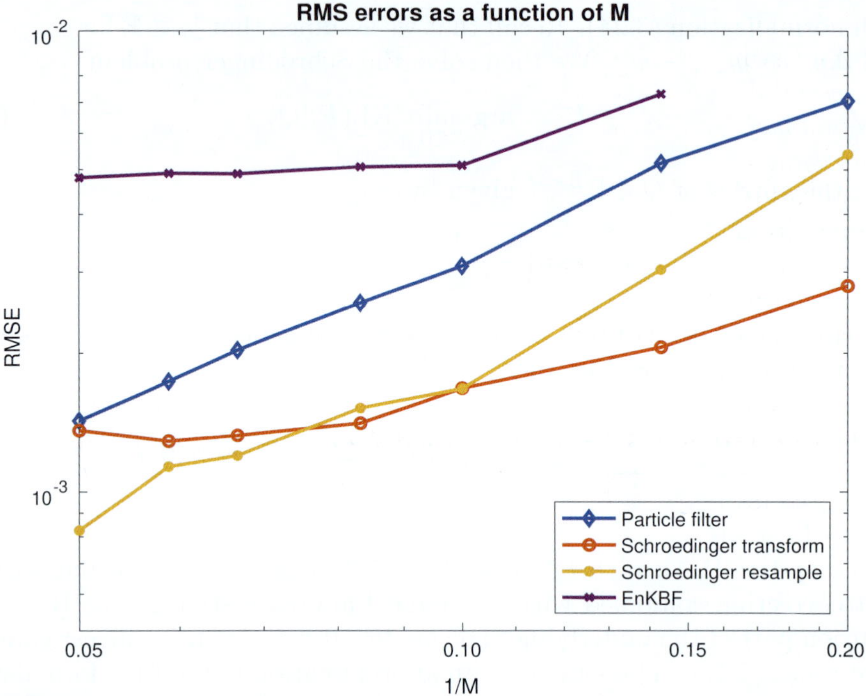

Figure 5.1. RMS errors as a function of sample size, M, for a standard particle filter, the EnKBF, and implementations of (5.4) (Schrödinger transform) and (5.5) (Schrödinger resample), respectively. Both Schrödinger-based methods outperform the standard particle filter for small ensemble sizes. The EnKBF diverged for the smallest ensemble size of $M = 5$ and performed worse than all other methods for this highly nonlinear problem.

(Schrödinger transform) and (5.5) (Schrödinger resample), respectively. See Figure 5.1. It can be seen that the Schrödinger-based methods outperform the standard particle filter in terms of RMS errors for small ensemble sizes. The Schrödinger transform method is particularly robust for very small ensemble sizes while Schrödinger resample performs better at larger sample sizes. We also implemented the EnKBF (5.1) and found that it diverged for the smallest ensemble size of $M = 5$ and performed worse than the other methods for larger ensemble sizes.

6. Conclusions

We have summarized sequential data assimilation techniques suitable for state estimation of discrete- and continuous-time stochastic processes. In addition to algorithmic approaches based on the standard filtering and smoothing framework of stochastic analysis, we have drawn a connection to a boundary value problem over joint probability measures first formulated by Erwin Schrödinger. We have argued that sequential data assimilation essentially needs to approximate such a boundary value problem with the boundary conditions given by the filtering distributions at consecutive observation times.

Application of these techniques to high-dimensional problems arising, for example, from the spatial discretization of PDEs requires further approximations in the form of localization and inflation, which we have not discussed in this survey. See, for example, Evensen (2006), Reich and Cotter (2015) and Asch *et al.* (2017) for further details. In particular, the localization framework for particle filters as introduced by Chen and Reich (2015) and Reich and Cotter (2015) in the context of scenario (A) could be generalized to scenarios (B) and (C) from Definition 2.4.

Finally, the approaches and computational techniques discussed in this paper are also relevant to combined state and parameter estimation.

Acknowledgement

This research has been partially funded by Deutsche Forschungsgemeinschaft (DFG) through grant CRC 1294 'Data Assimilation'. Feedback on earlier versions of this paper by Nikolas Kantas, Prashant Mehta and Tim Sullivan has been invaluable.

Appendices

A.1. Mesh-free approximations to Fokker–Planck and backward Kolmogorov equations

In this appendix we discuss two closely related approximations, first to the Fokker–Planck equation (2.26) with the (time-independent) operator (2.25) taking the special form

$$\mathcal{L}^\dagger \pi = -\nabla_z \cdot (\pi \nabla_z \log \pi^*) + \Delta_z \pi = \nabla_z \cdot \left(\pi^* \nabla_z \frac{\pi}{\pi^*} \right)$$

and, second, to its adjoint operator $\mathcal{L}$ given by (3.10).

The approximation to the Fokker–Planck equation (2.26) with drift term

$$f_t(z) = \nabla_z \log \pi^*(z) \tag{A.1}$$

can be used to transform samples x_0^i, $i = 1, \ldots, M$ from a (prior) PDF π_0 into samples from a target (posterior) PDF π^* using an evolution equation of the form

$$\frac{\mathrm{d}}{\mathrm{d}s} \check{Z}_s = F_s(\check{Z}_s), \tag{A.2}$$

with $\check{Z}_0 \sim \check{\pi}_0 := \pi_0$ such that

$$\lim_{s \to \infty} \check{Z}_s \sim \pi^*.$$

The evolution of the marginal PDFs $\check{\pi}_s$ is given by Liouville's equation

$$\partial_s \check{\pi}_s = -\nabla_z \cdot (\check{\pi}_s F_s). \tag{A.3}$$

We now choose F_s such that the Kullback–Leibler divergence $\mathrm{KL}\,(\check{\pi}_s \| \pi^*)$ is non-increasing in time, that is,

$$\frac{\mathrm{d}}{\mathrm{d}s} \mathrm{KL}\,(\check{\pi}_s \| \pi^*) = \int \check{\pi}_s \left\{ F_s \cdot \nabla_z \log \frac{\check{\pi}_s}{\pi^*} \right\} \mathrm{d}z \le 0. \tag{A.4}$$

A natural choice is

$$F_s(z) := -\nabla_z \log \frac{\check{\pi}_s}{\pi^*}(z),$$

which renders (A.3) formally equivalent to the Fokker–Planck equation (2.26) with drift term (A.1) (Reich and Cotter 2015, Peyre and Cuturi 2018).

Let us now approximate the evolution equation (A.2) over a reproducing kernel Hilbert space (RKHS) $\mathcal{H}$ with kernel $k(z - z')$ and inner product $\langle f, g \rangle_{\mathcal{H}}$, which satisfies the reproducing property

$$\langle k(\cdot - z'), f \rangle_{\mathcal{H}} = f(z'). \tag{A.5}$$

Following Russo (1990) and Degond and Mustieles (1990), we first introduce the approximation

$$\widetilde{\pi}_s(z) := \frac{1}{M} \sum_{i=1}^{M} k(z - z_s^i) \tag{A.6}$$

to the marginal densities $\check{\pi}_s$. Note that (A.5) implies that

$$\langle f, \widetilde{\pi}_s \rangle_{\mathcal{H}} = \frac{1}{M} \sum_{i=1}^{M} f(z_s^i).$$

Given some evolution equations

$$\frac{\mathrm{d}}{\mathrm{d}s} z_s^i = u_s^i$$

for the particles z_s^i, $i = 1, \ldots, M$, we find that (A.6) satisfies Liouville's equation, that is,

$$\partial_s \widetilde{\pi}_s = -\nabla_z \cdot (\widetilde{\pi}_s \widetilde{F}_s)$$

with

$$\widetilde{F}_s(z) = \frac{\sum_{i=1}^{M} k(z - z_s^i)\, u_s^i}{\sum_{i=1}^{M} k(z - z_s^i)}.$$

We finally introduce the functional

$$\mathcal{V}(\{z_s^l\}) := \left\langle \widetilde{\pi}_s, \log \frac{\widetilde{\pi}_s}{\pi^*} \right\rangle_{\mathcal{H}} = \frac{1}{M} \sum_{i=1}^{M} \log \frac{\frac{1}{M}\sum_{j=1}^{M} k(z_s^i - z_s^j)}{\pi^*(z_s^i)}$$

as an approximation to the Kullback–Leibler divergence in the RKHS $\mathcal{H}$ and set

$$u_s^i := -M\nabla_{z_s^i}\mathcal{V}(\{z_s^l\}), \tag{A.7}$$

which constitutes the desired particle approximation to the Fokker–Planck equation (2.26) with drift term (A.1). Time-stepping methods for such gradient flow systems have been discussed by Pathiraja and Reich (2019).

We also remark that an alternative interacting particle system, approximating the same asymptotic PDF π^* in the limit $s \to \infty$, has been proposed recently by Liu and Wang (2016) under the notion of Stein variational descent. See Lu, Lu and Nolen (2019) for a theoretical analysis of Stein variational descent, which implies in particular that Stein variational descent can be viewed as a Lagrangian particle approximation to the modified evolution equation

$$\partial_s \breve{\pi}_s = \nabla_z \cdot \left(\breve{\pi}_s^2 \nabla_z \log \frac{\breve{\pi}_s}{\pi^*}(z) \right) = \nabla_z \cdot (\breve{\pi}_s(\nabla_z \breve{\pi}_s - \breve{\pi}_s \nabla_z \log \pi^*))$$

in the marginal PDFs $\breve{\pi}_s$, that is, one uses

$$F_s(z) := -\breve{\pi}_s \nabla_z \log \frac{\breve{\pi}_s}{\pi^*}(z)$$

in (A.2). The Kullback–Leibler divergence is still non-increasing since (A.4) becomes

$$\frac{\mathrm{d}}{\mathrm{d}s}\mathrm{KL}\,(\breve{\pi}_s\|\pi^*) = -\int \|F_s\|^2\,\mathrm{d}z \leq 0.$$

A numerical discretization is obtained through the approximation

$$F_s(z') \approx \int F_s(z)\,k(z - z')\,\mathrm{d}z,$$

that is, one views the kernel $k(z - z')$ as a regularized Dirac delta function. This approximation leads to another vector field

$$\widehat{F}_s(z') := -\int \breve{\pi}_s(z)\{\nabla_z \log \breve{\pi}_s(z) - \nabla_z \log \pi^*(z)\}k(z - z')\,\mathrm{d}z$$

$$= \int \breve{\pi}_s(z)\{\nabla_z k(z - z') + k(z - z')\,\nabla_z \log \pi^*(z)\}\,\mathrm{d}z.$$

On extending the RKHS $\mathcal{H}$ and its reproducing property (A.5) component-wise to vector-valued functions, it follows that

$$\frac{\mathrm{d}}{\mathrm{d}s}\mathrm{KL}\,(\breve{\pi}_s\|\pi^*) = -\int F_s \cdot \widehat{F}_s\,\mathrm{d}z = -\langle \widehat{F}_s, \widehat{F}_s \rangle_{\mathcal{H}} \leq 0$$

along transformations induced by the vector field $\widehat{F}_s$. See Liu and Wang (2016) for more details.

We now turn our attention to the dual operator $\mathcal{L}_t$, defined by (3.10), which also arises from (4.6) and (4.20), respectively. More specifically, let us rewrite (4.20) in the form

$$\mathcal{A}_t \phi_t = -(h - \pi_t[h]) \tag{A.8}$$

with the operator $\mathcal{A}_t$ defined by

$$\mathcal{A}_t g := \frac{1}{\pi_t} \nabla_z \cdot (\pi_t \nabla_z g).$$

Then we find that $\mathcal{A}_t$ is of the form of $\mathcal{L}_t$ with π_t taking the role of π^*.

We also recall that (3.11) provides an approximation to $\mathcal{L}_t$ and hence to $\mathcal{A}_t$. This observation allows one to introduce a sample-based method for approximating the potential ϕ defined by the elliptic partial differential equation (A.8) for a given function $h(z)$.

Here we instead follow the presentation of Taghvaei and Mehta (2016) and Taghvaei *et al.* (2017) and assume that we have M samples z^i from a PDF π. The method is based on

$$\frac{\phi - e^{\epsilon \mathcal{A}} \phi}{\epsilon} \approx h - \pi[h] \tag{A.9}$$

for $\epsilon > 0$ sufficiently small and upon replacing $e^{\epsilon \mathcal{A}}$ with a diffusion map approximation (Harlim 2018) of the form

$$e^{\epsilon \mathcal{A}} \phi(z) \approx T_\epsilon \phi(z) := \sum_{i=1}^{M} k_\epsilon(z, z^i) \, \phi(z^i). \tag{A.10}$$

The required kernel functions $k_\epsilon(z, z^i)$ are defined as follows. Let

$$n_\epsilon(z) := \mathrm{n}(z; 0, 2\epsilon\, I)$$

and

$$p_\epsilon(z) := \frac{1}{M} \sum_{j=1}^{M} n_\epsilon(z - z^j) = \frac{1}{M} \sum_{j=1}^{M} \mathrm{n}(z; z^j, 2\epsilon\, I).$$

Then

$$k_\epsilon(z, z^i) := \frac{n_\epsilon(z - z^i)}{c_\epsilon(z)\, p_\epsilon(z^i)^{1/2}}$$

with normalization factor

$$c_\epsilon(z) := \sum_{l=1}^{M} \frac{n_\epsilon(z - z^l)}{p_\epsilon(z^l)^{1/2}}.$$

In other words, the operator T_ϵ reproduces constant functions.

The approximations (A.9) and (A.10) lead to the fixed-point problem[3]

$$\phi_j = \sum_{i=1}^{M} k_\epsilon(z^j, z^i)\,\phi_i + \epsilon \Delta h_i, \quad j = 1, \dots, M, \qquad (\text{A.11})$$

in the scalar coefficients ϕ_j, $j = 1, \dots, M$, for given

$$\Delta h_i := h(z^i) - \bar{h}, \quad \bar{h} := \frac{1}{M} \sum_{l=1}^{M} h(z^l).$$

Since T_ϵ reproduces constant functions, (A.11) determines ϕ_i up to a constant contribution, which we fix by requiring

$$\sum_{i=1}^{M} \phi_i = 0.$$

The desired functional approximation $\widetilde{\phi}$ to the potential ϕ is now provided by

$$\widetilde{\phi}(z) = \sum_{i=1}^{M} k_\epsilon(z, z^i)\{\phi_i + \epsilon \Delta h_i\}. \qquad (\text{A.12})$$

Furthermore, since

$$\nabla_z k_\epsilon(z, z^i) = \frac{-1}{2\epsilon} k_\epsilon(z, z^i)\left((z - z^i) - \sum_{l=1}^{M} k_\epsilon(z, z^l)(z - z^l)\right)$$
$$= \frac{1}{2\epsilon} k_\epsilon(z, z^i)\left(z^i - \sum_{l=1}^{M} k_\epsilon(z, z^l)z^l\right),$$

we obtain

$$\nabla_z \widetilde{\phi}(z^j) = \sum_{i=1}^{M} \nabla_z k_\epsilon(z^j, z^i)\, r_i = \sum_{i=1}^{M} z^i\, a_{ij},$$

with

$$r_i = \phi_i + \epsilon \, \Delta h_i$$

and

$$a_{ij} := \frac{1}{2\epsilon} k_\epsilon(z^j, z^i)\left(r_i - \sum_{l=1}^{M} k_\epsilon(z^j, z^l)\, r_l\right).$$

---

[3] It would also be possible to employ the approximation (3.11) in the fixed-point problem (A.11), that is, to replace $k_\epsilon(z^j, z^i)$ by $(Q_+)_{ji}$ in (3.11) with $\Delta t = \epsilon$ and $\pi^* = \pi_t$.

We note that

$$\sum_{i=1}^{M} a_{ij} = 0$$

and

$$\lim_{\epsilon \to \infty} a_{ij} = \frac{1}{M} \Delta h_i$$

since

$$\lim_{\epsilon \to \infty} k_\epsilon(z^j, z^i) = \frac{1}{M}.$$

In other words,

$$\lim_{\epsilon \to \infty} \nabla_z \widetilde{\phi}(z^j) = \frac{1}{M} \sum_{i=1}^{M} z^i \left(h(z^i) - \bar{h} \right) = K^M$$

independent of z^j, which is equal to an empirical estimator for the covariance between z and $h(z)$ and which, in the context of the FPF, leads to the EnKBF formulations (5.1) of Section 5.1. See Taghvaei *et al.* (2017) for more details and Taghvaei, Mehta and Meyn (2019) for a convergence analysis.

A.2. Regularized Störmer–Verlet for HMC

One is often faced with the task of sampling from a high-dimensional PDF of the form

$$\pi(x) \propto \exp(-V(x)), \quad V(x) := \frac{1}{2}(x - \bar{x})^{\mathrm{T}} B^{-1}(x - \bar{x}) + U(x),$$

for known $\bar{x} \in \mathbb{R}^{N_x}$, $B \in \mathbb{R}^{N_x \times N_x}$, and $U : \mathbb{R}^{N_x} \to \mathbb{R}$. The hybrid Monte Carlo (HMC) method (Neal 1996, Liu 2001, Bou-Rabee and Sanz-Serna 2018) has emerged as a popular Markov chain Monte Carlo (MCMC) method for tackling this problem. HMC relies on a symplectic discretization of the Hamiltonian equations of motion

$$\frac{\mathrm{d}}{\mathrm{d}\tau} x = M^{-1} p,$$

$$\frac{\mathrm{d}}{\mathrm{d}\tau} p = -\nabla_x V(x) = -B^{-1}(x - \bar{x}) - \nabla_x U(x)$$

in an artificial time τ (Leimkuhler and Reich 2005). The conserved energy (or Hamiltonian) is provided by

$$\mathcal{H}(x, p) = \frac{1}{2} p^{\mathrm{T}} M^{-1} p + V(x). \tag{A.13}$$

The symmetric positive definite mass matrix $M \in \mathbb{R}^{N_x \times N_x}$ can be chosen arbitrarily, and a natural choice in terms of sampling efficiency is $M = B^{-1}$

(Beskos *et al.* 2011). However, when also taking into account computational efficiency, a Störmer–Verlet discretization

$$p_{n+1/2} = p_n - \frac{\Delta\tau}{2}\nabla_x V(x_n), \tag{A.14}$$

$$q_{n+1} = q_n + \Delta\tau \widetilde{M}^{-1} p_{n+1/2}, \tag{A.15}$$

$$p_{n+1} = p_{n+1/2} - \frac{\Delta\tau}{2}\nabla_x V(x_{n+1}), \tag{A.16}$$

with step-size $\Delta\tau > 0$, mass matrix $M = I$ in (A.13) and modified mass matrix

$$\widetilde{M} = I + \frac{\Delta\tau^2}{4} B^{-1} \tag{A.17}$$

in (A.15) emerges as an attractive alternative, since it implies

$$\mathcal{H}(x_n, p_n) = \mathcal{H}(x_{n+1}, p_{n+1})$$

for all $\Delta\tau > 0$ provided $U(x) \equiv 0$. The Störmer–Verlet formulation (A.14)–(A.16) is based on a regularized formulation of Hamiltonian equations of motion for highly oscillatory systems as discussed, for example, by Reich and Hundertmark (2011).

Energy-conserving time-stepping methods for linear Hamiltonian systems have become an essential building block for applications of HMC to infinite-dimensional inference problems, where B^{-1} corresponds to the discretization of a positive, self-adjoint and trace-class operator $\mathcal{B}$. See, for example, Beskos *et al.* (2017).

Note that the Störmer–Verlet discretization (A.14)–(A.16) together with (A.17) can be easily extended to inference problems with constraints $g(x) = 0$ (Leimkuhler and Reich 2005) and that (A.14)–(A.16) conserves equilibria,[4] that is, points x_* with $\nabla V(x_*) = 0$, regardless of the step-size $\Delta\tau$.

HMC methods, based on (A.14)–(A.16) and (A.17), can be used to sample from the smoothing distribution of an SDE as considered in Sections 2.2 and 3.3.

A.3. Ensemble Kalman filter

We summarize the formulation of an ensemble Kalman filter in the form (3.20). We start with the stochastic ensemble Kalman filter (Evensen 2006), which is given by

$$\widehat{Z}_1^j = z_1^j - K(h(z_1^j) + \Theta^j - y_1), \quad \Theta^j \sim \mathrm{N}(0, R), \tag{A.18}$$

---

[4] Note that equilibria of the Hamiltonian equations of motion correspond to MAP estimators of the underlying Bayesian inference problem.

with Kalman gain matrix

$$K = P^{zh}(P^{hh} + R)^{-1} = \frac{1}{M-1} \sum_{i=1}^{M} z_1^i \, (h(z_1^i) - \bar{h})^{\mathrm{T}} (P^{hh} + R)^{-1}$$

and

$$P^{hh} := \frac{1}{M-1} \sum_{l=1}^{M} h(z_1^l) \, (h(z_1^l) - \bar{h})^{\mathrm{T}}, \quad \bar{h} := \frac{1}{M} \sum_{l=1}^{M} h(z_1^l).$$

Formulation (A.18) can be rewritten in the form (3.20) with

$$p_{ij}^* = \delta_{ij} - \frac{1}{M-1}(h(z_1^i) - \bar{h})^{\mathrm{T}}(P^{hh} + R)^{-1}(h(z_1^j) - y_1 + \Theta^j), \qquad \text{(A.19)}$$

where δ_{ij} denotes the Kronecker delta, that is, $\delta_{ij} = 0$ if $i \neq j$ and $\delta_{ii} = 1$.

More generally, one can think about ensemble Kalman filters and their generalizations (Anderson 2010) as first defining appropriate updates $\widehat{y}_1^i$ to the predicted $y_1^i = h(z_1^i)$ using the observed y_1, which is then extrapolated to the state variable z via linear regression, that is,

$$\widehat{z}_1^j = z_1^j + \frac{1}{M-1} \sum_{i=1}^{M} z_1^i (h(z_1^i) - \bar{h})^{\mathrm{T}} (P^{hh})^{-1} (\widehat{y}_1^j - y_1^j), \qquad \text{(A.20)}$$

which can be reformulated in the form (3.20) (Reich and Cotter 2015). Note that the consistency result

$$H\widehat{z}_1^i = \widehat{y}_1^i$$

follows from (A.20) for linear forward maps $h(z) = Hz$.

Within such a linear regression framework, one can easily derive ensemble transformations for the particles z_0^i at time $t = 0$. We simply take the coefficients p_{ij}^*, as defined for example by an ensemble Kalman filter (A.19), and apply them to z_0^i, that is,

$$\widehat{z}_0^j = \sum_{i=1}^{M} z_0^i \, p_{ij}^*.$$

These transformed particles can be used to approximate the smoothing distribution $\widehat{\pi}_0$. See, for example, Evensen (2006) and Kirchgessner *et al.* (2017) for more details.

Finally, one can also interpret the ensemble Kalman filter as a continuous update in artificial time $s \geq 0$ of the form

$$\mathrm{d}z_s^i = -P^{zh} R^{-1} \, \mathrm{d}I_s^i \qquad \text{(A.21)}$$

with the innovations I_s^i given either by

$$\mathrm{d}I_s^i = \frac{1}{2}(h(z_s^i) + \bar{h}_s) \, \mathrm{d}s - y_1 \, \mathrm{d}s \qquad \text{(A.22)}$$

or, alternatively, by

$$\mathrm{d}I_s^i = h(z_s^i)\,\mathrm{d}s + R^{1/2}\,\mathrm{d}V_s^i - y_1\,\mathrm{d}s,$$

where V_s^i stands for standard Brownian motion (Bergemann and Reich 2010, Reich 2011, Bergemann and Reich 2012). Equation (A.21) with innovation (A.22) can be given a gradient flow structure (Bergemann and Reich 2010, Reich and Cotter 2015) of the form

$$\frac{1}{\mathrm{d}s}\mathrm{d}z_s^i = -P^{zz}\nabla_{z^i}\mathcal{V}(\{z_s^j\}), \tag{A.23}$$

with potential

$$\mathcal{V}(\{z^j\}) := \frac{1-\alpha}{4}\sum_{j=1}^{M}(h(z^j) - y_1)^{\mathrm{T}}R^{-1}(h(z^j) - y_1)$$
$$+ \frac{(1+\alpha)M}{4}(\bar{h} - y_1)^{\mathrm{T}}R^{-1}(\bar{h} - y_1)$$

and $\alpha = 0$ for the standard ensemble Kalman filter, while $\alpha \in (0,1)$ can be seen as a form of variance inflation (Reich and Cotter 2015).

A theoretical study of such dynamic formulations in the limit of $s \to \infty$ and $\alpha = -1$ has been initiated by Schillings and Stuart (2017). There is an interesting link to stochastic gradient methods (Bottou, Curtis and Nocedal 2018), which find application in situations where the dimension of the data y_1 is very high and the computation of the complete gradient $\nabla_z h(z)$ becomes prohibitive. More specifically, the basic concepts of stochastic gradient methods can be extended to (A.23) if R is diagonal, in which case one would pick at random paired components of h and y_1 at the kth time-step of a discretization of (A.23), with the step-size Δs_k chosen appropriately. Finally, we also point to a link between natural gradient methods and Kalman filtering (Ollivier 2018) which can be explored further in the context of the continuous-time ensemble Kalman filter formulation (A.23).

A.4. Numerical treatment of forward–backward SDEs

We discuss a numerical approximation of the forward–backward SDE problem defined by the forward SDE (2.24) and the backward SDE (2.54) with initial condition $Z_0^+ \sim \pi_0$ at time $t = 0$ and final condition $Y_1(Z_1^+) = l(Z_1^+)/\beta$ at time $t = 1$. Discretization of the forward SDE (2.24) by the Euler–Maruyama method (3.5) leads to M numerical solution paths $z_{0:N}^i$, $i = 1,\ldots,M$, which, according to Definition 3.5, lead to N discrete Markov transition matrices $Q_n^+ \in \mathbb{R}^{M\times M}$, $n = 1,\ldots,N$.

The Euler–Maruyama method is now also applied to the backward SDE (2.54) and yields

$$Y_n = Y_{n+1} - \Delta t^{1/2}\Xi_n^{\mathrm{T}}V_n.$$

Upon taking conditional expectation we obtain

$$Y_n(Z_n^+) = \mathbb{E}[Y_{n+1}|Z_n^+] \tag{A.24}$$

and

$$\Delta t^{1/2}\mathbb{E}[\Xi_n\Xi_n^{\mathrm{T}}]V_n(Z_n^+) = \mathbb{E}[(Y_{n+1} - Y_n)\Xi_n|Z_n^+],$$

respectively. The last equation leads to

$$V_n(Z_n^+) = \Delta t^{-1/2}\mathbb{E}[(Y_{n+1} - Y_n)\Xi_n|Z_n^+]. \tag{A.25}$$

We also have $Y_N(Z_N^+) = l(Z_N^+)/\beta$ at final time $t = 1 = N\Delta t$. See page 45 in Carmona (2016) for more details.

We finally need to approximate the conditional expectation values in (A.24) and (A.25), for which we employ the discrete Markov transition matrix Q_{n+1}^+ and the discrete increments

$$\zeta_{ij} := \frac{1}{(\gamma\Delta t)^{1/2}}(z_{n+1}^i - z_n^j - \Delta t f_{t_n}(z_n^j)) \in \mathbb{R}^M.$$

Given $y_{n+1}^j \approx Y(z_{n+1}^j)$ at time level t_{n+1}, we then approximate (A.24) by

$$y_n^j := \sum_{i=1}^{M} y_{n+1}^i (Q_{n+1}^+)_{ij} \tag{A.26}$$

for $n = N - 1, \ldots, 0$. The backward iteration is initiated by setting $y_N^i = l(z_N^i)/\beta$, $i = 1, \ldots, M$. Furthermore, a Monte Carlo approximation to (A.25) at z_n^j is provided by

$$\Delta t^{1/2} \sum_{i=1}^{M} \{\xi_{ij}(\xi_{ij})^{\mathrm{T}}(Q_{n+1}^+)_{ij}\}v_n^j = \sum_{i=1}^{M} (y_{n+1}^i - y_n^j)\xi_{ij}(Q_{n+1}^+)_{ij}$$

and, upon assuming invertibility, we obtain the explicit expression

$$v_n^j := \Delta t^{-1/2} \left(\sum_{i=1}^{M} \xi_{ij}(\xi_{ij})^{\mathrm{T}}(Q_{n+1}^+)_{ij}\right)^{-1} \sum_{i=1}^{M} (y_{n+1}^i - y_n^j)\xi_{ij}(Q_{n+1}^+)_{ij} \tag{A.27}$$

for $n = N - 1, \ldots, 0$.

Recall from Remark 2.20 that $y_n^j \in \mathbb{R}$ provides an approximation to $\psi_{t_n}(z_n^j)$ and $v_n^j \in \mathbb{R}^{N_z}$ an approximation to $\gamma^{1/2}\nabla_z\psi_{t_n}(z_n^j)$, respectively, where ψ_t denotes the solution of the backward Kolmogorov equation (2.31) with final condition $\psi_1(z) = l(z)/\beta$. Hence, the forward solution paths $z_{0:N}^i$,

$i = 1, \ldots, M$, together with the backward approximations (A.26) and (A.27) provide a mesh-free approximation to the backward Kolmogorov equation (2.31). Furthermore, the associated control law (2.32) can be approximated by

$$u_{t_n}(z_n^i) \approx \frac{\gamma^{1/2}}{y_n^i} v_n^i.$$

The division by y_n^i can be avoided by means of the following alternative formulation. We introduce the potential

$$\phi_t := \log \psi_t, \tag{A.28}$$

which satisfies the modified backward Kolmogorov equation

$$0 = \partial_t \phi_t + \mathcal{L}_t \phi_t + \frac{\gamma}{2} \|\nabla_z \phi_t\|^2$$

with final condition $\phi_1(z) = \log l(z)$, where we have ignored the constant $\log \beta$. Hence Itô's formula applied to $\phi_t(Z_t^+)$ leads to

$$\mathrm{d}\phi_t = -\frac{\gamma}{2} \|\nabla_z \phi_t\|^2 \, \mathrm{d}t + \gamma^{1/2} \nabla_z \phi_t \cdot \mathrm{d}W_t^+ \tag{A.29}$$

along solutions Z_t^+ of the forward SDE (2.24). It follows from

$$\frac{\mathrm{d}\widehat{\mathbb{P}}}{\mathrm{d}\mathbb{Q}^u}\bigg|_{z_{[0,1]}} = \frac{l(z_1)}{\beta} \frac{\pi_0(z_0)}{q_0(z_0)} \exp\left(\frac{1}{2\gamma} \int_0^1 (\|u_t\|^2 \, \mathrm{d}t - 2\gamma^{1/2} u_t \cdot \mathrm{d}W_t^+) \right),$$

with $u_t = \gamma \nabla_z \phi_t$, $q_0 = \widehat{\pi}_0$, and

$$\log \frac{l(z_1)}{\beta} - \log \frac{\widehat{\pi}_0(z_0)}{\pi_0(z_0)} = \int_0^1 \mathrm{d}\phi_t$$

that $\widehat{\mathbb{P}} = \mathbb{Q}^u$, as desired.

The backward SDE associated with (A.29) becomes

$$\mathrm{d}Y_t = -\frac{1}{2} \|V_t\|^2 \, \mathrm{d}t + V_t \cdot \mathrm{d}W_t^+ \tag{A.30}$$

and its Euler–Maruyama discretization is

$$Y_n = Y_{n+1} + \frac{\Delta t}{2} \|V_n\|^2 - \Delta t^{1/2} \Xi_n^{\mathrm{T}} V_n.$$

Numerical values (y_n^i, v_n^i) can be obtained as before with (A.26) replaced by

$$y_n^j := \sum_{i=1}^M \left(y_{n+1}^i + \frac{\Delta t}{2} \|v_n^j\|^2 \right) (Q_{n+1}^+)_{ij}$$

and the control law (2.32) is now approximated by

$$u_{t_n}(z_n^i) \approx \gamma^{1/2} v_n^i.$$

We re-emphasize that the backward SDE (A.30) arises naturally from an optimal control perspective onto the smoothing problem. See Carmona (2016) for more details on the connection between optimal control and backward SDEs. In particular, this connection leads to the following alternative approximation

$$u_{t_n}(z_n^j) \approx \sum_{i=1}^{M} z_{n+1}^i \{ (\widehat{Q}_{n+1}^+)_{ij} - (Q_{n+1}^+)_{ij} \}$$

of the control law (2.32). Here $\widehat{Q}_{n+1}^+$ denotes the twisted Markov transition matrix defined by

$$\widehat{Q}_{n+1}^+ = D(y_{n+1})\, Q_{n+1}^+\, D(y_n)^{-1}, \quad y_n = (y_n^1, \ldots, y_n^M)^{\mathrm{T}}.$$

Remark. The backward SDE (A.30) can also be utilized to reformulate the Schrödinger system (2.58)–(2.61). More specifically, one seeks an initial π_0^ψ which evolves under the forward SDE (2.24) with $Z_0^+ \sim \pi_0^\psi$ such that the solution Y_t of the associated backward SDE (A.30) with final condition

$$Y_1(z) = \log \widehat{\pi}_1(z) - \log \pi_1^\psi(z)$$

implies $\pi_0 \propto \pi_0^\psi \exp(Y_0)$. The desired control law in (2.30) is provided by

$$u_t(z) = \gamma \nabla_z Y_t(z) = \gamma^{1/2} V_t(z).$$

REFERENCES[5]

W. Acevedo, J. de Wiljes and S. Reich (2017), 'Second-order accurate ensemble transform particle filters', *SIAM J. Sci. Comput.* **39**, A1834–A1850.

S. Agapiou, O. Papaspipliopoulos, D. Sanz-Alonso and A. Stuart (2017), 'Importance sampling: Computational complexity and intrinsic dimension', *Statist. Sci.* **32**, 405–431.

J. Amezcua, E. Kalnay, K. Ide and S. Reich (2014), 'Ensemble transform Kalman–Bucy filters', *Q. J. Roy. Meteorol. Soc.* **140**, 995–1004.

J. L. Anderson (2010), 'A non-Gaussian ensemble filter update for data assimilation', *Monthly Weather Rev.* **138**, 4186–4198.

M. S. Arulampalam, S. Maskell, N. Gordon and T. Clapp (2002), 'A tutorial on particle filters for online nonlinear/non-Gaussian Bayesian tracking', *IEEE Trans. Signal Process.* **50**, 174–188.

M. Asch, M. Bocquet and M. Nodet (2017), *Data Assimilation: Methods, Algorithms and Applications*, SIAM.

---

[5] The URLs cited in this work were correct at the time of going to press, but the publisher and the authors make no undertaking that the citations remain live or are accurate or appropriate.

A. Bain and D. Crisan (2008), *Fundamentals of Stochastic Filtering*, Vol. 60 of Stochastic Modelling and Applied Probability, Springer.

T. Bengtsson, P. Bickel and B. Li (2008), Curse of dimensionality revisited: Collapse of the particle filter in very large scale systems. In *Probability and Statistics: Essays in Honor of David F. Freedman*, Vol. 2 of IMA Collections, Institute of Mathematical Sciences, pp. 316–334.

K. Bergemann and S. Reich (2010), 'A mollified ensemble Kalman filter', *Q. J. Roy. Meteorol. Soc.* **136**, 1636–1643.

K. Bergemann and S. Reich (2012), 'An ensemble Kalman–Bucy filter for continuous data assimilation', *Meteorol. Z.* **21**, 213–219.

A. Beskos, M. Girolami, S. Lan, P. Farrell and A. Stuart (2017), 'Geometric MCMC for infinite-dimensional inverse problems', *J. Comput. Phys.* **335**, 327–351.

A. Beskos, F. J. Pinski, J. M. Sanz-Serna and A. M. Stuart (2011), 'Hybrid Monte Carlo on Hilbert spaces', *Stochastic Process. Appl.* **121**, 2201–2230.

D. Blömker, C. Schillings and P. Wacker (2018), 'A strongly convergent numerical scheme for ensemble Kalman inversion', *SIAM J. Numer. Anal.* **56**, 2537–2562.

L. Bottou, F. E. Curtis and J. Nocedal (2018), 'Optimization methods for large-scale machine learning', *SIAM Rev.* **60**, 223–311.

N. Bou-Rabee and J. M. Sanz-Serna (2018), Geometric integrators and the Hamiltonian Monte Carlo method. In *Acta Numerica*, Vol. 27, Cambridge University Press, pp. 113–206.

K. Burrage, P. M. Burrage and T. Tian (2004), 'Numerical methods for strong solutions of stochastic differential equations: An overview', *Proc. Roy. Soc. Lond. A* **460**, 373–402.

R. Carmona (2016), *Lectures on BSDEs, Stochastic Control, and Stochastic Differential Games with Financial Applications*, SIAM.

A. Carrassi, M. Bocquet, L. Bertino and G. Evensen (2018), 'Data assimilation in the geosciences: An overview of methods, issues, and perspectives', *WIREs Clim Change.* **9**, e535.

A. Carrassi, M. Bocquet, A. Hannart and M. Ghil (2017), 'Estimation model evidence using data assimilation', *Q. J. Roy. Meteorol. Soc.* **143**, 866–880.

Y. Chen and S. Reich (2015), Assimilating data into scientific models: An optimal coupling perspective. In *Frontiers in Applied Dynamical Systems: Reviews and Tutorials* (P. J. van Leeuwen *et al.* eds), Vol. 2, Springer, pp. 75–118.

Y. Chen, T. T. Georgiou and M. Pavon (2014), 'On the relation between optimal transport and Schrödinger bridges: A stochastic control viewpoint', *J. Optim. Theory Appl.* **169**, 671–691.

Y. Chen, T. T. Georgiou and M. Pavon (2016*a*), 'Entropic and displacement interpolation: A computational approach using the Hilbert metric', *SIAM J. Appl. Math.* **76**, 2375–2396.

Y. Chen, T. T. Georgiou and M. Pavon (2016*b*), 'Optimal steering of a linear stochastic system to a final probability distribution, Part I', *Trans. Automat. Control* **61**, 1158–1169.

N. Chustagulprom, S. Reich and M. Reinhardt (2016), 'A hybrid ensemble transform filter for nonlinear and spatially extended dynamical systems', *SIAM/ASA J. Uncertain. Quantif.* **4**, 592–608.

D. Crisan and J. Xiong (2010), 'Approximate McKean–Vlasov representation for a class of SPDEs', *Stochastics* **82**, 53–68.

M. Cuturi (2013), Sinkhorn distances: Lightspeed computation of optimal transport. In *Advances in Neural Information Processing Systems 26 (NIPS 2013)* (C. J. C. Burges et al., eds), pp. 2292–2300.

P. Dai Pra (1991), 'A stochastic control approach to reciprocal diffusion processes', *Appl. Math. Optim.* **23**, 313–329.

F. Daum and J. Huang (2011), Particle filter for nonlinear filters. In *2011 IEEE International Conference on Acoustics, Speech and Signal Processing (ICASSP)*, pp. 5920–5923.

J. de Wiljes, S. Reich and W. Stannat (2018), 'Long-time stability and accuracy of the ensemble Kalman–Bucy filter for fully observed processes and small measurement noise', *SIAM J. Appl. Dyn. Syst.* **17**, 1152–1181.

P. Degond and F.-J. Mustieles (1990), 'A deterministic approximation of diffusion equations using particles', *SIAM J. Sci. Comput.* **11**, 293–310.

P. del Moral (2004), *Feynman–Kac Formulae: Genealogical and Interacting Particle Systems with Applications*, Springer.

J. L. Doob (1984), *Classical Potential Theory and its Probabilistic Counterpart*, Springer.

R. Douc and O. Cappe (2005), Comparison of resampling schemes for particle filtering. In *4th International Symposium on Image and Signal Processing and Analysis (ISPA 2005)*, pp. 64–69.

A. Doucet, N. de Freitas and N. Gordon, eds (2001), *Sequential Monte Carlo Methods in Practice*, Springer.

T. A. El Moselhy and Y. M. Marzouk (2012), 'Bayesian inference with optimal maps', *J. Comput. Phys.* **231**, 7815–7850.

G. Evensen (2006), *Data Assimilation: The Ensemble Kalman Filter*, Springer.

P. Fearnhead and H. R. Künsch (2018), 'Particle filters and data assimilation', *Annu. Rev. Statist. Appl.* **5**, 421–449.

W. H. Fleming (1997), 'Deterministic nonlinear filtering', *Ann. Scuola Norm. Super. Pisa* **25**, 435–454.

H. Föllmer and N. Gantert (1997), 'Entropy minimization and Schrödinger processes in infinite dimensions', *Ann. Probab.* **25**, 901–926.

M. Frei and H. R. Künsch (2013), 'Bridging the ensemble Kalman and particle filters', *Biometrika* **100**, 781–800.

C. González-Tokman and B. R. Hunt (2013), 'Ensemble data assimilation for hyperbolic systems', *Phys. D* **243**, 128–142.

P. Guarniero, A. M. Johansen and A. Lee (2017), 'The iterated auxiliary particle filter', *J. Amer. Statist. Assoc.* **112**, 1636–1647.

J. Harlim (2018), *Data-Driven Computational Methods*, Cambridge University Press.

C. Hartmann, L. Richter, C. Schütte and W. Zhang (2017), 'Variational characterization of free energy: Theory and algorithms', *Entropy* **19**, 629.

J. Heng, A. N. Bishop, G. Deligiannidis and A. Doucet (2018), Controlled sequential Monte Carlo. Technical report, Harvard University. arXiv:1708.08396v2

A. H. Jazwinski (1970), *Stochastic Processes and Filtering Theory*, Academic Press.

N. Kantas, A. Doucet, S. S. Singh, J. Maciejowski and N. Chopin (2015), 'On particle methods for parameter estimation in state-space models', *Statist. Sci.* **30**, 328–351.

H. J. Kappen and H. C. Ruiz (2016), 'Adaptive importance sampling for control and inference', *J. Statist. Phys.* **162**, 1244–1266.

H. J. Kappen, V. Gomez and M. Opper (2012), 'Optimal control as a graphical model inference problem', *Machine Learning* **87**, 159–182.

D. Kelly, A. J. Majda and X. T. Tong (2015), 'Concrete ensemble Kalman filters with rigorous catastrophic filter divergence', *Proc. Natl Acad. Sci. USA* **112**, 10589–10594.

D. T. Kelly, K. J. H. Law and A. Stuart (2014), 'Well-posedness and accuracy of the ensemble Kalman filter in discrete and continuous time', *Nonlinearity* **27**, 2579–2604.

P. Kirchgessner, J. Tödter, B. Ahrens and L. Nerger (2017), 'The smoother extension of the nonlinear ensemble transform filter', *Tellus A* **69**, 1327766.

P. E. Kloeden and E. Platen (1992), *Numerical Solution of Stochastic Differential Equations*, Springer.

E. Kwiatowski and J. Mandel (2015), 'Convergence of the square root ensemble Kalman filter in the large ensemble limit', *SIAM/ASA J. Uncertain. Quantif.* **3**, 1–17.

R. S. Laugesen, P. G. Mehta, S. P. Meyn and M. Raginsky (2015), 'Poisson's equation in nonlinear filtering', *SIAM J. Control Optim.* **53**, 501–525.

K. Law, A. Stuart and K. Zygalakis (2015), *Data Assimilation: A Mathematical Introduction*, Springer.

F. Le Gland, V. Monbet and V.-D. Tran (2011), Large sample asymptotics for the ensemble Kalman filter. In *The Oxford Handbook of Nonlinear Filtering* D. Crisan and B. Rozovskii, eds), Oxford University Press, pp. 598–631.

B. Leimkuhler and S. Reich (2005), *Simulating Hamiltonian Dynamics*, Cambridge University Press.

C. Leonard (2014), 'A survey of the Schrödinger problem and some of its connections with optimal transportation', *Discrete Contin. Dyn. Syst. A* **34**, 1533–1574.

F. Lindsten and T. B. Schön (2013), 'Backward simulation methods for Monte Carlo statistical inference', *Found. Trends Machine Learning* **6**, 1–143.

J. S. Liu (2001), *Monte Carlo Strategies in Scientific Computing*, Springer.

Q. Liu and D. Wang (2016), Stein variational gradient descent: A general purpose Bayesian inference algorithm. In *Advances in Neural Information Processing Systems 29 (NIPS 2016)* (D. D. Lee *et al.*, eds), pp. 2378–2386.

E. N. Lorenz (1963), 'Deterministic non-periodic flows', *J. Atmos. Sci.* **20**, 130–141.

J. Lu, Y. Lu and J. Nolen (2019), 'Scaling limit of the Stein variational gradient descent: The mean field regime', *SIAM J. Math. Anal.* **51**, 648–671.

R. J. McCann (1995), 'Existence and uniqueness of monotone measure-preserving maps', *Duke Math. J.* **80**, 309–323.

S. K. Mitter and N. J. Newton (2003), 'A variational approach to nonlinear estimation', *SIAM J. Control Optim.* **42**, 1813–1833.

R. E. Mortensen (1968), 'Maximum-likelihood recursive nonlinear filtering', *J. Optim. Theory Appl.* **2**, 386–394.

M. Morzfeld, X. Tu, E. Atkins and A. J. Chorin (2012), 'A random map implementation of implicit filters', *J. Comput. Phys.* **231**, 2049–2066.

R. M. Neal (1996), *Bayesian Learning for Neural Networks*, Springer.

E. Nelson (1984), *Quantum Fluctuations*, Princeton University Press.

N. Nüsken, S. Reich and P. Rozdeba (2019), State and parameter estimation from observed signal increments. Technical report, University of Potsdam. arXiv:1903.10717

Y. Ollivier (2018), Online natural gradient as a Kalman filter. *Electron. J. Statist.* **12**, 2930–2961.

S. Pathiraja and S. Reich (2019), Discrete gradients for computational Bayesian inference. Technical report, University of Potsdam. arXiv:1903.00186

G. A. Pavliotis (2014), *Stochastic Processes and Applications*, Springer.

G. Peyre and M. Cuturi (2018), Computational optimal transport. Technical report, CNRS, ENS, CREST, ENSAE. arXiv:1803.00567

F. Pons Llopis, N. Kantas, A. Beskos and A. Jasra (2018), 'Particle filtering for stochastic Navier–Stokes signal observed with linear additive noise', *SIAM J. Sci. Comput.* **40**, A1544–A1565.

S. Reich (2011), 'A dynamical systems framework for intermittent data assimilation', *BIT Numer. Math.* **51**, 235–249.

S. Reich (2012), 'A Gaussian mixture ensemble transform filter', *Q. J. Roy. Meterol. Soc.* **138**, 222–233.

S. Reich (2013), 'A nonparametric ensemble transform method for Bayesian inference', *SIAM J. Sci. Comput.* **35**, A2013–A2024.

S. Reich and C. J. Cotter (2015), *Probabilistic Forecasting and Bayesian Data Assimilation*, Cambridge University Press.

S. Reich and T. Hundertmark (2011), 'On the use of constraints in molecular and geophysical fluid dynamics', *Eur. Phys. J. Spec. Top.* **200**, 259–270.

S. Robert, D. Leuenberger and H. R. Künsch (2018), 'A local ensemble transform Kalman particle filter for convective-scale data assimilation', *Q. J. Roy. Meteorol. Soc.* **144**, 1279–1296.

H. C. Ruiz and H. J. Kappen (2017), 'Particle smoothing for hidden diffusion processes: Adaptive path integral smoother', *IEEE Trans. Signal Process.* **62**, 3191–3203.

G. Russo (1990), 'Deterministic diffusion of particles', *Commun. Pure Appl. Math.* **43**, 697–733.

S. Särkkä (2013), *Bayesian Filtering and Smoothing*, Cambridge University Press.

C. Schillings and A. M. Stuart (2017), 'Analysis of the ensemble Kalman filter for inverse problems', *SIAM J. Numer. Anal.* **55**, 1264–1290.

E. Schrödinger (1931), 'Über die Umkehrung der Naturgesetze', *Sitzungsberichte der Preußischen Akademie der Wissenschaften, Physikalisch-Mathematische Klasse* **IX**, 144–153.

R. Sinkhorn (1967), 'Diagonal equivalence to matrices with prescribed row and column sums', *Amer. Math. Monthly* **74**, 402–405.

A. Taghvaei and P. G. Mehta (2016), Gain function approximation in the feedback particle filter. In *IEEE 55th Conference on Decision and Control (CDC)*, IEEE, pp. 5446–5452.

A. Taghvaei, J. de Wiljes, P. G. Mehta and S. Reich (2017), 'Kalman filter and its modern extensions for the continuous-time nonlinear filtering problem', *ASME. J. Dyn. Sys. Meas. Control.* **140**, 030904.

A. Taghvaei, P. Mehta and S. Meyn (2019), Gain function approximation in the feedback particle fitler. Technical report, University of Illinois at Urbana-Champaign. arXiv:1902.07263

S. Thijssen and H. J. Kappen (2015), 'Path integral control and state-dependent feedback', *Phys. Rev. E* **91**, 032104.

X. T. Tong, A. J. Majda and D. Kelly (2016), 'Nonlinear stability and ergodicity of ensemble based Kalman filters', *Nonlinearity* **29**, 657.

P. J. van Leeuwen (2015), Nonlinear data assimilation for high-dimensional systems. In *Frontiers in Applied Dynamical Systems: Reviews and Tutorials* (P. J. van Leeuwen *et al.* eds), Vol. 2, Springer, pp. 1–73.

P. J. van Leeuwen, H. R. Künsch, L. Nerger, R. Potthast and S. Reich (2018), Particle filter and applications in geosciences. Technical report, University of Reading. arXiv:1807.10434

E. Vanden-Eijnden and J. Weare (2012), 'Data assimilation in the low noise regime with application to the Kuroshio', *Monthly Weather Rev.* **141**, 1822–1841.

S. Vetra-Carvalho, P. J. van Leeuwen, L. Nerger, A. Barth, M. U. Altaf, P. Brasseur, P. Kirchgessner and J.-M. Beckers (2018), 'State-of-the-art stochastic data assimilation methods for high-dimensional non-Gaussian problems', *Tellus A* **70**, 1445364.

C. Villani (2003), *Topics in Optimal Transportation*, American Mathematical Society.

C. Villani (2009), *Optimal Transportation: Old and New*, Springer.

J. Xiong (2011), Particle approximations to the filtering problem in continuous time. In *The Oxford Handbook of Nonlinear Filtering* (D. Crisan and B. Rozovskii, eds), Oxford University Press, pp. 635–655.

T. Yang, P. G. Mehta and S. P. Meyn (2013), 'Feedback particle filter', *IEEE Trans. Automat. Control* **58**, 2465–2480.

C. Zhang, A. Taghvaei and P. G. Mehta (2019), 'A mean-field optimal control formulation for global optimization', *IEEE Trans. Automat. Control* **64**, 282–289.